ENVIRONMENTAL SCIENCE

For a Changing World

FOURTH EDITION

Susan Karr
Carson-Newman University

Macmillan is committed to lessening our company's impact on the environment especially in regard to carbon emissions. Based on the progress we have made thus far, along with the purchase of carbon offsets, Macmillan USA was carbon neutral for Scopes 1, 2, and 3 in the calendar years 2017 and 2019. For more information please visit: www.macmillansustainability.com

macmillan learning

Austin • Boston • New York • Plymouth

• A PARTNERSHIP BETWEEN •
macmillan learning & SCIENTIFIC AMERICAN.

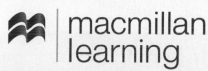

Senior Vice President, STEM: Daryl Fox
Program Director: Sandy Lindelof
Senior Program Manager: Jennifer Edwards
Marketing Manager: Leah Christians
Executive Content Development Manager, STEM: Debbie Hardin
Development Editor: Heather McCoy
Executive Project Manager, Content, STEM: Katrina Mangold
Editorial Project Manager: Michele Mangelli
Director of Content, Earth Sciences: Jennifer Driscoll Hollis
Executive Media Editor: Amy Thorne
Senior Media Editor: Emily Marino
Senior Media Editor: Alexandra Gordon
Editorial Assistant: Nathan Livingston
Marketing Assistant: Morgan Psiuk
Director, Content Management Enhancement: Tracey Kuehn
Senior Managing Editor: Lisa Kinne
Senior Content Project Manager: Won McIntosh
Senior Workflow Project Manager: Paul W. Rohloff
Production Supervisor: Robert Cherry
Director of Design, Content Management: Diana Blume
Design Services Manager: Natasha A. S. Wolfe
Cover Design Manager and Designer: John Callahan
Text Designer: Emiko Rose-Paul
Art Manager: Matthew McAdams
Illustrations: Troutt Visual Services
Director of Digital Production: Keri deManigold
Media Project Manager: Brian Nobile
Permissions Manager: Jennifer MacMillan
Photo Researchers: Krystyna Borgen, Sheena Goldstein
Composition: Lumina Datamatics, Inc.
Printing and Binding: LSC Communications
Cover Image: imagewerks/Imagewerks Japan/Getty Images

Library of Congress Control Number: 2020942134

ISBN-13: 978-1-319-24562-7
ISBN-10: 1-319-24562-5

Printed in the United States of America
1 2 3 4 5 6 25 24 23 22 21 20

Macmillan Learning
One New York Plaza
Suite 4600
New York, NY 10004-1562
www.macmillanlearning.com

w.h.freeman
Macmillan Learning

In 1946, William Freeman founded W. H. Freeman and Company and published Linus Pauling's *General Chemistry*, which revolutionized the chemistry curriculum and established the prototype for a Freeman text. W. H. Freeman quickly became a publishing house where leading researchers can make significant contributions to mathematics and science. In 1996, W. H. Freeman joined Macmillan and we have since proudly continued the legacy of providing revolutionary, quality educational tools for teaching and learning in STEM.

CSNafzger/Shutterstock

BRIEF CONTENTS

DETAILED CONTENTS

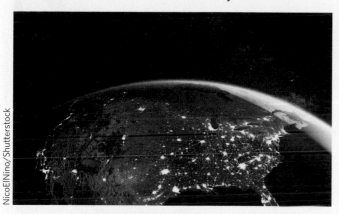

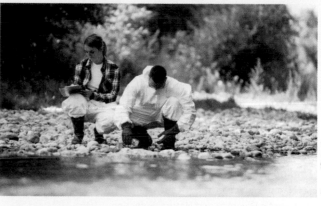

Roman Lukiw Photography/Moment/Getty Images

peeterv/Getty Images

CHAPTER 5 Managing Resources: Environmental Economics and Policy 197

Romolo Tavani/iStock/Getty Images

CHAPTER 6 Water Resources 252

Garry Solomon/EyeEm/Getty Images

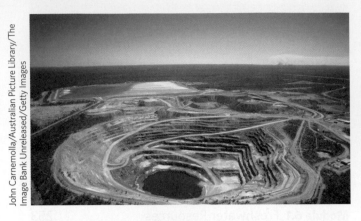

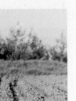

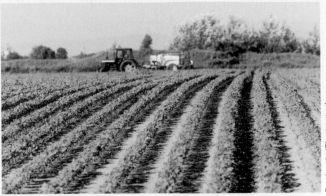

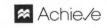

ping han/Alamy

Dennis MacDonald/AGE Fotostock

CHAPTER 11 Alternatives to Fossil Fuels 464

GIPhotoStock/Cultura/Getty Images

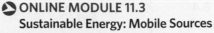

ABOUT THE AUTHOR

Stephen Karr

Susan Karr, MS, is an assistant professor in the Biology Department of Carson-Newman University in Jefferson City, Tennessee, and has been teaching for more than 25 years. She has served on environmental sustainability committees for both her university and her community. In addition to teaching non-majors courses in environmental science and zoology, she teaches an upper-level course in animal behavior in which she and her students train dogs from the local animal shelter in a program that improves the animals' chances of adoption. She received degrees in animal behavior and forestry from the University of Georgia.

Acknowledgments

I've discovered that an undertaking this big is truly a collaborative effort. It is amazing what you can accomplish when you work with talented and highly skilled people. I want to thank Jennifer Edwards, Senior Program Manager, for her support and guidance and Development Editor Heather McCoy and Media Editors Amy Thorne, Emily Marino, and Alexandra Gordon for their patience, insights, and outstanding editorial skills. I also want to thank Melinda Wenner Moyer, the accomplished writer and journalist whose research and writing have made these stories come to life. I gratefully acknowledge the entire team at Lumina Datamatics for their skills, vision, and patience in the production of the book. Thanks also go to Leah Christians, the marketing manager, and the sales force, who work tirelessly to see that the book is a success in the marketplace.

I've had tremendous support from my biology department colleagues at Carson-Newman University and from focus group participants, who offered advice, answered questions, and helped track down elusive information—thanks for sharing your expertise. I also owe a debt of gratitude to my environmental science students over the years for their questions, interests, demands, and passion for learning that have always challenged and inspired me. Thanks to my husband, Steve, for supporting me in so many different ways it is impossible to count them all. Finally, I want to thank my granddaughter, Chloe, with whom I roam the countryside rediscovering nature's wonders as often as I can.

A Word from the Author

For more than 25 years as an environmental science and biology instructor, I've found that "stories" capture the imagination of my students. Students are genuinely interested in environmental issues—using stories to teach these issues makes the science more relevant and meaningful to them. Many leave the class with an understanding that what they do really matters, and they are ready to act on that knowledge. This is why I am enthusiastic about our textbook **Environmental Science for a Changing World.**

The fourth edition breaks down the main topics into 11 chapters, each containing two or more modules that focus on specific topics relevant to the chapter. In each module of this text, students will be introduced to one or more major issues and concepts of environmental science through the context of a central case study—a story—threaded throughout the module. These stories are current, relevant, and captivating. Discrete sections and key concepts within each module help identify the environmental science content that students need to master, while the case study example helps put those concepts in context for better understanding. The most popular modules (those used by most instructors) are found in the print book, and additional modules are found online. A complete bibliography for all modules is also located online.

As with the previous editions, the text focuses on building core competencies for the nonmajor: environmental literacy, science literacy, and information literacy. Infographic questions, end-of-module questions and activities, and online exercises provide further opportunities to develop these competencies, as well as critical thinking skills.

Every person involved in this book—the writers, illustrators, editors, and fellow instructors—has one sincere objective: to help students become informed citizens able to analyze issues, evaluate arguments, discuss solutions, and recognize trade-offs as they make up their own minds about our most pressing environmental challenges.

Sincerely,

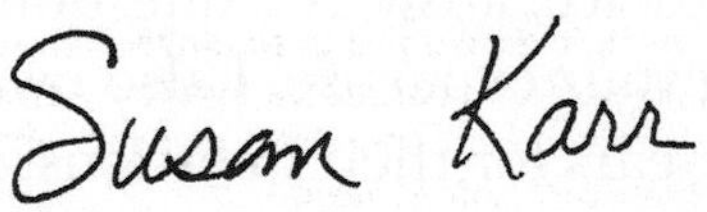

Susan Karr

Real people. Real stories. Real science.

Susan Karr's ***Environmental Science for a Changing World*** engages students to think critically about the environment as they learn how science is done through real stories about real people. From the text's journalistic approach to its dynamic online resources, students explore scientific processes and concepts in a real-world context. Now available together with Macmillan's new online learning platform **Achieve.**

EMPOWERING SCIENCE FOR EVERY STUDENT

Environmental Science for a Changing World *helps students create personal understanding of the natural world and their role in it.*

- The text offers a consistent methodology for teaching the field's essential scientific concepts, with each module centered around **Guiding Questions**.
- From the opening page to the end-of-module questions, each module follows an **engaging case study** featuring real scientific research that brings context and relevance to the module's subject.
- **Infographics** in the style of *Scientific American* magazine guide students step-by-step through essential processes and concepts.
- Supplemental studies from around the world are highlighted in the end-of-module **Global Case Studies activities**, which are also available online in Achieve, enabling students to learn more about each module's topic through additional real-world examples.
- **Bring It Home** boxes at the end of each module offer suggestions for ways students can make a difference individually or collectively.
- **Decision Point Simulations** give students the opportunity to choose different outcomes to see the environmental effects of their choices. These simulations are designed to help students understand the complexity of environmental decisions in a fun and interesting way.

CRITICAL THINKING AND ACTIVE LEARNING

Environmental Science for a Changing World focuses on building core competencies for the nonmajor, including environmental literacy, science literacy, information literacy, and critical thinking skills.

- **End-of-module questions and activities and online exercises** provide opportunities to develop these competencies.
- **Critical Thinking Questions** accompany each Infographic, encouraging students to engage with the illustrations, to explore concepts more deeply, and to build scientific literacy.
- Key concepts and skills are revisited and reinforced with **LearningCurve adaptive quizzing**.
- Each module is complemented by a collection of **in-class activities that can be adapted to any teaching modality (online only, hybrid, or face-to-face)** to help students develop a deeper understanding of the key concepts in environmental science.
- In addition, this edition features an expanded module on The Process of Science (Module 1.2), including new material on **information and media literacy**.

ROBUST RESOURCES THROUGH ACHIEVE

The Achieve learning platform supports educators and students throughout the full flexible range of instruction.

- The Achieve **design is guided by learning science research** through extensive collaboration and testing by both students and faculty, including two levels of Institutional Review Board approval for every study.
- Achieve features a **flexible suite of resources** to support learning core concepts, visualization, problem solving, and assessment. All student and instructor resources can be assigned or downloaded in Achieve.
- The **iClicker** classroom engagement system syncs with the Achieve **gradebook**.
- **Gradebook integration** can be set up with **your campus LMS** and with **Inclusive Access** programs.
- The Achieve gradebook provides insights for **just-in-time teaching** and for reporting on student and full-class achievement by learning objective. Powerful **analytics and instructor resources** in Achieve pair with exceptional environmental science content to provide an unrivaled learning experience.

Compelling, Integrated Stories within a Strong Pedagogical Framework

Students follow one **captivating story** through each module that illustrates core science concepts.

New coverage of timely topics such as water wars, global hunger, environmental health and infectious diseases, microplastic waste, regenerative farming, lithium mining, and sustainable energy exemplifies the book's contemporary approach to environmental science.

Maintaining the genetic and species diversity of our crops is vital to our ability to find crop varieties suitable for changing environmental conditions. *Helen Camacaro/Getty Images*

After reading this module, you should be able to answer these GUIDING QUESTIONS

1. How prevalent is hunger, and what are its causes?
2. What is malnutrition, and what problems can it cause?
3. What were the intent, scope, and outcome of the Green Revolution?
4. What is industrial agriculture, and what are the pros and cons of farming this way?
5. What are the trade-offs of fertilizer and pesticide use in agriculture?
6. How can genetic engineering be used in agriculture, and what are its trade-offs?
7. What are food self-sufficiency and food sovereignty, and why are they important?

Key Concept 6: Genetically engineering crops (or animals) to contain useful traits such as faster growth or an expanded growing range may increase food supplies but comes with environmental, economic, and ethical concerns.

Humans have been altering the genetic makeup of crops ever since the first budding farmers started cross-breeding plants to produce new varieties, and it was with these same artificial selection techniques that plant scientists created the high-yielding crops of the Green Revolution. But scientists now have new methods to directly alter the genetic makeup of an organism, and these techniques form the basis of what some farmers and scientists like to think of as Green Revolution 2.0, or the *Gene Revolution*—the next battle against hunger.

genetically modified organism (GMO) An organism that has had its genetic information modified to give it desirable characteristics such as pest or drought resistance.

transgenic organism An organism that contains genes from another species.

intragenic organism An organism whose own DNA has been edited.

cisgenic organism An organism that received DNA from a close relative, DNA that could have been acquired via traditional breeding.

Genetically modified organisms (GMOs) are defined as those organisms that have had their genetic information altered in a way that would not be possible by natural means. Scientists have been producing such organisms for decades, coaxing genetically modified bacteria to produce important medicines such as insulin—like tiny living drug factories. In the 1990s, researchers began applying the same technology to food crops. By transferring genes for desirable traits (like pest resistance or herbicide tolerance) from one species to another, they have created a new suite of genetically modified food crops—plants that can grow more plentifully and thrive in a wider range of habitats than they ordinarily would. Organisms that receive DNA (genetic material) from another species with which they could not naturally breed (a sexually incompatible species) are known as **transgenic organisms**.

As powerful as genetic engineering might be, it cannot create the traits we would like our crops to have—it can only transfer them. Seed banks are an important resource not only because they preserve seeds that might otherwise go extinct, but also because they are a repository for diverse genetic material. When scientists analyze the genes of seeds kept in banks, they can identify useful genes for plant breeding or the creation of GMOs. The best place to find these traits is among the populations with the most genetic diversity—those plants growing wild in their place of origin (wild-type plants)—the very plants Murray and others like her search out to collect seeds for safekeeping.

In recent years, new technologies have emerged that can create other types of genetically engineered (GE) organisms. Known as *genome editing*, such as the technique that goes by the acronym CRISPR, these methods allow scientists to more quickly and precisely edit the DNA of organisms. While these methods

Guiding Questions establish a clear pathway from the opening story to the end-of-module and online features. The Guiding Questions and **Key Concepts** bring the science to the forefront and reinforce fundamental concepts throughout the module.

Each module section includes an **Infographic**, where students will find the information they need to think critically about the Guiding Questions. Additional questions within the Infographics prompt students to reflect on the content and stretch their understanding.

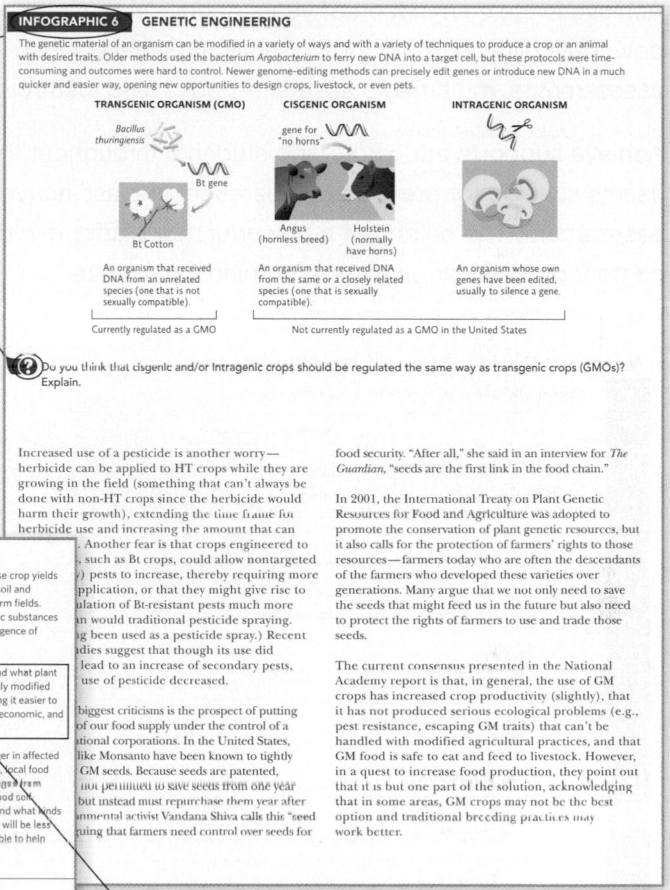

INFOGRAPHIC 6 GENETIC ENGINEERING

The genetic material of an organism can be modified in a variety of ways and with a variety of techniques to produce a crop or an animal with desired traits. Older methods used the bacterium *Argobacterium* to ferry new DNA into a target cell, but these protocols were time-consuming and outcomes were hard to control. Newer genome-editing methods can precisely edit genes or introduce new DNA in a much quicker and easier way, opening new opportunities to design crops, livestock, or even pets.

TRANSGENIC ORGANISM (GMO)
CISGENIC ORGANISM
INTRAGENIC ORGANISM
Bacillus thuringiensis
Bt gene
Bt Cotton
gene for "no horns"
Angus (hornless breed)
Holstein (normally have horns)
An organism that received DNA from an unrelated species (one that is not sexually compatible).
An organism that received DNA from the same or a closely related species (one that is sexually compatible).
An organism whose own genes have been edited, usually to silence a gene.
Currently regulated as a GMO
Not currently regulated as a GMO in the United States

Do you think that cisgenic and/or intragenic crops should be regulated the same way as transgenic crops (GMOs)? Explain.

Increased use of a pesticide is another worry—herbicide can be applied to HT crops while they are growing in the field (something that can't always be done with non-HT crops since the herbicide would harm their growth), extending the time frame for herbicide use and increasing the amount that can ... Another fear is that crops engineered to ..., such as Bt crops, could allow nontargeted ...) pests to increase, thereby requiring more ...pplication, or that they might give rise to ...lation of Bt-resistant pests much more ...n would traditional pesticide spraying. ...g been used as a pesticide spray.) Recent ...dies suggest that though its use did ... lead to an increase of secondary pests, ... use of pesticide decreased.

...biggest criticisms is the prospect of putting ...of our food supply under the control of a ...tional corporations. In the United States, ...like Monsanto have been known to tightly ... GM seeds. Because seeds are patented, ... not permitted to save seeds from one year ... but instead must repurchase them year after ...nmental activist Vandana Shiva calls this "seed ...uing that farmers need control over seeds for food security. "After all," she said in an interview for *The Guardian*, "seeds are the first link in the food chain."

In 2001, the International Treaty on Plant Genetic Resources for Food and Agriculture was adopted to promote the conservation of plant genetic resources, but it also calls for the protection of farmers' rights to those resources—farmers today who are often the descendants of the farmers who developed these varieties over generations. Many argue that we not only need to save the seeds that might feed us in the future but also need to protect the rights of farmers to use and trade those seeds.

The current consensus presented in the National Academy report is that, in general, the use of GM crops has increased crop productivity (slightly), that it has not produced serious ecological problems (e.g., pest resistance, escaping GM traits) that can't be handled with modified agricultural practices, and that GM food is safe to eat and feed to livestock. However, in a quest to increase food production, they point out that it is but one part of the solution, acknowledging that in some areas, GM crops may not be the best option and traditional breeding practices may work better.

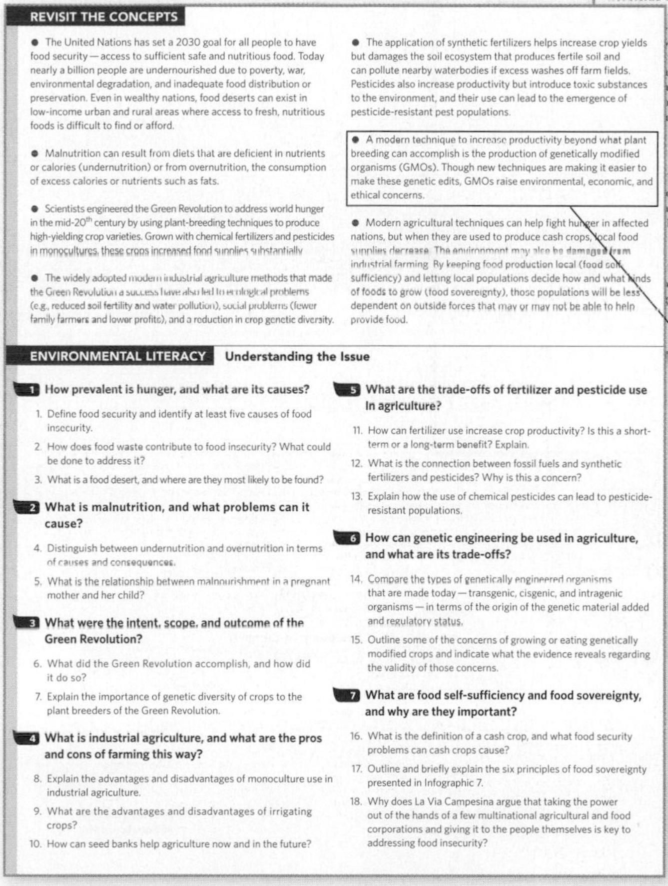

REVISIT THE CONCEPTS

- The United Nations has set a 2030 goal for all people to have food security—access to sufficient safe and nutritious food. Today nearly a billion people are undernourished due to poverty, war, environmental degradation, and inadequate food distribution or preservation. Even in wealthy nations, food deserts can exist in low-income urban and rural areas where access to fresh, nutritious foods is difficult to find or afford.
- Malnutrition can result from diets that are deficient in nutrients or calories (undernutrition) or from overnutrition, the consumption of excess calories or nutrients such as fats.
- Scientists engineered the Green Revolution to address world hunger in the mid-20th century by using plant-breeding techniques to produce high-yielding crop varieties. Grown with chemical fertilizers and pesticides in monocultures, these crops increased food supplies substantially.
- The widely adopted modern industrial agriculture methods that made the Green Revolution a success have also led to ecological problems (e.g., reduced soil fertility and water pollution), social problems (fewer family farmers and lower profits), and a reduction in crop genetic diversity.
- The application of synthetic fertilizers helps increase crop yields but damages the soil ecosystem that produces fertile soil and can pollute nearby waterbodies if excess washes off farm fields. Pesticides also increase productivity but introduce toxic substances to the environment, and their use can lead to the emergence of pesticide-resistant pest populations.
- A modern technique to increase productivity beyond what plant breeding can accomplish is the production of genetically modified organisms (GMOs). Though new techniques are making it easier to make these genetic edits, GMOs raise environmental, economic, and ethical concerns.
- Modern agricultural techniques can help fight hunger in affected nations, but when they are used to produce cash crops, local food supplies decrease. The environment may also be damaged from industrial farming. By keeping food production local (food self-sufficiency) and letting local populations decide how and what kinds of foods to grow (food sovereignty), those populations will be less dependent on outside forces that may or may not be able to help provide food.

ENVIRONMENTAL LITERACY Understanding the Issue

1 How prevalent is hunger, and what are its causes?

1. Define food security and identify at least five causes of food insecurity.
2. How does food waste contribute to food insecurity? What could be done to address it?
3. What is a food desert, and where are they most likely to be found?

2 What is malnutrition, and what problems can it cause?

4. Distinguish between undernutrition and overnutrition in terms of causes and consequences.
5. What is the relationship between malnourishment in a pregnant mother and her child?

3 What were the intent, scope, and outcome of the Green Revolution?

6. What did the Green Revolution accomplish, and how did it do so?
7. Explain the importance of genetic diversity of crops to the plant breeders of the Green Revolution.

4 What is industrial agriculture, and what are the pros and cons of farming this way?

8. Explain the advantages and disadvantages of monoculture use in industrial agriculture.
9. What are the advantages and disadvantages of irrigating crops?
10. How can seed banks help agriculture now and in the future?

5 What are the trade-offs of fertilizer and pesticide use in agriculture?

11. How can fertilizer use increase crop productivity? Is this a short-term or a long-term benefit? Explain.
12. What is the connection between fossil fuels and synthetic fertilizers and pesticides? Why is this a concern?
13. Explain how the use of chemical pesticides can lead to pesticide-resistant populations.

6 How can genetic engineering be used in agriculture, and what are its trade-offs?

14. Compare the types of genetically engineered organisms that are made today—transgenic, cisgenic, and intragenic organisms—in terms of the origin of the genetic material added and regulatory status.
15. Outline some of the concerns of growing or eating genetically modified crops and indicate what the evidence reveals regarding the validity of those concerns.

7 What are food self-sufficiency and food sovereignty, and why are they important?

16. What is the definition of a cash crop, and what food security problems can cash crops cause?
17. Outline and briefly explain the six principles of food sovereignty presented in Infographic 7.
18. Why does La Via Campesina argue that taking the power out of the hands of a few multinational agricultural and food corporations and giving it to the people themselves is key to addressing food insecurity?

The new **Revisit the Concepts** feature at the end of each module offers an excellent summary of main concepts useful for studying or as a pre-reading resource. Guiding Questions provide the pedagogical framework for the **end-of-module and online assessment** in Achieve, which includes references to relevant Infographics to tie material together.

Achieve

Achieve is the culmination of years of development work put toward creating the most powerful online learning tool for environmental science students. It houses all of our renowned assessments, multimedia assets, e-books, and instructor resources in a powerful new platform.

Achieve supports educators and students throughout the full range of instruction, including assets suitable for pre-class preparation, in-class active learning, and post-class study and assessment. The pairing of a powerful new platform with outstanding environmental science content provides an unrivaled learning experience.

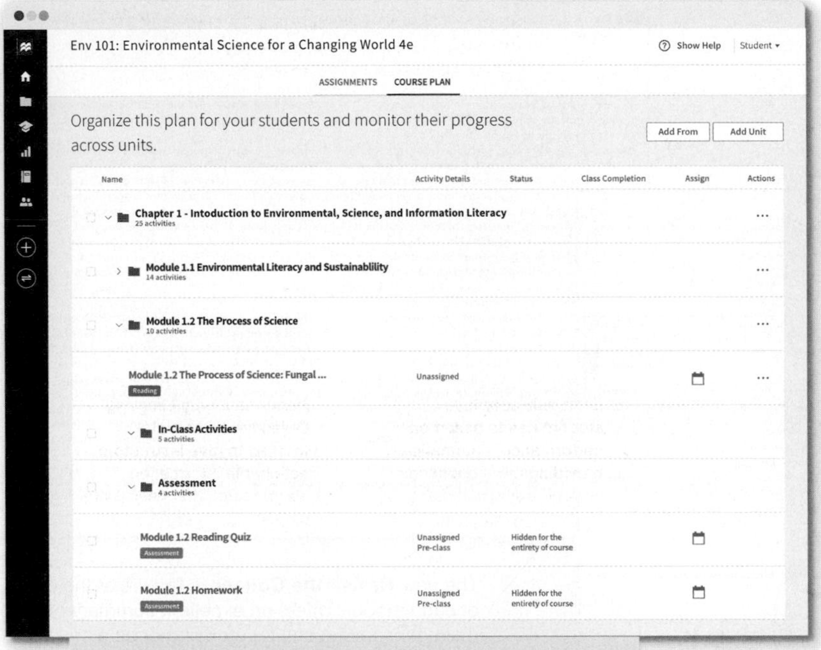

Highlights include:

- **A design guided by learning science research.** Co-designed through extensive collaboration and testing by both students and faculty, including two levels of Institutional Review Board approval for every study of Achieve.
- **A learning path of powerful content** including pre-class, in-class, and post-class activities and assessments. A detailed gradebook with insights for just-in-time teaching and reporting on student achievement by learning objective.
- **Easy integration and gradebook sync** with iClicker classroom engagement solutions.
- **Simple integration** with your campus LMS and availability through **Inclusive Access** programs.

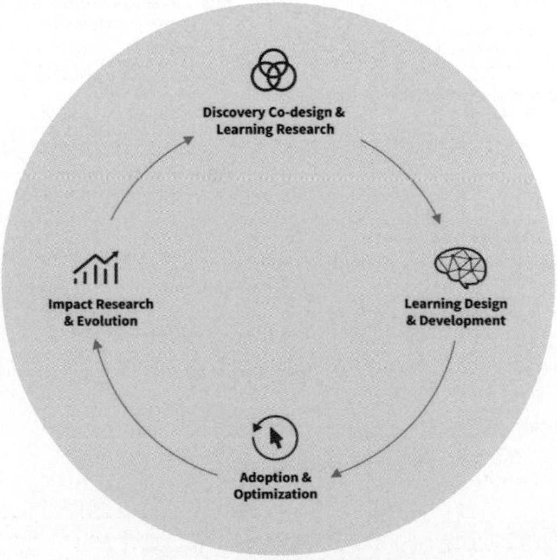

For more information or to sign up for a demonstration of Achieve, contact your local Macmillan representative or visit **macmillanlearning.com/achieve**

Full Learning Path

Achieve includes a full learning path of content for pre-class preparation, in-class active learning, and post-class engagement and assessment. It's the perfect solution for engaging all students in an inclusive teaching experience.

LearningCurve

LearningCurve's game-like quizzing motivates each student to engage with the course content, and reporting tools help instructors understand what their class needs.

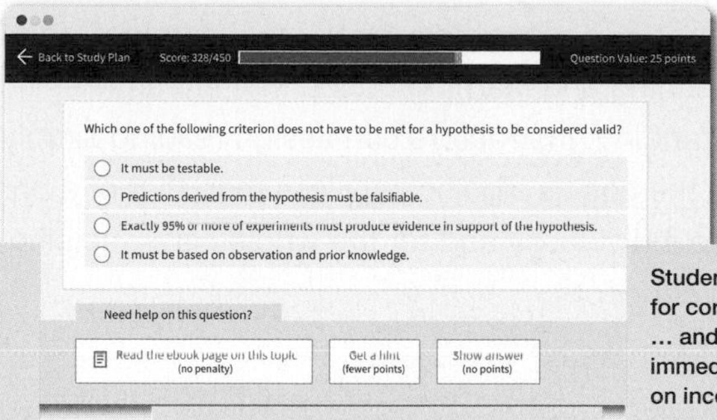

Module-specific quizzes adapt in difficulty based on individual student performance.

Students earn points for correct answers ... and receive immediate feedback on incorrect answers.

Questions are tagged to sections of the e-book to provide comprehensive coverage and easy-to-find help.

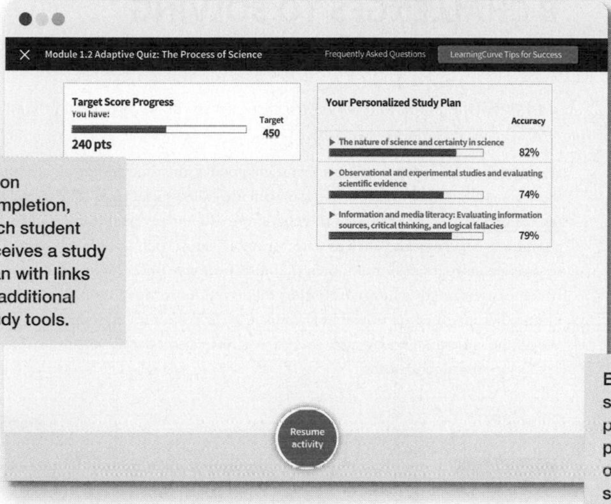

Upon completion, each student receives a study plan with links to additional study tools.

Each quiz adapts to student needs based on performance, automatically providing more questions on topics in which the student is struggling.

Performance by Learning Objective can also be tracked to give instructors actionable information about topics that need extra emphasis.

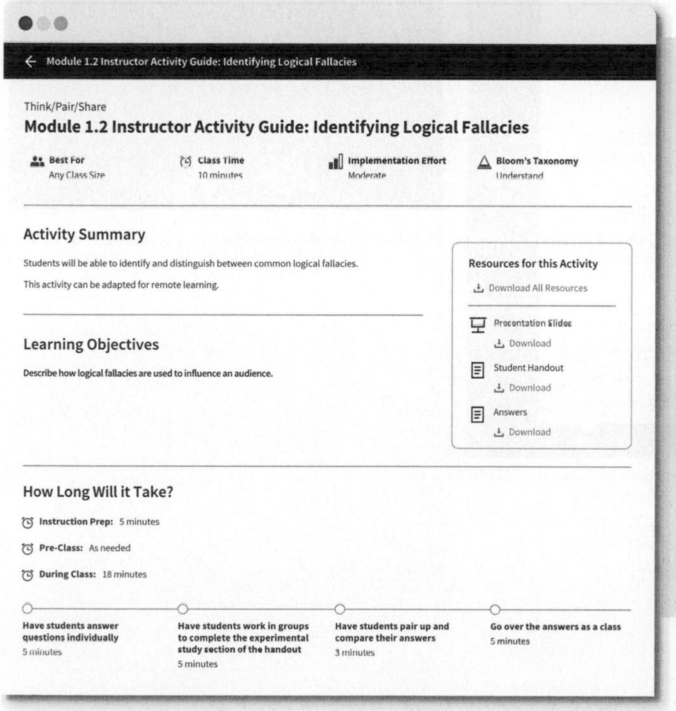

In-Class Activities

Because active learning is important to keeping students engaged in the material, each module includes in-class activities developed by Susan Karr that help students develop a deeper understanding of the key concepts in environmental science. These activities can be adapted for in-person, online, and hybrid use.

iClicker

Achieve includes gradebook integration with iClicker, allowing for a seamless pairing of the best classroom response system with the best in-class content.

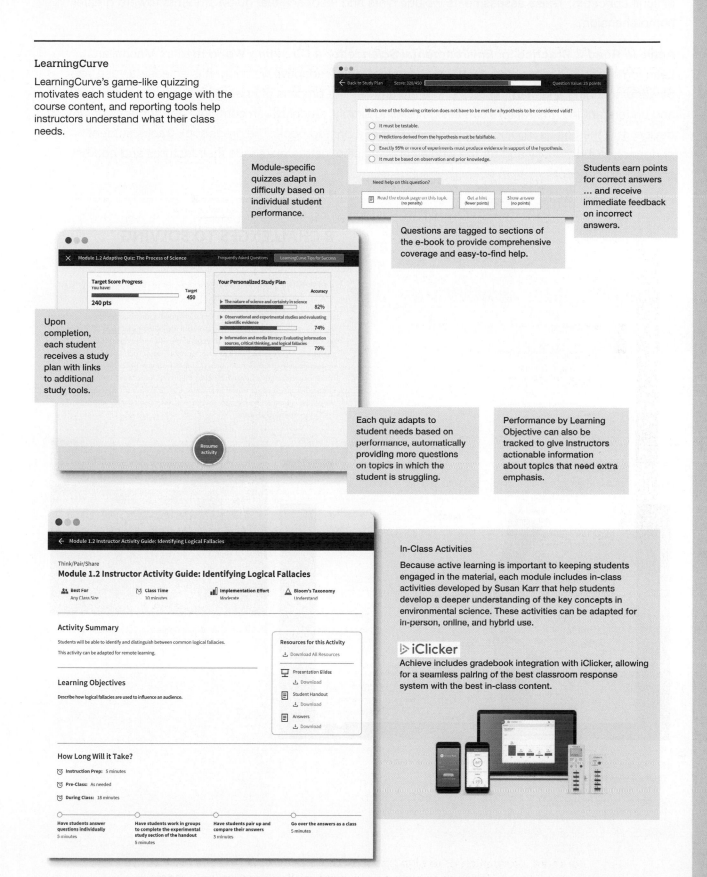

Achieve offers a diverse range of resources to target the most challenging concepts and skills in the course. **Conceptual animations** help students to visualize key concepts. In addition, **narrated lecture slides** offer a guided experience for students. **Tutorial Style Homework** questions help students learn difficult concepts. These assessments include hints and feedback that guide students toward greater comprehension.

Achieve Read & Practice for ***Environmental Science for a Changing World*** marries Macmillan Learning's mobile-accessible e-book and LearningCurve's adaptive quizzing. It is an easy-to-use yet powerful teaching and learning option that streamlines the process of increasing student engagement and understanding. Instructors can assign reading simply, students can complete assignments on any device, and the cost is significantly less than that of a printed book. The gradebook tracks student performance individually and for the whole class, helping instructors shape their lectures and address challenging concepts.

eBOOK
NOTEBOOK
15
8 CHALLENGES TO SOLVING ENVIRONMENTAL PROBLEMS
Key Concept 8: Impediments to solving environmental problems include short-term thinking and social traps.
Decisions by individuals or groups that seem good at the time and produce a short-term benefit but hurt society in the long run are called social traps. The tragedy of the commons is a social trap that often emerges when many people are using a commonly held resource, such as water or public land. Each person will act in a way to maximize his or her own benefit, but as everyone does this, the resource becomes overused or damaged. Herders might put more animals on a common pasture because they are driven by the idea that "if I don't use it, someone else will." We do the same thing today as we overharvest forests and oceans and as we release toxins into the air and water.
Scott Warren/Aurora/Getty Images
rch built in Brattahlid, Greenland—a Viking settlement founded by Erik the Red
ago.

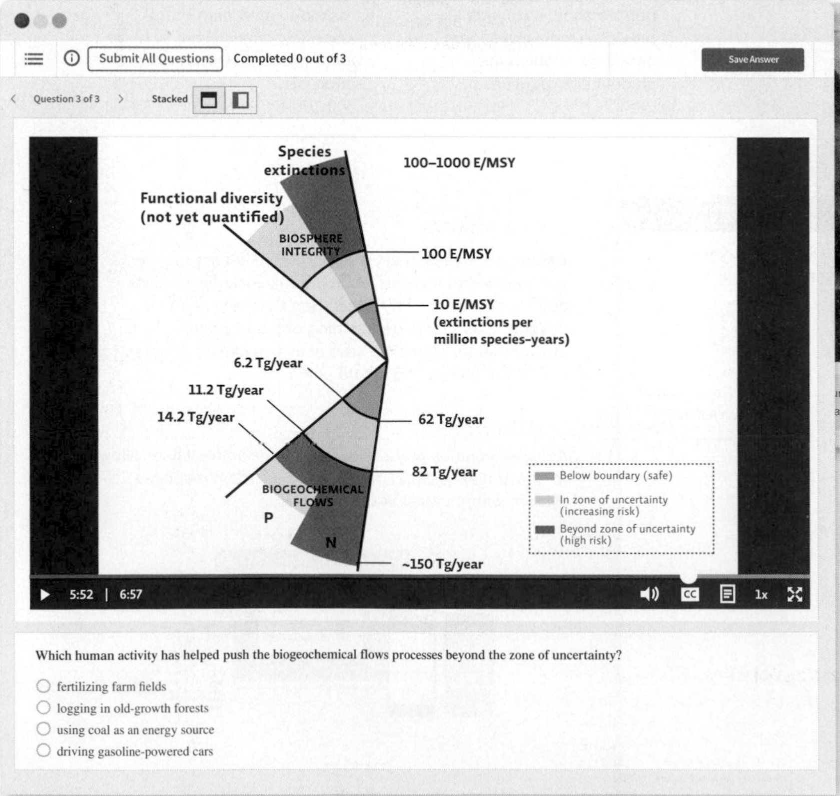
Submit All Questions
Completed 0 out of 3
Save Answer
Question 3 of 3
Stacked
Species extinctions
100–1000 E/MSY
Functional diversity (not yet quantified)
BIOSPHERE INTEGRITY
100 E/MSY
10 E/MSY (extinctions per million species-years)
6.2 Tg/year
11.2 Tg/year
14.2 Tg/year
62 Tg/year
82 Tg/year
BIOGEOCHEMICAL FLOWS
P
N
~150 Tg/year
Below boundary (safe)
In zone of uncertainty (increasing risk)
Beyond zone of uncertainty (high risk)
5:52 | 6:57
1x
Which human activity has helped push the biogeochemical flows processes beyond the zone of uncertainty?
fertilizing farm fields
logging in old-growth forests
using coal as an energy source
driving gasoline-powered cars

Powerful analytics, viewable in an elegant dashboard, offer instructors a window into student progress. Achieve gives you the insight to address students' weaknesses and misconceptions before they struggle on a test.

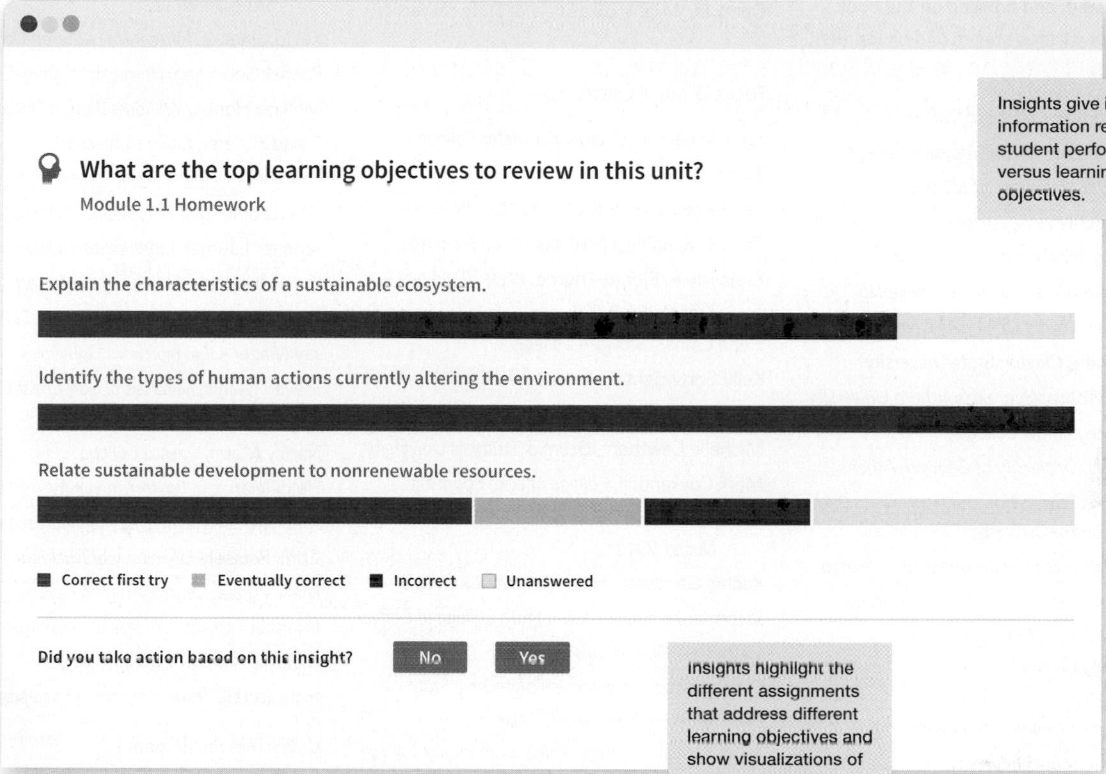

What are the top learning objectives to review in this unit?
Module 1.1 Homework
Explain the characteristics of a sustainable ecosystem.
Identify the types of human actions currently altering the environment.
Relate sustainable development to nonrenewable resources.
Correct first try
Eventually correct
Incorrect
Unanswered
Did you take action based on this insight?
No
Yes

Insights give instructors information regarding student performance versus learning objectives.

Insights highlight the different assignments that address different learning objectives and show visualizations of student performance.

Reports provide powerful views of student performance and can be viewed in a number of different ways.

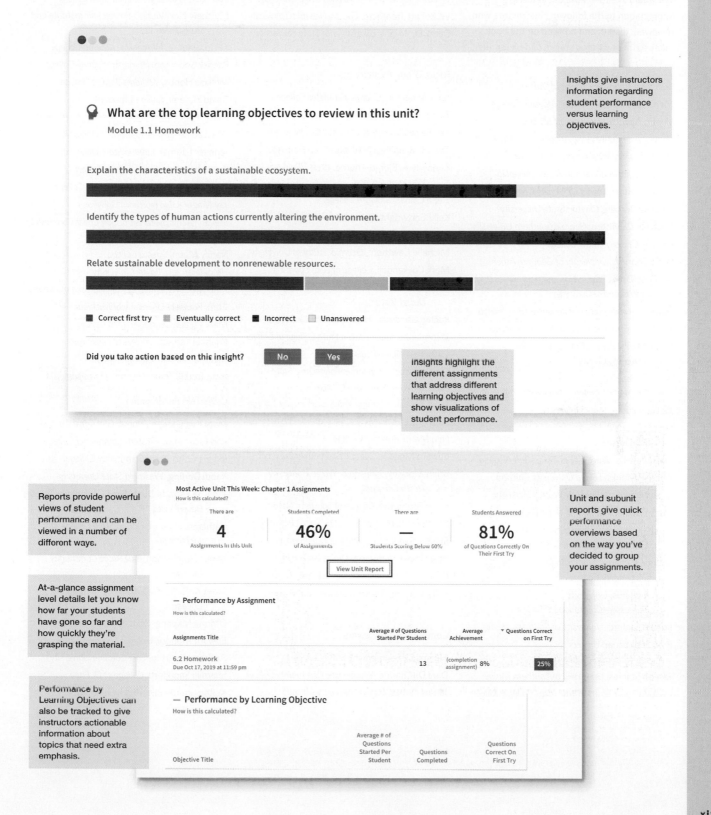

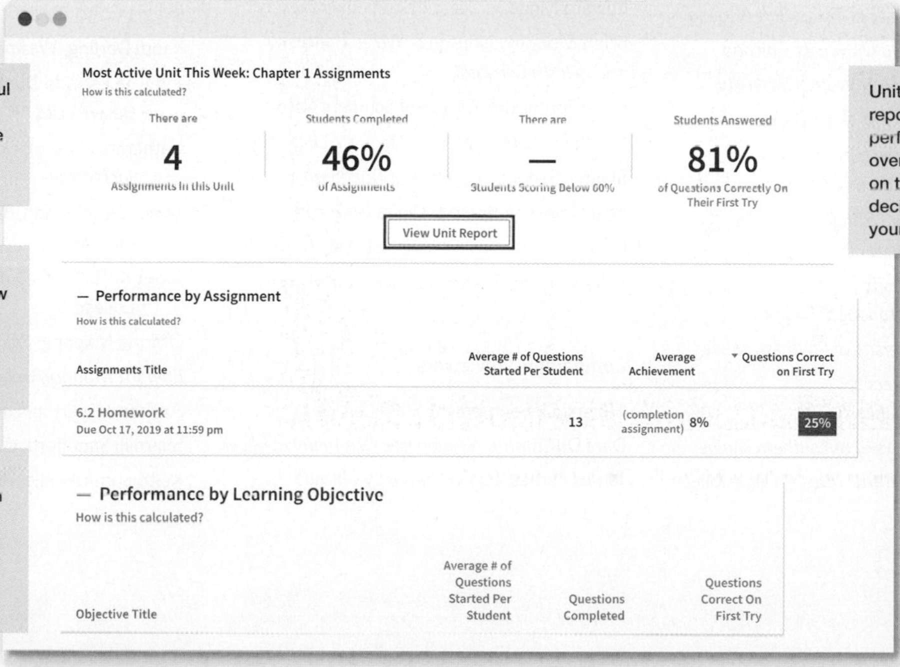

Unit and subunit reports give quick performance overviews based on the way you've decided to group your assignments.

At-a-glance assignment level details let you know how far your students have gone so far and how quickly they're grasping the material.

Performance by Learning Objectives can also be tracked to give instructors actionable information about topics that need extra emphasis.

REVIEWERS

We would like to extend our sincere appreciation to the following instructors who reviewed, tested, and advised on the book manuscript at various stages of development.

Shamili Ajgaonkar, *College of DuPage*
Merrilee Anderson, *Mount Aloysius College*
Martin Appold, *University of Missouri*
Shirley-Ann Behravesh, *Arizona State University*
Donna Bivans, *Pitt Community College*
Marnie Branfireun, *Simon Fraser University*
Laura Brentner, *Loyola University Chicago*
Stephen Burnett, *Clayton State University*
Michelle Cawthorn, *Georgia Southern University*
Lindsay Chaney, *Snow College*
Scott Connelly, *University of Georgia–Athens*
Shannon Davis-Foust, *University of Wisconsin–Oshkosh*
James DeFrancesco, *Loyola University Chicago*
Andrew Dutra, *Eastern Florida State College*
Megan Kelly, *Arrupe College of Loyola University Chicago*
Andrew Lapinski, *Reading Area Community College*
Jia Lu, *Valdosta State University*
Ian MacDonald, *Florida State University-Tallahassee*
Terri Matiella, *University of Texas–San Antonio*
Andrew Monks, *Loyola University Chicago*
Brian Mooney, *Johnson & Wales University*
Mark Olsen, *University of Notre Dame*
Sarah Praskievicz, *University of North Carolina–Greensboro*
Maggie Richards, *Front Range Community College*
Sean Richards, *University of Tennessee–Chattanooga*
Robert Sanford, *University of Southern Maine*
Brad Spanbauer, *University of Wisconsin–Oshkosh*
Joseph Staples, *University of Southern Maine*
Cristina Summers, *Central Texas College–Killeen*
Willetta Toole-Simms, *Azusa Pacific University*
W. Robert Trentham, *Carson-Newman University*
Derek D. Wright, *Lake Superior State University*

Focus Group Participants

John Anderson, *Georgia Perimeter College*
Tom L. Arsuffi, *Texas Tech University*
Teri Balser, *University of Wisconsin, Madison*
Tracy L. Benning, *University of San Francisco*
Kimberly A. Bjorgo-Thorne, *West Virginia Wesleyan College*
Elena Cainas, *Broward College*
Kelly Cartwright, *College of Lake County*
Mary Kay Cassani, *Florida Gulf Coast University*
Michelle Cawthorn, *Georgia Southern University*
Mark Coykendall, *College of Lake County*
JodyLee Estrada Duek, *Pima Community College, Desert Vista*
Rachel Goodman, *Hampden-Sydney College*
Jason Janke, *Metropolitan State College of Denver*
Catherine Kleier, *Regis University*
Charles Knight, *California Polytechnic State University, San Luis Obispo*
Janet Kotash, *Moraine Valley Community College*
Jean Kowal, *University of Wisconsin, Whitewater*
Nilo Marin, *Broward College*
Edward Mondor, *Georgia Southern University*
Brian Mooney, *Johnson & Wales University, North Carolina*
Barry Perlmutter, *College of Southern Nevada*
Matthew Rowe, *Sam Houston State University*
Shamili Sandiford, *College of DuPage*
Ryan Tainsh, *Johnson & Wales University*
Michelle Tremblay, *Georgia Southern University*
Kelly Watson, *Eastern Kentucky University*

Comparative Reviewers

Buffany DeBoer, *Wayne State College*
Dani DuCharme, *Waubonsee Community College*
James Eames, *Loyola University Chicago*
Bob East, *Washington & Jefferson College*
Matthew Eick, *Virginia Polytechnic Institute and State University*
Kevin Glaeske, *Wisconsin Lutheran College*
Rachel Goodman, *Hamdpen-Sydney College*
Melissa Hobbs, *Williams Baptist College*
David Hoferer, *Judson University*
Paul Klerks, *University of Louisiana at Lafayette*
Troy Ladine, *East Texas Baptist University*
Jennifer Latimer, *Indiana State University*
Kurt Leuschner, *College of the Desert*
Quent Lupton, *Craven Community College*
Jay Mager, *Ohio Northern University*
Steven Manis, *Mississippi Gulf Coast Community College*
Nancy Mann, *Cuesta College*
Heidi Marcum, *Baylor University*
John McCarty, *University of Nebraska at Omaha*
Chris Poulsen, *University of Michigan*
Mary Puglia, *Central Arizona College*
Michael Tarrant, *University of Georgia*
Melissa Terlecki, *Cabrini College*
Jody Terrell, *Texas Woman's University*

Class Test Participants

Mary Kay Cassani, *Florida Gulf Coast College*
Ron Cisar, *Iowa Western Community College*
Reggie Cobb, *Nash Community College*
Randi Darling, *Westfield State University*
JodyLee Estrada Duek, *Pima Community College, Desert Vista*
Catherine Hurlbut, *Florida State College at Jacksonville*
James Hutcherson, *Blue Ridge Community College*
Janet Kotash, *Moraine Valley Community College*
Offiong Mkpong, *Palm Beach State College*
Edward Mondor, *Georgia Southern University*
Anthony Overton, *East Carolina University*
Shamilli Sandiford, *College of DuPage*
Keith Summerville, *Drake University*

ENVIRONMENTAL SCIENCE
FOR A CHANGING WORLD

Danita Delimont/Gallo Images/Getty Images

CHAPTER 1

Introduction to Environmental, Science, and Information Literacy

Informed citizens have an understanding of how the environment works, how science is used to acquire that understanding, and how they can find and evaluate information. These tools can help us identify and pursue sustainable actions that benefit the environment and society.

Module 1.1

Environmental Literacy and Sustainability

An introduction to the scope and focus of environmental science and what it means to live sustainably

Module 1.2

The Process of Science

A primer on the scientific method and how it is used to investigate environmental science questions and an introduction to the critical evaluation of information

NicoElNino/Shutterstock

Module 1.1 Environmental Literacy and Sustainability

LESSONS FROM A VANISHED SOCIETY

What can we learn about sustainability from a vanished Viking society?

The remains of Hvalsey, a Viking settlement church, in southern Greenland. REDA&CO/Universal Images Group/Getty Images

After reading this module, you should be able to answer these GUIDING QUESTIONS

1. What are the purpose and scope of environmental science?
2. Why are both empirical and applied approaches useful in environmental science?
3. What characteristics make an environmental dilemma a "wicked problem"?
4. What does it mean to be sustainable?
5. Why do scientists think we are living in a new geologic epoch, the Anthropocene, and how might this move us beyond planetary boundaries we should not pass?
6. What are the characteristics of a sustainable ecosystem?
7. What can human societies and individuals do to pursue sustainability?
8. What challenges does humanity face in dealing with environmental issues?
9. Distinguish among anthropocentric, biocentric, and ecocentric worldviews.

Although not a tourist hotspot, Greenland offers some spectacular sights—colossal ice sheets, a lively seascape, rare and impressive wildlife (whales, seals, polar bears, eagles). But on his umpteenth trip to the island, Thomas McGovern was not interested in any of that. What he wanted to see was the garbage—specifically, the garbage that Viking settlers had left behind some seven centuries before.

McGovern, an archaeologist at the City University of New York, had been on countless expeditions to Greenland over the preceding 40 years. Digging through layers of peat and permafrost, he and his team had unearthed a museum's worth of artifacts that, when pieced together, were beginning to tell the story of the Greenland Vikings. But as thorough as their expeditions had been, that story was still maddeningly incomplete.

Here's what they knew so far: A thousand or so years ago, Norse settlers arrived in Greenland, possibly in search of new walrus populations to support the ivory trade. Ivory was a lucrative commodity, and walrus populations had been overhunted from outposts in Iceland, another colony of Norway. An infamous Norse Viking by the name of Erik the Red, exiled from Iceland as a sentence for a murder conviction, supposedly led a small group of followers to a vast expanse of snow and ice that he had dubbed Greenland—a name, legend says, he felt would attract more settlers.

Most of Greenland was not, in fact, green. It was a forbidding place marked by harsh winds and sparse vegetation. But tucked between two fjords along the southwestern coast, protected from the elements by jagged, imposing cliffs, the Norse settlers found a string of verdant meadows brimming with wildflowers. They set up camp here and proceeded to build a society similar to the one they had left behind in Iceland and their ancestral home of Norway. They farmed, hunted, and raised livestock. They built barns and churches as elaborate as the ones back home. They established an economy and a legal system, traded goods with mainland Europe, and, at their peak, reached a population of 3,000.

And then, after 450 years of prosperity, they disappeared, leaving little more than the beautiful, tragic ruins of a handful of barns and churches in their wake.

◉ WHERE IS THE VIKING SETTLEMENT IN GREENLAND?

Researchers suspected that disturbances in the natural environment—a cooling climate, loss of soil, problems with the food supply—may have been the deciding factors in their disappearance. While other researchers probed ice sheets and soil deposits in search of clues, McGovern stuck to the garbage heaps, or *middens*, as Vikings called them. Every farmstead had one, and every generation of the farmstead's owners threw their waste into it. The result was an archaeological jackpot: fine-grain details about what people ate, how they dressed, and the kinds of objects with which they filled their homes. It gave McGovern and his team a clear picture of how they lived.

If they dug deep enough, McGovern thought, it might also explain how they died.

1 ENVIRONMENTAL SCIENCE

Key Concept 1: Environmental science draws on science and nonscience disciplines to understand the natural world and our relationship to it.

From a modern developed society like the United States, it can be difficult to imagine a time and place when the natural world held such sway over our fate. Our food comes from a grocery store; our water, from a tap. Even our air is artificially heated and cooled to our liking. These days, it seems more logical to consider societal conflict, or even collapse, through the lens of politics or economics. But as we will see time and again throughout this book, the natural environment—and how we interact with it—plays a leading role in the sagas that shape human history; this is as true today as it was in the time of the Vikings.

environment The biological and physical surroundings in which any given living organism exists.

environmental science An interdisciplinary field of research that draws on the natural and social sciences and the humanities in order to understand the natural world and our relationship to it.

Environment is a broad term that describes the surroundings or conditions (including living and nonliving components) in which any given organism exists. **Environmental science**—a field of research that is used to understand the natural world and our relationship to it—is extremely interdisciplinary. It relies on a range of natural sciences (such as ecology, geology, chemistry, and engineering) to unlock the mystery of the natural world and to look at the role and impact of humans in the world. It also draws on social sciences (such as anthropology, psychology, and economics) and the humanities (such as art, literature, and music) to understand the ways that humans interact with, and thus impact and are impacted by, the ecosystems around them. **INFOGRAPHIC 1**

INFOGRAPHIC 1 ENVIRONMENTAL SCIENCE IS HIGHLY INTERDISCIPLINARY

Environmental science studies the natural world and how humans interact with and impact it. We must look to the natural and social sciences as well as to the humanities to help us understand our world and effectively address environmental issues and environmental questions such as "Why did the Vikings disappear from this region in Greenland, and how do humans live now in such a harsh environment?"

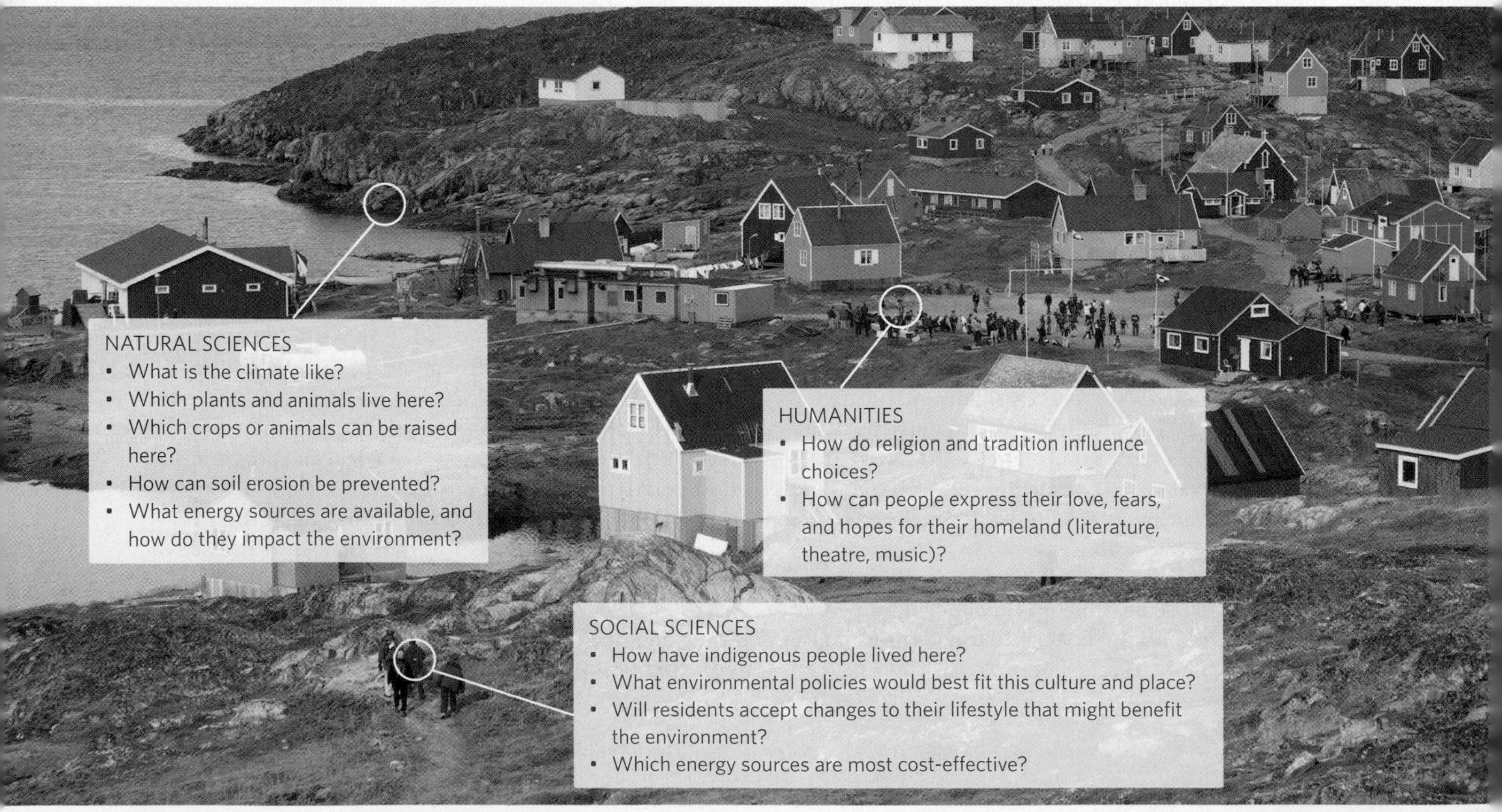

NATURAL SCIENCES
• What is the climate like?
• Which plants and animals live here?
• Which crops or animals can be raised here?
• How can soil erosion be prevented?
• What energy sources are available, and how do they impact the environment?
HUMANITIES
• How do religion and tradition influence choices?
• How can people express their love, fears, and hopes for their homeland (literature, theatre, music)?
SOCIAL SCIENCES
• How have indigenous people lived here?
• What environmental policies would best fit this culture and place?
• Will residents accept changes to their lifestyle that might benefit the environment?
• Which energy sources are most cost-effective?

Pete Ryan/National Geographic Creative

? How would you use your particular college major to help address an environmental problem?

2 EMPIRICAL AND APPLIED SCIENCE

Key Concept 2: Empirical investigations provide information about the natural world; applied science focuses on the practical application of scientific knowledge.

Environmental science is an **empirical science**: It scientifically investigates the natural world through systematic observation and experimentation. The studies of Viking middens and ice cores are examples of empirical science seeking to understand Greenland and its history. Today, soil studies might be done to reveal the composition and water-retention abilities of Greenland's soil. Environmental science is also an **applied science**: We use its findings to inform our actions and, in the best cases, to bring about positive change. For example, an understanding of the suitability of soil for agriculture could help farmers make choices about which crops to plant. **INFOGRAPHIC 2**

Early on, researchers thought that the Greenland colony's demise was largely due to the choices the settlers made—poor land management and rigid conservatism kept them from changing their ways. But like any good scientist, McGovern and other researchers continued to follow the data; recently clues have begun to surface that paint a different picture.

Greenland's interior is covered by ancient, vast ice sheets that stretch toward the horizon. To climate scientists these expanses of ice are a treasure trove. As snow falls, it absorbs particles from the atmosphere. As time passes, the snow and particles compact into ice, freezing in time perfect samples of the atmosphere as it existed when that snow first fell. By analyzing those ice-trapped particles, scientists can get a pretty good idea of what was happening to the climate at any given time. "It's like perfectly preserved slices of atmosphere from the past," says Lisa Barlow, a geologist and climate researcher at the University of Colorado at Boulder.

empirical science A scientific approach that investigates the natural world through systematic observation and experimentation.

applied science Research whose findings are used to help solve practical problems.

INFOGRAPHIC 2 DIFFERENT APPROACHES TO SCIENCE HAVE DIFFERENT GOALS AND OUTCOMES

Environmental science is used to systematically collect and analyze data to draw conclusions and use these conclusions to propose reasonable courses of action.

EMPIRICAL SCIENCE USED TO INVESTIGATE THE NATURAL WORLD

Ashley Cooper/Cavan Images

Through observation and experimentation (such as applying a dye to a glacier and measuring the distance it travels over time), glaciologists study the speed at which glaciers flow to shed light on glacier meltrates.

IN APPLIED SCIENCE, KNOWLEDGE IS USED TO ADDRESS PROBLEMS OR NEEDS

Verkis Consulting Engineers

Engineers use their understanding of flowing water's potential energy to harness its power; glacial meltwater can be diverted to produce hydroelectric power.

❓ If we don't have an application in mind for an empirical research topic, is it worth pursuing?

Ted Scambos & Rob Bauer, NSIDC

Scientist from the National Snow and Ice Data Center at the University of Colorado at Boulder working with an ice core drill.

To uncover those clues, a team of scientists drilled from the surface all the way down to the bedrock below and extracted a 10,000-foot-long cylinder of ice. Analysis of thin sequential segments of the ice core showed that when the Vikings first arrived in Greenland, the temperature had been higher than the average over the preceding 1,000 years. By the time the Vikings had vanished, temperatures had lowered so much that scientists call the period the Little Ice Age—a time when all the seasons were cooler than normal and winters were exceptionally cold.

In addition, analyses of mud cores taken from lake beds around the Viking settlements—samples that contain large amounts of soil that was blown into the lakes during Viking times—indicate that soil erosion had become a significant problem long before the region descended into a mini ice age. "This wasn't a climate problem," says Bent Fredskild, a Danish scientist who extracted and studied many of the mud cores. "This was self-inflicted. It happened the same way that soil erosion happens today—they overgrazed the land, and once it was denuded, there was nothing to anchor the soil in place. So the wind carried it away."

Overgrazing wasn't their only mistake. The Greenland Vikings also used 6-foot-thick slabs of grassland sod to insulate their houses; a typical home took about 10 acres of grassland to insulate. On top of that, they chopped down the birch forests, harvesting timber to provide fuel and build houses. Greenland's ecosystem was far too fragile to endure such pressures. The short, cool growing season meant that plants developed slowly, which in turn meant that the land could not recover quickly enough from the various assaults to protect the soil.

The Greenland settlers responded by changing some of their ways to better fit their new and, it turns out, changing environment. They irrigated fields and fertilized with manure. They depended less on livestock for food, whose numbers were dropping, and more on seals. But this wasn't easy. Hunting these animals was dangerous and required teamwork. All members of the community worked together to support these seal hunts, harvesting enough food to get them through the hard winters. But, it turns out, the path that they chose was only useful in the short term and may have prevented them from adapting further when things got even harder.

3 ENVIRONMENTAL ISSUES AS "WICKED PROBLEMS"

Key Concept 3: Environmental problems are difficult to solve because there are multiple causes and consequences, different stakeholders prefer different solutions, and potential solutions come with trade-offs.

The ability to understand environmental problems is referred to as **environmental literacy**. Such literacy is crucial to helping us become better stewards of Earth. This is especially important because many environmental problems are multifaceted and hard to solve. Scientists refer to them as *wicked problems*. Wicked problems can be extremely complex because they tend to have multiple causes, each one difficult to address. There are also multiple consequences, requiring that we fight a battle against the wicked problem on several fronts.

environmental literacy A basic understanding of how ecosystems function and of the impact of our choices on the environment.

Things are further complicated by the fact that different groups of people affected by the issue (stakeholders)

INFOGRAPHIC 3 WICKED PROBLEMS

Wicked problems are difficult to address because they have many causes and many consequences; and in many cases, each stakeholder hopes for a different solution. Solutions that address wicked problems usually involve trade-offs, so there is no clear "winner." One example of a wicked problem is climate change. Many causes exist for the current climate change we are experiencing, some of which are shown below. There are multiple consequences of climate change; these effects will be varied for different species and people, depending on where they live and their ability to adapt to the changes. While there are many solutions that can help address climate change, each brings new problems that must also be addressed.

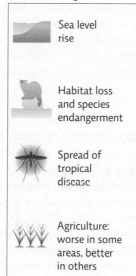

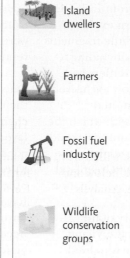

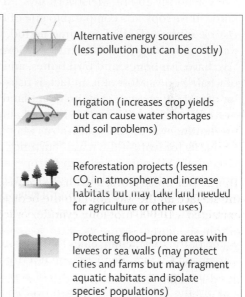

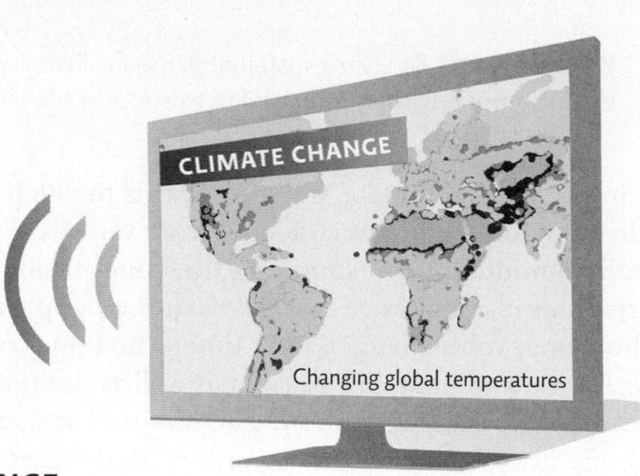

CLIMATE CHANGE

MULTIPLE CAUSES:	MULTIPLE CONSEQUENCES:	STAKEHOLDER DIFFERENCES:	SOLUTIONS COME WITH TRADE-OFFS:
Burning fossil fuels	Sea level rise	Island dwellers	Alternative energy sources (less pollution but can be costly)
Deforestation	Habitat loss and species endangerment	Farmers	Irrigation (increases crop yields but can cause water shortages and soil problems)
Methane from agriculture	Spread of tropical disease	Fossil fuel industry	Reforestation projects (lessen CO_2 in atmosphere and increase habitats but may take land needed for agriculture or other uses)
Overconsumption by modern society	Agriculture: worse in some areas, better in others	Wildlife conservation groups	Protecting flood-prone areas with levees or sea walls (may protect cities and farms but may fragment aquatic habitats and isolate species' populations)

? What are some other environmental "wicked problems" we face?

may prefer a different solution. For example, a citizen of a landlocked state or nation may feel the sea level rise expected with climate change is a minor problem, whereas a coastal resident may feel threatened by the sea level rise already occurring. These two individuals might differ in what they feel is a reasonable response to climate change.

Water and air pollution, the loss of biodiversity (the variety of species present), climate change, how to feed an ever-growing world: These are all crucial wicked problems that must be addressed. But because of their complexity, any given response to an environmental problem involves significant **trade-offs**. No single response is likely to present the ultimate solution, and each potential solution may create new problems that must be solved. In addition, when confronting wicked problems, we must consider not only their environmental but also their economic and social causes and consequences. Scientists refer to this trifecta as the **triple bottom line**: Solutions must be good for the environment, good for society, and affordable. **INFOGRAPHIC 3**

trade-offs The imperfect and sometimes problematic responses that we must at times choose between when addressing complex problems.

triple bottom line The combination of the environmental, social, and economic impacts of our choices.

4 SUSTAINABLE DEVELOPMENT

Key Concept 4: Living sustainably means living within the means of one's environment in a way that does not diminish the environment's ability to support life in the future.

Back in his Manhattan lab, McGovern sorts through hundreds of animal bones collected from various Greenland middens. By examining the bones from different layers, McGovern can tell what the people ate and how their diets changed over time. The bones of cattle, sheep, and goats are present as well as the remains of local wildlife such as caribou. The bones of seal and walrus are also present. As time went on, seal became an increasingly important part of the Greenlanders' diet.

Conspicuously absent, McGovern says, are fish of any kind. "If we look at a comparable pile of bones from [Norse settlers of] the same time period, from Iceland, we see something very different," McGovern explains. "We have fish bones, and bird bones, and little fragments of whale bones. Most of it, in fact, is fish—including a lot of cod." It appears that the Greenland settlers had moved away from fishing and focused almost exclusively on hunting walrus and seal. Walrus were hunted primarily for ivory—the main "currency" with which the Greenland Vikings traded with Europe—and seal could provide a larder of food that would last throughout the winter. However, this choice might have put Greenland colony on a path to decline and eventual extermination.

Some of the most telling clues to the mystery of the Greenland Vikings' demise come not from the Viking colonies, but from another group of people who lived nearby: the Inuit. Living in harsh northern climates has never been easy. The inhabitants of any region must learn how to "live in place"—learn to access and use available resources and to do so within the means of the environment. Sociologists call this *traditional ecological knowledge.* The Inuit had inhabited arctic regions such as this for millennia, building, over generations, a fine-tuned understanding of their environment and the animals on which they depended. They learned the seasonal cycles and exploited resources during seasons of plenty. The Inuit were expert hunters of ringed seal; they knew where, when, and how to capture this difficult-to-catch but very abundant food source. They also knew how to heat and light their homes with seal blubber (instead of firewood).

sustainable development Development that meets present needs without compromising the ability of future generations to do the same.

The Vikings might have survived the Little Ice Age if they had learned from their Inuit neighbors—replaced their own way of hunting and living with Inuit customs that better matched Greenland's environment. But excavations show that very few Inuit artifacts made their way into Viking settlements. The Viking's disdain for the Inuit, whom they called *skraelings* (Norse for "wretches"), likely prevented any meaningful contact or information sharing.

The Norse colonists of Greenland and Iceland brought their own traditional ecological knowledge, honed over millennia in Norway, to their new homes. This knowledge had served the Norse well in Norway, but it didn't always apply to their new environment. At some point in time, settlers in both Greenland and Iceland realized how profoundly different their new lands were from Norway: Growing seasons were shorter; thin soils were much more fragile; and because the land could not rebound quickly, rearing cattle was unsustainable.

Once they saw that their old ways would not work in their new environment, the Icelandic Vikings made changes. They not only switched from beef to fish and favored the hardier goat over sheep but also began conserving their wood and abandoned the highlands, where soil was especially fragile. As a result of these and other adaptations, they survived and prospered. The Icelanders responded to the limitations of their natural environment in a way that allowed them to meet present needs without compromising the ability of future generations to do the same—an approach known today as **sustainable development**.

The United Nations (UN) advocates sustainable development and offers guidance to achieve sustainability worldwide. In 2015, the UN published its *2030 Agenda for Sustainable Development,* presenting 17 goals that seek to protect the planet and ensure health and prosperity for all people. Goals focus on human concerns (e.g., poverty, hunger, equity, and justice) and environmental issues (e.g., resource use, pollution, ecosystem health, and climate change). **INFOGRAPHIC 4**

Like their Icelandic kinsfolk, Greenland settlers made changes: They abandoned the highlands and moved further up the fjords to the most protected areas they could find. They reduced their livestock herds and tried to improve grassland productivity by fertilizing pastures with animal manure—an action that worked in Norway, but in Greenland just made matters worse. The manure trapped water—this kept the ground frozen longer, essentially shortening an already short growing season.

INFOGRAPHIC 4 THE UNITED NATIONS' SUSTAINABLE DEVELOPMENT GOALS

In 2015, the United Nations released its 2030 Sustainable Development Agenda and put forth 17 Sustainable Development Goals in the pursuit of its overarching goal to "end poverty, protect the planet, and ensure prosperity for all people." Specific 15-year targets are set for each goal.

Goals:
Eliminate poverty by 2030; implement social protection systems for everyone and create sound regional, national, and international policies that are "pro-poor and gender-sensitive."

Goals:
End hunger and all forms of malnutrition; double the agricultural productivity of small-scale farmers, the main source of food and income for many of the world's poor and hungry.

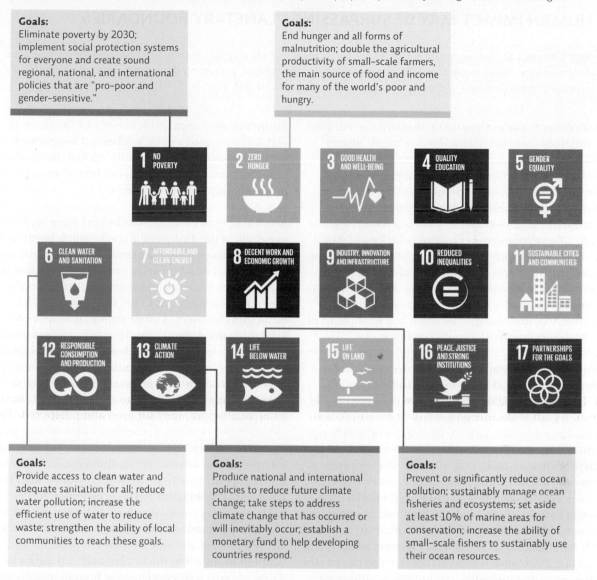

Goals:
Provide access to clean water and adequate sanitation for all; reduce water pollution; increase the efficient use of water to reduce waste; strengthen the ability of local communities to reach these goals.

Goals:
Produce national and international policies to reduce future climate change; take steps to address climate change that has occurred or will inevitably occur; establish a monetary fund to help developing countries respond.

Goals:
Prevent or significantly reduce ocean pollution; sustainably manage ocean fisheries and ecosystems; set aside at least 10% of marine areas for conservation; increase the ability of small-scale fishers to sustainably use their ocean resources.

Which of the sustainability goals above do you feel will be the hardest to reach? What kinds of things can be done to help achieve that goal?

As farm productivity declined, the Greenlanders depended more and more on seal for food—an enterprise made less efficient by their move further from the open ocean. These dangerous hunts required most of the men (and boats) in the settlement for a large part of the summer. One bad storm or attack by hostile Inuit could seriously deplete the number of able-bodied men in the settlement, and time saw their numbers slowly erode. At a time when the Little Ice Age was making life exceedingly difficult and agriculture all but impossible, the Greenlanders may have become so specialized as walrus and seal hunters that they lacked the skills or equipment needed to pursue fishing on a large scale (as their Icelandic kinfolk did) or to adopt the more appropriate and less dangerous Inuit hunting methods. Changing course may have proved impossible.

Adding to their plight, they soon found themselves increasingly isolated. Europe was dealing with the Little Ice Age too and found little incentive to attempt the dangerous journey to Greenland—a colony now with little to offer in needed trade goods. As time wore on, ships that had visited every year came less and less often. After a while, they did not come at all. For the Greenland Vikings, who depended on the Europeans for iron, timber, and other essential supplies, this loss proved devastating.

5 HUMAN IMPACT MAY BE SURPASSING PLANETARY BOUNDARIES

Key Concept 5: Human impact on Earth may be so great that we may be ushering in a new geological epoch: the Anthropocene. These impacts may be approaching or, in some cases, exceeding the boundaries that delineate safe operating zones for some of the physical and biological processes that allow Earth to support life as we know it.

When it comes to the environment, modern societies are not as different from the Vikings as one might assume. The Vikings initially tried to use farming methods that were ill-suited to Greenland's climate and natural environment. We, too, use farming practices that strip away topsoil and diminish the land's fertility. We have overharvested our forests and in so doing have triggered a cascade of environmental consequences: loss of vital habitat and biodiversity, soil erosion, and water pollution. We have overfished and overhunted and have allowed invasive species to devastate ecosystems. And whereas the Greenlanders showed signs of trying to adjust to their environment amidst changes, we seem less inclined to do so as these environmental assaults continue.

The Iceland Vikings learned to live in their new home, adjusting their way of life and society itself to meet the challenges they faced. What about modern humans? Do we have the ecological knowledge needed to live sustainably? Many of our own problems stem from a disconnect in our understanding of the relationship between our actions and their environmental consequences. For example, many people in the United States don't realize that entire mountains are being leveled to produce their electricity, destroying thousands of acres of habitat and polluting the air and water of local residents—all to access thin coal seams beneath the Appalachian Mountains. Likewise, we are often ignorant of the connection between the burning of that coal and mercury-contaminated fish or increased asthma rates.

We also face a suite of new problems that did not trouble the Vikings. A global population poised to top 11 billion by 2100 will strain Earth's resources like never before. Air and water pollution threaten all species, and climate change, caused largely by our use of greenhouse gas–emitting fossil fuels, endangers ecosystems and human communities worldwide.

In fact, the impact of humans on Earth is so great that in 2000, scientists Paul Crutzen and Eugene Stoermer proposed that we have entered a new geologic time interval, suggesting the name **Anthropocene** (the age of humans). They argued that distinctive geological evidence of our existence is accumulating and will be left behind long after we are gone—evidence such as the global distribution of radioactive elements from nuclear weapons or the profusion of plastic pollution.

Anthropocene The newly proposed geologic epoch that is marked by modern human impact.

According to the current Geological Time Scale, we are presently in the Holocene Epoch (which began at the end of the last ice age, almost 12,000 years ago), but in 2019, a panel of scientists commissioned to examine the evidence formally endorsed the Anthropocene as a new epoch designation to the International Union of Geological Sciences. The group will now move forward to identify the defining geologic marker present in Earth's rock layers that will designate the transition from the Holocene to the Anthropocene Epoch. This transition to a new epoch has significance, not so much for which marker is chosen to mark the changeover, but for what it might mean for our future. **INFOGRAPHIC 5A**

For all of human history, humanity has enjoyed the relatively stable planetary conditions and climate of the Holocene Epoch. Johan Rockström of Stockholm University's Stockholm Resilience Center argues that human actions are so altering Earth's environment on a global scale that they threaten the stability of the Earth systems that we and other species depend on and are adapted to. In 2009, Rockström and colleagues proposed a new approach to evaluate how human actions were affecting the ability of the planet to sustainably support humanity—the *Planetary Boundaries framework.*

The Planetary Boundaries framework quantitatively identifies "safe operating zones" for nine important Earth systems or processes on which we depend. When our actions take us past the boundaries that delineate those safe zones, we enter a zone of uncertainty that represents increasing risk of unacceptable environmental change. Beyond that, we enter high-risk territory where "tipping points" might be surpassed from which the ecosystems and Earth processes on which we depend cannot recover.

A 2015 evaluation suggests we have surpassed the safe operating zone for four of the planetary boundaries;

INFOGRAPHIC 5A THE ANTHROPOCENE: A NEW EPOCH

Human impact increased dramatically after the Industrial Revolution and again after World War II. This second surge in resource use and pollution generation, dubbed the *Great Acceleration*, marks the beginning of a new geologic epoch—the Anthropocene. Research is underway to determine which geologic marker will serve as the definitive physical evidence that separates the Holocene (our current epoch designation) from this new geologic time period.

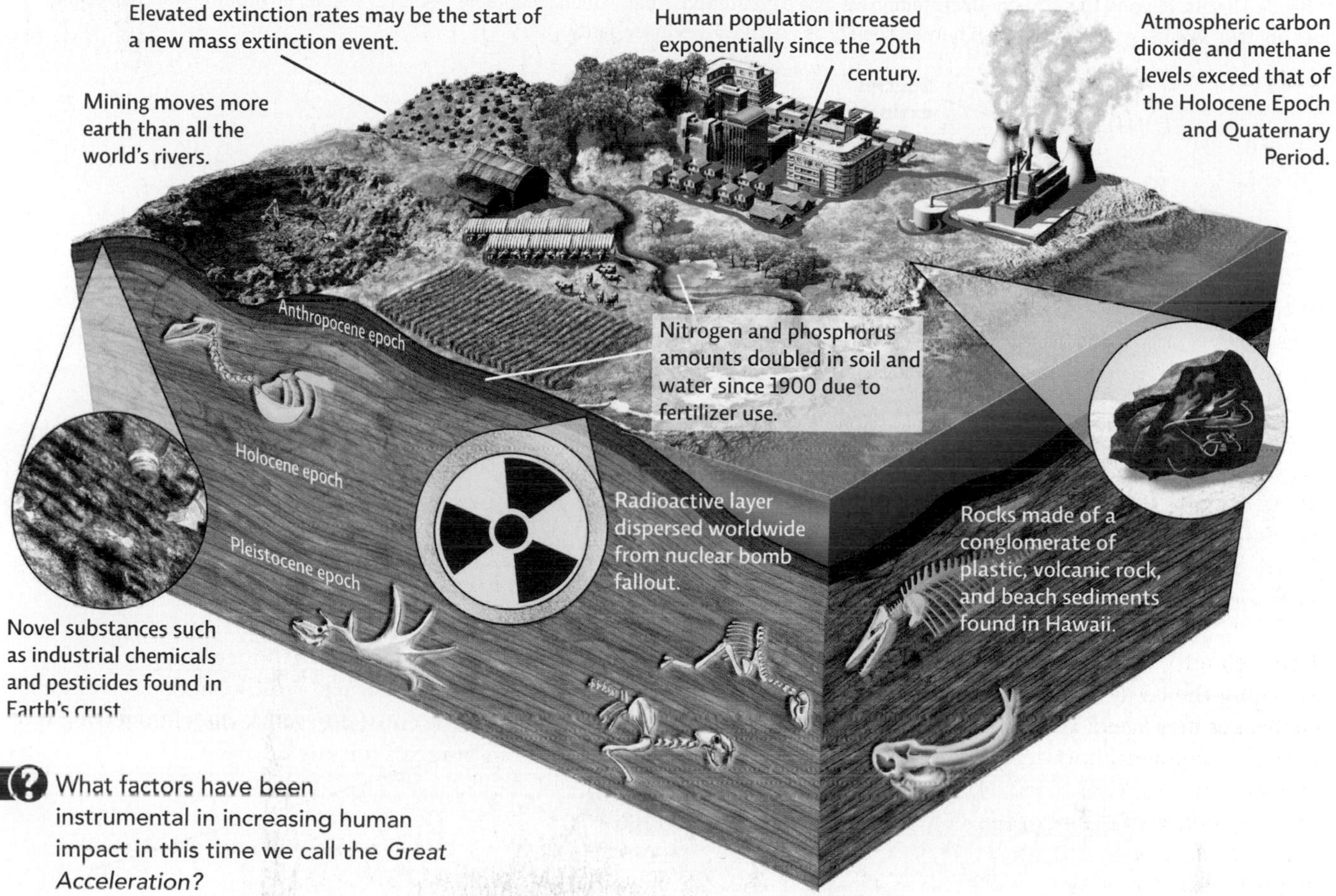

Elevated extinction rates may be the start of a new mass extinction event.
Human population increased exponentially since the 20th century.
Atmospheric carbon dioxide and methane levels exceed that of the Holocene Epoch and Quaternary Period.
Mining moves more earth than all the world's rivers.
Anthropocene epoch
Nitrogen and phosphorus amounts doubled in soil and water since 1900 due to fertilizer use.
Holocene epoch
Radioactive layer dispersed worldwide from nuclear bomb fallout.
Rocks made of a conglomerate of plastic, volcanic rock, and beach sediments found in Hawaii.
Pleistocene epoch
Novel substances such as industrial chemicals and pesticides found in Earth's crust
What factors have been instrumental in increasing human impact in this time we call the *Great Acceleration?*

for two of them, we are already in high-risk territory. And climate change is, and will continue to be, one of the leading threats to sustainability, impacting many of the vital Earth systems on which we depend. Planetary Boundaries research is beginning to influence sustainable development policy, but Rockström's group stresses that the focus for sustainability policy should be on staying within those safe-operating boundaries, not pushing up against the boundaries. **INFOGRAPHIC 5B**

The Millennium Ecosystem Assessment, a UN report that evaluates environmental problems and makes recommendations about addressing those problems, concludes that human actions are straining the ability of the planet's ecosystems to sustain future generations. But there is hope: If we act now, the report's authors write, we can still reverse much of the damage. Some of the best lessons about how we can do this come from the natural environment itself.

jrphoto6/Getty Images

Current climate change is linked to human activities, most notably burning fossil fuels. The effects on ecosystems, species, and human societies are far reaching and will continue to escalate if we don't take steps to address them.

INFOGRAPHIC 5B PLANETARY BOUNDARIES

Based on the assumption that the conditions of the relatively stable Holocene Epoch represent the only planetary state that we know can support modern human society, researchers have developed the Planetary Boundaries framework to identify levels of human impact that keep us within the safe operating zone for vital Earth processes. Values outside these quantitatively determined boundaries (e.g., atmospheric CO_2 levels greater than 350 ppm) put us in a zone of uncertainty as to whether Earth's systems can recover from damage or continue to operate at needed levels. Beyond this zone of uncertainty, we face a greater possibility of unacceptable, and irreversible, environmental change that puts our own wellbeing — indeed, even human society as we know it — in jeopardy.

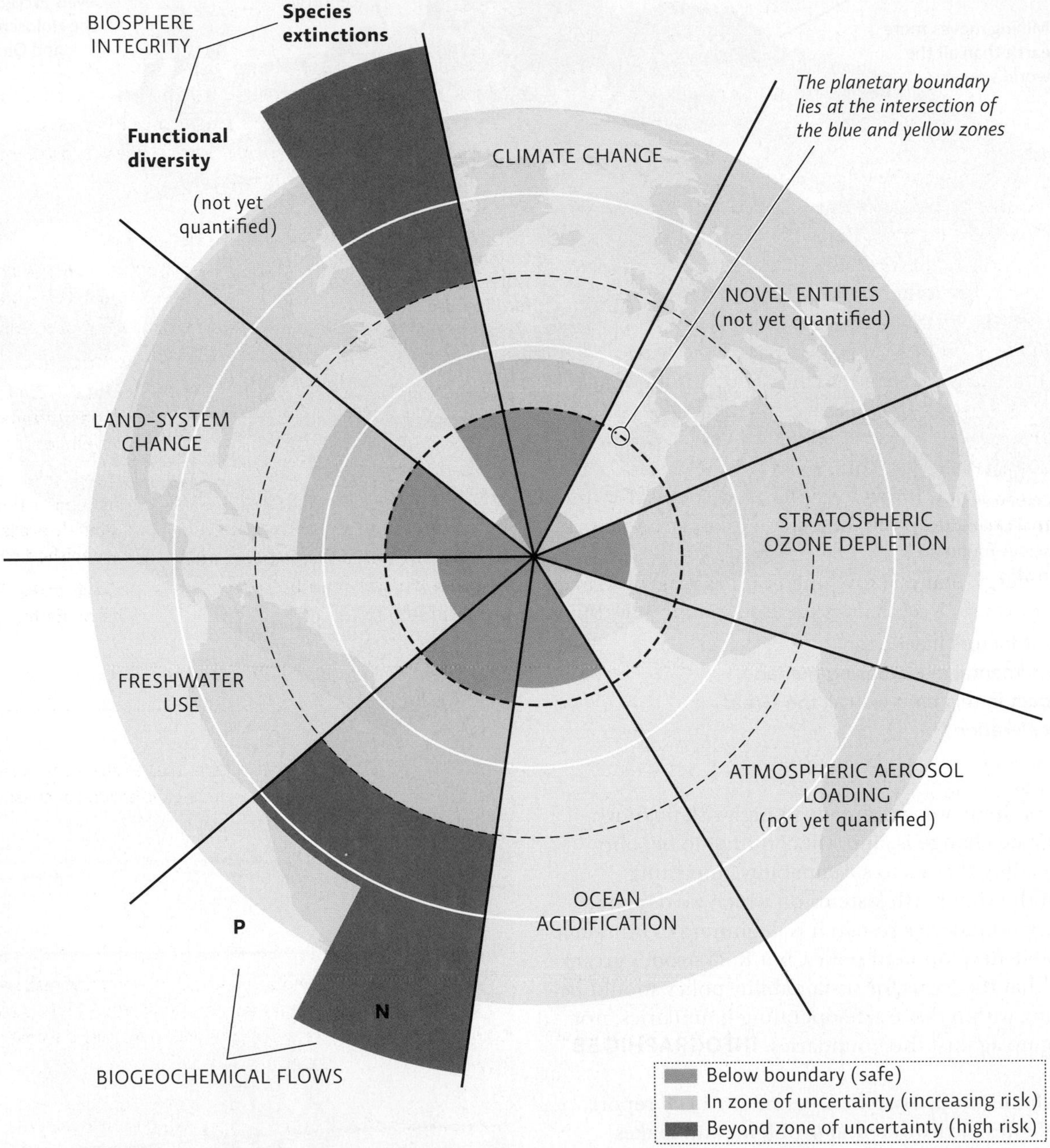

Understanding that there is uncertainty in quantitatively determining the safe operational zones for these parameters, how close to these proposed planetary boundaries do you think we ought to allow ourselves to go in our quest to keep the planet operating in a way that supports us?

6 THE CHARACTERISTICS OF A SUSTAINABLE ECOSYSTEM

Key Concept 6: Natural ecosystems are sustainable because of the way they acquire energy, use matter, control population sizes, and depend on local biodiversity to meet their needs.

Compared to their counterparts in Greenland, the Icelandic Vikings were more successful at responding to their environment as they adopted more **sustainable** practices. They didn't have to look far for a model of sustainability: Natural ecosystems are sustainable. This means they use resources—namely, energy and matter—in a way that ensures that those resources continue to be available.

To survive, organisms need a constant, dependable source of energy. But as energy passes from one part of an ecosystem to the next, the usable amount declines; therefore, new inputs of energy are always needed. A sustainable ecosystem is one that relies on **renewable energy**—energy that comes from an infinitely available or easily replenished source. (In fact, *replenishable energy* may be a better descriptor since energy, once it is used by living things, cannot be renewed and used again.) For the vast majority of ecosystems on Earth, that energy source is the Sun. Photosynthetic organisms such as plants trap solar energy and convert it to a form that they can readily use or that can be passed up the food chain to other organisms.

Unlike energy, matter (anything that has mass and takes up space) can be recycled and reused indefinitely; the key is not using it faster than it is recycled. Sustainable ecosystems recycle matter so that the waste from one organism ultimately becomes a resource for another.

It is also important to keep populations in check so that resources are not overused and there is enough food, water, and shelter for all. In sustainable ecosystems, predators, competitors for resources, and disease all provide population control.

Finally, sustainable ecosystems depend on local **biodiversity** to perform many of the jobs just mentioned. Different species aid in population control; others have different ways of trapping and using energy and matter that boost productivity and efficiency (see Module 2.1). Higher biodiversity (greater number of species and more variation between individuals within a species) generally means more energy can be captured, more matter can be recycled and at a faster pace, and population sizes can be better controlled. **INFOGRAPHIC 6**

sustainable Capable of being continued indefinitely.

renewable energy Energy that comes from an infinitely available or easily replenished source.

biodiversity The variety of life on Earth.

INFOGRAPHIC 6 FOUR CHARACTERISTICS OF A SUSTAINABLE ECOSYSTEM

Ecosystems found on Earth today have the capacity to be naturally sustainable—those that did not die out long ago. They all share characteristics that allow the capture of energy and use of matter in a way that allows them to persist over time, all without degrading the environment itself.

SUSTAINABLE ECOSYSTEMS

RELY ON RENEWABLE ENERGY

Ecosystems must rely on sources of energy that are replenished daily because energy that is used by one organism is "used up" and cannot be used by another; energy is NOT recycled. This means new inputs are constantly needed.

RECYCLE MATTER

No new matter arrives on Earth, so ecosystems must make do with what they have. Fortunately, matter can be recycled, and organisms in ecosystems use matter resources over and over again; the waste of one becomes resource for the next.

HAVE POPULATION CONTROL

The sizes of the various populations in an ecosystem are kept in check by disease, predators, and competitors. This prevents a population from getting too large and damaging the ecosystem it, and others, depends on.

DEPEND ON LOCAL BIODIVERSITY

Ecosystems access energy, recycle matter, and control population sizes largely through the actions of their resident species. The more species present, the better these jobs get done.

? Identify an example of each of these sustainability characteristics from the ecosystem where you live.

Thus, natural ecosystems live within their means, and each organism contributes to the ecosystem's overall function. This is not to imply that natural ecosystems are perfect places of total harmony, but those that are sustainable meet all four of these characteristics.

7 NATURE AS A MODEL FOR SUSTAINABLE ACTIONS

Key Concept 7: Human societies can become more sustainable by mimicking the way natural ecosystems operate.

Human ecosystems are another story. Humans tend to rely on **nonrenewable resources**—those whose supply is finite or is not replenished in a timely fashion. The most obvious example of this is our reliance on fossil fuels such as coal, natural gas, and oil to power our society. Fossil fuels are replenished only over vast geologic time—far too slowly to keep pace with our rampant consumption of them. On top of that, we have a hard time keeping our population size under control, despite (or maybe because of) all the advances of modern technology. We generate volumes of waste, much of it toxic, and have yet to fully master the art of recycling. Our actions are decreasing biodiversity at an alarming rate, and we completely miss the point that diversity in general is good—we depend on just a few energy sources, and three crops (rice, wheat, and corn) predominate in our agricultural systems.

nonrenewable resources A resource that is formed more slowly than it is used or that is present in a finite supply.

Increasingly, however, we humans are looking to nature to help us learn how to change our ways. By studying and emulating nature, we learn what is reasonable to do and how to do it. Scientists are applying this knowledge for all kinds of things: for example, to design better products, to find more sustainable methods of growing crops, and to trap and use sustainable energy. **INFOGRAPHIC 7**

Despite such efforts, some concerned scientists and environmentalists say that modern global societies are not acting nearly as quickly as they could or should. What prevents us from changing our ways, even in the face of brewing calamity?

INFOGRAPHIC 7 SUSTAINABLE ECOSYSTEMS CAN BE A USEFUL MODEL FOR HUMAN SOCIETIES

SUSTAINABLE ECOSYSTEMS:	WE CAN MIMIC THIS IN OUR OWN SOCIETIES BY:
RELY ON RENEWABLE ENERGY	**USING SUSTAINABLE ENERGY SOURCES** We can move away from nonrenewable fuels such as fossil fuels by turning to sustainable energy sources such as solar, wind, geothermal, and biomass (harvested at sustainable rates).
RECYCLE MATTER	**USING MATTER CONSERVATIVELY AND SUSTAINABLY** We can reduce our waste by reducing our use of resources and by recovering, reusing, and recycling matter that we do use; we would also benefit from minimizing the toxins we create or release into the environment that degrade our natural resources.
HAVE POPULATION CONTROL	**GETTING HUMAN POPULATION GROWTH UNDER CONTROL** While predation controls many natural populations, there are many ways to reduce human birth rates without increasing the death rate through war or disease.
DEPEND ON LOCAL BIODIVERSITY	**DEPENDING ON LOCAL HUMAN CONTRIBUTIONS AND BIODIVERSITY** Protecting biodiversity will help us achieve the three preceding goals; we can also regard the use of diversity as a metaphor and emulate nature by using a variety of local energy sources, building materials, and crops and by exploring the many ideas and innovations that come from a diverse human community.

? What obstacles stand in society's way in pursuing the actions mentioned in this graphic?

8 CHALLENGES TO SOLVING ENVIRONMENTAL PROBLEMS

Key Concept 8: Impediments to solving environmental problems include short-term thinking and social traps.

Decisions by individuals or groups that seem good at the time and produce a short-term benefit but hurt society in the long run are called **social traps**. The **tragedy of the commons** is a social trap that often emerges when many people are using a commonly held resource, such as water or public land. Each person will act in a way to maximize his or her own benefit, but as everyone does this, the resource becomes overused or damaged. Herders might put more animals on a common pasture because they are driven by the idea that "if I don't use it, someone else will." We do the same thing today as we overharvest forests and oceans and as we release toxins into the air and water.

Other social traps emerge when beneficial actions set into motion events that build over time and create problems later. An example is using a pesticide to kill pests. Initially the pest population is reduced and crop productivity increases, but over time, the individuals that survive the pesticide can give rise to a pest population that is resistant to the pesticide. The pest population can even grow bigger if competing or predatory organisms are also killed by the pesticide. What started out as a beneficial action—reducing a pest population with pesticides—actually made the problem worse later on.

Addressing social traps requires a consideration of the long-term effects of our actions. In some cases, *education* about the consequences of our actions may help, but only if there is cooperation among all users. If the resource can be *privatized*—sold to users so that they have ownership of the resource—the owners may be more likely to protect that resource and use it wisely. If a resource cannot be owned (e.g., the oceans, the atmosphere), then *regulation*—binding laws that dictate how it can be used—may be needed to sustainably manage the resource. **INFOGRAPHIC 8**

Environmental literacy—understanding how ecosystems operate and how our actions impact society and the environment—is our best hope for avoiding such traps. When people are aware of the consequences of their decisions, they can make better choices. Addressing social traps might include policies (regulation) or education about the issue, but any actions should consider long-term costs and benefits, not just short-term ones. For example, Flint, Michigan, is experiencing a serious problem of lead in drinking water. The health problems this is causing are the result of the decision to save money by not investing in appropriate water infrastructure (see Module 6.1). Short-term gains were favored, and now the people of Flint are faced with the long-term consequences of that decision.

social traps Decisions by individuals or groups that seem good at the time and produce a short-term benefit but that hurt society in the long run.

tragedy of the commons The tendency of an individual to abuse commonly held resources in order to maximize his or her own personal interest.

Scott Warren/Cavan Images

A replica of the first church built in Brattahlid, Greenland—a Viking settlement founded by Erik the Red more than 1,000 years ago.

INFOGRAPHIC 8 THE TRAGEDY OF THE COMMONS

Social traps, such as the tragedy of the commons, are decisions that seem good at the time and produce a short-term benefit but that hurt society (usually in the long run).

The tragedy of the commons can occur when resources aren't "owned" by anyone (they are commonly held). Individuals who try to maximize their own benefit can end up harming the resource itself, as illustrated with this example of a "common pasture."

This common resource (pasture) can support four animals sustainably. Each of the four farmers benefits equally.

If one farmer adds two more cows, the commons becomes degraded. All four farmers share in the degradation (less milk produced per cow), but the farmer with three cows gets all the benefit from adding the extra cows.

The other farmers also must add cows to return to the same level of production as before. **Over time, the commons degrades even more, and at some point, it will no longer support any animals.**

Addressing social traps requires "long-term thinking." Here are three approaches that can be used.

EDUCATION

If farmers understand that their long-term success depends on proper land management and cooperation, they may refrain from adding too many cows to the pasture.

PRIVATIZATION

If farmers own their patch of land, they may be less likely to abuse it since they have already invested in it and will lose money if it becomes degraded.

REGULATION

If farmers are required by law to restrict the number of cows they bring to a common pasture, the resource will be protected.

? Why do you suppose humans are so prone to being caught in social traps like the tragedy of the commons?

9 WORLDVIEWS AND ENVIRONMENTAL ETHICS

Key Concept 9: Our worldview reflects how we value the natural world and influences the ethical choices we make regarding the environment.

environmental ethic The personal philosophy that influences how a person interacts with his or her natural environment and thus affects how one responds to environmental problems.

Conflicting worldviews present another challenge to sustainable living. Because our *worldviews*—the windows through which we view our world and existence—are influenced by cultural, religious, and personal experiences, they vary across countries and geographic regions, even within a society. People's worldviews determine their **environmental ethic**, or how they interact with their natural environment and respond to environmental problems. When people or groups with different worldviews approach environmental problems, they often draw different conclusions about how best to proceed.

The Vikings may have had an **anthropocentric worldview**—one where human lives and interests are most important. They may have viewed other species as having *instrumental value*—meaning these species were valued only for their benefit to humans. Forests were a source of timber; grasslands, a source of home insulation and a feeding ground for cattle.

A **biocentric worldview** values all life. From a biocentric standpoint, every organism has an inherent right to exist, regardless of its benefit (or harm) to humans; each organism has *intrinsic value*. This worldview would lead us to be mindful of our choices and avoid actions that indiscriminately harm other organisms or put entire species in danger of extinction.

An **ecocentric worldview** values the ecosystem as an intact whole, including all of the ecosystem's organisms and the nonliving processes that occur within the ecosystem. Considering the same forests and grasslands from an ecocentric worldview, the Vikings might have decided to protect both not only to harvest the resources they could but also to protect the complex processes that could produce those resources only when they remained intact. **INFOGRAPHIC 9**

Andreas Strauss/LOOK-foto/Getty Images

This lovely meadow in the Sassalb mountain range in Switzerland could be valued for many reasons. Some people might value it solely for its usefulness as a site for recreation or harvesting medicinal plants (anthropocentric worldview). Others might feel that every organism living there has a right to exist and would want to keep it intact as a way to protect each one (biocentric worldview). Still others might value the functioning meadow ecosystem as a whole and so would want to protect the living and nonliving components found there for the services they provide (ecocentric worldview).

Back in Greenland, in the silt-covered ruins of a Viking farmhouse dating back several centuries, archaeologists found the bones of a hunting dog and a newborn calf. The knife marks covering both indicate that the animals were butchered and eaten. "It shows how desperate they were," says McGovern. "They would not have eaten a baby calf, or a hunting dog, unless they were starving."

The loss of the Norse settlement on Greenland had once been seen as a failure of the community to adapt to their surroundings—a collapse caused by unfortunate circumstances and bad choices. However, McGovern's research now suggests it was more of a *decline*, which came about due to an inability of the Norse community to "anticipate an unknowable future." Although they had some success adapting to their new surroundings and to the changes that came their way in the first few centuries of their settlement of Greenland—an admirable feat considering the harsh conditions of their environment—in the end, they may have become too specialized to adapt further (most of their time, energy, and technology devoted to seal hunts) and too isolated to survive.

Throughout human history, societies have come and gone. For some, their environments had changed for the worse—sometimes at their own hands—contributing to their disappearance. The initial evidence that suggested the Greenland colony vanished due to poor choices—self-inflicted environmental damage and failure to adjust to a changing environment—made a tidy, cautionary tale about how we should avoid making the same mistakes and seek to live sustainably. But the reality—the Greenlanders did try to make changes and live within the means of their environment but failed—is a more sobering reminder that even if we try to adapt and adjust to a changing environment, there are no guarantees. This suggests we should not take our environment for granted or degrade it in ways that diminish its ability to support us. To do this, we must

anthropocentric worldview A human-centered view that assigns intrinsic value only to humans.

biocentric worldview A life-centered approach that views all life as having intrinsic value, regardless of its usefulness to humans.

ecocentric worldview A system-centered view that values intact ecosystems, not just the individual parts.

INFOGRAPHIC 9 WORLDVIEWS AND ENVIRONMENTAL ETHICS

People's environmental worldviews describe how they see themselves in relation to the world around them. Their worldview influences their environmental ethic, which in turn influences how they interact with the natural world. We present three common worldviews here (there are others).

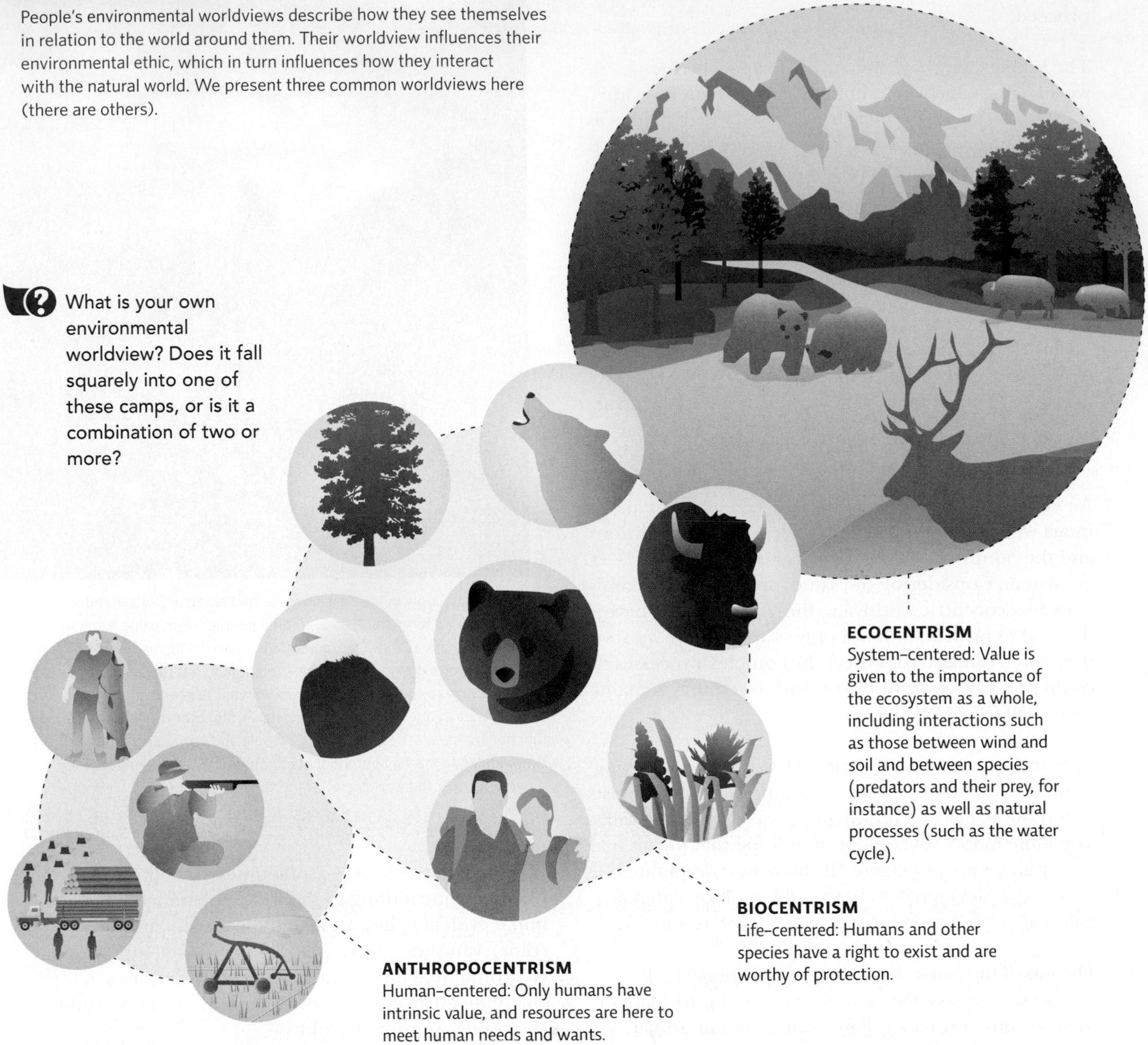
What is your own environmental worldview? Does it fall squarely into one of these camps, or is it a combination of two or more?
ECOCENTRISM
System-centered: Value is given to the importance of the ecosystem as a whole, including interactions such as those between wind and soil and between species (predators and their prey, for instance) as well as natural processes (such as the water cycle).
BIOCENTRISM
Life-centered: Humans and other species have a right to exist and are worthy of protection.
ANTHROPOCENTRISM
Human-centered: Only humans have intrinsic value, and resources are here to meet human needs and wants.

understand our environment—how it works and how our actions affect it. We cannot assume there will always be time "later" to change our ways.

Our own story is being written now in much the same way, but hopefully, we still have enough time to shape our own narrative. Will it be dug up 1,000 years hence from the ruins of what we leave behind? Or will it be passed down by the voices of our successors, who continue to thrive long after we are gone? Ultimately, the answer is up to us.

Select References:

Dugmore, A. J., et al. (2012). Cultural adaptation, compounding vulnerabilities and conjunctures in Norse Greenland. *Proceedings of the National Academy of Sciences, 109*(10), 3658–3663.

Jackson, R., et al. (2018). Disequilibrium, adaptation, and the Norse settlement of Greenland. *Human Ecology, 46*(5), 665–684.

Kintisch, E. (2016). The lost Norse. *Science, 354*(6313), 696–701.

Lewis, S. L., & Maslin, M. A. (2015). Defining the Anthropocene. *Nature, 519*(7542), 171–180.

Rockström, J., et al. (2009). A safe operating space for humanity. *Nature, 461*(7263), 472–475.

GLOBAL CASE STUDIES **SOCIETAL COLLAPSE** Achieve

Greenland colony was not the only once-thriving society to collapse due, at least in part, to environmental problems. Read the online case studies provided with this map for a look at some other notable examples from history.

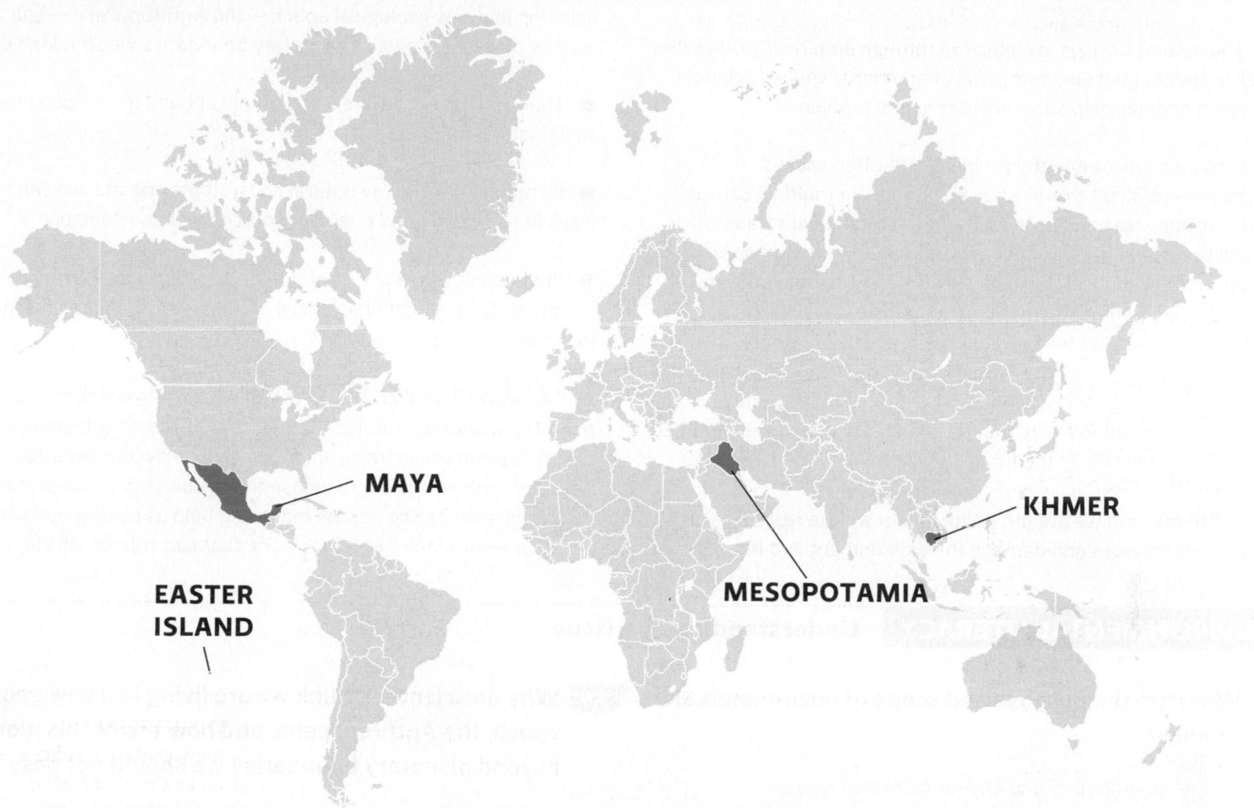

BRING IT HOME

PERSONAL CHOICES THAT HELP

The concept of sustainability unites three main goals: environmental health, economic profitability, and social and economic equity. All sorts of people, philosophies, policies, and practices contribute to these goals; concepts of sustainable living apply on every level, from the individual to the society as a whole. In other words, every one of us participates. Throughout this book, you will have the opportunity to learn about personal actions that can help address environmental issues, but a good starting place is to learn about your own environment and the place you call home.

Individual Steps

- Discover your local environment. What parks or natural areas are close by? Does your campus have any natural areas? Visit one and spend a little time observing nature and your own reactions. Write down your thoughts or share your experiences with a friend.
- Are there restaurants, grocery stores, or other retail venues accessible through public transportation or within walking distance of your campus or home? For a week or two, try walking or riding a bike or bus to these businesses instead of driving to others farther away. Is this a reasonable option for you? Why or why not?
- Discover what's happening in your community. Read newspapers and monitor blogs covering environmental and quality-of-life issues. Alert your local, regional, and national representatives about the issues you care about, and vote for government officials who support the causes you support.

Group Action

- Discover your own interests. There is a group for every interest — from outdoor recreation, wildlife viewing and preservation, and environmental education to transportation and air quality issues. Get involved with organizations working to improve environmental issues or address social change and human rights. One person can make a difference, but a group of people can cause a sea of change.

REVISIT THE CONCEPTS

- Environmental science draws on science and nonscience disciplines to understand the natural world and our relationship to it.
- The natural world is investigated through empirical studies that seek to understand environmental phenomena; applied scientific research findings help solve environmental problems.
- However, environmental problems are often wicked problems — difficult to solve because there are multiple causes and consequences and potential solutions come with trade-offs. In addition, different stakeholders often prefer different solutions. Solutions being considered need to address the triple bottom line — the environmental problem at hand as well as social and economic impacts of the problem or proposed solutions.
- The ultimate goal is to live sustainably — within the means of one's environment in a way that does not diminish the environment's ability to support life in the future.
- But at present, we are not doing that — we use resources at unsustainable rates and damage the environment and its species in ways that diminish nature's ability to replace the resources we take. In fact, humanity's impact on Earth is so great that our actions are ushering in a new geological epoch — the Anthropocene — and we may be pushing up against planetary boundaries we shouldn't cross.
- Thankfully, we have a ready example at hand in our quest to live sustainably — nature itself.
- By mimicking the way natural ecosystems operate, we can find ways to use matter and energy resources more sustainably.
- But it won't be easy. We tend to operate on short-term time frames and get caught up in social traps that cause problems in the long run.
- Our worldview also influences how we view a problem and potential solutions. Shifting our focus to long-term outcomes, taking action to avoid social traps, and recognizing how our personal worldview influences our response to environmental issues may help us understand the choices we make and help us pursue options that keep our environment a vibrant place that can support all life.

ENVIRONMENTAL LITERACY Understanding the Issue

1 What are the purpose and scope of environmental science?

1. Define environment and environmental science.
2. Why is environmental science considered an interdisciplinary field that includes the natural and social sciences and the humanities?

2 Why are both empirical and applied approaches useful in environmental science?

3. Distinguish between empirical and applied science.
4. Some scientists study peatlands to look for clues about how ancient civilizations lived and died. Others work with peatlands to manage these ecosystems as sources of fuel or soil additives (peat moss). Which of the activities above describes empirical science, and which describes applied science? Explain.

3 What characteristics make an environmental dilemma a "wicked problem"?

5. Consider the wicked problem of air pollution. Identify some of the causes and consequences of air pollution. What are some possible solutions, and what are the trade-offs that come with those solutions?
6. What is the triple bottom line, and why does it need to be considered when trying to identify solutions to environmental problems?

4 What does it mean to be sustainable?

7. Why does short-term thinking often lead to actions that are not sustainable? Give an example.
8. Identify three Sustainable Development Goals, and explain how accomplishing each would bring the world population closer to sustainability.

5 Why do scientists think we are living in a new geologic epoch, the Anthropocene, and how might this move us beyond planetary boundaries we should not pass?

9. What are some of the proposed geologic markers in Earth's strata that would distinguish the Anthropocene from the Holocene?
10. Explain the Planetary Boundaries concept, and identify the status of the Earth systems that have been evaluated.

6 What are the characteristics of a sustainable ecosystem?

11. Why is it important that sustainable ecosystems only rely on a renewable energy source such as the Sun?
12. Why must ecosystems recycle and reuse matter?

7 What can human societies and individuals do to pursue sustainability?

13. Identify at least one action we could take as a society to follow each of the four characteristics of a sustainable ecosystem.

8 What challenges does humanity face in dealing with environmental issues?

14. Explain the tragedy of the commons using the example of driving your car — you know the pollution that it releases to the air harms the environment, but you continue to do it anyway.
15. Why do we fall prey to social traps, and how can we avoid them?

9 Distinguish between anthropocentric, biocentric, and ecocentric worldviews.

16. A classmate says that she is anthropocentric because human well-being is her top priority. Explain how she could also be ecocentric.

SCIENCE LITERACY Working with Data

Greenland's climate is a challenging one in which to grow crops. The Norse Vikings first settled Greenland during a warmer-than-average period (similar to today's temperatures), but farming there still would have been challenging. One of the crops that the Vikings grew was barley — a cold-tolerant grain. Barley can grow in a cool climate, but it will stop growing at temperatures below 40°F. The crop needs around 15 inches of rain over the growing season; production will decline with less rain. Use the graphs below that show current climate data for Norsaq, Greenland, to answer the questions that follow.

A: GREENLAND: Average monthly precipitation

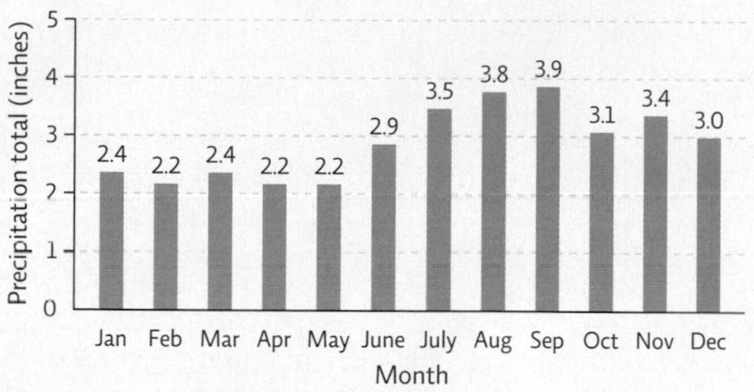

B: GREENLAND: Average monthly temperature

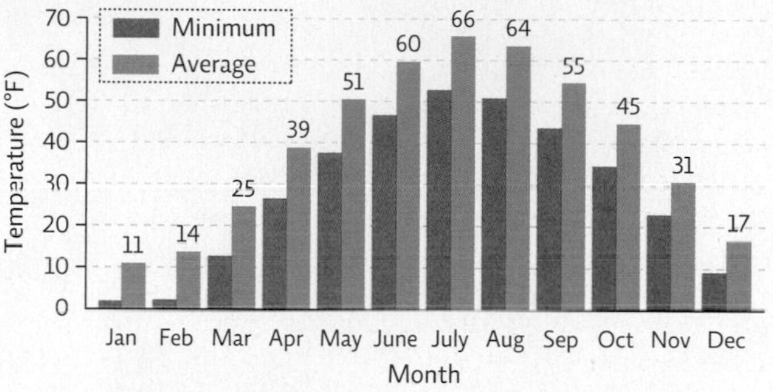

Interpretation

1. Which season is the driest? wettest?
2. Based on the average for each month, what is the total precipitation in a year?
3. For how many months of the year does the average temperature stay below freezing?

Advance Your Thinking

4. Consider barley's temperature needs for growth. In what month could a Greenland farmer plant a barley crop? In what month would it be best to harvest barley?
5. Barley needs around 90 days to mature and be ready for harvest. Is the Greenland growing season long enough to support this growth?
6. Calculate the total precipitation that would fall, on average, during the growing season for barley. Does enough rain fall to support this crop?

INFORMATION LITERACY Evaluating Information

The Lorax, a children's book by Dr. Seuss, tells the story of the Lorax, a fictional character who speaks for the trees against the greedy Once-ler who represents industry. Written in 1971, *The Lorax* was banned in parts of the United States for being an allegorical political commentary. Today the book is used for educating children about environmental concerns (see www.seussville.com/educators/the-lorax-project/). Even so, some people consider the book inappropriate for young children due to its "doom-and-gloom" environmentalism.

The book *The Truax*, by Terri Birkett, follows a forest industry representative offering a logging-friendly perspective to an anthropomorphic tree, known as the Guardbark. This story was criticized for containing skewed arguments and, in particular, a nonchalant attitude toward endangered species. About 400,000 copies of the book have been distributed to elementary schools nationwide.

Read both books. You can find *The Lorax* at your local public library, and you can download *The Truax* as a PDF from http://woodfloors.org/truax.pdf.

Evaluate the stories and work with the information to answer the following questions:

1. What are the credentials of the author of each book? In each case, do the person's credentials make him or her reliable/unreliable as a storyteller? Explain.
2. Connect each story to the key concepts in the module:
   a. What are the underlying attitudes and worldviews of each story?
   b. Does each story reflect social traps and, if so, in what way?
   c. How might each story contribute to environmental literacy? Explain.
   d. What does each story have to say about sustainability? Explain.
3. What supporting evidence can you find for the main message in each story? In the story itself? From doing some research?
4. What is your response to each story? What do you agree and disagree with in each case? Explain.

Additional study questions are available at Achieve.macmillanlearning.com. Achieve

FUNGAL ATTACKER THREATENS BATS

Unraveling the mystery behind bat deaths

Little brown bat (*Myotis lucifugus*) in flight. Edward Kinsman/Getty Images

After reading this module, you should be able to answer these GUIDING QUESTIONS

1. How do scientists study the natural world?
2. What influences the degree of certainty in a scientific explanation?
3. Why are observational and experimental studies needed to investigate the natural world?
4. What is the difference between a correlation and a cause-and-effect relationship, and how can science help policy makers address environmental problems?
5. What is information literacy, and why is it important in environmental science?
6. What is critical thinking, and how can it counter logical fallacies used in arguments?

In the winter of 2007, David Blehert, chief of the Wildlife Disease Diagnostic Labs at the National Wildlife Health Center, received a very troubling phone call from his colleague, Al Hicks. Hicks is a biologist tasked with surveying, among other things, bat populations in New York State. This year, they found a big problem: thousands of bats dead at several different sites and near-total colony collapse at some of them.

But the bats hadn't merely died; they had suffered, badly—their skin was dry and flaking, and their bodies were so emaciated that bones pushed easily through skin. Many of them were found at the cave opening, a sign that they had been roused from hibernation and had been looking for insects to replenish their fat stores. Of course, because it was the dead of winter, there were no insects to be found. "In thirty years of doing this work, he said he had never seen anything like it," Blehert recalls.

Before long, surveyors across the region were reporting similar troubles—including cases where entire colonies had died. Nobody knew what was killing the bats, but because most biologists noted a peculiar white fuzz on the victims' muzzles, the mystery affliction earned the moniker *white-nose syndrome* (WNS).

In the next years, WNS spread rapidly, reaching 37 states and 5 Canadian provinces. It has infected at least seven different species of bat and is driving at least two of them to the edge of extinction. Researchers estimate that more than 6 million bats have died from the disease. For scientists like Blehert, those grim numbers would fuel a race against time to answer two urgent questions: What is causing the disease? And can it be stopped?

WHERE IS WHITE-NOSE SYNDROME FOUND?

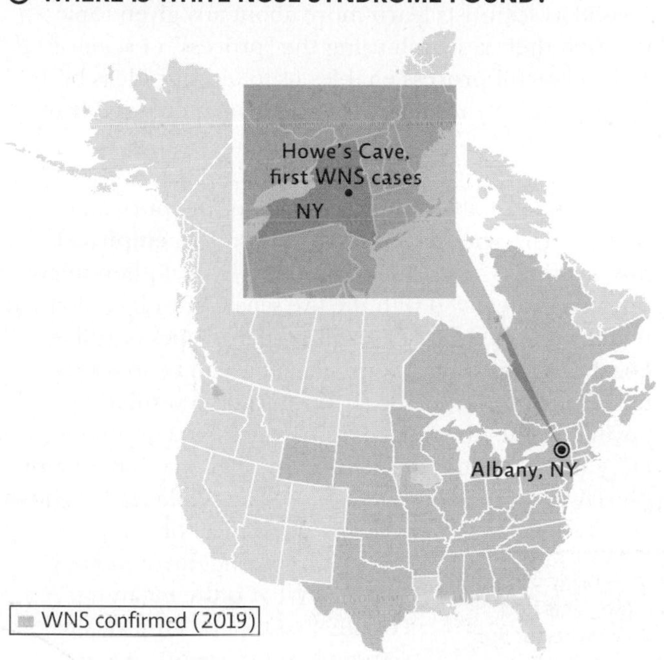

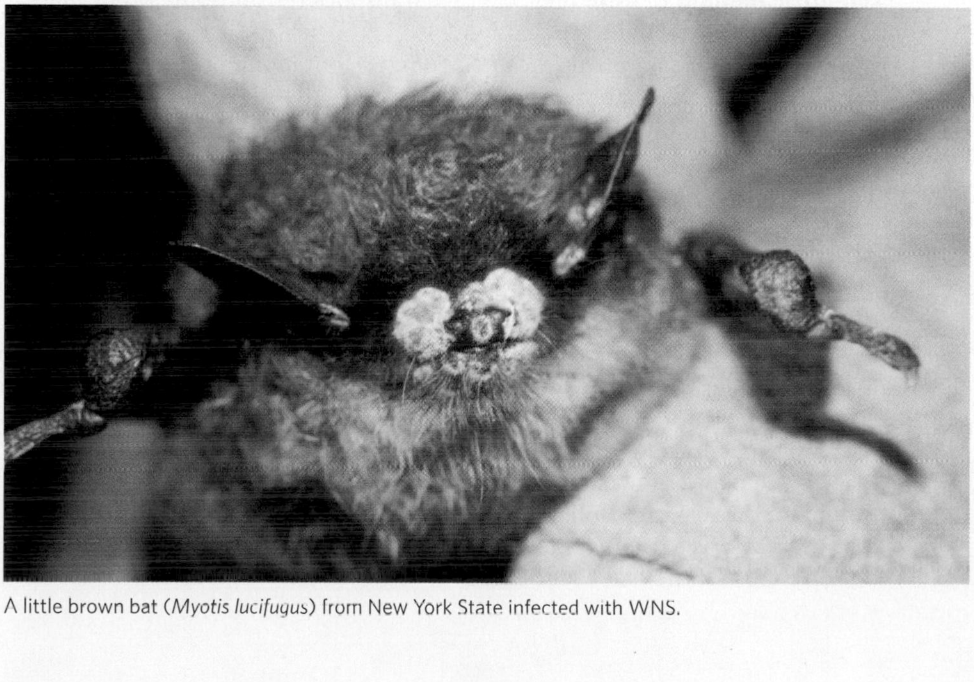

Thomas Deerinck, NCMIR/Science Source

A little brown bat (*Myotis lucifugus*) from New York State infected with WNS.

1 THE NATURE OF SCIENCE AND THE SCIENTIFIC METHOD

Key Concept 1: Scientists investigate the natural world using a transparent method of inquiry, and their findings are evaluated by other scientists.

Science is both a body of knowledge (basic facts and complex explanations) and the process used to get that knowledge. The "body of knowledge" part of science is ever changing; facts, figures, and understandings are revised as scientists learn more about any given topic through their research using the "process" of science. This powerful process enables us to test our ideas by gathering evidence and then to evaluate the quality of that evidence.

Of course, not all questions fall under the purview of science. The scientific method is based on **empirical observations**—information about physical phenomena that can be detected with the five senses or with equipment used to extend those senses (like microscopes or sonar). For that reason, only phenomena that can be *objectively* observed (that is, the observation can be verified by anyone in the same place, using the same equipment, etc.) are fair game for science. Phenomena that are not objectively observable (What is my dog thinking? Do ghosts exist?) and ethical or religious questions (Is the death penalty wrong? What is the meaning of life?) cannot be empirically studied and therefore are not within the purview of science.

science A body of knowledge (facts and explanations) about the natural world and the process used to get that knowledge.

empirical observations Information detected with the senses—or with equipment that extends our senses.

scientific method The procedure scientists use to empirically test a hypothesis.

inferences Conclusions drawn based on observations.

hypothesis A possible explanation for what we have observed that is based on some previous knowledge.

experimental prediction A statement that identifies what is expected to happen if the hypothesis being tested is correct.

Like all good scientists, Blehert and his colleagues applied the **scientific method** to solve the mystery of the dying bats. They worked logically from previous knowledge and observations of the natural world toward new questions and possible answers. They started with observations: thousands of dead bats and a mysterious white fuzz on most of their noses. They combined these observations with existing knowledge to make some **inferences** (explanations of what else might be true): Because bats hibernate in cold, dark, wet places—the kind of places where fungi also thrive—they inferred that the fuzz was a fungus. And because most of the bats that died had it on their noses and bodies, they inferred that it was somehow related to the deaths.

Next, they gathered more evidence to investigate these inferences. Samples of the white fuzz were collected and identified in the lab as a fungus. Scientists eventually came to identify it as *Pseudogymnoascus destructans* (Pd).

"It was an obscure little fungus, that hardly anyone even knew existed," Blehert says. "And it was part of a family that was not known to be pathogenic at all."

Was this fungus causing the bat deaths? If so, where did it come from? To begin answering those questions, researchers studying the problem would need to generate some hypotheses. A **hypothesis** is a proposed answer to the research question—an inference—based on previous knowledge and current observations.

Researchers had no trouble coming up with hypotheses, including these two:

1. The white-nose infection caused by the fungus was secondary and opportunistic; it had always been present but was only able to grow on the bats because their defenses had been weakened by another pathogen.

2. The fungus was new to the region; it had been transported from some other place.

To figure out if one of these ideas was correct, they would have to test each one.

To test a hypothesis, a researcher designs a scientific study and makes an **experimental prediction**, a statement that identifies what is expected to happen if the hypothesis being tested is correct. A scientific study must be designed to be a "fair test"—this means the results of the test could support *or* falsify (prove wrong) the prediction. If evidence repeatedly falsifies an experimental prediction, the hypothesis is rejected and alternative hypotheses can be tested. If a hypothesis is supported, researchers repeat the study to validate the data and generate new predictions that test the same hypothesis from different angles. As supporting evidence mounts from replicate studies and from multiple predictions, we become more confident in our data and conclusions.

INFOGRAPHIC 1 INVESTIGATING SCIENTIFIC QUESTIONS

Scientists work from observations and previous knowledge to ask new questions and pose possible answers (hypotheses) to those questions. They test these hypotheses by generating a testable prediction and designing a study to test that prediction. Evidence is collected, evaluated, and shared, generating new predictions and sometimes new hypotheses, making this an ongoing process that allows researchers to delve deeper and deeper, building on previous studies to gain new knowledge.

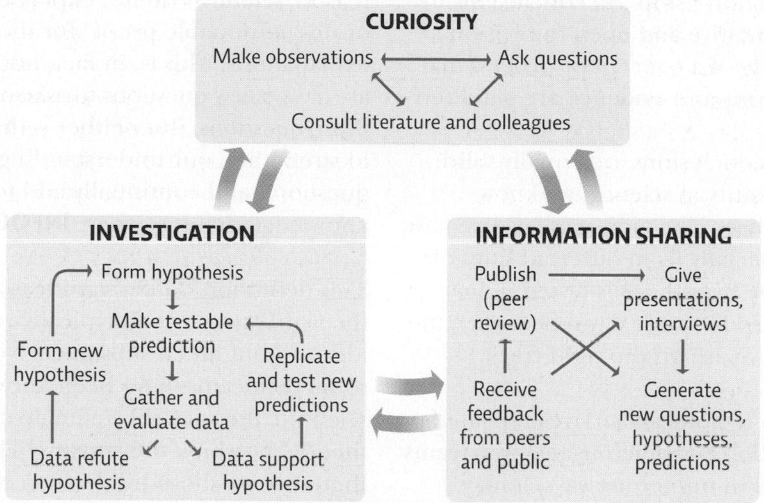

If the data support a hypothesis, new predictions are still tested. Why is this useful?

Before being published, scientific reports are subjected to **peer review**—they are reviewed by a group of third-party experts. Studies that are not well designed or conducted are not accepted for publication. Therefore, peer-reviewed published research represents high-quality scholarship in the field. This process might slow down the dissemination of information but adds a needed check on the validity of that information. There are even formal processes for the retraction of a study that has come into question after publication. **INFOGRAPHIC 1**

2 CERTAINTY IN SCIENCE: FROM HYPOTHESIS TO THEORY

Key Concept 2: Scientific certainty increases as evidence mounts and can be quite strong, but science never claims or even expects it will reach a level of absolute, unquestionable proof because there are always new questions to ask and new experiments to perform.

Hypothesis testing is at the heart of science. For example, Blehert's team investigated the hypothesis that the fungal infection was opportunistic and the bats were already weakened from another infection. Close examination of internal organs (heart, liver, lungs, and kidneys) did not reveal any abnormalities—no sign of a problem or previously undetected infection that would make the bats more susceptible to the fungus. The skin infections caused by WNS were occurring in the absence of other infections. The *opportunistic hypothesis* was rejected, and scientists moved on.

Meanwhile, as word of the colossal bat die-offs spread throughout the scientific community, researchers in Europe and China identified the same fungus in their bat caves, only the bats there were perfectly healthy. It appears that Pd had been native to those regions for a very long time. Because the bats in those areas had evolved with the fungus, they likely had adaptations to help them survive the infections (see Module 3.1 for an explanation of coevolution). Scientists are currently investigating the nature of these adaptations to determine why European and Asian bats do not succumb to the disease.

This new information lent some weight to the second hypothesis about the origin of the fungus: It had been brought to North

peer review A process whereby researchers' work is evaluated by outside experts to determine whether it is of a high enough quality to publish.

America from elsewhere (the *novel pathogen hypothesis*). The fact that the fungus is found in the dirt of infected caves and on clothing and equipment from cavers who visit these caves suggests a means of transport from cave to cave. When evidence mounts like this to support a given hypothesis, scientists suspect they are on the right track. But even then, they don't stop. All conclusions in science are considered tentative and open to revision because our understanding of a concept or process may change as more observations and evidence are gathered.

That is not to say that all conclusions are equally valid. There are degrees of certainty in science; we know some things better than others. The more evidence we have in support of an idea, especially from different lines of investigation (in which the hypothesis is tested in many different ways, not just tested the same way repeatedly), the more certain we are that we are on the right track.

theory A widely accepted explanation of a natural phenomenon that has been extensively and rigorously tested scientifically.

If a hypothesis survives repeated testing by numerous research teams and in numerous ways, it may be incorporated into a **theory**: a widely accepted explanation that has been extensively and rigorously tested. Theories represent the highest level of certainty a scientific explanation can attain. But in keeping with the tentative nature of science, even a well-substantiated theory is open to further study and can be revised, sometimes significantly, if new data strongly support a new conclusion. For this reason, scientists do not expect or require "absolute or unquestionable proof" for their conclusions and explanations. This is, in fact, not possible, because there are always new questions to ask and new ways to investigate those questions. But neither is this a weakness—it is a way to strengthen our understanding of a topic, uncover new questions, and continually add to and revise the body of knowledge that is science. **INFOGRAPHIC 2**

This definition differs significantly from the lay use of the word *theory*, which typically refers to a speculative idea without much substance. Nonscientists, including some politicians, have been known to dismiss some scientific theories (like climate change) as "just a theory," by which they mean, "This is unproven and therefore we shouldn't have to consider it." But that line of thinking represents a serious misunderstanding of what a scientific theory really is. Theories in science

Many bat species live together by the hundreds or even thousands in caves or other protective sites. This communal living greatly increases the chance of spreading infection. Here, a colony of hibernating gray bats is being inspected for signs of WNS.

STEPHEN ALVAREZ/National Geographic Creative

INFOGRAPHIC 2 CERTAINTY IN SCIENCE

There are degrees of certainty in science; we know some ideas are better than others. The more evidence we have in support of an idea, especially when the evidence comes from different lines of inquiry, the more certain we are that we are on the right track. But since all scientific information is open to further evaluation, we do not expect or require "absolute" proof.

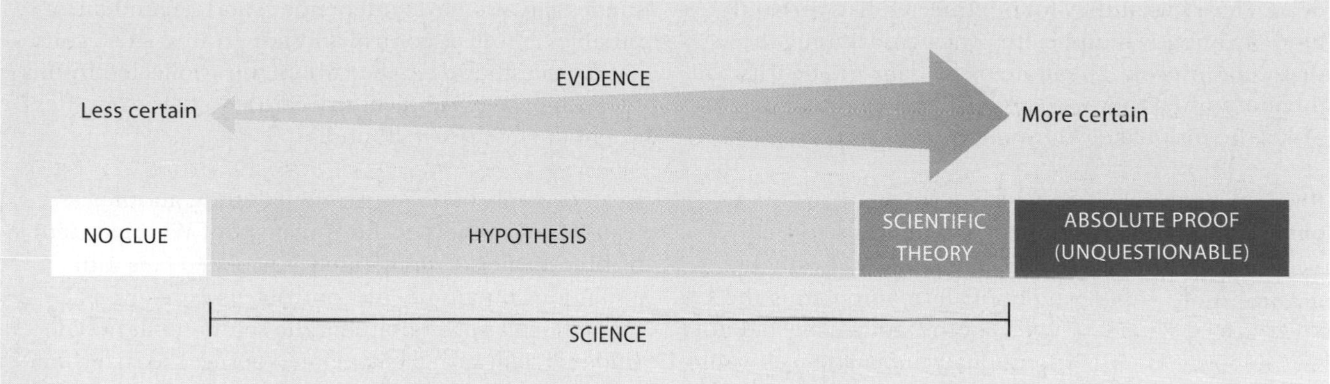

? Why do scientists say that a hypothesis can be disproved but never proven?

represent the highest level of certainty assigned to an explanation. They should not be rejected without substantial scientific evidence to the contrary.

When it came to the bat die-offs, scientists were nowhere near a theory. But they did have enough information to hypothesize further: Since bats don't travel across oceans, the fungus was probably introduced to U.S. caves by humans—most likely cave explorers who carried fungal spores overseas on their clothing and gear—and then spread to multiple colonies by bat-to-bat contact.

Scientists collected more data from bat caves and bat carcasses. The cold-adapted fungus seemed to thrive on hibernating bats, but it did not appear to affect active bats. Researchers suspected bats were easier prey for the fungus during hibernation because in that state, their body temperatures were lower than normal and their immune systems were suppressed. In active (awake) bats, the fungus's growth would be slower because those bats were warmer, and their immune systems were fully active and could fight off the infection.

During hibernation, bats are in a state of inactivity, conserving energy by reducing their metabolic rate and body temperature. Bats will occasionally rouse from this inactive state to drink water and excrete waste. These arousals take a lot of energy, but bats normally have just enough energy (fat stores) saved up to get them through 8 to 12 arousal periods. Examination of infected bats revealed that the fungus appeared to attack the hairless areas, namely their wings, ears, and noses—areas vital to bats' ability to maintain body temperatures during hibernation.

Because bats with WNS appeared to die of starvation, Blehert and others hypothesized that the fungal infection was causing them to rouse more often and thus burn up their fat reserves too early in the season. "They would scurry to the cave openings in search of insects to eat," Blehert explains. "But in the dead of winter, none would come." Without insect meals to replenish their stores, they would starve to death before spring arrived.

3 OBSERVATIONAL AND EXPERIMENTAL STUDIES

Key Concept 3: Experimental studies manipulate conditions, whereas observational studies collect data without intentionally altering any conditions. Each supplies different lines of evidence for analysis.

With evidence mounting that an introduced fungus from overseas was indeed responsible for the bat die-offs, scientists still needed to test the hypothesis that death was caused by arousing too often and using up fat stores before the end of winter (and the return of insects).

To do this, a research team led by Bucknell University biologist DeeAnn Reeder attached temperature-sensitive dataloggers to the backs of 504 free-ranging little brown bats (*Myotis lucifugu*). By recording body temperature, these dataloggers would tell the researchers how often a bat aroused during hibernation and how long it "slept" between arousals. At the end of the hibernation season, bats were recaptured (or collected if dead) and dataloggers retrieved. For each bat, researchers noted if it was infected with WNS and whether it was alive or dead. Here's what they found: Infected bats aroused more frequently than healthy ones. And among those that were infected, the more tissue damage the bats had, the more often they woke. And the more often they woke, the more likely they were to die.

Reeder's datalogger study is an example of an **observational study**—one where scientists collect data in the real world without intentionally manipulating the subject of the study. They did not intentionally inoculate bats with WNS; they simply followed bats, measured a parameter of interest, and noted which bats were infected and which were not. In observational studies, researchers may simply gather data to learn about a system or phenomenon (like measuring the amount of CO_2 in the atmosphere over time to see if it is rising or falling), or they may be comparing different groups or conditions found in nature (like monitoring asthma cases in areas with high and low air pollution). These types of studies are very useful in environmental science because they enable us to look at whole systems that cannot be manipulated in a lab or field setting.

observational study Research that gathers data in a real-world setting without intentionally manipulating any variable.

experimental study Research that manipulates a variable in a test group and compares the response to that of a control group that was not exposed to the same variable.

independent variable The variable in an experiment that a researcher manipulates or changes to see if the change produces an effect.

dependent variable The variable in an experiment that is evaluated to see if it changes due to the conditions of the experiment.

test group The group in an experimental study that is manipulated such that it differs from the control group in only one way.

control group The group in an experimental study to which the test group's results are compared; ideally, the control group will differ from the test group in only one way.

statistics The mathematical evaluation of experimental data to determine how likely any difference observed is due to the variable being tested.

Experimental studies, by contrast, involve intentional manipulation of experimental conditions.

A factor that is changed is called a variable (such as whether test subjects are exposed to a pathogen). Specifically, scientists *manipulate* the **independent variable** and *measure* the **dependent variable** to see if it is affected.

These types of studies rely on test groups and control groups. The **test group** subjects are exposed to a variable that scientists want to study, and their results are compared to a **control group** that was not exposed to the variable (or was exposed to the variable to a different degree). Ideally, all other factors remain the same (such as temperature or light exposure)—these are called constants. (Observational studies can also have independent and dependent variables as well as control and test groups—i.e., control bats in the study described above were collected from a site with no WNS and compared to test bats from sites where the disease was found.)

In an example of an experimental study, another team of researchers at the University of Winnipeg led by Lisa Warnecke deliberately inoculated bats with a solution containing Pd; control bats were given a sham inoculation containing the solution but no Pd (independent variable: type of exposure) to determine whether the fungus affected the length of the bats' sleep periods during hibernation (dependent variable: length of sleep periods). Warnecke's results showed that bats that had been intentionally exposed to Pd experienced shorter sleeping bouts because they woke up more often than control bats, providing evidence that Pd infections did interfere with hibernation. **INFOGRAPHIC 3**

Studies of this type must pass ethical review before beginning in order to ensure animals do not experience undue pain or stress; however, animal studies of any type raise ethical concerns. When scientists choose between experimental studies like this one or observational studies that do not impose a potentially harmful manipulation, they weigh the ethics of intentionally making animals (or people) sick against the benefits that research might bring, such as finding a cure for a disease or its cause.

As mentioned earlier, in science there are degrees of certainty; we know some things better than others. These degrees of certainty are expressed in terms of probabilities with a branch of mathematics called **statistics**. Once data are collected from an observational or experimental study, statistical analysis is done to calculate the probability that the variable being tested influenced the results. Specifically, it calculates the likelihood that any difference observed between two groups is just a result of natural variation (i.e., not *statistically significant*) or if it represents genuine differences between the groups (i.e., the differences are *statistically significant*).

In Reeder's study, for example, average sleep periods were longer in the groups of bats without WNS compared to survivors who had WNS, but these differences were not different enough to be statistically significant, so we conclude that there is no difference in

INFOGRAPHIC 3 HOW DO SCIENTISTS COLLECT EVIDENCE TO ANSWER QUESTIONS?

Scientists collect evidence to test ideas. Experimental studies are used when the test subjects can be intentionally manipulated; observational studies allow scientists to look at entire ecosystems or other complex systems. Both can be used to test the same hypothesis.

BACKGROUND KNOWLEDGE
Bats burn most of their winter calories during brief arousals from sleep periods.

QUESTION
Do bats with WNS wake up more often during hibernation?

HYPOTHESIS
WNS causes bats to awaken more frequently, causing them to burn calories, depleting their fat stores.

SCIENTIFICALLY TEST THE HYPOTHESIS.

OBSERVATIONAL STUDY (Reeder et al.)

Prediction: Bats with WNS will have shorter sleep periods than bats without WNS due to more frequent arousal.

Procedure: Researchers tracked bats in six different caves, three caves with WNS and three without. They determined how long each animal was awake by monitoring the bats' body temperature. The length of sleep periods for uninfected bats was compared to that of bats with WNS.

RESULTS

SLEEP PERIOD LENGTH AND WNS STATUS

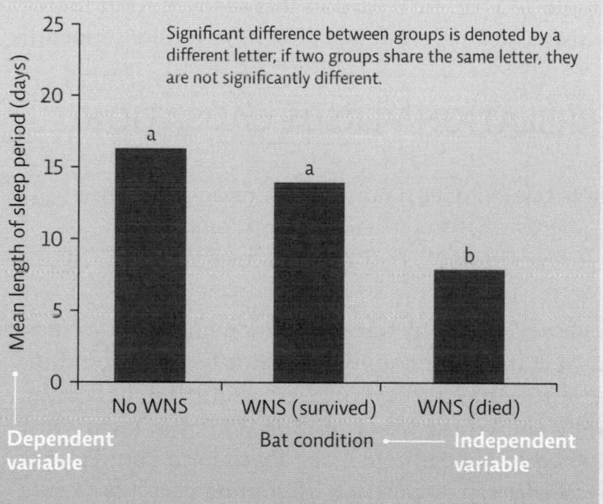

EXPERIMENTERS' CONCLUSION: Increased mortality/disease state is associated with shortened sleep periods due to frequent arousal episodes.

EXPERIMENTAL STUDY (Warnecke et al.)

Prediction: Bats inoculated with WNS will awaken from sleep periods more frequently than bats given a sham inoculation.

Procedure: Bats were randomly divided into two groups. The control group was given a sham inoculation; the test group was inoculated with WNS fungus (Pd). The length of sleep periods was monitored at different stages of hibernation.

RESULTS

AROUSAL IN HIBERNATING BATS WITH OR WITHOUT WNS

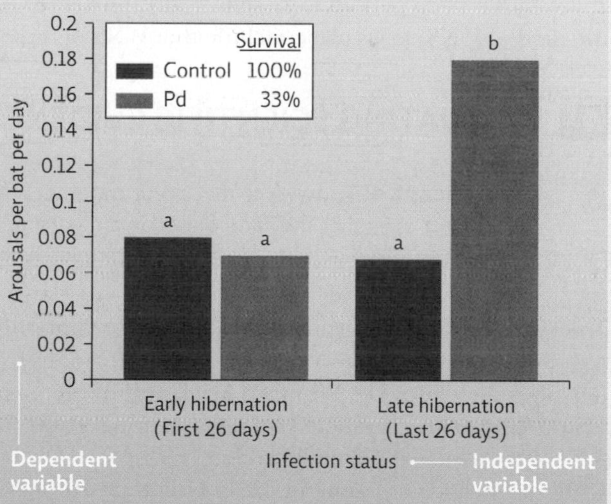

EXPERIMENTERS' CONCLUSION: Bats with WNS have increased arousal frequencies later in hibernation, contributing to death by starvation.

Both studies provide evidence that infection with WNS reduces sleep period length and is linked to higher rates of death. Future studies could build on these results. For example, future observational studies might look more closely at factors that affect sleep period length or contribute to the spread of the disease in natural settings; future experimental studies could examine the physiological mechanisms that lead to death.

Why are both observational and experimental studies useful when trying to answer the original question, "Do bats with WNS wake up more often during hibernation?"

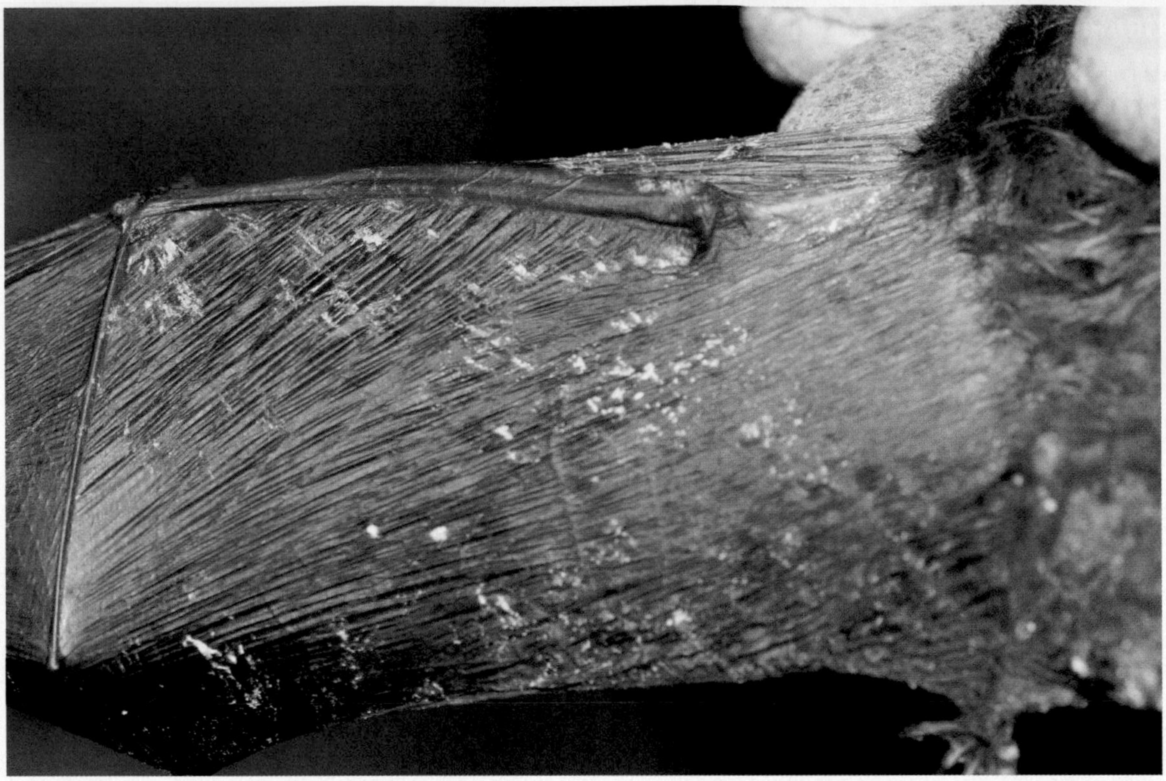

Thomas Deerinck, NCMIR/Science Source

A close-up of the wing of a little brown bat infected with WNS. The fungus infiltrates hairless skin such as that found on the muzzle, ears, and wings of bats, damaging the tissue.

the duration of sleep periods between these two groups. However, the sleep periods were statistically shorter for bats who died of WNS, so we can conclude that WNS in these bats did increase arousals later in hibernation compared to uninfected bats or WNS bats that survived. (For more on statistics and experimental design, see online Appendix 3.)

4 EVALUATING SCIENTIFIC EVIDENCE: CORRELATION VERSUS CAUSATION

Key Concept 4: Two events that occur together are correlated, but this does not necessarily mean one caused the other or even that they are related in a meaningful way. A scientific analysis of an environmental problem provides the information necessary to determine what policy, if any, is needed to address that issue.

Observational and experimental studies differ not only in their approach, but also in the type of information they ultimately provide. Winifred Frick and her team at Boston University conducted an observational study that compared bat colony sizes over time in areas with and without WNS. They observed a sharp decline in colony size, but only in the colony experiencing WNS infections.

correlation Two things occurring together but not necessarily having a cause-and-effect relationship.

cause-and-effect relationship An association between two variables that identifies one (the effect) occurring as a result of or in response to the other (the cause).

Observational studies like this (and the Reeder study) reveal **correlations**. That is, they can show that two things occur together: A colony was somehow exposed to WNS, and most bats in that colony died shortly thereafter. But observational studies can't necessarily tell us if one thing caused the other or even if the two factors are related in a meaningful way (because it could just be a coincidence—perhaps something else that wasn't evaluated caused the die-off). So, although scientists suspected from the very start that the white fuzz was related to the bat die-offs, they needed more evidence to determine whether the two events were connected in a **cause-and-effect relationship**. For that, they needed experimental studies like Warnecke's that intentionally exposed bats to the fungus. If all other factors are the same between two groups except the one variable that is manipulated (in this case, exposure to the fungus)

INFOGRAPHIC 4 CORRELATION VERSUS CAUSATION

Because they cannot control the extraneous variables in a research study, observational studies can only provide correlational evidence that two or more factors are related. Experimental studies that can manipulate and control variables can establish whether a cause-and-effect relationship exists.

A: POPULATION SIZE IN COLONIES WITH AND WITHOUT WNS

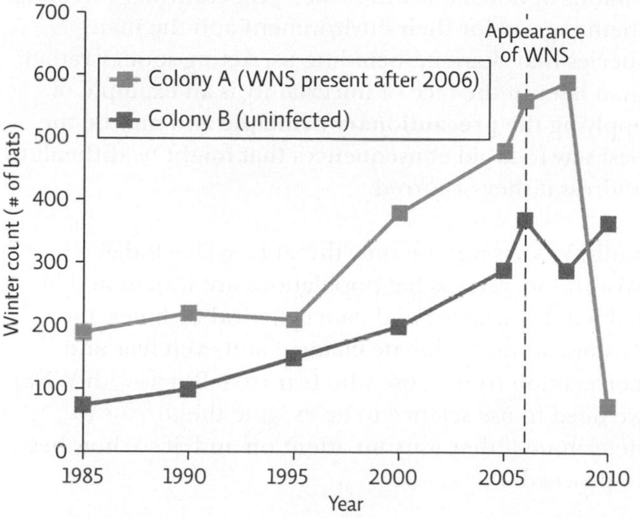

Study A: Observational study that followed two colonies in years before and after WNS affected one of the colonies.

This study provides correlational evidence that suggests WNS caused the population crash. However, it does not provide cause-and-effect evidence since something else might have caused population size to fall in that colony.

B: EXPOSURE TO WNS FUNGUS AND BAT SURVIVAL

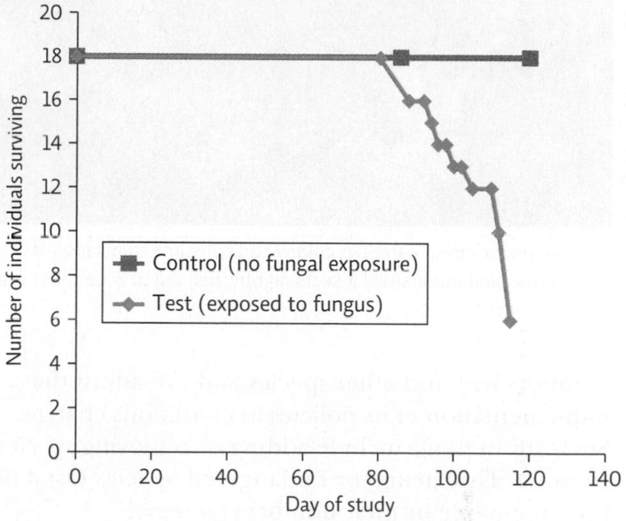

Study B: Experimental study that monitored survival in bats exposed to the fungus that causes WNS (Pd) compared to bats with no Pd fungal exposure.

In this study, a cause-and-effect relationship can be established since the only difference between the groups was Pd exposure.

Other than WNS, what else could have decreased the number of bats in Colony A after 2008?

and we see differences in the dependent variable we are monitoring (in this case, bat survival), then it is reasonable to conclude that the manipulation caused the effect. **INFOGRAPHIC 4**

It's important to understand that while experimental evidence is generally considered stronger, precisely because it can determine cause and effect, observational studies that amass a lot of evidence in support of a given correlation are also quite powerful. For example, we have only correlational evidence that smoking causes lung cancer in humans, but this relationship is not in question. In fact, observational studies that produce the same results as experimental studies offer particularly strong corroborating evidence. This is because they demonstrate the correlation between the variables even when the populations are exposed to many other factors that might influence the outcome. Conversely, when the results from observational studies do not agree with those from controlled experimental studies, we can conclude that other factors are probably at play.

A scientific analysis of environmental problems provides the information needed to take the next step—to determine whether that problem needs to be addressed and how to go about doing so.

When it is established that action is needed, the next step is to create a **policy** to deal with the problem. These policies need to be informed by science. To deal with WNS, the U.S. Fish and Wildlife Service (USFWS) drew from available research to create a national plan that addresses the disease. The plan lays out guidelines for a wide variety of actions, including disease surveillance and management procedures as well as outreach programs to educate the public about best practices and the need for containment. It has also closed some hibernation caves to the public in an effort to prevent humans from carrying fungal spores from one cave to another.

Policies must also be flexible and able to adapt to new scientific findings. The USFWS

policy A formalized plan that addresses a desired outcome or goal.

National Geographic Creative

Bats are major insect eaters. A colony can consume thousands of mosquitoes and other small insects nightly, helping to keep these insect pests in check.

monitors bats and other species and can adjust the implementation of its policies as conditions change. Such adjustments include adding or removing species from the Threatened or Endangered Species List if new threats emerge or their numbers increase.

One dilemma faced by policy makers is the fact that gathering strong evidence about the cause of an environmental problem takes time, sometimes longer than we can wait if we are to take meaningful action. This means policies often are based on incomplete data. For example, we don't know yet what the loss of bats would mean for the areas threatened, but it could be quite severe—especially in the face of global warming that is expanding the range of disease-carrying mosquitoes. Like most predators, bats are vital to their ecosystem. They are important predators of nocturnal insects like mosquitoes. In fact, every year bats provide billions of dollars' worth of free pest control. Protecting them is good for their environment and the many species that share it, including us. Acting sooner rather than later in the face of uncertainty is an example of applying the **precautionary principle** and may be the best way to avoid consequences that might be difficult to address if they occurred.

Sadly, WNS is not the only threat bats face today. Worldwide, various bat populations are threatened by habitat disturbance and damage, wind turbines, the bushmeat trade, climate change, and even fear and persecution from those who fear bats. But as with WNS, we need to use science to investigate the threats to determine if they warrant attention and, if so, how best to proceed.

Scientists will continue to study the disease, possible treatment and prevention options, and the ecological impacts of bat declines, building on previous studies and pursuing new avenues of research that appear promising. Each new piece of information raises new questions—questions that are addressed through the process of science.

5 INFORMATION AND MEDIA LITERACY

Key Concept 5: Published information about scientific topics abounds in our modern world. Information literacy skills help individuals determine the reliability of that information.

Understanding how scientists study the natural world and ask questions allows us to apply the scientific method to investigate a question. But most often we will be evaluating the research of others, and this means we need to be able to distinguish between reliable and unreliable sources of information—a skill known as **information literacy**. It's the key to drawing reasonable evidence-based conclusions about any given issue or topic; it is especially important in public health and science issues, because when it comes to these issues, misinformation abounds, especially digital information—readily available to most of us on our computers and smart phones. One of the first steps in information literacy is identifying what type of information source one is consulting—is it a firsthand account from the original person or group that created or experienced it, or is it a retelling?

Our first choice for reliable information would be those that are **primary sources**, sources that present original data or information, including novel scientific experiments and firsthand accounts of any given observation. Journal entries and interviews are primary information sources, as are original research articles published in scientific journals. The research papers published by Reeder and Warnecke are primary sources; this means you can view the data (the evidence) on which the researchers based their conclusions and evaluate it yourself.

precautionary principle Acting in a way that leaves a safety margin when the data are uncertain or severe consequences are possible.

information literacy The ability to find and evaluate the quality of information.

primary source Information source that presents original data or firsthand information.

Satirical news outlets like "The Onion" deliberately present news in an outrageous fashion, using humor, irony, and exaggeration. The story may be for entertainment only, or its intention may be to expose what the author sees as corruption or irresponsible actions. Political satire is seen as an important part of democracy and a way to broach difficult and even taboo topics.

Most of us start a step or two lower than primary sources—perhaps it is a newspaper article about some recent research or event or a video we see on television or online. This information may be coming from a **secondary source**, one that presents and interprets information solely from primary sources, such as a science journalist who reviews a recent research article.

But most likely, our initial information comes from **tertiary sources**, those that use one or more secondary sources. Most books, including textbooks, and reports from the popular press are tertiary sources. Most blogs, websites, and news shows that provide additional commentary or foster debate also qualify as tertiary sources. But because tertiary references do not consult the primary information source, they may perpetuate errors made by a secondary source. For this reason, they are less dependable than secondary sources or primary sources. That is not to say secondary and tertiary sources are not useful—they are a good starting place. But if you have unanswered questions or if there are inconsistencies in the reports you encounter, going back to the primary source is the best way to reconcile those differences or find answers to unanswered questions. For any claim you encounter, you should be asking, "What's the evidence for that claim?" Where do you find that evidence? Primary sources.

Information literacy in the digital age must be expanded to include **media literacy**—the ability to evaluate digital sources of information. The ease of publishing information digitally and the lack of oversight for many pieces of digital information mean that we all need to have the skills to determine the veracity of a piece of information.

Information and media literacy skills allow us to identify information presented under the guise of news that is misleading in some way. Questionable "news" has been around a long time; today it comes in a variety of forms:

- *Fake news* is false information. In a digital format, it may be *clickbait*—shocking titles to get you to click on a story (common

secondary source Information source that presents and interprets information solely from primary sources.

tertiary source Information source that uses information from at least one secondary source.

media literacy The ability to evaluate digital sources of information.

INFOGRAPHIC 5 MEDIA LITERACY: EVALUATING DIGITAL "NEWS" STORIES

More and more people are getting their news from digital sources, including social media, but this news may not be authentic or reliable; it may even be intentionally fabricated but presented in a way to fool the reader. To evaluate the reliability of any news story, there are some basic questions you can ask to help navigate the world of digital news.

QUESTIONABLE "NEWS" COMES IN A VARIETY OF FORMS:

- **Fake news:** False; often clickbait—shocking titles to get you to click on a story
- **Misleading news:** Some truth but is incomplete, is out of context, and/or contains false statements; often supports an agenda
- **Satire:** Intentionally outrageous; often political commentary

CONSIDER THE SOURCE
Is the author a credible source of information? What is the mission or purpose of the website?

DETERMINE IF IT'S REASONABLE
Does the information seem reasonable, or is it outrageous? Might it be satire?

EVALUATE THE ACCURACY
Are sources cited for statistics, facts, quotes, or images? Is the story current?

CORROBORATE THE INFORMATION
Can you find the same story on other reputable websites? Has it been verified or debunked on a fact-checking website?

IDENTIFY AUTHOR BIAS
Does the author or website have a position it is promoting? Does it stand to profit in some way if you accept the claims?

ACKNOWLEDGE PERSONAL BIAS
Might your own biases affect your willingness to believe what you are reading?

Which type of "questionable" news story do you think is the hardest to identify? Why?

on websites where authors make money based on how often the link to that story is opened).

- *Misleading news* contains some truth but may be incomplete or out of context and/or contain false information; it is often presented to support an agenda. Political propaganda often falls under this category.
- *Satire* is intentionally outrageous. It may take some investigation to determine if the story is offered as legitimate news or as satirical commentary.

A reader must be diligent when trying to determine the credibility of a news story. Tips for ferreting out questionable news are offered in Infographic 5. **INFOGRAPHIC 5**

6 INFORMATION LITERACY: CRITICAL THINKING AND LOGICAL FALLACIES

Key Concept 6: Evaluating a claim requires critical-thinking skills: Always ask for the evidence and evaluate it, look out for logical fallacies, and keep an open mind—be willing to go where the evidence leads you.

Information on environmental issues abounds, and it is our responsibility to determine which information sources and arguments are sound and which are faulty or out of date. While how best to deal with WNS in bats may not be a hot topic on the minds of most people, many other environmental problems make headlines daily and are the subject of heated debate. Should coal be phased out, or is it still a vital lifeline needed by workers in coal-producing states? How should we respond to climate change on an individual or national level? Is the added expense of water or air pollution prevention worth its costs?

Search long enough, and you will find arguments on many sides of each of these (and other) environmental questions. Sorting through the rhetoric of any complex environmental problem requires diligence and the use of critical-thinking skills that allow one to logically evaluate the evidence—the same skills used by scientists to evaluate scientific studies.

An evaluation of any argument or claim should include an evaluation of the logic that underlies that claim—is it supported by evidence? In some cases, the way a claim is stated can be misleading—this may be intentional or the result of a poorly constructed argument. Literary devices used to confuse or sway the audience to accept a claim or position in the absence of evidence, or by twisting the evidence, are referred to as **logical fallacies**. For example, if bat conservationists had called for the immediate closure of all U.S. caves that housed bats after the 2006 discovery of so many dead bats at Howe's Cave, they would have been guilty of a *hasty generalization*. At that point in time, it was too early to know what was causing the deaths or if it could be transmitted to other caves.

Logical fallacies can persuade others to accept a conclusion—or to doubt one. Industry, for example, has been found to deliberately mislead consumers in order to make sales. Perhaps that comes as no surprise—after all, they are in the business of making money—but it is particularly disturbing when the products they sell are dangerous. For example, the tobacco industry argued for years that smoking cigarettes was safe, even though they had ample evidence from internal and external investigations that it was quite dangerous.

To combat the growing mountain of evidence that smoking was linked to a number of serious health problems, the tobacco industry developed what science historian Naomi Oreskes calls a "playbook"—definitive steps that could be taken to confuse consumers. This included hiring scientists who would question studies that showed harm (*appeal to authority*) and claiming that diseases like cancer and heart disease were too complex to pin any particular individual's ailment on smoking (*appeal to complexity*). *Ad hominem* attacks (discrediting the scientist or physician personally), *red herring* statements (extraneous information that is true but not relevant), and other logical fallacies were also used to sway the public and instill doubt that tobacco was dangerous. In one of their own internal memos, a public relations firm working for the tobacco industry stated that "doubt is our product." If they could cast doubt on the science, they could delay regulation. **INFOGRAPHIC 6**

Critical thinking is the antidote to logical fallacy. It enables individuals to logically assess and reflect on information and reach their own conclusions. Critical thinking is a skill that can be broken down into a handful of tenets:

Be skeptical. Just like a good scientist, don't accept claims without evidence, even from an expert. This doesn't mean refusing to believe anything; it simply means requiring evidence before accepting a claim as reasonable, especially if it is counter to well-established scientific consensus.

Evaluate the evidence. Is the claim derived from anecdotal evidence (unscientific observations, single occurrences) or from scientific studies? If it is based on studies, how relevant are those studies to the claim? (Were they done on human subjects or other species? Intact organisms or cells in a petri dish?)

Watch out for author biases. Is the person or group making the claim trying to promote a position? Is it financially tied to one conclusion or another? Is it trying to use evidence to support a predetermined conclusion?

Be open-minded. Try to identify your own biases or preconceived notions ("The government can't be trusted," "Most people overreact to things like this," etc.), and be willing to follow the evidence where it takes you.

Humanity faces many environmental problems today, from rising levels of greenhouse gases to water and air pollution to biodiversity loss (and many others), that must be addressed if we are to reach the goal of living sustainably. (See Module 1.1 for an introduction to living sustainably.) It is vital that we base our policies on sound science rather than on speculation or what we'd like to be true.

Unfortunately, making logical decisions based on evidence, even when we have compelling evidence, may not be as easy as it sounds. Cognitive and social science research is revealing some interesting aspects regarding how the human brain operates. We would like to think we are rational creatures, weighing evidence to reach logical conclusions. But research shows that when presented with evidence that supports our current beliefs, this evidence tends to strengthen those beliefs. However, when presented with evidence that contradicts what we believe to be true, no matter how compelling, many of us simply reject it out of hand.

To counter this "confirmation bias," we must be willing to consider other viewpoints and, in doing so, look at the evidence and evaluate its veracity. If we approach new information knowing we may be biased to our current position, we may be more open to the evidence and be willing to go where the evidence takes us—just like a scientist.

logical fallacies Arguments that attempt to sway the reader without using reasonable evidence.

critical thinking Skills that enable individuals to logically assess information, reflect on that information, and reach their own conclusions.

INFOGRAPHIC 6 LOGICAL FALLACIES

Logical fallacies are arguments used to confuse or sway someone to accept a claim/position in the absence of evidence.

COMMON LOGICAL FALLACIES	EXAMPLES
Hasty generalization: Draws a broad conclusion on too little evidence.	***Hundreds of cave explorers annually visit a cave where there are many dead bats. The visitors must be responsible for the problem.*** Just because a cave with bat fatalities is visited by people is not evidence that the visitors are related to the deaths. More evidence is needed.
Red herring: Presents extraneous information that does not directly support the claim but that might confuse the reader/listener.	***Asian bats also harbor the fungus but are rarely harmed by it.*** The implication is that what is true for Asian bats will also be true for other bats. The fact that Asian bats are not affected is not evidence that North American bats would also be safe from exposure to the same fungus.
Ad hominem attack: Attacks the person/group that is presenting the opposite view rather than addressing the evidence.	***Conservationists are alarmists who just want to restrict recreational access to caves. Or: Cave explorers cannot be trusted to take proper precautions to avoid cave contamination.*** These statements are attacking the group, not the evidence.
Appeal to authority: Does not present evidence directly but instead makes the case that an "expert" agrees with the position or claim.	***WNS should (or should not) be addressed because "Dr. Smith," an expert in the field, advocates that position.*** Naming a group or person in support of a position is not evidence in itself. The evidence in support of that position should be evaluated.
Appeal to complexity: A statement or an implication that an issue is too complicated to understand.	***So many factors threaten bats that it is impossible to determine the role WNS is playing in their decline.*** This statement is assuming there is no way to examine each factor. The situation might be complex, but that doesn't mean it cannot be analyzed or addressed.
False dichotomy: The argument sets up an either/or choice that is not valid. Issues in environmental science are rarely black and white, so easy answers (it is "this" or "that") are rarely accurate.	***The only way to save bats is to close visitor access to all caves.*** An argument that presents only two options regarding cave access — either close all caves or close none — is a false dichotomy. It is likely that access to some caves should be limited or prohibited to protect bats but that others could remain open for recreation with little or no impact on bats.

Which logical fallacy do you feel is the hardest to recognize and might slip by the average person? Why?

Katie Gillies/Bat Conservation International

Wildlife biologist Sybill Amelon holds a bat from a study that tested exposure to bacterial antifungal compounds as a treatment for the disease in infected bats. All the bats survived and were released into the wild after recovery.

As information accumulates on the cause of the effects of WNS, a shift in the focus of researchers is taking place. Longtime bat researcher Daniel Lindener says research now focuses on breaking the cycle of infection and finding ways to stem the spread of the disease. There is even some evidence that the bats themselves may be showing signs of adapting to the fungus.

One closely monitored population of Indiana bats (*Myotis sodalist*) has been observed to wake up together. (Normally, arousal from hibernation is not a "group" behavior.) This adaptation might help the bats expend less energy during arousal periods. More research is needed to elucidate how this behavior might help and if it is found in other populations or species of bats.

Research on treatments is also underway. Studies revealing that the Pd fungus is very sensitive to ultraviolet (UV) radiation have led researchers to begin testing its effectiveness as a potential treatment. And the results of experimental tests of a vaccine to WSN, administered by injection or orally to treat bats, are encouraging—88% of vaccinated bats survived, compared to only 30% of unvaccinated bats. Researchers hope to develop a spray-on version of the oral vaccine; bats would ingest the vaccine during grooming.

Another possible treatment came from microbiologist Chris Cornelison, then of Georgia State University. Cornelison knew that organic compounds produced by the bacterium *Rhodococcus rhodochrous* reduced the amount of fungus that grows on bananas. He had read about WNS and wondered if it might do the same for bats. He teamed up with U.S. Forest Service scientists to test his idea in the field. Bats already infected with WNS were collected early in hibernation and briefly exposed (48 hours) to petri dishes containing the bacterium. At the end of the hibernation period, all 75 test bats were alive, although some still had wing damage from the fungus.

Although it is too soon to know if the bacterial chemicals or any of the other treatments being tested will prove to be an effective treatment, teams of scientists are back in the lab and in the field trying to find out while, as Cornelison says, "there are still bats to treat."

Select References:

Blehert, D. S., et al. (2009). Bat white-nose syndrome: An emerging fungal pathogen? *Science, 323*(5911), 227.

Cornelison, C. T., et al. (2014). A preliminary report on the contact-independent antagonism of *Pseudogymnoascus destructans* by *Rhodococcus rhodochrous* strain DAP96253. *BMC Microbiology, 14*(1), 246.

Frick, W. F., et al. (2016). White-nose syndrome in bats. In *Bats in the Anthropocene: Conservation of bats in a changing world* (pp. 245–262). Cham (Zug), Switzerland: Springer International.

Hayman, D. T., et al. (2017). Long-term video surveillance and automated analyses reveal arousal patterns in groups of hibernating bats. *Methods in Ecology and Evolution, 8*(12), 1813–1821.

LaGarde, J., & Hudgins, D. (2018). *Fact vs. fiction: Teaching critical thinking skills in the age of fake news.* Portland, OR: International Society for Technology in Education.

Palmer, J. M., et al. (2018). Extreme sensitivity to ultraviolet light in the fungal pathogen causing white-nose syndrome of bats. *Nature Communications, 9*(1), e00148.

Reeder, D. M., et al. (2012). Frequent arousal from hibernation linked to severity of infection and mortality in bats with white-nose syndrome. *PLoS One, 7*(6), e38920.

Warnecke, L., et al. (2012). Inoculation of bats with European *Geomyces destructans* supports the novel pathogen hypothesis for the origin of white-nose syndrome. *Proceedings of the National Academy of Sciences, 109*(18), 6999–7003.

GLOBAL CASE STUDIES FUNGAL INVADERS

Achieve

Non-native pathogenic invaders can wreak havoc on species that don't possess adequate adaptations to fight off the invader or if the invader has no natural predators in its new environment. The way we plant crops (in large monocultures) also makes them vulnerable to attack. Here are some notable fungal invaders affecting plants and animals around the world.

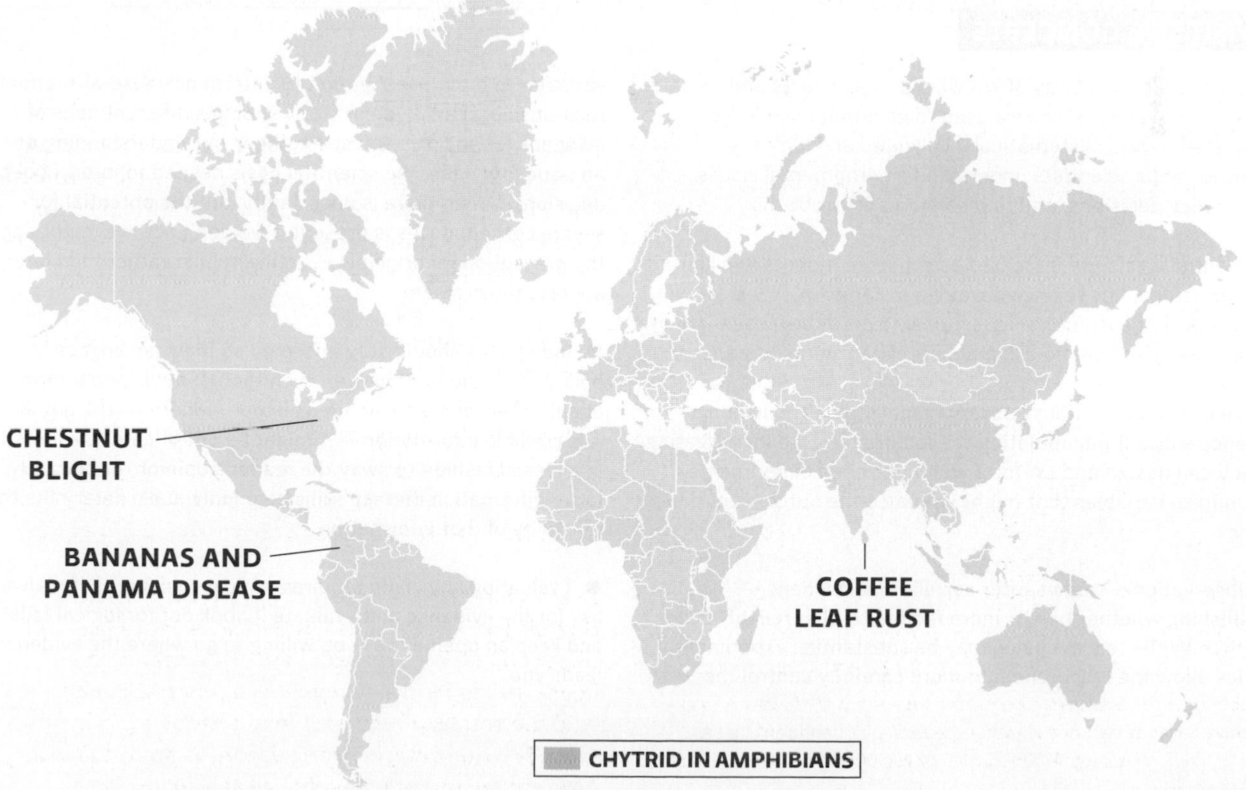

BRING IT HOME

PERSONAL CHOICES THAT HELP

The story of how scientists unraveled the mystery of what was killing bats is a great example of how science documented a problem, uncovered its cause while discarding unsupported hypotheses, and informed public policy to address the problem. Anyone can be a scientist by logically and systematically collecting evidence to answer scientific questions. How scientific information is or is not put into action has far-reaching consequences, making science literacy a matter of importance for every individual.

Individual Steps

- Practice thinking like a scientist. Go outside for 10 minutes, and observe the world around you. Make observations of what you see or hear. What predictions could you make from your observations? How could you test them?
- Stay informed. Read or watch a science-related article or show once a month.
- Attend a city council or county board meeting to see how policy issues are addressed in your area.
- Make knowledgeable voting decisions on ballot initiatives.

Group Action

- Demonstrate the importance of scientific literacy to your friends and family. Develop three additional questions from the material in the module, and discuss them over dinner.
- Support science education: Find out about public lectures and programs in your area, and attend one with your friends.
- Are there caves with bats in your area? Contact local wildlife officials to see if there are any projects that can be done to assist in the protection of the bats or caves.
- Serve on local civic committees that address environmental issues in your community.

BartCo/iStock/Getty Images

REVISIT THE CONCEPTS

- Science is both a body of knowledge (basic facts and complex explanations) and the *process* used to gain that knowledge. Physical evidence, systematically collected and logically analyzed, helps scientists understand environmental issues, guide policy decisions, and find solutions to problems.

- Scientific certainty increases as evidence mounts and can be quite strong, but science never claims or even expects it will reach a level of absolute, unquestionable proof because there are always new questions to ask and new experiments to perform.

- Scientists may conduct observational studies, collecting evidence without manipulating the factors they are investigating, or they can design and perform well-controlled experiments to minimize variables that might influence the outcome of the study.

- Observational studies offer correlational evidence, establishing whether two or more factors tend to reliably occur together. While this evidence may be substantial, experimental studies allow the researchers to more carefully control the variables in their investigation to determine cause-and-effect relationships. Each type of study supplies different lines of evidence for analysis and strengthens our understanding of an issue, providing the scientific basis needed to inform policy decisions. When there is uncertainty, but the potential for severe consequences is present if we don't act, we might apply the precautionary principle — acting sooner rather than later to address the problem.

- Individuals should stay informed so they can engage in policy discussions. However, even though published information about scientific topics abounds in our modern world, not all of it is credible information — it might be out of date, presented in a biased fashion to sway the reader's opinion, or blatantly false. Information literacy skills help individuals determine the reliability of that information.

- Evaluating any claim requires critical-thinking skills: Always ask for the evidence and evaluate it, look out for logical fallacies, and keep an open mind — be willing to go where the evidence leads you.

ENVIRONMENTAL LITERACY Understanding the Issue

1 How do scientists study the natural world?

1. Identify some questions that ARE scientifically testable and some that are NOT scientifically testable.
2. Outline the steps of the scientific method. Why is this more like a cycle than a linear process?
3. If an experiment produces results that confirm an experimental prediction and support the hypothesis, what are the next steps? What should be done if the results fail to support the hypothesis?
4. Distinguish between a hypothesis and an experimental prediction. Why is it useful to test a hypothesis using many different experimental predictions?

2 What influences the degree of certainty in a scientific explanation?

5. Why does the realm of scientific certainty span a spectrum of understanding from hypothesis to theory but excludes "no clue" and "absolute proof"?
6. What increases our certainty that our interpretation of results and conclusions are correct?
7. Why is it unreasonable to reject an accepted scientific explanation on the basis that it is "just a theory"?

3 Why are observational and experimental studies needed to investigate the natural world?

8. Scientists want to know whether a pesticide from a farm field is harming fish in a nearby stream. Identify an observational study and an experimental study that could address this question.
9. Why is it necessary to have a control group in a scientific study?
10. Identify a circumstance in which it would be more acceptable or reasonable to conduct an observational study than to conduct an experimental one.

4 What is the difference between a correlation and a cause-and-effect relationship, and how can science help policy makers address environmental problems?

11. We have only correlational evidence that smoking causes lung cancer. Why don't we have cause-and-effect evidence for this relationship? Why do we consider the correlational evidence we do have for the smoking-cancer connection to be strong?
12. What did it take to shift the conclusion that Pd and WNS were correlated to the conclusion that Pd causes WNS?
13. When might policy makers not have the option of waiting until they have definitive scientific evidence regarding the best course of action before drafting policies to deal with a problem?
14. The best policies are informed by science. What scientific evidence led to the policy of closing infected and uninfected caves to visitors?

5 What is information literacy, and why is it important in environmental science?

15. Distinguish among primary, secondary, and tertiary sources of information, and rank them in terms of reliability.
16. What is media literacy, and why is it needed in this age of abundant information?

6 What is critical thinking, and how can it counter logical fallacies used in arguments?

17. Compose an argument against taking actions to address white-nose syndrome that contains at least three logical fallacies, and identify each logical fallacy you employed.
18. Identify and explain four tenets in critical thinking that can help you logically assess and reflect on information in order to draw your own conclusions.

SCIENCE LITERACY Working with Data

To determine how bats contract WNS in the wild, J. M. Lorch and colleagues used different methods to expose bats to *Pseudogymnoascus destructans* (Pd), the fungus that causes WNS. Healthy bats were collected from the wild and the following groups set up:

- **Inoculation group:** 29 bats were directly treated with Pd from a laboratory culture of the fungus.
- **Contact exposure group:** 18 bats were housed in an enclosure with infected bats.
- **Airborne exposure group:** 36 bats were kept in cages close to, but not in direct contact with, WNS bats.
- Two control groups were used:
  - **Negative control group:** 34 bats with no exposure to Pd
  - **Positive control group:** 25 bats collected from the wild that already had WNS (compares the disease in the wild to that of experimental bats)

HOW DO BATS GET INFECTED WITH PD?

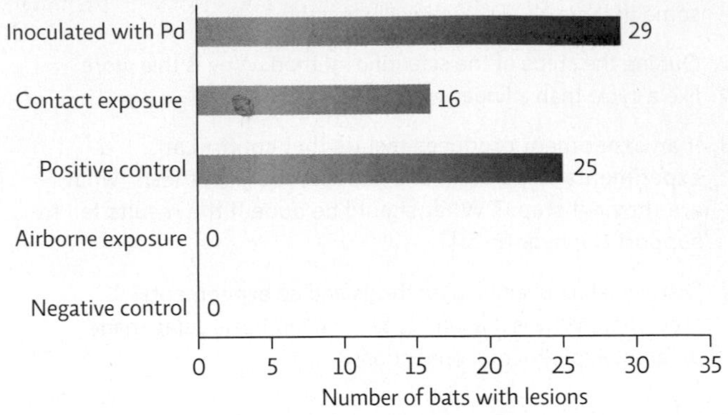

Interpretation

1. Look at the bars on the graph. Which exposure methods appear to transmit Pd and cause WNS lesions?
2. Compare the effectiveness of direct treatment with Pd (inoculation) to contact exposure with infected bats for causing WNS lesions. Does one route appear to be more effective at transmitting the disease than the other? How did you determine this?
3. Look at the sample size (number of bats) of each group. Now go back and look at your answer to question #2. Do you need to revise your answer? Explain.
4. Determine the percentage of each group that developed lesions, and redraw the graph using "Percentage of bats with WNS lesions present" as your *x*-axis title. Do you think this is a better way to show these data? Explain.

Advance Your Thinking

5. Researchers also collected data on mortality and found that the airborne exposure group had a slightly higher death rate than the contact exposure group, but a statistical analysis of these groups found that this difference was not statistically significant. Can you conclude that airborne exposure is more likely to lead to death than contact exposure? Explain.
6. What is the purpose of the positive and negative controls?
7. Why was this research done? Why might it be useful to know how the infection spreads?

INFORMATION LITERACY Evaluating Information

Use your understanding of the process of science to evaluate the threats and programs aimed at protecting bats. Go to the National Park Service (NPS) webpage "Threats to Bats," found at https://www.nps.gov/subjects/bats/threats-to-baats.htm.

Investigate the four categories of threats listed there: habitat loss, wind energy, white-nose syndrome, and climate change.

1. For each category, answer questions a and b:
   a. Summarize the threat posed by this category.
   b. What can be done to decrease the threat it poses?
2. Now, go back to the "Wind Energy" webpage, select a link for one of the "papers" or "reports" posted there, and answer the following questions:
   a. What is the title of the document you chose?
   b. What type of document is this (scientific report, general information, government report)?
   c. Look over the report. Read the abstract or introduction if one is provided. What is the focus of the report?
   d. Scan the document. Does the report offer scientific evidence in support of its claims or conclusions? Explain.
   e. Does this document appear to be a credible source of information? Explain.
3. Drawing from your reading of the report on the "Wind Energy" webpage and your understanding of WNS (from Module 1.2 and the NPS webpage on WNS), compare the threat to bats posed by wind turbines to that of white-nose syndrome in terms of severity of threat and our ability to respond. Which do you think we have a better chance of successfully addressing, and why?

Additional study questions are available at Achieve.macmillanlearning.com. Achieve

CHAPTER 2

Ecology

Ecology is the study of interactions organisms have with their environment and other organisms. Understanding these interactions and how species respond to changing conditions can help us as we try to prevent human-imposed ecosystem damage or attempt ecosystem restoration where needed.

Module 2.1

Ecosystems and Nutrient Cycling

A look at the living and nonliving components of ecosystems and the ways energy and matter are captured and moved through ecosystems

Module 2.2

Population Ecology

An examination of ecological populations and how ecologists study them and the biotic and abiotic factors that influence whether they thrive or decline

Module 2.3

Community Ecology

A look at the relationships between populations in a community, the impact of human actions that break community connections, and ways to protect or even restore damaged ecosystems

vitranc/Getty Images

Module 2.1 Ecosystems and Nutrient Cycling

ENGINEERING EARTH

An ambitious attempt to replicate Earth's life support systems falls short

Biosphere 2 in Tucson, Arizona. *Buyenlarge/Archive Photos/Getty Images*

After reading this module, you should be able to answer these GUIDING QUESTIONS

1. What is the hierarchy of organization recognized by ecologists, and why might it be useful to recognize such distinctions?
2. Why do ecosystems need a constant input of energy yet can handle the fact that they do not receive appreciable new inputs of matter?
3. How do environmental factors affect the distribution and makeup of biomes?
4. What is a population's range of tolerance, and how does it affect the distribution of a population or its ability to adapt to changing conditions?
5. How does carbon cycle through ecosystems, how have humans disrupted this cycle, and what problems can this disruption cause?
6. How does nitrogen cycle through ecosystems, how have humans disrupted this cycle, and what problems can this disruption cause?
7. How does phosphorus cycle through ecosystems, how have humans disrupted this cycle, and what problems can this disruption cause?

On September 26, 1993, with their first mission complete, four men and four women emerged from Biosphere 2—a hulking dome of custom-made glass and steel—back into the Arizona desert, where throngs of spectators stood cheering. They had been sealed inside the facility, along with 3,000 other plant and animal species, for exactly 2 years and 20 minutes; it was the longest anyone had ever survived in an enclosed structure.

WHERE IS BIOSPHERE 2?

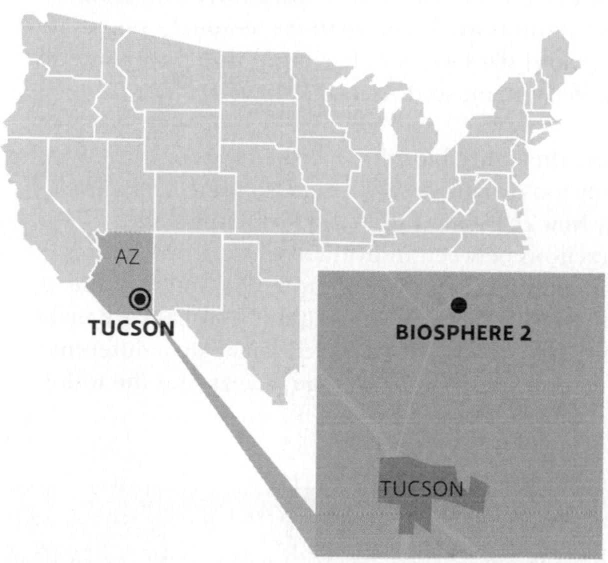

The feat was part of a grand experiment with two main goals. First, scientists hoped that by studying this mini Earth, which could be controlled and manipulated in ways the real Earth could not, they might better understand our own planet's delicate balance and how best to protect it. Second, they wanted to create an entirely self-contained enclosure that could sustain life—the kind that might one day be used to colonize the Moon or Mars.

By the time the biospherians emerged from the structure, they had learned a lot about the complexities of ecosystems—one of the key lessons was the realization that Earth's natural and self-sustaining ecosystems were much harder to replicate than they had hoped. More than one-third of the flora and fauna had gone extinct, including most of the vertebrates and all of the pollinating insects. Morning glory vines had overrun food crops. Cockroaches and ants were thriving. And eating too many sweet potatoes had turned the biospherians themselves bright orange. (A string of plant diseases had decimated other crops.)

On top of that, nitrous oxide (laughing gas) had grown concentrated enough to "reduce vitamin B_{12} synthesis to a level that could impair or damage the brain," according to one interim report. And oxygen levels had plummeted from 21% to 14% (just barely enough to sustain human life). Project engineers had been forced to pump in outside air, violating the facility's sanctity as a closed system, in order to ensure adequate oxygen levels.

To be sure, the eight biospherians had survived and the facility still held enormous potential as a scientific tool—an opportunity to conduct experiments on ecosystem scales. In a recent memoir, biospherian Mark Nelson points out that humans are already conducting "unplanned planetary experiments" as we release pollutants and extract resources at record rates. "What happens when tens of thousands of new synthetic chemicals produced in vast quantities are introduced into basic air, water, food, and soil cycles?" Nelson asks.

This small replica of Earth would be a far better place to test the environmental impacts of our actions

TIM ROBERTS/AFP/Getty Images

The eight biospherians emerge from Biosphere 2 after living there for 2 years.

than our own life support system. But would it work—could we create a functioning biosphere on our own? To answer that question, we need to answer a few others first: What exactly is a biosphere, and just how did Biosphere 2's creators set about building one?

1 THE ECOLOGICAL HIERARCHY: FROM BIOSPHERE TO INDIVIDUAL

Key Concept 1: Ecologists identify a nesting organization of life from individuals to the biosphere, often focusing on the levels of population, community, and ecosystem to examine how species respond to and affect the natural world.

The field of ecology focuses on how **species** interact with other components in their environment. In other words, it is about relationships. These relationships can be examined at different levels. The term **biosphere** refers to the total area on Earth where living things are found—the sum total of all its **biomes** (regions identified by their distinctive collection of vegetation, such as boreal forest or tallgrass prairie). An **ecosystem** includes all the organisms in a given area (the **biotic** components) in addition to the nonliving (**abiotic**) components of the physical environment in which they interact—things like wind, rain, and sunlight. In the natural world, ecosystems assume a range of shapes and sizes—a single simple tide pool qualifies as an ecosystem; so does the Mojave Desert.

Having the entire biosphere as the focus of study is usually too expansive to manage, so some ecologists study how ecosystems function by focusing on interactions between individuals of the same species within a **population**. For example, the wolves of the Northern Rockies and Yellowstone National Park make up one population there; the elk make up a different population, as do the bison, the beaver, and the willow

species A group of plants or animals that have a high degree of similarity and can generally only interbreed among themselves.

biosphere The sum total of all of Earth's ecosystems.

biome One of many distinctive types of ecosystems determined by climate and identified by the predominant vegetation and organisms that have adapted to live there.

ecosystem All of the organisms in a given area plus the physical environment in which, and with which, they interact.

biotic The living (organic) components of an ecosystem, such as the plants and animals and their waste (dead leaves, feces).

abiotic The nonliving components of an ecosystem, such as rainfall and mineral composition of the soil.

population All the individuals of a species that live in the same geographic area and are able to interact and interbreed.

community All the populations (plants, animals, and other species) living and interacting in an area.

Visitors can tour the facility. Here, they view the desert biome.

INFOGRAPHIC 1 ORGANIZATION OF LIFE: FROM BIOSPHERE TO INDIVIDUAL

Ecologists recognize a nesting hierarchy of organization from the biosphere down to the individual organism. Each category is made up of the smaller ones.

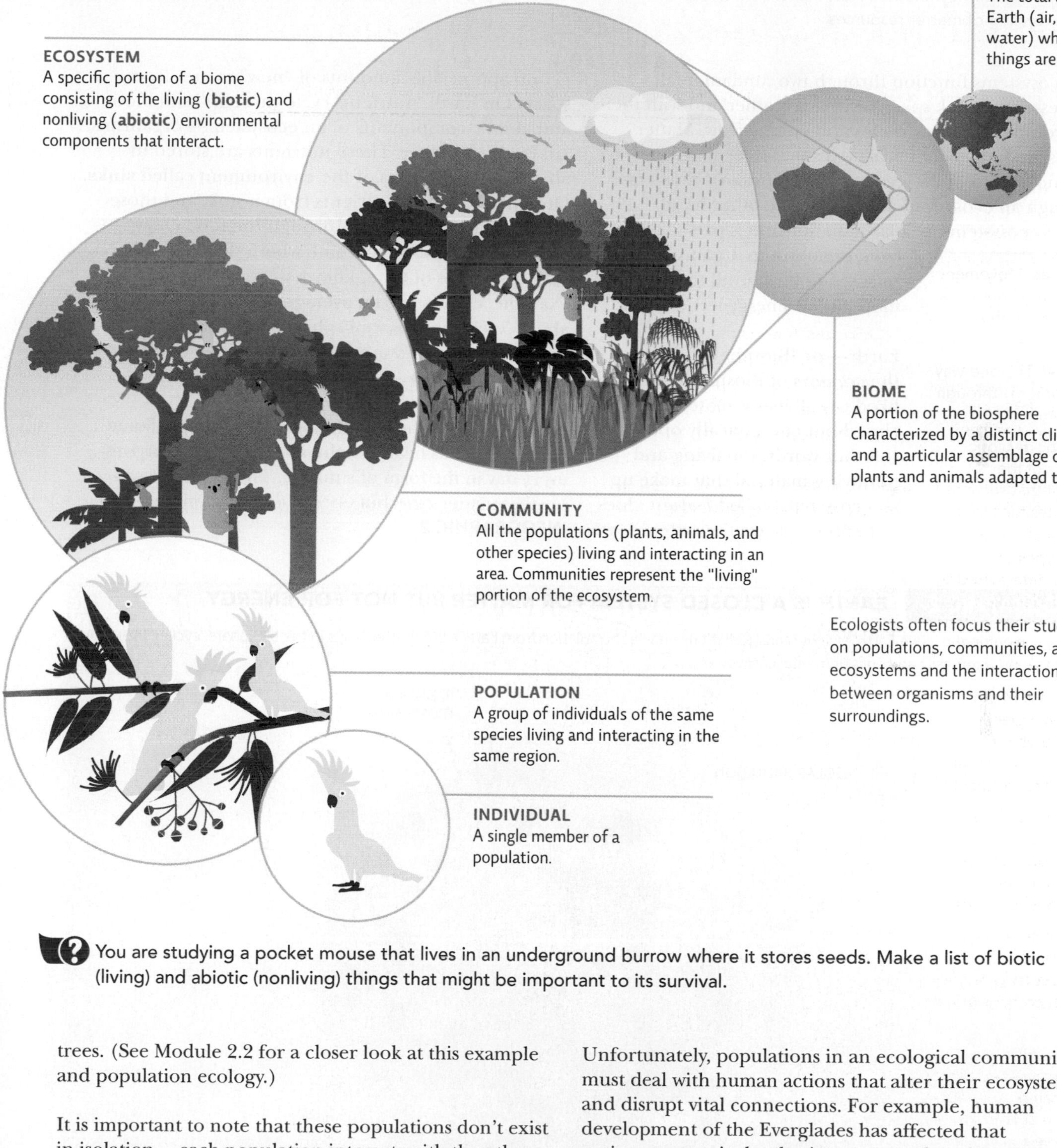

BIOSPHERE
The total area on Earth (air, land, or water) where living things are found.

ECOSYSTEM
A specific portion of a biome consisting of the living (**biotic**) and nonliving (**abiotic**) environmental components that interact.

BIOME
A portion of the biosphere characterized by a distinct climate and a particular assemblage of plants and animals adapted to it.

COMMUNITY
All the populations (plants, animals, and other species) living and interacting in an area. Communities represent the "living" portion of the ecosystem.

Ecologists often focus their study on populations, communities, and ecosystems and the interactions between organisms and their surroundings.

POPULATION
A group of individuals of the same species living and interacting in the same region.

INDIVIDUAL
A single member of a population.

? **You are studying a pocket mouse that lives in an underground burrow where it stores seeds. Make a list of biotic (living) and abiotic (nonliving) things that might be important to its survival.**

trees. (See Module 2.2 for a closer look at this example and population ecology.)

It is important to note that these populations don't exist in isolation—each population interacts with the other species around it. When ecologists study the collection of populations that live together and interact within a given area, they are focusing on the ecological **community**. **INFOGRAPHIC 1**

Unfortunately, populations in an ecological community must deal with human actions that alter their ecosystem and disrupt vital connections. For example, human development of the Everglades has affected that entire community by altering water cycles, changing the availability of nutrients and resources for every population in the ecosystem, and changing the way they interact with each other—an example covered in depth in Module 2.3.

2 ENERGY AND MATTER IN ECOSYSTEMS

Key Concept 2: A constant input of energy is always needed because once it is used by one organism, it cannot be reused by another. New inputs of matter to Earth, however, are negligible, so life also depends on the constant cycling of matter resources.

All ecosystems function through two fundamental processes that link species with each other and with their environment: **matter cycles** and **energy flow**. Matter cycles are biogeochemical cycles that refer specifically to the movement of life's essential chemicals or nutrients through an ecosystem. Energy, on the other hand, enters ecosystems as solar radiation and is passed along from organism to organism, some released as heat, until there is no more usable energy left.

matter cycles Movement of life's essential chemicals or nutrients through an ecosystem.

energy flow The one-way passage of energy through an ecosystem.

sinks Abiotic or biotic components of the environment that serve as storage places for cycling nutrients.

Earth—or "Biosphere 1," as the creators of Biosphere 2 liked to call it—is materially closed but energetically open. In other words, the living and nonliving material that make up an ecosystem, called *biomass*, does not enter or leave the system—we get no appreciable amounts of "new" matter from space. On Earth, nutrients cycle through both biotic and abiotic components of an ecosystem—organisms, air, land, and water. These nutrients are stored in abiotic or biotic parts of the environment called **sinks**. Organisms acquire nutrients from a sink, and those nutrients are then cycled through the food chain, eventually returning to a sink where they linger for various lengths of time, known as *residence times*. For example, carbon has an average residence time of about 5 years in the atmosphere, about 50 years in terrestrial soils, and more than 3,000 years in deep ocean sediments.

Energy, on the other hand, does come and go: Some energy leaves as heat or light, and new energy arrives every day in the form of sunlight. Therefore, we can say that matter *cycles* but energy *flows* in a one-way trip. **INFOGRAPHIC 2**

INFOGRAPHIC 2 EARTH IS A CLOSED SYSTEM FOR MATTER BUT NOT FOR ENERGY

Energy can enter and leave Earth as light (solar radiation) and heat (radiation from Earth), but matter stays in the biosphere, cycling in and out of organisms (biomass) and environmental components.

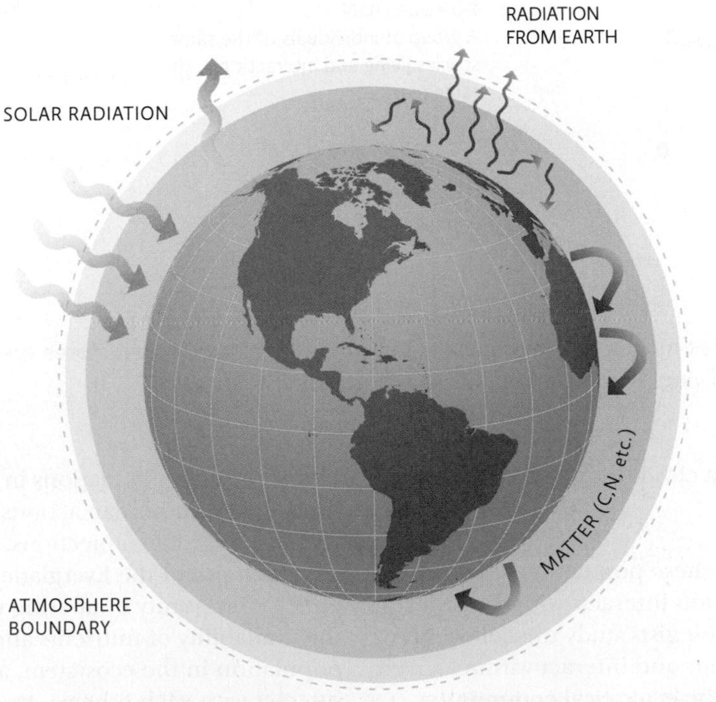

Why is it so important that species recycle matter and that they depend on a source of energy that is readily replenished (renewable)?

3 BIOMES

Key Concept 3: The biome that is present in a given area is influenced by the physical and climatic characteristics of its environment, particularly precipitation amount and temperature for terrestrial biomes.

Biomes are specific portions of the biosphere determined by climate and identified by the predominant vegetation and organisms adapted to live there. Biomes can be divided into three broad categories—terrestrial, marine, and freshwater. Within those three categories are several narrower groups, and within those are a variety of subgroups. For example, forests, deserts, and grasslands are the three main types of terrestrial biomes. Within the forest biome category are different types of forests, such as tropical, temperate, and boreal forests, and within each of those groups are subgroups (e.g., dry tropical forest and tropical rainforest). **INFOGRAPHIC 3**

When ecologists study entire ecosystems, their ability to manipulate such a broad study area is limited; they usually focus on making observations and trying to discern cause and effect from those observations. This is no small challenge; precise measurement of every relevant ecological factor is simply not possible, and even the simplest factor (e.g., a change in rainfall, the loss of a single species, variations in solar radiation reaching the ground) can impact many of the other factors and affect the ecosystem as a whole. But this doesn't mean we can't gain insights from ecosystem-scale studies. Scientists measure important parameters of natural ecosystems to gather evidence to try to determine cause and effect; they often compare those that are affected by natural disasters or human impact to less disturbed ecosystems in a kind of "natural experiment." From this, ecologists make their best estimations of how multiple factors affect one another (often via mathematical modeling). The more parameters that are properly measured and linked to one another, the better, but because all parameters and relationships cannot possibly be included, there may be considerable uncertainty when it comes to understanding what is happening at the ecosystem level.

Biosphere 2 offered ecologists an unprecedented research tool: a mini planet where a variety of environmental variables—from temperature and water availability to the relative proportions of oxygen and carbon dioxide (CO_2) at any given moment—could be tightly controlled and precisely measured. "Manipulating these variables and tracking the outcomes could greatly advance our understanding of natural ecosystems and all the minute, complex interactions that make them work," says Kevin Griffin, a Columbia University plant ecologist who conducted research at the Biosphere 2 facility. "The plan was to use that knowledge to figure out how to repair degraded ecosystems in the real world, so that they continue to provide the services so essential to our survival."

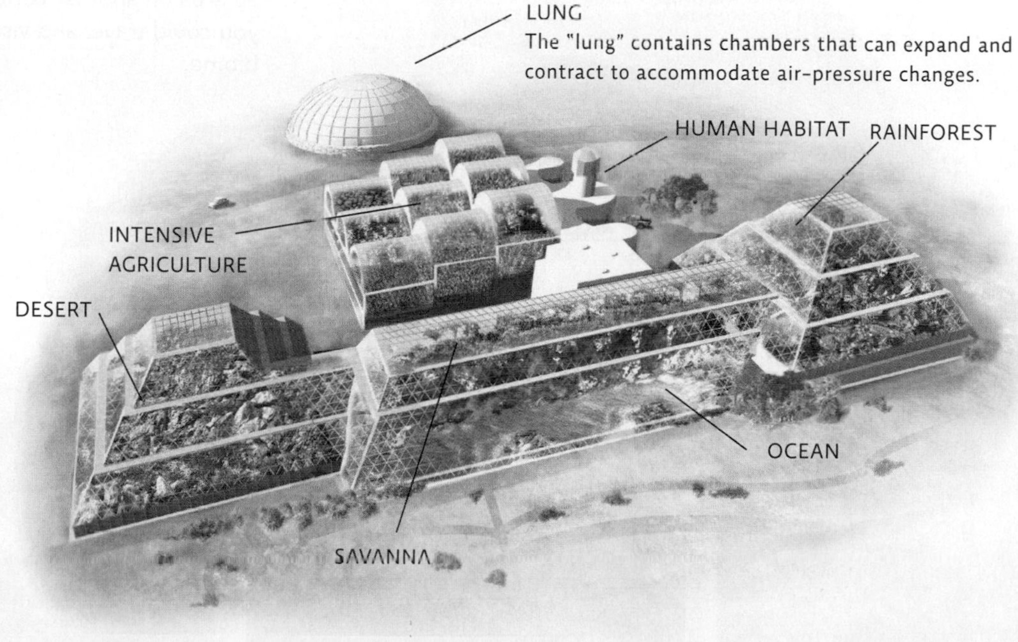

Biosphere 2 houses several biomes under one roof, each contributing to overall function. One of the challenges faced by designers was how to include a variety of biomes in the close quarters of the 3-acre Biosphere 2 structure. For example, in nature, a tropical rainforest, whose temperature is fairly constant, would not be next to an arid desert, which drops in temperature at night. To deal with this, an ocean was placed between the desert and rainforest to serve as a temperature buffer.

INFOGRAPHIC 3 GLOBAL TERRESTRIAL BIOMES

Achieve

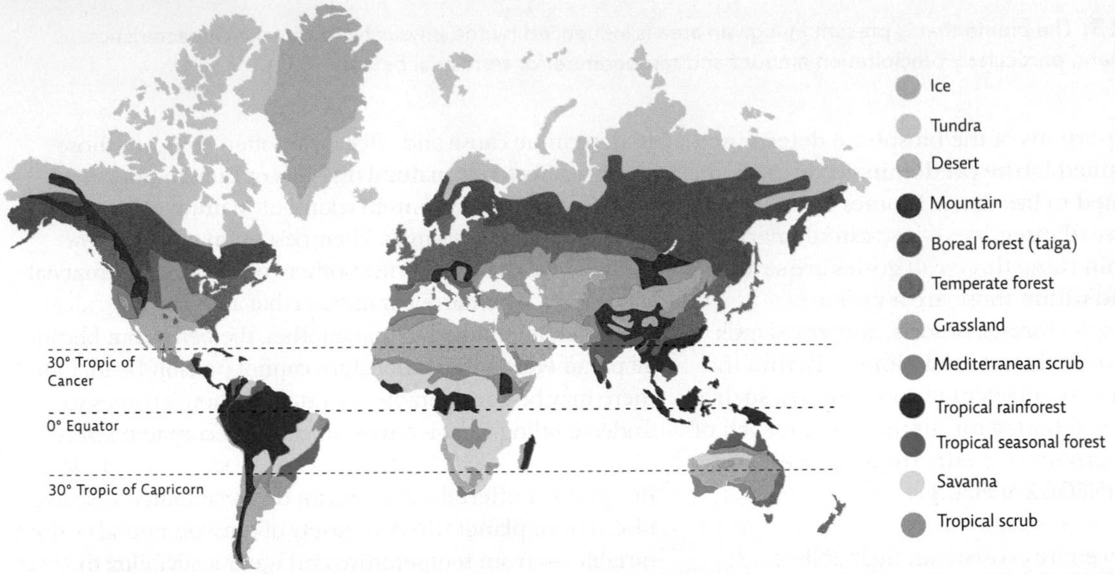

Terrestrial biomes are specific types of terrestrial ecosystems with characteristic temperature and precipitation conditions. Temperature varies with latitude (decreasing as one moves away from the equator) and altitude (decreasing as elevation increases); thus, a cold climate can be found above 60° north and south latitudes as well as on an equatorial mountaintop. Latitude also affects precipitation, with wet areas occurring at the equator and around 60° north and south and dry areas occurring around 30° north and south (due to global air circulation patterns).

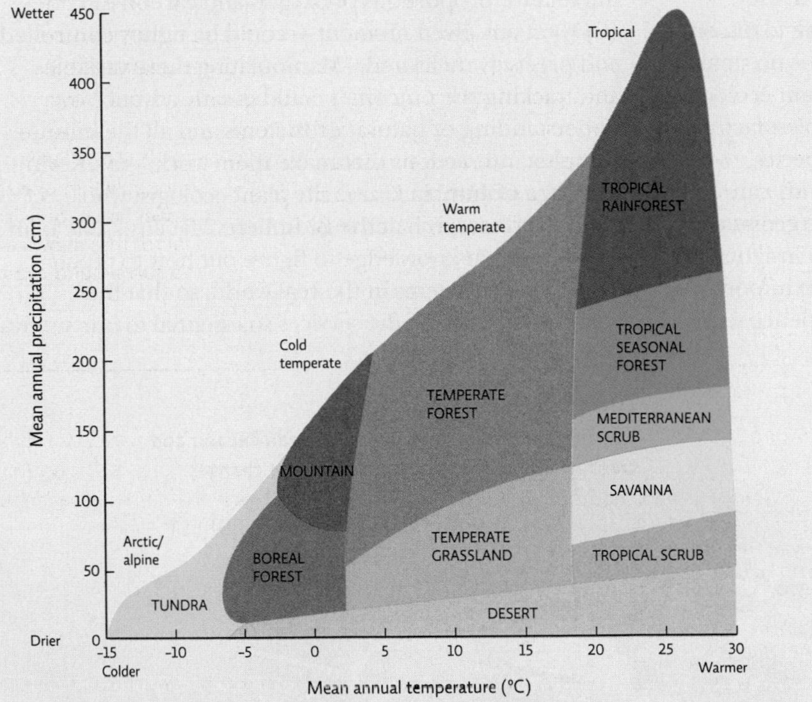

This biome climograph shows the approximate distribution of terrestrial biomes with regard to annual precipitation and temperature. As precipitation increases and more plant life is able to be supported, deserts, scrublands, or grasslands give way to forests. The temperature also influences what type of desert, grassland, or forest is present (for example, temperate forest versus boreal forest).

What biome do you live in? Identify an area on another continent where you could travel and visit that same biome.

DESERT

tonda/Getty Images

BOREAL FOREST

CSNafzger/Shutterstock

SAVANNA

Kondrachov Vladimir/Shutterstock

4 RANGE OF TOLERANCE AND ITS IMPACT ON SPECIES DISTRIBUTION

Key Concept 4: Limiting factors determine the distribution and size of populations. Variability within a population increases its range of tolerance, expanding its distribution and increasing the chance it will be able to adapt to changing conditions.

Planning each biome of Biosphere 2 required a mind-boggling array of considerations—not only how diverse plant and animal species would interact within and across biomes but also what nutrient requirements of each organism they planned to include. Termites, for example, would need enough dead wood at the beginning of closure to sustain them until some of the larger plants began dying off. Termites live in the soil, stirring it and allowing air to penetrate soil particles. If the termites ran out of dead wood and starved to death, organisms living in the soil would not get enough oxygen, and the entire desert would be jeopardized. Hummingbirds, on the other hand, would need nectar-filled flowers. "Try figuring out how many flowers a day a hummingbird needs," says Tony Burgess, a University of Arizona ecologist who helped design the biomes in Biosphere 2. "From there you need to know what the blooming season is, and then what the nectar load per flower is. And then you have to translate all of that into units of hummingbird support. Now imagine doing that sort of thing about 3,000 times."

Every species depends on a variety of needed resources; when one or more of these resources is in short supply, growth slows. In addition, living things can only survive and reproduce within a certain range (between the upper and lower limits that they can tolerate) for a given critical resource or environmental condition, referred to as their **range of tolerance** for that factor.

The more individual variability there is among members of a population, the wider the range of tolerance might be. Take temperature, for example. Some individuals might tolerate warmer or colder temperatures than others because they were "born that way"—it is encoded in their genes. Genetic differences that allow some individuals to tolerate or even thrive at the edges of the population's tolerance offer the population a chance to adapt to changing conditions (such as a warmer climate) if needed. The less genetically diverse the population is, the more narrow its range of tolerance and the less likely it will survive a change in conditions. **INFOGRAPHIC 4**

range of tolerance The range, within upper and lower limits, of a limiting factor that allows a species to survive and reproduce.

INFOGRAPHIC 4 RANGE OF TOLERANCE FOR LIFE

Populations have a range of tolerance for any given environmental factor—an upper and a lower limit beyond which it cannot survive (in this example, temperatures that are too cool or too hot). Most individuals, like the butterflies in this population, can be found around the optimum temperature, although what is "optimum" for each individual may differ slightly because of genetic variability. These differences expand the range of tolerance for a population as a whole, increasing the potential for it to adapt to changing conditions.

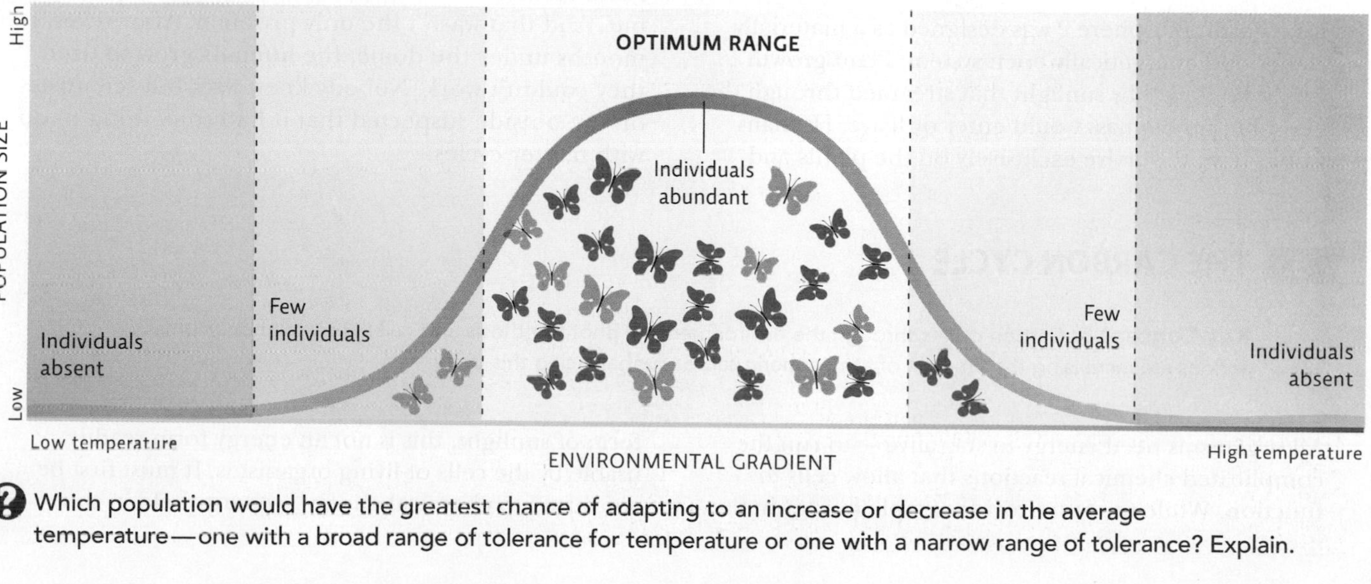

Which population would have the greatest chance of adapting to an increase or decrease in the average temperature—one with a broad range of tolerance for temperature or one with a narrow range of tolerance? Explain.

Ecologists routinely monitor a variety of limiting factors as a way of assessing ecosystem health, but anticipating what each individual organism would need to survive in Biosphere 2 before the fact proved daunting.

A dedicated team of scientists—oceanographers, forest ecologists, plant physiologists, and others—spent 2 years sorting through these challenges. Drawing on their combined expertise, they set about choosing the combination of soils, plants, and animals that seemed most capable of working together to re-create the delicate balances that had made Biosphere 1 such a spectacular success. Each biome was created from a carefully selected array of species: For example, the marsh biome was composed of intact chunks of swampland harvested from the Florida Everglades, and the savanna was composed of grasses from Australia, South America, and Africa.

C. Kurt Holter/Shutterstock

Organisms like these mushrooms (a fungus) growing on a decaying log are important consumers that help recycle nutrients in an ecosystem.

But it wasn't long before the rigor and pragmatism of good science began to clash with the idealism of Biosphere 2 financiers. And when that happened, critics say, science lost out. Some scientists worried that the ocean wouldn't get enough sunlight to support plant and animal life. Others opposed the use of soil high in organic matter; soil microbes decompose the soil's organic carbon and release it into the air as CO_2. Although this organic soil might eliminate the need for chemical fertilizers, there were concerns that it would provide too much fuel for the soil microbes and would thus send atmospheric CO_2 concentrations through the roof. Despite these concerns, scientific advisers to the project were overruled.

The first few months of Mission 1 went smoothly enough, but eventually, plants and animals started dying. Humans grew hungry and mysteriously sleepy.

Like Earth, Biosphere 2 was designed as a materially closed and energetically open system: Plant growth would be fueled by sunlight that streamed through the glass, but no biomass would enter or leave. Humans would have to survive exclusively on the plants and animals they could grow and harvest under the dome. The system as a whole would have to continuously recycle every last bit of nutrient that was in the soil on day 1.

At first, the carefully constructed agricultural biome seemed well suited to the challenge. Carrots, broccoli, peanuts, kale, lettuce, and sweet potatoes were grown on broad half-acre terraces that sat adjacent to the sprawling six-story human habitat. A bevy of domestic animals also provided sustenance—goats for milk, chickens for eggs, and pigs for pork. Indeed, eating only what they could grow made the biospherians healthier.

But the glazed glass of the dome admitted less sunlight than had been anticipated. Less sunlight meant less biomass production. And that meant less food. Now, a few months in, food was starting to run out. And that wasn't the only problem. After several months under the dome, the humans grew so tired they couldn't work. Nobody knew why, but scientists on the outside suspected that it had something to do with matter cycles.

5 THE CARBON CYCLE

Key Concept 5: Carbon cycles through the environment via photosynthesis and cellular respiration. Human actions are increasing the amount of atmospheric carbon, unbalancing this cycle.

All organisms need energy to stay alive—to run the complicated chemical reactions that allow cells to function. While the Sun provides ample energy in the form of sunlight, this is not an energy form readily usable by the cells of living organisms. It must first be converted to chemical energy—energy held in the chemical bonds of molecules. When those bonds are broken, the energy can be released for use by cells. So life on Earth needs a way to convert abundant solar energy into chemical energy. That process is **photosynthesis**.

Using solar energy, plants and other photosynthesizers combine carbon dioxide (CO_2) and water (H_2O) to produce a sugar molecule ($C_6H_{12}O_6$); oxygen (O_2) is a by-product of this reaction. For terrestrial plants, the most important sink or source of carbon is the atmosphere. (Oceans and soil are also abiotic sinks for carbon. Oceans absorb CO_2 directly from the atmosphere, and soils accumulate it during decomposition.) Because plants and other photosynthesizers "produce" sugar (an organic molecule) from inorganic atmospheric CO_2, they are called **producers**.

This sugar molecule represents stored chemical energy that the producer can use. A **consumer**, the organism that eats the plant (or that eats the organism that eats the plant), also uses the chemical energy of sugar. This energy is released to the cell via **cellular respiration**—a process in which the sugar molecule is broken down and the energy in the chemical bonds that held the atoms of the molecule together is released. This energy can be captured by cells to run cellular reactions needed by the organism. All organisms—producers and consumers—perform cellular respiration. (For more information on producers, consumers, and the food chain, see Module 2.3.)

From its initial incorporation into living tissue via photosynthesis to its ultimate return to the atmosphere through cellular respiration, carbon moves in and out of various molecular forms (CO_2 or sugar) and in and out of living things as it moves through the **carbon cycle**.

As mentioned earlier, human activity is changing the amount of carbon in the atmosphere, unbalancing the carbon cycle. When we burn fossil fuels for energy, we add extra CO_2 to the atmosphere—about 25% of this is absorbed by the oceans and another 25% has been taken up by terrestrial plants (due to a boost in photosynthesis). We also add extra CO_2 to the atmosphere when we burn large swaths of forest, most often to clear land for agriculture. In addition, deforestation has the net effect of increasing atmospheric CO_2—not just because more is added if the trees are burned, but because we remove the largest photosynthetic organisms on the planet that would be pulling CO_2 out of the atmosphere: trees. **INFOGRAPHIC 5**

In Biosphere 2, carbon cycled just like it does on Earth; carbon moved from living tissue to the atmosphere and back in the same predictable manner. But as the biospherians' energy waned, it became clear that something was going wrong.

It turned out that oxygen levels had fallen steadily—from 21% down to 14%. At such low concentrations, the Biosphere 2 residents were unable to convert the food they consumed into usable energy. "We were just dragging ourselves around the place," Jayne Poynter, one of the biospherians, says. "And we had sleep apnea at night. So we'd wake up gasping for air because our blood chemistry had changed."

In just a few months, some 7 tons of oxygen—enough to keep six people breathing for 6 months—had gone missing. As scientists from Columbia University later discovered, it had to do with that rich organic soil that the Biosphere 2 planners had insisted be used. The abundant soil microbes (consumers) were gobbling up all that O_2 and converting it into CO_2 as they decomposed the organic matter in the soil. The biospherians responded by filling all unused planting areas with morning glory vines, a pretty and fast-growing species they hoped would maximize the amount of CO_2 converted back into O_2 by photosynthesis. But even with an abundance of plants and enough CO_2, photosynthesis was still limited by the availability of sunlight; even the morning glories couldn't keep up with the soil microbes in their warm, well-watered, highly organic soil.

Adding to the confusion, concrete used to build parts of Biosphere 2 was absorbing some of the CO_2 and converting it into calcium carbonate, trapping some of the carbon and oxygen in this unexpected sink.

photosynthesis The chemical reaction performed by producers that uses the energy of the Sun to convert carbon dioxide and water into sugar and oxygen.

producer An organism that converts solar energy to chemical energy via photosynthesis.

consumer An organism that obtains energy and nutrients by feeding on another organism.

cellular respiration The process in which all organisms break down sugar to release its energy, using oxygen and giving off CO_2 as a waste product.

carbon cycle Movement of carbon through biotic and abiotic parts of an ecosystem via photosynthesis and cellular respiration as well as in and out of other reservoirs, such as oceans, soil, rock, and atmosphere.

INFOGRAPHIC 5 THE CARBON CYCLE

In photosynthesis, producers use solar energy to combine CO_2 and H_2O to make sugar ($C_6H_{12}O_6$), releasing O_2 in the process. When an organism (a producer or any consumer that eats another organism) needs energy, it breaks apart the sugar via the reverse reaction, cellular respiration. Oxygen is required for this reaction (which is why it is called "respiration").

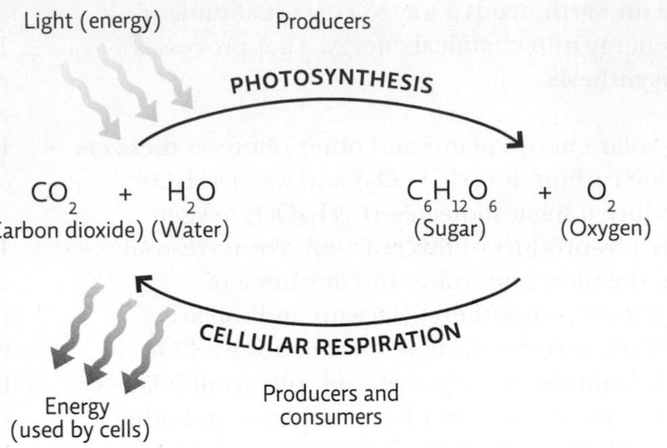

Light (energy)
Producers
PHOTOSYNTHESIS
CO_2 + H_2O
(Carbon dioxide) (Water)
$C_6H_{12}O_6$ + O_2
(Sugar) (Oxygen)
CELLULAR RESPIRATION
Energy (used by cells)
Producers and consumers

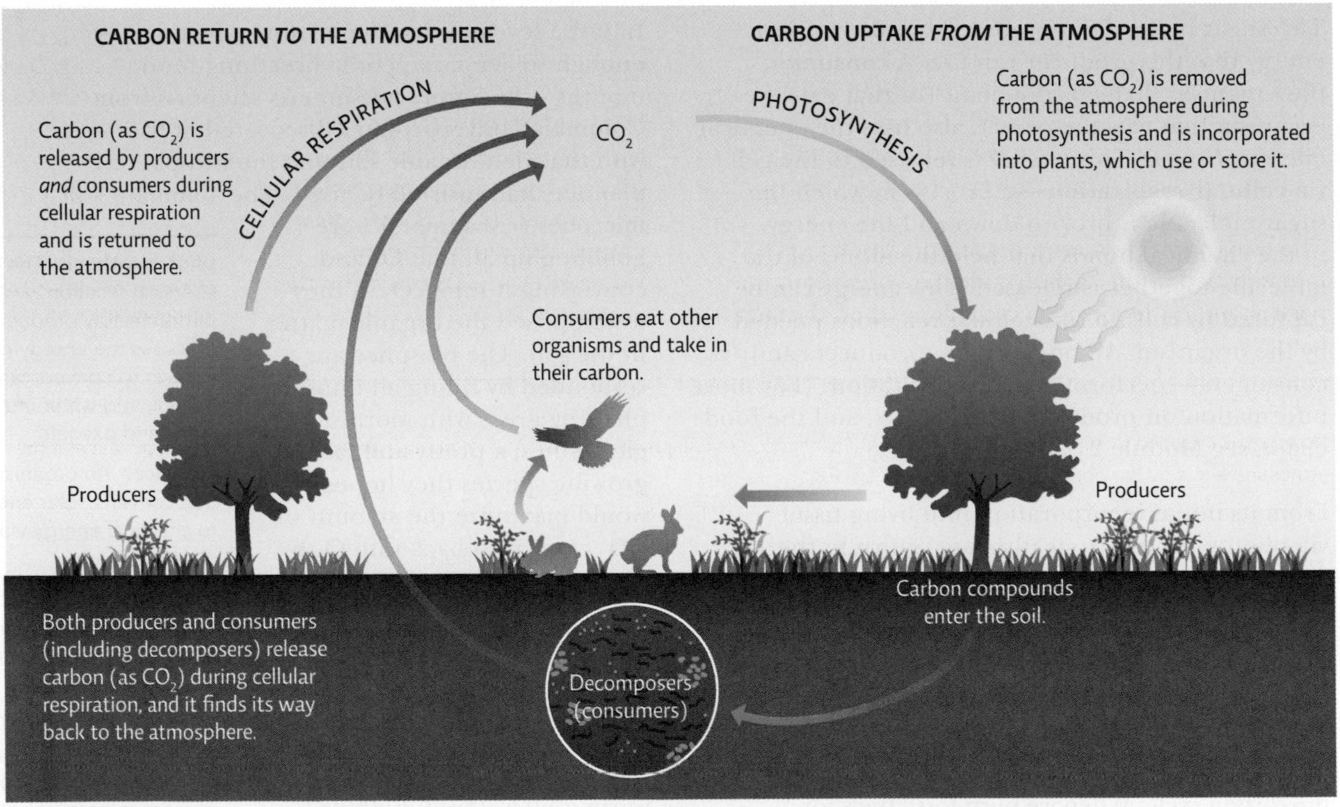

CARBON RETURN *TO* THE ATMOSPHERE
Carbon (as CO_2) is released by producers *and* consumers during cellular respiration and is returned to the atmosphere.
CELLULAR RESPIRATION
CO_2
CARBON UPTAKE *FROM* THE ATMOSPHERE
Carbon (as CO_2) is removed from the atmosphere during photosynthesis and is incorporated into plants, which use or store it.
PHOTOSYNTHESIS
Consumers eat other organisms and take in their carbon.
Producers
Producers
Carbon compounds enter the soil.
Both producers and consumers (including decomposers) release carbon (as CO_2) during cellular respiration, and it finds its way back to the atmosphere.
Decomposers (consumers)

Carbon cycles in and out of living things during photosynthesis and cellular respiration. As consumers eat other organisms, carbon is transferred. Most of Earth's carbon is actually stored in rocks or dissolved in the planet's oceans, but some carbon is stored in the bodies of organisms and in soil.

The balance of exchange between carbon sinks is affected by human activities that increase the amount of CO_2 in the atmosphere.

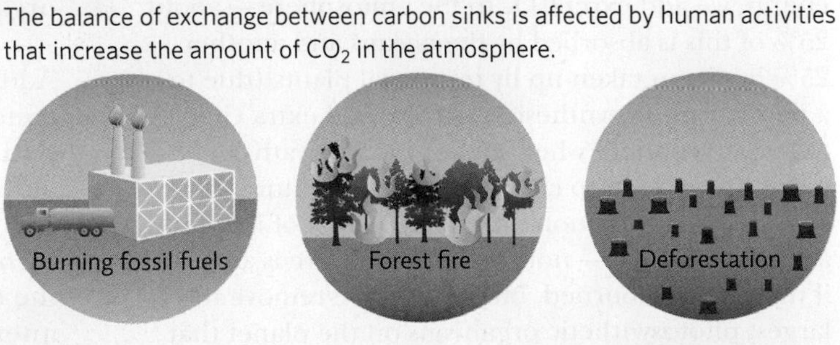

Burning fossil fuels
Forest fire
Deforestation

Explain how each of these human impacts (burning fossil fuels, forest fires, and deforestation) can result in a net increase in atmospheric carbon.

6 THE NITROGEN CYCLE

Key Concept 6: Nitrogen cycles through the environment in steps that depend on a wide variety of bacteria. Human actions are increasing the amount of usable nitrogen in the environment.

Besides carbon, other chemicals essential for life, such as nitrogen and phosphorus, cycle through ecosystems. Nitrogen, the most abundant element in Earth's atmosphere, is needed to make proteins and nucleic acids, but plants cannot utilize nitrogen in its atmospheric form (N_2). All plant life, and ultimately all animal life, too, depends on microbes (bacteria) to convert atmospheric nitrogen into usable forms as part of the **nitrogen cycle**.

In a process called **nitrogen fixation**, bacteria convert atmospheric nitrogen (N_2) into ammonia (NH_3), which plants subsequently take up through their roots; consumers take in nitrogen via their diet. In other steps of the nitrogen cycle, different species of bacteria are responsible for **nitrification**, the conversion of ammonia to nitrate (NO_3^-)—another form that can be directly used by the bacteria or taken up by plants. Still other bacteria convert nitrate back to molecular nitrogen (N_2), which finds its way back into the atmosphere in the process of **denitrification**.

The nitrogen cycle does not operate in isolation—it is closely tied to the carbon cycle. Soil organisms that participate in the nitrogen cycle depend on the sugars produced by plants for their main source of carbon. In return, they provide the plants with much-needed nitrogen. (The same relationship is seen with bacteria of the phosphorus cycle.)

Like the carbon cycle, human impact is also disrupting the nitrogen cycle. In fact, our use of fertilizers, the release of animal or human waste (sewage) into the environment, and vehicle emissions are adding as much fixed nitrogen to the environment as the natural nitrogen cycle does—a doubling of available nitrogen! This excess means nitrogen may no longer be a limiting factor for plant growth, but different plants respond to this in different ways. So the makeup of an ecological community can change, with ripple effects felt throughout the ecosystem. **INFOGRAPHIC 6**

nitrogen cycle A continuous series of natural processes by which nitrogen passes from the air to the soil to organisms and then returns back to the air or soil.

nitrogen fixation Conversion of atmospheric nitrogen into a biologically usable form, carried out by bacteria found in soil or via lightning.

nitrification Conversion of ammonia (NH_3) to nitrate (NO_3^-).

denitrification Conversion of nitrate to molecular nitrogen (N_2).

INFOGRAPHIC 6 THE NITROGEN CYCLE

Achieve

Nitrogen, needed by all living things to make biological molecules like protein and DNA, continuously moves in and out of organisms and the atmosphere in a cycle dependent on a variety of bacteria.

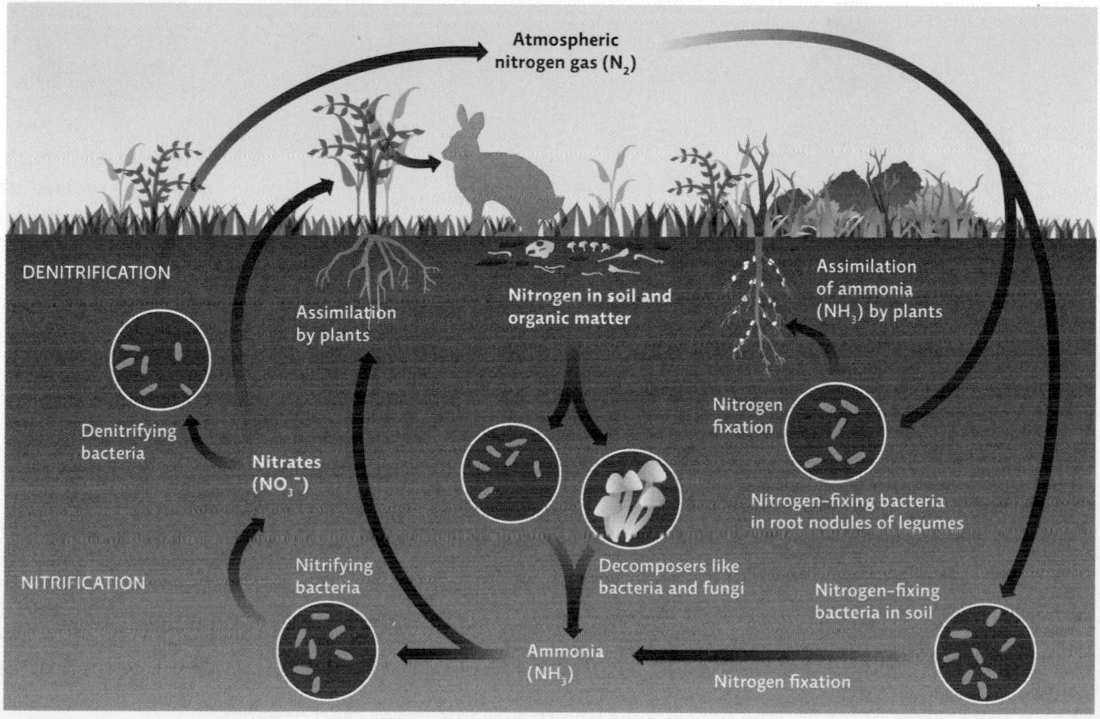

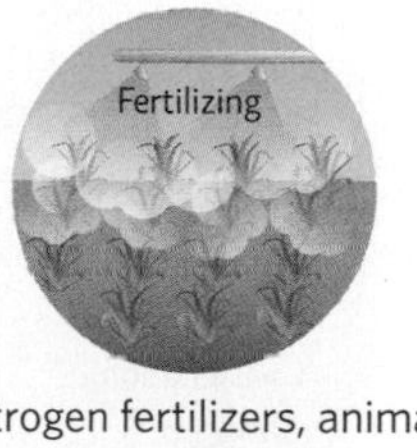

Nitrogen fertilizers, animal waste, and sewage promote plant growth, but this can deplete other soil nutrients and pollute aquatic ecosystems.

Burning fossil fuels contributes to nitrogen pollution such as smog and acid rain.

? Look closely at the nitrogen cycle. How many different types of microbes are needed to complete the entire cycle?

The biospherians were not immune to nitrogen cycle disruptions. Thanks to the overabundance of soil microbes in Biosphere 2, levels of nitrous oxide, a normal by-product of denitrification, got high enough to interfere with vitamin B_{12} metabolism, affecting the biospherians' nervous systems.

7 THE PHOSPHORUS CYCLE

Key Concept 7: Phosphorus moves slowly through the environment, depending on physical and biological processes. Human impact has also unbalanced this cycle.

The phosphorous cycle was also disrupted in Biosphere 2. Unlike nitrogen and carbon, phosphorus—which is needed to make important biological molecules like DNA and RNA—is found only in solid or liquid form on Earth, so the **phosphorus cycle** does not move through the atmosphere. Instead, phosphorus passes from inorganic to organic form through a series of interactions with water and organisms. Like the nitrogen cycle, the phosphorus cycle has been disrupted by the release of additional phosphorus into the environment by human activities such as mining, the use of fertilizers, and the release of sewage or animal waste. (See Module 8.1 for more on the environmental impacts of the agricultural use of nitrogen and phosphorus.) **INFOGRAPHIC 7**

phosphorus cycle A series of natural processes by which the nutrient phosphorus moves from rock to soil or water to living organisms and back to soil.

In Biosphere 2, phosphorus got trapped in the water system, polluting aquatic habitats, because the underwater and terrestrial plants there were dying off too quickly to complete this cycle. Biospherians removed excess nutrients from their water supply by passing the water over algal mats that would absorb the phosphorus and could then be harvested, dried, and stored.

As food reserves dwindled, the eight biospherians split into two factions. One group felt that scientific research

INFOGRAPHIC 7 THE PHOSPHORUS CYCLE

Achieve

Phosphorus, needed by all organisms to make DNA, cycles very slowly. It has no atmospheric component but instead depends on the weathering of rock to release new supplies of phosphate (PO_4) into bodies of water or the soil, where it dissolves in water and can be taken up by organisms. Microbes also play a role when they break down organic material and release the phosphate to the soil.

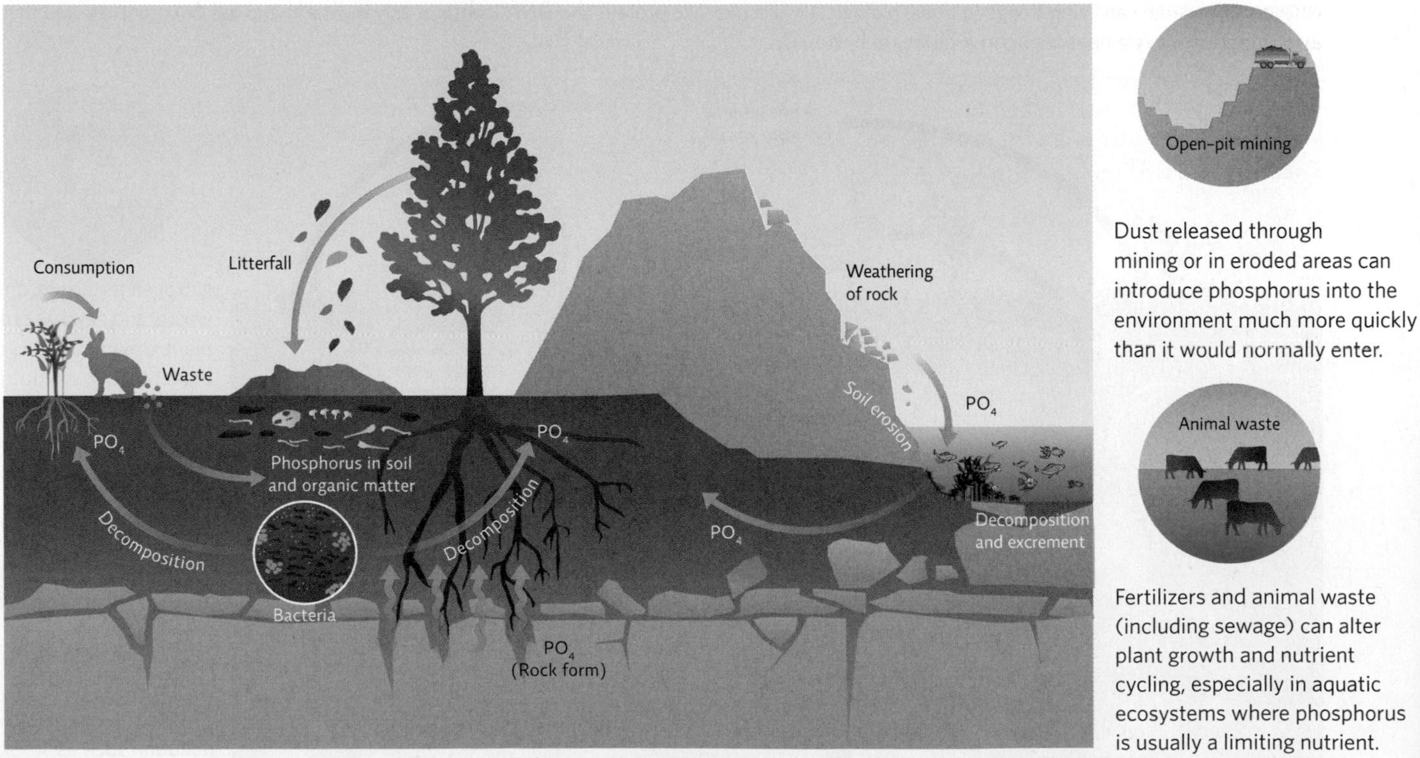

Dust released through mining or in eroded areas can introduce phosphorus into the environment much more quickly than it would normally enter.

Fertilizers and animal waste (including sewage) can alter plant growth and nutrient cycling, especially in aquatic ecosystems where phosphorus is usually a limiting nutrient.

? How might phosphorus from farms enter aquatic ecosystems (rivers, streams, lakes, and oceans)?

was the top priority and wanted to import food so that they would have enough energy to continue with their experiments. The other group felt that maintaining a truly closed system—one where no biomass was allowed to enter or leave—was the project's most important goal; proving that humans could survive exclusively on what the dome provided would be essential to one day colonizing the Moon or Mars. To them, importing food would amount to a mission failure.

"It was a heartbreaking split," Poynter says. "Just 6 months into the mission, and two people on the other side of the divide had been my closest friends going in." Eventually, Poynter snuck in food. That wasn't the only breach. To solve the various nutrient cycle conundrums, the project's engineers installed a CO_2 scrubber and pumped in some 4 million gallons of oxygen.

Three years after the first mission was completed, the editors of the respected journal *Science* deemed the entire project a failure. "Isolating small pieces of large biomes and juxtaposing them in an artificial enclosure changed their functioning and interactions, rather than creating a small working Earth as originally intended," they wrote. For the $200 million dome to survive as a scientific enterprise, they concluded, it would need dramatic retooling.

Biosphere 2 might not have met its original goal of creating a self-supporting human habitat, but it was not a failure as some reports claimed. Negative results can be just as informative as positive ones—in some cases even more so because they uncover gaps in our knowledge and help us decide how to move forward. Biosphere 2 taught scientists that Earth is an exceptionally complex and dynamic place—ecosystem components intertwine and respond to one another in many complicated ways. Each is governed by a countless array of interacting factors, and a change in one can set off a whole chain of events that degrade the system's capacity to sustain life. This complexity will make it difficult, if not impossible, for humans to re-create a fully functional and self-sustaining biosphere habitat capable of supporting human life long term.

In fact, Biosphere 2's greatest liability—its skyrocketing CO_2 levels—is proving to be a valuable asset. "Now it's like a time machine," says Griffin, who points out that Biosphere 2 is allowing us a look at the consequences of elevated atmospheric CO_2 levels, the main contributor to climate change today. Recent research by Griffin has uncovered some of the complexities of carbon cycling. His group saw unexpected fluctuations in carbon release at various levels in the tree canopy, telling him there is much we still don't know about how carbon cycles—data that could only be gathered in an enclosed forest such as that found in Biosphere 2.

Today, school groups and tourists can visit Biosphere 2, and scientists from all over the world still use the facility to study a variety of research topics at scales not possible outside of the facility. For example, the ambitious Landscape Evolution Observatory Project has constructed three large identical artificial landscapes that will allow scientists to precisely monitor how these landscapes change over time as climate parameters such as temperature, humidity, and wind speed are manipulated—the triplicate landscapes allow researchers to assess inherent variability in the complex process of landscape evolution.

Ultimately, though, the most valuable lesson Biosphere 2 has provided might just be how irreplaceable Biosphere 1 is.

Justin Peterson, a Biosphere 2 undergraduate intern, assists PhD candidate Henry Adams with his Pinon Pine Tree Drought Experiment. The experiment's goal is to predict the effects of climate shifts on the trees.

Select References:

Cohen, J. E., & Tilman, D. (1996). Biosphere 2 and biodiversity: The lessons so far. *Science, 274*(5290), 1150–1151.

Griffin, K. L., et al. (2002). Canopy position affects the temperature response of leaf respiration in *Populus deltoides. New Phytologist, 154*(3), 609–619.

Marino, B., et al. (1999). The agricultural biome of Biosphere 2: Structure, composition and function. *Ecological Engineering, 13*(1), 199–234.

Nelson, M. (2018). *Pushing our limits: Insights from Biosphere 2.* Tucson, AZ: University of Arizona Press.

Pangle, L. A., et al. (2015). The Landscape Evolution Observatory: A large-scale controllable infrastructure to study coupled Earth-surface processes. *Geomorphology, 244,* 190–203.

Poynter, J. (2009). TED Talk: *Life in Biosphere 2.* Retrieved from www.ted.com/talks/jane_poynter_life_in_biosphere_2?language=en

GLOBAL CASE STUDIES BIOMES

Achieve

The different biomes of Earth reflect different climatic and physical conditions of the places where they are found. These case studies present more details about the physical and biological makeup of four very different biomes.

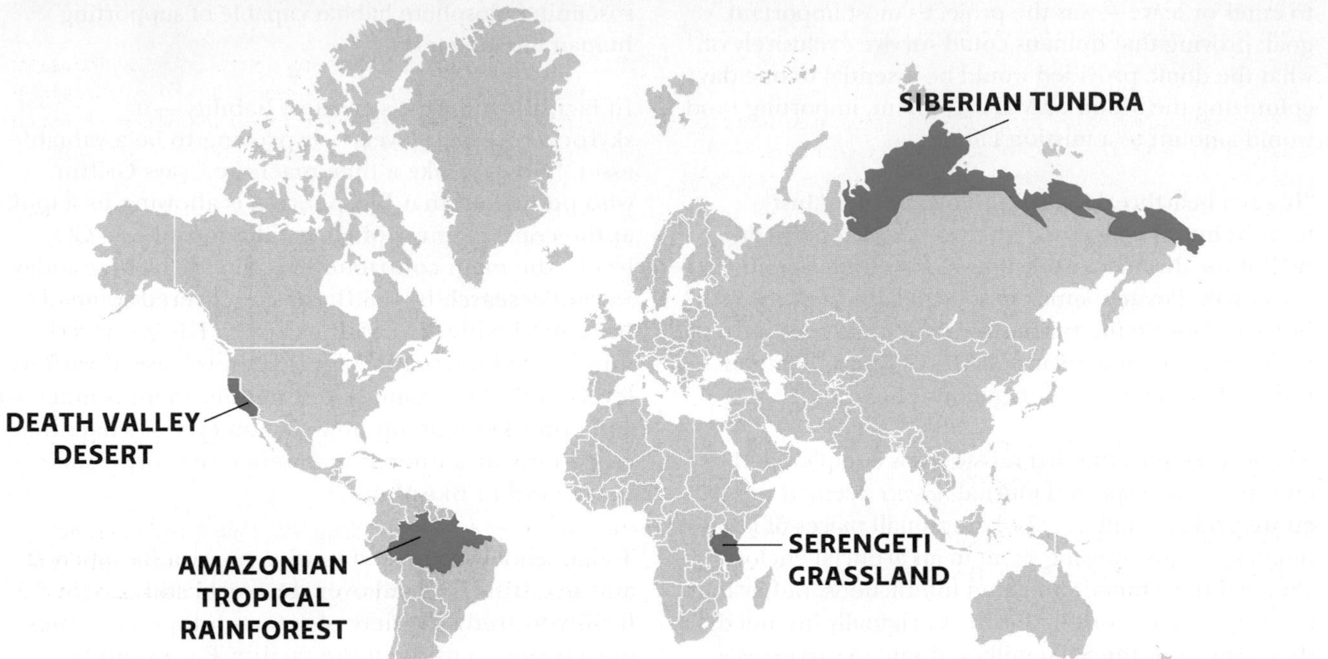

BRING IT HOME

PERSONAL CHOICES THAT HELP

Nutrient cycling is critical for maintaining Earth's ecosystems, but we interfere with nutrient cycles through our daily activities. Driving a car interferes with the carbon cycle by releasing carbon from fossil fuel reservoirs. Applying synthetic chemical fertilizers to food crops interferes with the nitrogen cycle by adding soluble nitrogen compounds to aquatic ecosystems through runoff. The challenge is to figure out ways to work with nutrient cycles rather than against them — in other words, to return nutrients to the sinks from which they come. How might this be done in our daily lives and in our own communities?

Individual Steps

• Reduce your fossil fuel use to curtail carbon, nitrogen, and sulfur emissions. Take public transportation, walk, ride a bike, and drive a fuel-efficient vehicle.

• Compost food and yard waste. Then use this material to fertilize flower beds, trees, and garden plots. Composting will reduce or eliminate the need for synthetic fertilizers in your yard or garden.

• Support legislation to increase fuel efficiency of vehicles and subsidies for clean, renewable energy. More fuel-efficient cars and cleaner energy sources will reduce our carbon outputs from fossil fuel use.

Group Action

• Participate in or organize an event to plant trees or native grasses. By doing so, you can help recapture the carbon put into the atmosphere by driving a car.

• Many urban areas are grateful for individuals being willing to plant native trees, shrubs, and wildflowers along roadways or in parks and other public spaces.

• Public policy currently prevents large-scale composting of municipal wastes in many areas. Working to change these policies will extend the life of our landfills and make use of valuable nutrient-rich materials.

Peter Essick/Cavan/Alamy

REVISIT THE CONCEPTS

- Ecologists identify a nesting organization of life. From small to large, this is individual, population, community, ecosystem, biome, and biosphere. Ecologists often focus on the levels of population, community, and ecosystem to examine how species respond to and affect the natural world.

- Ecosystems need a constant input of energy because once energy is used by one organism, it cannot be reused by another. However, species do recycle matter, so the fact that Earth gets no appreciable amounts of new matter is not a problem.

- The biome that is present in a given area is influenced by the physical and climatic characteristics of its environment, such as precipitation and temperature. Rainfall and temperature are examples of limiting factors that determine the distribution and size of populations within their environment.

- Genetic variability within a population increases its range of tolerance, expanding its distribution and increasing the chance it will be able to adapt to changing conditions.

- Carbon cycles through the environment via photosynthesis (the process by which plants trap and convert solar energy into sugar, a form of chemical energy that cells can use) and cellular respiration (the release of usable energy when the sugar is broken down).

- Nitrogen, needed to make important biomolecules like protein, cycles in steps that depend on a wide variety of soil bacteria.

- Phosphorus moves slowly through the environment, depending on physical and biological processes. Human impact has unbalanced all three of these important nutrient cycles, altering community composition worldwide.

ENVIRONMENTAL LITERACY Understanding the Issue

1 What is the hierarchy of organization recognized by ecologists, and why might it be useful to recognize such distinctions?

1. List the six ecological organizational categories from small to large, starting with *individual*.
2. Why do ecologists focus mainly on the study of populations, communities, and ecosystems?
3. Give examples of biotic and abiotic factors important to ecosystems.

2 Why do ecosystems need a constant input of energy yet can handle the fact that they do not receive appreciable new inputs of matter?

4. Explain what it means to say that Earth is energetically open but a closed system with regard to matter.
5. Relate the concept of matter sinks and residence time to matter cycling.

3 How do environmental factors affect the distribution and makeup of biomes?

6. What two main climatic factors determine which biome is present at a given location?
7. Look at the location of temperate forests on Earth in the biome climograph in Infographic 3. What can you surmise about the climatic conditions in North America and Europe/Asia where these forests are found?

4 What is a population's range of tolerance, and how does it affect the distribution of a population or its ability to adapt to changing conditions?

8. Using the example of spring wildflowers and the critical factor of rainfall, explain the term *range of tolerance* (in terms of the distribution of wildflowers within their range).
9. Human actions are shrinking populations and reducing their genetic variation. How might this affect the ability of a population to respond to an environmental alteration like climate change?

5 How does carbon cycle through ecosystems, how have humans disrupted this cycle, and what problems can this disruption cause?

10. Draw the equation for photosynthesis and then insert an arrow to show the direction of cellular respiration.
11. What is the purpose of photosynthesis and cellular respiration? Who performs these reactions: producers, consumers, or both?

6 How does nitrogen cycle through ecosystems, how have humans disrupted this cycle, and what problems can this disruption cause?

12. What effect would a wildfire that burned so hot that it sterilized the soil, killing all the microbes, have on the nitrogen cycle?

7 How does phosphorus cycle through ecosystems, how have humans disrupted this cycle, and what problems can this disruption cause?

13. What steps did the biospherians take to address the imbalance in the phosphorus cycle in Biosphere 2? Why was this an effective solution?
14. Consider the two human impacts that unbalance the nitrogen cycle identified in Section 7. Which one also impacts the phosphorus cycle?

SCIENCE LITERACY Working with Data

An experiment was done at Biosphere 2 to test the effect of drought and higher temperatures on pinyon pines. Estimates of photosynthetic activity and cellular respiration were made by measuring CO_2 uptake or output (measured in terms of micromoles [μmoles] of carbon exchange per square meter of leaf area per second).

Look at Graphs A and B, and answer the following questions.

A: MIDDAY NET PHOTOSYNTHESIS

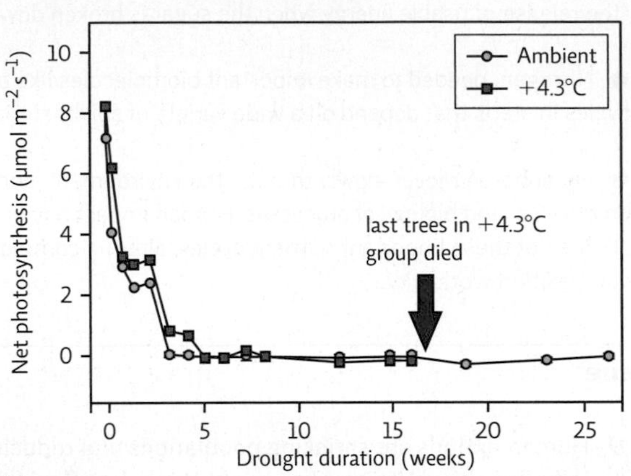

B: PREDAWN CELLULAR RESPIRATION

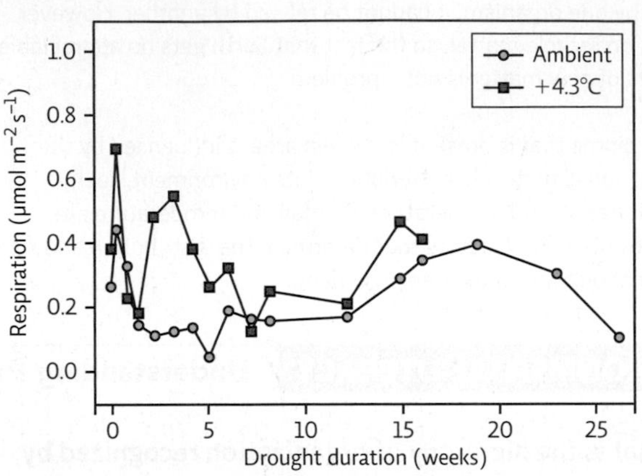

Interpretation

1. Look at Graph A. What is the general trend for photosynthetic activity over the course of the experiment for both the ambient and elevated temperature groups?
2. Look at Graph B. What is the general trend for cellular respiration activity over the course of the experiment for both the ambient and elevated temperature groups?

Advance Your Thinking

3. Why did cellular respiration also fall off toward the end of the experiment? (**Hint:** Link this back to the trees' photosynthetic activity.)
4. What can you conclude about the cause of death of the trees?
5. The researchers also measured the ability of the trees to take up and distribute water to their tissues in all the test trees and saw no difference — no breakdown in the ability to do this in the drought-stricken trees at either test temperature. Why did they collect these data?

INFORMATION LITERACY Evaluating Information

Earth is vastly larger than Biosphere 2, but its resources are just as finite. People must decide, for each biome, whether to leave some places untouched or use all of an area and its resources for human purposes. One example would be plowing under an entire prairie and growing wheat, displacing all of the native plants and animals. The rainforests are another such biome. Left alone, they produce huge amounts of oxygen for the whole planet and also support millions of species. Many humans want to use the trees for lumber and the land to grow crops or raise cattle.

Can we use rainforests in sustainable ways? Go to the Rainforest Alliance's website (www.rainforest-alliance.org) and read about "Sustainable Agriculture Certification" and Rainforest Alliance-certified products such as coffee, tea, and cocoa.

Evaluate the website and work with the information to answer the following questions:

1. Evaluate the reliability of this information source:
   a. Identify a claim the organization makes about sustainable agriculture and the evidence it gives in support of this claim. Is it convincing?
   b. Does it give sources for its evidence?
2. Search the website and read about Rainforest Alliance Certification. What does it mean to be Rainforest Alliance Certified^{TM}?
3. Using the search tool on this website, enter the key word *cocoa* and read about Rainforest Alliance Certified cocoa.
   a. In your opinion, is certification an adequate tool to ensure sustainability in cocoa farming? Explain.
   b. Would you be willing to buy cocoa (or another food product) — and pay more for this product — if you knew it was sustainably produced? Explain.

Additional study questions are available at Achieve.macmillanlearning.com. Achieve

Module 2.2 Population Ecology

GRAY WOLVES RETURN TO YELLOWSTONE

Endangered gray wolves return to the American West

Biologists track wolves from a helicopter. Once they spot a wolf, they dart it with a tranquilizer so that they can fit it with a radio collar. The wolf is unharmed.

BARRETT HEDGES/National Geographic Creative

After reading this module, you should be able to answer these GUIDING QUESTIONS

1. What is a population, and why do ecologists study them?
2. What population distributions are seen in nature?
3. What is the importance of population size and density?
4. What is exponential growth, and when does it occur in a population?
5. What is logistic growth, and when does it occur in a population?
6. How do density-dependent and density-independent factors affect population growth?
7. What are the life-history strategies of r- and K-selected species, and how do they relate to population growth patterns and their ability to respond to environmental changes?
8. What are top-down and bottom-up regulation, and how do they contribute to trophic cascades?

At least half a dozen times each winter, Doug Smith climbs into a helicopter, gun in tow, and hunts wolves (*Canis lupus*) in Yellowstone National Park. He's not looking to kill them—just to put them to sleep for a little while so that he can outfit them with radio collars to track the sizes of their packs, what they eat, and where they go over the course of the following year. Smith, a biologist, spends about 200 hours per year on these "hunting" expeditions, which are part of the Yellowstone Gray Wolf Restoration Project that he leads. The project has been responsible for reintroducing a total of 41 wolves to Yellowstone since 1995, after their disappearance as a result of predator control programs implemented by the U.S. government in the early 20th century.

Sometimes, though, things go awry on Smith's radio collar missions. For instance, the tranquilizer dart doesn't fully sedate big wolves—some of which can reach 175 pounds. Smith is forced to approach the animals while they're still awake. "I have to grab them on the scruff of the neck and manhandle them until I get the collar on," he explains. "They're typically not dangerous then—they've had enough drug to be kind of out of it—but they're still able to walk around. It's a wild experience." Sometimes the wolves—who are typically frightened of the helicopter—try to attack it while it's hovering with the doors open, just out of their reach. "I've had two females turn and run at the helicopter, teeth gnashing, jumping up trying to get me," Smith recalls. "I'm hovering above it, going back and forth, and I can't get a shot because all I'm seeing is face and teeth." Despite these adventures, Smith says a wolf has never actually bitten him—and he has tranquilized and collared more than 300 of them.

WHERE IS YELLOWSTONE?

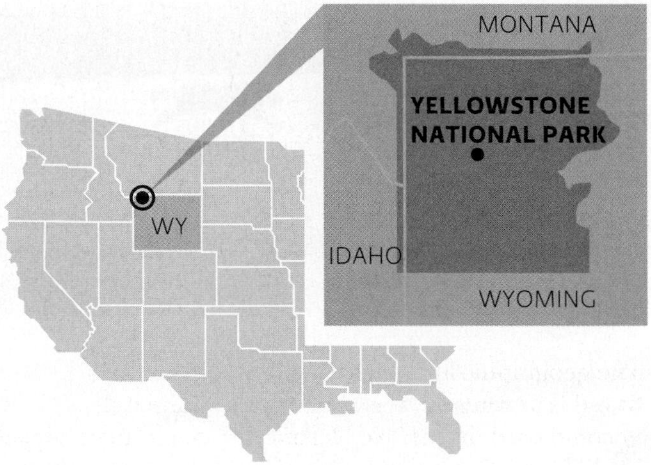

1 POPULATION DYNAMICS

Key Concept 1: Ecologists study populations to better understand what makes them thrive, decline, or become overpopulated in an effort to manage them and the other populations they impact.

population All the individuals of a species that live in the same geographic area and are able to interact and interbreed.

Ecologists monitor **populations** of organisms in ecosystems around the world for a variety of reasons, whether to protect endangered species, to manage economically valuable species such as commercial fisheries or timber, or to control pests. (See Module 2.1 for an introduction to the concept of an ecological population.) In Yellowstone, Smith and others monitor wolf populations. Elk (*Cervus elaphus*) populations, a popular game species, are also closely monitored, as are the populations of aspen, willow, and cottonwood trees, the most important winter food for elk. In some cases, elk numbers and aspen growth are closely tied to the size of the wolf population—but not in all areas. It turns out that many factors play a role in the sizes of these populations.

National Park Service

Doug Smith and Yellowstone Delta wolf 487M, the largest wolf collared in the winter of 2005.

The geographic area where a given population is found is its **range**. The historic range of the wolf encompassed most of North America, but human impact has reduced that range tremendously. Smith and his colleagues monitor the range of the Yellowstone wolves to understand how well the population is doing. As the wolf population grew and new packs formed, the wolves expanded their range into other areas of the Northern Rockies outside of the park boundaries. If their range begins to shrink, this might be a warning that habitats are declining in quality or that other pressures are restricting the movement of wolves. **INFOGRAPHIC 1**

range The geographic area where a species or one of its populations can be found.

2 POPULATION DISTRIBUTIONS

Key Concept 2: A population's distribution within its range is influenced by behavioral and ecological factors and is a reflection of how individuals interact with each other and their environment.

In addition to the geographic range occupied by a population, another important feature is **population distribution**, or the location and spacing of individuals within their range. A number of factors affect distribution, including species characteristics, topography, and habitat makeup. Ecologists typically speak of three types of distribution. In a **clumped distribution**, individuals are found in groups or patches within the habitat. Yellowstone examples include social species like the prairie dog or beaver that are clustered around a necessary resource, like water. Wolves travel in packs and therefore have a clumped distribution. Elk, one of their prey species, congregate as well; living in herds offers some protection against the wolves.

population distribution The location and spacing of individuals within their range.

clumped distribution A distribution in which individuals are found in groups or patches within the habitat.

INFOGRAPHIC 1 **GEOGRAPHIC RANGE OF THE GRAY WOLF**

The range of a species represents its geographic distribution. Distinct populations may exist within the range, especially if the species is broadly distributed. A small population of the Mexican wolf, a subspecies of the gray wolf that once ranged widely from central Mexico throughout the Southwest, is being restored to Arizona and New Mexico in a reintroduction program similar to the one that brought wolves back to the Northern Rockies.

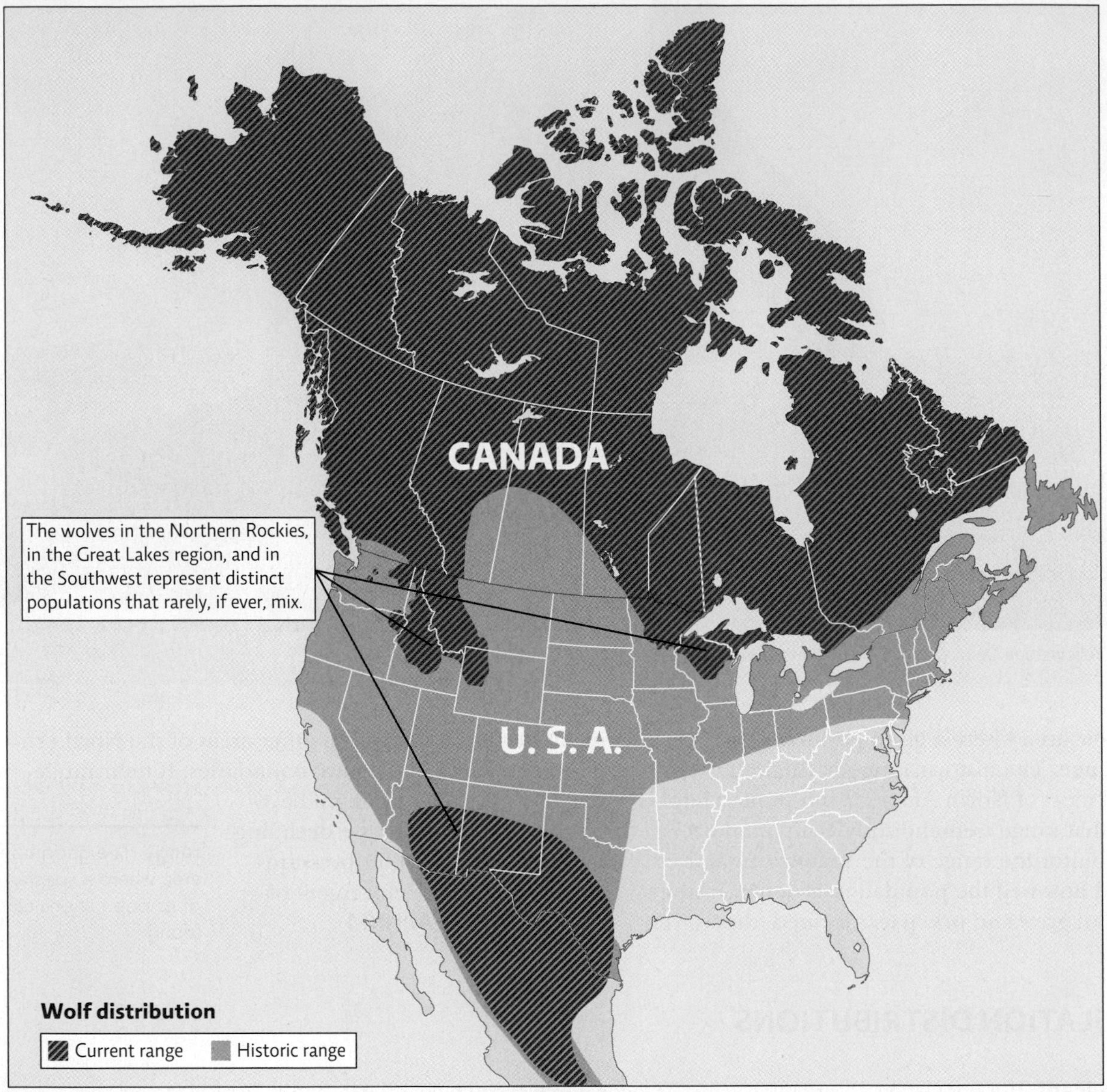

CANADA
U. S. A.
The wolves in the Northern Rockies, in the Great Lakes region, and in the Southwest represent distinct populations that rarely, if ever, mix.
Wolf distribution
Current range
Historic range

What are some factors that might expand the range of the Mexican wolf? What might cause it to shrink?

random distribution
A distribution in which individuals are spread out over the environment irregularly, with no discernible pattern.

uniform distribution
A distribution in which individuals are spaced evenly, perhaps due to territorial behavior or mechanisms for suppressing the growth of nearby individuals.

In a **random distribution**, individuals are spread out in the environment irregularly, with no discernible pattern. Random distributions are sometimes seen in homogeneous environments, in part because no particular spot is considered better than another. Species that rely on wind and water to disperse their offspring—like wind-blown seeds or the free-floating larvae of coral—also often have a random distribution.

Uniform distributions, rare in nature, include individuals that are spaced evenly, perhaps due to territorial behavior (e.g., gulls nesting on a crowded beach will fill the available space, but each breeding pair keeps neighbors from getting too close) or mechanisms for suppressing growth of nearby individuals (e.g., some plant species produce toxins that inhibit growth of would-be neighbors). **INFOGRAPHIC 2**

If population ecologists understand the normal distribution pattern for the species they are studying, they can better track the species and

INFOGRAPHIC 2 POPULATION DISTRIBUTION PATTERNS

The distribution of individuals in populations varies from species to species and is influenced by biotic and/or abiotic factors.

Tim Graham/Getty Images

Elk stay in herds, which offers some protection against predators.

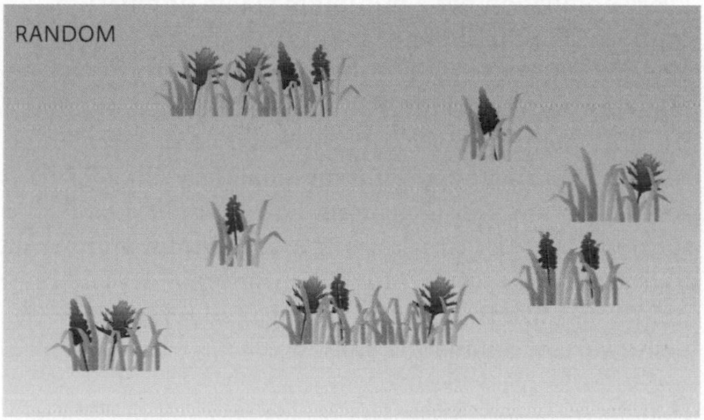

John Elk/Getty Images

The seeds of many Yellowstone flower species are distributed randomly and germinate where they fall.

Jack Dykinga/Nature Picture Library

The creosote bush of the desert Southwest produces toxins that prevent other bushes from growing close by.

In which population distribution pattern would individuals within the population experience the most competition with other individuals in their population? Why, then, is this distribution pattern ever seen?

determine if a problem is emerging. For example, one would not expect to find wolves everywhere in Yellowstone — they will be in areas with elk or other prey species. However, if a wildflower that normally has a random distribution over a sunny hillside shows a clumped distribution one season, this might be an indication that something is causing that species to be lost in areas where it should be found — perhaps pollution has contaminated the area or a new competitor has moved in. Knowing what to expect allows a scientist to look for the cause if a population's distribution changes. In addition, an understanding of a species' distribution within its range provides information about its habitat needs, which can be used when designing wildlife preserves or migration corridors that allow animals to move from one region to another.

3 POPULATION SIZE AND DENSITY

Key Concept 3: Populations require minimal sizes and densities to reproduce successfully and maintain social ties, but high population density can lead to problems such as disease and overuse of resources.

Understanding how any population interacts with biotic and abiotic forces in its environment, through programs like the Wolf Restoration Project, is key to preventing species from disappearing from their ecosystem forever. That's because many factors influence the livelihood of a species and its ability to survive and reproduce. For wolves, the main threat was humans.

In the early 20th century, humans set out to exterminate wolves from the American West. (They were already gone from most areas east of the Mississippi River.) As part of a government-sponsored program, Congress allocated $125,000 for the Predator Control Program, which employed poison—and later hunting—to eradicate predators and rodents that might harm crops or livestock. Wolves were one of the targeted species because wolves preyed on livestock. Wolf eradication was also supported because it boosted deer and elk populations for hunting. "It was park policy to kill all predators, and wolves were their biggest objective," Smith explains. Between 1914 and 1926, at least 136 wolves were killed in Yellowstone. In 1944, the last known wolf in the Yellowstone area was killed.

Humans not only threatened wolves by hunting them but also destroyed the animals' natural habitat (and that of their prey) when they cut down forests to build farms and ranches. They starved the animals by hunting elk, deer, and bison, wolves' usual sources of food. At the time, people didn't think that this combination of changes would very nearly cause wolves to go extinct or result in exploding elk and deer populations. But now that scientists like Smith have spent years watching how the animals live, they have a much better understanding of what needs to be done to keep them alive.

population dynamics Changes over time in population size and composition.

minimum viable population The smallest number of individuals that would still allow a population to be able to persist or grow, ensuring long-term survival.

carrying capacity (*K*) The maximum population size that a particular environment can support indefinitely.

population density The number of individuals per unit area.

Before humans started killing wolves in the early 1900s, the exact numbers of gray wolves living in Yellowstone were unknown, but estimates range from 300 to 400. Elk populations at that time hovered around 10,000 animals. The size of a population in a given geographic area is determined by the simultaneous interplay of factors that increase the number of individuals in a population (births and immigration) and those that decrease numbers (deaths and emigration). An understanding of these factors helps us predict population size at any given time.

Ecologists who study changes in population size and makeup (e.g., the average age of the individuals in the population or the proportion of males to females) are studying its **population dynamics**. They find that the population size of some species increases and decreases rather predictably (barring a catastrophic event), while others tend to fluctuate more randomly, affected by a variety of factors.

Every population has a **minimum viable population**, or the smallest number of individuals that would still allow a population to be able to persist or grow, ensuring long-term survival. This is an important concept when considering how to conserve endangered or threatened species. A population that is too small may fail to recover for a variety of reasons. For example, some species' courtship rituals require a minimum number of individuals for success. Other activities that depend on numbers—like flocking, schooling, and foraging—fail below certain population sizes.

Genetic diversity (inherited variety between individuals in a population) is also important: A population with little genetic diversity is less able to adapt to changes and is therefore more vulnerable to environmental change. A small population is also subject to inbreeding, which allows harmful genetic traits to spread and weaken the population. (See Modules 3.1 and 3.2 for more on genetic diversity.) Conversely, a population that has exceeded **carrying capacity**—the population size that a particular environment can support indefinitely without long-term damage to the environment—also causes problems.

Another important metric is **population density**—the number of individuals per unit area. Population density varies enormously among species or even among populations of the same species in different ecosystems. Similar to problems encountered in a small population, if a population's density is too low, individuals may have difficulty finding mates, or the only potential mates may be closely related individuals, which can lead to inbreeding, loss of genetic variability, and, potentially, extinction. Density that is too high can also cause problems, such as increased competition, fighting, and spread of disease. Deer, elk, and moose populations in the United States, whose density has increased in recent years because of exploding numbers combined with shrinking habitats, now frequently suffer from an infectious disease known as chronic wasting disease. **INFOGRAPHIC 3**

INFOGRAPHIC 3 POPULATION SIZE AND DENSITY INFLUENCE LONG-TERM POPULATION SUCCESS

Population density influences how well a population thrives. When population density is too low, a population may suffer social or reproductive problems; if the size falls below the minimum viable population size for that species, the long-term viability of the population is in doubt. But when populations grow too large for an area, problems associated with high densities may also negatively impact the population.

PROBLEMS IF POPULATION SIZE AND DENSITY ARE:

TOO LOW (below minimum population size)	TOO HIGH (above carrying capacity)
Normal social behaviors are deficient (e.g., group foraging or defense).	Social behaviors break down with overcrowding.
Unable to find mates.	Spread of disease increases.
Normal courtship and mating behaviors don't occur.	Food supplies are insufficient.
Genetic diversity falls (inbreeding).	Increased chance of conflict with humans.
Important community connections may be lost, affecting other species.	Damage to environment from overuse of resources.

? A major problem many species face today is a shrinking habitat due to habitat destruction and fragmentation by human actions. How would a shrinking habitat affect the density of a population, and what problems might this lead to for that population?

4 EXPONENTIAL POPULATION GROWTH

Key Concept 4: Growth and resistance factors influence population growth. Exponential growth occurs when population growth is unrestricted; however, it will not continue indefinitely.

Critics began raising concerns about the Predator Control Program in the 1920s, recognizing that wolves played a key role in their ecosystem as a top predator. Without the wolf, elk and deer populations could grow unchecked, leading to a damaged and overgrazed habitat.

Scientists can monitor elk populations to determine whether they are growing, and how fast, using some simple mathematical models that describe population growth over time. The annual **population growth rate** is determined by *birth rate* (the number of births per 1,000 individuals per year) minus the population *death rate* (the number of deaths per 1,000 individuals per year). Immigration and emigration rates might also be factored into the equation, if appropriate.

Population growth rates can also be determined by taking a simple census of population size at the two time points in question. For example, in 1920, the elk population was 10,000; the next year, it was 10,500, an increase of 500 animals. Using the simple equation that divides the change in population size by the original population size, we find that this population increased by 5% (500 / 10,000 = 0.05).

Assuming growth remains constant, we can predict future population size using the growth rate calculated. For example, in 1932, the elk population was 16,000. If it maintained an annual growth rate of 5%, we would

population growth rate The change in population size over time that takes into account the number of births and deaths as well as immigration and emigration numbers.

predict that the population would be 16,800 in 1933 ($16{,}000 \times 0.05 = 800$).

growth factors Resources individuals need to survive and reproduce that allow a population to grow in number.

resistance factors Things that directly (predators, disease) or indirectly (competitors) reduce population size.

biotic potential (*r*) The maximum rate at which the population can grow due to births if each member of the population survives and reproduces.

exponential growth The kind of growth in which a population becomes proportionally larger each breeding cycle; produces a J curve when plotted over time.

Growth rates can also be negative, reflecting a shrinking population. Between 2009 and 2010, the elk population decreased from 6,070 to 4,635, giving this population a growth rate of −24% ($4{,}635 - 6{,}070 = -1{,}435$; $-1{,}435 / 6{,}070 = -0.24$).

Population growth is dependent on the presence of **growth factors** (resources individuals need to survive or reproduce). Conversely, **resistance factors** (things that reduce population size by directly or indirectly killing individuals or prompting emigration), such as predators, competitors, diseases, or pollution, will decrease population size.

When there are no environmental limits to survival or reproduction, a population will reach its maximum per capita rate of increase (r), called its **biotic potential**. This occurs, theoretically, when every female reproduces to her maximal potential and every offspring survives. A population increasing in this manner will quickly grow to fill its environment. This period of growth, which can't go on indefinitely, is referred to as **exponential growth**, named for the mathematical function it represents.

Populations that have a high biotic potential have high *fecundity* (females typically produce many offspring, reach reproductive maturity quickly, and produce many "clutches" per year). The higher the biotic potential, the faster the population of a given species will grow under ideal conditions. Yellowstone species such as deer mice and the problematic non-native weed known as spotted knapweed have higher biotic potential than species such as grizzly bears and spruce trees.

In nature, exponential growth is typically seen when a species first enters a new environment or when there is an influx of new resources. The population has a high

Terry Whittaker Wildlife/Alamy

Rodents have a high biotic potential. A population that moves into a habitat with a good source of food and few predators, like a barn without a resident cat or rat snake, can quickly multiply.

birthrate—most individuals must have access to enough food, water, and habitat in which to reproduce—and a low death rate. The loss of predators can also lead to exponential growth among their prey species. For instance, without wolves to thin their ranks, elk numbers in Yellowstone doubled between 1914 and 1932, after the Predator Control Program had been implemented. This led to the need to cull the herd (in spite of increased hunting pressure), a standard practice for many years that reduced the herd to around 6,000 in 1968. When this practice ended, elk population sizes climbed to a high of around 19,000 in the 1980s before wolves were reintroduced.

Population growth is often evaluated in terms of its *doubling time*—the time it takes for the population to double in size if the growth rate is constant. For a population growing exponentially, a simple mathematical equation can estimate doubling time: $70 / r =$ doubling time in years, where r is the growth rate, entered as a whole number (e.g., if the growth rate is 2%, $r = 2$).

A population that is growing exponentially will have a *J-shaped curve* if plotted on a graph with time on the x axis and population size on the y axis. The J curve shows a slight lag at first and then a rapid increase. This is due to the fact that the larger the population, the faster it grows, even at the same growth rate. Think of it this way: Doubling a small number yields a number that is still small. Doubling a large number, on the other hand, produces a very large number. **INFOGRAPHIC 4**

As an example of how profound exponential growth can be, imagine if someone offered to give you a penny one day and then, each subsequent day for a month, doubled the amount given to you the previous day. On day 1, you would have 1 cent; on day 2, you would be given 2 cents; on day 3, you'd be given 4 cents; and on day 4, you would be given 8 cents. On day 31, you would receive almost $11 million, giving you a 31-day total of more than $21 million. Exponential growth can create large populations quickly.

INFOGRAPHIC 4 **EXPONENTIAL GROWTH OCCURS WHEN THERE ARE NO LIMITS TO GROWTH**

Because deer mice have a high biotic potential, even a single pair could produce thousands of descendants in their lifetime.

BIOTIC POTENTIAL OF DEER MICE

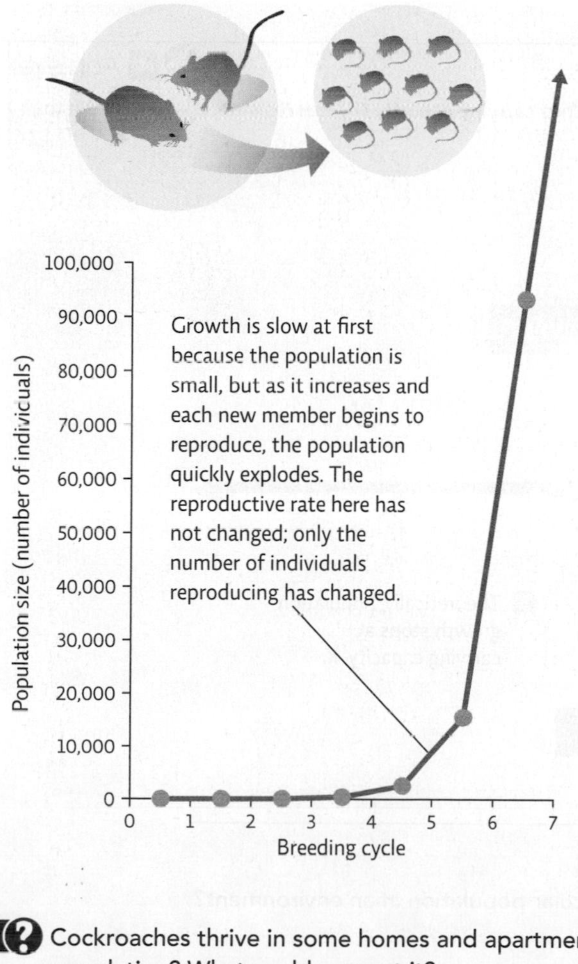

100,000
90,000
80,000
70,000
60,000
50,000
40,000
30,000
20,000
10,000
0
Population size (number of individuals)
0 1 2 3 4 5 6 7
Breeding cycle
Growth is slow at first because the population is small, but as it increases and each new member begins to reproduce, the population quickly explodes. The reproductive rate here has not changed; only the number of individuals reproducing has changed.

REPRODUCTIVE RATE

Assume each pair produces 10 pups/litter, none of the pups die, and each adult female continues to reproduce. The per capita rate of increase would be:

$$r = \frac{\text{births} - \text{deaths}}{\text{original population size}}$$

For this example:

$$r = \frac{10 - 0}{2} = 5$$

This represents a 500% growth rate!

The generation time of a deer mouse is about 10 weeks (3 weeks gestation, 7 weeks to reach reproductive age). That means this single pair could produce more than 9,000 descendants in just 60 weeks (15 months)!

DOUBLING TIME

The doubling time of a population experiencing exponential growth can be calculated using the "Rule of 70": Divide 70 by the per capita growth rate (r) over a period of time such as per year or per breeding cycle. This equation is most useful when the population growth rate is small (but constant). A population increasing at 3.5% per generation would double in 20 generations.

$$\frac{70}{3.5} = 20$$

? Cockroaches thrive in some homes and apartments. What could trigger exponential growth in a cockroach population? What could prevent it?

5 LOGISTIC POPULATION GROWTH

Key Concept 5: As a population's size approaches carrying capacity, exponential growth may transition to logistic growth, slowing population growth rates.

Exponential growth can't last forever. As a population begins to fill its environment, resistance factors begin to slow its growth rate. Available resources become scarce as more individuals use them. This can lead to starvation for some; crowding may bring about an increase in disease and aggression for others. Predation pressure may also increase as the more numerous prey are easier to track and capture or simply because the predator population itself has increased. All of these stressors can increase death rates.

logistic growth The kind of growth in which population size increases rapidly at first but then slows down as the population becomes larger; produces an S-shaped curve when plotted over time.

Birth rates may also decline. Underfed individuals may not be able to successfully reproduce; others may be unable to find a suitable area for rearing young in an overcrowded habitat. This kind of growth—in which as population size increases, growth rate decreases—is called **logistic growth.** A population that grows logistically will produce an *S-shaped curve* if plotted on a graph with time on the *x* axis and population size on the *y* axis. The S is created by initial exponential growth that produces the J-shaped curve, followed by decelerating growth as the species approaches its carrying capacity (signified as *K* in population mathematical models)—its maximum sustainable population size, where it levels off.

Carrying capacity is determined by the presence of growth factors and varies between species; the same environment can support many more elk than wolves, for example. Over time, a population's carrying capacity can change. If resources are diminished at a faster rate than they are replenished, the carrying capacity will drop. If, on the other hand, new resources are added or become available, perhaps due to the loss of a competitor, the carrying capacity for a given species will rise. **INFOGRAPHIC 5**

INFOGRAPHIC 5 LOGISTIC POPULATION GROWTH Achieve

Exponential growth turns into logistic growth (S curve) as population size approaches carrying capacity (*K*) and resistance factors begin to limit survival.

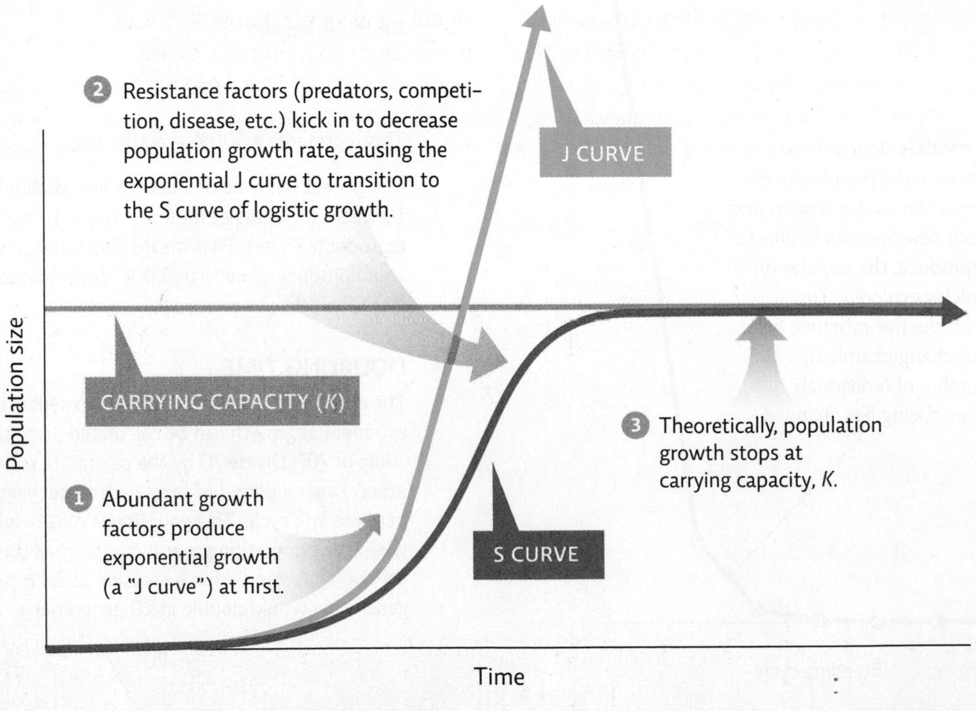

2 Resistance factors (predators, competition, disease, etc.) kick in to decrease population growth rate, causing the exponential J curve to transition to the S curve of logistic growth.
J CURVE
CARRYING CAPACITY (K)
Population size
1 Abundant growth factors produce exponential growth (a "J curve") at first.
S CURVE
3 Theoretically, population growth stops at carrying capacity, K.
Time

? What could increase or decrease the carrying capacity for a particular population in an environment?

6 DENSITY-DEPENDENT AND DENSITY-INDEPENDENT GROWTH FACTORS

Key Concept 6: Some factors that influence population size have more of an impact when that population is large (density-dependent factors), whereas others will have the same effect on large and small populations alike (density-independent factors).

Populations will grow as long as growth factors are available, but as the population gets larger, resources start to decline. The *limiting factor*, the resource that is most scarce, tends to determine carrying capacity. The effects that predators (a resistance factor) have on populations can vary widely, in part because predators are **density-dependent factors**—their effects on a prey's population go up as the size of that prey population goes up.

Density-dependent factors are usually biotic—imposed by other species. In the case of wolves preying on elk, the denser an elk population is, the easier it is for wolves to successfully find and take a member of that population. In the same way, elk are density-dependent factors on the ability of young aspens to grow after sprouting (aspen is an important winter food for elk; they eat the young shoots).

Competition for scarce resources among members of a population, or between members of two different populations vying for the same resource, also increases as the density of one or both populations increases. For the elk, this means less food for each member of the population when competing mule deer are present. Smith also found evidence that competition impacts the wolf population in a density-dependent fashion. As the size of wolf packs in Yellowstone increased, so did the incidence of wolf-on-wolf aggression, contributing to a subsequent decline in the wolf population. Disease, too, spreads more easily in larger populations with higher densities.

On the other hand, **density-independent factors** (usually abiotic rather than biotic factors) affect a population no matter how large or small it is, such as natural disasters like droughts, storms, and fire. Human impact such as toxic pollution could also affect members of a population regardless of its size. In these cases, the chance of one individual dying has nothing to do with how many others are in the area. For example, a population of elk trapped in a flood or a mudslide after an intense rain event will die regardless of how many individuals there are in the affected population.

Determining how these factors contribute to a population's increase or decrease gives scientists tools they can use to monitor and manage populations. For example, an elk herd that is decreasing in number would lead a manager to investigate potential causes such as greater predation or spread of a disease. An elk herd that is increasing in size might signal a loss of predators and lead to decisions to help control that population to avoid problems with overpopulation of the elk. **INFOGRAPHIC 6**

density-dependent factors Factors, such as predation or disease, whose impact on a population is influenced by the size of that population.

density-independent factors Factors, such as a storm or an avalanche, whose impact on a population is not related to population size.

Wild Horizon/Getty Images

The bare trunks of these aspen trees show the winter browse line from herbivores such as elk and deer, which eat leaves as high as they can reach during winter. As forage becomes more and more scarce, these animals will even strip off the bark, damaging the trees.

INFOGRAPHIC 6 DENSITY-DEPENDENT AND DENSITY-INDEPENDENT FACTORS AFFECT POPULATION SIZE

Density-dependent factors exert more of an effect as population size increases. On the other hand, density-independent factors have the same effect regardless of population size.

DENSITY DEPENDENT

COMPETITION

James Balog/Cavan Images

Mule deer compete with elk for food; the larger the deer population, the greater the competition.

DISEASE

Karl Gehring/Denver Post/Getty Images

Infectious diseases, such as chronic wasting disease, which weakens and eventually kills the animal, spread more easily in large populations of elk, deer, or moose.

PREDATORS

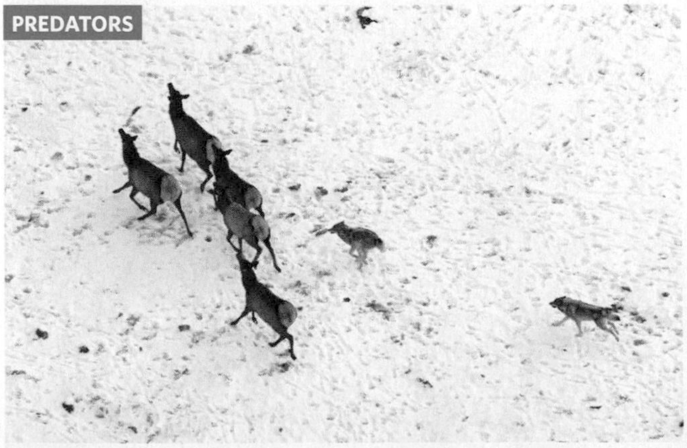

NPS Photo/M. Metz

Predators are more successful at capturing prey from larger populations than from smaller ones.

DENSITY INDEPENDENT

FIRES

David R. Frazier/DanitaDelimont.com/Danita Delimont Photography/Newscom

A forest fire sweeping through an area will drastically alter population sizes.

FLOODS

Debbie Noda/ZUMA Press/Modesto/California/U.S./Newscom

Floods can decimate some populations; others are increased when new organisms are swept into the area.

AVALANCHE AND OTHER NATURAL DISASTERS

Johner Images/AGE Fotostock

An avalanche can kill any animal in its path.

? Identify the following as either density-dependent or density-independent factors for an elk population: a tick infestation, building a dam that floods a valley in elk habitat, drought, bison, and a blizzard.

7 LIFE-HISTORY STRATEGIES: *r*- AND *K*-SPECIES

Key Concept 7: The population size of *r*-selected species can increase or decrease quickly if the environment changes. *K*-selected species' populations don't fluctuate as widely, but they are less able to respond to environment changes.

The biology of a species (which reflects its adaptations for growth, reproduction, and survival) also affects its populations' growth potential. For instance, ecologists recognize a continuum of **life-history strategies** among species. Species whose members mature early, have high fecundity, and have relatively short life spans are known as ***r*-selected species**—so named because of their high rate of population increase (*r*). Yellowstone *r*-selected species, such as deer mice and spotted knapweed, are well adapted to exploit unpredictable environments and are able to increase quickly (exponential growth) if resources suddenly become available.

On the other hand, ***K*-selected species**, which in Yellowstone include bears, wolves, and slow-growing trees like spruce, are found at the other end of the continuum. Individuals in these species have longer life spans, are slow to mature, and have lower fecundity. Because of this, their reproductive rates are lower and they are more likely to exhibit logistic growth. This means their population growth rates are responsive to minor changes in environmental conditions; they decrease or increase slowly if resources show a gradual decrease or increase in availability. For example, females may have fewer offspring or reproduce less often when food resources are in short supply. This responsiveness tends to keep population sizes close to carrying capacity (*K*). **INFOGRAPHIC 7**

Some species, like elk and deer, have characteristics of both

life-history strategies Biological characteristics of a species, such as life span and fecundity, that influence how quickly a population can potentially increase in number.

***r*-selected species** Species that have a high biotic potential and that share other characteristics, such as short life span, early maturity, and high fecundity.

***K*-selected species** Species that have a low biotic potential and that share characteristics such as long life span, late maturity, and low fecundity; generally show logistic population growth.

A gray wolf on the prowl at Yellowstone National Park.

KenCanning/Getty Images

INFOGRAPHIC 7 LIFE-HISTORY STRATEGIES

Different species have different potentials for population growth, known as life-history strategies. A species' biology may place it anywhere along a continuum between two extremes — the *r*- and *K*-selected species.

CHARACTERISTICS OF *r*-SELECTED SPECIES

1. Short life
2. Rapid growth of individual
3. Early maturity
4. Many small offspring
5. Little parental care
6. Adapted to unstable environment
7. Prey
8. Uses many habitats and resources

CHARACTERISTICS OF *K*-SELECTED SPECIES

1. Long life
2. Slower growth of individual
3. Late maturity
4. Few large offspring
5. High parental care
6. Adapted to stable environment
7. Predators
8. Needs specific habitat and resources

r-SELECTED SPECIES — *K*-SELECTED SPECIES

NORBERT ROSING/National Geographic Creative

Wayne Lynch/AGE Fotostock

Juniors Bildarchiv/F314/Alamy

Tom Reichner/Shutterstock

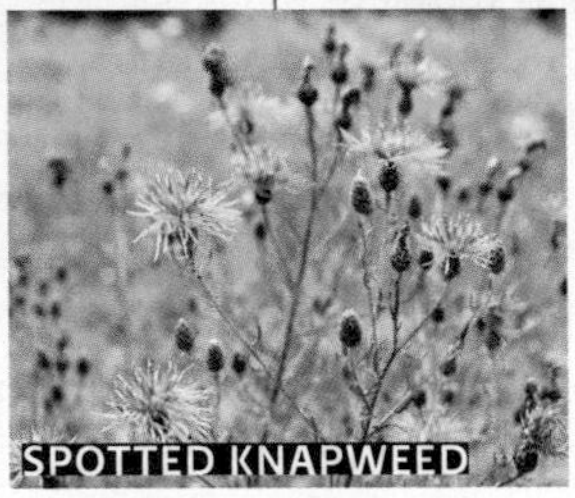

John W. Bova/Science Source

TAYLOR KENNEDY-SITKA PRODUCTIONS/ National Geographic Creative

Consider an aquatic ecosystem. Where would you place the following organisms on a life-history continuum: tuna, sperm whale, plankton, and jellyfish? (Hint: Identify the most extreme *r*- and *K*-species and then place the others between those on the continuum.)

r- and *K*-species; they fall somewhere in the middle of the continuum. They are large organisms that have one or two offspring per year and provide parental care (*K* characteristics), but their population sizes can increase rapidly if conditions are favorable for growth and survival (*r* characteristics).

K-species and *r*-species often experience different types of population change. For instance, population sizes tend to be stable, especially for *K*-species, in undisturbed, mature areas. However, it may take a long time for a *K*-species to bounce back if sudden or major changes in their environment decrease their population size. For this reason, many of our endangered species are *K*-species.

On the other hand, *r*-species, with their high reproductive potential, sometimes have sudden, rapid population growth characterized by occasional surges to very high population numbers, which may overshoot carrying capacity, followed by sudden crashes, especially in response to seasonal availability of food or temperature changes (also known as a boom-and-bust growth cycle); their high rate of reproduction does not allow the population the time to adjust and produce fewer offspring as resources become scarce or conditions change. When this occurs, the population that exceeds carrying capacity will drop below carrying capacity and then increase again; some populations will eventually level off close to carrying capacity, while others continue to overshoot and crash. But this tendency to overshoot and crash is not a weakness — it is a strength that allows *r*-species to inhabit unpredictable or seasonally changing environments.

8 TOP-DOWN AND BOTTOM-UP REGULATION

Key Concept 8: Population size is influenced by the presence of predators (top-down regulation) as well as the availability of needed resources (bottom-up regulation). Which one has the greatest impact varies from population to population, but both can trigger trophic cascades, having effects throughout the food chain.

Thanks in part to Smith's determination, the Yellowstone Gray Wolf Restoration project is going strong. Wolf numbers steadily grew after reintroduction to a high of 174 in 2003, then steadily decreased, leveling off in the last decade to between 80 and 100 wolves, a number Smith says may reflect the current carrying capacity for wolves. But one chief lesson hammered home by observing wolves in Yellowstone is that populations do not exist in isolation—they live in communities with other populations and influence each other greatly.

In the many decades when wolves were absent, elk populations crashed periodically when food was scarce due to drought, extreme winters, or overgrazing. This is known as **bottom-up regulation**—regulation by the presence or absence of growth factors that affect population numbers, such as the availability of water, food, or sunlight. Even physical disturbances like fire are considered bottom-up regulators when they free up nutrients, boosting plant growth. For decades, ecologists felt this was the major regulating force at work on populations throughout an ecosystem.

Recently, evidence is emerging that predators, too, play a major role in determining population sizes—that their impact is not just incidental, only killing weak or old individuals that would soon die anyway. The ability of a predator to control the size of its prey populations is an example of **top-down regulation**. In 1995, when the wolf reintroduction program began, the winter Yellowstone elk population was around 17,000. This number has steadily declined to around 5,000 today, a decline linked to the return of the wolf—beginning around 2008, the leading cause of death for young elk changed from starvation to predation. The wolf, it seems, has shifted control of the elk population to top-down regulation. Aspen and willow also appear to be controlled from the top down by elk in many areas. Because elk like to eat the tender young shoots, few saplings are seen in areas where elk are abundant and not deterred from grazing by predators.

The top-down effect of a predator is not limited to the prey that the predator actively takes. As with bottom-up regulation, effects can be felt throughout the food chain in what is known as a **trophic cascade**. (*Trophic* refers to feeding levels on a food chain; see Module 2.3 for an introduction to trophic levels.) In Yellowstone, aspen and willow owe their recovery to wolves. These trees, whose growth had been severely restricted when wolves were absent, are recovering thanks to lower elk grazing pressure—due both to lower elk population density and to behavioral changes; elk are warier when wolves are around and tend to stay in more open areas, away from aspen and willow stands.

Ecologists have long debated the relative importance of top-down regulation versus bottom-up regulation. In Yellowstone, the picture that seems to be emerging is that both top-down and bottom-up regulation are important in controlling elk populations and the plant species on which they feed. John Terborgh of Duke University explains it this way: Bottom-up factors determine how much energy enters the ecological community, and top-down factors influence how it is distributed up and down the food chain. **INFOGRAPHIC 8**

The return of the wolves is even affecting the way rivers run in Yellowstone National Park, according to research by Robert Beschta and William Ripple. Constant grazing from the overabundant elk in the 1990s kept willow saplings so short (most were under 2 feet tall) that they could no longer help shore up the stream banks, resulting in a gradual widening of the streambeds. This reduced the frequency of flooding after heavy rains or spring snowmelt—historically an important part of these streamside ecosystems. But with fewer elk feeding on them, by 2017, willows averaged just over 10 feet tall; this improved the streamside habitat by providing shade to the stream and promoting the re-vegetation of the streambanks. Stabilized streambanks held the stream channel in place, allowing a return to the occasional flooding vital for these habitats.

bottom-up regulation Population sizes in a community are limited primarily by availability of resources that enhance growth and survival of organisms lower on the food chain.

top-down regulation Population sizes in a community are limited primarily by predation from organisms at the top of the food chain.

trophic cascade Top-down effects from the presence or absence of a top predator that propagate all the way down a food chain to the ecosystem's plant communities.

INFOGRAPHIC 8 TOP-DOWN AND BOTTOM-UP REGULATION

Population size is affected by the presence of both resistance factors (things that reduce population size) and growth factors (the availability of resources that allow the population to grow). In most cases, both impact population size, although which plays the greater role may vary from species to species or even within a species. Top-down regulation can lead to trophic cascades that affect all levels of the food chain.

TOP-DOWN REGULATION

According to this model, population sizes are primarily determined by the control exerted by top predators, which can set into motion a trophic cascade that affects all levels of the food chain — predation decreases herbivore populations, which in turn increases the population size of plants in the ecosystem.

BOTTOM-UP REGULATION

Conversely, population sizes might be more heavily influenced by the availability of resources needed for plant growth. In this model, the size of plant community determines the population size of animals that eat plants (herbivores), which in turn determines the population size of animals that eat the herbivores.

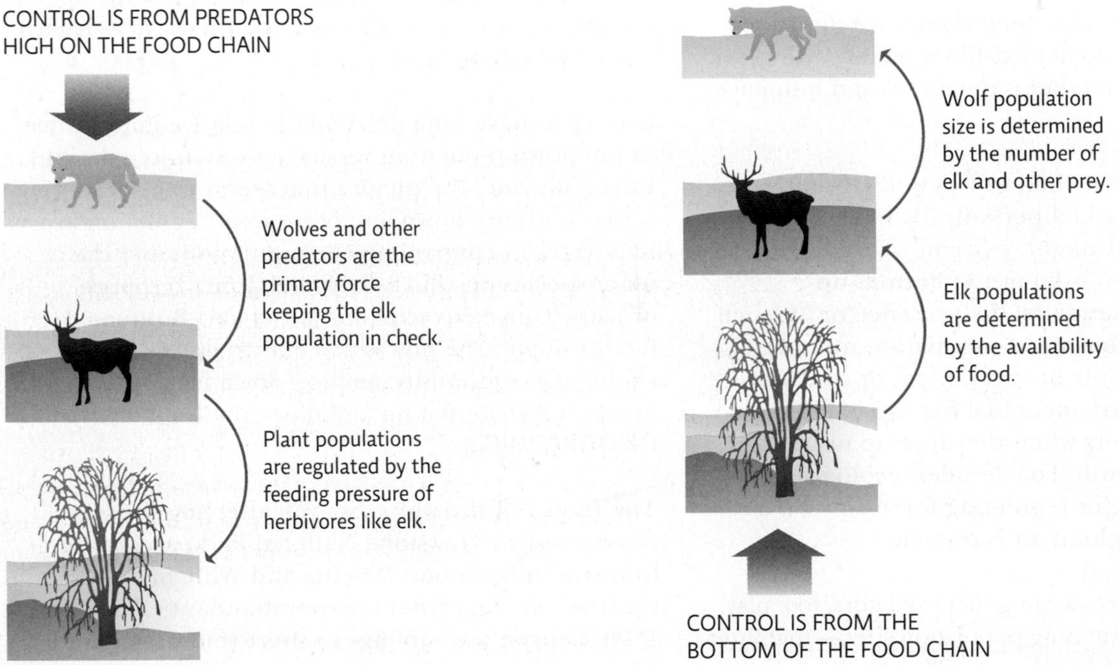

CONTROL IS FROM PREDATORS HIGH ON THE FOOD CHAIN
Wolves and other predators are the primary force keeping the elk population in check.
Plant populations are regulated by the feeding pressure of herbivores like elk.
Wolf population size is determined by the number of elk and other prey.
Elk populations are determined by the availability of food.
CONTROL IS FROM THE BOTTOM OF THE FOOD CHAIN

In most unprotected areas of the United States, wolf population densities are much lower than they are inside the protected Yellowstone National Park. Do you think wolves exert top-down control on elk populations in these unprotected areas? Explain.

Wolves, they say, have set in motion a *landscape-scale trophic cascade.*

The success of the Yellowstone reintroduction program led to the wolf being "delisted" in all the northern Rocky Mountain states by 2012, although not without opposition from some scientists and conservation groups. (A delisted species is no longer protected by the federal Endangered Species Act, and management authority is returned to the state.) In 2019, citing stable populations, the U.S. Fish and Wildlife Service proposed to delist the gray wolf in all U.S. states. (This does not apply to the Mexican wolf, which has a separate listing under the Endangered Species Act.) This too is being met with opposition from scientists who say it is too soon to remove protection.

The Northern Rockies delisting has allowed wolf hunts. In most states, there are quotas set for trophy hunting; there are no limits set for killing wolves deemed dangerous to humans or livestock. To Smith, the management of wolves includes the protection of some packs as well as policies that allow hunting of others. He reasons that hunting wolves only in areas where they conflict with humans may be one of the best ways to protect wolves in wild places like Yellowstone. If people know they can protect their animals and livelihoods, they may be more amenable to allowing the wolves to remain in wilderness areas.

To ensure that the wolves reintroduced to Yellowstone are given a chance to really flourish, Smith and his colleagues diligently stay on the wolves' trails, studying their population dynamics. "We want to know their population size, what they are eating, where they are denning, how many pups they have and how many survive, and how the wolves interact with each other," he explains. Why is it so important to ensure that the wolves do well? Simply put: "They were here first," he says. "We want to restore the original inhabitants to the Park."

Select References:

Beschta, R. L., & Ripple, W. J. (2019). Can large carnivores change streams via a trophic cascade? *Ecohydrology, 12*(1), e2048.

Painter, L. E., et al. (2018). Aspen recruitment in the Yellowstone region linked to reduced herbivory after large carnivore restoration. *Ecosphere, 9*(8), e02376.

Ripple, W. J., et al. (2016). What is a trophic cascade? *Trends in Ecology & Evolution, 31*(11), 842–849.

Smith, D., et al. (2018). *Yellowstone National Park Wolf Project Annual Report 2017* (YCR-2018-03). Yellowstone National Park, WY: National Park Service, Yellowstone Center for Resources.

Terborgh, J. W. (2015). Toward a trophic theory of species diversity. *Proceedings of the National Academy of Sciences, 112*(37), 11415–11422.

Lynx and snowshoe hares go through boom-and-bust population cycles. Lynx populations are controlled by the size of the hare populations (bottom-up regulation), whereas hare populations are controlled by lynx predation (top-down regulation).

Tom & Pat Leeson/AGE Fotostock

GLOBAL CASE STUDIES POPULATION GROWTH PATTERNS

Achieve

Different species exhibit different population growth patterns, a reflection of their biology and the species with which they interact. Read about the species presented here for a closer look at four examples of population growth.

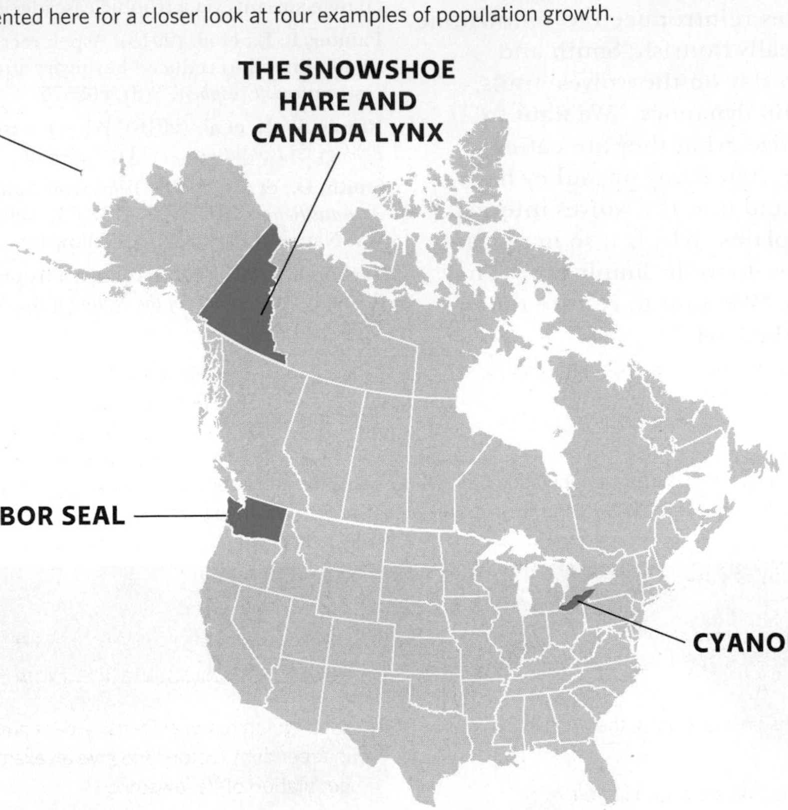

BRING IT HOME

PERSONAL CHOICES THAT HELP

Understanding the factors that influence how populations change can help us manage species that are facing extinction or help us control (or even eliminate) non-native species that are causing problems. How people view species and their connection to our world has a large impact on how management plays out.

Individual Steps

- Learn more about wolves at the International Wolf Center (www.wolf.org).
- Use the Internet and books on wildlife to research what your area might have been like prior to human settlement. Which species have been extirpated, and which ones have been introduced? How have wildlife populations changed as a result of human action?
- Even if you don't hunt or fish, consider purchasing a hunting or fishing license; proceeds help fund wildlife management programs in your state.
- Visit www.regulations.gov to submit comments on proposed rules, such as the removal or addition of a species to the Endangered Species List.

Group Action

- Explore organizations that support predator preservation, such as Defenders of Wildlife and Keystone Conservation, for suggestions on how you can help educate others about the importance of predators.
- Join a local, regional, or national group that works to monitor, protect, and restore wildlife habitats, such as the Defenders of Wildlife Volunteer Corps.
- Investigate predator compensation funds such as the Defenders of Wildlife Wolf Compensation Trust and the Maasailand Preservation Trust (for livestock losses due to lion predation). Do you think this is a worthwhile approach?

John E Marriott/All Canada Photos

REVISIT THE CONCEPTS

- Ecologists study populations to better understand what makes them thrive, decline, or become overpopulated.
- Within their range, populations may be distributed in a clumped, random, or uniform pattern, a reflection of how individuals interact with each other and their environment.
- Populations require minimal sizes and densities to reproduce successfully and maintain social ties, but high population density can lead to problems such as disease and food shortages.
- When growth factors (e.g., food, habitat) are abundant and resistance factors (e.g., predators) are low, populations can grow at an exponential rate (at or near their biotic potential).
- Exponential growth cannot continue indefinitely. As a population's size approaches carrying capacity, exponential growth may transition to logistic growth, slowing population growth rates.
- Some factors that influence population size have more of an impact when that population is large (density-dependent factors), whereas others will have the same effect on large and small populations alike (density-independent factors).
- The life history of a species also influences its potential for growth; *r*-selected species have the potential for boom-and-bust growth cycles, whereas *K*-selected species tend to show logistic growth that is responsive to environmental conditions.
- Population size is influenced by the presence of predators (top-down regulation) as well as the availability of needed resources (bottom-up regulation). Which one has the greatest impact on a given population varies, but both can trigger trophic cascades, affecting the entire food chain.

ENVIRONMENTAL LITERACY Understanding the Issue

1 What is a population, and why do ecologists study them?

1. Why might ecologists focus on monitoring the population of a single species?
2. Identify several populations that are found in your area.

2 What population distributions are seen in nature?

3. Why is it useful for ecologists to understand how a species they are studying is distributed within its ecosystem?
4. What population distribution is more common for a social species such as elk or prairie dogs? Explain.

3 What is the importance of population size and density?

5. What is the minimum viable population, and why might it be important to wildlife managers?
6. Explain what problems can emerge if a population's density is too low or if it is too high.

4 What is exponential growth, and when does it occur in a population?

7. What effect might the sudden influx of a new food source or the removal of a predator have on a population?
8. What is doubling time? What would be the doubling time for a population whose annual growth rate is 5%?
9. Why can't exponential growth continue indefinitely?

5 What is logistic growth, and when does it occur in a population?

10. Kangaroo rats eat seeds and are eaten by coyotes. Under what conditions might the kangaroo rat population increase logistically rather than exponentially?

6 How do density-dependent and density-independent factors affect population growth?

11. Distinguish between density-dependent and density-independent factors and give an example of each for the wolf population of Yellowstone.
12. Lesser goldfinches are small, seed-eating birds. In cities, both wild hawks and domestic cats eat these birds. Discuss several density-dependent and density-independent factors that could affect their carrying capacity.

7 What are the life-history strategies of *r*- and *K*-selected species, and how do they relate to population growth patterns and their ability to respond to environmental changes?

13. Why are *r*-selected species more likely to experience "overshoot and crash" population growth cycles compared with *K*-selected species?
14. Compare the life-history strategy of a deer mouse with that of a bear and identify each as either an *r*- or *K*-selected species.

8 What are top-down and bottom-up regulation, and how do they contribute to trophic cascades?

15. In a bottom-up regulation scenario, what kinds of things would determine the size of an elk herd?
16. Explain the concept of a trophic cascade using the example of wolves, elk, willow, and streams in Yellowstone.

SCIENCE LITERACY Working with Data

To understand the effect of the increased wolf population on the abundance of elk in the Yellowstone area, researchers have been careful to track a variety of factors that might also be influencing the size of the elk population. The following graphs show the size of the elk population since the introduction of the wolf in 1995 relative to the wolf population (Graph A) and relative to the area's "dryness" for that area over that same time period (Graph B) as measured by the Palmer Drought Severity Index. Values range from +10 (very wet) to −10 (very dry), with zero representing the long-term average.

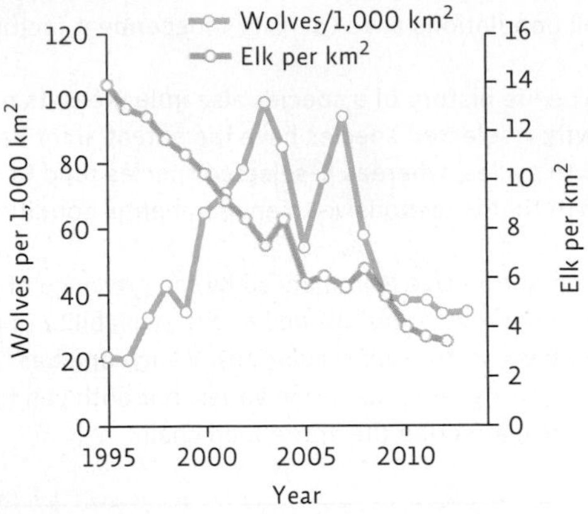

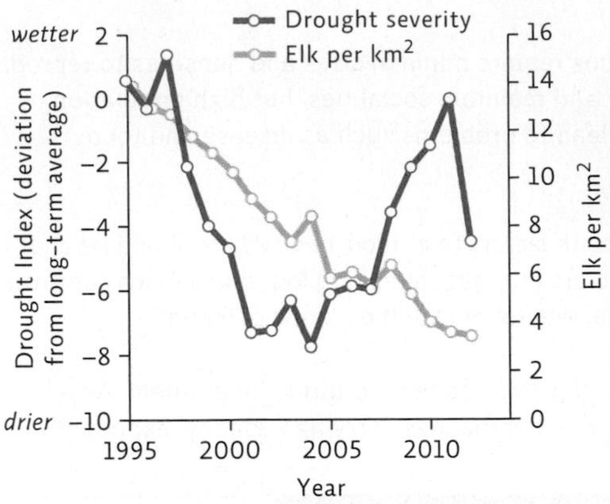

Interpretation

1. Look at Graph A. What is the relationship between the two population sizes over the time frame shown?
2. Using only the data from Graph A, what conclusion could you draw regarding the effect of the wolf reintroduction on the elk population?
3. Now look at Graph B to determine if drought severity is correlated with elk population size. Using only these data, what conclusion could you draw regarding the effect of drought on the elk population?

Advance Your Thinking

4. Records show that elk harvests by hunters increased in the 1990s and 2000s. Additionally, the population size of grizzly bears, a major predator of elk calves, tripled in that time frame. How might this have impacted overall elk abundance?
5. An evaluation of the predation rate (the proportion of the elk population killed by wolves) reveals wolves killed less than 4% of the elk population in the 7 years after introduction, but they now kill around 12–16% of the elk population. Does this offer evidence that the wolves are exerting predation pressure on the elk?
6. Considering all the data provided here, what is your conclusion regarding the effect of the wolf introduction on the abundance of elk?

INFORMATION LITERACY Evaluating Information

For over 100 years, up until the 1960s, wolves were lost from their historic range in most of the United States. Over the past few decades, studies have indicated that wolves are quite important to the functioning of an ecosystem. The U.S. Fish and Wildlife Service has worked to reintroduce wolves in various U.S. states, but they often encounter fierce resistance from other groups concerned with the safety of humans and livestock.

Search the Internet for information on the reintroduction of the gray wolf to Yellowstone. Read two articles that support the program or focus on the advantages and two articles that are opposed to the program or focus on the disadvantages.

Evaluate the websites and work with the information to answer the following questions:

1. Answer the following questions for each of the four articles you read to determine if the websites you visited are reliable information sources:
   a. What is the title of the article? Who are the authors?
   b. Summarize the main points of the article and identify the author's position regarding wolf reintroduction programs.
   c. What supporting evidence do the authors give for their claims?
   d. Do they give sources for their evidence? If so, evaluate the reliability of these sources.
   e. Do the authors or organization appear to be biased for or against wolves? Explain.
   f. Based on your answers to the above questions, is this a reliable source of information? Explain.
2. Write an essay that presents your own position on wolf reintroduction programs.
   a. Do you believe that wolves should be reintroduced in a few isolated areas; in many areas, including those where human contact is frequent; or not at all? Justify your decision.
   b. Do you believe that the hunting of wolves outside protected areas like Yellowstone National Park should be allowed? Justify your answer.

Additional study questions are available at Achieve.macmillanlearning.com. Achieve

Module 2.3 Community Ecology

THE FLORIDA EVERGLADES: A COMMUNITY IN CRISIS

A bird species in the Everglades reveals the intricacies of a threatened ecosystem

Wood storks gathering nesting material. *cnx4004/iStock/Getty Images*

After reading this module, you should be able to answer these GUIDING QUESTIONS

1. What is an ecological community?
2. What can we learn about a community by identifying its food web?
3. What are trophic levels, how are they classified, and why are these illustrated as a pyramid?
4. How is species diversity measured, and why is it important to a community?
5. What are core and edge habitats, and how can habitat fragmentation affect the species that inhabit them?
6. What is a keystone species and why is it important?
7. What species interactions are seen in a community, and what problems can emerge if these interactions are disrupted?
8. What is the goal of restoration ecology, and what obstacles do restoration ecologists face?
9. What is ecological succession, and how can we use this knowledge to assist in ecosystem restoration?

James Rodgers steered his canoe toward a large cypress tree as sunlight trickled through the dizzy pattern of leaves overhead. The tree had several wood stork nests in it, and Rodgers and his assistant wanted to get a closer look at all of them. They were in the thick of a dense swamp near the northwestern edge of the Florida Everglades, and it was the height of breeding season for the storks—eggs had hatched, and nestlings everywhere were crying, loudly, for food. Rodgers was silent. He knew from experience that too much human disturbance could "flush" the wary adult storks—forcing them to flee in a hurry, which would leave their babies vulnerable to aerial predators, such as hawks or crows, that would feed on the chicks in the nest.

The wood stork is an unassuming sort of bird: more than 3 feet tall, yes, but also covered with a mottled black-and-white coat of feathers—bland compared to some of its tropical neighbors. Despite the lack of majesty of the wood storks, however, Rodgers and others at the Florida Fish and Wildlife Conservation Commission keep close tabs on their ranks.

Here's why: In the late 1970s, the number of nesting pairs in Florida began to plummet, falling from an estimated 10,000 in 1975 to around 2,600 in 1978. By the early 1980s, the bird had earned a spot on the Endangered Species List and numbers continued to drop. It was then that Rodgers and his colleagues were first tasked with determining which of several factors (reduced nesting habitat? health of females? damaged feeding grounds?) was most responsible for the decline of this particular bird. And it was through those research efforts—focused intently on the wood stork—that they found an entire ecosystem on the brink.

WHERE ARE THE FLORIDA EVERGLADES?

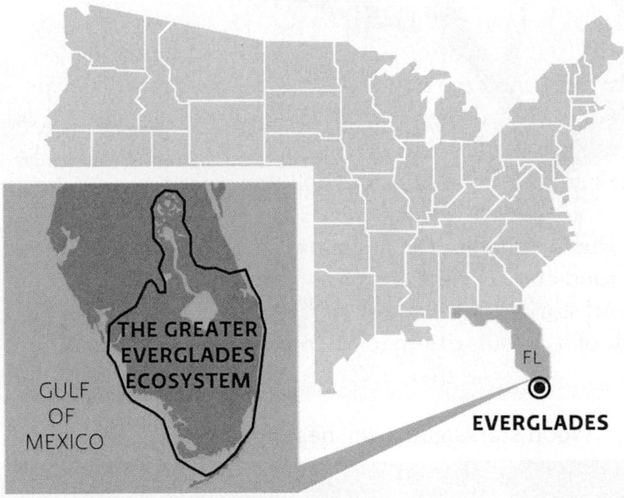

IMBODEN, OTIS/National Geographic Creative

The inlets of Everglades National Park contain a unique mix of tropical and temperate plants and animals, including more than 700 plant and 300 bird species.

1 COMMUNITY ECOLOGY

Key Concept 1: Community ecologists study the many populations that live and interact with each other and their environment within a given area.

Community ecology is the study of how all the populations in a given ecosystem function—how space is structured, why certain species thrive in certain areas, and how individual species in the same community interact with one another and with their **habitat**. Each species there occupies a unique **niche**—that is, its role and set of interactions in the community, how it gets its energy and nutrients, and its preferred habitat.

Some species will be **niche specialists**—they have very specific habitat or resource requirements, and this restricts where they can live. However, specialists are often highly adapted to acquire these resources and can outcompete others who might be trying to utilize the same resources. On the other end of a spectrum are the **niche generalists**. These species can use a wide variety of resources. The advantage to this adaptation is flexibility—if one food resource dwindles, for example, the generalist can simply switch to another.

In the Everglades, the alligator is a generalist with regard to food choices; it will eat just about anything. The wood stork is more of a specialist feeder and prefers fish of a certain size that are found in shallow, murky water. As long as these fish are present, the stork's exceptional fishing skills give it an advantage over other fish-predators. But if these fish are few in number, the stork has few other options, turning to the less abundant salamanders and frogs or trying to catch larger or smaller fish.

Ecological communities are complex places defined by their many interconnected species. The Everglades is made up of a wide variety of ecosystems—marshes, prairies, swamps, and forests—each with an array of species that are adapted to the unique conditions of their habitat and that interact closely with each other. The wood stork is one of many different Everglades wading birds that feast on fish, crustaceans, frogs, and insects. Sawgrass, periphyton (a mix of algae and bacteria), and an assortment of small plants and water-tolerant trees support a wide variety of birds, reptiles, mammals, fish, and countless insects. The alligator serves as an important apex predator. All these populations together make up the communities of the Everglades. **INFOGRAPHIC 1**

Community ecologists investigate how various species contribute to ecosystem services like pollination, water purification, and nutrient cycling (see Modules 3.2 and 5.1 for more on ecosystem services). For example, wetlands like the Florida Everglades help capture contaminants and excess nutrients, preventing them from reaching downstream fresh- and saltwater ecosystems. The slow-moving water is better able to soak into the ground, which helps refill groundwater supplies, and the wetland provides flood control, vital ecosystem services for the people of South Florida. But human impact has drained much of the wetland and reordered the landscape, and pollution has disrupted nutrient cycles—actions that are altering the ability of the Everglades communities to continue to provide these services.

Before giving way to a hodgepodge of resorts, sugar plantations, and dense urban centers, the region was defined by an uninterrupted web of natural ecosystems, collectively known as the Everglades, that stretched from Lake Okeechobee to the Florida Bay. It was here that wood storks flourished. In the 1930s, an estimated 15,000 to 20,000 pairs nested throughout the southeastern United States—largely in South Florida, where foraging grounds were ideal.

But it was not long after they first discovered the Everglades that American explorers began hatching plans to drain and then develop them. Swamps and muddy rivers choked with grass were seen as having no inherent value. "From the middle of the 19th century to the middle of the 20th, the United States went through a period in which wetland removal was not questioned," says University of Florida geographer and historian Christopher Meindl. "In fact, it was considered the proper thing to do." The Central and Southern Florida Project, authorized by the U.S. Congress in 1948, set out to systematically drain the Everglades.

As the human population swelled in the region, water that once fed swamps and marshes was drained away or rerouted to the faucets of

community ecology The study of all the populations (plants, animals, and other species) living and interacting in an area.

habitat The physical environment in which individuals of a particular species can be found.

niche The role a species plays in its community, including how it gets its energy and nutrients, what habitat requirements it has, and with which other species and parts of the ecosystem it interacts.

niche specialist A species with very specific habitat or resource requirements that restrict where it can live.

niche generalist A species that occupies a broad niche because it can utilize a wide variety of resources.

INFOGRAPHIC 1 **THE EVERGLADES COMMUNITY**

An ecological community consists of all the populations of species that live and interact in one area. In the Florida Everglades, different communities can be found in the lagoons, out in the sawgrass marshes, in the cypress swamps, and in the many other unique habitats in this large wetland ecosystem — unique communities that are all interdependent and connected. The species that make up a given community fill specific niches, and each contributes to the overall health of its ecosystem.

? Think of an ecological community in an ecosystem near you. Identify some of the populations that make up that community.

burgeoning developments. And as water levels changed, becoming deeper in some areas and completely disappearing in others, the total wading bird population plummeted—by 90% between the 1930s and 1990s. As ecologists would soon discover, these changes disrupted the entire ecosystem—from the health of giant wading birds right down to the movement of matter and energy.

2 THE FOOD WEB

Key Concept 2: The flow of matter and energy through a community is represented by a food chain that shows who eats whom and always begins with producers. The combination of all the food chains in one area makes up a community's food web.

food chain A simple, linear path starting with a plant (or other photosynthetic organism) that identifies what each organism in the path eats.

Energy is the foundation of every ecosystem; it is captured by photosynthetic organisms and then passed from organism to organism via the **food chain**—a simple, linear path that shows who eats whom. Any given ecosystem might have dozens of individual food chains. Linked together, they create a **food web**, which shows all the many connections in the community. Both food chains and webs help ecologists track energy and matter through a given community. They can vary greatly in length and complexity between different types of ecosystems. But most share a few common features, and all are made up of the same basic building blocks—namely, producers and consumers.

Florida wood storks sit near the top of a food web that begins with sawgrass, other plants like cypress and mangrove trees, and periphyton. These photosynthetic organisms are all known as **producers**. Producers capture energy directly from the Sun and convert it to food (sugar) via photosynthesis. (See Infographic 5 in Module 2.1.) They are then eaten by a wide range of **consumers**—organisms that gain energy and nutrients by eating other organisms. Animals, fungi, and most bacteria and protozoa are consumers. **INFOGRAPHIC 2**

The wood stork is a specialist feeder that hunts almost exclusively in shallow, muddy, plant-filled water that is so cloudy that fish cannot be seen. It walks through the water snatching up fish it feels with its long bill with a reflex that snaps its bill shut in just 25 milliseconds. It's the fastest reflex known in all vertebrates, and it enables the wood stork to capture prey that no other wading birds can access. But for this to work, the prey must be densely concentrated. Even a small drop in feeding success can impact the ability of these

food web A linkage of all the food chains together that shows the many connections in the community.

producer An organism that converts solar energy to chemical energy via photosynthesis.

consumer An organism that obtains energy and nutrients by feeding on another organism.

INFOGRAPHIC 2 EVERGLADES FOOD WEB

The food web of the Everglades is very complex and varies among the different ecosystems found there. Periphyton algae mats form the base of the food web and may be the most important producers in the ecosystem. The American alligator is the main apex predator, although when young is also the prey of various birds, fish, mammals, and even other alligators.

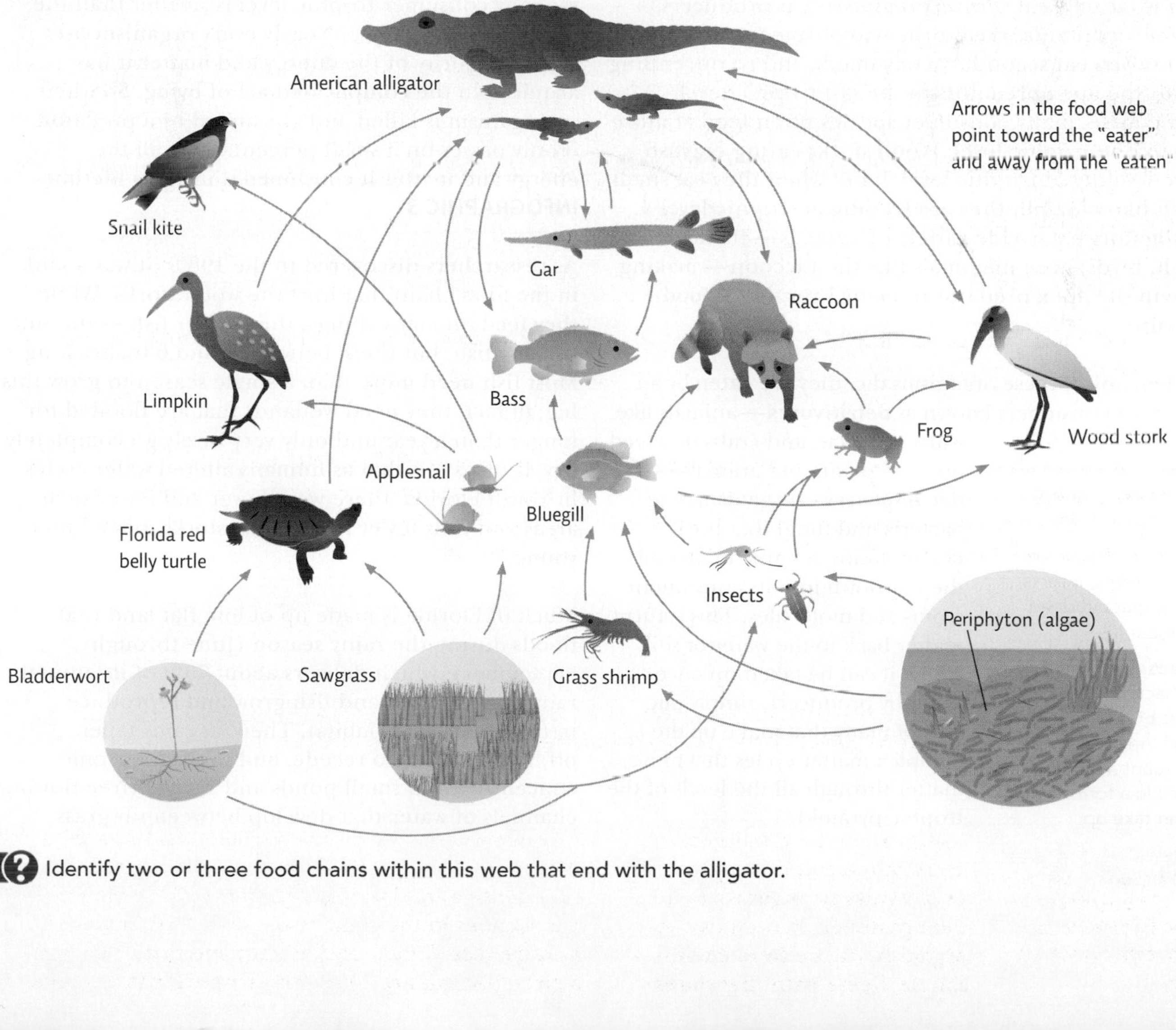

Identify two or three food chains within this web that end with the alligator.

birds to successfully rear their young, a factor that can have serious repercussions for this *K*-selected species. Storks mature slowly and may not reproduce until they are 4 years old. They lay a single clutch of eggs per year, and it is only in years of abundant food that all the young (perhaps three or four) survive. In lean years, only one chick may survive, or none. (See Module 2.2 for more on *K*- and *r*-selected species.)

indicator species A species that is particularly vulnerable to ecosystem perturbations, and that, when we monitor it, can give us advance warning of a problem.

It's this sensitivity that makes the wood stork such a good **indicator species** for the Everglades. An indicator species is one that's particularly vulnerable to ecosystem perturbations. Because even minor environmental changes can affect them dramatically, they can warn ecologists of a problem before it grows. "It is much easier to follow one or two species than to try and monitor an entire ecosystem," says Rodgers, who is a wood stork specialist. "So if an indicator species can be identified, this makes it much easier to keep tabs on the health of the ecosystem."

3 THE TROPHIC PYRAMID

Key Concept 3: The trophic levels of the food chain are shown as a pyramid; lower levels are larger than upper ones. Because energy only enters at the producer level and the organisms at each level use most of the energy they have taken in, only a small percentage is passed on to support the next level.

The different levels at which organisms feed or capture energy are known as **trophic levels**. Consumers are organized into trophic levels based on what they eat. *Primary consumers* eat producers, *secondary consumers* eat primary consumers, *tertiary consumers* eat secondary consumers, and so on, ending with the apex predators in the last trophic level. Of course, many consumer species often feed at more than one trophic level: Wood storks eating crayfish are feeding at trophic level 3, but when they eat small fish like bluegill, they are feeding at trophic level 4. Alligators eat a wide variety of animals—turtles, fish, birds, even mammals like the raccoon—making them the apex predator in many Everglades food chains.

When any of these organisms die, they are eaten by an army of consumers known as **detritivores**—animals like worms, insects, and crabs that feed on dead plants and animals—and **decomposers**—organisms like bacteria and fungi that break decomposing organic matter all the way down into its constituent atoms and molecules. This returns matter back to the water or soil where it can be taken up once again by producers, simply one step of many that make up the complex matter cycles that move matter through all the levels of the trophic pyramid.

trophic levels Feeding levels in a food chain.

detritivores Consumers (including worms, insects, and crabs) that eat dead organic material.

decomposers Organisms such as bacteria and fungi that break organic matter all the way down to constituent atoms or molecules in a form that plants can take up.

As one moves up the food chain, energy and biomass (matter comprising all the organisms at that level) decrease, creating a trophic pyramid. The reason that any consumer trophic level is smaller than the one below it is simple: Nearly every organism uses up the majority of the energy and matter it has acquired in the complicated act of living. So when an organism is killed and consumed by a predator, it only passes on a small percentage of all the energy and matter it consumed during its lifetime. **INFOGRAPHIC 3**

As researchers discovered in the 1980s, it was a kink in the food chain that hurt the wood storks. While they feed on many things, they prefer fish—and not just any fish, but those between 1 and 6 inches long. Most fish need more than a single season to grow this big; in fact, they need wetlands that are flooded for longer than a year and only very rarely go completely dry. It turns out that as humans altered water cycles in South Florida, there were fewer and fewer such areas and thus fewer fish for the storks to feed their young.

Much of Florida is made up of low, flat land that floods during the rainy season (June through September), which delivers about 75% of its annual rainfall. Many wetland fish grow and reproduce in this expanding habitat. Then, as rains taper off, water begins to recede, and the fish become concentrated in small ponds and *sloughs* (free-flowing channels of water that develop between sawgrass

INFOGRAPHIC 3 TROPHIC PYRAMID

Energy enters at the base of the food chain in the first trophic level (TL) via photosynthesis and is passed on to higher levels as consumers feed on other organisms. This is shown as a pyramid (smaller on top) because only a small percentage of the energy is passed on to each higher level, with the majority being "lost" to the environment (usually as heat from the energy that the organism burns in day-to-day life before it is eaten). Most food chains have only four or five levels due to this progressive loss of energy. Although the actual amount varies from ecosystem to ecosystem, for illustration purposes, we show 10% passing on to the next higher level.

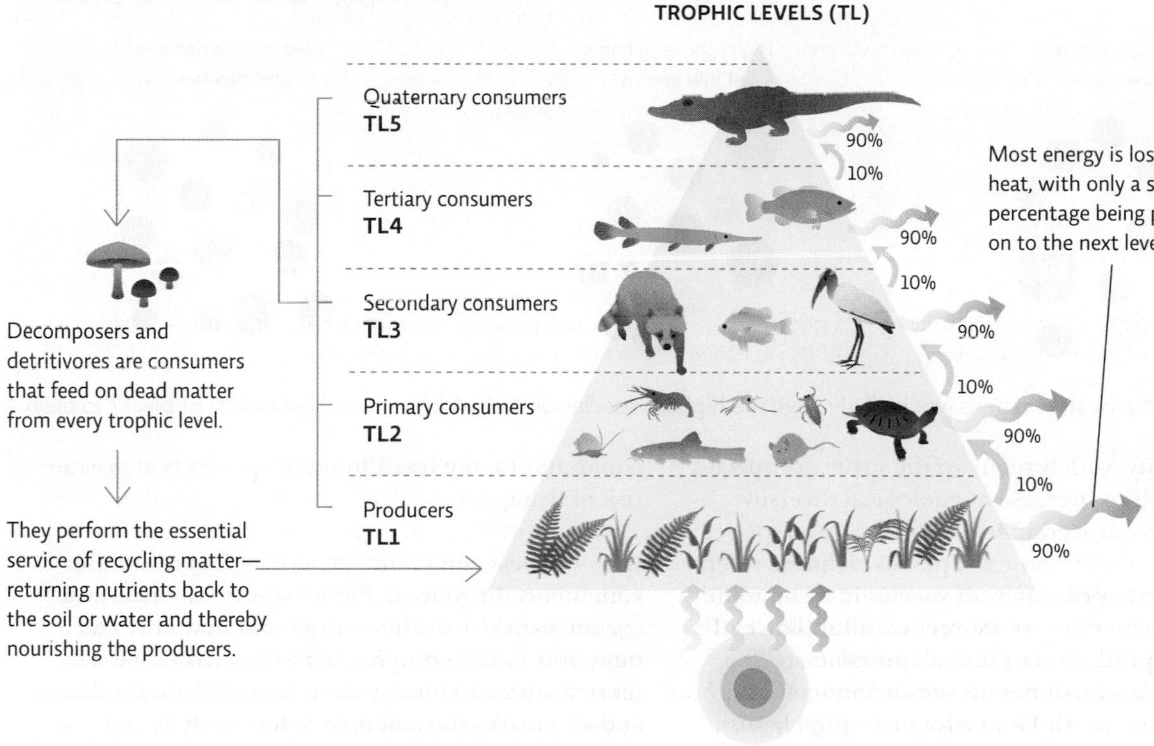
TROPHIC LEVELS (TL)
Quaternary consumers
TL5
Tertiary consumers
TL4
Secondary consumers
TL3
Primary consumers
TL2
Producers
TL1
90%
10%
90%
10%
90%
10%
90%
10%
90%
Most energy is lost as heat, with only a small percentage being passed on to the next level.
Decomposers and detritivores are consumers that feed on dead matter from every trophic level.
They perform the essential service of recycling matter—returning nutrients back to the soil or water and thereby nourishing the producers.

Why are there seldom more than five trophic levels?

prairies; pronounced "slew"). Foraging storks follow these receding waters, from upland ponds to lowland coastal areas, feeding on fish. They are so dependent on this water cycle that their breeding cycle is regulated by water levels. Such profound connectedness — between landscape and life — is common in the Everglades.

4 SPECIES DIVERSITY

Key Concept 4: Species diversity is measured in terms of the number of different species present (richness) and the relative abundance of each species (evenness). High richness and evenness tend to make a community more resilient to environmental changes.

Strong connections between species and their environment, and among species, give rise to ecological diversity, a measure of the number of species at each trophic level as well as the total number of trophic levels and available niches. (See Module 3.2 for more on ecological diversity.) Greater ecological diversity means more niches and thus more ways for matter and energy to be accessed and exchanged. This generally increases a community's **resilience** — its ability to adjust to changes in the environment and return to its original state rather quickly.

resilience The ability of an ecosystem to recover when it is damaged or perturbed.

INFOGRAPHIC 4 SPECIES DIVERSITY INCLUDES RICHNESS AND EVENNESS

The species diversity in an area is a measure of species richness (the total number of species) and species evenness (a comparison of the population size of each species). The Everglades contains forested areas known as hardwood hammocks. Each forest plot shown here contains 15 trees, but they differ in terms of species richness and evenness.

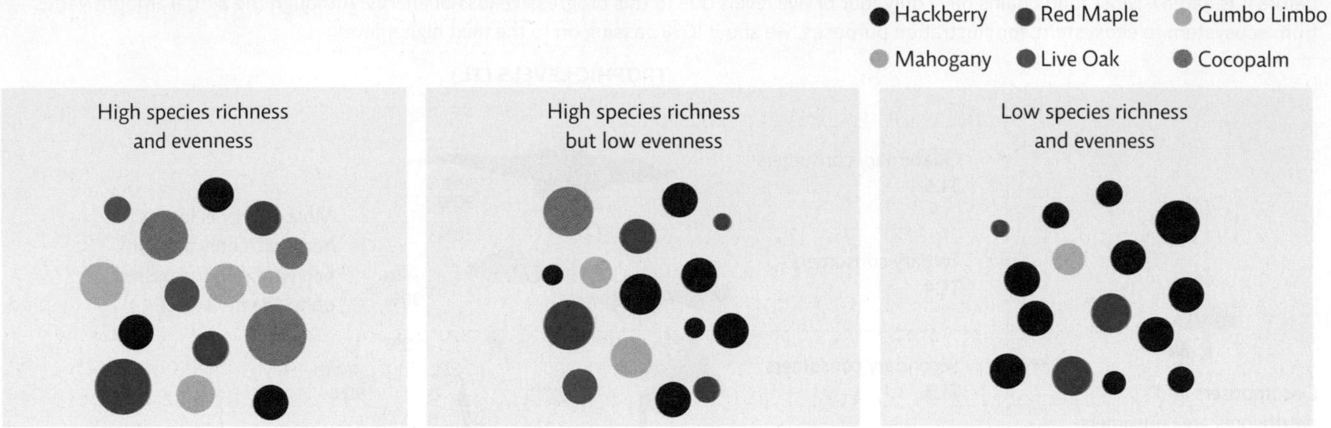

Which forest plot shown here would likely have the highest species diversity (richness and evenness) of birds? Explain.

Species diversity, which refers to the variety of species in an area, tends to increase as ecological diversity increases. It is measured in two different ways: species richness and species evenness. **Species richness** refers to the total number of different species in a community. **Species evenness** refers to the relative abundance of each individual species. In general, populations at a higher trophic level will have fewer members than those at a lower trophic level, but populations within the same trophic level should have relatively similar numbers. If they do, the community is said to have high species evenness. If, on the other hand, one or two species dominate any given trophic level and there are few members of other species, then the community is said to have low species evenness. In such uneven communities, the less abundant species is at greater risk of dying out.

species diversity The variety of species in an area; includes measures of species richness and evenness.

species richness The total number of different species in a community.

species evenness The relative abundance of each species in a community.

Both richness and evenness have an impact on the community. In general, higher species richness and evenness make for a more diverse community and a more intricate food web. Greater intricacy enables more matter and energy to be brought into the system and also makes the community less likely to collapse in the face of calamity. Of course, it isn't always as simple as "more species are better." When non-native species are introduced to an area, species richness may initially increase, but these non-natives may disrupt the community in ways that ultimately diminish diversity. The Everglades is facing a formidable invasive species that is even challenging alligators for the role of apex predator and changing the food web in the area—the Burmese python. (See Module 3.1 and its Global Case Studies map for more on invasive species and the Burmese python.) **INFOGRAPHIC 4**

5 HABITAT STRUCTURE: EDGE AND CORE REGIONS

Key Concept 5: Community composition is affected by the physical structure of the habitat, with some species preferring to inhabit ecotone regions where one habitat meets another (the edge) and others staying deep within one habitat (the core). Human impact that fragments habitats may increase edge and decrease core areas, harming core species and disrupting community connections.

ecotones Regions of distinctly different physical areas that serve as boundaries between different communities.

A community's composition and diversity are also heavily influenced by its physical features. As abiotic physical features like temperature and moisture change, so does community composition. This often happens in **ecotones**, places where two different ecosystems meet—like the edge between a forest and field or between a river and shore. The different physical makeup of these edges creates different conditions, known as **edge effects**, which either attract or repel certain species. For example, it is drier, warmer, and more open at the edge between a forest and field than it is further into the forest. This difference produces conditions favorable to some species but not others.

Ecotones may also attract some species that use different aspects of the two adjacent communities; fish such as young snapper or grunts, for example, prefer to live in areas where seagrass beds are fairly close to a shoreline populated by mangrove trees. The mangrove "prop" roots, which anchor the trees into the wet, sandy ground below, offer the fish safety from predators during the day but are close enough to the seagrass beds where the snapper and grunts feed at night for easy "commuting." These fish are not found in coastal areas without the combination of protective coastal mangrove trees and close-by offshore seagrass beds.

Species that thrive in edge habitats like this are called *edge species.* Other species, those that can only be found deep within a given habitat, are called *core species.* Some of the many species that find food and refuge in the seagrass, such as crustaceans, sea urchins, and worms, prefer to stay in core areas, where they are better hidden and protected from wave action or can make use of deeper sediment buildup in these inner areas.

Unfortunately, human actions often fragment large expanses of intact core habitat, increasing edge habitat. This fragmentation scenario plays out in ecosystems around the world, especially terrestrial ones such as forests or grasslands where urban/suburban, agricultural, and industrial development create patchworks of formerly expansive habitats. Edge species may thrive in these patchy habitats, but because core species may not venture out across the edge in search of new habitat, they are easily trapped by habitat fragmentation; we may eliminate these core species altogether if we don't leave enough core area behind. In other instances, edge habitats are totally destroyed and replaced with other habitats, such as the removal of mangrove forests in areas undergoing coastal development, eliminating the edge species that resided there. (See Module 3.2 for more on habitat fragmentation.) **INFOGRAPHIC 5**

edge effect The change in species diversity that occurs due to the different conditions that either attract or repel certain species at an ecotone.

Nicholas Reuss/Getty Images

Deer are an edge species — they feed in fields but need the cover of a forest when they bed down to rest.

INFOGRAPHIC 5 EDGE EFFECTS

MANGROVE EDGES

The mangrove–seagrass ecotone provides an example of an edge effect. Fish such as immature gray snapper and bluestriped grunt "commute" between the mangrove trees and the seagrass beds. The proximity of these two areas is vital to provide both the protection during the day and feeding opportunities at night that these young fish need.

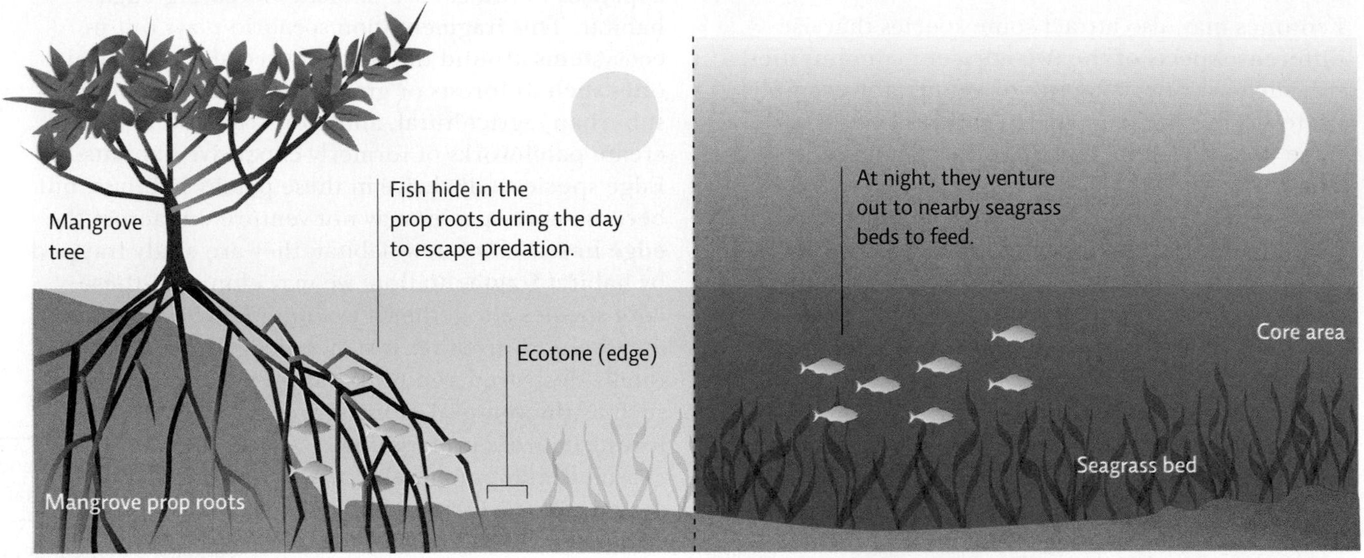

Many coastal mangrove areas are being fragmented as stretches of mangrove are removed for residential or commercial development. What might happen to the gray snapper and bluestriped grunt populations if the mangrove trees are removed from part of the shoreline? What impact would this have on the seagrass beds?

FOREST EDGES

Habitat structure influences where species live. Edge species, like white-tailed deer, prefer habitats with forest and field edges, whereas core species, like some warblers (small birds), prefer the inner areas of forest and do not readily venture into edge regions.

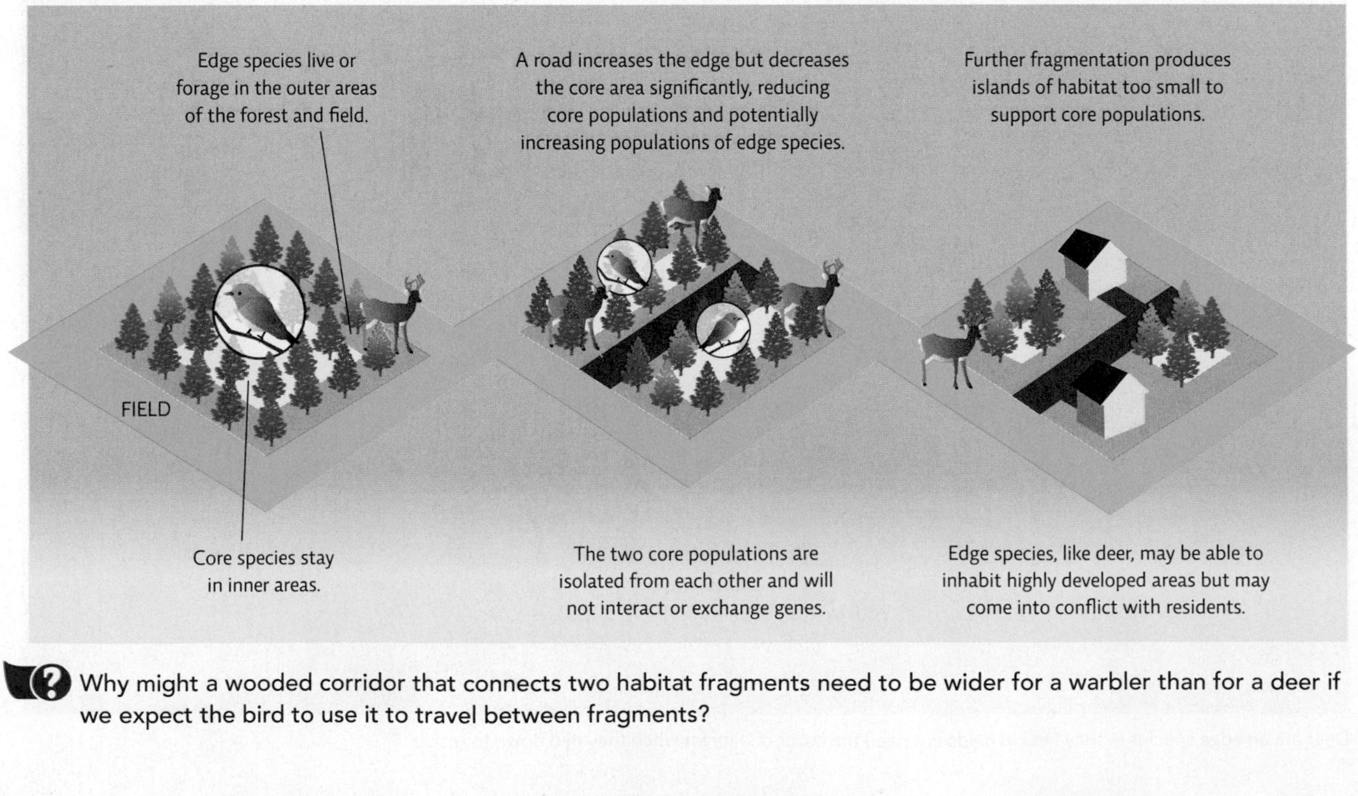

Edge species live or forage in the outer areas of the forest and field.
A road increases the edge but decreases the core area significantly, reducing core populations and potentially increasing populations of edge species.
Further fragmentation produces islands of habitat too small to support core populations.
FIELD
Core species stay in inner areas.
The two core populations are isolated from each other and will not interact or exchange genes.
Edge species, like deer, may be able to inhabit highly developed areas but may come into conflict with residents.

Why might a wooded corridor that connects two habitat fragments need to be wider for a warbler than for a deer if we expect the bird to use it to travel between fragments?

6 KEYSTONE SPECIES

Key Concept 6: Keystone species are particularly important to other members of their community, and if their numbers decline, many other species may be negatively affected.

Wood storks are spectacular fliers. They can reach altitudes as high as 5,000 feet and can glide for miles without flapping their wings. When foraging grounds dry up, flood, or are converted into human developments, these aerial skills are pushed to the limit—flying as much as 75 miles in some cases in search of food.

But they weren't the only species to struggle in the newly developed region. As cities replaced swamps and roads replaced rivers in the Florida Everglades, the flow of water was disrupted like never before. As sloughs ran dry, key detritivores and decomposers like worms, grass shrimp, and microbial communities that had thrived there were decimated. With each species lost, connections were broken, leading to the decline of the snakes, fish, alligators, turtles, and wading birds that fed on them. All species contribute to their ecosystem, but it turns out that some are particularly important.

The replacement of mangrove forests with oceanfront resorts illustrates this point. It turns out that mangrove trees are a **keystone species**—one that provides a unique service that impacts its community more than its mere abundance would predict. It's a species that many other species depend on and one whose loss creates a substantial ripple effect, disrupting interactions for many other species and, ultimately, altering food webs. From their natural habitat at the water's edge, mangrove "prop" roots stabilize the shoreline and provide shelter for a wide variety of fish. So when the mangrove forests are cleared, many other species suffer: the fish that hide among their roots, the fish that feed on those fish, and so on.

Alligators are also a keystone species in the Everglades, one

keystone species A species that impacts its community more than its mere abundance would predict, often altering ecosystem structure.

The gopher tortoise is a keystone species in the longleaf pine forests of Georgia and Florida. Its burrows provide refuge from weather and fire to more than 400 different species.

Shellphoto/iStock/Getty Images

that a great many species depend on during the dry season. As the waters recede, depressions made by alligators (gator holes) are some of the few places that still hold standing water. These holes, which range from about 5 to 50 feet in diameter and from half a foot to 5 feet deep, become refuges for fish, invertebrates, and aquatic plants; they also become very attractive to the animals who feed on these aquatic creatures. Without gator holes, many species would not survive the dry season.

As an apex predator, alligators also help control the population of a wide variety of species. This generalist predator tends to prey on the most abundant populations at any given time. As one population declines in abundance, the alligator turns to other, more abundant prey. This allows the declining population to recover and keeps the abundant population in check. (For more examples of keystone species, see the Global Case Studies map at the end of this module.) **INFOGRAPHIC 6**

Wood storks also depend on the presence of alligators, but not just for the dry-season gator holes. In the 1980s, Rodgers and his colleagues embarked on a comprehensive study of stork nests in an effort to see in which types of trees the storks preferred to nest and whether the availability of those trees was impacting their ability to breed. "We went to 20 stork colonies," Rodgers remembers. "We measured every tree, recorded its species, size, cored it for age, noted its branching structure." The conclusion, reached after 5 years of painstaking work, can be summed up in a single sentence, Rodgers says: Wood storks will nest in just about anything, as long as it's surrounded by water that is patrolled by alligators. "Without the alligators, raccoons swim across, and climb up and destroy everything," Rodgers says. "Without the alligators, when predators get in, we've seen [the storks] abandon entire colonies."

INFOGRAPHIC 6 **KEYSTONE SPECIES SUPPORT ENTIRE ECOSYSTEMS**

Some species are especially important to their ecosystem. If a keystone species is lost or declines in number, the ecosystem could change drastically, and other species that depend on it may suffer or be lost.

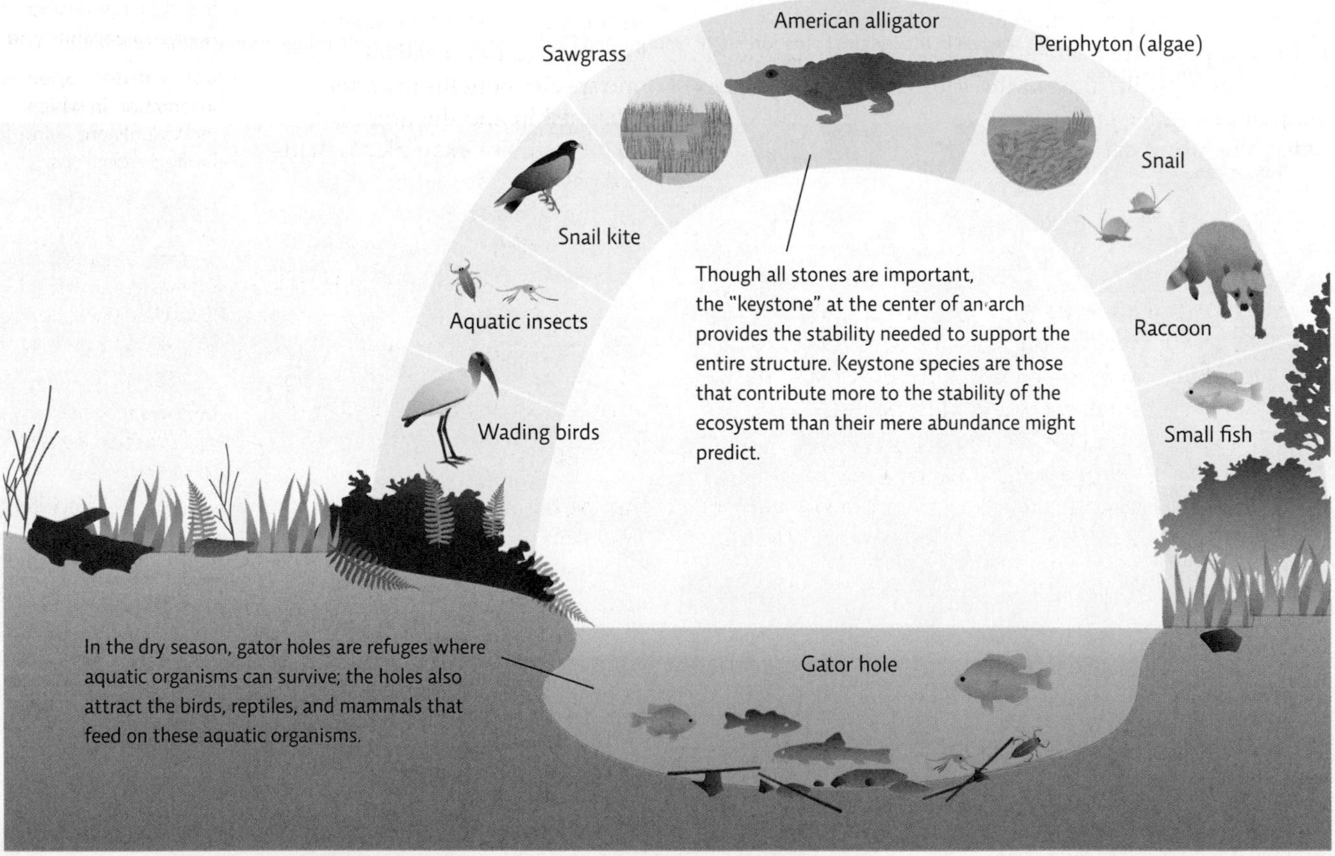

Keystone species whose actions alter the habitat in a way that benefits other species are also called ecosystem engineers. Other than the alligator, identify a species that acts as an ecosystem engineer in its ecosystem and explain how its actions benefit other species.

7 SPECIES INTERACTIONS

Key Concept 7: Interactions within and between species may be beneficial, neutral, or harmful to participants, but all are critical to energy capture and flow and to matter cycling. Changes that interfere with these interactions can imperil many others and decrease the overall functioning of the ecosystem and the services it provides.

Communities are all about relationships. Successful communities are those where a balance has evolved between all the organisms living there. Species interactions serve many purposes; for example, they control populations and affect carrying capacity. Biodiversity (many species, a great deal of variety within a species) is important because more diversity means more ways to capture, store, and exchange energy and matter (see Module 3.2). But it is not sheer numbers that matter most; it is all the connections among species—how they help or hurt one another—that determine how and how well an ecosystem works. Each species is unique and thus interacts in its own unique ways with all the species around it.

Many species have adaptations that bind them to others, that allow them to coexist, or that facilitate **predation** or escape from predators. In the Everglades, for example, alligators have adaptations—like sharp teeth and powerful jaws—that allow them to stalk and capture prey, while most of the fish they prey on have adaptations—like camouflage and a wary nature—that help them avoid capture.

Competition—the vying among organisms for limited resources—is another way that species interact. In general, it is subtle rather than outright fighting. *Intraspecific competition* (that which occurs among members of the same species) is generally stronger than *interspecific competition* (that which occurs among members of different species). This is because members of the same species share exactly the same niche and thus compete for all resources in that niche, whereas members of different species may compete for only a single resource, like water.

Other neighbors—those that prey on the same food or inhabit similar niches—find a way to partition resources. That is, they divvy up the goods in a way that reduces competition and allows several species to coexist. For example, limpkins and snail kites (two niche specialist birds that feed almost exclusively on apple snails) hunt in different regions of the Everglades. This strategy—known as **resource partitioning**—increases the ecosystem's overall capture of matter and energy and thus benefits the entire community.

There are other strategies, too, that keep an ecosystem functioning and strong. Some of these interactions show a tremendous interdependency on the part of the participants. Known as **symbiosis**, these relationships can take one of three forms. The most commonly recognized form of symbiosis is **mutualism**, where both species benefit from the relationship. Research by Lucas Nell has revealed that a surprising mutualistic relationship exists between storks and alligators. As mentioned earlier, nesting near patrolling alligators increases the breeding success of storks. But the alligators also benefit by feeding on nestlings that drop out of stork nests. Nell's research showed that alligators that patrolled the sloughs that surround breeding bird colonies were better fed and had larger fat reserves than those feeding in other areas.

In the symbiotic relationship known as **commensalism**, one species benefits from the relationship and the other is unaffected. For example, in fields and pastures throughout the Southeast, cattle egrets follow grazing cattle, snatching up the bugs disturbed by the hooves of the passing cows. This has no effect on the cows—they don't eat the bugs and are not disturbed by the bugs—but it greatly enhances the foraging efficiency of the egrets.

Parasitism, where one species benefits from the relationship and the other is negatively affected, is also a form of symbiosis due to the close biological tie between the parasite and its host. Parasites may be external, such as ticks, fleas, and leeches, or internal, such as intestinal worms. Guinea worm disease, highlighted in Module 4.3, is just one example of a wide variety of parasitic diseases that afflict human populations. **INFOGRAPHIC 7**

predation Species interaction in which one individual (the predator) feeds on another (the prey).

competition Species interaction in which individuals are vying for limited resources.

resource partitioning The use of different parts or aspects of a resource by different species rather than direct competition for exactly the same resource.

symbiosis A close biological or ecological relationship between two species.

mutualism A symbiotic relationship among individuals of two species in which both parties benefit.

commensalism A symbiotic relationship among individuals of two species in which one benefits from the presence of the other but the other is unaffected.

parasitism A symbiotic relationship among individuals of two species in which one benefits and the other is negatively affected.

INFOGRAPHIC 7 SPECIES INTERACTIONS

The heart of a functioning community is its species interactions. Some interactions are beneficial and others cause conflict, but all are important in keeping matter and energy flowing through an ecosystem.

MUTUALISM Both species benefit: The moth gains nutrition while the flower gets pollinated.
COMMENSALISM One species benefits, and the other is unaffected: The heron can catch twice as many fish when foraging alongside the ibis; this doesn't impact the ibis's ability to forage.
PARASITISM One species (the parasite) benefits, and the other (the host) is harmed: An animal that has too many leeches will be weakened from the loss of blood.
PREDATION One species benefits (predator), and the other is harmed (prey): Alligators prey on a variety of animals and are prey themselves when young.
RESOURCE PARTITIONING Both species benefit by partitioning a resource rather than competing for it: Though they both eat apple snails, snail kites and limpkins don't directly compete for them since each predator feeds in a different region of the Everglades.
COMPETITION All participants are negatively affected: Apple snails, a variety of fish, and crustaceans all compete for periphyton algae as a food source.

Choose an ecosystem other than the Everglades and give examples of mutualism, commensalism, parasitism, predation, competition, and resource partitioning.

By ensuring that all populations persist, even as individuals die, these delicate checks and balances allow more energy to be captured and exchanged and thus increase the amount of biomass the ecosystem is able to produce. When individual species are lost or when a landscape is physically altered, the balance is tipped. And when that happens, things can fall apart. Fast.

8 RESTORATION ECOLOGY

Key Concept 8: Human impact often reduces species diversity. It may be difficult to restore all the species and their connections when we try to repair ecosystem damage; therefore, our best course of action is to avoid the damage in the first place.

Indicator species like the wood stork gave ecologists an early warning that the Everglades ecosystem was suffering from the drastic changes imposed by human impact. The U.S. Army Corps of Engineers was called on to investigate the damage to the Everglades that resulted from nearly 50 years of unchecked expansion. They found that the size of the Everglades had been reduced by 50% since the late 1800s. Constructed canals and levees had dramatically altered water levels, leaving some areas parched and others flooded. And poorly timed water releases were further starving ecosystems that had already been affected by salt levels that were too high, excessive nutrients (from agricultural runoff), and an ever-growing list of non-native species.

Ignoring these problems any longer could greatly imperil the 10 million people who had made their home in the region. "What folks finally realized when we reexamined the area was that the wetlands were this essential filter—they cleaned the water of pollutants," says Kim Taplin, a restoration ecologist who works for the U.S. Army Corps of Engineers. "So as the ecosystems have suffered, water quality has declined considerably. We're going to have millions of people with no clean water, unless we fix it."

Avoiding the ecosystem damage in the first place is always our best option, but when damage is done, fixing it is the work of restoration ecologists. **Restoration ecology** is the science that deals with the repair of damaged or disturbed ecosystems. Indicator species like the stork not only alert ecologists to a problem but also are monitored as a gauge of success in the restoration efforts: Is the population recovering as expected? It requires a special blend of skills—not only biology and chemistry but also engineering and a heavy dose of politics.

In 2000, the U.S. Congress enacted the most comprehensive—and expensive—ecological repair project in history. The Comprehensive Everglades Restoration Plan, or CERP, included more than 60 construction projects to be completed over a 30-year period. The idea was to restore some of the natural flow of water through the Everglades, starting with the Tamiami Trail.

The Tamiami Trail, a 150-mile stretch of U.S. highway that connects the South Florida cities of Tampa and Miami, is a serious barrier to freshwater flow in the region. It's also heavily traveled. That means the U.S. Army Corps must not only tear down the road but also build something in its place. "Most of our restoration projects involve building even more structures," says Tim Brown, project manager for the U.S. Army Corps of Engineers' Tamiami Trail project. "It's a delicate balance. We of course want to restore as much of the natural system as possible. But we are also charged with protecting lives and property, and in this case, that means building bridges." **INFOGRAPHIC 8**

For the wood stork, things are looking up. The population is slowly increasing across the southeastern United States, and in 2014, the species was "downlisted"

restoration ecology The science that deals with the repair of damaged or disturbed ecosystems.

Joe Cavaretta/Sun Sentinel/TNS/Getty Images

Canals that drained (and still drain) the Everglades allowed communities like Sunrise, Florida, to be built where wetlands once existed.

INFOGRAPHIC 8 THE COMPREHENSIVE EVERGLADES RESTORATION PLAN (CERP)

Restoration efforts attempt to return an ecosystem back to its original condition, or as close as possible. Earlier drainage projects in the Everglades moved water that would normally flow slowly through South Florida into canals that quickly sent it to the Atlantic Ocean or Gulf of Mexico. The Everglades Restoration Plan focuses on redirecting some of that water to flow slowly back along its historic path through the Everglades. This will help restore some of the wetland areas and replenish groundwater supplies in South Florida.

On the maps below, darker green areas represent wetland areas or river floodplains; white arrows show overland water flow.

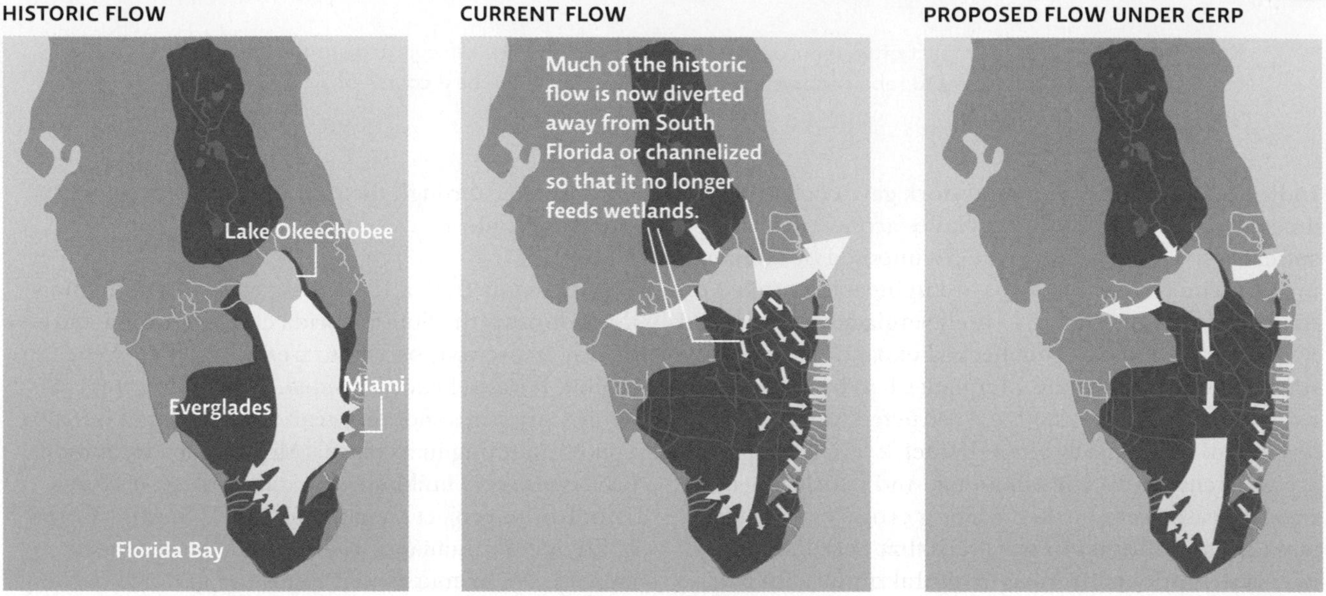

Historically, the Everglades covered most of South Florida—more than 10,000 square kilometers (4,000 square miles). Water flowed slowly through the landscape, keeping much of the area flooded year-round.

Projects to drain the wetlands and divert water to agricultural lands disrupted normal flow, drained about half of the wetlands, and resulted in water shortages for the downstream ecosystems and for people as well.

The goal of CERP is to restore some water flow back to historic wetland areas through various methods, including the removal of more than 240 miles of canals. This will allow more freshwater to soak into the ground, benefiting both the ecosystems and the residents of South Florida.

? Explain how rerouting some of South Florida's water flow back through the center of the state, through restored wetlands, will help provide residents with more drinking water.

from endangered to threatened. The number of breeding pairs in the Everglades sometimes exceeds the restoration target minimum of 1,500 breeding pairs, but numbers fall lower than that during poor-weather years.

Even before humans stepped in to drain the Everglades, wood stork nesting success has always fluctuated with the weather—sometimes wildly. The ideal conditions include a wet summer and fall (a boost to fish populations) followed by a gradual drying during winter and an even drier spring, conditions that concentrate fish into shallow pools. Just such a year was 2107, and wood stork breeding pairs increased dramatically with 3,894 nests recorded in that year's *South Florida Wading Bird Survey*. Similar numbers were seen in 2018. Looking beyond the yearly fluctuations that weather brings, researchers report slow but steady improvements, but the species is still falling short of restoration targets over the long term.

"The bottom line, though, is that there's only so much we can do," says Rodgers. "It's a lot just to figure out what the baseline was or should be. Some plant species have probably gone extinct, and some non-natives are virtually impossible to remove. What we can do is figure out what some of the big obstacles to recovery are, remove them, and after that, let nature take its course."

A new problem is making the timely restoration of the Everglades even more vital—climate change. Rising sea level and saltwater intrusion into freshwater habitats are already occurring and are serving as an added incentive to restore the wetland so that it can capture more freshwater for the ecosystems, farms, and cities of South Florida. (See Module 10.2 for more on climate change and its effects on Florida.)

9 ECOLOGICAL SUCCESSION

Key Concept 9: Over time, ecosystems naturally transition from one community to another in response to changing environmental conditions. An understanding of this ecological succession process can guide our restoration efforts.

Although the changes the Everglades have experienced are extreme, changes to ecological communities are really the norm; nature is not static. Predictable transitions can sometimes be observed in which one community replaces another, a process known as **ecological succession**.

Primary succession begins when **pioneer species** move into new areas that have not yet been colonized. In terrestrial ecosystems, the earliest pioneer species are usually lichens—a symbiotic combination of algae and fungus. Lichens can tolerate the barren conditions. As time goes by and lichens live, die, and decompose, they produce soil. As soil accumulates, other small plants move in—typically sun-tolerant annual plants that live 1 year, produce seed, and then die—and the plant community grows. Gradually, the plant growth itself changes the physical conditions of the area—covering sun-drenched regions with broad, shady leaves, for example. Since these conditions are no longer suitable for the plants that created them, new species move in, and those changes beget even more changes until the pioneers have been completely replaced by a succession of new species and communities.

Secondary succession describes a similar process that occurs in an area that once held life but has been damaged somehow; the level of damage the ecosystem has suffered determines which stage of plant community moves in. For example, a forest completely obliterated by fire may start close to the beginning with small herbs and grasses, whereas one that has suffered only moderate losses may start midway through the process with shrubs or sun-tolerant trees moving in. The stages are roughly the same for any terrestrial area that can support a forest: first annual species, then shrubs, then sun-tolerant trees, then shade-tolerant trees. Grasslands follow a similar pattern, with different species of grasses and forbs (small leafy plants) moving in over time.

Wetland areas also go through succession, responding to the presence of water and sediment depth. In the Everglades, each ecosystem is guided along this path by its own constellation of forces. Some, like the iconic sawgrass ecosystems, are fire-adapted; fire returns them to early stages again and again, where the underwater roots of the emergent plants (those that are rooted underwater but grow above the waterline) survive and quickly regrow. Others, if left undisturbed, would pass through successional stages of pioneers (grasses) to shrubs or small trees to larger species of trees, depending on the deposition of soil and proximity of the water table (the top of the groundwater in the area) to the land surface. **INFOGRAPHIC 9**

The progression of ecological succession in damaged areas might be viewed by some as a "repair" sequence. While we can certainly step in to assist in this natural progression to help a damaged ecosystem recover to a former state, the ecosystem is simply doing what comes naturally—responding to changing conditions.

Intact ecosystems have a better chance at recovering from, and thus surviving, perturbations. More complex communities are also more resilient because it is less likely that the loss of one or two species will be felt by the community at large—even if some links in the food web are lost, other species are there to fill the void.

Of course, if keystone species are lost, the community will feel the effect. The loss of the alligator from the complex Everglades community would impact many species and change the face of the ecosystem.

However precarious their recovery might be, wood storks have rebounded in some areas. Some say this rebound is the result of careful conservation efforts. Others insist that it is merely the result of above-average rainfall in some years. For his part, Rodgers sees another trend at work. Once again, he says, the storks are trying to tell us something. "They have shifted their center of distribution from South Florida to Central and North Florida," he says. "They're now spilling into Georgia and

ecological succession Progressive replacement of plant (and then animal) species in a community over time due to the changing conditions that the plants themselves create (more soil, shade, etc.).

primary succession Ecological succession that occurs in an area where no ecosystem existed before (e.g., on bare rock with no soil).

pioneer species Species that move into an area during early stages of succession.

secondary succession Ecological succession that occurs in an ecosystem that has been disturbed but not rendered lifeless.

INFOGRAPHIC 9 ECOLOGICAL SUCCESSION

FOREST ECOLOGICAL SUCCESSION DEPENDS ON SOIL AND LIGHT AVAILABILITY

Lichens can attach to bare, uncolonized rock surfaces; as they grow and die over the years, soil begins to form and accumulate, allowing mosses to move in.

Soil builds up over time, allowing larger plants to move in—first, small herbs and grasses and, later, small shrubs.

Sun-tolerant trees, such as pine and cedar, move in when there is enough soil to support large root systems.

Taller plants produce shade, changing conditions and favoring different plant species.

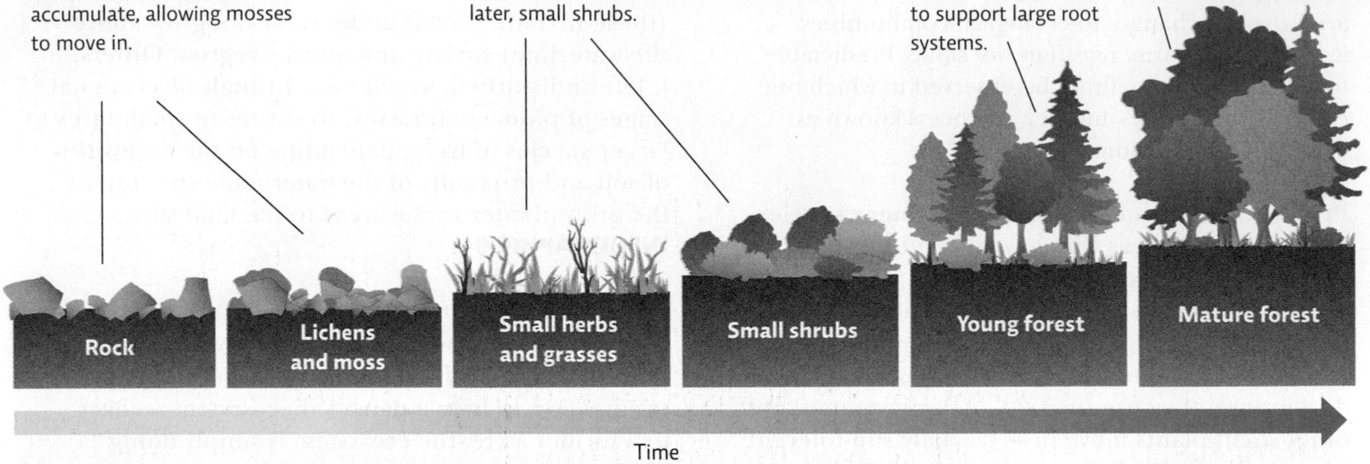

In terrestrial ecosystems, we see natural stages of succession occur whenever a new area is colonized or an established area is damaged. Sun-tolerant species give way to shade-tolerant ones as more soil is built up, supporting larger plant species.

EVERGLADES ECOLOGICAL SUCCESSION DEPENDS ON THE WATER LEVEL

When water is more than 20 inches deep, floating and submerged vegetation grow. As sediments collect, emergent grasses that are rooted underwater—but grow tall enough to emerge above the water surface—move in and establish the sawgrass marsh.

Water-tolerant cypress and willow move in and replace sawgrass as sediment becomes deeper and more stabilized.

Other species of trees, such as the pond apple, move in as sediment gets closer to the water surface.

If dry land emerges from the water, upland species, such as oak, move in and replace the water-tolerant species of the mixed swamp forest.

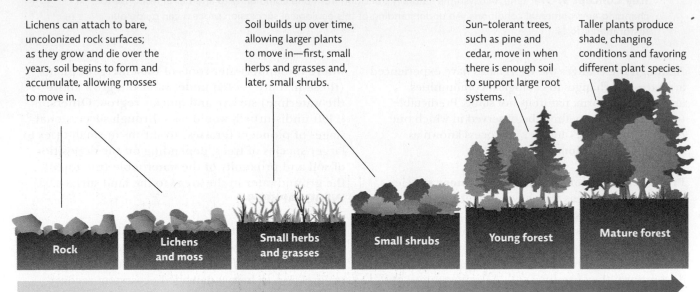

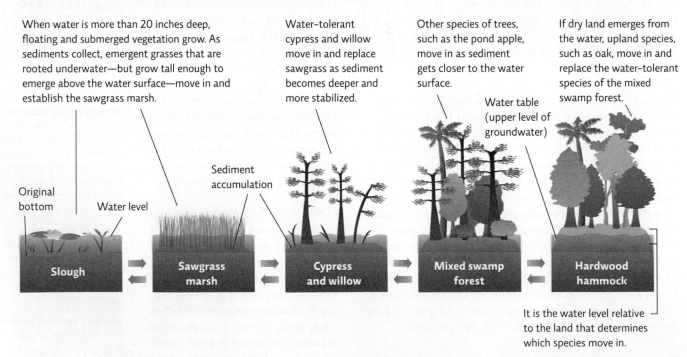

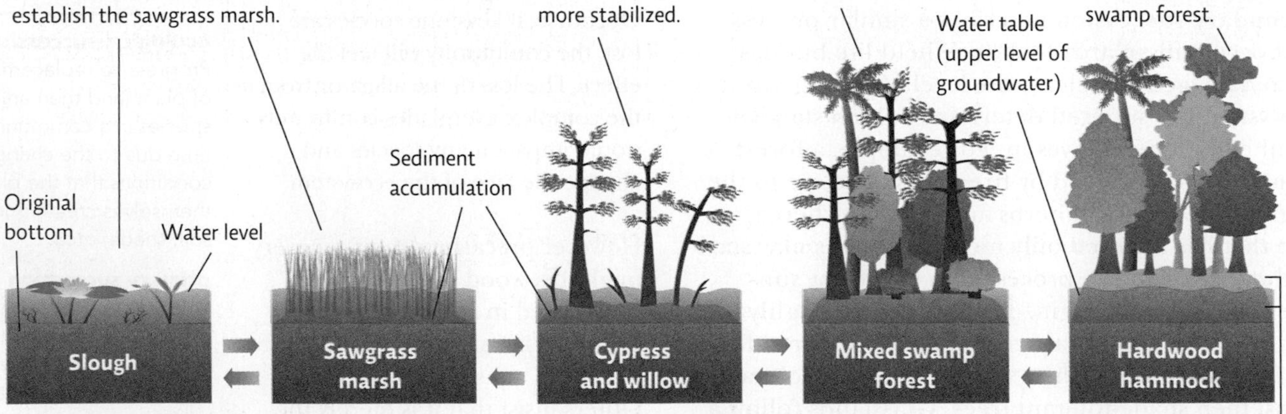

It is the water level relative to the land that determines which species move in.

Succession in wetlands can actually go both ways due to periodic fires, cyclic rainfall patterns, and variable water levels. Rising water might flood an area anew, or sediment buildup might continue to raise the land relative to the water level. In the absence of disturbance, succession will progress to the hardwood hammock forest when sediment builds up enough to expose dry land.

? Look at the forest successional stages shown in the top part of this diagram. Why can't small shrubs or young trees grow on the land shown supporting small herbs and grasses?

Ron Erwin/All Canada Photos/Alamy

Pileated woodpeckers are a keystone species that many other species depend on. Many different species commandeer abandoned woodpeckers' holes in a commensal relationship that benefits the new hole occupant but does not affect the woodpecker that originally constructed the nest hole.

North Carolina—something we've never seen before." Rodgers suspects that the shift has something to do with the way climate is changing in the region, although he says much more research is needed before anyone can say for certain. "We're still trying to figure out what that means," he says. "But we know it's a clue to something."

Select References:

Brandt, L. A., et al. (2016). *System-wide indicators for Everglades restoration.* Unpublished Technical Report. Retrieved from http://www.evergladesrestoration.gov/content/documents/system_wide_ecological_indicators/2016_system_wide_ecological_indicators.pdf

Frederick, P., et al. (2009). The white ibis and wood stork as indicators for restoration of the Everglades ecosystem. *Ecological Indicators, 9*(6), S83–S95.

Nell, L. A., et al. (2016). Presence of breeding birds improves body condition for a crocodilian nest protector. *PloS One, 11*(3), e0149572.

Rodgers, J. A., et al. (1996). Nesting habitat of wood storks in North and Central Florida, USA. *Colonial Waterbirds, 19*(1), 1–21.

Siddig, A. A., et al. (2016). How do ecologists select and use indicator species to monitor ecological change? Insights from 14 years of publication in *Ecological Indicators. Ecological Indicators, 60,* 223–230.

South Florida Water Management District. (2018). *South Florida wading bird report* (Vol. 23; M. I. Cook & M. Baranski, Eds.). Retrieved from https://www.sfwmd.gov/sites/default/files/documents/sfwbr_2017.pdf

GLOBAL CASE STUDIES KEYSTONE SPECIES

Achieve

Just like the alligator in the Everglades, ecological communities everywhere have keystone species, each contributing to its ecosystem in unique ways. Go online to read about these additional examples.

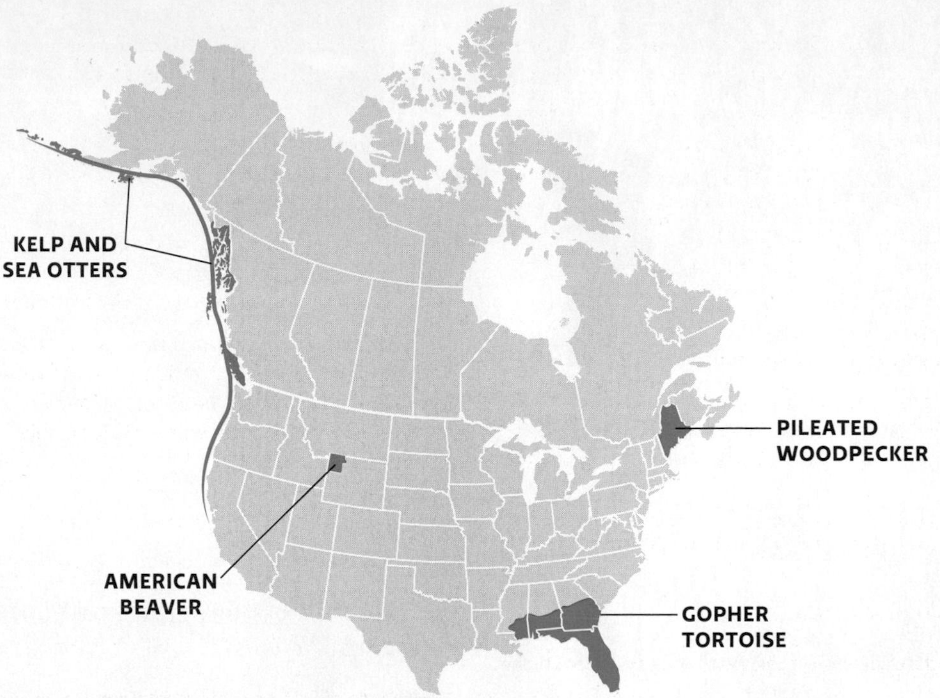

BRING IT HOME

PERSONAL CHOICES THAT HELP

The world is full of weird and wonderful species. Every year we discover new information about how intricate our biological communities are. By restoring habitats and increasing our understanding of the relationships between species, we can better ensure their long-term survival.

Individual Steps

- Visit a park or nature preserve and watch for signs of species interactions. Do you hear animals or birds? Can you see signs of predation or herbivory?
- Buy a Duck Stamp. Usually purchased by waterfowl hunters for license purposes, nonhunters can purchase a stamp, which supports wetland conservation in the National Wildlife Refuge System.
- Follow the U.S. Fish and Wildlife Service Open Space blog to learn more about wildlife and issues facing conservation (www.fws.gov/news/blog).

Group Action

- The Everglades case study is an example of a very extensive restoration project. Call your local park district or nature preserve or visit www.epa.gov/wetlands to learn about restoration work that is happening in your area and how you can become involved.
- Several U.S. laws protect wetlands and the ecosystem services they provide; these include the Clean Water Act and the Coastal Barrier Resources Act. Keep abreast of proposed policy changes to weaken or strengthen these acts and let your elected representatives know how you feel about those changes. Encourage your friends and family to do the same.

Chris Matula/ZUMA Press/Newscom

REVISIT THE CONCEPTS

- Community ecologists study the many populations that live and interact with one another and with their environment within a given area.
- Matter and energy flow through a community via a food chain that begins with producers and has consumers at each higher trophic level.
- These trophic levels are shown as a pyramid — lower levels are larger than upper ones because energy only enters at the producer level, and the organisms at each level use most of the energy they have taken in (therefore, only a small percentage is passed on to support the next level).
- Community ecologists monitor species diversity in terms of the number of different species present (richness) and the relative abundance of each species (evenness). High richness and evenness tend to make a community more resilient to environmental changes.
- Some species prefer to inhabit ecotone regions where one habitat meets another (the edge), and others stay deep within one habitat (the core); anything that fragments habitats may increase edge and decrease core areas, harming core species and disrupting community connections.
- Keystone species are particularly important to their community. If their numbers decline, many other species may be negatively affected.
- Countless interactions take place within and between species. These may be beneficial, neutral, or harmful to participants, but all are critical to energy capture and flow and to matter cycling.
- It may be difficult to restore all the species and their connections when we try to repair ecosystem damage; therefore, our best course of action is to avoid the damage in the first place.
- In nature, ecosystems and their communities are constantly changing: Communities transition from one to another over time in response to changing environmental conditions. An understanding of this ecological succession process can guide our restoration efforts.

ENVIRONMENTAL LITERACY Understanding the Issue

1 What is an ecological community?

1. Define habitat and explain how it relates to the concept of niche.
2. Distinguish between niche generalists and niche specialists in a community. What are the advantages and disadvantages of each of these strategies?

2 What can we learn about a community by identifying its food web?

3. Explain how matter and energy flow through a food chain using the terms *producer* and *consumer*.
4. Relate the concept of a food chain to a food web. Which concept might be more useful to a community ecologist?

3 What are trophic levels, how are they classified, and why are these illustrated as a pyramid?

5. Identify the levels of the trophic pyramid and explain the connection between adjacent levels.
6. The levels on a trophic pyramid are shown as a pyramid because each level is smaller than the one below it. Why is this relationship seen?

4 How is species diversity measured, and why is it important to a community?

7. How does species diversity generally change as the variety of habitat and available niches increases?
8. Explain why both species richness and species evenness are important for a healthy ecosystem.

5 What are core and edge habitats, and how can habitat fragmentation affect the species that inhabit them?

9. Which species are more negatively affected by habitat fragmentation: edge or core species? Explain.

6 What is a keystone species, and why is it important?

10. How do alligators fit the definition of a keystone species?

7 What species interactions are seen in a community, and what problems can emerge if these interactions are disrupted?

11. Think about the many roles played by the alligator in the Everglades. For each of the relationships listed here, identify an example of the alligator and another species that illustrates that relationship: predation, competition, mutualism, and commensalism.
12. What is resource partitioning, and how does it reduce competition?

8 What is the goal of restoration ecology, and what obstacles do restoration ecologists face?

13. Why might it be hard to restore damaged ecosystems like the Everglades?
14. What is an indicator species, and why is it useful when monitoring the health of an ecosystem?

9 What is ecological succession, and how can we use this knowledge to assist in ecosystem restoration?

15. What is the importance of ecological succession in nature?
16. Distinguish between primary and secondary ecological succession.
17. How might a restoration ecologist use an understanding of ecological succession to help repair a damaged area?

SCIENCE LITERACY Working with Data

The number of breeding pairs of wading birds has been closely tracked in the Everglades since the 1970s. Here we show the annual number of wood stork and snowy egret nests in the Everglades ecosystem since 2000, as reported in the *2017 South Florida Wading Bird Report*.

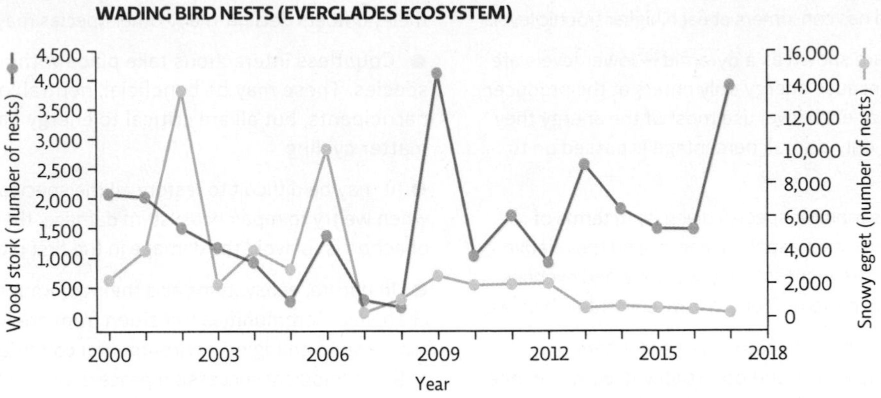

Interpretation

1. Which year had the greatest number of stork nests? Egret nests? For each species, which year had the fewest number of nests?
2. Over what 2-year period did the greatest change (increase or decrease) in nest number occur for storks? For egrets?
3. Compare the number of nests in the year 2000 to later years for each species. From 2001 to 2017, which was greater: the number of years in which nest totals exceeded that of the year 2000 or the number of years when there were fewer nests?

Advance Your Thinking

4. The best weather conditions for storks are a wet summer and fall followed by a dry winter and spring, conditions that concentrate fish into shallow pools. Which year(s) most likely produced this weather pattern?
5. Do you think the snowy egret needs the same seasonal weather patterns as the wood stork? Explain.
6. In some cases, we can extrapolate from trend lines to estimate future values. What about these data? How successful do you think you would be if you tried to predict how many stork and egret nests would be found in subsequent years? Explain.

INFORMATION LITERACY Evaluating Information

Woodlands are some of the most important communities in the United States. They are rich with wildlife, and we see them as places worth preserving. The United States began fighting wildfires in a systematic way around the turn of the 20th century. Recently, however, there have been increased discussions about the automatic response of immediately quelling all wildfires; some ecologists argue that some wildfires are helpful and should be allowed to burn. After all, change from storms, fires, and floods is part of the natural cycle of an ecosystem.

Learn about wildfires and decide for yourself what the best response should be.

1. Search the Internet for information about fire ecology to help develop an informed opinion about how we should respond to wildfires on wildlands. Topics to research include fire-adapted ecosystems, wildfire suppression, the cost of fighting wildfires, and prescribed fires. You may visit as many websites as you need, but you must visit at least four different websites for your research. Answer the following questions to evaluate whether each of the websites you visit is a reliable information source:
   a. Who are the authors of the information given in this article or on this webpage?
   b. What supporting evidence do the authors give for their claims?
   c. Do they give sources for their evidence? If so, evaluate the reliability of these sources.
   d. Do you detect any strong biases for or against fighting wildfires? Explain.
   e. Summarize the position of this source regarding how we should respond to wildfires or what could be done to prevent wildfires.
   f. Based on your answers to the above questions, is this a reliable source of information? Explain.
2. Based on the information you obtained from the websites that you deemed to be reliable information sources, write an essay that addresses the following two questions. Provide evidence from your sources in support of your position.
   a. Should all wildfires on wildlands be fought immediately?
   b. What, if anything, should be done in an area to prevent or lessen a possible wildland wildfire?

Additional study questions are available at Achieve.macmillanlearning.com. Achieve

CHAPTER 3

Evolution and Biodiversity

The incredible variety of life on Earth is the result of evolution of species over time; however, human impact is causing a disturbing loss of biodiversity. We can take steps to protect and preserve species, but this requires good science, adequate funding, and political will for success.

Module 3.1

Evolution and Extinction

A look at evolution by natural selection and the problems that result when drastic changes are imposed on populations that exceed their ability to adapt

Module 3.2

Biodiversity

An evaluation of the extent and importance of biodiversity and the impact of human actions on current biodiversity

Online Module 3.3

Preserving Biodiversity

A survey of the approaches being pursued to protect species

Online Modules are available at Achieve.macmillanlearning.com.

Roman Lukiw Photography/Moment/ Getty Images

Module 3.1 Evolution and Extinction

A TROPICAL MURDER MYSTERY

Finding the missing birds of Guam

The brown tree snake (*Boiga irregularis*). *Clement Carbillet/BIOSPHOTO/Alamy*

After reading this module, you should be able to answer these GUIDING QUESTIONS

1. What is biological evolution, and how do populations adapt to changes via natural selection?
2. Why is genetic diversity important to natural selection?
3. What is coevolution, and what problems can emerge when species that did not coevolve together suddenly share a habitat?
4. How do random events influence the evolution of a population?
5. How do humans, intentionally or accidentally, affect the evolution of a population?
6. What factors affect the pace of evolution and extinction, and why are extinctions that occur quickly more of a concern than those that take a long time to unfold?
7. How do the mass extinction events of the past compare to extinctions during intervening times and today?

On a crisp December morning in 2013, representatives of several federal agencies met on Anderson Air Force Base in Guam—a South Pacific island and U.S. territory—to watch an experiment that sounded more like science fiction than science. As they looked on with binoculars, military personnel in a small fleet of helicopters dropped dead baby mice, one by one, into the surrounding jungle. The mice had been laced with acetaminophen and fitted with lightweight streamers that served as mini parachutes.

The mice could be thought of as paratroopers in a war that the island has been fighting for half a century against a most elusive, yet devastating enemy: brown tree snakes. Introduced accidentally from other Pacific islands to Guam through ships sometime back in the 1940s, they have driven almost all of the island's native bird species to extinction.

The mouse airdrop was part of the effort to decrease the snake population. Acetaminophen is lethal to the snakes. The parachutes would ensure that the mice would catch in the trees, where the snakes live and eat. Operation Mouse Drop is just one of many tools currently employed to address a problem that has plagued wildlife biologists for half a century and that, for years, was shrouded in mystery.

WHERE IS GUAM?

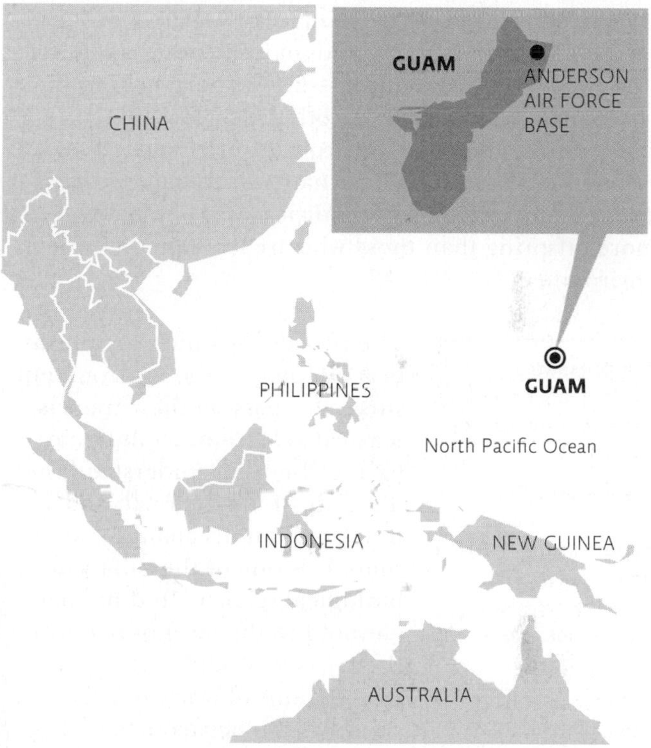

Dr. Julie Savidge holding a Mariana Fruit-Dove. This species only occurs on certain islands within the Mariana Islands, and the last sighting on Guam was in 1985. This bird was caught as part of an early blood-sampling effort to see if exotic diseases might be causing the bird decline on Guam.

It was the late 1960s when the birds of Guam began dying off with disturbing speed. By the early 1980s, four species had gone extinct, ten others were in danger of joining them, and wildlife experts had no clue why. Biologist Julie Savidge, a PhD student at the University of Illinois, took on the project and headed to Guam. Early hypotheses (diseases or pesticides) didn't pan out, but the locals were certain that brown tree snakes (*Boiga irregularis*), non-native snakes up to 6 feet long, were responsible for the birds' demise. Savidge began to investigate whether these reptiles might be causing the **extinctions**. Surely it was something else; could a few snakes really obliterate a whole island's worth of birds?

extinction The complete loss of a species from an area; may be local (gone from an area) or global (gone for good).

1 NATURAL SELECTION AS A MECHANISM FOR EVOLUTION

Key Concept 1: Populations can adapt to a changing environment when individuals whose inherited traits make them better suited to survive or reproduce leave more offspring with those traits on average than other less suited individuals—a process known as natural selection.

Before they started disappearing, Guam was home to 18 native species of birds, each specially suited to life on the island.

Populations usually contain individuals that are genetically different from one another. According to the evolutionary theory first put forth by Charles Darwin and Alfred Russel Wallace, a **selective pressure** on a population—a nonrandom influence that affects who survives or reproduces—favors individuals with certain inherited traits over others (such as better camouflage, tolerance for drought, or enhanced sense of smell). These individuals have *differential reproductive success* compared to other individuals: They leave more offspring than those who are less suited for their environment.

selective pressure A nonrandom influence that affects who survives or reproduces.

natural selection The process by which organisms best adapted to the environment survive to reproduce, leaving more offspring than less well-adapted individuals.

genes Stretches of DNA, the hereditary material of cells, that each direct the production of a particular protein and influence an individual's traits.

alleles Variants of genes that account for the diversity of traits seen in a population.

evolution Differences in the gene frequencies within a population from one generation to the next.

gene frequencies The assortment and abundance of particular variants of genes relative to each other within a population.

The process by which organisms best adapted to the environment survive to pass on their traits is **natural selection**. Evolutionary biology helps us understand the diversity of life on Earth and how populations change over time. It is one of the pillars of biological sciences and has been elevated to the level of scientific theory (see Module 1.2) by the vast amount of evidence that supports the occurrence of evolution and the mechanisms by which it happens.

For most populations, more offspring are born than can survive, since resources are limited and many species produce large numbers of young. Since only some individuals will survive, over time, the population will contain more and more of these better-adapted individuals and their offspring. Ultimately, this changes how common certain variants of **genes** are in the population (these variants are called **alleles**): The frequency (percentage in the population) of some alleles increases and that of others decreases. When this occurs, the population has experienced **evolution**, or changes in the **gene frequencies** within a population from one generation to the next. Natural selection may be *stabilizing*, *directional*, or *disruptive*, depending on which genetic traits are favored or selected against. **INFOGRAPHIC 1**

It is important to note that *individuals* are selected but *populations* evolve; individuals do not change their own genetic makeup to produce new necessary adaptations, such as bigger size or pesticide resistance. If they get the opportunity to reproduce, they pass on their traits to the next generation. If they cannot tolerate environmental changes, as was the case with the first bird species to disappear from Guam (the bridled white-eye), they die or fail to reproduce and do not pass on their genes. Individuals may be able to adjust their behavior to accommodate environmental changes, but if a trait is not genetically controlled, and therefore is not heritable, it cannot be passed on to the next generation.

Bill Gozansky/Alamy

Green anoles in Florida are evolving in response to the presence of the non-native brown anole from Cuba. These two lizard species compete for the same food in the same area of the trees in which they hunt. The larger brown anole is winning out, pushing the green anole higher up in the tree—a more difficult hunting ground. Directional selection is favoring green anoles with larger toepads, a trait that allows them to better grip the thinner branches higher up in the trees.

INFOGRAPHIC 1 NATURAL SELECTION AT WORK

Achieve

When the environment presents a selective force (e.g., a new predator, changing temperatures, change in food supply), natural selection is the primary force by which populations adapt. The survivors are those who were lucky enough to have genetic traits that allowed them to survive in their changing environment. (Others who did not possess the trait were not as likely to survive to reproduce.) Because survivors pass on those adaptations to their offspring, the gene frequencies of the population change in the next generation, which means some traits are more common and others are less common than they used to be. When this happens, the population is said to have evolved.

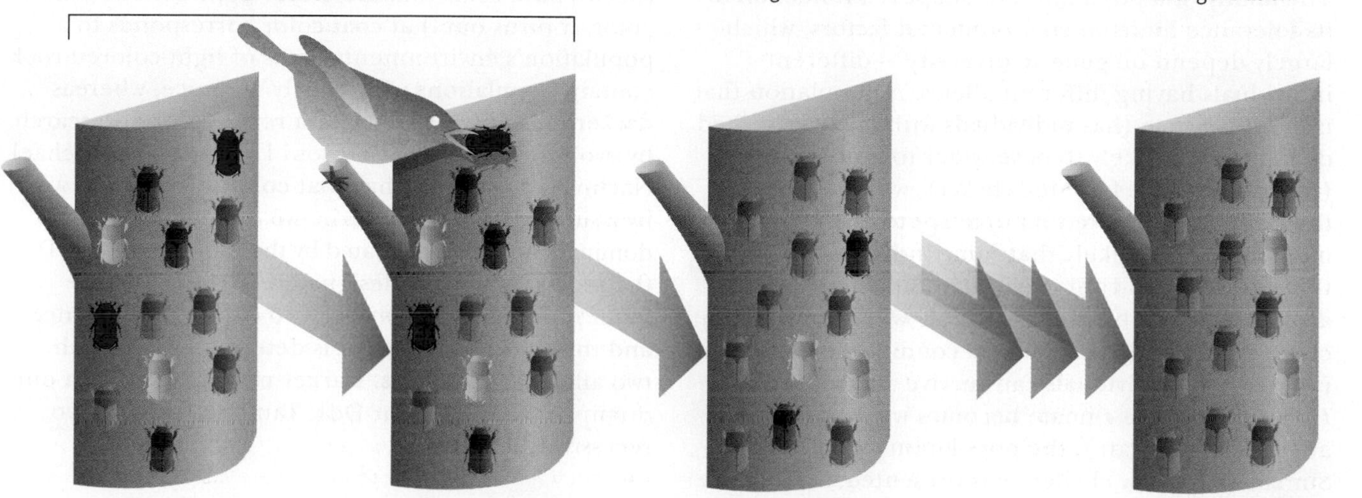

Beetles resting on this tree vary in color.

Genetic variation exists in the population: Individuals possess inherited differences.

Individuals with the less favorable trait (coloration that makes them stand out on a tree trunk) are more likely to be eaten.

Differential reproductive success: Not everyone will survive to reproduce.

Fewer dark individuals are born (though recombination might produce some from light or tan parents).

Gene frequencies have changed: The population is evolving.

Over time, the population may be mostly or solely made up of tan individuals.

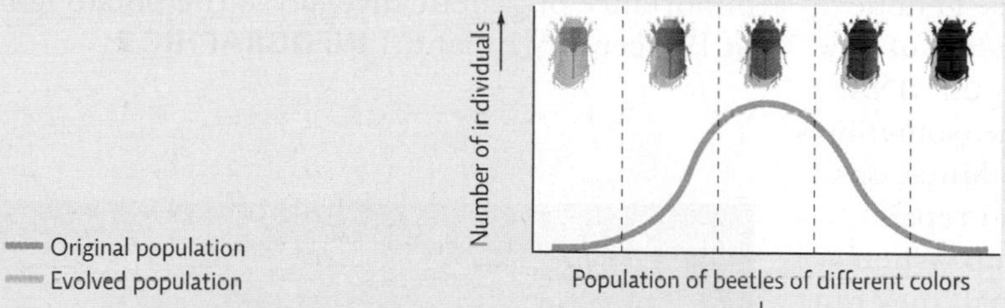

Natural selection can have different outcomes, depending on what varieties in the population the environment favors or selects against.

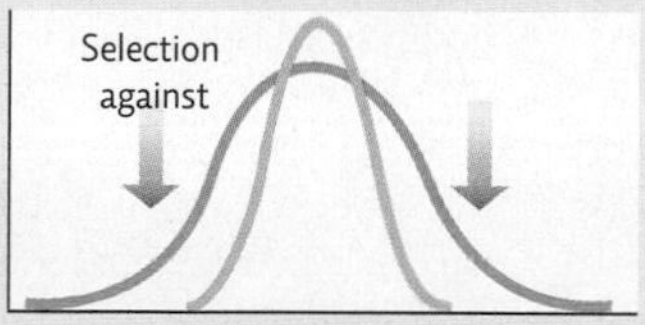

Stabilizing selection favors the norm and selects against extremes.

All trees are tan; tan beetles are favored.

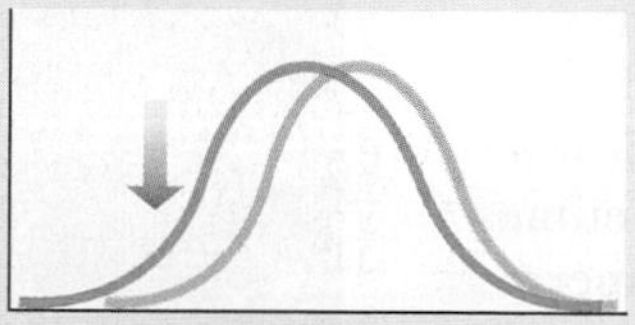

Directional selection continually favors a particular extreme of the trait (bigger, darker, etc.).

In areas with trees darkened from pollution, darker beetles are favored.

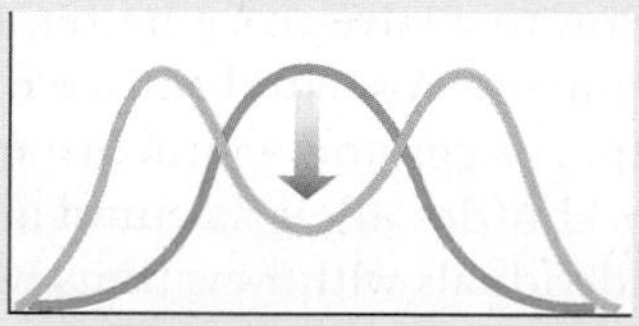

Disruptive selection favors the extremes but selects against the intermediate forms.

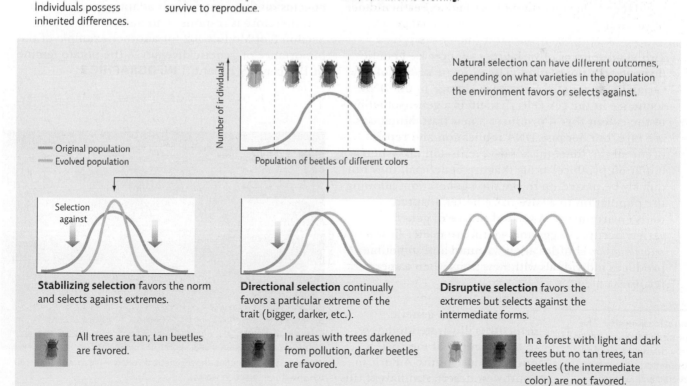

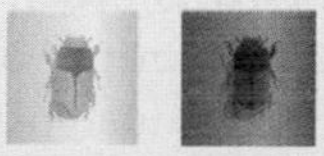

In a forest with light and dark trees but no tan trees, tan beetles (the intermediate color) are not favored.

Identify the gene frequencies of the "original population" for each color morph (dark gray, dark brown, light brown, dark tan, and light tan) by counting the number of each and expressing it as a percentage of the whole. Now do the same for the "later generation." Has evolution occurred? Explain.

2 GENETIC DIVERSITY AND NATURAL SELECTION

Key Concept 2: Genetic diversity in a population is the raw material on which natural selection operates. The more diverse a population, the more likely there will be individuals present who can withstand or even thrive if environmental conditions change.

The ability of a population to adapt is a reflection of its tolerance limits to environmental factors, which largely depend on **genetic diversity**—different individuals having different alleles. A population that is highly diverse (has individuals with many inherited differences) is likely to have wider tolerance limits (see Infographic 4 in Module 2.1), which increases the population's potential to adapt to changes. This means it is more likely that some individuals will exist that can withstand (or even thrive in) the changes and that the population as a whole will survive. If a change occurs that produces a condition outside the range where individuals can survive and reproduce (for instance, the climate becomes warmer than anyone can tolerate), the population will die out. Similarly, if a new challenge is presented, such as the introduction of a new predator or competitor, the survival of the population will depend on whether there are any individuals in the population who can effectively deal with the new species. If the snakes on Guam were indeed responsible for killing the birds, any birds that happened to have effective snake-avoidance behaviors would have had a greater chance of survival.

Two main sources of variation can increase genetic diversity in a population. The ultimate source of new variability is genetic *mutation*, a change in the DNA sequence in the sex cells that alters a gene, sometimes to the extent that it produces a new trait. Mutations are rare, but because DNA replication and repair occur all the time, these rare events add up. When a mutation produces traits that are beneficial, they can quickly be passed on to the next generation, allowing the population to evolve to be better adapted to its environment. A second source of genetic variety occurs as eggs and sperm are made: *Genetic recombination* shuffles alleles around and sometimes produces individuals with new traits when a sperm fertilizes an egg.

genetic diversity The heritable variation among individuals of a single population or within a species as a whole.

The value of this genetic diversity is illustrated today in the example of the rock pocket mouse of the American Southwest desert. Animals of this species have coats that are either light tan or a darker color. It turns out that coat color corresponds to a population's environment: Areas of light-colored rock contain populations with mostly tan mice, whereas darker mice inhabit black lava rock regions. Research by evolutionary biologists Hopi Hoekstra and Michael Nachman has shown that coat color is determined by a single gene that comes in two different alleles. The dominant allele is designated by the uppercase letter D; the recessive allele is designated by the lowercase letter d. All individuals have two copies of the gene, and the color of their coat is determined by which two alleles they possess. Darker mice have at least one dominant allele (DD or Dd). Tan mice possess two recessive alleles (dd).

It is likely that coat color provides camouflage and protection from visual hunters, but only if the mouse is on a background of the same color. A study on deer mice (a similar species) showed that predatory owls are more successful at capturing mice on a contrasting background. This gives support to the conclusion that coat color is adaptive as camouflage and therefore is responsive to natural selection. (See Section 3 of Module 3.2 for another example of the importance of genetic diversity—the potato famine in 19th-century Ireland.) **INFOGRAPHIC 2**

Robert Hamilton/Alamy/Diomedia

These Asian lady beetles show genetic diversity—the raw material on which natural selection works.

INFOGRAPHIC 2 EVOLUTION IN ACTION

Natural selection produces populations with different gene frequencies (more or less of a particular gene variant or allele). For this to occur, there must be genetic variation (more than one allele for a given trait) and a selective pressure (a reason one variant is better than another in a given situation).

Different color morphs of the rock pocket mouse (*Chaetodipus intermedius*) are found on different-colored rocky outcroppings in the desert Southwest. An evaluation of the mice living on or near the Pinacate lava flow in southern Arizona represents the first documentation of the genetic basis (in this case, a single gene) for a naturally favored trait. The well-known peppered moth is another example in which different color variants are favored in different habitats, but the genes responsible for that trait have not yet been identified.

Even though there is gene flow between dark and light populations that are close to each other, populations on tan rock have mostly tan individuals, and populations on dark rock have mostly dark individuals, suggesting a strong selective pressure that favors one color over the other.

Predatory owls are likely the selective pressure that favors different coat colors in different habitats.

Rick & Nora Bowers/Alamy

Tan mice (the recessive trait, dd) predominate in light-colored rocky outcroppings.

Dr. Hopi Hoekstra

Darker mice (the dominant trait, DD or Dd) predominate on darker lava rocks.

Sean Locke/Getty Images

Lyn Alweis/Denver Post/Getty Images

A study done with dark and light varieties of deer mice revealed that owls caught twice as many opposite-colored mice (dark mice on a light background or tan mice on a dark background) as mice whose coloration matched their background, even in almost total darkness. Owl predation is therefore likely to be a strong selective pressure on coat color, driving directional selection that produces either light or dark populations of mice, depending on the background.

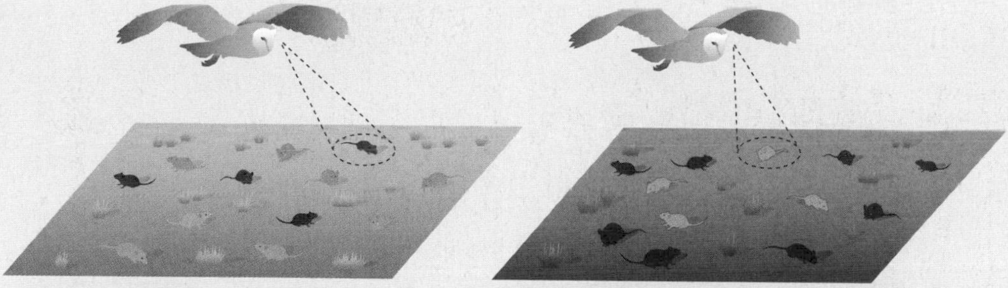

Dr. Hopi Hoekstra

Dr. Hopi Hoekstra

If this mouse population migrated to an area with a red rock habitat with visual predators like hawks or owls, what could prevent the population from evolving into one with red coats?

3 COEVOLUTION

Key Concept 3: Two species can become highly adapted to each other when each becomes the selective pressure that favors certain traits in the other, a process known as coevolution. Species that never coevolved with a particular predator or competitor may not have the traits needed to survive if that species invades their habitat.

A special type of natural selection, known as **coevolution**, occurs when two species each provide the selective pressure that determines which of the other's traits are favored by natural selection. Predator and prey species usually evolve together, each exerting selective pressures that shape the other. As predators get better at catching prey, the only prey to survive are those a little better at escaping, and it is those individuals that reproduce and populate the next generation. This game of one-upmanship continues generation after generation, with each species affecting the differential survival and reproductive success of the other. The result can be a predator extremely well equipped to capture prey and prey extremely well equipped to escape. **INFOGRAPHIC 3**

coevolution A special type of natural selection in which two species each provide the selective pressure that determines which traits are favored by natural selection in the other.

invasive species A non-native species (a species outside its range) whose introduction causes or is likely to cause economic or environmental harm or harm to human health.

endemic Describes a species that is native to a particular area and is not naturally found elsewhere.

Non-native species that cause ecological, economic, or human health problems and are hard to eradicate are considered **invasive species**, and they can cause significant damage in areas they invade. In fact, invasive species are one of the leading causes of species endangerment worldwide (see Module 3.2).

If the birds on Guam were indeed eradicated by the invasive snake species, it was because the speed at which the eradication happened prevented the bird populations from potentially coevolving survival strategies to deal with the new snake population. The brown tree snake was already well adapted to preying on birds. But Guam's bird populations had never faced such a predator and had no natural defenses. It was an unfair fight. And like populations worldwide that are isolated on islands or mountaintops or by fragmented habitats, Guam's populations were further handicapped by their isolation; they rarely, if ever, received new individuals from populations elsewhere. Indeed, some of Guam's species were **endemic**—found nowhere else—so no other populations even existed to contribute new members that might be better adapted to the snake.

Many ground-nesting birds and turtles in Hawaii have no defenses against this invasive mongoose, a skilled predator that eats their eggs and hatchlings.

Malcolm Schuyl/Alamy

INFOGRAPHIC 3 **COEVOLUTION ALLOWS POPULATIONS TO ADAPT TO EACH OTHER**

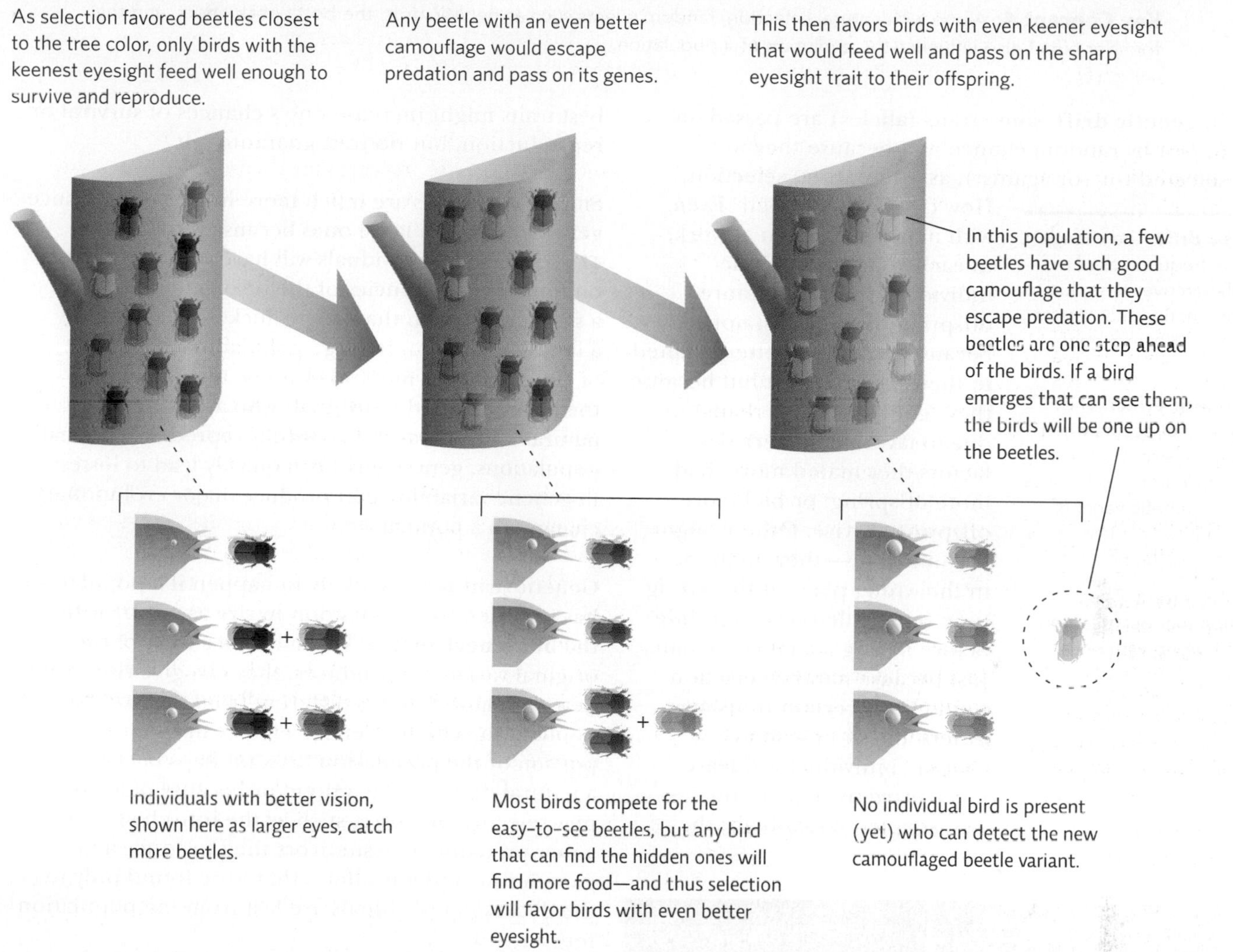

Achieve
As selection favored beetles closest to the tree color, only birds with the keenest eyesight feed well enough to survive and reproduce.
Any beetle with an even better camouflage would escape predation and pass on its genes.
This then favors birds with even keener eyesight that would feed well and pass on the sharp eyesight trait to their offspring.
In this population, a few beetles have such good camouflage that they escape predation. These beetles are one step ahead of the birds. If a bird emerges that can see them, the birds will be one up on the beetles.
Individuals with better vision, shown here as larger eyes, catch more beetles.
Most birds compete for the easy-to-see beetles, but any bird that can find the hidden ones will find more food—and thus selection will favor birds with even better eyesight.
No individual bird is present (yet) who can detect the new camouflaged beetle variant.

What do you predict would happen to the original beetle population if a species of bird with extremely keen eyesight (like those shown on the right side of the diagram) were accidentally introduced into the beetle's habitat?

Some of Guam's bird species went extinct sooner than others. For instance, the endemic bridled white-eye, the gregarious bird species that was extinguished first, was very small, raising the possibility that the small size of these birds might have put them at a disadvantage. Larger species like flycatchers survived longer, though most of them, too, eventually disappeared. Other bird species experienced **extirpation**; the Guam rail, for instance, is gone from Guam, but other populations still live on the nearby island of Rota. If some individuals of the bridled white-eye or other extinct species had been able to avoid the snake (perhaps due to a heritable trait that made them more wary of the predator), their descendants might have produced new populations that could cohabitate with the snake.

When populations diverge because of isolation, food availability, new predators, or habitat fragmentation that prevents the ability of population members to freely interbreed, new species may arise (*speciation*). This increases the number of species in a community and sometimes produces specialists that can exploit open niches. This separation may be physical (e.g., geographic boundaries the individuals won't cross) or may arise when something prevents some individuals from choosing others as mates, as may happen when individuals spend their time in different parts of their habitat. However, not all evolution is driven in this manner. Random events play a role, too, typically by decreasing genetic diversity rather than increasing it.

extirpation Locally extinct in one geographic area but still found elsewhere.

4 RANDOM EVENTS AND EVOLUTION

Key Concept 4: Along with natural selection, random events such as genetic drift, the bottleneck effect, and the founder effect also influence the evolution of a population.

In **genetic drift**, some traits (alleles) are passed on or lost by random chance, not because they were selected for (or against), as with natural selection. How could this happen? Even with natural selection at work, in each generation, some individuals may leave more offspring than others, not because they were better adapted to their environment but because they "got lucky"—perhaps due to favorable external factors they mated more, had more offspring, or had more offspring survive. Others might be "unlucky"—they might be in the wrong place at the wrong time (e.g., killed by a mudslide) before having a chance to mate. Just because natural selection would favor certain traits over others, doesn't mean every "better" individual will leave more progeny than all the less well-adapted individuals; the best traits might increase one's chances of survival or reproduction, but no trait guarantees it.

genetic drift The change in gene frequencies of a population over time due to random mating that results in the loss of some gene variants.

bottleneck effect The situation that occurs when population size is drastically reduced, leading to the loss of some genetic variants and resulting in a less diverse population.

founder effect The situation that occurs when a small group with only a subset of the larger population's genetic diversity becomes isolated and evolves into a different population, missing some of the traits of the original group.

Small populations are much more likely to experience genetic drift than large ones because the offspring of a few "lucky" individuals will have a greater impact on the gene frequencies of the next generation in a small population than a few "lucky" individuals in a large population. In large populations, the effect of genetic drift is more likely to be masked by all the better-adapted individuals who are favored by natural selection and successfully reproduce. In small populations, genetic drift can quickly lead to losses in genetic variability that produce major evolutionary changes in a population.

Genetic drift is more likely to happen if a population has experienced a reduction in size as occurs with the **bottleneck effect**. When only a subset of the original variants reproduces, they can give rise to a new population that is different from the original population. The bottleneck effect can occur when a portion of the population dies, perhaps because of a natural disaster like a flood or because of a strong new selective pressure, such as the introduction of a new predator. The survivors then produce a new generation, and any alleles that were found only in the deceased individuals are lost from the population forever.

Susan Karr

The descendants of the giant tortoises of the Galápagos Islands, which might have arrived at these isolated islands on floating vegetation, have evolved into distinct populations. The larger domed-shelled tortoises, like this one living on Santa Cruz, are found in areas with abundant plant life. The saddle-backed tortoises are adapted to arid regions with less food; their shell is angled upward in the front, allowing the animal to stretch its neck higher in search of food.

The **founder effect** also reduces population size and therefore available alleles. Consider the situation in which a small subset of a population colonizes a new area. If this subset (the founding population) happens to be less genetically diverse than the original and if the subset becomes completely isolated from the original group such that there is no mixing of the two populations (and no chance to reintroduce those missing alleles), the founding population will produce a population that has different gene frequencies than the original population. (See Module 3.2 for more on the effect of isolation.)

Today human impact increases instances of both the founder effect and the bottleneck effect. Much of what we do isolates populations into smaller groups, forcing them into these situations. **INFOGRAPHIC 4**

INFOGRAPHIC 4 RANDOM EVENTS CAN ALTER POPULATIONS THROUGH GENETIC DRIFT

GENETIC DRIFT

Genetic drift occurs when random events eliminate some gene variants (alleles) from a population. This happens because, in addition to natural selection, chance also influences who survives or reproduces. Genetic drift is more likely to accumulate and have major effects in small populations.

BOTTLENECK

If something causes a large part of the population to die, leaving the survivors with only a portion of the original genetic diversity, the population may recover in size but will not be as genetically diverse as the original population.

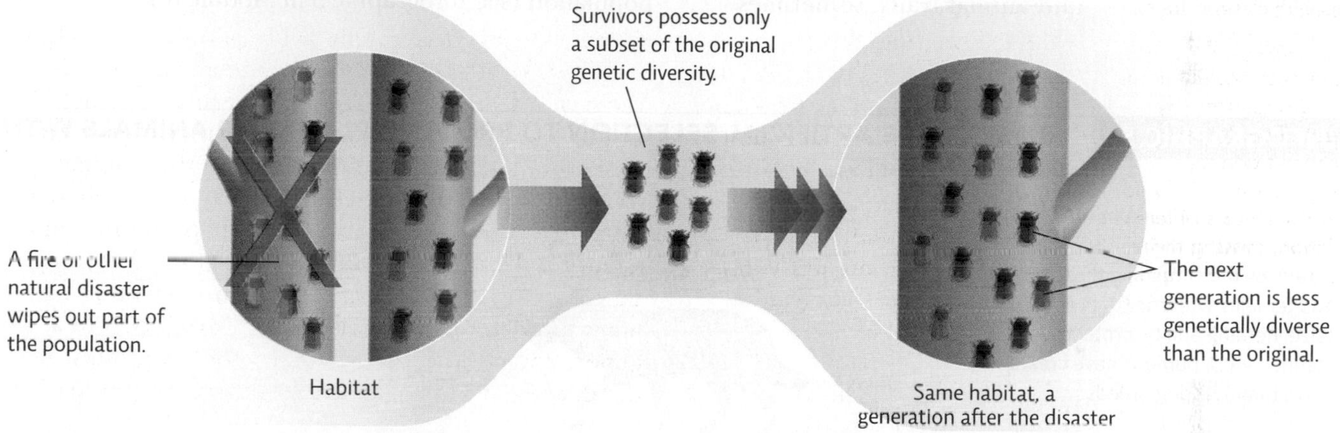

FOUNDER EFFECT

If a small subset of a population that possesses only a fraction of the genetic variability of the original population colonizes a new area and the subset becomes completely isolated from the original group, the new population will likely produce descendant populations that have different gene frequencies than the original population.

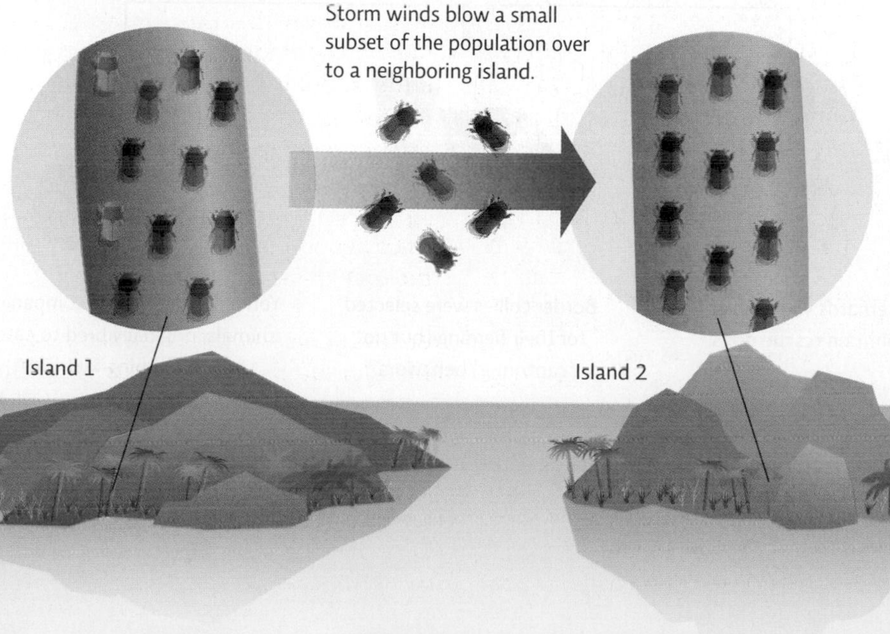

Storm winds blow a small subset of the population over to a neighboring island.
Island 1
Island 2

How is the bottleneck effect similar to and different from the founder effect?

5 ARTIFICIAL SELECTION

Key Concept 5: In artificial selection, humans choose which traits to keep and which to eliminate from a population through selective breeding. Our actions have also inadvertently led to the evolution of antibiotic- or pesticide-resistant populations.

The introduction of the brown tree snake to Guam was an accident: A snake hitchhiker crossed the ocean on a ship and landed in a veritable bird buffet. But humans also directly affect the evolution of a population through **artificial selection**. Artificial selection works the same way as natural selection, but the difference is that the selective pressure is us (humans). For many animal and plant species, humans choose who breeds with whom in an attempt to produce new individuals with desired traits. By doing this over many generations, people have accentuated certain plant and animal traits, sometimes to extremes. For instance, artificial selection created domestic dogs from their wolf ancestors. **INFOGRAPHIC 5**

artificial selection
A process in which humans decide which individuals breed and which do not in an attempt to produce a population of plants or animals with desired traits.

But evolution is ever at work. Pesticide- and antibiotic-resistant populations can emerge as an inadvertent human-influenced selection. When we apply a chemical that kills a pest or pathogen, some individuals survive because of their natural genetic resistance; that is, the individuals were already resistant even though they had never encountered the chemical. These survivors are then the only individuals who reproduce, producing the next generation that is also pesticide-resistant, ultimately changing the frequency of resistant genes in the population (see Infographic 5 in Module 8.1).

INFOGRAPHIC 5 **HUMANS USE ARTIFICIAL SELECTION TO PRODUCE PLANTS OR ANIMALS WITH DESIRED TRAITS**

All dogs (*Canis lupus familiaris*) are descendants of the wolf (*Canis lupus*). By only breeding those males and females with the traits desired (size, herding ability, protective instinct, etc.), humans have created more than 170 dog breeds.

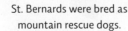

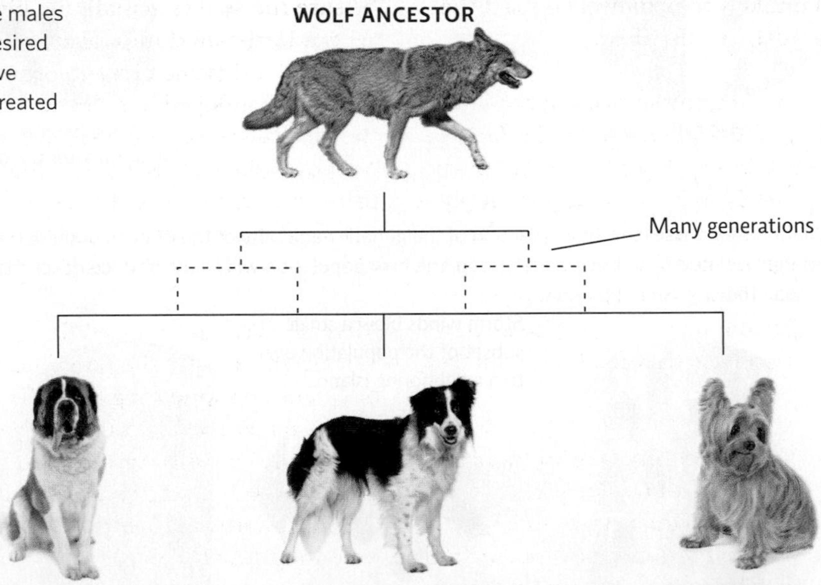

St. Bernards were bred as mountain rescue dogs.

Border collies were selected for their herding (but not capturing) behaviors.

Yorkshire terriers are companion animals, originally bred to catch rats in clothing factories.

Wolf: DaddyBit/Getty Images; St. Bernard: GlobalP/Getty Images; Border collie: GlobalP/Getty Images; Yorkshire terrier: jimmyjamesbond/Getty Images

? **Why can we say that artificial selection is goal-directed but natural selection is not?**

6 THE PACE OF EVOLUTION AND EXTINCTION

Key Concept 6: The pace of evolution and extinction is generally slow and is affected by population size and genetic diversity, reproductive rate, generation time, and the strength of the selective pressures at play. When extinctions unfold over long periods of time, better-adapted species tend to replace their predecessors and the niche remains filled; rapid extinction events may eliminate well-adapted species and break important community connections.

The pace of evolution by natural selection is not constant or the same for all species, but in general, it is slow—changes accumulate over generations, and speciation events can take thousands or millions of years. Evolution's pace is affected by a variety of factors, including population characteristics such as the genetic diversity and size of the population, aspects of the species' biology such as its biotic potential (maximum reproductive rate) and generation time, and the strength of the selective pressure a population is facing.

As mentioned earlier, genetic diversity is important because it provides options for natural selection to favor or select against. If a new selective pressure favors a less common trait in the population or if a new favorable trait arises, it can quickly displace other variants in subsequent generations, resulting in evolution of the population. If this results in a speciation event, it can eliminate an ancestral species. When this happens—when a new species replaces an ancestral one—the outcome is unlikely to negatively impact the ecosystem because the niche is still filled.

In addition to genetic diversity, the size of the population also makes a difference in how quickly natural selection can produce a change in a population: Beneficial traits can spread more quickly in smaller populations simply because it is more likely that the individuals with the trait will find each other and mate (as long as it is not a population that is widely dispersed). Of course, as mentioned earlier, smaller populations can also be at a higher risk of extinction because they likely contain less genetic diversity.

Reproductive rate and generation time also influence how quickly a population can adapt to changes. Populations of *r*-selected species (those with high reproductive rates and fast generation times) decimated by a disturbance or a depleted resource can quickly bounce back and repopulate the area because the remaining individuals can produce so many offspring in a single breeding cycle. Many *endangered species* (species in high danger of becoming extinct; see Module 3.2), on the other hand, are *K*-selected species, with slower reproductive rates and longer generation times; therefore, they take longer to recover if population numbers fall. Selective pressures that change over the course of just a few years can eliminate a species with a generation time of many years—there is simply not enough time for those able to withstand the stressor to grow up and produce progeny who can also withstand the stressor. (For more on *r*- and *K*-selected species, see Module 2.2.)

The strength of the selective pressure also affects how quickly natural selection might produce a change in a population. One of the reasons the demise of birds in Guam was so stupefying was that it happened so quickly—particularly for the small birds, which were easiest for the snakes to eat. Larger birds disappeared later, when the snakes started eating their nestlings and eggs. **INFOGRAPHIC 6**

To determine whether the snakes were a strong selective pressure on the birds of Guam, Savidge first had to be sure the snakes actually ate birds. To test this, she set out bird-baited traps around Guam and on nearby Cocos Island (which has no brown tree snakes). What she found shocked her: "In one area [on Guam] where the birds were extinct, 75% of my traps got hit within 4 nights," she recalls. On Cocos, all the birds used as bait survived. Surveys of Guam's abundance of small mammals also revealed heavy predation pressure by the snakes—mice and shrews had declined by 94%.

All in all, Savidge amassed three lines of evidence that convinced her that she had finally solved the mystery of Guam's disappearing birds: The geographic location of the snakes correlated strongly with the birds' disappearance, brown tree snakes were willing to eat birds, and other small mammals also went missing after the snakes' arrival. Brown tree snakes, she concluded, were the culprit.

The hunting skill of the snake and the lack of antipredator behaviors in the birds made this new predator a strong selective pressure; a less proficient predator or one whose hunting style was familiar to the birds would have been a weaker selective pressure. Indeed, one lesson of the brown tree snake invasion

INFOGRAPHIC 6 THE PACE OF EVOLUTION

The speed at which evolution can occur is influenced by a variety of factors. While the relationships below are not guaranteed to affect the rate of evolution as shown (e.g., even a population with high genetic diversity would not be able to evolve [or survive] if *no one* in the population could withstand the environmental disturbance experienced), in general, the following correlations are seen.

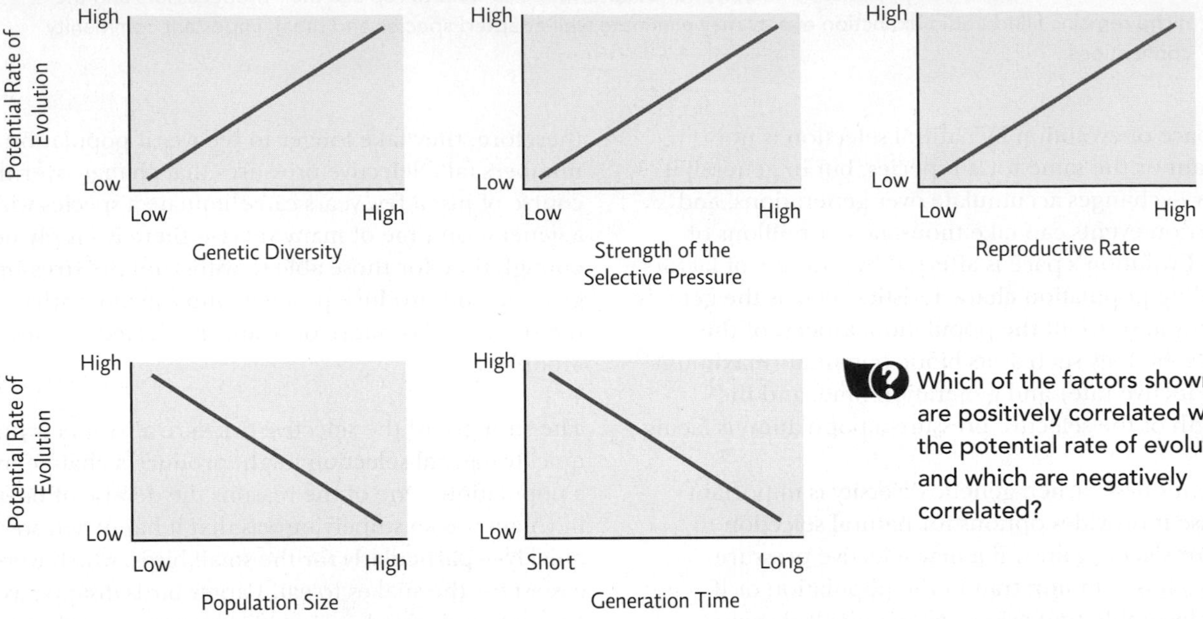

Which of the factors shown are positively correlated with the potential rate of evolution, and which are negatively correlated?

is that while speciation typically occurs at a slow pace, extinction can occur much more quickly if the rate of change exceeds the ability of the population to adapt. These rapid extinction events are a concern because they can break community connections and leave unfilled niches, negatively impacting other species. This is, in fact, already occurring.

Twelve native bird species are now extinct or extirpated on Guam, including all fruit-eating birds. This is having dire consequences for the island's trees, according to a 2017 study by Iowa State University researcher Haldre Rogers. Around 70% of Guam's trees are fruit producers and depend on mutualistic partners for seed dispersal; unfortunately, in Guam, no other species have stepped in to fill this role. Rogers and her team estimate the growth of new trees could be down by as much as 90% due to failed germination of seeds. On the other hand, Rogers's research has also revealed that spiders are flourishing without enough birds around to keep their numbers in check. Guam's forests are changing drastically, all because a few snakes found their way to this island.

7 MASS EXTINCTIONS: PAST AND PRESENT

Key Concept 7: Extinction rates were much higher in mass extinction events than at other times. Past mass extinctions are linked to natural causes; today human impact appears to be causing another mass extinction.

Extinction is nothing new on Earth. By most estimates, more than 99% of all species that ever lived on the planet have gone extinct. Based on a critical analysis of the fossil record, scientists agree that there have been five *mass extinction events*—when species have gone extinct at much greater rates than during intervening times, each event leading to the loss of 75% or more of the species present on Earth. The most infamous of these was the extinction event that occurred at the transition from the Cretaceous to the Tertiary period, 65 million years ago. Most scientists agree this event was set off by an asteroid impact in the Gulf of Mexico; more than 75% of all species, including the dinosaurs, were wiped out. **INFOGRAPHIC 7A**

Earth's past mass extinctions were due to catastrophic events or physical changes to the atmosphere or oceans, which altered the environment faster than species' ability to adapt. Though most species loss

INFOGRAPHIC 7A EARTH'S MASS EXTINCTIONS

There have been five mass extinctions in Earth's history, defined as extinction events that eliminated a large number of species in a short period of time (geologically speaking). Each is believed to have been caused by major environmental changes such as the meteor that struck Earth at the end of the Cretaceous period.

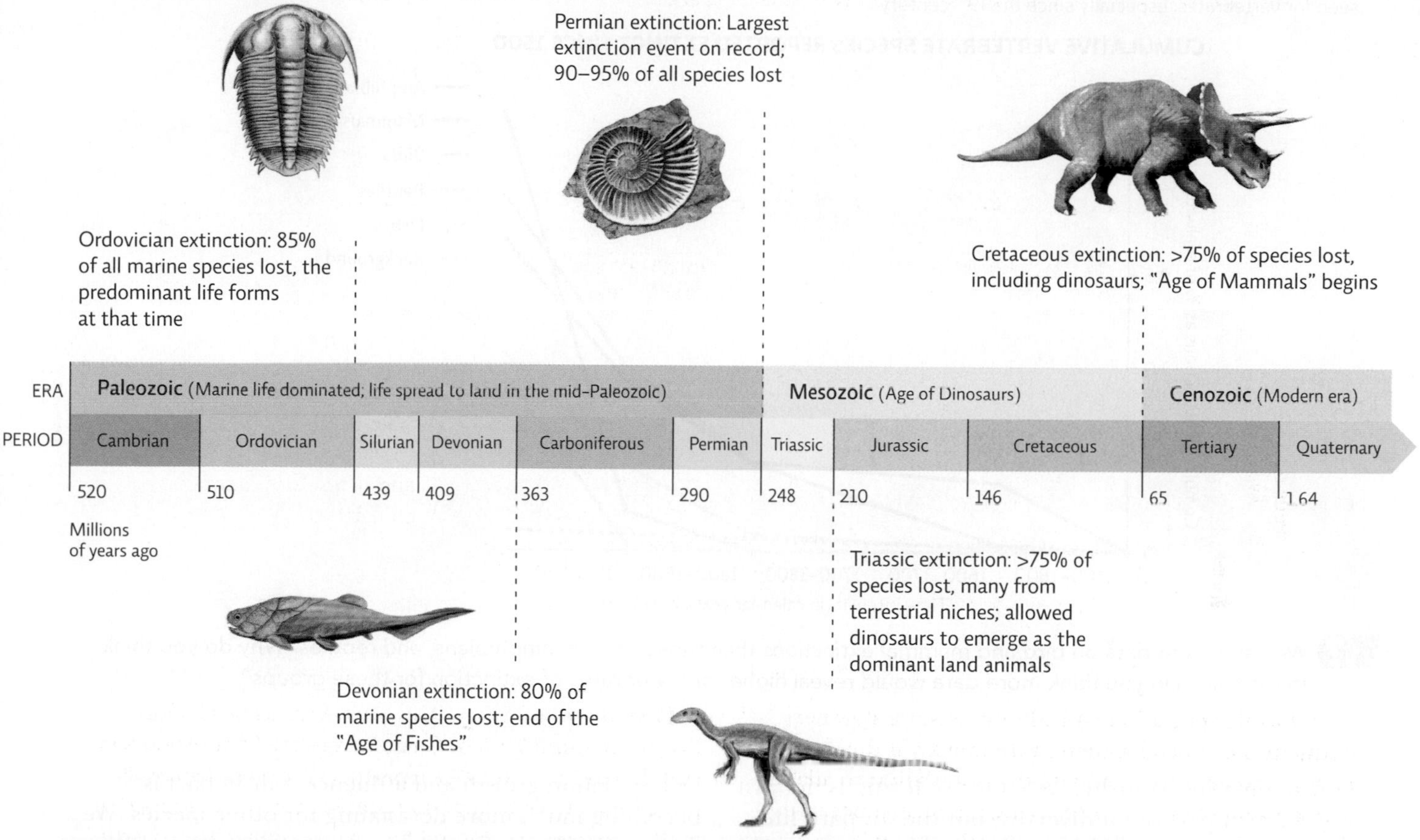

Why do we see mass extinction events occurring at transitions from one geologic time period to the next?

wasn't overnight, taking from hundreds to millions of years to unfold, these kinds of events eventually led to the emergence of new species as surviving populations adapted to the available niches (a process that took millions of years). Cycles of extinction and evolution ultimately gave rise to the diversity of life we see on Earth today—estimates range from 3 to 100 million species.

Throughout most of Earth's history, the **background rate of extinction**—the average rate of extinction that occurs between mass extinction events—has been slow. The fossil record tells us that, on average, 1 or 2 species out of every 1 million species goes extinct each year. In a world with 3 million species, this would be 3 to 6 species per year; if Earth is home to 100 million species, that would be 100 to 200 species per year.

A 2015 study led by Gerardo Ceballos compared the number of vertebrate species extinctions that have been documented since the year 1500 to what would have been expected if extinction had occurred at a background rate that was determined by an extensive examination of the mammalian fossil record. The data revealed a sharp increase in the number of extinctions beginning about 200 years ago. At background rates, Earth should have seen 9 vertebrate extinctions since 1900; 477 have occurred—that's 53 times higher than expected. **INFOGRAPHIC 7B**

Unfortunately, it's not just vertebrates that are in trouble—many invertebrate groups appear to be declining at precipitous rates. Scientists are particularly concerned at the decline of insect species. Insects, the largest species group by far, are crucial to ecosystem function, providing services such as pollination, matter recycling,

background rate of extinction The average rate of extinction that occurred before the appearance of humans or that occurs between mass extinction events.

INFOGRAPHIC 7B **THE RATE OF EXTINCTION IS ACCELERATING**

Recent rates of species extinctions are well above that expected by the background rate, a trend attributed to human impact such as habitat destruction, the introduction of invasive species, and climate change. This graph, based on analysis techniques developed by Ceballos, shows estimate of extinctions since the year 1500 for vertebrates in comparison to the background rate of extinction. Accelerated extinction rates are seen for vertebrates, especially since the 19th century.

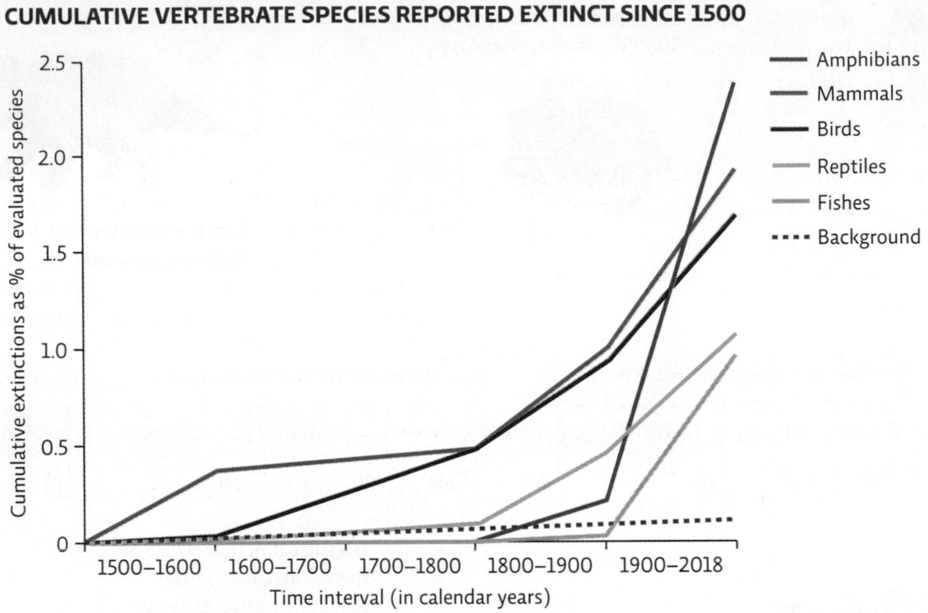

? **We have more data on bird and mammal extinctions than those for fish, amphibians, and reptiles. Why do you think this is true? Do you think more data would reveal higher or lower rates of extinction for these groups?**

and population control. "What we're losing is not just the *diversity* part of biodiversity, but the *bio* part: life in sheer quantity," wrote Brooke Jarvis in a recent *New York Times* article.

In the most comprehensive evaluation of biodiversity ever completed, a 2019 global assessment by the Intergovernmental Science-Policy Platform on Biodiversity and Ecosystem Services estimates 25% of evaluated plant and animal species are threatened; some face imminent extinction without immediate action. Extrapolating from this to include all groups, the authors estimate that 1 million species are at risk. The current rate of extinction is estimated to be 100 to 1,000 times higher than the background rate, and this rate is accelerating.

Most scientists agree that the accelerated extinction we are witnessing can be considered a sixth major extinction event, and that it is largely driven by human actions. As our use of resources increases, driven by population growth and affluence, our impact is becoming much more devastating for other species. We remove the resources they need to survive, minimize their habitat ranges, harvest them at unsustainable rates, introduce new predators or competitors, and strip them of their genetic diversity, all of which slowly eliminate them. British researchers Ian Owens and Peter Bennett analyzed the extinction risk for 1,012 *threatened* bird species (those at risk for becoming endangered) and found that habitat destruction was cited as a risk factor in 70% of the cases. Other human interventions, such as the introduction of non-native species or overharvesting, were implicated in 35% of the cases. (See Module 3.2 for more on species endangerment.)

Some might wonder why scientists are so concerned about species extinctions—after all, extinction is a natural event. But it's the pace of extinctions that is concerning. Our changes to the environment can be so rapid or so great that natural selection simply cannot keep up—perhaps because a new needed trait is not present in the population, or it cannot spread quickly enough to prevent a population collapse. Module 3.2 will examine the value of biodiversity and why its loss can threaten us all. Indeed, an understanding of how populations evolve, and how we can affect that process, can help us avoid actions that create problems such as the emergence of antibiotic-resistant bacteria or untimely species extinctions.

In Guam, the near-total disappearance of birds between the 1960s and 1980s was a biological murder mystery. The brown tree snake was the cause of their demise, but the inadvertent introduction of this species by humans puts the blame squarely on us.

Efforts to control the brown tree snake on Guam are ongoing and aggressive; as many as 15,000 snakes are captured annually. Control methods include baited traps, canine tracking teams, and toxic "mouse-drops"; there are even rapid response teams that are deployed when a snake sighting is reported on a nearby island. The goal is not to eradicate the snakes (which probably isn't possible) but to control and contain them so that they don't find their way to other islands, explains Diane Vice, a wildlife biologist with the Guam Department of Agriculture. "The hope is to create safe habitat," Vice says, "so that these beautiful native species can once again thrive."

Select References:

Ceballos, G., et al. (2015). Accelerated modern human–induced species losses: Entering the sixth mass extinction. *Science Advances, 1*(5), e1400253.

Hockstra, H. E., et al. (2005). Local adaptation in the rock pocket mouse (*Chaetodipus intermedius*): Natural selection and phylogenetic history of populations. *Heredity, 94*(2), 217–228.

Rogers, H. S., et al. (2017). Effects of an invasive predator cascade to plants via mutualism disruption. *Nature Communications, 8*, 14557.

Savidge, J. A. (1987). Extinction of an island forest avifauna by an introduced snake. *Ecology, 68*(3), 660–668.

GLOBAL CASE STUDIES INVASIVE SPECIES

Achieve

Non-native species that find their way to other ecosystems can wreak havoc on native species unprepared to deal with them. Here are a few notable invasive species that are taking their toll on U.S. ecosystems.

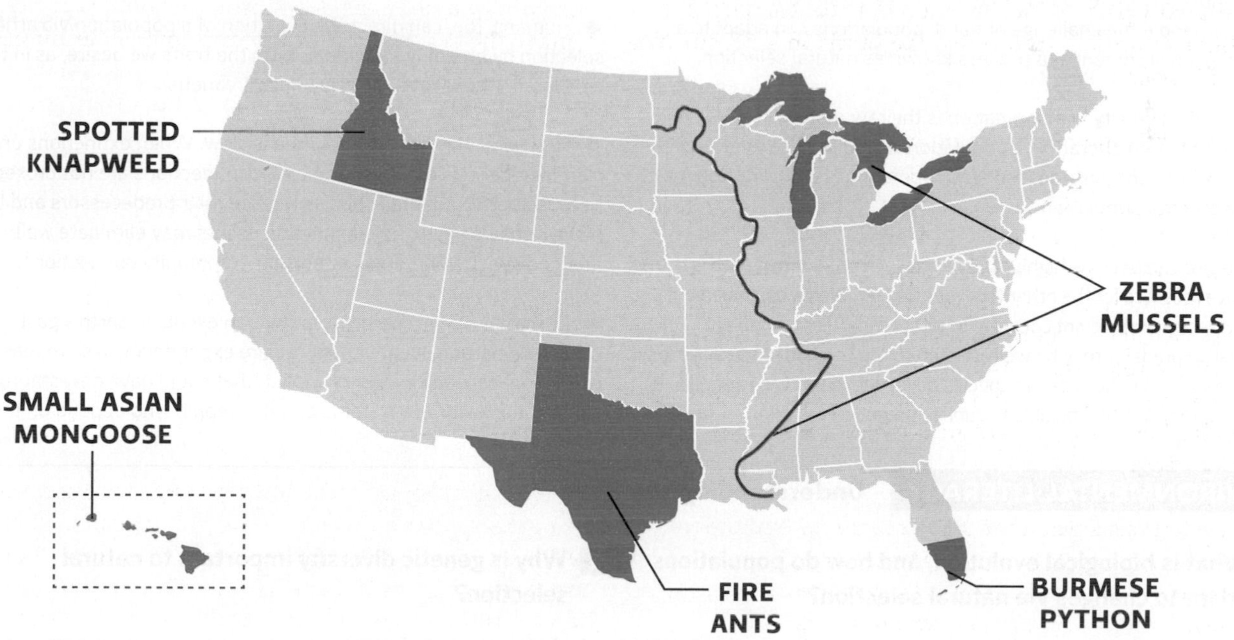

BRING IT HOME

PERSONAL CHOICES THAT HELP

The astonishing variety of life found on Earth is a result of natural selection favoring those individuals within populations that are best able to survive in their particular environment. Given enough time, some populations may be able to adapt to environmental changes. However, human activities may disrupt natural ecosystems so that organisms cannot adapt fast enough to survive, and those organisms may go extinct. Conservation actions can help protect vulnerable organisms and ecosystems.

Individual Steps

- When buying a home, consider an older established area of your community or a location close to work. Suburban sprawl reduces habitat for wildlife, and reliance on cars causes greenhouse gas emissions that contribute to species-threatening climate change.
- Save your pocket change and, at the end of every year, donate the money to a land, marine, or wildlife protection agency.
- Create a personal blog that includes photographs of wildlife, facts about current threats to plants and animals, and articles about conservation.

Group Action

- Throw a party in support of wildlife conservation. Take a collection at the door and donate the money to an organization that supports conservation.
- "Adopt an Organism." The U.S. Fish and Wildlife Service maintains a database of endangered plants and animals in every state. Research this database to find wildlife that interests you. Determine what agency, conservation group, or legislator you could contact and then start your own protection campaign. See what meetings, petitions, and legislation could impact your organism and get involved.

JOEL SARTORE/National Geographic Creative

REVISIT THE CONCEPTS

- Evolutionary biology helps us understand the diversity of life on Earth and how populations change over time. In response to selective pressures that favor the survival or reproductive success of some individuals over others due to inherited traits that better equip them to handle the challenge at hand, populations can adapt to a changing environment — a process known as natural selection.
- Genetic diversity in a population is the raw material on which natural selection operates. A population is said to have evolved if the frequency of genes in a descendant population is different from that in its ancestral population.
- Species can become highly adapted to each other when each is the selective pressure for the other (coevolution). Problems can emerge when species that did not coevolve together meet. For example, a non-native predator may be well adapted to prey on native species in its new habitat but if those native species never coevolved with a similar predator, they may not be able to survive predation.
- Along with natural selection, random events such as genetic drift, the bottleneck effect, and the founder effect also influence the evolution of a population.
- Humans, too, can direct the evolution of a population via artificial selection by breeding individuals with the traits we desire, as in the development of animal breeds or plant varieties.
- In general, the pace of evolution is slow. When extinctions unfold over long periods of time, community connections are not broken because better-adapted species replace their predecessors and the niche remains filled; rapid extinction events may eliminate well-adapted species and break important community connections.
- There have been five mass extinction events in Earth's past, caused by natural events; today we are experiencing a sixth mass extinction, caused by human impact, that could have devastating consequences for ecosystems and the people who depend on them.

ENVIRONMENTAL LITERACY Understanding the Issue

1 What is biological evolution, and how do populations adapt to changes via natural selection?

1. A butterfly population adapts to a warming environment — explain how this could happen via natural selection.
2. Use the example of a bird population that most commonly feeds on medium-sized insects to distinguish between *stabilizing*, *directional*, and *disruptive selection*.
3. Explain this statement: Individuals are selected but populations evolve.

2 Why is genetic diversity important to natural selection?

4. As the amount of genetic diversity in a population increases, how does this affect the likelihood that the population will be able to adapt to environmental changes?
5. Does the presence of genetic diversity in a population guarantee it will be able to adapt to changes? Explain.
6. Using the example of the rock pocket mouse, explain the importance of genetic diversity to a population.

3 What is coevolution, and what problems can emerge when species that did not coevolve together suddenly share a habitat?

7. Use the concept of coevolution to explain why the brown tree snake is a much more devastating predator on Guam than it is in its native habitat.
8. Suppose a native Guam bird species began to show evasive behaviors that allowed it to avoid the brown tree snake. Describe a potential coevolution scenario that might have allowed this adaptation to eventually predominate in the population.

4 How do random events influence the evolution of a population?

9. Why is genetic drift considered random whereas natural selection is considered nonrandom?
10. Why might a population, isolated by the founder effect, be more vulnerable to extinction than the original population from which it came?

5 How do humans, intentionally or accidentally, affect the evolution of a population?

11. Give some examples of artificial selection by humans that have been beneficial to human society.
12. Explain how pesticide resistance might evolve in an insect pest.
13. Suppose you wanted to use artificial selection to produce a breed of hairless dogs for people who are allergic to dog hair. How would you go about doing this?

6 What factors affect the pace of evolution and extinction, and why are extinctions that occur quickly more of a concern than those that take a long time to unfold?

14. What factors influence the pace of evolution in the absence of a mass extinction event?
15. Why are *K*-selected species more vulnerable to extinction than *r*-selected species?

7 How do the mass extinction events of the past compare to extinctions during intervening times and today?

16. Why do most scientists think that we are in the midst of a sixth mass extinction?
17. Why are the changes being imposed by humans difficult for species to adjust to?

SCIENCE LITERACY Working with Data

The following graph depicts the relationship between numbers of extinctions and human population size since the 19th century.

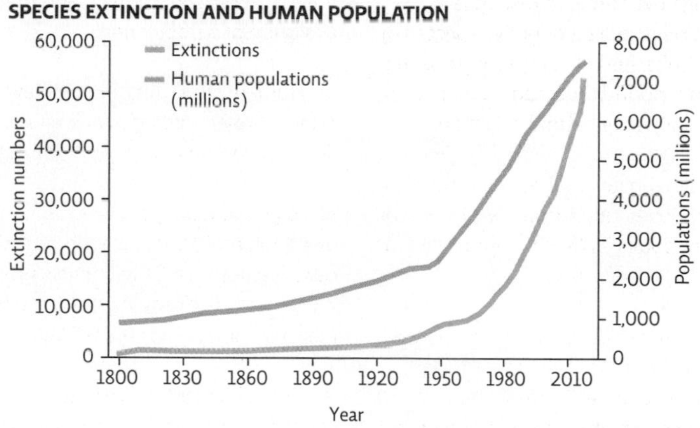

Interpretation

1. Describe what is happening to:
   a. the extinction rate over time.
   b. human population growth over time.
2. Around what point in time do we see a change in the rate of species extinctions?
3. The two curves have been graphed together. What is the implication of presenting these data in this manner?

Advance Your Thinking

4. The *y* axis is labeled "Extinction numbers." What taxonomic units are being measured? What if the taxonomic unit being evaluated here had been genus (a taxonomic group that can contain more than one species) or family (a taxonomic group that can contain more than one genus)? Would you be more or less concerned about the trend of extinctions? Explain.
5. What type of relationship is suggested by the figure: correlational or causal? What additional data would you like to see to support the graph's main point?

INFORMATION LITERACY Evaluating Information

Life on Earth as we know it is a result of millions of years of evolutionary processes. By most accounts, we are currently witnessing a mass extinction event at our own hands, one that will result in changes in biodiversity, which will necessarily affect life on Earth, including humans.

Investigate this Problem

1. Visit the website of the Center for Biological Diversity at www.biologicaldiversity.org and search the website for information using the key words *extinction crisis*. Read a few articles and then select one article about extinction to evaluate. Identify it (title, author and date [if provided], and URL) and then answer the questions that follow.

Evaluate the Article

2. In your own words, summarize the main point(s) or position of the article you are evaluating.
   a. Identify three claims made in the article. (If there are not three claims in the article you chose, chose a different one to evaluate.)
   b. Identify the evidence presented to back up each claim and evaluate that evidence. Is it sufficient? (If no evidence is given, where might you go to find that evidence?)
   c. Are references given for the evidence? Explain.

Evaluate the Organization

3. Determine if this a reliable information source:
   a. Does the organization have a clear and transparent agenda? Explain.
   b. Who runs this website? Do the credentials of the staff make this source reliable/unreliable? Explain.

Dig Deeper

4. Go to the International Union for Conservation of Nature (IUCN) website (www.iucnredlist.org).
   a. Choose five species at random that are identified as vulnerable, endangered, or critically endangered. Read about each species, and list the threats each faces.
   b. Create one master list of all the threats you encountered for your five species, and categorize them as either *human caused* or *caused by natural events*. Which list is longer?

Draw Conclusions

5. Based on your research, does human impact play a role in the endangerment of species? What actions would be most useful to address these threats? Do you think these actions should be pursued? Support your conclusions with evidence from your research.

Additional study questions are available at Achieve.macmillanlearning.com. Achieve

Module 3.2 Biodiversity

OIL PALM PLANTATIONS THREATEN TROPICAL FORESTS

Can we have tropical forests and our palm oil too?

Oil palm plantation at the forest edge. *Rich Carey/Shutterstock*

After reading this module, you should be able to answer these GUIDING QUESTIONS

1. What is biodiversity?
2. Why is biodiversity important?
3. How do genetic, species, and ecological diversity each contribute to ecosystem function and services?
4. What are biodiversity hotspots, and why are they important?
5. What role does isolation play in a species' vulnerability to extinction?
6. What are the causes and consequences of biodiversity loss?

Even before his plane landed on the Indonesian island of Sumatra, Laurel Sutherlin could tell the scene awaiting him would be worse than expected. They were still 30,000 feet over the Java Sea, and already a sickly haze of yellow smoke seemed to be enveloping the aircraft. The 30-something naturalist and environmental crusader had traveled all the way from California to see firsthand the impact the burgeoning palm oil industry was having on this South Pacific island.

Palm oil rose to prominence as a dietary replacement for *trans fats*, a type of fat linked to heart disease and other health problems. The oil comes from the waxy orange fruits that sprout from oil palm trees. It can be extracted from both the flesh and the seed of the fruits and makes up roughly half of the planet's edible oil production.

In recent years, it has made its way into a mind-boggling list of consumer products—foods like chocolate, peanut butter, cereal, and biscuits as well as cosmetics, shampoos, diaper cream, and toilet cleaner. By some accounts, nearly half of all U.S. packaged food and household products contain palm oil and our palm oil consumption has increased more than eightfold since 2000, reaching 1.4 million tons in 2018. Almost all that palm oil comes from Southeast Asia.

Like other areas where oil palm plantations are established, the island of Sumatra, where Sutherlin's plane was now landing, suffers from high rates of deforestation. His eyes and nose were assaulted with smoke the instant he stepped off the plane. "I actually had to suppress an initial panic that I would

WHERE IS SUMATRA?

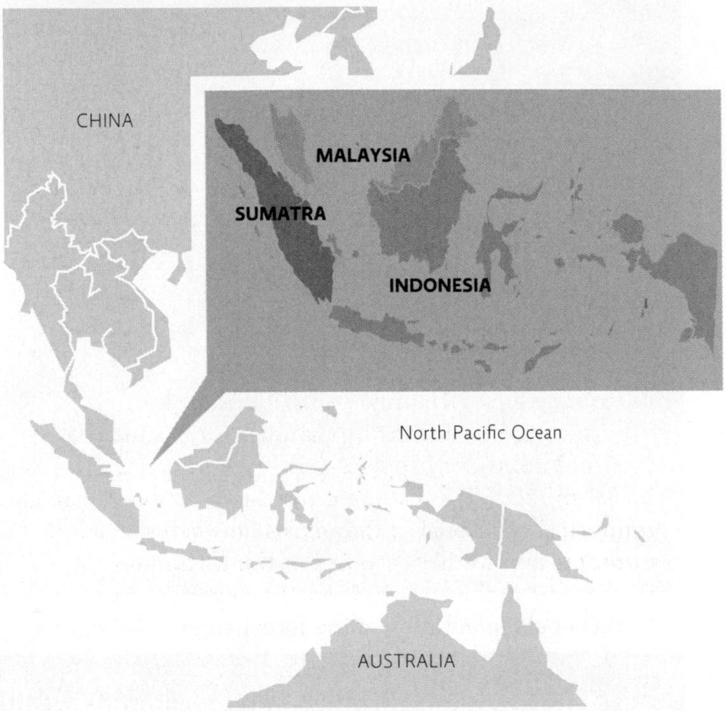

A young worker collects palm fruit clusters at an oil palm plantation in Sumatra. Oil can be extracted from both the flesh and the seed of the fruit.

Dimas Ardian/Bloomberg/Getty Images

suffocate from the smoke," he would later recall. In fact, he had been lucky to land at all; air traffic would be canceled later that day due to poor visibility. Sutherlin had heard stories about the devastation wrought by deforestation. But he was not prepared for what he was about to see.

1 BIODIVERSITY: THE VARIETY OF LIFE

Key Concept 1: We have identified only a fraction of the species that make up the tremendous variety of life (biodiversity) on Earth. We know much more about smaller groups such as plants and vertebrates than more diverse groups like invertebrates.

Rainforests, including those being cleared in Indonesia and Malaysia, contain the greatest concentration and variety of plant and animal terrestrial life-forms on Earth. This variety is called **biodiversity**, and it is the most unique and extraordinary feature of our planet. So great is the diversity of life on Earth that it is virtually impossible to know just how many species exist; in fact, many believe that the vast majority of all living species have yet to be discovered or identified by humans.

So far, science has identified around 1.5 million species, and common estimates for the total number of species on Earth range from 3 to 11 million. A highly regarded study by Camilo Mora estimated the total number of species (excluding bacteria) to be 8.7 million. This estimate was based on the reliable numerical relationship that exists between species numbers and numbers found in the taxonomic groups above the species level (from genera to phyla)—categories that are easier to track and count. A 2017 study by Brendan Larsen and colleagues that predicts a much larger number of invertebrate species and includes an estimate of bacterial species (1.75 billion species) raises the total number of species to an astonishing 2 billion.

"The numbers are mind boggling," says Jim Miller, a scientist at the Missouri Botanical Garden in St. Louis. "It's almost unfathomable." What we do know is that some life-forms are far more diverse than others; there are far more insects than there are vertebrates, for example, and there are relatively few mammals (fewer than 5,500 species) overall. **INFOGRAPHIC 1**

We also know that some regions of the world contain far greater concentrations of biodiversity—megadiversity—than other regions. Indonesia is one such region; it has more species of mammals, parrots, palms, swallowtail butterflies, and coral than any other country on the planet. Roughly 10% of all flowering plant species are found there.

A recent census of a 1.3-square-mile plot on the Indonesian island of Siberut, for example, revealed 139 different species of trees. Indeed, tropical forests (and tropical coral reef ecosystems) are the most biodiverse in the world. John Terbough of the Duke University Center for Tropical Conservation reports that a 2.5-acre research plot in an Amazonian rainforest contained close to 300 tree species. Compare this to the tree species diversity of the boreal forest that spans all of northern Canada—about 20 species. But habitat destruction and other pressures are threatening this biodiversity on a global scale.

Since 1950, more than 560 million acres of tropical forestland have been cleared—that's an area larger than Alaska and California combined. Indonesia alone lost roughly 23 million acres between 2000 and 2017. Much of this deforestation has been to make way for oil palm plantations. Tropical forests in Africa and South America are also being lost to oil palms. Rubber plantations, too, account for significant deforestation in these areas—globally, around 25 million acres have been converted to rubber plantations; 85% of that is in Southeast Asia, and the extent of rubber plantations is expected to increase fourfold by 2024.

The environmental costs of the clearing of tropical forests are expected to be far-reaching. For one thing, such rapid and colossal loss of rainforest contributes to regional and global climate change and is almost certain to accelerate global warming. (For much of 2015, Indonesia's greenhouse gas emissions exceeded those of the United States due to the massive forest clearing and fires.) For another, forest loss contributes to soil erosion, water pollution, and flooding—not to mention a scarcity of resources on which local human communities depend. Rampant air pollution from forest clearing poses a significant threat to human health in the region.

But the most immediate consequences, by far, will be visited upon the region's wildlife. Fewer than 50% of

biodiversity The variety of life on Earth; it includes species, genetic, and ecological diversity.

INFOGRAPHIC 1 BIODIVERSITY ON EARTH

We have identified about 1.5 million species so far, but our knowledge of Earth's total biodiversity is scant. Even our accounting of "known species" — those described in the scientific literature — is an estimate, because some may be erroneously identified as separate species while others go unrecognized as distinct species. Of the species we have identified, insects far outnumber any other group of organisms; beetles alone account for about 25% of all known species — more than all known or estimated plant species. Vertebrates (the group to which humans belong) likely make up only 1% of all organisms on Earth.

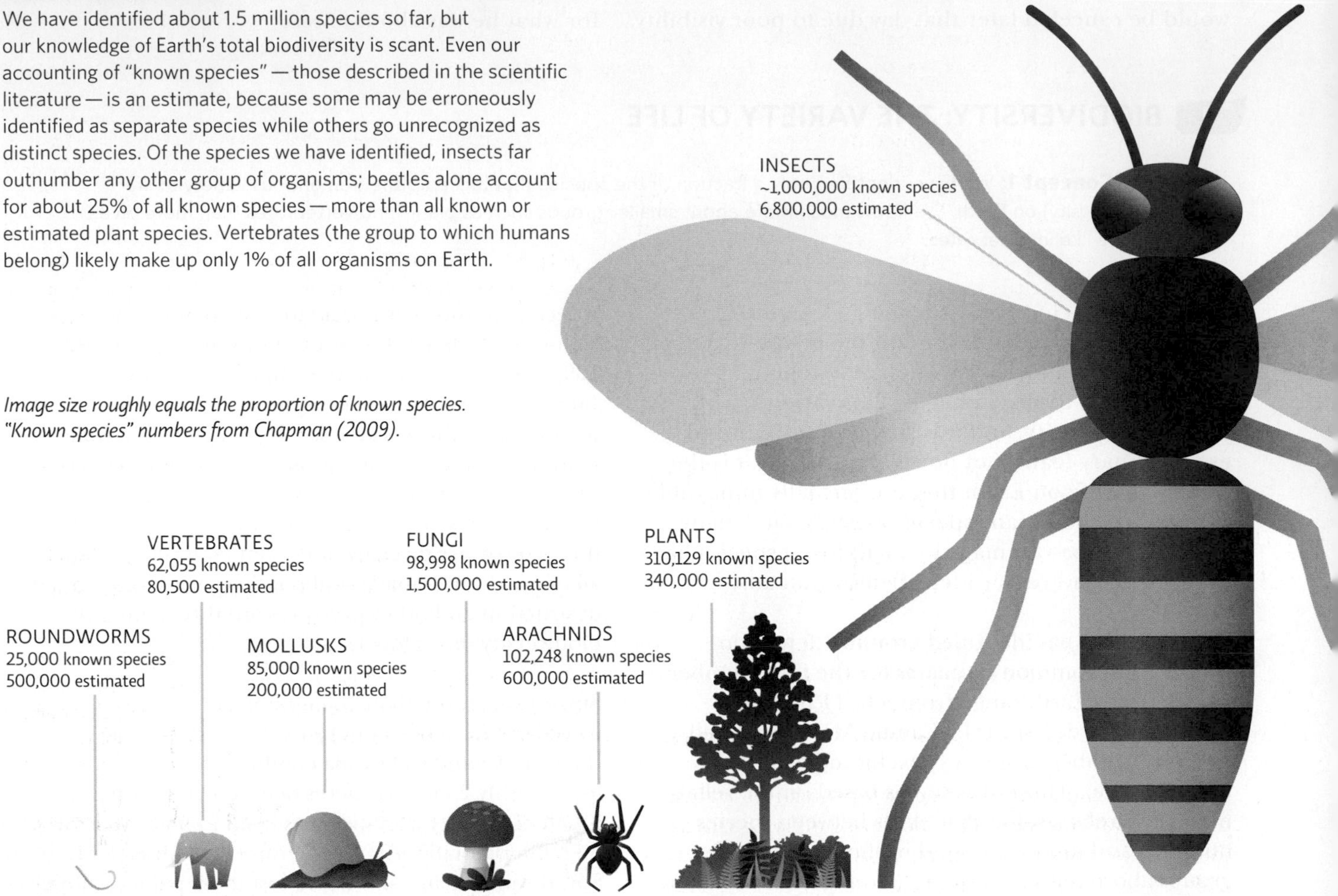

How important do you think it is to get an accurate accounting of the number of species on Earth?

the species normally found in the natural forest will be found in oil palm plantations. Mammals, in particular, tend to avoid the plantations; only about 10% of the original mammalian fauna are typically present. Scientists estimate that converting a forest into an industrial oil palm plantation results in the death or displacement of more than 95% of the orangutans living there. The habitat loss also impacts countless bird, reptile, insect, and plant species, many of which fail to thrive or even return at all because other species on which they depend are gone.

"When you look at a candy bar or package of crackers in a grocery store, or a jar of peanut butter in your kitchen, it's difficult to imagine that it has anything to do with orangutans going extinct," Sutherlin says. "But the link is actually quite direct."

Vier Pfoten/Four Paws/Rhoi/Shutterstock

Rescue workers from the animal charity Four Paws found this orangutan mother holding her daughter tightly as the pair was being surrounded by a group of young men paid to hunt and kill orangutans in the area. The orangutans were rescued by the charity workers and relocated to a remote area in the rainforest.

2 THE VALUE OF BIODIVERSITY

Key Concept 2: Biodiversity contributes to the health and wellbeing of ecosystems, which in turn benefits human populations by providing ecosystem resources and services as well as cultural and health benefits.

Biodiversity is responsible for much more than the majesty of nature. It provides a wide array of **ecosystem services**—important to all species, including humans. For starters, it provides the key connections between individual species and between species and their environment. These connections help regulate the ecosystem as a whole: Insects pollinate flowers, for example. Photosynthetic organisms (plants on land, algae and phytoplankton in the sea) bring in energy, produce oxygen, and sequester carbon. Other organisms capture and pass along important nutrients like nitrogen and phosphorus. Still others help purify the air and water, and all species eventually become food for other creatures. Water also cycles through living things; in a forest, thousands of gallons a year are captured and passed along by each tree, releasing enough water vapor into the atmosphere to affect local rainfall. Life-giving soil forms as organisms decompose: Soil supports plant life whose roots, in turn, hold it in place, keeping it from being washed away in rains or floods. Meanwhile, predators and competitors keep each other in check so that no single species grows too populous or gobbles up too many needed resources.

Biodiversity can also provide direct protection against disease in various ways. The incidence of Lyme disease in the United States, for example, is often lower in areas with higher biodiversity (especially areas with squirrels and opossums, two species that effectively remove disease-carrying ticks, an action that prevents the spread of the disease to other organisms). Likewise, in recent research, John Swaddle showed that fewer human cases of West Nile virus occur in eastern U.S. counties with higher bird biodiversity, possibly because some species are less effective at transmitting the virus than others; the more these species are present, the less likely the virus will be transmitted to humans.

Biodiverse ecosystems have economic value as well. Forests provide not only food, fuel, and building materials but also pharmaceuticals. People use the chemicals they extract from plants and animals not only as a way to attend to their individual human health but also as a source of income. Roughly half of all modern medicines—including medicine cabinet staples like aspirin and codeine, as well as most hypertension drugs and some cancer-fighting superstars—were originally derived from plants, many of them traditional remedies used for centuries. Over generations, rural forest-based communities in Sumatra have perfected a sophisticated management of forest resources that produces hillsides of coffee, elegant groves of cinnamon trees, acres of terraced rice paddies, and a healthy mixture of fruit trees, vegetables, tobacco, and other useful edible, commercial, and medicinal plants.

Biodiversity supplies cultural benefits, too—whether it is the enjoyment of a natural area for recreation or aesthetic appreciation or a societal tradition rooted in nature. Both cultural and economic benefits of biodiversity are ultimately tied to the ecosystem regulation and support services that a rich community of species provides—without the many species that keep ecosystems functioning, our ability to extract resources or enjoy nature would be much reduced.

Ever since Robert Constanza's 1997 study that estimated the annual value of ecosystem services to be worth almost twice the annual gross domestic product (GDP) of the entire world, environmental economists have set out to quantify the monetary worth of these services to bring

ecosystem services Essential ecological processes that make life on Earth possible.

Ron Boardman/Life Science Image/FLPA/Science Source

One ecosystem service of biodiversity is the provision of food, both through the food we eat and the other species that help that food grow, such as pollinators and soil decomposers.

attention to this often underappreciated value. The most recent estimate puts that value at $125 trillion—an amount roughly equal to the GDP of the entire world economy. Even nature-based recreation is a multibillion-dollar business worldwide. Whale watching, a pastime in more than 80 countries, brings in $2 billion annually. The U.S. Department of Fish and Wildlife Service estimated the economic value of hunting, fishing, and wildlife viewing in 2016 to be more than $157 billion. (For more on valuing ecosystem services, see Module 5.1.)

instrumental value An object's or species' worth, based on its usefulness to humans.

Unfortunately, the ability of nature to provide these services and meet human needs (and wants) is declining in almost all of the ecosystem service categories evaluated—only agricultural resources (food, fiber, and biofuels) have increased in recent decades, but these gains may be lost with further ecosystem damage. **INFOGRAPHIC 2**

INFOGRAPHIC 2 ECOSYSTEM SERVICES

We depend on genetically diverse, species-rich communities to provide the goods and services we use every day. When biodiversity declines in an ecosystem, its ability to meet the needs of humans and other species is diminished.

1. HUMAN PROVISIONS
- Food
- Fiber products, such as cotton and wool
- Fuel
- Pharmaceuticals

2. CULTURAL BENEFITS
- Aesthetic
- Spiritual
- Educational
- Recreational

Ecosystem regulation and support make the services of human provisions and cultural benefits possible.

3. ECOSYSTEM REGULATION AND SUPPORT

Nutrient cycling	Air/water purification	Soil formation	Wildlife habitat
Pollination	Flood control	Erosion prevention	Genetic resources
Seed dispersal	Climate regulation	Pest control	Waste treatment

1.
Many forest products can be harvested.
2.
Many people enjoy spending time in nature.
3.
Shoreline vegetation protects the bank from erosion.

Danita Delimont/Gallo Images/Getty Images

In terms of reasons to value biodiversity, which of the beneficial ecosystem services listed here are the most important to you personally? Which ones do you think might be the most influential to society?

Of course, many people feel that the value of any given species goes beyond these **instrumental values** (i.e., the ecological, cultural, and economic benefits it brings). Many say that a species like the orangutan has **intrinsic value** (an inherent right to exist) and is therefore worth preserving, regardless of what benefits it might provide.

intrinsic value An object's or species' worth based on its mere existence; it has an inherent right to exist.

3 TYPES OF BIODIVERSITY: GENETIC, SPECIES, AND ECOLOGICAL

Key Concept 3: Genetic, species, and ecological diversity all enhance the ability of an ecosystem to function and for its members to adapt to changing conditions.

Biodiversity can be broken down into three types of diversity that really represent three different levels of diversity, from the population up to the ecosystem. **Genetic diversity** is the heritable variation among individuals of a single population or within the species as a whole. This is diversity at the population level. Genetic diversity provides the raw material that allows populations to adapt to their environment (potentially leading to the emergence of new species). Without genetic diversity, even the most well-adapted species could be in danger of facing extinction if conditions change. (See Module 3.1.)

The importance of genetic diversity was tragically illustrated in 19th-century Ireland. At the time, potatoes were the main food source for one-third or more of the population. Potatoes grew well in the Irish climate, despite the plant not being native to Ireland — it hails from South America. On the slopes of the Andes Mountains, thousands of different genetic varieties of potato plants grew, each adapted to the slightly different niche found at different latitudes and elevations, traits that were selected for by nature and later enhanced with artificial selection by hundreds of generations of Andean farmers. But only a few varieties of potatoes were imported to Ireland, and by the 1840s, one type was predominately planted — the "lumper." This variety grew well and fed millions until a fungus arrived in the mid-1800s that attacked potatoes. In the Andes, this fungus damaged some crops, but because all the potato varieties were so different, even when some plants died, others survived. But in Ireland, without genetic diversity in the potato "population," what killed one, killed them all. The vulnerable lumper succumbed to the fungus, and the potato crop was wiped out for several years. Between 1845 and 1851, more than 1 million people died of starvation and 1 million more emigrated (many to the United States).

We can also find biodiversity at the community level — that is, **species diversity**. This is an accounting of the number of species living in an area (richness) and their distribution (evenness) (see Module 2.3).

Some ecosystems naturally have higher species diversity than others. Areas with high species diversity often owe that variety to their **ecological diversity** — the variety of habitats, niches, and ecological communities in an ecosystem. Rainforests contain a wide variety of habitats and considerable physical complexity, creating many niches. Each of the many species living in a tropical rainforest occupies an individual niche, and they all contribute to the functioning of the ecosystem (including vital services like energy capture, nutrient cycling, and decomposition). For example, a wide variety of plant species occupy every level of the rainforest, from the forest floor to the canopy, each adapted to capture the sunlight available in the forest level where they are found. This large producer base supports many trophic levels and many species within those trophic levels, increasing the efficient use and transfer of nutrients in the ecosystem. Few matter resources go unused. The loss of species (extinction) in an ecosystem removes contributing members from that ecosystem, potentially impacting other species and the ability of that ecosystem to function and provide ecosystem services. **INFOGRAPHIC 3**

genetic diversity The heritable variation among individuals of a single population or within the species as a whole.

species diversity The variety of species, including how many are present (richness) and their abundance relative to each other (evenness).

ecological diversity The variety within an ecosystem's structure, including many communities, habitats, niches, and trophic levels.

INFOGRAPHIC 3 **BIODIVERSITY INCLUDES GENETIC, SPECIES, AND ECOSYSTEM DIVERSITY**

There is individual variation between these *Delia* butterflies.
BIODIVERSITY
GENETIC DIVERSITY
Variations in the genes among individuals of the same species
Indonesia's rainforest ecosystems contain varied habitats and multiple niches that support a complex community.
SPECIES DIVERSITY
The variety of species present in an area, including the number of different species that are present as well as their relative abundance
Tropical forests have some of the highest plant and animal species diversity in the world.
ECOLOGICAL DIVERSITY
The variety of habitats, niches, trophic levels, and community interactions

? Give an example of genetic diversity and species diversity from the grocery store produce section.

4 BIODIVERSITY HOTSPOTS

Key Concept 4: Protecting biodiversity hotspots, areas with high numbers of endangered endemic species, can be a cost-effective way to protect many endangered species.

Tropical forests tend to be particularly flush with both ecological and species diversity, thanks largely to the abundant sunlight and climatic conditions conducive to growth. The forests that are currently being laid low in Sumatra are no exception. They have been left unhampered for so many millennia that these steamy ecosystems swarm with a cornucopia of life: elephants, orangutans, tapirs, tigers, and every manner of bird and beetle the human imagination can fathom. "The truth is, no one has any idea how many species used to live here," Sutherlin says. "Half the species in these forests have yet to be described to science."

Since areas with high ecological diversity offer so many unique habitats and niches, they often have a large number of **endemic** species that are specially adapted to that locale and naturally found nowhere else on Earth. They are most commonly found in small isolated ecosystems (e.g., islands, mountaintops) because the species' population members cannot easily disperse and share genes with other populations.

endemic A species that is native to a particular area and is not naturally found elsewhere.

biodiversity hotspot An area that contains a large number of endangered endemic species.

endangered species Species at high risk of becoming extinct.

Biodiversity hotspots are areas that have high endemism (greater than 1,500 endemic plant species) and have lost at least 70% of their original habitat. These areas contain a large number of **endangered species** (those at risk of becoming extinct).

Thirty-six land and aquatic regions have been designated as hotspots, and many are found in the tropics. The latest biodiversity hotspot to be added to the list is the North American Coastal Plain — the coastal region from Massachusetts down to the Gulf Coast of Texas and northern Mexico and extending up the Mississippi Basin into southern Missouri and Kentucky. With the number of endemic plant species topping 1,800 and having lost 85% of its original natural habitat, it more than qualifies as a hotspot.

This region is especially rich in turtle species (39% of its reptile species are endemic) and amphibians (47% are endemic). **INFOGRAPHIC 4**

Because they are pockets of high biodiversity and contain many unique species found nowhere else, biodiversity hotspots are conservation priorities. Investing time and

INFOGRAPHIC 4 BIODIVERSITY HOTSPOTS

Biodiversity hotspots, areas with high numbers of endemic but endangered species, cover a small percentage of land and water areas but hold more than 40—50% of all plant and vertebrate endemic animal species. Most hotspots are located in tropical biomes or in isolated terrestrial ecosystems, such as mountains or islands. Even small disturbances, such as a small farm plot or road that cuts through the area, can threaten endemic species that populate specialized niches in these hotspots.

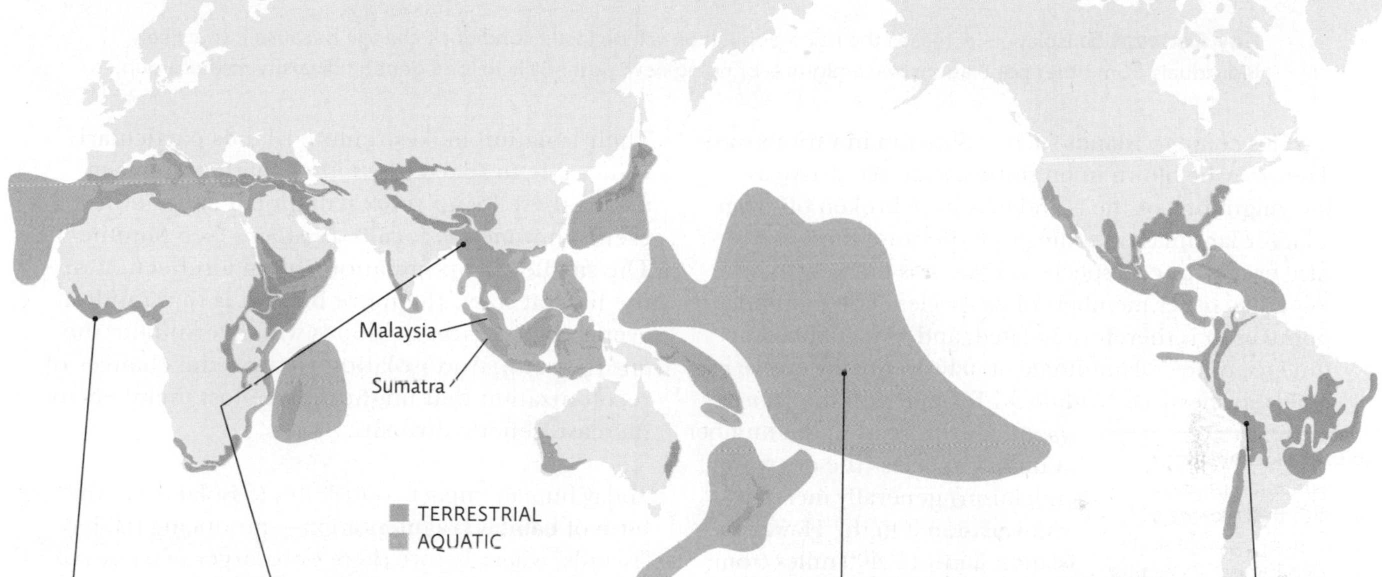

INDO-BURMA 12,000 endemic species; 754 threatened species; at least 9 extinct

POLYNESIA-MICRONESIA 3,278 endemic species; 476 threatened species; at least 235 extinct

GUINEAN FORESTS 2,050 endemic species; 936 threatened species; number of extinctions unknown

TROPICAL ANDES 16,733 endemic species; 814 threatened species; at least 2 extinct

Many hotspots are in tropical areas, but some are also found in areas further north and south of the equator. Why do you suppose so many nontropical coastal areas are biodiversity hotspots?

money in protecting these areas can potentially provide a lot of return on that investment—many endangered species, or species vulnerable for endangerment, will be protected.

The island of Sumatra is one such hotspot; it is home, for example, to the endemic Sumatran tiger. The pygmy elephant is likewise endemic to the nearby island of Borneo. And the orangutan, one of humans' closest relatives on the evolutionary tree, is endemic to the region, with different types of orangutan found on the islands of Sumatra and Borneo. All of these species are being driven to the edge of extinction by our pursuit of palm oil.

5 ISOLATION AND EXTINCTION RISK

Key Concept 5: Isolation increases the risk a population will be lost if conditions change because it is unlikely individuals from other populations will emigrate, bringing new, potentially helpful genetic diversity to the group.

Species come to islands such as Sumatra in various ways: They may be blown in on storms, they may arrive as lost migrators, or the island may have broken off from a larger landmass at some point. Because these are rare events, once a species arrives, it is unlikely to be joined by other members of its species. The founding population is therefore isolated, and as it adapts over time to its new island home, it may eventually evolve into a new species (see Module 3.1 for more on the *founder effect*). For this reason, the number of unique species (the degree of endemism) generally increases with isolation. On the Hawaiian Islands, almost 2,400 miles from the nearest mainland, 90% of native species are endemic.

habitat fragmentation The destruction of part of an area that creates a patchwork of suitable and unsuitable habitat areas that may exclude some species altogether.

Their isolation makes remote islands particularly vulnerable to species loss; their high endemism means these losses represent global extinction events, not merely local *extirpations* (see Module 3.1). The smaller the population (often a reflection of the habitat size), the more likely it is that random events such as fires or floods will exterminate the entire group. And isolation reduces the chances of recolonization that might replace lost members or increase genetic diversity.

Today human impact contributes to isolation in the form of **habitat fragmentation**—producing habitat "islands" where before there were larger expanses of uninterrupted habitat. Deforestation that leaves patches of forest may not be suitable for species that need large

Wildlife corridors like this one over the Trans-Canada Highway in Banff National Park in Alberta, Canada, allow animals safe passage over roads.

Robert McGouey/Wildlife/Alamy

expanses of forest. Some species, including orangutans, simply will not cross over the deforested patch to disperse from one side of the patch to the other, so the patches effectively isolate them from other parts of their population. Even a narrow or little-used road may isolate individuals on one side or the other.

Small, completely isolated ecological communities are at the highest risk for species extinction. Large-scale experimental studies show that habitat fragmentation reduces biodiversity by as much as 75%, with significant reductions in biomass and nutrient cycling. Research by Luke Gibson and colleagues found that after dam construction flooded a Thailand forest, leaving behind some small islands covered in forest, almost all small mammal species were lost in as little as 5 years on the smallest islands. Isolation can be a recipe for disaster in the face of rapid environmental change. (See the *Science Literacy* activity at the end of this module for more on Gibson's research.)

Stuart Pimm of Duke University and his colleagues analyzed the fossil record and the living species on Pacific islands colonized by humans about 4,000 years ago and concluded that as many as half of the species might have been driven to extinction after humans arrived. In fact, even though they make up a tiny percentage of the land mass on Earth, islands have accounted for about half of all recorded extinctions in the past 400 years. In Hawaii alone, fully one-half of the indigenous flora faces immediate extinction. Such a high extinction rate threatens the fragile tapestry of life on these islands, from the soil and freshwater supplies to the health and economic future of the islands' residents. **INFOGRAPHIC 5**

INFOGRAPHIC 5 ISOLATION CAN AFFECT POPULATIONS

Achieve

Isolation can increase the number of endemic species in an area because local populations do not "share" genes with other populations. Over time, an isolated population may diverge from its ancestral population as it becomes adapted to its immediate environment.

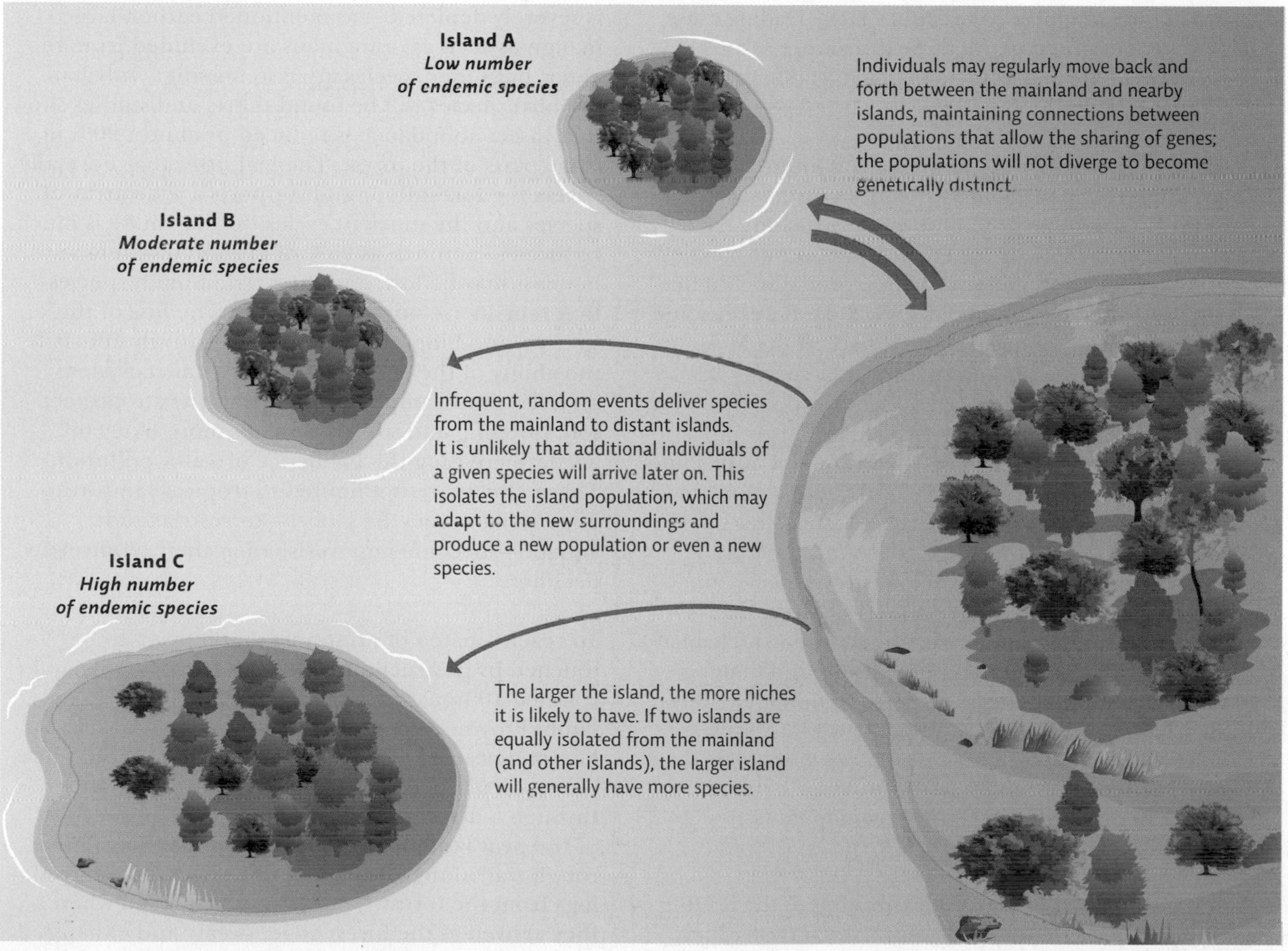

Island A
Low number of endemic species
Individuals may regularly move back and forth between the mainland and nearby islands, maintaining connections between populations that allow the sharing of genes; the populations will not diverge to become genetically distinct.
Island B
Moderate number of endemic species
Infrequent, random events deliver species from the mainland to distant islands. It is unlikely that additional individuals of a given species will arrive later on. This isolates the island population, which may adapt to the new surroundings and produce a new population or even a new species.
Island C
High number of endemic species
The larger the island, the more niches it is likely to have. If two islands are equally isolated from the mainland (and other islands), the larger island will generally have more species.

? Why might populations of small animals, like salamanders that live only at high elevations on adjacent mountains, remain isolated from one another, even though there are no barriers between the two mountains?

Wildlife corridors that connect fragmented habitats can be effective conservation tools. Specially constructed road overpasses or underpasses or ribbons of natural habitat that connect two habitat fragments can allow individuals to travel more freely and are vitally important in maintaining genetic diversity in the populations—especially for species that range widely and live at low population densities, such as large carnivores like grizzly bears and wolves, or those that live in small family groups like the orangutan. For oil palm plantations, preserving natural corridors around rivers has the added benefits of protecting the water from sediment pollution and reducing the loss of floodplain land by decreasing riverbank erosion.

6 THREATS TO BIODIVERSITY

Key Concept 6: The leading threats to biodiversity today come from human activities. Biodiversity loss endangers ecosystems and the species (including humans) that depend on the ecosystem services these areas provide.

The world's biodiversity is threatened from many angles. The scientific consensus is that human actions are behind the high extinction rates of the recent past, especially since 1900. These threats include overexploitation (legal and illegal), pollution, anthropogenic climate change, the introduction of invasive species (see Module 3.1 for a detailed example), and the number one cause—habitat destruction and fragmentation. Globally and in many regions of the Indonesian rainforest, agriculture is the leading cause of habitat destruction. All these threats are related to overpopulation and affluence, which lead to greater resource use.

In recent decades, despite increasing international attention to the problems of biodiversity loss, each of these factors has gotten worse. Overfishing has claimed some 85% of oyster reefs and as much as 90% of the world's populations of large predatory fish like tuna and cod (see Online Module 8.4). Pesticides and mercury pollution have all but killed off the Mekong River dolphin and imperil many other species of insects, insect-eating birds, and frogs and salamanders (amphibians are a particularly vulnerable group). Climate change has claimed the Bramble Cay melomys, a small rodent from Australia, and is threatening to bring the iconic polar bear in the Arctic Circle, not to mention many other species around the world, to disastrous ends. Meanwhile, invasive species are running amok from Alabama to Zimbabwe, thanks to intentional and accidental introductions. And habitat fragmentation is isolating populations and limiting their ability to migrate or disperse to new areas. It is also making commercially valuable wildlife, such as elephants with their ivory tusks, more vulnerable than ever to human poaching. (See Online Module 3.3 for more on this threat to African elephants.) **INFOGRAPHIC 6**

Our palm oil obsession is contributing to the leading cause of species endangerment: habitat destruction. By obliterating certain parts of an ecosystem, habitat destruction makes that ecosystem physically less diverse, which then makes it less biodiverse. The loss or even decline of a species can have significant effects on its ecological community, which, in turn, can lead to a reduction of ecosystem services.

The differences between an oil palm plantation and the natural forest it replaced are stark. Biodiversity is severely depleted—as mentioned earlier, large mammals such as orangutans are excluded from the areas, but other species are also missing: Only half the bird species will be found there, and studies show that insect abundance is reduced by almost 90% in some parts of the forest. The leaf litter that normally covers the forest floor and supports a wide array of species and the nutrient cycles they facilitate is much reduced—as much as 95% of ground-dwelling ant biomass may be lost, and the predominant species that remain are not forest natives. The loss of these species, and biomass in general, negatively affects the ability of the ecosystem to perform ecosystem services. For example, without leaf litter to protect soil, flooding and soil erosion are more likely to occur, increasing the incidence of water pollution. These effects are not limited to tropical rainforests. Ecosystems across the globe—terrestrial and aquatic—become impoverished as their biodiversity declines.

To reach a patch of rainforest that had yet to be touched by destruction, Sutherlin and his colleagues traveled 10 hours through the night from Riau to Jambi Province, then another 4 hours by car over horrendous dirt roads to South Sumatra. From there, they took motorcycles over winding-ribbon trails through a desolate oil palm plantation to the edge of the peat lands and then continued by foot "on a rough trail along a canal dug by loggers to remove logs from the forest." Sutherlin was thrilled when they arrived at the forest edge, sweaty and exhausted,

INFOGRAPHIC 6 **THE MAIN THREATS TO BIODIVERSITY TODAY COME FROM HUMANS**

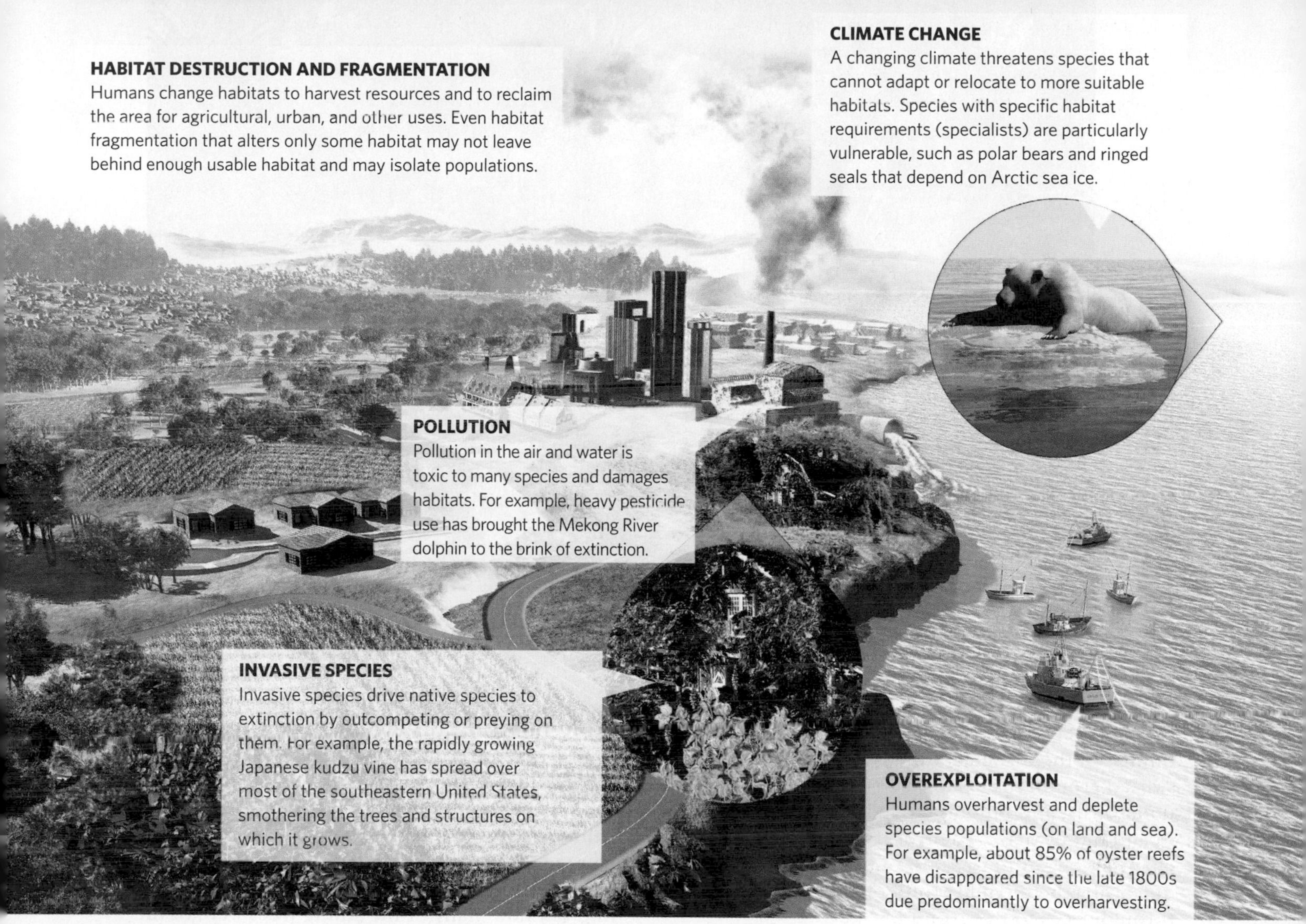

HABITAT DESTRUCTION AND FRAGMENTATION
Humans change habitats to harvest resources and to reclaim the area for agricultural, urban, and other uses. Even habitat fragmentation that alters only some habitat may not leave behind enough usable habitat and may isolate populations.
CLIMATE CHANGE
A changing climate threatens species that cannot adapt or relocate to more suitable habitats. Species with specific habitat requirements (specialists) are particularly vulnerable, such as polar bears and ringed seals that depend on Arctic sea ice.
POLLUTION
Pollution in the air and water is toxic to many species and damages habitats. For example, heavy pesticide use has brought the Mekong River dolphin to the brink of extinction.
INVASIVE SPECIES
Invasive species drive native species to extinction by outcompeting or preying on them. For example, the rapidly growing Japanese kudzu vine has spread over most of the southeastern United States, smothering the trees and structures on which it grows.
OVEREXPLOITATION
Humans overharvest and deplete species populations (on land and sea). For example, about 85% of oyster reefs have disappeared since the late 1800s due predominantly to overharvesting.

HUMAN ACTIVITIES THAT THREATEN SPECIES

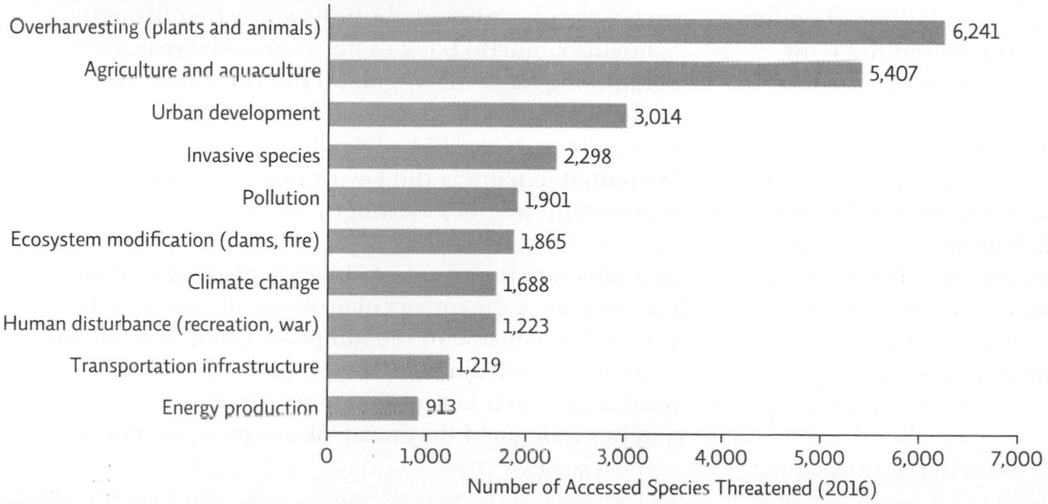

In a more detailed breakdown of the threats, overexploitation from hunting, fishing, and gathering wild plants is the main threat to species that have been accessed so far by the International Union for Conservation of Nature, with agricultural activities a close second. Some of these categories are interconnected (e.g., some pollution threats are agricultural in origin, and many of these contribute to habitat destruction).

? Which categories on this graph represent habitat destruction/fragmentation? Do these data support the claim that habitat destruction/fragmentation is the leading cause of species endangerment?

Compared to the natural forests they replace, oil palm plantations have less than half the bird species and as little as 10% of the insect species. This decline in biodiversity reduces the effectiveness of ecosystem services such as flood control and carbon storage.

to finally see some tall trees still standing. Monkeys howled in the distance. An electric blue butterfly swirled around his head. And spider-hunters, dollarbirds, and bulbuls circled overhead, flitting in and out of view. "Then, as if on cue," he says, "a chainsaw began to roar just out of sight, followed quickly by the terrible sound of trees crashing through trees to the ground."

Virtually all of the forest's inhabitants are facing annihilation. "The remaining populations of endemic Sumatran rhinos are widely considered to be the living dead," says Sutherlin. "Their habitat is too sparse, too fragmented, and too disturbed, their numbers too few."

In Sumatra and elsewhere in Southeast Asia, the solution to protecting biodiversity will have to involve palm oil production. It's the most productive oil-bearing crop at our disposal, producing around 1.6 tons per acre. Soybeans, by comparison, produce only 0.17 ton per acre. "To get the same amount of soybean oil, you would have to use ten times as much land," geneticist Raviga Sambanthamurthi explains.

In an effort to increase that productivity even more, Sambanthamurthi has developed a seed-screening technique that allows farmers to identify the most productive strains of oil palm for planting—more palm oil per tree means fewer trees need to the planted. Before this screening method was developed, farmers had to plant seeds and wait 5 or 6 years to find out how many of the young trees were of the most productive variety. "That puts years back on the clock," she says. It was the dream of making such a significant contribution as this that brought Sambanthamurthi back to her native Malaysia after completing her PhD in genetics at the University College of London. "I knew if I was going to make a difference, it would be in palm oil," she said. "Sustainable palm is the key to protecting our biodiversity, and our heritage."

In addition to increased productivity, there are other ways to reduce the impact of growing oil palms. To be certified sustainable by the nonprofit group Roundtable on Sustainable Palm Oil (RSPO), palm oil producers must adhere to practices that minimize the use of fire and pesticides, and they must take steps to prevent soil erosion and water pollution. The rights of workers (fair wages and good working conditions) are a high priority, and the local community is given an opportunity to weigh in on proposed oil palm plantations. In addition, growers cannot clear old-growth forests or those with high biodiversity or cultural value for conversion to plantations.

Sustainable supplies of palm oil began to increase steadily around 2008. In 2019, about 19% of the global supply was sustainably grown—up from only 6% in 2013. However, sustainable production plateaued around 2015, even as the use of palm oil continues to increase. By 2050, demand is expected to quadruple. Reducing demand and growing oil palms more sustainably will be critical if we want to protect these forests and their species.

One thing is certain: There is still much value in saving these forests. Orangutans still swing freely through the canopies, and new species of lizards and birds continue to be discovered. Despite the destruction plaguing Indonesia's forests, all hope is not yet lost.

Select References:

Chapman, A. D. (2009). Number of living species in Australia and the world. *Report for the Australian Biological Resources Study, Canberra, Australia.*

Gibson, L., et al. (2013). Near-complete extinction of native small mammal fauna 25 years after forest fragmentation. *Science, 341*(6153), 1508–1510.

Larsen, B. B., et al. (2017). Inordinate fondness multiplied and redistributed: The number of species on Earth and the new pie of life. *The Quarterly Review of Biology, 92*(3), 229–265.

Newbold, T., et al. (2016). Has land use pushed terrestrial biodiversity beyond the planetary boundary? A global assessment. *Science, 353*(6296), 288–291.

Pimm, S., et al. (2006). Human impacts on the rates of recent, present, and future bird extinctions. *Proceedings of the National Academy of Sciences, 103*(29), 10941–10946.

Singh, R., et al. (2013). The oil palm SHELL gene controls oil yield and encodes a homologue of SEEDSTICK. *Nature, 500*(7462), 340–344.

Stork, N. E., et al. (2015). New approaches narrow global species estimates for beetles, insects, and terrestrial arthropods. *Proceedings of the National Academy of Sciences, 112*(24), 7519–7523.

Swaddle, J. P., & Calos, S. E. (2008). Increased avian diversity is associated with lower incidence of human West Nile infection: Observation of the dilution effect. *PLoS One, 3*(6), e2488.

The biodiverse rainforests of Indonesia are the last place on Earth where tigers, rhinos, and elephants exist together in the wild, but those populations are critically endangered from habitat destruction.

Edo Schmidt/Alamy

GLOBAL CASE STUDIES BIODIVERSITY HOTSPOTS

Achieve

Species in each of the world's biodiversity hotspots are threatened by a unique constellation of forces, but most of these threats are human in origin. Read about four of these hotspots in more detail to see what threats they face and what is being done to protect them.

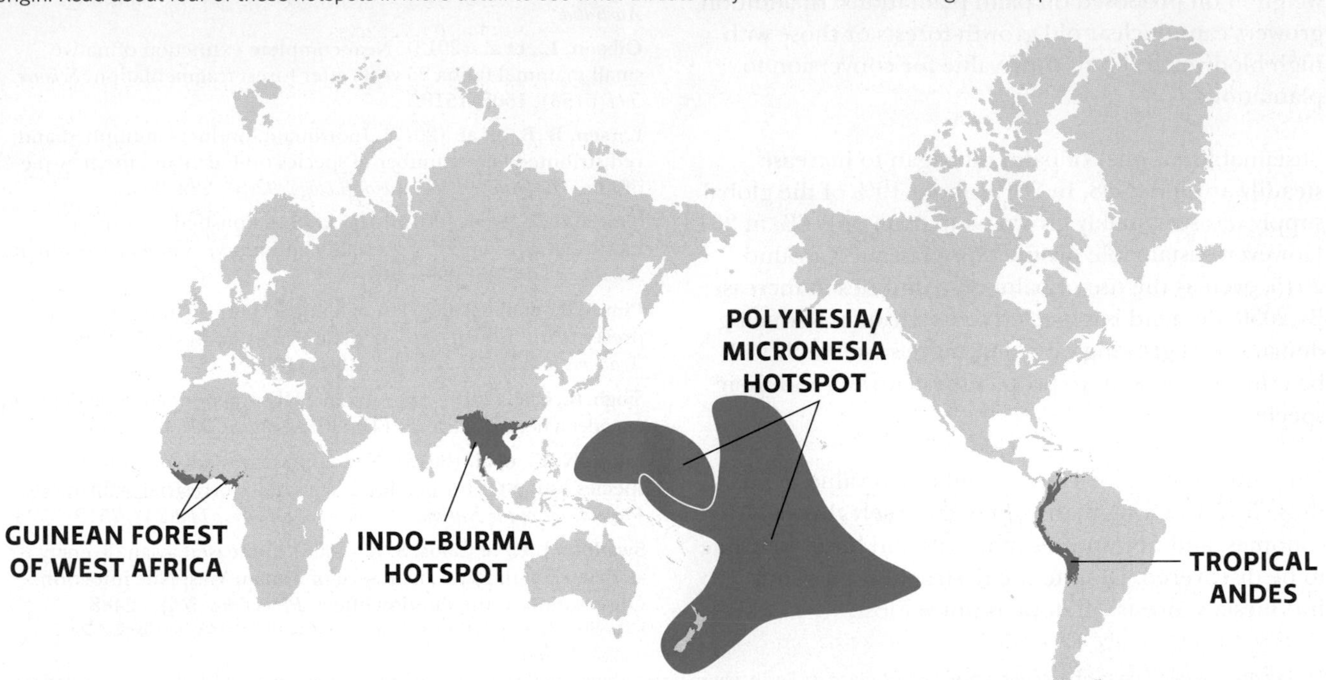

BRING IT HOME

PERSONAL CHOICES THAT HELP

Species and habitats provide numerous benefits to people, including water and air purification, food sources, recreation, and medicine. Unfortunately, many species are facing threats at ever-increasing levels. The good news is that we as a society have a direct impact on these threats and can make changes to ensure the survival of many at-risk species.

Individual Steps

- Don't buy products made from wild animal parts such as horns, fur, shells, or bones. Only buy captive-bred tropical aquarium fish, not wild-caught fish.
- Visit the Rainforest Action Network website at www.ran.org/issue/palm_oil to read about its Snack Food 20 campaign. Also visit www.ran.org/issue/snackfood20 for an evaluation of how the top 20 snack food producers are doing in their quest to source sustainable palm oil.
- Look for labels on products that identify the palm oil it contains as "Certified Organic" or "Certified Sustainable Palm Oil RSPO" with its trademarked logo, a globe-shaped palm.
- Make your backyard friendly to wildlife, using suggestions from the National Wildlife Federation (www.nwf.org).
- Install an Audubon Guide or iNaturalist app on your smartphone or buy a field guide to learn the plant and animal species in your area.
- The Endangered Species Act was the first U.S. legislation established to protect species diversity. To learn more about current challenges and updates to the program, visit the U.S. Environmental Protection Agency (www.epa.gov/espp).

Group Action

- Work with faculty and other students to organize a bioblitz for a protected area in your region. A bioblitz, which is an intensive survey of all the biodiversity in the area, can generate a large amount of data to be used for habitat management and species protection.
- Join a citizen science program monitoring wildlife. Many regional conservation groups have monitoring opportunities and provide training. For national programs, see the Cornell Lab of Ornithology website at www.birds.cornell.edu and the Izaak Walton League of America website at www.iwla.org.

REVISIT THE CONCEPTS

- The variety of life (biodiversity) on Earth is tremendous—perhaps as many as 11 million species, even more when bacteria diversity is considered—and there is also great diversity within most of those species. We have identified only a fraction of existing species, and we know much more about smaller groups such as plants and vertebrates than more diverse groups like invertebrates.

- Biodiversity contributes to the health and wellbeing of ecosystems, which in turn benefits human populations by providing ecosystem resources and services as well as cultural and health benefits.

- Three types of biodiversity are genetic, species, and ecological diversity. Each of these enhance the ability of an ecosystem to function and its members to adapt to changing conditions.

- Unfortunately, Earth's biodiversity is declining rapidly. Many species are endangered, and others have already been lost to extinction. Protecting biodiversity hotspots, areas with high numbers of endangered endemic species, can be a cost-effective way to protect many endangered species.

- Many endangered species (though certainly not all) live in isolated areas — islands or mountaintops or other areas that are disconnected from similar habitats. Isolation increases the risk a population will be lost if conditions change because it is unlikely individuals from other populations might emigrate, bringing new, potentially helpful genetic diversity to the group.

- The leading threats to biodiversity today come from human activities. In order of threat, they are habitat destruction, overexploitation, climate change, pollution, and invasive species. Biodiversity loss endangers ecosystems and the species (including humans) that depend on the ecosystem services these areas provide.

ENVIRONMENTAL LITERACY Understanding the Issue

1 What is biodiversity?

1. How many species are estimated to live on Earth, and which group contains the most species (other than bacteria)?
2. Where are the most biodiverse terrestrial and ocean ecosystems in the world found?
3. Why do you suppose we know more about smaller species groups such as vertebrates and plants than larger groups like arthropods and worms?

2 Why is biodiversity important?

4. Identify the three categories of ecosystem services and give examples of each.
5. Distinguish between intrinsic and instrumental value.
6. How might biodiversity loss impact the ecosystem service of human provisions?

3 How do genetic, species, and ecological diversity each contribute to ecosystem function and services?

7. How does the amount of ecological diversity affect species diversity?
8. Explain why genetic diversity is considered a population concept, whereas species diversity is a community concept.
9. Define genetic diversity and use the example of the potato blight in Ireland to explain the importance of genetic diversity to a population.

4 What are biodiversity hotspots, and why are they important?

10. What is an endemic species, and how does that relate to biodiversity hotspots?
11. Where are most biodiversity hotspots located?
12. What role do biodiversity hotspots play in efforts to protect biodiversity?

5 What role does isolation play in a species' vulnerability to extinction?

13. Consider three islands: a small island close to the mainland, a large island close to the mainland, and a large island far from the mainland or other islands. Which island is most likely to have the highest number of endemic species? Explain.
14. Why are isolated populations more vulnerable to extinction than populations that are not isolated from each other?
15. Other than living in naturally isolated locations such as islands or mountaintops, how else might a population become isolated from other populations of its species?

6 What are the causes and consequences of biodiversity loss?

16. List the five leading causes of species endangerment. Which is the leading threat?
17. There are no easy answers to protecting biodiversity, especially when human actions such as agriculture — clearly needed to feed the world — harm species. We know palm oil production harms species, but if we stop growing it, threats to species might get worse. Why?
18. How can you, as an individual, help maintain biodiversity worldwide? Justify your choices.

SCIENCE LITERACY Working with Data

Habitat loss is currently the main driver of species endangerment and extinction, but habitat loss need not be complete to cause a problem; habitat fragmentation may also be an insurmountable problem for some species. Islands that are created when a river is dammed to form a reservoir provide instant habitat fragments. Luke Gibson and his team evaluated the number of small mammal species in large (10–56 hectares [25-140 acres]) and small (<10 hectares [>25 acres]) forested islands in Chiew Larn Reservoir of Thailand. Island sampling was done shortly after the reservoir was formed (about 6 years after isolation); the islands were sampled again about 26 years after isolation. Their results are below. (For comparison, on average, nine species were found on mainland [pre-reservoir] plots; the richness did not change in this mainland forest over the study period.)

Number of Small Mammal Species (Richness) Found on Islands 6 and 26 Years After Isolation

	Island (ID #)	*Area of Island (hectares)*	*Species Richness (6 years)*	*Species Richness (26 years)*
Large Islands	6	56.3	12	5
	5	12.1	9	3
	9	10.4	7	1
Small Islands	28	4.7	2	2
	7	1.9	3	2
	33	1.7	1	1
	3	1.4	2	1
	41	1.1	3	1
	39	1.0	3	1
	40	0.8	2	1
	2	0.4	2	1
	16	0.3	2	1

Interpretation

1. Before evaluating the data, draw a graph that compares species richness of large islands 6 years after isolation versus 26 years after isolation and that shows the same for small islands. (*Hint*: Calculate the average species richness values for each group and draw a bar graph that allows you to directly compare the richness of large islands after both sampling periods to the richness of small islands after both sampling periods.)
2. Consider that species richness before isolation was 9. How does the species richness compare in large islands before isolation, 6 years after isolation, and 26 years after isolation? How does it compare for the small islands over those sampling periods?

Advance Your Thinking

3. What might lead to the difference in species richness losses in large islands compared to small islands?
4. On all of the islands, the most common (and sometimes only) small mammal 26 years after isolation was a non-native rat. Could this have had an influence on the loss of the other native species? Explain.
5. What conclusion can be drawn regarding the value of leaving behind fragmented forest landscapes for protecting species that live in the habitat fragments?

INFORMATION LITERACY Evaluating Information

Visit the World Wildlife Fund (WWF) website (www.worldwildlife.org). Explore the website to learn about the organization and their work to protect endangered species.

1. What is the mission of the WWF? How do you know this?
2. Read about several of the species that WWF is protecting: What conservation techniques do they use to protect these species? Does their approach support their stated mission?
3. What scientific evidence does the organization provide to indicate that their work is effective at protecting species?
4. Read about the WWF's work in Borneo and Sumatra. What threats do they identify for this region, and how are they helping species and the local communities there? Do you think their work is helping the region? Explain.
5. What is your overall opinion of the WWF? Would you do anything differently if you were in charge? If so, what?
6. Is this an organization you would support financially? Explain.

Additional study questions are available at Achieve.macmillanlearning.com. Achieve

CHAPTER 4

Human Populations and Environmental Health

Many environmental problems arise from the sheer number of people, as well as high individual impacts of the human population. Some solutions lie in reducing population growth rates, making our cities more sustainable, and taking steps to create healthy environments.

Module 4.1

Human Populations

An examination of past and present human population growth, ways to reduce its current rate of increase, and the ecological footprint of populations

Module 4.2

Urbanization and Sustainable Communities

An introduction to smart urban development and ways to reduce the per capita impact of city residents while improving their quality of life

Module 4.3

Environmental Health and Toxicology

An assessment of the impact of biological, chemical, and physical environmental factors on human health and ways to address environmental health issues

peeterv/Getty Images

Module 4.1 Human Populations

THE KERALA MODEL

India's path to population control

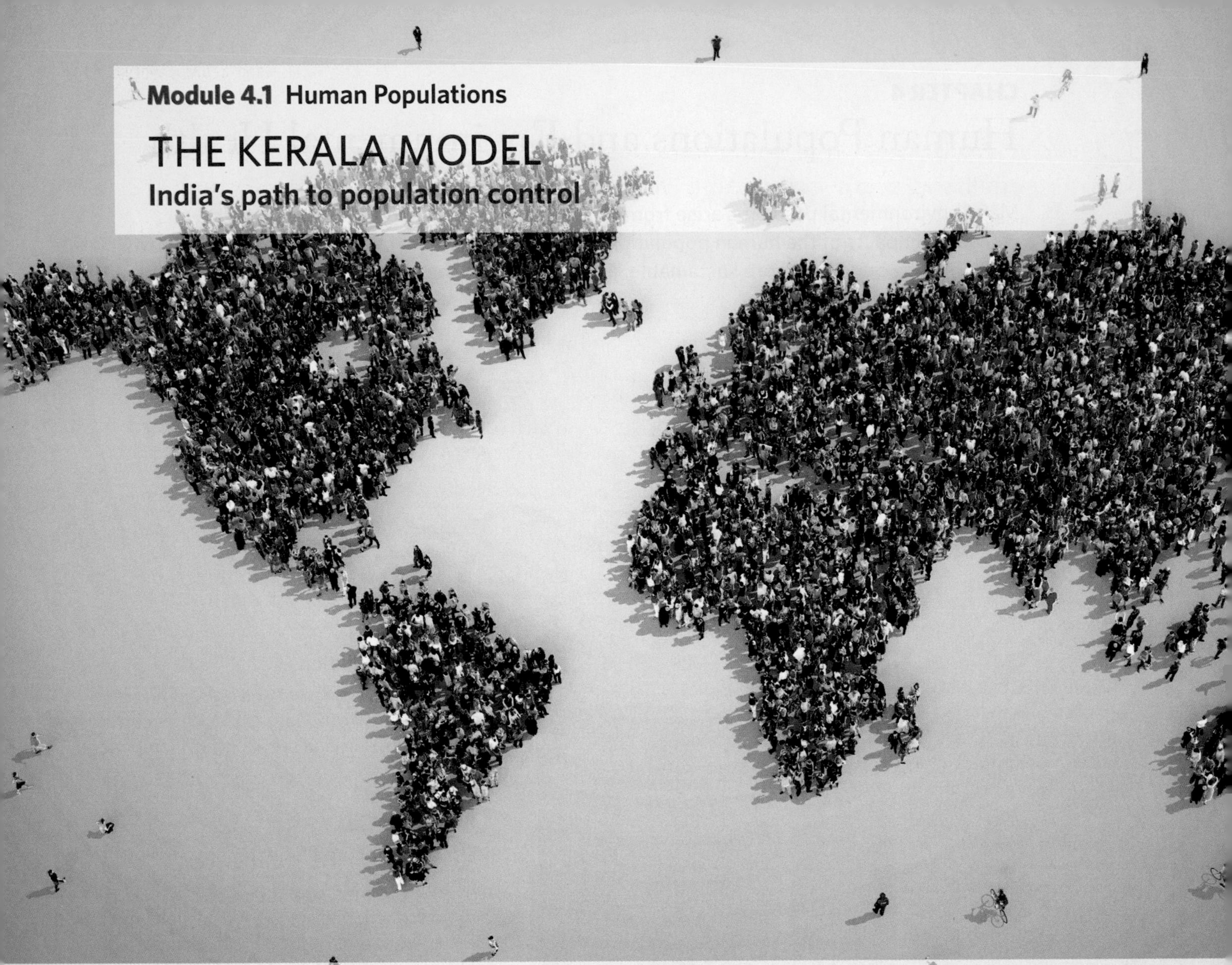

More than 7.7 billion people live on Earth and the human population is still growing, increasingly straining the resources needed to support us as well as other species. *Arthimedes/Shutterstock*

After reading this module, you should be able to answer these GUIDING QUESTIONS

1. How and why have human population size and growth rate changed over time?
2. How big is the human population today, and where do most people live?
3. What factors affect human population growth?
4. How does the age structure of a population influence its growth?
5. What is the demographic transition, and why is it important?
6. How can addressing social justice issues help achieve zero population growth?
7. What determines Earth's carrying capacity for humans, and can the planet support the current or future (projected) human population long term?

Back in 1988, a swarm of volunteers descended on Ernakulam, a city of about three million people in the state of Kerala, India. They were on a mission to teach as many people as possible to read and write, and they were not leaving any stones unturned. "Classes were held in cowsheds, in the open air, in courtyards," an organizer told *The New York Times*. "For fishermen, we went to the seashore. In the hills, tribal groups sat on rocks. Leprosy patients were taught to hold a pencil in stumps of hands with rubber bands."

India is a developing country on track to surpass China as the world's most populous nation in the near future. Its huge and still rapidly growing population must find ways to stem population growth in a world and a country quickly running low on resources. Like much of India, Kerala, a small, bean-shaped state tucked into the southwest corner of India, is both very poor and very crowded. The average Keralan earns just $1/70^{th}$ of what the average U.S. citizen earns and lives in just half the space of the average resident of New Jersey (the most densely populated U.S. state with about 1,200 people per square mile). But Kerala has some unique features that set it apart from other, equally poor and crowded places. Women tend to be more educated and have more power, for example. And decent health care is the norm for all citizens.

The Ernakulam campaign made Kerala the first fully literate state in all of India. But it also brought about another success—one that stretched far beyond the volunteers' makeshift classrooms: As literacy rates went up in Kerala, the state's population stopped growing so quickly. The achievement was so unprecedented that economists came up with a new term to describe it: the Kerala Model. To understand the relationship between literacy and population growth, we must first take a closer look at the nature of human populations and the forces that cause them to grow and shrink.

WHERE IS KERALA?

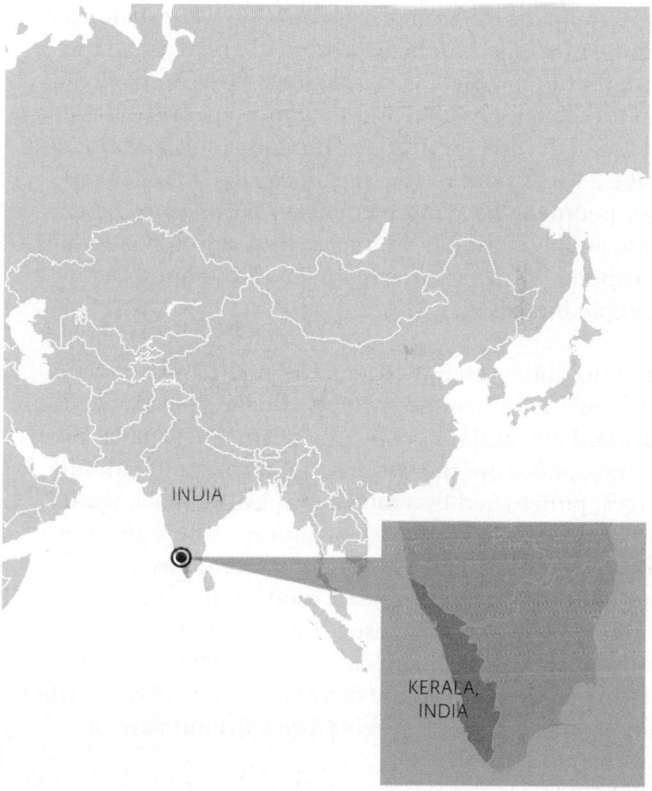

Educational opportunities, especially for women and young girls, are highly correlated with decreased population growth rates.

Robert van der Hilst/Corbis Documentary/Getty Images

1 HUMAN POPULATION: PAST, PRESENT, AND FUTURE

Key Concept 1: The human population grew very slowly for most of human history. Only recently have accelerated growth rates sent our population soaring past 7 billion. World population is projected to stabilize near 11 billion around 2100.

Population growth rate is a measure of how quickly a population will increase in size. Right now, global population size is the largest it has ever been, and the growth rate is very high. For most of human history, our population size was orders of magnitude smaller than it is now. The Agricultural Revolution—the prehistoric transition from hunter-gatherer societies that were small and nomadic to farming societies that were stationary and larger—led to the first dramatic increase in the number of people, for two reasons. First, it enabled us to produce more food and feed more people. Second, it created a greater need for human labor. Because they take a lot of work to run, farms created an incentive for people to have more children. With farms, people also ate better and lived longer. As a result, death rates dropped, birth rates stayed high, and the number of people increased.

But not indefinitely. Before long, population growth slowed and human population size held steady at around 100 million people for centuries. Population grew slowly from the Bronze Age through the Middle Ages, punctuated by a dip caused by the Black Death in Europe, until the next revolution—the Industrial Revolution—led to dramatic improvements in agriculture (which increased food supplies) and sanitation (which decreased the spread of diseases). Advancements in health care, too, improved survival. Those changes drove the death rates down even further and spurred the next, even more extreme wave of population growth.

More recently, Earth has witnessed yet another surge in human population growth as the improvements in the provision of food, health care, and sanitation brought on by the Industrial Revolution have spread from developed countries like the United States to developing ones like India and China. These advances have enabled people in all countries to live longer than ever before—life expectancy at birth has more than doubled since 1800, when it was less than 30 years of age. (According to the United Nations, in 2019, life expectancy at birth was age 72 worldwide, and age 65 in less developed countries.) Many countries have life expectancies in the low 80s—the highest 2019 value was that of Japan at age 84; the United States came in at age 79. Even the country with the lowest life expectancy today (Central African Republic at age 53) exceeds that of countries with the highest life expectancies in 1800! Still, we shouldn't lose sight of the fact that some countries have life expectancies considerably lower than others, highlighting the need to actively work toward addressing the health threats each faces. (See Module 4.3 for more on environmental health.)

population growth rate The change in population size over time that takes into account the number of births and deaths as well as immigration and emigration numbers.

zero population growth The absence of population growth; occurs when birth rates equal death rates.

In 1950, world population was 2.5 billion; in 1972, it reached 3.85 billion; and by 2019, it had doubled, topping 7.7 billion people. We are adding about a half billion people every 6 years!

Though still increasing—and very quickly in some areas—overall, human population is growing at a slower rate than it was in the mid-20th century, thanks to decreases in birth rates. Annual global growth rates peaked around 1963 at 2.2% and slowly decreased to the current level of just over 1% (1.1% in 2019). Though a growth rate of barely more than 1% doesn't sound like much, if our growth rate stayed the same, by 2100, there would be more than 21 billion people on Earth.

Future projections that assume lesser growth rates predict human population size will stabilize somewhere between 7.3 billion (note this is a decline) and 15.6 billion by 2100—the mid-range projection of 10.9 billion is believed to be the most likely scenario. (See the *Science Literacy* activity at the end of this module for a look at these projections.) Where we eventually stabilize will depend largely on how quickly we reach a stasis, where death and birth rates even out (a growth rate of 0%). That stasis is referred to as **zero population growth**. **INFOGRAPHIC 1**

Because of concerns about how many people the planet can support and what will happen if population growth goes unchecked, zero population growth has long been the goal of many individual countries around the world, including India.

INFOGRAPHIC 1 HUMAN POPULATION THROUGH HISTORY

Population size in 2050 and beyond will largely depend on future growth rates and may stabilize around 11 billion; if population growth rates do not decline as quickly as expected, a high-end estimate puts our population at more than 15.6 billion by 2100.

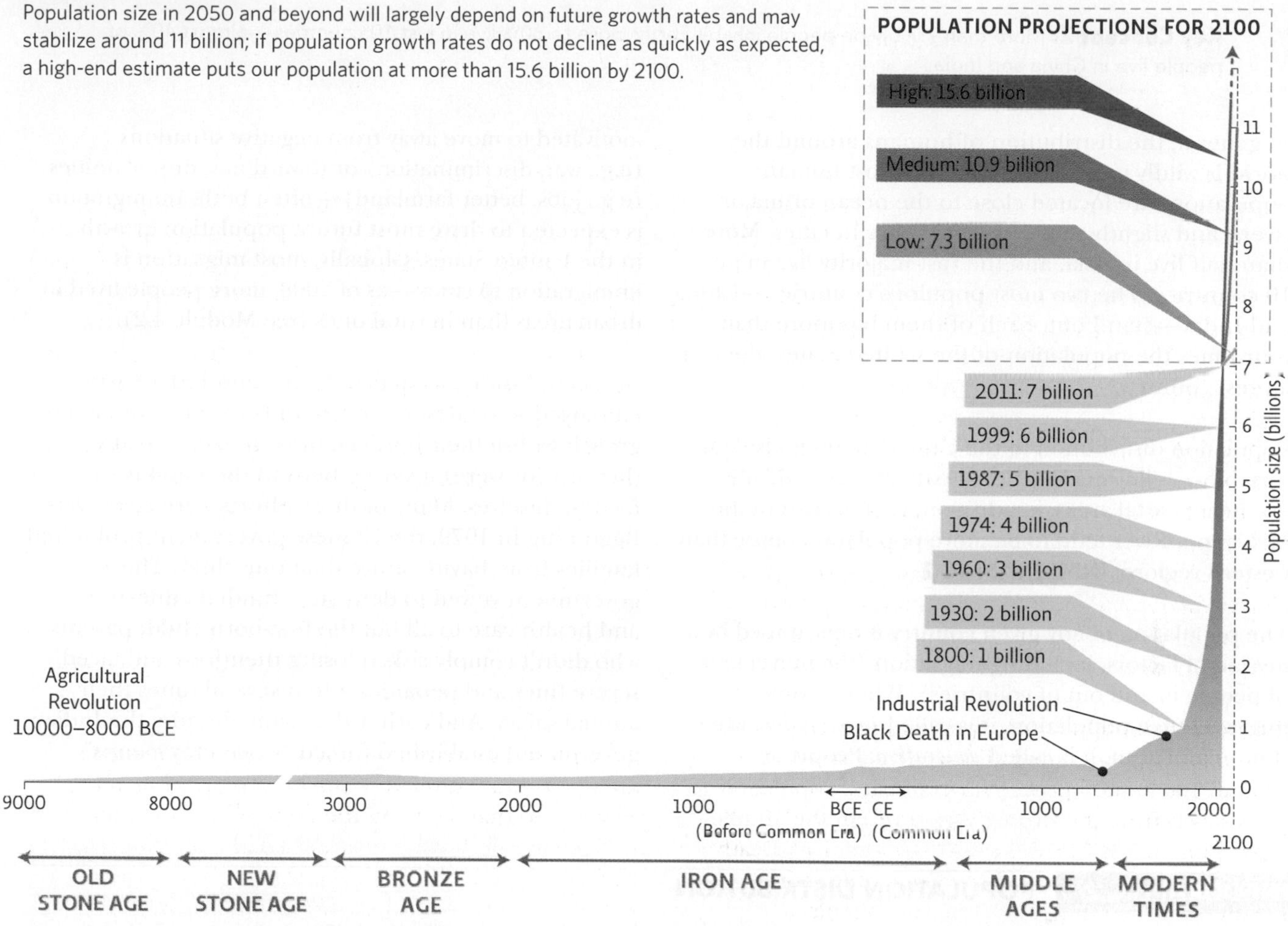

What kinds of factors might result in a population of 15.6 billion by 2100 (the high-variant estimate), and what might prevent our population from growing this large? For the medium and low estimates, what circumstances might lead to populations of those sizes?

A crowded city street in Kolkata, India. Global human population growth began to surge in the late 20th century and continues to grow at a fast pace, adding a billion people every 12 years or so.

Steve Raymer/National Geographic Image Collection/Alamy

2 HUMAN POPULATION: SIZE AND DISTRIBUTION

Key Concept 2: More than 7.7 billion people inhabit Earth; close to 60% live in just 10 countries—almost 40% of people live in China and India.

In general, the distribution of humans around the world is wildly uneven. Worldwide, most human populations are located close to the ocean or major rivers, and slightly more than 55% live in cities. More than half live in Asia, and the vast majority live in just 10 countries. The two most populous countries—China and India—stand out. Each of them has more than four times the population of the United States, the next largest country.

Population distribution in the United States is also quite variable—following the global pattern, around 50% live near coastal areas; in addition, regions east of the Mississippi River tend to be more population dense than western regions. **INFOGRAPHIC 2**

The population of any given country is determined by a myriad of factors, including migration (the movement of people in and out of countries). When people migrate into a population, it is called *immigration*; when they migrate out, it is called *emigration*. People are motivated to move away from negative situations (e.g., war, discrimination) or toward new opportunities (e.g., jobs, better farmland)—often both. Immigration is expected to drive most future population growth in the United States. Globally, most migration is immigration to cities—as of 2008, more people lived in urban areas than in rural ones (see Module 4.2).

In recent history, leaders of China and India have employed some desperate attempts to curb population growth within their borders, in large part because they are (or were) growing beyond their ability to feed themselves. Many of those efforts were coercive: Beginning in 1979, the Chinese government prohibited families from having more than one child. The government vowed to deny state-funded education and health care to all but the first-born child; parents who didn't comply risked losing their jobs and faced severe fines and penalties, often several times their annual salary. And earlier that same decade, the Indian government established forced "vasectomy camps"

INFOGRAPHIC 2 POPULATION DISTRIBUTION

Human population is not evenly distributed around the globe. Around 60% of the world's population live in only 10 countries; around 35% of all people live in China or India. Population growth in China is slowing, and the population size of India should surpass it around 2027. Future population size increases in the United States are expected to be due to immigration.

WORLD POPULATION BY CONTINENT (2019)

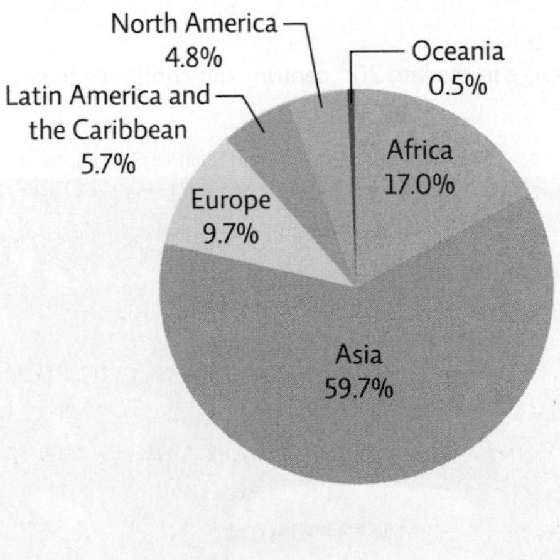

THE TEN MOST POPULOUS COUNTRIES (2019)

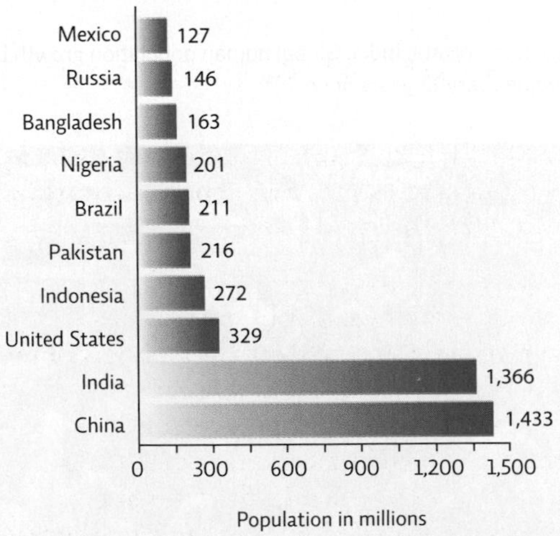

Given a total population size of 7.7 billion in 2019, what percentage of people lived in China and India that year? What might explain why these two Asian countries are so much more heavily populated than other countries with similar or larger geographic sizes, such as the United States and Russia?

REUTERS/K. K. Arora/Newscom

In many regions of Asia, especially India, Pakistan, and Bangladesh, infrastructure such as trains and roadways cannot accommodate the growing population.

across the country. They withheld food ration cards, business licenses, and even water for crop irrigation if men didn't "volunteer" for the sterilization surgery. Women too were coerced or paid (a form of coercion to those in poverty) to have sterilization surgery, and this practice continues today, sometimes with tragic consequences when hurried procedures done on women in poor health result in death.

Regardless of how well or poorly these efforts succeed in reducing population growth, most experts agree that they represent human rights violations and that it would be better to find a noncompulsory way to address human reproduction and population growth.

It turned out that in Kerala, a densely packed society with a unique culture was doing just that.

3 FACTORS THAT AFFECT POPULATION GROWTH

Key Concept 3: A variety of pronatalist pressures, such as high infant mortality, lead to higher desired family size, which correlates well with the high population growth rates seen in less developed countries. It is in these areas where most future growth will occur.

Demography is the statistical analysis of populations. Demographers evaluate metrics known as **demographic factors** in an effort to understand how human populations change over time. In simple terms, populations grow when births outnumber deaths. As we saw in Module 2.2, factors that favor births and survival are considered population *growth factors*, whereas factors that lead to death or reduced reproduction are termed *resistance factors*. And the factors that affect births and deaths aren't all that different between human populations and those of other species: disease, famine, and predators (or war in our case) are resistance factors; access to clean water, nutritious food, and protection from the elements are examples of growth factors. For humans, a decrease in resistance factors and an increase in growth factors have contributed to our phenomenal population growth over the past century or so.

demographic factors Population characteristics such as birth rate that influence changes in population size and composition.

Demographic factors tend to differ substantially between more developed countries (those with a moderate to high standard of living on average) and less developed countries (those with a weak economy and a lower standard of living than a developed country). For example, the *crude birth rate* (the number of offspring per 1,000 individuals) is, on average, more than three times as high in less developed countries (34.5 births per 1,000 people) compared to developed ones (10.8 births per 1,000 people), but in some countries, it is much higher than this average: Niger has the highest crude birth rate in the world (46.3)—almost four times higher than that of the United States (12.0). Meanwhile, the *crude death rate* (the number of deaths per 1,000 individuals per year) is less variable worldwide. In 2019, it was 7.5 per 1,000 people for the planet as a whole with an average crude death rate of 7.0 in less developed countries and 10.2 in more developed countries. (A higher death rate is found in those developed countries that have an aging population [more elderly people in their population].) Because of their high birth rates, by far, most current and future population growth is occurring, and will continue to occur, in developing nations.

But as much as demographic factors can differ between locations, they do tend to be influenced by the same forces everywhere. Health (especially of children), education (especially of females), economic conditions, culture, and religion all work together to determine a country's **desired fertility** (the number of children the average couple says they want to have). These same factors strongly affect the **total fertility rate (TFR)**—the number of children a woman actually has in her lifetime. As of 2019, the global TFR was 2.5, though like birth rates, these TFRs are highly variable from place to place.

Desired fertility and TFR tend to closely track one another, meaning that women tend to have the number of children they say they want to have. And both numbers are linked to population growth: As desired fertility increases, TFR increases; and as TFR increases, so does population size.

Forces that increase desired fertility (meaning they drive people to want more children) are called **pronatalist pressures**—things like cultural or religious views that favor large families, or the need for children to work the farm or care for elderly parents. In many regions where birth rates are high, women may not have access to contraception (called "unmet need") due to the inability to afford birth control, cultural taboos against its use, or pressure from patriarchal societies that view children as a sign of male fertility.

desired fertility The ideal number of children an individual indicates he or she would like to have.

total fertility rate (TFR) The number of children the average woman has in her lifetime.

pronatalist pressure Factor that increases the desire to have children

childhood mortality rate The number of children under 5 years of age who die per every 1,000 live births in that year.

replacement fertility The rate at which children must be born to replace the previous generation.

One of the most significant pronatalist pressures in less developed countries like India is the **childhood mortality rate**—the number of children under 5 years of age who die per every 1,000 live births in that year. (Many of these deaths occur before age 1, so demographers often speak in terms of the *infant mortality rate.*) Statistics on childhood mortality vary greatly between more and less developed nations from a high of 123 deaths per 1,000 births in Chad to a low of 2 deaths per 1,000 births in Japan and the countries of Scandinavia. (The world average is 40.)

Higher childhood mortality rates tend to lead to higher desired fertility and thus higher TFRs, because couples will have more children to account for the likelihood that some of those children will die young. By the same token, decreasing childhood mortality will decrease birth rates because people tend to have fewer children when those children can be expected to survive. The bottom line: When childhood mortality decreases, so does desired fertility. And as desired fertility falls, so does TFR. **INFOGRAPHIC 3**

In the 1970s, the consensus among development experts was that the best way to reduce a given country's TFR was to grow the economy: Raising a country's (or state's) gross domestic product (GDP), the thinking went, would foster a drive among people toward material wealth, which would in turn improve living conditions, and lead to lower childhood mortality rates and lower TFRs. A rising tide, the reasoning went, would raise all boats.

But in the beginning of that decade, Indian economist Amartya Sen and his colleagues began arguing for a different approach: Build development policies around quality of life, not economic growth. Sen (who would go on to win a Nobel Prize for his work) used Kerala as a prime example. The state had a low GDP but high literacy rates, and even its poorest citizens enjoyed wide access to child and maternal health clinics (which offered birth control and nutritional assistance). Even as the economy stagnated, this unprecedented access to health care and education (especially for women) was driving population growth down and quality of life up.

In the 1950s, the population growth rate in Kerala had been one of the highest in India. But in the 1970s, it was beginning to fall. The rate at which children must be born to replace the previous generation, **replacement fertility**, was reached around 1987, and within two

INFOGRAPHIC 3 PRONATALIST PRESSURES: FACTORS THAT INFLUENCE POPULATION GROWTH

Most population growth in the recent past has occurred in developing countries, and it is there that most future growth will occur. The pronatalist pressures that increase population growth rates can be economic, social, cultural, or related to family health. Because actual fertility closely matches desired family size, steps that decrease desired family size are key to reducing high birth rates in many developing countries.

DESIRED FERTILITY CORRELATES WITH TOTAL FERTILITY

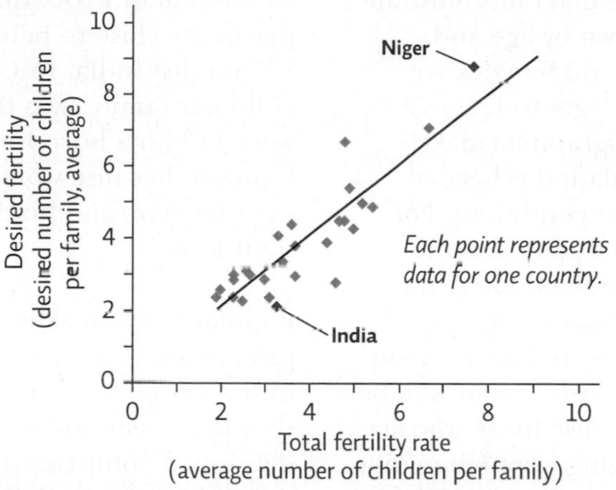

PRONATALIST PRESSURES

High infant mortality rates, which are closely tied to poverty

Lack of education or job opportunities for women

Valuing children (cultural and religious views); practical help (e.g., farms, care of elderly)

Unmet need for contraception

Why does high infant mortality increase population growth rather than decrease it?

decades the state had achieved sub-replacement fertility. (Global replacement fertility is 2.1 children per woman, rather than 2, because not all children survive and not all couples have children.) Kerala's achievements are impressive: the lowest birth and death rates, lowest childhood mortality rate, and highest life expectancy at birth (age 75.1) in all of India.

Sen might be on to something. More and more social scientists are arguing that monetary wealth or economic growth is the wrong gauge of societal progress and wellbeing. (The GDP of a nation goes up, for example, when air pollution increases health costs for the population—this may be good for the economy, but it is not good for society.) The goal of striving for a high quality of life for all citizens was formalized by Bhutan in the 1970s—they seek to increase "Gross National Happiness" with a balanced approach that integrates economic development and cultural and environmental protections. The United Nations recognized the importance of this by signing a resolution that calls for the pursuit of a more equitable approach to economic growth that considers the "happiness and wellbeing of all people"; the resolution established March 20 as International Happiness Day.

Of course, reducing TFR did not immediately solve Kerala's overpopulation problem. Two decades after reaching replacement fertility, the state is still waiting to reach zero population growth; the reasons have to do with the breakdown of the population by age and its sex ratio (the proportion of males to females).

4 AGE STRUCTURE AND POPULATION MOMENTUM

Key Concept 4: Youthful populations have a great deal of population momentum and will continue to grow even if these young people reproduce at replacement rates.

To predict a population's future growth, demographers use **age structure** diagrams. These diagrams illustrate how a given population breaks down by age and gender, showing how many males and females are found in specific age groups (e.g., ages 0–4, ages 5–9, and so on). These data help demographers make predictions about where the population is headed and make inferences about current conditions. For example, a population with mostly young people, like that of Niger, has a greater potential for growth, because many people are of childbearing age or younger — this population cohort is still, or will soon be, having children. A population with similar numbers in all age classes is fairly stable because those who are born replace those who die. An aging population, like that of Japan, may even be shrinking if more of the population is found in older, post-reproductive age classes than in younger ones.

When a population is poised to continue growing for some time, even as TFR falls — usually because it has a high proportion of younger people compared to older people — that society is said to have **population momentum**. If there are a lot of young women in the population who have not yet given birth to any children, they have the potential to contribute to a population size increase for another generation, even if these individuals reproduce at replacement rates. In other words, there will be far more births than deaths. Many countries in Africa have very young populations, and population momentum will contribute greatly to the expected high population growth in these regions.

Kerala has population momentum for two reasons: lots of younger people and lots of females. In fact, Kerala is one of the only states in all of India where women outnumber men. This feature is also a result of its unique approach to health and education: In many patriarchal societies where male children are preferred, girls are aborted or given up for adoption at much higher rates than boys. For example, a study of one Bombay (now Mumbai) clinic found that of 8,000 abortions performed in the early 1990s, only one was of a male fetus. (In one report, India's Department of Women and Child Development wrote that "to be born female comes perilously close to being born less than human.") And it's not just India. In China, where until recently "one child per family" was the strictly enforced rule, there were 113 boys born for every 100 girls. In Kerala, however, because women can receive education, earn income, and hold some power, families keep their girls.

age structure The percentage of the population that is distributed into various age groups

population momentum The tendency of a young population to continue to grow even after birth rates drop to replacement fertility.

Populations that skew older may grow slowly, not grow at all, or even shrink, since many or most of the people are past childbearing age. Once they have been industrialized for a while, more developed countries tend to transition to older age distributions. Currently, many industrialized nations — including the United States and many European countries — have top-heavy age structure diagrams with many older people. In many cases, countries struggle to provide for these rapidly aging populations. Japan, home to the world's fastest aging population, is encouraging workers to stay at their jobs well past retirement age to lessen the surge of government spending on retirees. In 2010, to relieve retirement age problems (and thus government pension payouts), France raised the retirement age from 60 to 62, sparking riots throughout the nation. In the United States, the prospect of baby boomers' retirement bankrupting Social Security has spurred intense and vitriolic debate. **INFOGRAPHIC 4**

The workforce, too, is being affected in nations with an aging population. In the United States, for example, worries about retirement funding are keeping many older citizens in the workforce, making it harder for younger workers to get started in their careers. At the opposite end of the spectrum, in some areas, there are not enough young people to take over jobs left behind by retirees. China's one-child policy has led economists to predict that the annual size of the labor force aged 20–24 would shrink substantially by 2020 — perhaps by as much as 50%. This shortfall prompted the Chinese government to end the one-child policy in 2015; families are now allowed to have two children.

Most population growth is occurring and is expected to continue to occur in less developed countries, with African countries leading the way. China will likely peak

INFOGRAPHIC 4 AGE STRUCTURE AFFECTS FUTURE POPULATION GROWTH

The fastest-growing regions are those with a youthful or very young population.

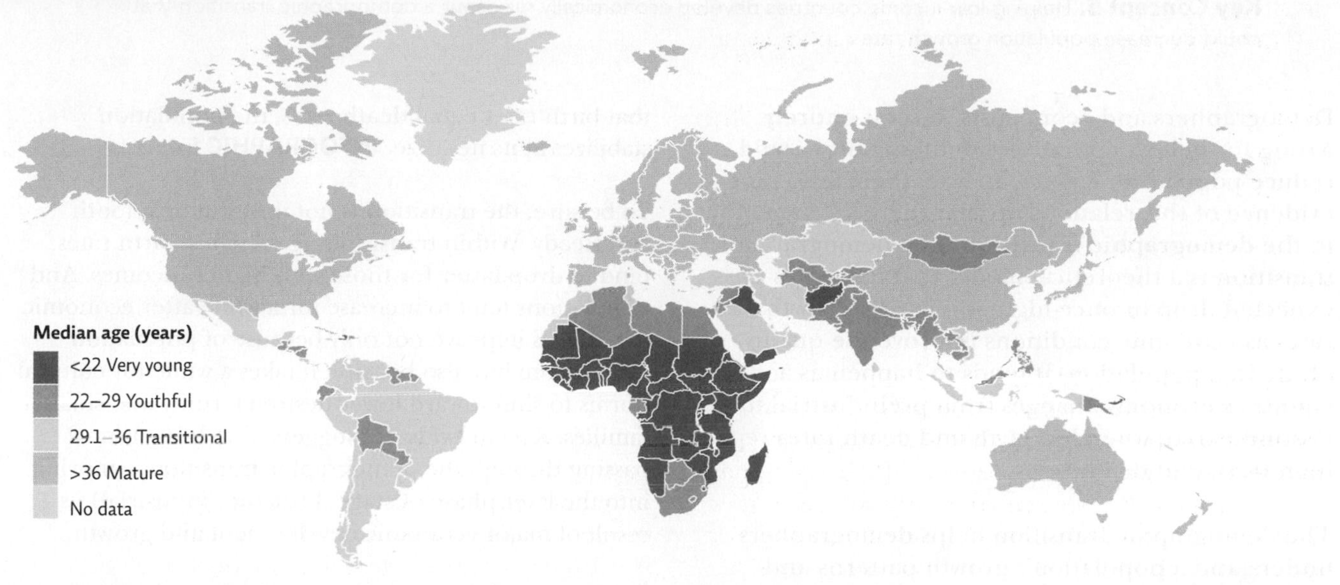

Age structure diagrams show the distribution of males and females of a population in various age classes. The width of a bar shows the percentage of the total population that is in each gender and age class. The more young people in a population, the more population momentum it has; it will continue to grow for some time. The more people of reproductive age, the higher the growth rate, but this measure is also influenced by income; growth rates decline as income increases.

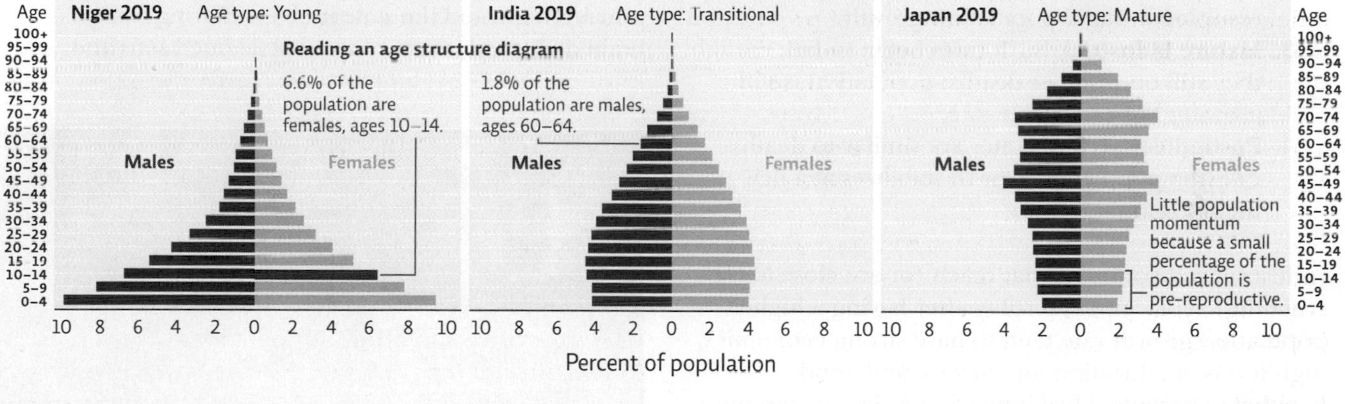

In Niger's young population, 76% of the people are under 30, and there is a very high capacity for growth (high population momentum).

India's transitional population is growing more slowly than Niger's because India's pre-reproductive age cohorts make up a smaller part of the population.

Japan's mature population has a fairly even distribution among age classes, with two slight bulges seen around 40 and 65. This population is fairly stable or may even be decreasing slowly as deaths start to outnumber births.

What percentage of the total population (males and females) is between 0 and 4 years of age in each of the three countries shown here (Niger, India, and Japan)?

close to 1.46 billion around 2030 (it was 1.43 billion in 2019) and is expected to shrink slightly after that due to an aging population. Demographers predict India will peak around 2060 at 1.65 billion.

Despite reaching replacement fertility in the 1990s, Kerala's population is still growing at a natural rate of 6.7 per 1,000 people, and experts say it will take several years more to reach zero population growth.

5 ADDRESSING POPULATION GROWTH: THE DEMOGRAPHIC TRANSITION

Key Concept 5: Helping low-income countries develop economically may spur a demographic transition that could decrease population growth rates.

Demographers and economists weren't entirely wrong for believing that economic stability would reduce population growth. In fact, there is so much evidence of this relationship that there's a name for it: the **demographic transition**. The demographic transition is a theoretical model that describes the expected drop in once-high population growth rates as economic conditions improve the quality of life in a population. It tends to happen as a country's economy changes from preindustrial to postindustrial, when low birth and death rates replace high birth and death rates.

The demographic transition helps demographers understand a population's growth patterns and predict future ones. It has four stages:

1. **Preindustrial**: Birth and death rates are high but similar, so population growth is slow or stationary; population size is low.
2. **Industrializing**: Better conditions lead to lower death rates; however, birth rates remain high, resulting in rapid population growth.
3. **Mature industrial**: Birth rates begin to fall, though they still outnumber deaths; population is still growing but at a slower rate.
4. **Postindustrial**: Birth rates are similar to death rates, so population growth stabilizes at a new higher population size.

Industrialized countries that reach (or are close to reaching) replacement fertility after having a higher population growth rate tend to have strong economies, high levels of education for citizens, and good healthcare systems. The United States, Japan, and most of the European Union are examples of countries that are postindustrial. The reasons that economic stability begets population stability are believed to be tied to better health: Economic stability improves access to health care and prenatal care, so that death rates (including childhood mortality) decline. As it becomes clear that more children will make it to adulthood, desired fertility declines, and TFR along with it. The lag between the drop in death rates and the subsequent drop in birth rates results in a higher overall population, but once birth rates decline, such that birth rates equal death rates, the population stabilizes at its new size. **INFOGRAPHIC 5**

demographic transition A theoretical model that describes the expected drop in once-high population growth rates as economic conditions improve the quality of life in a population.

To be sure, the transition is not uniform or smooth and steady. Within transitioning societies, birth rates tend to drop faster for those with higher incomes. And populations tend to increase for a while after economic conditions improve not only because of population momentum but also because it takes a while for cultural norms to shift toward lower desired fertility and smaller families. Recent evidence suggests that India is now passing through the demographic transition—moving into the later phase of stage 3 (mature industrial) as a result of major economic development and growth.

Kerala made it through the demographic transition in the early 1990s—well ahead of the rest of India and without first ushering in an era of economic growth. This was an amazing feat, to be sure, but it came with some serious side effects. Because the economy remains stagnant, university graduates have a hard time finding work that befits their educational status. As a result, unemployment remains high, and the state itself is suffering from brain drain, with many of its most talented sons and

AP Photo/Rafiq Maqbool

Education in India is producing many new graduates, but the job market in some areas lags behind, leaving many qualified individuals competing for scarce jobs like these hopeful candidates at a job fair in Hyderabad, India.

INFOGRAPHIC 5 DEMOGRAPHIC TRANSITION

Some industrialized nations have gone through the demographic transition from high birth and death rates to low birth and death rates, giving them a stable population. Demographers are considering adding a fifth stage, *Declining,* for those populations experiencing more deaths than births. It is unclear whether industrialization will produce the same pattern in all developing countries. Still, steps to improve the quality of life and decrease death rates are important worldwide, even if other measures are needed to slow birth rates.

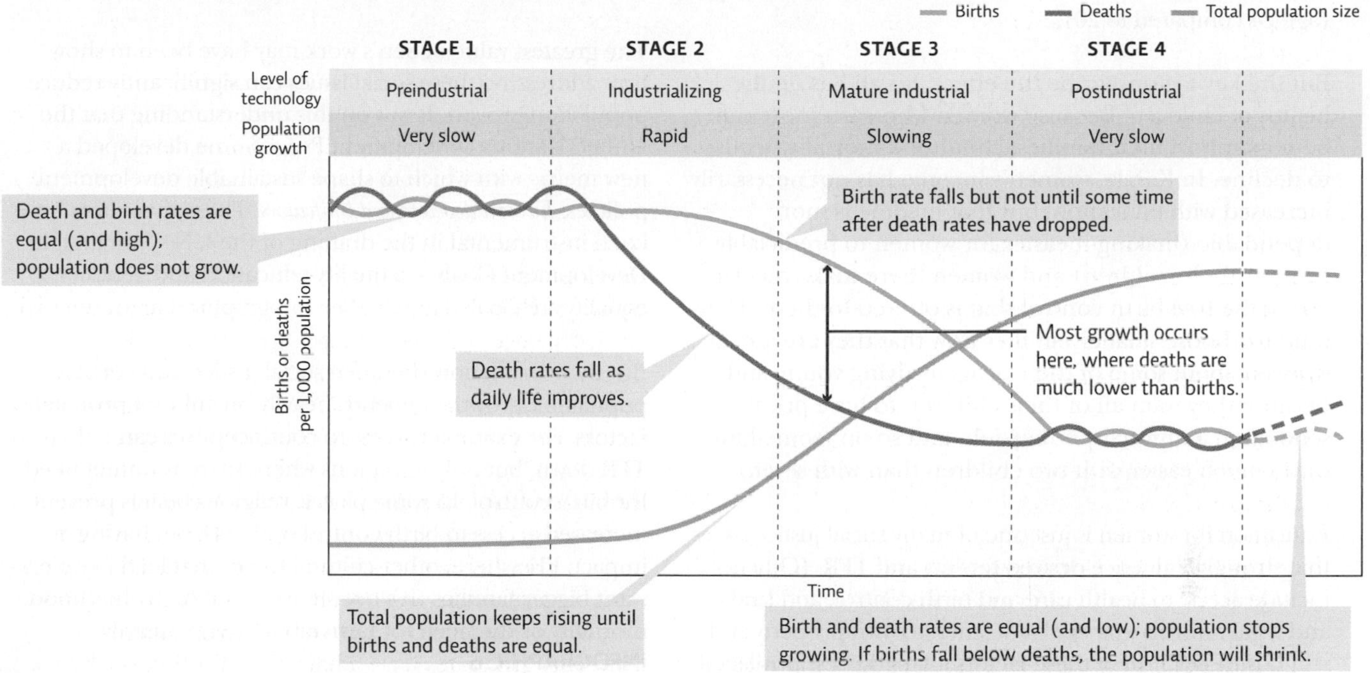

Why might industrialization *not* lead to a demographic transition in all countries?

daughters seeking employment elsewhere. Meanwhile, the state's commitment to health and education for all has left it with significant budget deficits. Indeed, even as development experts around the world look to Kerala as a model of how to do more with less, some wonder if the state's victory over population growth will prove sustainable. "Can the Kerala Model survive even in Kerala," asks noted environmentalist Bill McKibben, "or will it be remembered chiefly as an isolated and short-term outbreak from a prison of poverty?"

6 ADDRESSING POPULATION GROWTH: SOCIAL JUSTICE

Key Concept 6: In many cases, addressing social justice issues, especially improving the wellbeing of women and children, is the key to reducing total fertility and reaching zero population growth.

When analyzing a population's dynamics, demographers look to see if a relationship, or correlation, exists between various demographic factors (e.g., between income and TFR). Some correlations, like that of TFR and desired fertility, are positive—both factors track together (as one increases or decreases, so does the other). TFR and the use of contraception show a negative correlation—as one increases, the other decreases. (See Section 4 of Module 1.2 for more on correlations.)

Demographers have put forth many explanations for Kerala's unprecedented decline in population growth rate in the absence of substantial economic growth, but one factor stands out among the others: Kerala also had India's highest overall literacy rate (94%) and the highest literacy rate for women, by far. In fact, of all the factors that Sen and his colleagues cited for Kerala's high quality of life and decreasing population rates (land reform, educational access, child welfare, health care), education ranked as the most significant. And not just education: education of women in particular.

It turns out that education of girls and economic opportunities for women correlate with lower population growth in many regions, and it's not difficult to see why: Education empowers women to take more control over their lives and their fertility—by demanding and using birth control, for example, or by marrying later or delaying childbirth while pursuing a career. The typical Keralan woman marries somewhat later than her Indian counterpart (age 24 compared to 20).

But the key factor may be the effect this all has on the health of children. Because women who earn more can better support their families, childhood mortality tends to decline. In Kerala, women's income has not necessarily increased with education, but that income is more dependable (making it easier for women to predictably care for their children) and women there make regular use of the free birth control that is offered to them. They tend to choose smaller families now that they are less worried about some of their progeny dying young and because they want all of their children to have private schooling, a university education, and so on (something that's much easier with two children than with seven).

Education for women is just one of many social justice issues that strongly influence desired fertility and TFR. (Others include access to health care and birth control, and land and legal reforms that enable women to own property and run businesses.) All of these factors affect the less privileged members of society the most. In the United States, the high infant mortality rate (deaths to children under one year of age) of 6 deaths per 1,000 births—relative to other developed countries (between 1 and 4 in most developed countries)—manifests itself mainly in those living in poverty: Babies born into poverty are much more likely to die than those born to U.S. families that can afford decent health care. (Most of the countries with lower infant mortality rates have some form of national healthcare system.)

The greatest value of Sen's work may have been to show how addressing these social issues can significantly reduce population growth. It was on this understanding that the United Nations Development Programme developed a new metric with which to shape sustainable development policies: *human development indicators*. These indicators have been instrumental in the drafting of the UN's Sustainable Development Goals—a quality education and gender equality are Goals 4 and 5. (See Infographic 4 in Module 1.1.)

But that correlation (between social justice and curbing population growth) depends heavily on cultural pronatalist factors. For example, access to contraceptives can help drive TFR down, but only in regions where there is unmet need for birth control. In some places, religious beliefs prevent increased access to birth control options from having an impact. Elsewhere, other cultural forces that lead people to want bigger families (it's not always about high childhood mortality or the need for farm labor) weigh heavily. **INFOGRAPHIC 6**

Kerala's culture gave it three distinct advantages in spreading education and health care. First, the state was

Programs that provide microloans, such as this one offered by Bharat Financial Inclusion Ltd. (BFIL), allow women to start their own businesses and rise out of poverty. BFIL works with groups of five women who support each other and hold joint liability for the loans. Training and support from BFIL are strong, and loan repayment is greater than 99%.

Adeel Halim/Bloomberg/Getty Images

INFOGRAPHIC 6 ACHIEVING ZERO POPULATION GROWTH IS LINKED TO SOCIAL JUSTICE

Achieve

Addressing pronatalist pressures can help reduce population growth rates. This often involves bringing social justice to women and families so they can better care for themselves and their children. The graphs below show the correlation between total fertility rate (the average number of children per female) and three important factors: childhood mortality, education for women, and access to birth control. Each point on a graph represents the data for one country.

CHILDHOOD MORTALITY AND TOTAL FERTILITY RATE

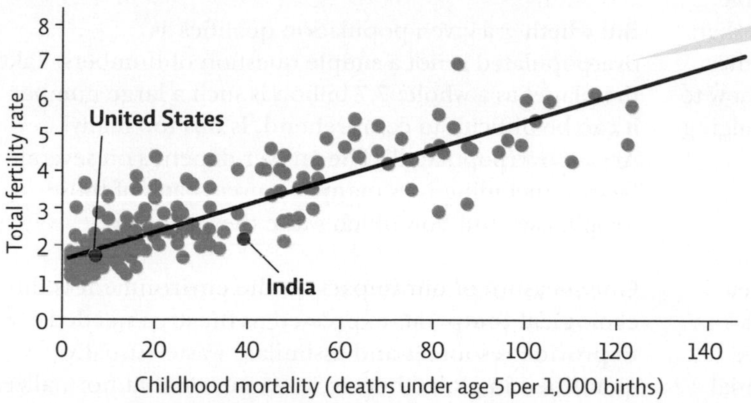

Infant mortality is closely correlated to poverty; therefore, actions that decrease poverty in a population should also reduce infant mortality and that population's growth rate.

EDUCATION OF WOMEN AND TOTAL FERTILITY RATE

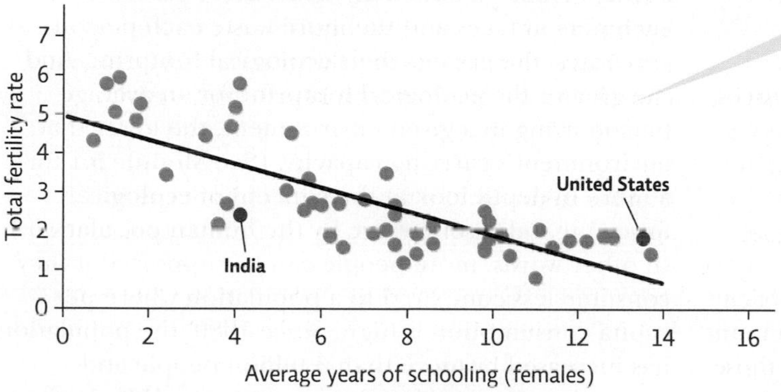

Fertility declines as educational opportunities for girls and women increase. This means that funding for education and job opportunities for women can be very effective at lowering TFR.

CONTRACEPTIVE USE AND TOTAL FERTILITY RATE

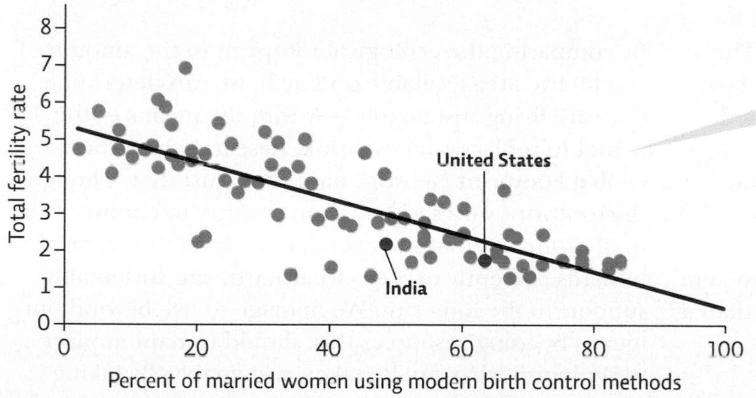

Access to contraceptives and family planning have been effective in many areas of the world. However, in countries with high desired fertility, such as Niger, providing contraceptives may have little impact on TFR.

? **Which of the graphs shown here depict positive correlations? Which are negative correlations? Are positive correlations "good" and negative correlations "bad"? Explain.**

always matrilineal. That is, power, land, and money passed down through daughters rather than sons, and women (especially those of higher castes) were routinely educated. So when progressive leaders began extending that right to poorer communities, the culture at large was primed for it. Second, there is no real divide between urban and rural areas in Kerala. So unlike other states in India (and across the developing world), schools and health clinics didn't get concentrated in cities, but instead were accessible to both groups. And third, the state boasted an above-average concentration of high-quality missionary schools that catered to people of all faiths.

7 HOW MANY PEOPLE CAN EARTH SUPPORT?

Key Concept 7: The number of people Earth can support depends on how many resources each uses. We may have already exceeded our planet's carrying capacity by using more resources than it can replace in the long term.

In the decades since Kerala achieved near 100% literacy and sub-replacement fertility, the state has been heralded far and wide as a model of sustainable development. Demographers and development workers the world over have been studying and discussing the Kerala Model for what it might teach them about how to achieve similar successes in other countries with bulging populations and small economies.

But some experts say that Kerala's most valuable lessons may actually be for wealthy countries, not poor ones. "Kerala suggests a way out of two problems simultaneously," McKibben writes. "Not only the classic development goal of more food in bellies and more shoes on feet, but also the emerging, equally essential task of living lightly on the Earth, using fewer resources, creating less waste."

In other words, not only has Kerala succeeded at reducing the number of people who consume resources, it did so without increasing the consumption of resources by each member of its society. In so doing, it did not exceed its carrying capacity.

Carrying capacity is the maximum population size that a given environment can support. The size of that capacity is determined by a range of forces, including the amount of crucial resources available and the rate at which those resources can be replenished (if they are renewable), as well as the rate of resource consumption by the population. (See Module 2.2 for more on carrying capacity.) The amount of waste generated plays a role, too; if waste is generated faster than the ecosystem can assimilate it, the ecosystem will suffer and the rate of resource replenishment will decrease. In other words, a degraded ecosystem can sustain far fewer people than a well-maintained healthy one.

When a population's size exceeds the carrying capacity of its environment, that environment is said to be **overpopulated**. Some parts of the world are certainly overpopulated today; they have exceeded the carrying capacity for their region, and as a result they face food and water scarcity issues as well as problems managing their waste.

But whether a given population qualifies as overpopulated is not a simple question of numbers. Take the planet as a whole: 7.7 billion is such a large number it can be difficult to comprehend. Is this too many? Are we overpopulated? The answer depends on several factors, including how many resources each of those people uses and how much waste they produce.

One measure of our impact on the environment is the **ecological footprint**, expressed as the area needed to provide resources and assimilate waste (usually quantified in global hectares—a metric that normalizes different types of land and aquatic areas in terms of average global productivity). The more resources each person takes and the more waste each person generates, the greater their ecological footprint. And the greater the ecological footprint for an average person living in a given environment, the lower that environment's carrying capacity. (See Module 5.1 for a more in-depth look at the concept of ecological footprint and resource use by the human population.) In other words, more people can be supported if they consume less compared to a population where per capita consumption is high. Since 1980, the population has increased by more than 3 billion people and per capita consumption has increased by 15%; both contribute to our rising footprint.

By comparing the ecological footprint to the amount of productive area available on Earth, we can determine if we are living sustainably—within the means of the planet to replace what we took. Researchers at the Global Footprint Network have done just that. Though the footprint varies widely from country to country, they have estimated that the overall ecological footprint of humans currently exceeds what Earth can sustainably support in the long run. We manage to live beyond our means by using resources that should remain in place to replenish what can be taken sustainably. By taking more now than Earth can replace at its own ecological pace, we are reducing Earth's carrying capacity for the future. To give an example, if you cut down all the trees in your forest faster than they regrow, before long you no longer have a forest. Can you wait for it to recover? Have you left it in a condition that allows its recovery? **INFOGRAPHIC 7**

carrying capacity The maximum population size that a particular environment can support indefinitely; for human populations, it depends on resource availability and the rate of per capita resource use by the population.

overpopulated The number of individuals in an area exceeds the carrying capacity of that area.

ecological footprint The land area needed to provide the resources for, and assimilate the waste of, a person or population.

INFOGRAPHIC 7 HUMAN POPULATION AND EARTH'S CARRYING CAPACITY

A comparison of humanity's ecological footprint (what we take) to Earth's productive capacity (what is available) can reveal whether our population has exceeded the long-term carrying capacity of Earth. Current assessments suggest we currently use resources faster than Earth's systems can replace — a 2018 estimate puts humanity's footprint at 1.7 Earths — and our footprint is increasing. This overshoot is caused by both our sheer numbers and the rise in per capita resource use, especially in developed countries.

GLOBAL ECOLOGICAL FOOTPRINT

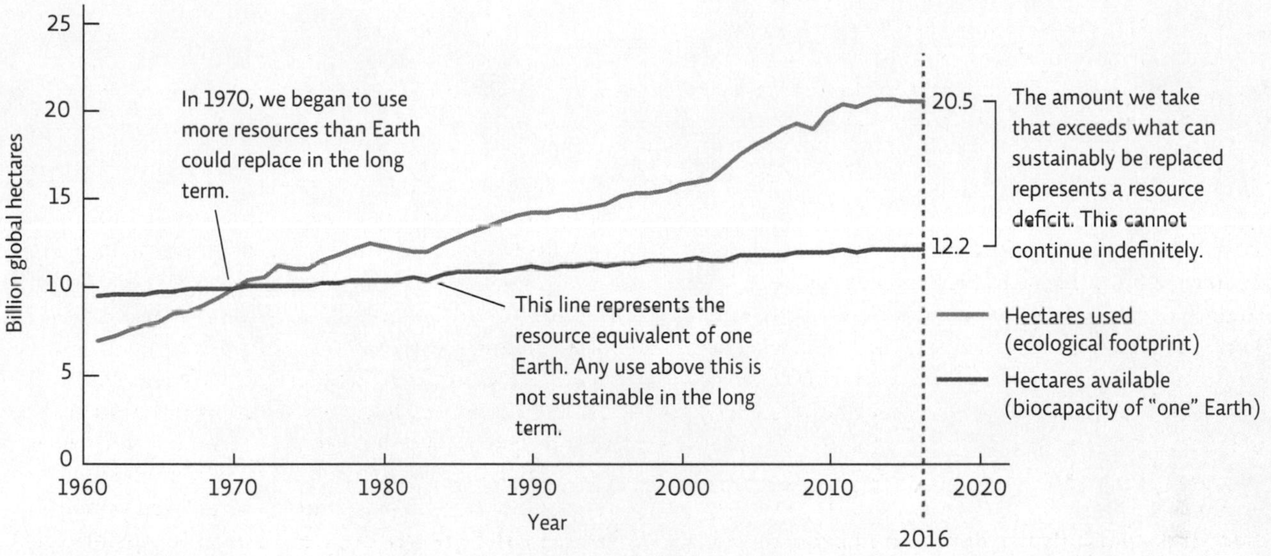

25
20
15
10
5
0
Billion global hectares
1960
1970
1980
1990
2000
2010
2020
Year
2016
In 1970, we began to use more resources than Earth could replace in the long term.
This line represents the resource equivalent of one Earth. Any use above this is not sustainable in the long term.
20.5
12.2
The amount we take that exceeds what can sustainably be replaced represents a resource deficit. This cannot continue indefinitely.
Hectares used (ecological footprint)
Hectares available (biocapacity of "one" Earth)

PER CAPITA FOOTPRINT COMPARISON

The graph above suggests the human population as a whole has overshot carrying capacity, but there is great disparity in the resources used and waste produced among the countries of the world. The per capita footprint of developing countries is much lower, in general, than that of developed ones.

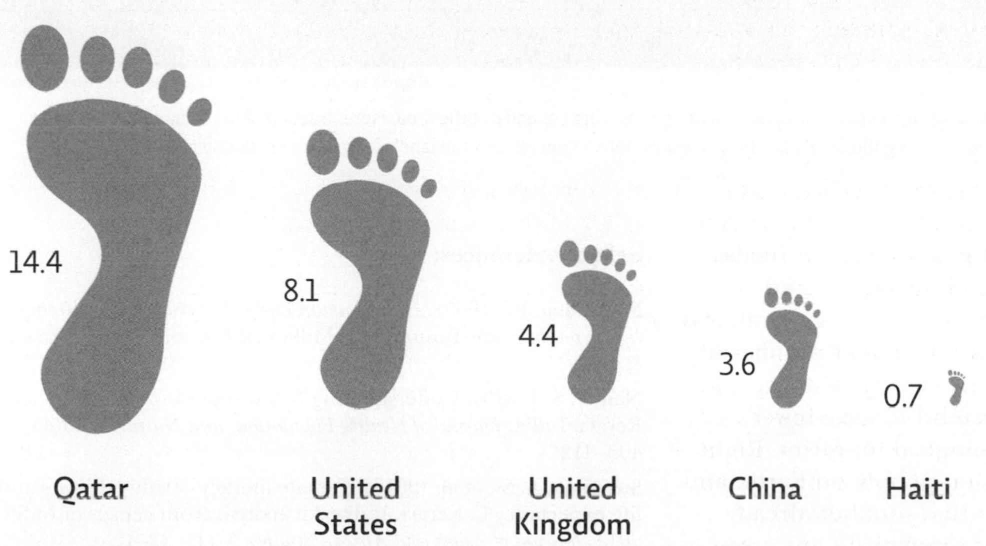

14.4
8.1
4.4
3.6
0.7
Qatar
United States
United Kingdom
China
Haiti

In 2016, Qatar had the largest per capita footprint, but as a small country, it contributes little to the overall impact. The large U.S. footprint has a greater impact due to its large population size, but China's total footprint (around 5 billion global hectares [gha]) exceeds that of the United States, which is estimated to be around 2.7 billion gha. (The world footprint totals around 20.5 billion gha.)

Values shown are global hectares per person (2013).

? Using the data values shown for 2016 on the graph, calculate how much we exceeded Earth's capacity to support us and use this value to calculate how may "Earths" we were using in that year.

Ricardo Funari/Brazil Photos/LightRocket via Getty Images

When humans overharvest resources or damage ecosystems, we reduce the carrying capacity of the area. Here, tracts of Amazonian rainforest are systematically destroyed for agriculture, reducing the area's ability to support native species and the land's long-term productivity.

Carrying capacity is itself a moving target. It tends to increase with technological innovations and decrease with environmental damage. In general, less developed regions tend to have a greater number of people but a smaller per capita ecological footprint, and more developed regions tend to have fewer people, each with a large ecological footprint. Right now, there are close to 8 billion people on Earth, and at least some experts say that that number already exceeds the planet's carrying capacity. In any case, whether we stabilize at 10 or 16 billion, or somewhere in between, will depend on how quickly we achieve replacement fertility worldwide.

Select References:

McKibben, B. (1995). *Hope, human and wild: True stories of living lightly on the Earth.* Boston, MA: Milkweed Editions, Little, Brown & Co.

Nair, P. S. (2010). Understanding below-replacement fertility in Kerala, India. *Journal of Health, Population, and Nutrition, 28*(4), 405–412.

Susuman, A. S., et al. (2016). Female literacy, fertility decline and life expectancy in Kerala, India: An analysis from census of India 2011. *Journal of Asian and African Studies, 51*(1), 32–42.

United Nations, Department of Economic and Social Affairs, Population Division. (2019). *World population prospects 2019: Highlights* (ST/ESA/SER.A/423). New York, NY: Author.

GLOBAL CASE STUDIES POPULATION GROWTH OR CONTROL IN OTHER REGIONS

Achieve

There are many reasons why populations grow and many ways to curb that growth. See the online case studies for five examples of population growth or control in different regions of the world.

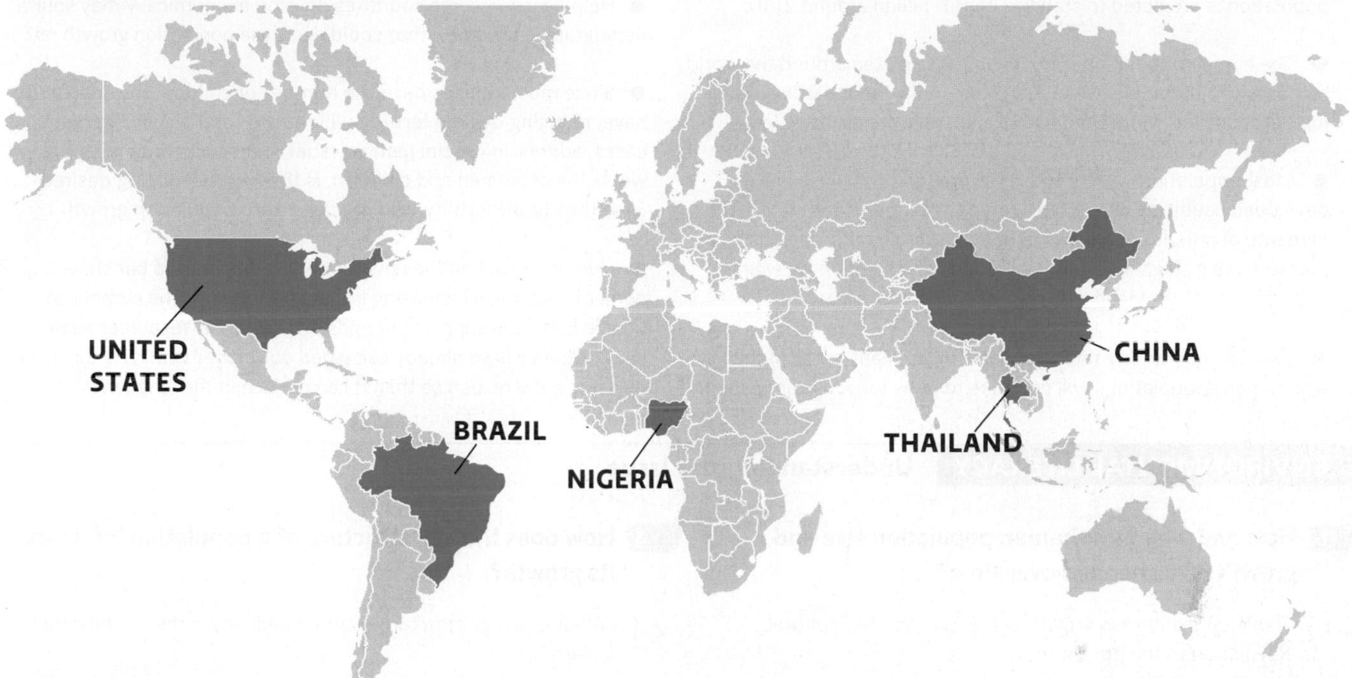

BRING IT HOME

PERSONAL CHOICES THAT HELP

The impact of humans on the planet is created by a combination of population size and per capita resource use. The issue of population and carrying capacity is complex. We cannot have a truly sustainable society until key components such as poverty, lack of education, and basic human rights are addressed.

Individual Steps

- Buy Fair Trade Certified–labeled products. These products provide a livable wage to workers and are often linked to education and community development.
- Research the products you buy to make sure that you are not supporting child slave labor, sweatshop facilities, or environmentally destructive actions.
- Volunteer to teach an adult to read in a literacy program or offer your services as a tutor to children in your community.

Group Action

- Raise money and invest it in a socially responsible project. The Foundation for International Community Assistance (www.finca.org) and Kiva (www.kiva.org) are nonprofit organizations that provide microloans to help people start small businesses in less developed countries.
- Join an organization such as Habitat for Humanity, which builds housing for low-income families.
- We all know that our tendency as humans is to build up and out; however, revitalizing our current older downtown areas is important to prevent new habitat destruction as well as to make use of current infrastructure. Urge your local government to keep shopping and dining establishments in historic downtowns. If needed, work to develop a community partnership that starts clean-up programs and community gardens in abandoned lots.

REVISIT THE CONCEPTS

- Human population is increasing rapidly and will soon top 8 billion. This rapid growth is recent — our population grew very slowly for most of human history. Population growth is slowing, and world population is projected to stabilize near 11 billion around 2100.

- The human population is not evenly distributed around the world. Almost 40% of people live in China and India; around 60% live in just 10 countries. More than half of all people are city dwellers.

- Most population growth today occurs (and will occur) in less developed countries where birth rates still exceed death rates. This high rate of population growth is due to a variety of pronatalist pressures (e.g., high infant mortality, lack of education for women, cultural norms) that lead to higher desired family size.

- Even though fertility rates are declining in many areas of the world, many populations will continue to grow for some time due to their predominantly youthful populations (population momentum). For a time, births will far outnumber deaths in these populations.

- Helping low-income countries develop economically may spur a demographic transition that could decrease population growth rates.

- Since most families have the number of children they desire to have, reducing desired fertility will reduce total fertility. In many cases, addressing social justice issues, especially improving the wellbeing of women and children, is the key to reducing desired (and thus total) fertility, and reaching zero population growth.

- Human impact on the environment is due both to our sheer numbers and to an increasing impact per person. The number of people Earth can support depends on how many resources each uses. We may have already exceeded our planet's carrying capacity by using more resources than it can replace in the long term.

ENVIRONMENTAL LITERACY Understanding the Issue

1 How and why have human population size and growth rate changed over time?

1. Describe the human growth curve from the Agricultural Revolution to the present day.
2. What factors led to the increase in population growth rate during and after the Industrial Revolution?

2 How big is the human population today, and where do most people live?

3. Describe the distribution of people around the world.
4. What are immigration and emigration? Do they impact worldwide population or just local population sizes?
5. Where does the United States rank globally in terms of population size? What will contribute to future growth in this country?

3 What factors affect human population growth?

6. Where is most future population growth expected to occur?
7. Describe the link between desired fertility and the total fertility rate of a population, and explain how they affect population growth.
8. Identify several factors that contribute to the high growth rate in developing nations.
9. How does infant or child mortality correlate to population growth rate, and why is this relationship generally seen?

4 How does the age structure of a population influence its growth?

10. What is an age structure diagram and why is this useful data to collect?
11. Why will a youthful population still be growing in size even after it has reached replacement fertility?

5 What is the demographic transition, and why is it important?

12. Why do birth rates tend to fall some time after death rates have dropped in the Industrializing stage of the demographic transition?

6 How can addressing social justice issues help achieve zero population growth?

13. How do job opportunities and education for women correlate to total fertility and population growth rates?
14. Why must culture and religious traditions be taken into consideration when designing programs to reduce population growth?

7 What determines Earth's carrying capacity for humans, and can the planet support the current or future (projected) human population long term?

15. What is carrying capacity, and how does it relate to the per capita ecological footprint of an area?
16. How could it be possible that we are already living beyond the long-term carrying capacity of Earth?

SCIENCE LITERACY Working with Data

The United Nations (UN) Population Division predicts future population growth based on scenarios of different rates of fertility (fertility variants). The constant fertility projection shows population growth at 2015–2020 growth rates; the medium variant is based on the median of thousands of modeled trajectories, and the high and low variants represent a fertility rate ±0.5 births from the medium-variant TFR.

A: ESTIMATED AND PROJECTED WORLD POPULATION ACCORDING TO DIFFERENT SCENARIOS

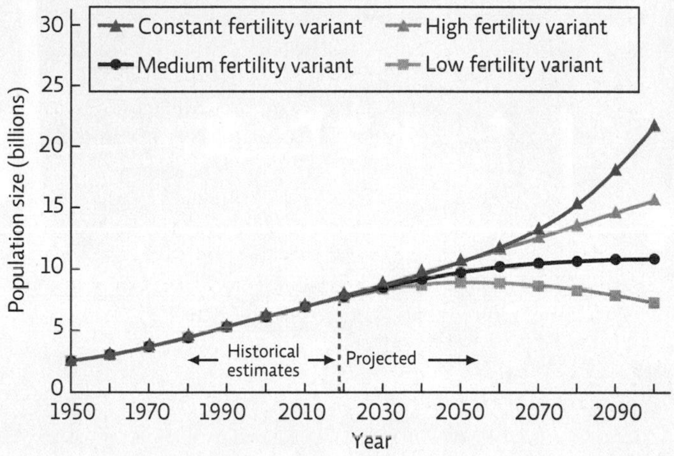

B: PROJECTED POPULATION CHANGE BETWEEN 2020 AND 2100 (MEDIUM VARIANT)

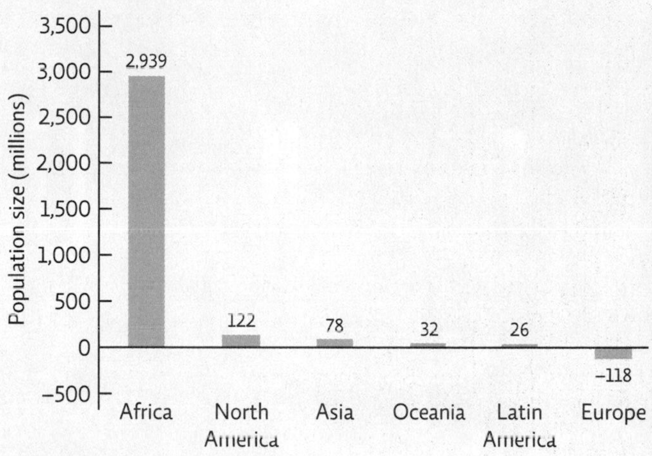

Interpretation

1. What are the population projections for each fertility variant in 2050? In 2100?
2. Describe the trend in population growth for each line seen in Graph A.
3. How does the information in Graph A relate to that in Graph B?

Advance Your Thinking

4. Why does the UN produce population projections based on different fertility rates?
5. Why is Africa projected to have such large growth if the medium-variant population projections assume that global fertility rates will be at replacement? Use information from the module to support your answer.

INFORMATION LITERACY Evaluating Information

Many people consider providing education and job opportunities for women to be important not only for their effects on fertility but also as a matter of social justice. Two organizations that work on this are the Foundation for International Community Assistance (FINCA) and Kiva, which provide microfinance services to low-income people, mostly women. Explore the FINCA website (www.finca.org) and the Kiva website (www.kiva.org).

Evaluate the websites and work with the information to answer the following questions for each organization:

1. Evaluate each organization to determine if it is a trustworthy organization with a clear and transparent agenda:
   a. Who runs this organization? Do their credentials make the work of the organization and the information presented reliable or unreliable? Explain.
   b. What are the mission and vision of this organization? What are its underlying values? How do you know this?
2. Explore the "About" links.
   a. How well does each explain the principle of microfinance? Give your own definition of *microfinance.*
   b. Does the website provide supporting evidence that its programs can help the poor? Is the evidence reliable? Explain.
   c. Do you think the business model of FINCA or Kiva is valid and effective? Explain.
3. How might the FINCA or Kiva model influence cultural, economic, and demographic factors that influence population growth?
4. Explore each website to see how interested people can participate.
   a. Who can be a part of FINCA's or Kiva's solution? How can they get involved?
   b. If you decided to contribute to a microloan program, which of these organizations would you choose? Why?

Additional study questions are available at Achieve.macmillanlearning.com. Achieve

CREATING GREEN CITIES

Building a better backyard in the Bronx

Hunts Point Riverside Park in the Bronx was built on abandoned riverside property and provides much-needed green space to this urban area.

Majora Carter

After reading this module, you should be able to answer these GUIDING QUESTIONS

1. What is the pattern of global urbanization in recent decades?
2. What are the trade-offs associated with urban areas?
3. What is environmental justice and what common problems does it address?
4. What environmental problems does suburban sprawl generate?
5. What is the value of urban green space?
6. How does smart growth help make a city "green"?
7. What are some of the design features of a green building?

As she walked her dog Xena, Majora Carter considered her options. She had moved back home to save money while she attended graduate school. Initially, she had wanted as little to do with the decaying neighborhood as possible. But then she'd taken work at a community development center, and now a colleague at the city parks department was offering her a $10,000 grant to come up with a waterfront development project for her neighborhood. Carter was balking.

At the moment, she and her neighbors were fighting the city's attempt to move a new waste facility to the East River waterfront. With 30 transfer stations in the South Bronx, their tiny parcel of New York City already handled 40% of the entire city's commercial waste, not to mention four power plants, two sludge-processing plants, and the largest food distribution center in the world. All told, some 60,000 diesel trucks passed through the neighborhood every week. In exchange for this burden, area residents boasted the highest asthma and obesity rates in the country, along with some of the poorest air quality. Another waste facility would only make matters worse. Consumed with this battle, Carter wasn't sure she had the time or energy to take on another project.

Besides, the idea of developing waterfront property in the South Bronx seemed a bit naïve to her. Like most of her neighbors, Carter had lived in the neighborhood most of her life; she knew full well how inaccessible the surrounding river was to residents. The waterfront—all of it—had long been claimed by industry. There was simply nothing left to develop, she thought, as she and Xena made their way along their usual route—past the transfer station, roaring with diesel-powered, garbage-filled 18-wheelers, along a winding string of garages filled with auto glass shops, metal work, and produce shipments.

And then Xena began pulling her toward an abandoned lot, one they had passed a million times without bothering to notice. After a futile effort to resist, Carter allowed herself to be led down a garbage-strewn path, through a ramble of towering weeds—the kind of place one would never venture alone at night. There at the end, sparkling in the early morning light, was the East River. Carter stood in awe. How many other forgotten patches of waterfront were there, she wondered? Maybe the river wasn't so inaccessible after all.

WHERE IS THE BRONX, NEW YORK?

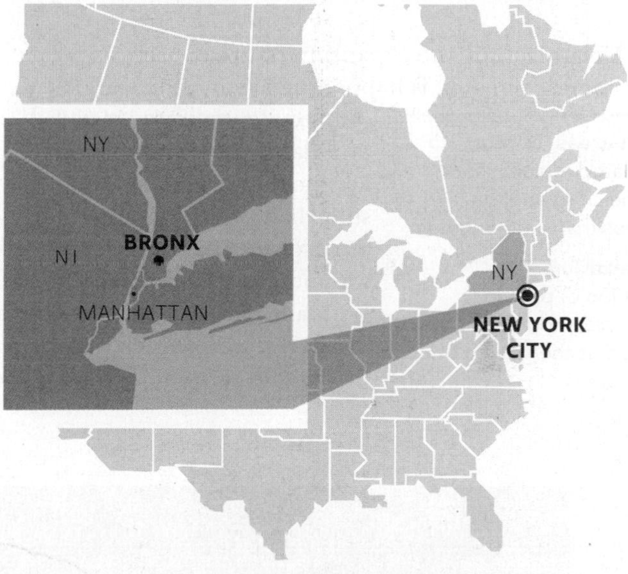

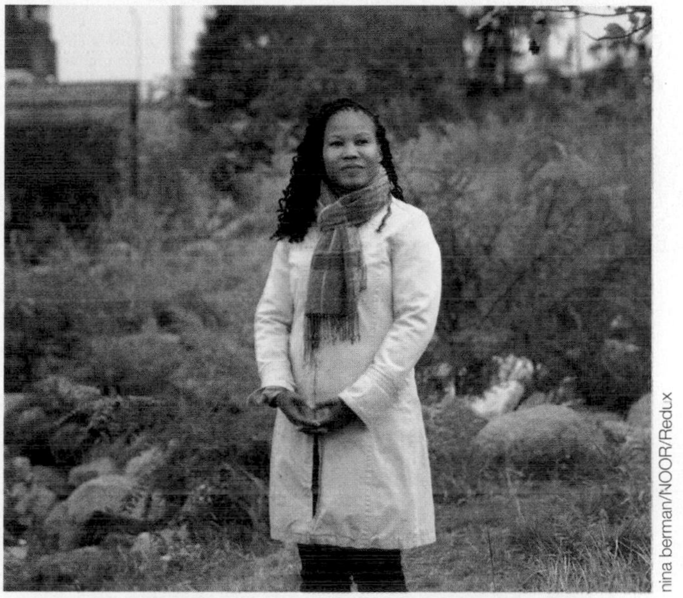

Majora Carter received a MacArthur "Genius Grant" for her work in revitalizing the South Bronx, improving the environment of an underserved population.

1 GLOBAL URBANIZATION PATTERNS

Key Concept 1: Today, more people live in urban than in rural areas. The number of megacities has greatly increased since 1950 and will continue to do so in the future.

Ever since 2008, and for the first time in human history, more than half the world's population (55% in 2018) lives in **urban areas**—densely populated regions that include both cities and the suburbs that invariably surround them. In the United States the proportion is even higher: More than 80% of Americans are urban dwellers.

Urbanization, the expansion of urban areas as people migrate to cities, is happening around the world at an unprecedented rate. As global population swells, rural lands are morphing into urban and suburban ones, and ordinary cities are growing into *megacities*—those with at least 10 million residents. With almost 19 million inhabitants in 2018, the New York City metropolitan area qualifies as the largest city in the United States and one of the world's 33 megacities.

urban areas Densely populated regions that include cities and the suburbs that surround them.

urbanization The migration of people to large cities; sometimes also defined as the growth of urban areas.

In the coming decades, the world's share of urban dwellers will continue to grow. In 2018 about 23% of the world population, or around 1.7 billion people, lived in a city of at least 1 million. Human population continues to grow at a rapid rate (see Module 4.1), and the United Nations predicts that the next billion people added to the planet (by 2030) will live in these big cities—the rural population is expected to slowly but steadily decrease over time. Cities everywhere will continue to grow, including megacities. By 2035, it is predicted that there will be 43 megacities—compared to only five in 1980 and 33 in 2019—and more of them will be amazingly large: 12 will have over 20 million residents. By 2035, New York City, USA, Mexico City, Mexico, and São Paulo, Brazil, will likely be the only 20 million+ cities that are not in Asia. **INFOGRAPHIC 1**

Bloomberg/Getty Images

Megacities, those with at least 10 million inhabitants, are increasing in number. The largest city in the world, Tokyo, Japan, is home to more than 37 million people—about twice as many as in the largest U.S. city (New York City).

INFOGRAPHIC 1 URBANIZATION AND THE GROWTH OF MEGACITIES

The world's population is becoming more urban. The growth of megacities, those with at least 10 million people, has increased dramatically over the last half century. In 1950, there were only two megacities: New York City with 12.3 million and Tokyo, Japan, with 11.3 million. In 2018, there were 33 megacities, with Tokyo's population of more than 37 million making it the largest city in the world. This map shows of the number of megacities in 2018 and the number that are expected to be megacities in 2030.

MEGACITIES OF THE WORLD IN 2018 AND 2030

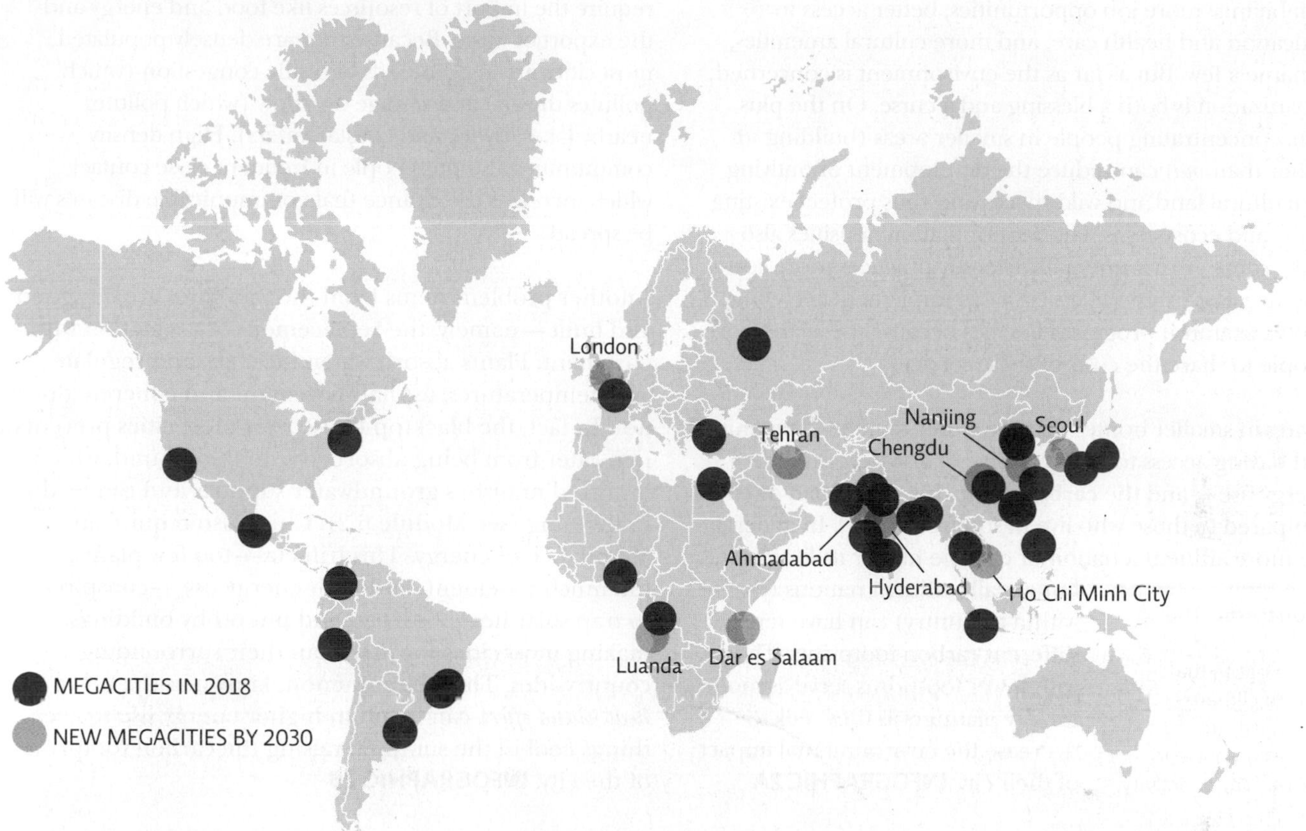

World population is continuing to increase, though at a slower rate than in the 20th century. Most future growth will be in urban areas.

WORLD POPULATION GROWTH BY LOCATION

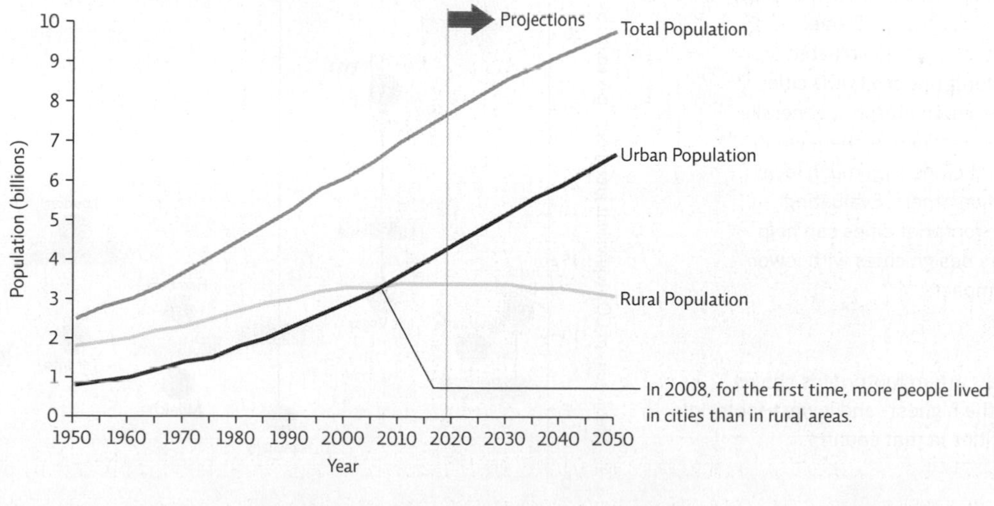

What are the advantages and disadvantages of living in a large urban area?

2 THE TRADE-OFFS OF URBAN LIVING

Key Concept 2: Cities offer many services and opportunities, and can have a lower than average per capita carbon footprint, but have higher rates of crime and disease and problems dealing with waste and stormwater.

To be sure, cities bring some obvious advantages to their inhabitants: more job opportunities, better access to education and health care, and more cultural amenities, to name a few. But as far as the environment is concerned, urbanization is both a blessing and a curse. On the plus side, concentrating people in smaller areas (building *up* rather than *out*) can reduce the development of outlying agricultural land and wild spaces and thus protect existing farms and ecosystems. Higher population densities also make some environmentally friendly practices more cost-effective. For example, it's easier to implement recycling and mass transit programs in cities because there are more people to share the costs of these services.

Living in smaller homes that are closer to needed amenities and having access to mass transit can also decrease the energy use—and the **carbon footprint**—of urban dwellers compared to those who live in suburban areas. In general, the more affluent a nation or city, the higher its footprint, but equally affluent nations (or cities within a country) can have very different carbon footprints. Those with lower footprints serve as models for city planners as they seek to decrease the environmental impact of their city. **INFOGRAPHIC 2A**

carbon footprint The amount of CO_2 (and other greenhouse gases that contribute to climate change) released to the atmosphere by a person, company, nation, or activity.

On the minus side, cities are *locally* unsustainable: They require the import of resources like food and energy and the export of waste. Because they are densely populated, most cities are also hotbeds of traffic congestion (which pollutes the air) and sewage overflow (which pollutes nearby lakes, rivers, and coastal waters). High-density communities also put people in frequent close contact, which increases the chance that communicable diseases will be spread.

Another problem stems from the way cities are designed and built—namely, the replacement of vegetation with pavement. Plants absorb water, filter air, and regulate area temperatures; asphalt pavement and concrete do not. In fact, the blacktop that covers most cities prevents rainwater from being absorbed into the ground, which in turn diminishes groundwater supplies and can lead to flooding (see Module 6.2). Cities also require an abundance of energy. This trifecta—too few plants, too much pavement, and high energy use—conspires to trap solar heat absorbed and put off by buildings, making most cities warmer than their surrounding countrysides. This phenomenon, known as the *urban heat island effect*, can result in higher energy use to keep things cool in the summer, raising the carbon footprint of the city. **INFOGRAPHIC 2B**

INFOGRAPHIC 2A URBAN LIVING CAN REDUCE ONE'S ENVIRONMENTAL IMPACT

Many large cities have lower ecological footprints than their suburban neighbors due to things like less personal vehicle travel, smaller homes, and efficiencies of scale. Daniel Moran led a study that evaluated the carbon footprint of 13,000 cities. Though the carbon footprint generally increases with affluence, there were many affluent cities with much lower footprints than others. Evaluating these lower-footprint cities can help city planners design cities with lower ecological impacts.

The national carbon footprint is shown along with the highest- and lowest-footprint evaluated cities in that country.

ANNUAL PER CAPITA CARBON FOOTPRINT

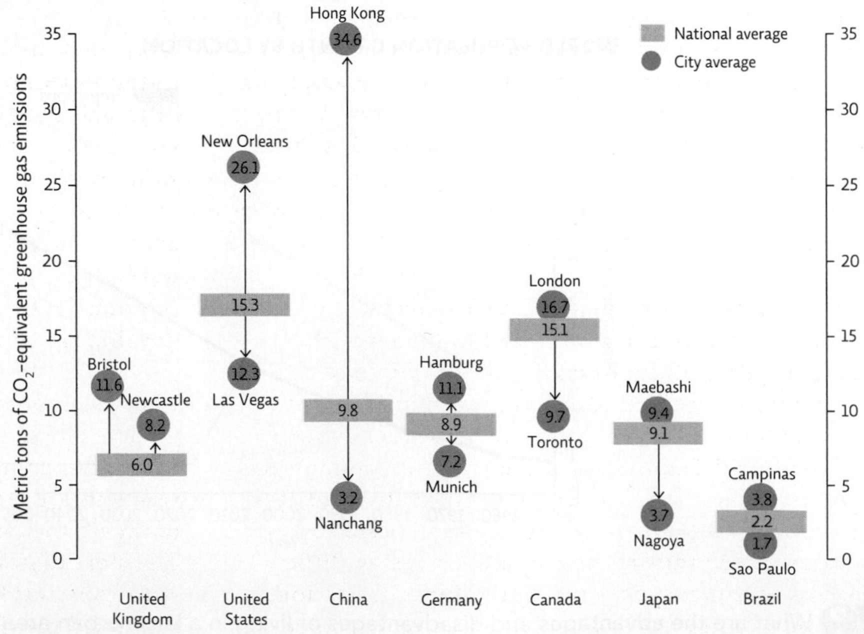

Why might the carbon footprint in Hong Kong be so different from Nanchang?

INFOGRAPHIC 2B TRADE-OFFS OF URBANIZATION

ADVANTAGES	DISADVANTAGES
• Lower impact per person due to smaller homes and less traveling	• Dependence on food and resource inputs from outside the city • Concentrated wastes that have to be transported away
• Higher energy efficiency in stacked housing than in freestanding buildings	• Urban heat island effect, which increases energy needs and can have health consequences
• More transportation options, which lessens the need for personal vehicles • Closer proximity to destinations, which makes walking and mass transit viable options	• Traffic congestion and its associated air pollution due to high population densities
• Easier-to-implement zoning ordinances	• Possibly higher disease and violence in concentrated inner-city areas
• More job opportunities because local collaboration from a diverse community fosters innovation and ingenuity	• Higher cost of living, which limits who can afford to live in the city
• More services for citizens, including more educational and cultural opportunities and better healthcare options	• Less green space, which leads to stormwater problems

? How might the disadvantages of urbanization be addressed to lessen their impact?

3 ENVIRONMENTAL JUSTICE ISSUES

Key Concept 3: Environmental justice recognizes the right of everyone to a safe and healthy environment and a voice in policies that impact their environment. Minorities and low-income communities are more likely to suffer environmental injustice than other communities.

All urban dwellers are vulnerable to the health effects associated with pollutants. But in most cities, the pros and cons of city living are unevenly realized. For example, New York City is one of the wealthiest, most populous cities in the world, but most of the cultural amenities, top-notch healthcare facilities, and job opportunities are concentrated in Manhattan, while most of the garbage, sewage, and power plants are located in the Bronx. This imbalance has spawned a whole new area of activism known as **environmental justice**, based on the idea that no community should be saddled with more environmental burdens and fewer environmental benefits than any other. (See Module 4.3 for more on environmental justice issues related to chemical hazard exposure.)

Violations of environmental justice occur everywhere—in the Bronx, in low-income communities of the Deep South, in minority communities of Midwestern cities—and inequities can be found in every U.S state and every country of the world. Minority and low-income communities are more likely to bear the burden of an unhealthy environment (e.g., close proximity to polluting or undesirable industry, poor-quality housing, little access to green space, and limited transportation options). They are also less likely to have a voice in decision-making processes that affect their community. In the United States, the Environmental Protection Agency (EPA) and other federal agencies seek to address issues of environmental injustice and offer programs and avenues for individuals or communities to address problems they face. (See Module 5.2 for more on environmental policy.) **INFOGRAPHIC 3**

Environmental justice issues are particularly relevant in the most impoverished cities in the world. In Mumbai, the largest city in India, with just over 20 million people, more than 8.5 million are slum dwellers who live in appalling, overcrowded conditions—as many as 18,000 people per acre—without adequate sanitation or running water. Globally, more than 1 billion of the world's population lives in slums, mostly in large cities in developing countries.

With most future population growth predicted to occur in large cities, urban planners are desperately searching for ways to create cities where the basic needs of residents are met and where the environmental benefits outweigh the environmental costs. The story of how the South Bronx waterfront was lost and then reclaimed provides important lessons about how to do this.

environmental justice The concept that access to a clean, healthy environment is a basic human right.

INFOGRAPHIC 3 ENVIRONMENTAL JUSTICE

All communities should have equal access to a clean, safe, and healthy environment but traditionally minorities and low-income communities are saddled with more environmental burdens than their neighboring communities. The EPA and other federal agencies identify goals (shown in this diagram) and provide programs that work to bring about equity in environmental quality and participation in decision-making processes for all communities. Many nonprofit organizations, such as the Environmental Justice Foundation, also work toward those same goals.

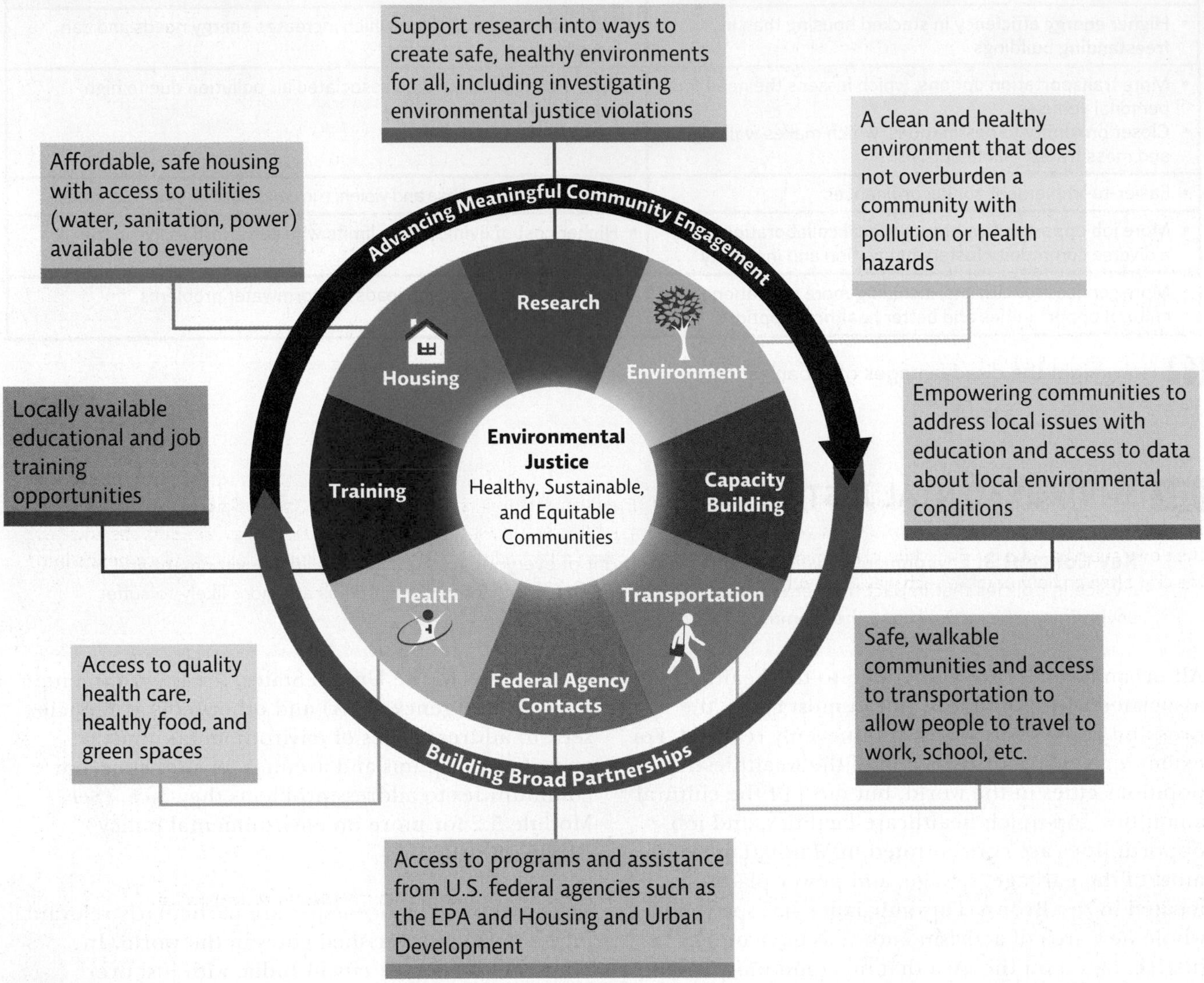
Support research into ways to create safe, healthy environments for all, including investigating environmental justice violations
Affordable, safe housing with access to utilities (water, sanitation, power) available to everyone
A clean and healthy environment that does not overburden a community with pollution or health hazards
Locally available educational and job training opportunities
Empowering communities to address local issues with education and access to data about local environmental conditions
Access to quality health care, healthy food, and green spaces
Safe, walkable communities and access to transportation to allow people to travel to work, school, etc.
Access to programs and assistance from U.S. federal agencies such as the EPA and Housing and Urban Development
Advancing Meaningful Community Engagement
Building Broad Partnerships
Research
Housing
Environment
Training
Capacity Building
Health
Transportation
Federal Agency Contacts
Environmental Justice
Healthy, Sustainable, and Equitable Communities

❓ The federal government has an ambitious plan to erase environmental injustice. Why, then, is it still prevalent in our society?

4 SUBURBAN SPRAWL

Key Concept 4: Suburban sprawl displaces farmland and natural areas; suburban residents often have a higher per capita environmental impact than urban dwellers.

urban flight The process of people leaving an inner-city area to live in surrounding areas.

In the late 1940s, the South Bronx was a mostly European-descended, white, working-class suburb of Manhattan. But as more Hispanic and black Americans moved to the area, whites moved to nearby commuter towns. Today, **urban flight**—the process of people leaving a city center for surrounding areas—is triggered by a variety of forces, including overcrowding, noise and air pollution, the high cost of city living, and, in some cases, racial tensions.

No matter what the cause, urban flight results in **suburban sprawl**—a slow conversion of rural areas outside of a city into suburban and exurban ones. **Exurbs** are more sparsely populated towns beyond the immediate suburbs whose residents also commute into the city for work.

As its name suggests, suburban sprawl tends to spread out over long corridors in an unplanned and often inefficient manner. By covering ever-greater swaths of terrain with concrete and pavement, sprawl reduces the amount of land available for farming, wildlife, and the provision of *ecosystem services* (see Modules 3.2 and 5.1). Because of the haphazard way in which they are developed, the resulting communities are heavily dependent on driving; unlike cities, which are densely populated and can accommodate mass transit systems, suburbs and exurbs require residents to drive almost everywhere they need to go. And because suburban homes are typically larger than urban ones (and exurban homes are often even larger than suburban ones), they tend to have a greater environmental impact. **INFOGRAPHIC 4**

Urban planners today are well aware of these perils and often work to mitigate them. But in the 1960s, urban planning tended to accommodate urban flight and the sprawl that came with it. The Cross Bronx Expressway, completed in 1972, enabled suburban commuters to completely bypass the Bronx as they traveled in and out of Manhattan each day. But the expressway segregated the ailing borough from the rest of New York City. "The South Bronx was utterly cut off," says Marta Rodriguez, a lifelong Bronx resident and a colleague of Carter's. "We didn't stand a chance."

suburban sprawl Low-population-density developments that are built outside of a city.

exurbs Towns beyond the immediate suburbs whose residents commute into the city for work

INFOGRAPHIC 4 SUBURBAN SPRAWL

Urban flight often leads to suburban sprawl—low population density in developments outside a city. Homes typically get larger the farther they are from the city, and residents have a larger environmental impact (larger homes and more time spent driving). The suburbs now have their own suburbs—the exurbs, which are commuter towns that are beyond the traditional suburbs but whose residents still commute into the city, often an hour or more each way. Both suburbs and exurbs often displace farmland and wildlands.

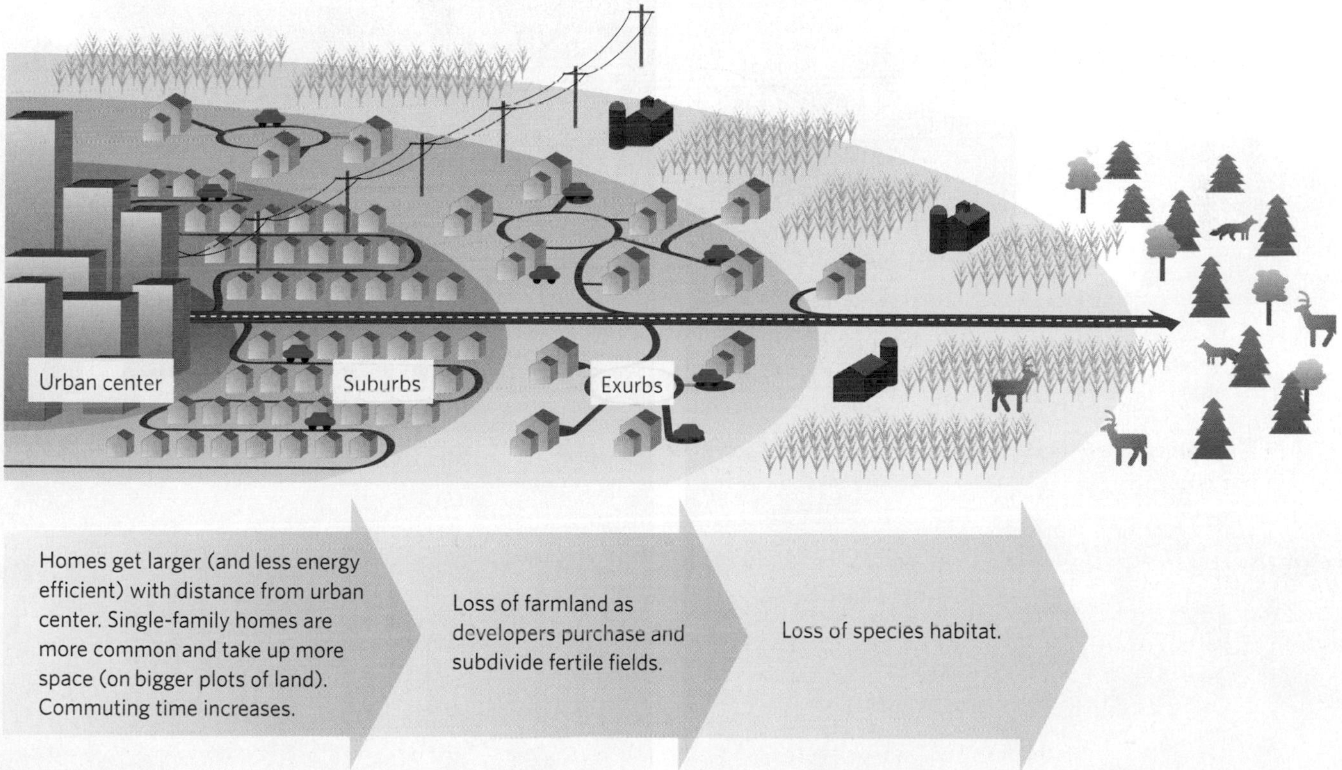

Urban center
Suburbs
Exurbs
Homes get larger (and less energy efficient) with distance from urban center. Single-family homes are more common and take up more space (on bigger plots of land). Commuting time increases.
Loss of farmland as developers purchase and subdivide fertile fields.
Loss of species habitat.

If you worked in an urban center, would you rather live there, in the suburbs, or in the exurbs? Explain.

5 THE VALUE OF GREEN SPACE

Key Concept 5: Green spaces improve the local environment and mental and physical health, and have been linked to decreased crime rates.

Carter knew that replacing the abandoned lot with a park would be a big first step toward righting some of the wrongs that her community had endured. The trees and plants would trap pollutants from the air, preventing them from infiltrating people's lungs. The grass and soil would absorb rainwater so that it could no longer carry trash and detritus from the streets into the river. And claiming even a small patch of waterfront for themselves would give Carter and her neighbors a sense of ownership, not to mention a connection to nature and a place to stretch their legs.

green space A natural area such as a park or undeveloped landscape containing grass, trees, or other vegetation in an urban area, usually set aside for recreational use.

Indeed, studies have shown that **green spaces** in a city improve both the physical and psychological health of people who live near them by providing more opportunities for physical activity, relaxation, and social interaction. Community gardens offer the chance for fresh produce — a healthy food option often lacking in low-income urban areas.

There is even evidence that the presence of green space makes an urban area safer. In Bogotá,

Hunts Point Riverside Park before construction began and after completion. The park reclaimed a degraded area of the Bronx, giving the residents access to the river for fishing, kayaking, and canoeing. The green space provides opportunities for recreation and community gatherings and is part of a greater vision, the Bronx River Greenway, which will connect communities all along the river.

Majora Carter

Before

After

Colombia, in the late 1990s, for example, a particularly environmentally conscious mayor noticed that while his city was designed to accommodate heavy automobile traffic, the vast majority of his electorate did not drive. So he narrowed municipal thoroughfares from five lanes to three, expanded bike lanes and pedestrian walkways, and established a string of parks and public plazas throughout the city. The result? People stopped littering. Crimes rates dropped. And slowly but surely, city residents reclaimed their streets.

Green spaces also provide other environmental and societal benefits. They provide habitat for wildlife, help mitigate against the urban heat island effect, reduce stormwater runoff after a rainfall, absorb air pollution, and help offset greenhouse gas emissions. Property values also tend to go up in urban areas close to parks or other green spaces, and community members are more likely to take pride in their surroundings and participate in community events in these areas. **INFOGRAPHIC 5**

Because of the growing list of green space benefits, many cities are actively putting in more parks, greenbelts, and other open spaces. For example, Vancouver, Canada, has a goal of ensuring that all residents are within a 5-minute walk of a green space and has committed to planting 150,000 trees by 2020.

INFOGRAPHIC 5 THE VALUE OF GREEN SPACES

Urban green spaces provide a wide variety of benefits to the residents and their community.

SOCIETAL BENEFITS
• Community pride and participation
• Higher property values
• Lower crime rate
HEALTH BENEFITS
• Physical activity opportunities
• Relaxation and rejuvenation
• Social interactions
ENVIRONMENTAL BENEFITS
• Trees absorb CO_2 and offset greenhouse gas emissions
• Unpaved areas reduce stormwater runoff and flooding
• Wildlife habitat
• Reduced urban heat island effect

Explain how more green space might lead to a lower crime rate.

6 GREEN CITIES AND SMART GROWTH

Key Concept 6: Green cities actively pursue sustainability through smart growth strategies that improve quality of life while decreasing per capita impact.

Carter and those who joined her project were on track to reclaim their slice of the Bronx. Starting with the $10,000 seed grant from the city parks department, they leveraged a small fortune in additional grants, donations, and private investment, until they had finalized plans to build a $3 million park, complete with gardens, grassy knolls, and East River kayaking. Hunts Point Riverside Park—the spot that Carter stumbled upon—would be the borough's first waterfront park in more than 60 years. But that was just the beginning.

Energized by their successful riverside park project, Carter and her neighbors formed a nonprofit called the Sustainable South Bronx (SSBx). The group immediately set its sights on an even grander vision: They would create a 1.5-mile greenbelt around the entire community, including a waterfront greenway and other open spaces, all connected by an interlinking system of bike and pedestrian pathways. They would also disassemble the Sheridan Expressway, a stretch of abandoned highway originally meant to cut across the entire northeast Bronx, and turn it into 28 acres of additional parkland.

But a green city is more than just a city with green spaces. A **green city** is one that pursues sustainable options that make them more environmentally friendly and socially equitable. *New Urbanism,* as the movement is called, maintains that cities (both now and in the future) have the capacity to reduce our per capita environmental impact, even as they improve the quality of life for people, provided they are designed properly.

The idea of a green city is admirable, but how is it sustainability measured? One method, the *Green City Index,* evaluates indicators from several categories to determine how far along on a path to sustainability a city might be. The analysis examines the progress a city has already made and its plans for future improvements. For example, water and energy use are evaluated, noting progress in metrics such as water use efficiency and the percentage of energy that comes from clean energy sources. Solid waste management and environmental governance (the quality of an environmental strategic plan, if one exists) are analyzed. Air pollution and the city's carbon footprint (CO_2 emissions) are measured, as are the availability of public transportation and the amount of green space. The value of a green city analysis is the identification of sustainability pursuits that are well done (as an example other cities can emulate) as well as areas where city planners need to focus their attention for continued progress toward sustainability.

green city A city designed to improve environmental quality and social equity while reducing its overall environmental impact.

smart growth Strategies that help create walkable communities with lower environmental impacts.

infill development The development of empty lots within a city.

Greening our cities becomes ever more vital as urban growth in many areas, especially the developing world, is increasing at unprecedented rates. And though we have a long way to go to make cities an environmental solution rather than one of the environment's biggest problems, the consensus emerging from environmentalists, sociologists, and economists is that the future lies in cities—where most people will live and perhaps where most people should live. Cities promote interaction among a diverse group of people. This in turn promotes the exchange of ideas and lessens cultural and economic barriers. Many cities around the world are pursuing sustainable development and paving the way for others to do the same.

Sustainable cities are cities where the environmental pros outweigh the cons—where sprawl is minimized, walkability is maximized, and the needs of inhabitants are met locally. Strategies that help create green cities—those with walkable communities with lower environmental impacts—are known as **smart growth**. To achieve self-sufficiency, for example, a sustainable city might maintain a mixture of open and agricultural land along its outskirts to provide a large part of the local food, fiber, and fuel crops, along with recreational opportunities and ecological services. Waste and recycling facilities could also be located nearby, along with other enterprises aimed at producing resources needed by area residents.

To stave off sprawl, the same city might establish *urban growth boundaries*—outer city limits beyond which major development would be prohibited. Keeping any outward growth that does occur as close to mass transit as possible minimizes the impacts of transportation. **Infill development**—developing empty lots within a city—and building "up" (a parking garage) rather than "out" (an expansive parking lot) also minimizes the amount of land used. Plano, Texas, did this when it created its Legacy Town Center, a mixed-use community with a range of housing options alongside shops, parks, and restaurants in an existing office park. The area is now attracting other industries, bringing more jobs to Plano.

nina berman/NOOR/Redux

A volunteer gardener at Finca Del Sur, a garden in the South Bronx, tends the corn stalks while a passenger train goes by in the background. The garden was created on an empty plot of land bordered by a highway exit ramp and a commuter train line.

Smart growth also provides a range of options for reducing traffic congestion and the air pollution that comes with it: reliable public transportation, car-sharing programs that allow residents to use cars when needed for a monthly fee, and sidewalks and overhead passageways that allow pedestrians to safely cross busy roads. To encourage more walking and less driving, zoning laws might allow for mixed land uses, where residential areas are located reasonably close to commercial and light industrial ones.

A final smart growth strategy is to seek out community and stakeholder input so that decisions can be made in a collaborative fashion. For example, as part of their standard protocol, city planners in Chicago, Illinois, engage a wide variety of stakeholders as they pursue a growth and economic development path that links local planning efforts to the greater regional vision. **INFOGRAPHIC 6**

Of course, building an ideal city from scratch is easy compared with the task of overhauling an existing city. Upgrading decaying infrastructure like roads, public places, and sewage and water lines can be more expensive than new construction, and the process is disruptive to residents. Even so, there are plenty of ways that American cities can push themselves into the environmental plus column.

Persuading people to support smart growth, as Carter and her colleagues soon discovered, is a matter of showing them that the benefits could be economic as well as environmental. "You need to show them what we call the *triple bottom line,*" says James Chase, vice president of SSBx (and Carter's husband), referring to the economic, social, and environmental impacts of any decision. "Developers, government, and residents all need some tangible, positive return." A major park project would surely be a boon for all three. Developers would be guaranteed millions in waterfront development contracts. Residents could look forward to cleaner air and water, a prettier neighborhood, and better health. Local businesses and state and city governments would save a bundle in healthcare costs. The greenbelt would also spur the local economy: Such a vast stretch of public space would attract street vendors, food stands, bicycle shops, and sporting goods stores. (See Modules 1.1 and 5.1 for more on the triple bottom line.) Many of these benefits are being realized—the area has seen an increase in the number of small businesses and jobs in recent years.

INFOGRAPHIC 6 SUSTAINABLE CITIES AND SMART GROWTH

Smart growth can be applied to large cities or to smaller communities. It employs strategies that make efficient use of land to create pleasant livable communities with less environmental impact than current suburban areas.

Which of these smart growth principles would be most appealing to you if you were looking for a place to live in a city?

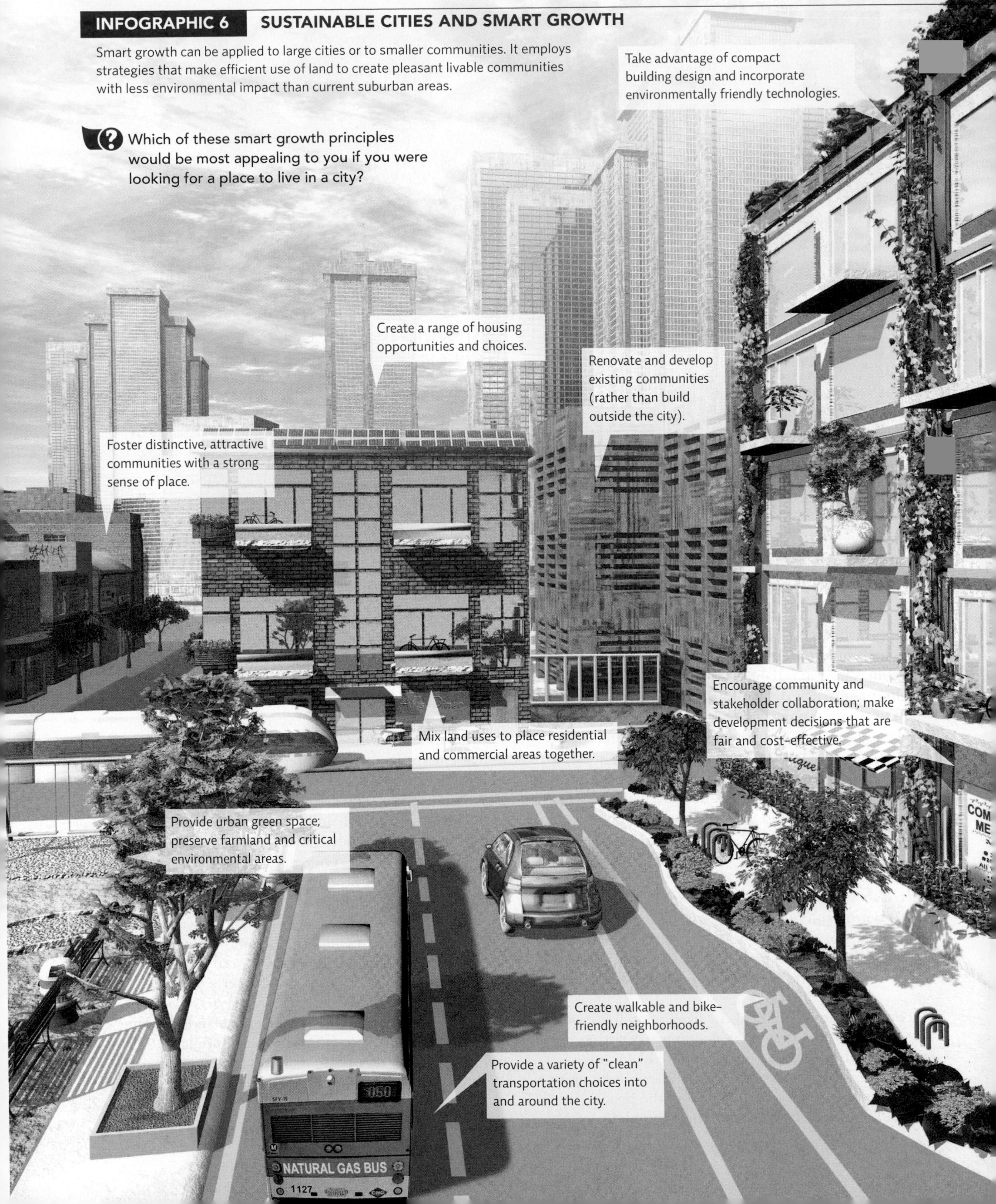

Take advantage of compact building design and incorporate environmentally friendly technologies.
Create a range of housing opportunities and choices.
Renovate and develop existing communities (rather than build outside the city).
Foster distinctive, attractive communities with a strong sense of place.
Encourage community and stakeholder collaboration; make development decisions that are fair and cost-effective.
Mix land uses to place residential and commercial areas together.
Provide urban green space; preserve farmland and critical environmental areas.
Create walkable and bike-friendly neighborhoods.
Provide a variety of "clean" transportation choices into and around the city.
NATURAL GAS BUS
050

7 GREEN BUILDING

Key Concept 7: Green building design focuses on efficient use of energy and water and on building materials with low environmental and health impacts.

Buildings in cities are major energy users—they consume an estimated 40% of global energy and generate around 30% of the world's greenhouse gases—making them key targets in the quest to reduce our overall environmental impact. Addressing this opportunity is a movement known as **green building**—that is, the construction of buildings that are better for the environment and the health of those who use them.

In one program implemented by the U.S. Green Building Council, buildings that meet a minimum standard are awarded a **Leadership in Energy and Environmental Design (LEED)** certification. Points are awarded based on how well the building meets certain criteria, such as energy and water efficiency. Of course, buildings can be built or retrofitted without seeking LEED certification (a costly procedure) but for those organizations that choose to participate, it offers publicity and an accountability that may encourage the pursuit of loftier goals.

The Bronx Library Center is a silver-certified LEED building. It earned the silver certification by recycling 90% of the waste materials created during the construction of the building, using architectural design and efficient heating and cooling systems to save 20% of energy costs, and using sustainably grown wood in 80% of the construction lumber. In its own green-building initiative, Carter's team launched Smart Roofs, LLC, a green-roof installation company. Green roofs are one type of rain garden—an area with plants suited to local temperature and rainfall conditions (see Module 6.2). Research shows that more than $5 million in annual cooling costs could be saved if green roofs were installed on just 5% of the city's buildings. That amount of green roofing could achieve an annual reduction of 350,000 metric tons of greenhouse gases. Riverkeep, an environmental nonprofit, found that green roofs can retain 800 gallons of stormwater for every $1,000 of investment—easing pressure on the city's overburdened sewer systems and mitigating water pollution from stormwater runoff. **INFOGRAPHIC 7**

green building Construction and operational designs that promote resource and energy efficiency and provide a better environment for occupants.

Leadership in Energy and Environmental Design (LEED) A certification program that awards a rating (standard, silver, gold, or platinum) to buildings that include environmentally sound design features.

Robert Stolarik/The New York Times/Redux

Via Verde, a subsidized housing development, earned LEED gold certification thanks to the incorporation of a variety of sustainability features that address environmental and social challenges. It has become a model for similar projects throughout the city and elsewhere.

INFOGRAPHIC 7 **GREEN BUILDING**

Many steps can be taken to build or retrofit a building so that it has less environmental impact and is a healthier environment for those who live, work, or go to school there. The nonprofit group Green Building Council certifies buildings through its LEED (Leadership in Energy and Environmental Design) program. A building receives a standard, silver, gold, or platinum rating based on a variety of criteria that include energy efficiency, sustainable building material use, and innovative design.

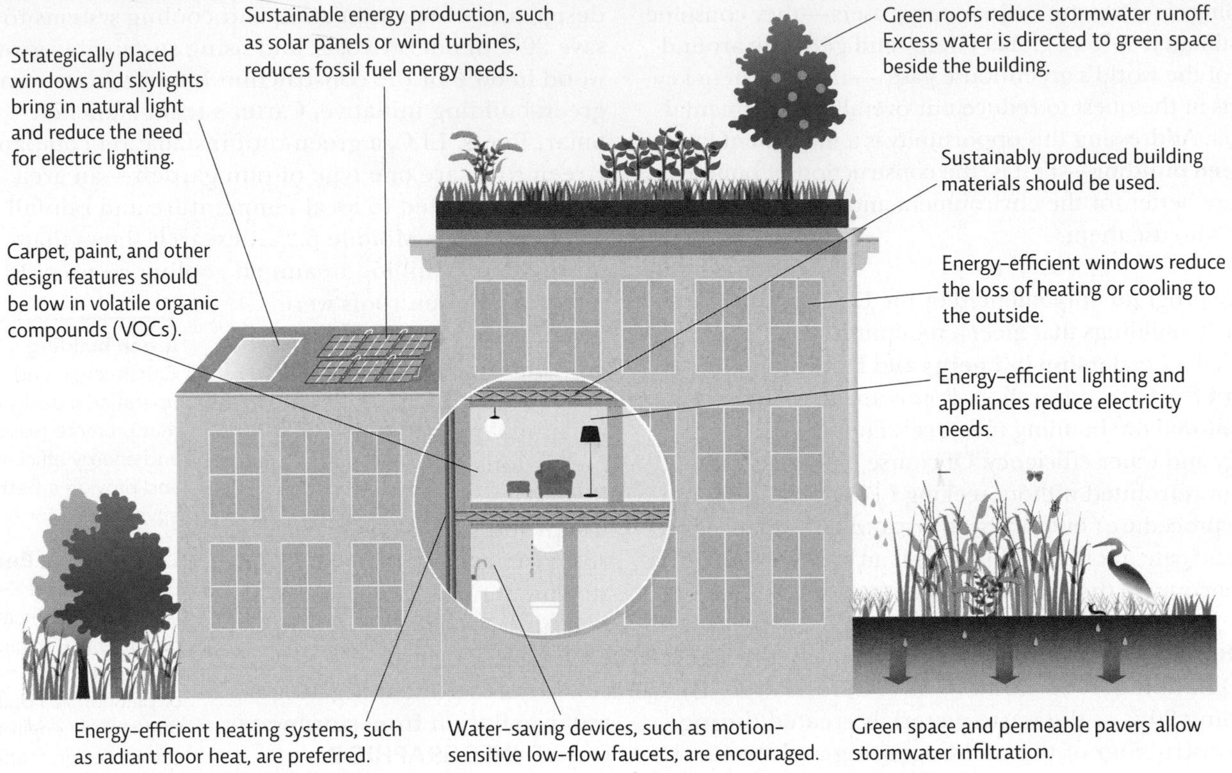

Strategically placed windows and skylights bring in natural light and reduce the need for electric lighting.
Sustainable energy production, such as solar panels or wind turbines, reduces fossil fuel energy needs.
Green roofs reduce stormwater runoff. Excess water is directed to green space beside the building.
Sustainably produced building materials should be used.
Carpet, paint, and other design features should be low in volatile organic compounds (VOCs).
Energy-efficient windows reduce the loss of heating or cooling to the outside.
Energy-efficient lighting and appliances reduce electricity needs.
Energy-efficient heating systems, such as radiant floor heat, are preferred.
Water-saving devices, such as motion-sensitive low-flow faucets, are encouraged.
Green space and permeable pavers allow stormwater infiltration.

How could a LEED-certified building help address urban problems such as the urban heat island effect and stormwater issues?

By the time the Hunts Point Riverside Park opened, dozens of cities across the country—from Madison, Wisconsin, to Miami, Florida—had taken up the mantle of sustainability and smart growth. The United States leads the world in green building, with more than 38,000 LEED-certified buildings or projects under construction in 2018. The Bronx received an award for Via Verde, a LEED gold mixed-income housing complex that incorporates a wide variety of green building features such as solar panels, natural lighting, and green roofs. It also has easy access to transportation and a neighborhood medical clinic on site. The project was so successful it prompted New York City to change its green zoning rules to make it easier to implement similar projects throughout the city.

Select References:

Beckett, K., & Godoy, A. (2010). A tale of two cities: A comparative analysis of quality of life initiatives in New York and Bogotá. *Urban Studies, 47*(2), 277–301.

Carter, M. (2006, February). *Greening the ghetto.* Retrieved from http://www.ted.com/talks/majora_carter_s_tale_of_urban_renewal

Lee, A. C. K., et al. (2015). Value of urban green spaces in promoting healthy living and wellbeing: Prospects for planning. *Risk Management and Healthcare Policy, 8,* 131–137.

Moran, D., et al. (2018). Carbon footprints of 13,000 cities. *Environmental Research Letters, 13*(6), 064041.

United Nations, Department of Economic and Social Affairs, Population Division. (2018). *The world's cities in 2018—data booklet* (ST/ESA/ SER.A/417). New York, NY: Author.

GLOBAL CASE STUDIES GREEN CITIES

Many cities of the world are taking steps to develop more sustainably in an effort to improve their local environments and their standards of living. Green cities have many things in common, such as recognizing the importance of civic involvement, having a government commitment to sustainable development, and pursuing a holistic approach that looks at all the ways the city can reduce its ecological footprint while still developing and growing. While there are many others, here are some notable examples.

BRING IT HOME

PERSONAL CHOICES THAT HELP

A sustainable community is one that promotes economic and environmental health and social equity. It is one in which the health and wellbeing of all citizens are considered, while those citizens help implement and maintain the community.

Individual Steps

- Investigate and support sustainable businesses in your area (see www.sustainablebusiness.com). Research products before you purchase them to understand the impact of your consumption choices (see www.goodguide.com).
- If you have a balcony or yard, plant flowers, vegetables, or trees.
- Support local businesses by shopping and dining close to home.
- See how well you can plan for a sustainable community. Download and play the strategy game Fate of the World (www.gamesforchange.org/play/fate-of-the-world/), and see how policies you put in place impact global climate change, rainforest preservation, and resource use.

Group Action

- Join neighborhood clean-up days. If you can't find one, organize one.
- Reduce reliance on cars. Start a petition to get more bike lanes in your city. Ride public transit more often.
- Find out how colleges and universities are working toward sustainable practices at www.AASHE.org.
- Attend a meeting of your city council or county commission, and ask members to look into smart growth opportunities.

REVISIT THE CONCEPTS

- As of 2008, more people live in cities than in rural areas, and those cities are getting bigger. The number of megacities is increasing; most are, and will be, in Asia and Africa.
- Urban living offers many services and opportunities not readily available in smaller municipalities or in rural areas, and city dwellers can have a lower carbon footprint, but cities also may have higher rates of crime and disease, problems dealing with waste and stormwater, and the urban heat island effect.
- Violations to environmental justice, the right of everyone to a safe and healthy environment, are more likely to occur in minority and low-income communities where less preferred industries (e.g., waste facilities) or infrastructure (highways) are often sited.
- To escape some of the negative aspects of cities, many residents have moved into suburban and exurban areas, though they often have a higher per capita environmental impact than urban dwellers.
- Within a city, green spaces can improve the local environment, have positive effects on the mental and physical health of residents, and are linked to decreased crime rates.
- The creation of green cities — that actively pursue sustainability through design and city services that improve quality of life while decreasing per capita impact — can lessen the overall environmental impact of cities. Smart growth describes strategies that allow cities to develop in a way that minimizes environmental impact while enhancing community living.
- One tool in the smart growth toolbox is green building — design that focuses on efficient use of energy and water and on building materials with low environmental and health impacts.

ENVIRONMENTAL LITERACY Understanding the Issue

1 What is the pattern of global urbanization in recent decades?

1. How has the distribution of human population between urban and rural dwellers changed over time?
2. What is a megacity, where are most of the world's megacities today, and where will most be in the future?

2 What are the trade-offs associated with urban areas?

3. Point out several reasons why the typical city dweller has a lower environmental impact than a suburban dweller. What are some problems that might be greater for a large city than a smaller suburban one?
4. In general, as affluence increases, so too does the environmental impact of a city. How might increasing affluence help decrease that impact?

3 What is environmental justice and what common problems does it address?

5. If urban centers have better health care and job opportunities than outlying areas, why are some city residents exposed to worse, not better, living conditions?
6. How can cases of environmental injustice be addressed?

4 What environmental problems does suburban sprawl generate?

7. What are suburbs and exurbs? Why might one choose to live outside rather than inside the city?
8. Compare the environmental impact of residents in urban centers, suburban areas, and exurban areas.

5 What is the value of urban green space?

9. Why might living in areas with green spaces be beneficial to the health and wellbeing of residents?
10. Identify some of the environmental benefits of urban green spaces.

6 How does smart growth help make a city "green"?

11. Why is it so important that we build or transition our cities into "green cities," especially the cities of the future?
12. Explain the triple bottom line, using the Bronx waterfront restoration as an example.
13. What are the major strategies of smart growth?
14. What is infill development, and how can it help revitalize an urban area and help a city pursue sustainable development?

7 What are some of the design features of a green building?

15. Describe a LEED-certified building. What criteria are used to evaluate a building for LEED certification?
16. How does green building fit into a smart growth plan?

SCIENCE LITERACY Working with Data

One common measurement of how well a building performs with regard to energy usage is its energy use intensity (EUI). EUI is expressed as energy used (in gigajoules) relative to a building's size so that structures of different sizes can be compared. This graph plots data for 36 LEED-certified office buildings, comparing their actual performances to expected performances, based on their designs. (Each data point represents a different building's EUI.) The line shown is not a trend line for the actual data; it is a line that bisects the graph at a 45-degree angle. This line allows us to see how closely the expected values match the observed values.

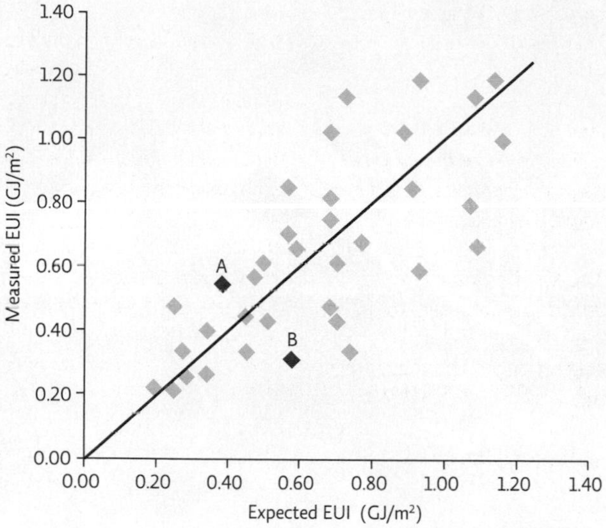

Interpretation

1. Look at the points labeled A and B. For each, what was the expected EUI predicted by the design plan? What was the actual measured EUI for this building?
2. Which of these two buildings, A or B, is performing better than expected?
3. Give the coordinate points for the one building that performed exactly as expected.
4. What does the line in the scatterplot tell you about those buildings above or below it?

Advance Your Thinking

5. About how well do LEED-certified buildings meet or exceed their predicted EUI?
6. The average EUI for non-LEED-certified office buildings is 2.19 GJ/m^2. How do LEED-certified buildings, even those that do not perform as well as expected, compare to this average?

INFORMATION LITERACY Evaluating Information

More and more people around the world live in cities, and the number of megacities is increasing. There are costs and benefits to living in cities, and in an effort to increase the benefits, there is a growing movement toward "greening" cities. Environmental and social scientists have published a number of studies documenting the effects of greener cities on environmental and human health.

Go to the website Green Cities: Good Health (http://depts.washington.edu/hhwb/). Read the introduction to the site.

Evaluate the website and work with the information to answer the following questions:

1. Determine if this is a reliable information source with a clear and transparent agenda:
   a. Who runs this website? Do the organization's credentials make it reliable or unreliable? Explain.
   b. Who are the authors? What are their credentials? Do they have the scientific background and expertise to lend credibility to the website?

Choose one of the RESEARCH THEMES links, such as "Social Strengths" or "Local Economies," then answer the following questions:

2. Which page did you choose and what type of information is provided there? What is the source of the information?
3. List a few of the fast facts provided. Then scroll down the page and read each section. For the topic you chose, what is the primary claim? What data are provided to support the claim?
4. Do you find the data convincing? Why or why not? How does what you read relate to your own life? Give specific examples.

Additional study questions are available at Achieve.macmillanlearning.com. Achieve

Module 4.3 Environmental Health and Toxicology

ERADICATING A PARASITIC NIGHTMARE

Human health is intricately linked to the environment

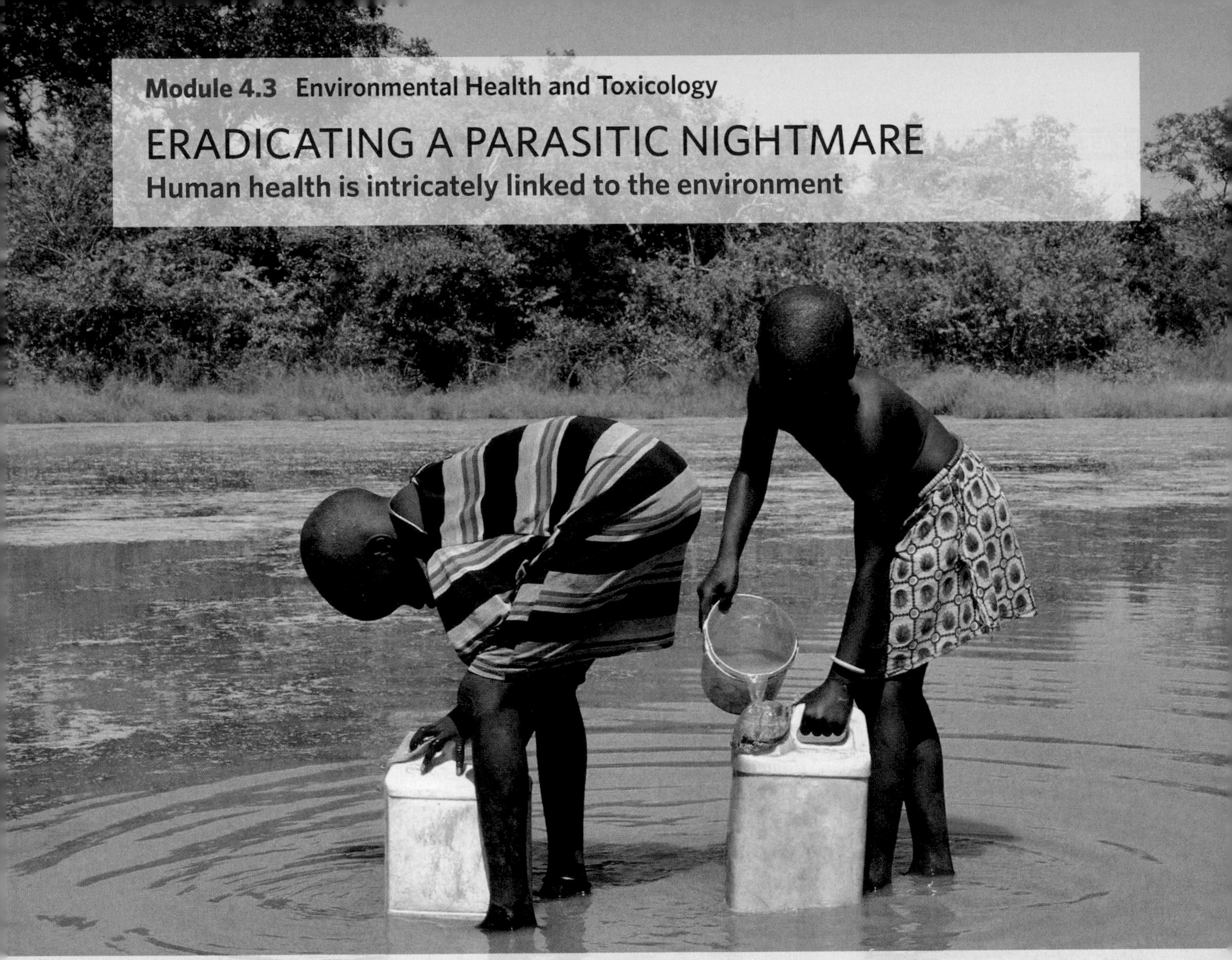

Children collect drinking water from a pond using filters that remove the water fleas that carry Guinea worm larvae.

AP Images/Maggie Fick

After reading this module, you should be able to answer these GUIDING QUESTIONS

1. What environmental hazards do humans face?
2. How do environmental factors and human actions contribute to the spread of infectious diseases?
3. What is the field of public health and how does it work to improve the health of human populations?
4. What health, societal, and environmental problems result from exposure to chemical hazards?
5. What factors influence the toxicity of a substance?
6. How is toxicity determined and what role does risk assessment play in determining the "safe dose" of a toxic substance?
7. What actions and regulations reduce the risk of environmentally mediated health problems?

Ernesto Ruiz-Tiben shook the tube of water and held it up to the sunlight so that the women who were gathered around him could see the tiny black flecks that had settled out. There was a soft, collective gasp at the spectacle. The black flecks—tiny "water fleas" known as copepods—offered the Nigerian women the first visible proof of what Ruiz-Tiben, director of the Carter Center's Guinea Worm Eradication Program, had been trying to explain to them: The water they drank and bathed in was contaminated with tiny crustaceans, and these creatures were solely responsible for the searing worm infections that seemed to sweep through the village every year or so, usually right around harvest time.

Guinea worm infection begins when a person ingests water contaminated with copepods infected with Guinea worm larvae. Once digested, the copepods release the larvae, which burrows into the victim's abdominal tissue. There, the larvae mature into adult males and females who then mate. Male worms die, but the females remain and grow—up to 3 feet long—while migrating through the victim's tissue. About 1 year later, the females force their way out by releasing acid just beneath the skin, creating a painful blister that drives victims to plunge the blister into water; the female worm then squirts out a dense cloud of milky white larvae, starting the cycle over again.

The disease is not fatal, but recovery is very slow and debilitating. Even after she releases her larvae, the mother worm can take as long as 3 months to fully emerge from the skin, during which time the victim is often completely laid up—unable to work or even walk around much, depending on which limbs are affected and with how many worms. There are no medications or vaccines for *dracunculiasis*, or Guinea worm disease (GWD) as it is more commonly known, and the infection itself does not confer immunity; that means the same people can fall prey to the worms over and over again.

WHERE IS NIGERIA?

The emergence of a Guinea worm is excruciatingly painful. Once the worm appears, it can be wound around a stick or piece of gauze and pulled out, a few centimeters per day. Complete removal can take several weeks.

Louise Gubb/Getty Images

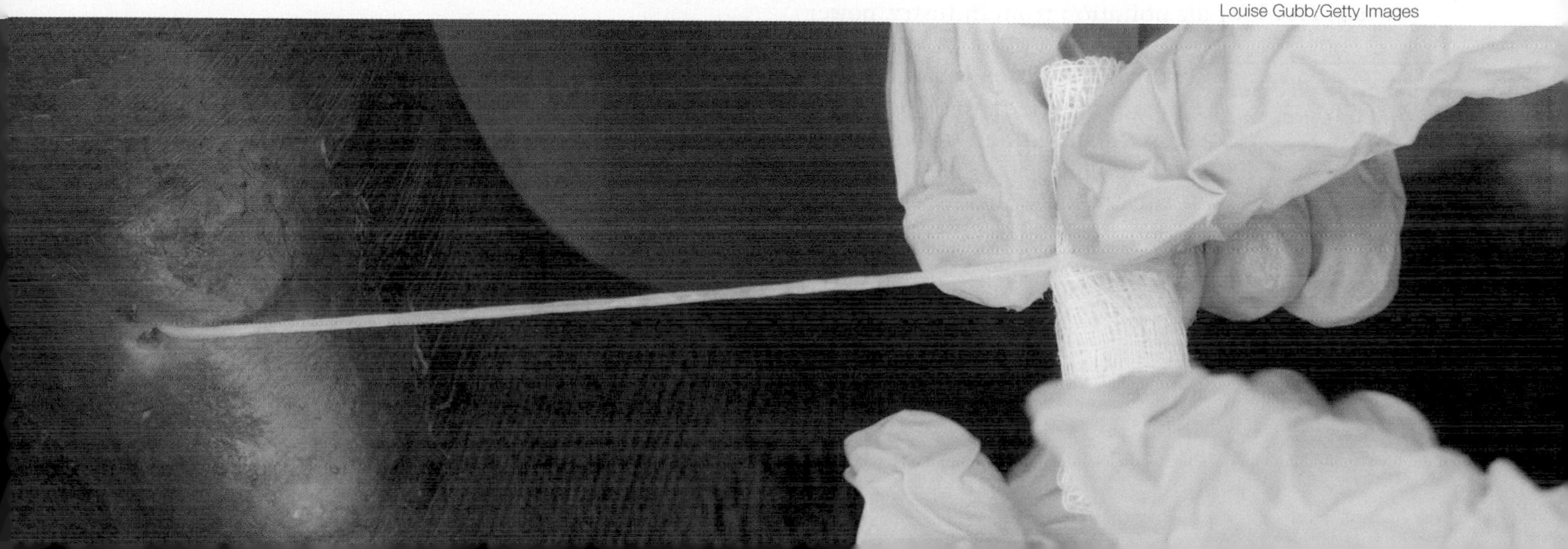

1 ENVIRONMENTAL HAZARDS

Key Concept 1: Humans face a variety of environmental factors that negatively affect health. Air and water pollution are the leading environmental health threats globally, but the spread of infectious diseases can also be environmentally mediated.

Environmental factors contribute significantly to disease, injury, and death; worldwide, they account for about a quarter of deaths and disease. Human action creates or contributes to many of these environmental problems, but that also means that many of these factors can be addressed to reduce the incidence of environmentally mediated illness and death.

Some environmental hazards are *physical hazards*—things that are harmful upon exposure (ultraviolet radiation that causes sunburn, floodwaters that increase the risk of drowning, etc.) or that cause damage without direct contact (e.g., extreme heat and cold, loud noises, or strong vibrations). Even the human-built environment—with its speeding cars and overcrowded urban areas—has physical hazards. *Chemical hazards* cause damage by virtue of their chemical makeup; air and water pollution, and occupational or household exposure to hazardous chemicals are examples. *Biological hazards* include infectious agents as well as plants, animals, and fungi that produce dangerous substances (e.g., venom from snakes and jellyfish; poisons from mushrooms); many infectious diseases have a strong environmental link and many are **communicable diseases**—they can be spread from one victim to another.

The leading environmental threat to human health today is air pollution. The small particles and chemicals found in air pollution can damage lungs and blood vessels. In addition, impaired lungs that cannot bring in enough oxygen will stress the heart (which is tasked with delivering oxygen to the body's tissues), further contributing to cardiovascular disease. Outdoor air pollution from industry, power plants, and vehicles is a problem throughout the world and the main source of air pollution in the developed world. In developing countries, indoor air pollution produced from burning fuel inside poorly ventilated homes is especially problematic. (See Module 10.1 for more on the health effects of air pollution.)

communicable disease A disease that can spread from one person to another.

noncommunicable disease (NCD) An illness that is not transmissible between people.

In fact, in terms of the threats they face and the health problems most commonly seen in the population, the differences between more developed and less developed countries are vast. In more developed countries, **noncommunicable diseases (NCDs)**, such as cardiovascular disease and cancer, represent the bulk of the disease burden. (Many NCDs are lifestyle diseases—those that are largely determined by choices about things like diet and exercise—though air pollution is also a contributing factor.) In less developed countries, lifestyle NCDs like these are ticking upward, but infectious diseases, linked to environmental factors such as lack of clean water, poor sanitation, and the burning of solid fuels indoors for heat and energy, are still the leading cause of death in many areas. It is here that death rates due to infections and parasitic diseases, like GWD, are the highest.

Of course, we are never exposed to any hazard—be it physical, chemical, or biological—in isolation. Rather, different types of hazards interact with one another and with other elements of the human environment in ways that can make it tricky for healthcare workers to map cause-and-effect relationships. For example, someone negatively affected by a chemical hazard such as air pollution may be more susceptible to a biological hazard such as a lung infection. An environment can also be contaminated with multiple hazards at the same time, such as when drinking water is tainted with both chemical pollutants and disease-causing bacteria. **INFOGRAPHIC 1**

While the environment can present many hazards, a healthy environment can mitigate many of these hazards and even improve health. Trees help purify air and wetlands do the same for water. Well-ventilated buildings reduce our exposure to pollution and pathogens. And living near green spaces (natural areas) can improve mental and physical health (see Module 4.2).

INFOGRAPHIC 1 TYPES OF ENVIRONMENTAL HAZARDS

A variety of environmental hazards impact human health. Though some of these threats are natural in origin, anthropogenic causes account for most of the hazards we currently face today. In addition, hazards can interact to create new threats. For example, floodwaters bring the risk of drowning but may also deliver pathogens and toxic chemicals to victims. Exposure to one threat, such as contracting an infectious disease, may make one more vulnerable to health complications from chemical exposure.

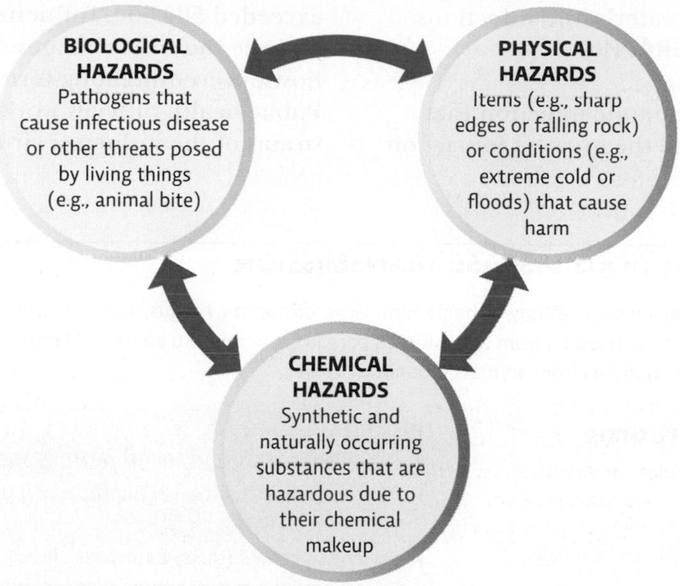
BIOLOGICAL HAZARDS
Pathogens that cause infectious disease or other threats posed by living things (e.g., animal bite)
PHYSICAL HAZARDS
Items (e.g., sharp edges or falling rock) or conditions (e.g., extreme cold or floods) that cause harm
CHEMICAL HAZARDS
Synthetic and naturally occurring substances that are hazardous due to their chemical makeup

? Earthquakes are natural disasters that can unleash a variety of environmental hazards. Identify physical, chemical, and biological hazards that might result from an earthquake.

2 BIOLOGICAL HAZARDS

Key Concept 2: Infectious diseases—including zoonotic diseases, which are spread between animals and people—threaten many human populations. Human modification of the environment can facilitate the spread of these diseases.

Infectious diseases—those caused by a **pathogen**—can simply be annoying (e.g., a cold) or they can be serious and even life-threatening (e.g., meningitis, Ebola). Pathogens can be acquired in a variety of ways such as direct contact with infected individuals or surfaces, or by ingesting contaminated food or water. Airborne exposure, too, can transmit pathogens in dust or air droplets.

While some diseases, such as smallpox and HIV, are only found in humans and only transmitted between humans, many infectious diseases are **zoonotic**—that is, they can spread between infected animals and humans. In fact, 75% of all *emerging infectious diseases*—those that are new to humans or have rapidly increased their range or incidence in recent years—are zoonotic. The novel coronavirus that caused the COVID-19 pandemic of 2020 likely arose in an animal and spread to humans, possibly in an Asian meat market that handled live animals. Many zoonotic diseases are spread through **vectors**—agents that carry a pathogen and transmit it to another organism. Vector organisms, such as mosquitoes or ticks, are adapted to carry the pathogen and do not succumb to infection, but they can pass it along from host to host. In the case of GWD, copepods serve as the vector, transmitting guinea worm larvae to humans through contaminated water. (Inanimate objects such as dust or food can also transmit pathogens and as such are considered vectors.)

Efforts to reduce the incidence of GWD began as part of an

pathogen An infectious agent that causes illness or disease.

zoonotic disease An infectious disease of animals that can be transmitted to humans.

vector An agent that transmits a pathogen to an organism.

international program to provide safe drinking water to all people. According to the World Health Organization (WHO), well over 1 billion people per year fall victim to **waterborne diseases**. By providing safe drinking water, Ruiz-Tiben and his colleagues hoped that they might eradicate GWD and eliminate other waterborne infections. **INFOGRAPHIC 2**

waterborne disease An infectious disease acquired through contact with contaminated water.

The zoonotic infection that has had the greatest impact on human health over the past several hundred years is influenza, which can spread from birds and pigs to people. The Spanish Flu epidemic of 1918 sickened one-third of the world's population and killed an estimated 50 million people; deaths attributed to the more recent H_1N_1 flu epidemic of 2009 may have exceeded 500,000. Influenza is very hard to control because the virus changes every year in its animal hosts, preventing long-term immune protection. Public health officials work every year to predict which strains of the influenza virus are most likely to become

INFOGRAPHIC 2 INFECTIOUS DISEASE TRANSMISSION

Achieve

Infectious diseases are illnesses caused by a pathogen (bacterium, virus, protozoan, fungus, worm, or prion). Most of these are communicable diseases—they can be passed on from one infected person (or animal) to another. Many of the emerging infectious diseases we face today can be transmitted to humans from animals (zoonotic).

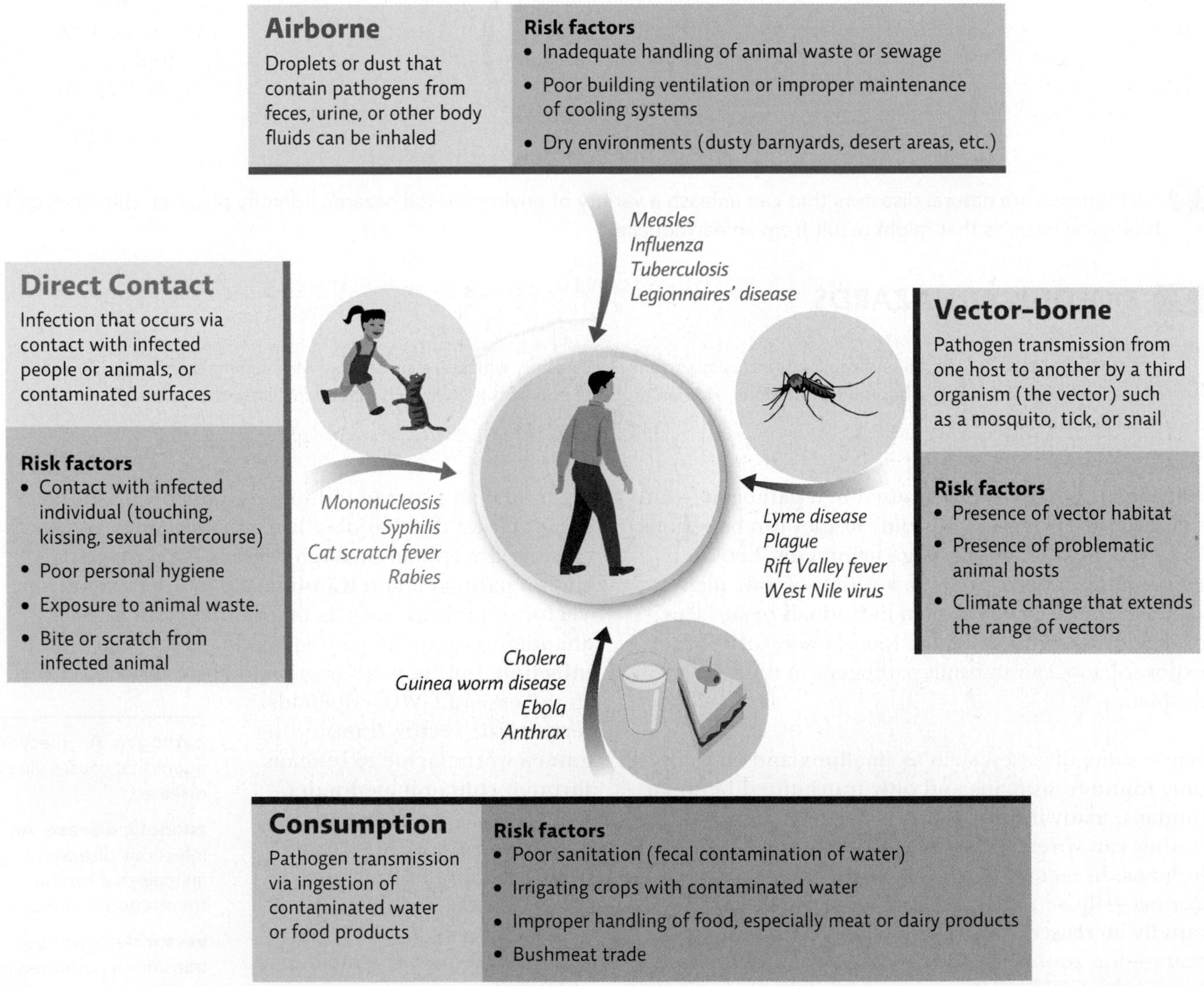
Airborne
Droplets or dust that contain pathogens from feces, urine, or other body fluids can be inhaled
Risk factors
• Inadequate handling of animal waste or sewage
• Poor building ventilation or improper maintenance of cooling systems
• Dry environments (dusty barnyards, desert areas, etc.)
Measles
Influenza
Tuberculosis
Legionnaires' disease
Direct Contact
Infection that occurs via contact with infected people or animals, or contaminated surfaces
Risk factors
• Contact with infected individual (touching, kissing, sexual intercourse)
• Poor personal hygiene
• Exposure to animal waste.
• Bite or scratch from infected animal
Mononucleosis
Syphilis
Cat scratch fever
Rabies
Vector-borne
Pathogen transmission from one host to another by a third organism (the vector) such as a mosquito, tick, or snail
Risk factors
• Presence of vector habitat
• Presence of problematic animal hosts
• Climate change that extends the range of vectors
Lyme disease
Plague
Rift Valley fever
West Nile virus
Cholera
Guinea worm disease
Ebola
Anthrax
Consumption
Pathogen transmission via ingestion of contaminated water or food products
Risk factors
• Poor sanitation (fecal contamination of water)
• Irrigating crops with contaminated water
• Improper handling of food, especially meat or dairy products
• Bushmeat trade

Identify at least three reasons why waterborne and vector-borne diseases are especially problematic in less developed areas and for those in poverty.

pandemic (spread around the world), and they produce a vaccine that protects against those strains. Some years the vaccine misses one or more strains, but even in those years the vaccine offers some level of protection.

Humans have been altering their environment for millennia and these changes have made life better for millions. However, our actions have also increased our exposure and vulnerability to natural disasters and disease, especially in modern times. We burn fossil fuels and pursue farming practices that increase air pollution. In crowded cities (and college dorms!) where people live in close quarters, the chance of contracting a communicable disease is increased.

Throughout the developing world, urbanization has increased the incidence of a wide range of waterborne pathogens, thanks in part to a wide variety of human-made objects, such as old tires and discarded plastic food containers, that find second life as vessels for rainwater. This water provides an excellent habitat for a suite of vectors that transmit a host of mosquito-borne diseases such as dengue and malaria as well as other debilitating parasitic worm diseases like river blindness (onchocerciasis), transmitted by black flies. Dams—especially those in tropical regions—do something similar: They create large bodies of standing water that have been associated with an uptick in the same cadre of diseases. Climate change, too, is expanding the range of many of these vectors.

And it's not just developing countries that are feeling the effects of human impact on environmentally mediated diseases. In the northeastern United States, for example, habitat fragmentation has contributed to an increase in the incidence of Lyme disease by increasing the populations of field mice (which carry but don't effectively remove disease-carrying ticks) while decreasing the number of squirrels and opossums (which are much more successful at removing ticks before becoming infected, thus reducing the incidence of disease transmission).

Cary Institute of Ecosystem Studies/J.Brunner

Ticks that carry the pathogen that causes Lyme disease can pass that on to their hosts, but field mice are more likely to become infected by the pathogen than other hosts such as squirrels or opossums. In environments with low species diversity and a lot of field mice, Lyme disease incidence in humans goes up because there is a greater chance that a tick that bites a person previously bit an infected field mouse and picked up the pathogen.

3 PUBLIC HEALTH PROGRAMS

Key Concept 3: Public health officials work to improve the health of a population as a whole by identifying health risks to the populace and recommending steps to reduce that risk.

Solving community health problems like GWD is the job of **public health**, a field that deals with the health of human populations as a whole, often focusing on disease prevention and understanding how diseases spread through a community. From international organizations like the WHO to national programs such as the U.S. Centers for Disease Control and Prevention (which is supported by state and local public health departments), public health **epidemiologists** like Ruiz-Tiben work to gauge the overall health of a target group by examining patterns of disease in the population. They use statistical analysis (e.g., rates of infant mortality, incidence of various diseases) to identify specific health threats to groups of people; they then recommend ways to mitigate those threats. Public health successes, like programs that reduce the incidence of childhood diseases, have contributed to humanity's rapid post–Industrial Revolution population growth (see Module 4.1).

Environmental health is the branch of public health that focuses on potential health hazards in the world around us; such hazards include not only things like contaminated water, air, and soil but also human behaviors—hand washing and water drinking, for example—that help determine whether those factors become hazards. Fortunately, many environmental hazards (e.g., poor hygiene or contaminated drinking water) are *modifiable*—that is, we can take action to change them. It is on these modifiable hazards that environmental public health workers like Ruiz-Tiben focus their efforts.

public health The science that deals with the health of human populations.

epidemiologist A scientist who studies the causes and patterns of disease in human populations.

environmental health The branch of public health that focuses on factors in the natural world and the human-built environment that impact the health of populations.

When trying to stop an outbreak like GWD, epidemiologists have four basic steps to follow: Identify the source of the problem (how are people getting sick?), determine how the disease progresses, identify actions to decrease the spread of infections, and, finally, implement those actions and track their success.

For GWD, Ruiz-Tiben already knew the source and method of transmission of the disease, and he knew how the disease progressed in the victim. His work was also aided in another way—when the Carter Center began its quest to eradicate GWD, humans were the only known reservoir for adult worms (only larvae can survive in copepods, and just for a few weeks). That meant that the third step in controlling the Nigerian outbreak—determining what to do to break the Guinea worm's life cycle and thus obliterate the disease—was straightforward, too. Ruiz-Tiben and his crew simply needed to change human behavior.

Knowing all this, Ruiz-Tiben could then focus on the fourth and last epidemiological step needed to end the outbreak—provide needed resources and education to populations affected by GWD. He knew that convincing villagers to filter their water before drinking it would stop new infections. Getting the villagers to apply a mild pesticide called Abate would decontaminate area ponds. And teaching infected individuals to avoid communal swimming or bathing while worms were emerging from the skin, and treating new infections as soon as blisters emerge, would break the worm's life cycle once and for all. Their goal was total eradication of GWD; if they succeeded, it would be only the second disease in human history (after smallpox) to be completely wiped off the face of Earth. **INFOGRAPHIC 3**

As straightforward as it all sounded, Ruiz-Tiben found it difficult to convince villagers that water they consumed a year prior had caused their disease. He thought he might make some headway by showing them the dead copepods in the water, but they stood firm in their resistance.

As many of us already know, it can be difficult to change human behavior. The United States has recently been dealing with fallout from this problem as it pertains to vaccines. Many dangerous childhood infections can be prevented with childhood immunizations, yet in recent years, a small proportion of Americans has been choosing not to vaccinate their children over mistaken fears that vaccines are unsafe, a decision with devastating consequences.

Consider measles—a serious disease that can lead to deafness, brain damage, and, rarely, death. It is so contagious that a person can be infected by walking into a room where an infected person coughed *hours*

INFOGRAPHIC 3 **PUBLIC HEALTH PROGRAMS CAN HELP ADDRESS ENVIRONMENTAL HEALTH PROBLEMS**

An examination of the life cycle of the Guinea worm demonstrates steps environmental public health officials can use to address environmentally mediated infectious disease.

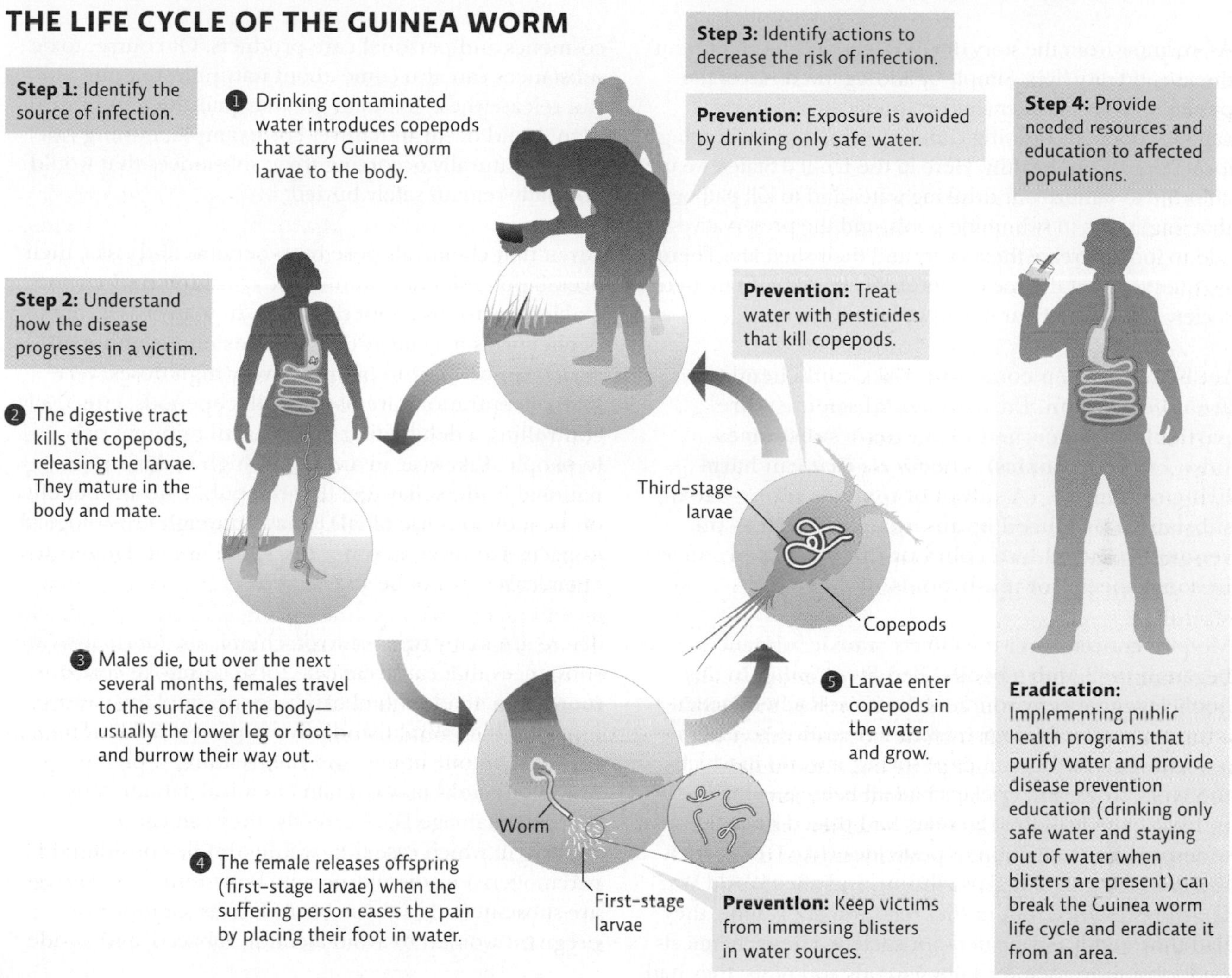
Achieve
THE LIFE CYCLE OF THE GUINEA WORM
Step 1: Identify the source of infection.
1 Drinking contaminated water introduces copepods that carry Guinea worm larvae to the body.
Step 3: Identify actions to decrease the risk of infection.
Prevention: Exposure is avoided by drinking only safe water.
Step 4: Provide needed resources and education to affected populations.
Step 2: Understand how the disease progresses in a victim.
2 The digestive tract kills the copepods, releasing the larvae. They mature in the body and mate.
Prevention: Treat water with pesticides that kill copepods.
Third-stage larvae
Copepods
3 Males die, but over the next several months, females travel to the surface of the body—usually the lower leg or foot—and burrow their way out.
5 Larvae enter copepods in the water and grow.
Eradication: Implementing public health programs that purify water and provide disease prevention education (drinking only safe water and staying out of water when blisters are present) can break the Guinea worm life cycle and eradicate it from an area.
Worm
4 The female releases offspring (first-stage larvae) when the suffering person eases the pain by placing their foot in water.
First-stage larvae
Prevention: Keep victims from immersing blisters in water sources.

Why is the management of GWD considered a public health issue rather than just a personal health issue?

before. The United States had more cases during the first five months of 2019 than it had any year since 1994, even though the infection was declared eradicated in 2000. These new outbreaks are largely due to falling vaccination rates: A single infected person can quickly infect others who might not have been vaccinated or who are too young or ill to receive the vaccine.

This outbreak highlights the importance of education in public health. Some people still believe that vaccines cause autism, a claim that has been thoroughly debunked numerous times in large, carefully conducted studies. In fact, the small study that initially linked vaccines with autism was later shown to be fraudulent. Yet fear and misinformation linger, in part because of a lack of education and the proliferation of erroneous information. Like smallpox, measles has no animal vector, so the disease could be globally eradicated with a strong vaccination program and enough public support and education.

4 CHEMICAL HAZARDS

Key Concept 4: The use of industrial chemicals in modern society has greatly improved the quality of life, but the environment is also now full of natural and synthetic toxic substances that have the potential to harm living things.

As we know from the story of GWD, chemicals can prevent disease and save lives: Simply by adding low doses of the pesticide Abate to water sources, public health workers can kill copepods carrying Guinea worm, thereby keeping local populations healthy. Here in the United States, we use chlorine to sanitize our drinking water and to kill pathogens that might lurk in swimming pools, and the preservatives we add to food increase their safety and their shelf life. There is no question that the use of industrial chemicals in modern society has boosted our quality of life.

Yet benefits often come with risks, and chemicals are no exception. Environmental scientists are particularly concerned about **toxic substances** (also known as **toxics**), chemicals that can harm living organisms. (A subset of toxics is *toxins*—toxic substances produced by an organism, such as the venom produced by a cobra or the poisons produced by some species of mushrooms.)

Modern concern over the effects of toxic substances began in 1962, with a book called *Silent Spring*. In this book, legendary environmental activist Rachel Carson, a former marine biologist, asked her readers to imagine a world without the sounds of spring, a world in which the birds, frogs, and crickets had all been poisoned by toxic chemicals. Just 20 years had passed since the widespread introduction of pesticides (like DDT, which was used to control malaria during and after World War II), she explained, but in that relatively short time, they had thoroughly permeated our society. These chemicals were obviously great for killing weeds and pests; they had done an amazing job keeping mosquito-borne diseases like malaria at bay during World War II and combating hunger by boosting global food production. But no one had paid attention to the effects they might have on nontarget species or on their (or our) ecosystems. *Silent Spring* created an uproar, which led to much stricter regulations for chemical pesticides in general, and, in the United States, a complete ban on DDT in particular. But half a century later, we are still struggling to effectively regulate the chemicals in our world—both synthetic (human-made) and natural.

Carson's worries were not unfounded. Toxic substances pervade the human-built and natural environments—in addition to industrial chemicals and pesticides, toxics are found in building materials, fabrics and carpets, food containers, cleaning supplies, paints—even in cosmetics and personal care products. Of course, toxic substances can also come about naturally, but our actions can release these toxics in greater quantities and rates than would occur in nature. For example, mining can release naturally occurring toxic substances that would normally remain safely buried.

toxic substance/toxic A substance that causes damage when it contacts, or enters, the body.

Given that chemicals pose both benefits and risks, their overall potential impact must be considered when making decisions about their use. In many cases, the use of chemicals may be justified. For example, although Abate can be toxic to people at very high doses, very low concentrations are used to kill copepods, effectively controlling a debilitating disease with minimal risk to people. Likewise, in areas with high malaria rates, national leaders may feel that the public health benefits of the judicious use of DDT may outweigh its ecological impacts. However, in other cases, the use of dangerous chemicals may not be so justified.

There are many types of toxic chemicals. *Carcinogens* are substances that cause cancer. Tobacco smoke contains more than 40 identified carcinogens; smoking causes lung, bladder, throat, and other types of cancer. The furniture in our homes and the cleaning products under our sinks may contain chemical carcinogens. *Mutagens* damage DNA directly; they can cause cancer (in which case they would also be considered carcinogens) or disrupt normal body function. *Teratogens* are substances that cause birth defects; cautions to pregnant women to avoid alcohol, tobacco, and a wide

Bettmann/Getty Images

The first public test of an insecticidal fogging machine at Jones Beach State Park, New York, in July 1945. As part of the testing, a 4-mile area was blanketed with the DDT fog.

variety of pharmaceuticals reflect the need to protect the developing child from teratogens.

Some chemicals, known as *poisons*, damage or kill cells or disrupt biochemical processes (the rat poison warfarin, for example, disrupts blood clotting, causing the animal to bleed internally). Other chemicals that cause local tissue damage, such as chlorine in pool water making your eyes red and itchy, are called *irritants*. *Sensitizers* are substances that can lead to the development of an allergic response after repeated exposure. The urushiol oil in poison ivy is a natural example, but many chemical solvents in cleaners, degreasers, and glues contain sensitizing chemicals. Some chemicals have multiple effects: Formaldehyde is an irritant, sensitizer, mutagen, carcinogen, teratogen, and poison.

Another class of hazardous chemical is what's known as an *endocrine disruptor*—a chemical that interferes with the body's system of hormones. Endocrine disruptors often mimic hormones and interfere with development or bodily functions. Many are estrogenic—they mimic the female hormone estrogen and thus have a feminizing effect on the body. Petroleum-based chemicals such as those in many plastics (bisphenol A [BPA], phthalate) and pesticides, and some pharmaceuticals are endocrine disruptors in humans and other species. **INFOGRAPHIC 4**

Exposure to high enough doses of poisons and irritants, and even a small dose of a chemical to which one has developed an allergy, leads to **acute effects**—like headache or breathing difficulty that comes from inhaling paint or solvent fumes. Chemicals that are considered safe in low doses—such as Abate in drinking water—can nevertheless cause acute effects if a person mishandles a concentrated solution. **Chronic effects** can emerge over time after long-term exposure to low doses of a toxic. Examples of chronic toxicity abound: the health problems that emerge after years of breathing polluted air, the gradual decrease in sperm counts in men exposed to estrogen mimics.

Some people and groups are more likely to be exposed to toxic substances than others; it might depend on where they live or their occupations, but it is often linked to their race or economic status. This fact challenges the idea of environmental justice—that access to a clean, healthy environment is a basic human right. In 2014 and 2015 in Flint, Michigan, a city in which 42% of the population lives in poverty, families were being exposed to lead—a metal that affects children's brain development and can cause serious cognitive deficits—in their drinking water. As is often true with environmental issues, it is minority and lower-income groups who most often bear the brunt of negative impacts of societal choices, and Flint was no exception. Most of the affected residents lacked the economic or political clout to advocate for change. Scientists and activists joined the fight and helped uncover the truth about how Flint's water had become contaminated and demand justice. (See Infographic 3 in Module 4.2 for more on environmental justice.)

acute effect Adverse reaction that occurs very rapidly after exposure to a toxic substance has occurred.

chronic effect Adverse reaction that happens only after repeated long-term exposure to low doses of a toxic substance.

INFOGRAPHIC 4 TOXIC SUBSTANCES

We are exposed to a variety of toxic substances in our daily lives—at home, in our food, at our workplace, and the outdoor environment. Here, we show some of the different types of toxic substances, based on their harmful effects.

TYPE OF TOXIC SUBSTANCE	EFFECT	EXAMPLES
Poison	Causes direct damage upon exposure at a high enough dose	Pesticides, cleaning solutions, drain cleaner, pharmaceuticals, antifreeze
Irritant	Causes localized damage to tissue such as skin or eyes	Household cleaners, chlorine, fabric softeners
Sensitizer	Can cause an allergic reaction to develop	Formaldehyde, latex
Carcinogen	Causes cancer by causing mutations in DNA	Components in tobacco smoke, paints, perchloroethylene (dry cleaning solvent)
Teratogen	Causes birth defects	Alcohol, components in tobacco smoke, some pharmaceuticals, heavy metals (e.g., mercury and lead)
Endocrine disruptor	Interferes with the hormones of the body	Many chemicals in plastics such as bisphenol A (BPA) and phthalates; some pesticides; glycol ethers (in paints, cosmetics)

? Could a toxic substance be classified into more than one category?

5 FACTORS THAT AFFECT TOXICITY

Key Concept 5: Chemical characteristics (e.g., potency, persistence, and solubility) of a substance will affect its toxicity. The route of exposure, personal traits of individuals, and interactions with other chemicals also influence its potential for harm.

Several key factors determine how dangerous a given toxic will be; some of those factors have to do with the chemical's own characteristics. The first such characteristic is a chemical's **potency**—how much or little of it is needed to cause harm. In general, the more potent a chemical is, the less of it is needed to damage an organism. For example, sodium cyanide is very potent; a tiny amount will kill a person. Sodium chloride (table salt) is far less potent; it would take a great deal more of it to do the same amount of harm. But that doesn't mean sodium chloride is always harmless. As toxicologists are fond of saying, "The dose makes the poison." In other words, all chemicals are dangerous if the dose is large enough—even water. Conversely, most substances are safe, and some are even helpful, in low enough doses. Consider the rat poison mentioned earlier, warfarin. It is used as a blood-thinning drug in patients with high blood pressure (but in concentrations much, much smaller than found in the rat poison). However, there are some substances so toxic that even minuscule amounts will cause problems.

Another factor that affects a chemical's toxicity is its **persistence** or degradability—how easily it breaks down into its constituent parts. Because a persistent chemical is chemically stable, it lasts a long time, affecting ecosystems and the organisms that live there well after its initial release. One reason why Abate is not considered as worrisome as some other pesticides is because it is not persistent—bacteria and light degrade it within days or weeks. It also binds easily to sediment and plastic containers, so it doesn't persist in the water very long. Many synthetic chemicals are, on the other hand, quite persistent, and sometimes that is by design (the chemicals in plastics are formulated to last a long time so the products made from them are durable).

potency The dose size required for a chemical to cause harm.

persistence A measure of how resistant a chemical is to degradation.

solubility The ability of a substance to dissolve in a water- or fat-based liquid or gas.

bioaccumulation The buildup of a substance in the tissues of an organism over the course of its lifetime.

biomagnification The increased concentration of substances in the tissue of animals at successively higher levels of the food chain.

How long a chemical remains in our bodies is another factor that affects its toxicity, and that "residence time" is affected by the chemical's **solubility**—that is, whether it can dissolve in fat or water. Its solubility is important because it impacts whether or how readily we can excrete it. Water-soluble chemicals have the advantage of being a substance we can excrete—our kidneys (our main organs of excretion) can extract them from our bloodstream and expel them via urine. This means they don't linger in our bodies for very long. But water-soluble toxics are a particular problem for aquatic organisms that can easily take up these substances from their watery environment. This exposure can wreak slow havoc on aquatic environments and, by extension, on the ecosystems that surround them.

Fat-soluble chemicals present an extra level of complexity. Our cells can readily absorb these chemicals and once they're inside, our bodies have a hard time expelling them—kidneys can only excrete water-soluble substances. The liver can convert some fat-soluble molecules into a water-soluble by-product, allowing it to be excreted in urine. (This is the case for Abate.) But when our livers can't make this conversion, fat-soluble chemicals are stored in our fatty tissue where they can build up in a process known as bioaccumulation.

Bioaccumulation refers to the buildup of substances in the tissue of an individual organism over the course of its lifetime. Most of the chemicals in plastics, for example, are fat-soluble and most if not all of us harbor these and other chemicals in our tissues. Bioaccumulation also leads to a phenomenon known as **biomagnification**, the fact that animals higher on the food chain bioaccumulate more chemicals than the organisms they eat. Here's why: When animals that are higher up on the food chain eat other animals that have bioaccumulated toxics, they consume their preys' lifetime dose of those toxics—and they do this every time they eat. Top predators, such as tuna, can have more than a million times the amount of a toxic as an organism at the bottom of the food chain. This impacts us as well: When we eat them, we consume all the toxics that they have picked up in a lifetime of preying on smaller animals. **INFOGRAPHIC 5**

Other factors that affect toxicity pertain not to the chemical itself, but to the context of the exposure or the individuals affected. For example, both the route of exposure (whether it was inhaled, consumed, or injected, or just touched the surface of the skin) and the dose of that exposure (a little? a lot?) play a role in determining how serious the effects of toxic exposure

INFOGRAPHIC 5 BIOACCUMULATION AND BIOMAGNIFICATION

Animals can acquire fat-soluble toxic substances through air, water, or food sources. The substances build up in the tissue of the animal over its lifetime if it has continued exposure. These fat-soluble substances are passed on to predators.

BIOACCUMULATION: (*occurs in the individual*)

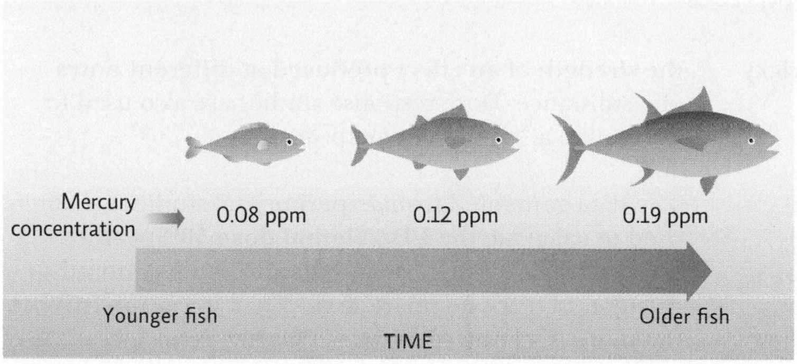

The fish accumulates some mercury every day; the longer it lives, the more it will accumulate and store in its tissues.

BIOMAGNIFICATION: (*a food chain phenomenon*)

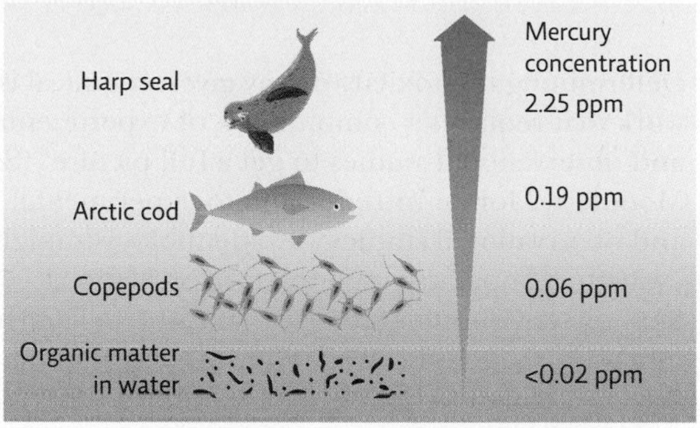

The harp seal ends up with a much higher dose of mercury than other organisms that are lower on the food chain because its prey have higher levels of mercury than prey eaten by consumers lower on the food chain.

Why don't animals bioaccumulate or biomagnify water-soluble substances?

will be. Personal characteristics of the individual being exposed to the toxic also influence its effect. Some people are genetically predisposed to be more (or less) sensitive to a chemical; illness or age may also make a person more vulnerable. The body is particularly sensitive to chemical exposures during the stages of embryonic and child development, putting the very young at higher risk than adults.

The type and amount of chemicals already in one's system can also influence how the newly introduced toxic behaves. Some chemicals in the body might combine to increase overall toxicity in a predictable manner; such effects are referred to as **additive effects**. For example, ampicillin and imipenem are two antibiotics that can be taken for a bacterial infection. Each works in a slightly different way and, taken together, they boost the effectiveness of their treatment.

Other chemicals may reduce toxicity due to interactions between the toxics that "cancel each other out," or at least lessen the damage; these are known as **antagonistic effects**. Antivenom used to treat a snake bite is an example. Drugs, too, may lessen the effect of each other—if taken with aspirin, ibuprofen can reduce the effectiveness of aspirin as a blood thinner. And others still may work together to produce **synergistic effects**, which are effects even bigger than either chemical would be expected to produce on its own. Synergistic effects can be beneficial: Laboratory studies have shown, for instance, that mixing Abate with a plant extract called devil's hair increases its effectiveness nearly fivefold. Mixing Abate with devil's hair could allow public health workers to use much less Abate and get the same disease-protecting effects.

But synergistic effects can be dangerous, too: It is for this reason that we are warned not to mix drugs (or drugs and alcohol). For example, acetaminophen (Tylenol) is a safe, over-the-counter pain medicine. It is processed in the liver by two different enzyme pathways; the primary pathway breaks down acetaminophen to safe by-products that are then excreted, but the other pathway converts a small amount of the drug to a toxic by-product, even at the recommended dose. Normally, this is a very small amount and it doesn't harm the person. If a person drinks alcohol while taking acetaminophen, however, the primary pathway gets tied up processing the alcohol and the acetaminophen gets processed by the secondary pathway, potentially producing enough of the toxic by-product to cause death. (An overdose of acetaminophen will also do this, so always take the recommended amount and no more!)

Because we are never exposed to only one hazardous chemical, predicting the effect of exposure to any of the chemicals to which we are exposed, or may have bioaccumulated, is difficult to do. Still, scientists try, and they use a variety of methods to evaluate a chemical's potential for harm.

additive effects Exposure to two or more chemicals that has an effect equivalent to the sum of their individual effects.

antagonistic effects Exposure to two or more chemicals that has a lesser effect than the sum of their individual effects would predict.

synergistic effects Exposure to two or more chemicals that has a greater effect than the sum of their individual effects would predict.

6 STUDYING TOXIC SUBSTANCES

Key Concept 6: A variety of scientific studies are needed to determine toxicity and safe doses for chemical exposure. Risk assessments allow us to weigh the pros and cons of a chemical; applying the precautionary principle is needed when there is uncertainty about the risk but a chance for significant harm.

Determining the toxicity of any given chemical is tricky work that requires a combination of experimental and observational studies to get a full picture. (See Module 1.2 for an introduction to experimental and observational studies.) Epidemiologists might evaluate a population to look for correlations between environmental factors and the incidence of a disease—such as the availability of proper sanitation and outbreaks of cholera. These observational studies (known as *epidemiology studies*) don't manipulate any factors—they just look for commonalities among those who have a particular health problem.

dose-response study An experiment that tests the strength of the effect produced at different doses of a substance.

LD_{50} (lethal dose 50%) The dose of a substance that would kill 50% of the test population.

NOAEL (no-observed-adverse-effect level) The highest dose where no adverse effect is seen.

Animal studies are particularly useful because they enable scientists to see the impact of a chemical in an intact organism with fully functioning systems. One common experimental study, a **dose-response study**, exposes test subjects (living organisms or cells in a Petri dish) to various doses of the chemical to evaluate the strength of an effect produced at different doses of a substance. Dose-response studies are also used to determine a "safe level of exposure."

The data generated from experimental studies are often used to calculate the **LD_{50} (lethal dose 50%)**—the dose that kills 50% of the population. This standard calculation helps compare toxics: The lower the LD_{50}, the more toxic the substance. The LD_{50} typically varies from species to species; Abate has a low LD_{50} for copepods (0.1 ppm), so it can kill them in very small concentrations, but it has a high LD_{50} for most vertebrates, which means it is much less toxic to this class of organisms.

To identify a "safe dose," scientists often calculate the **NOAEL (no-observed-adverse-effect level)**, the highest dose where there is no adverse effect. The actual safety standard (i.e., how much of a chemical is acceptable for humans, or for environmental entities like waterways) is set much lower than the NOAEL, to account for uncertainty arising from the limitations of experimental studies (e.g., can't experiment on humans, can't account for all chemical interactions, etc.). **INFOGRAPHIC 6**

INFOGRAPHIC 6 DOSE-RESPONSE STUDIES

Achieve

Dose-response studies evaluate the effect a toxic substance has on test subjects at various doses. Charting the change in the response being measured (e.g., appearance of a rash, impaired kidney function, death) as the dose increases is one step in trying to determine a safe dose or level of exposure.

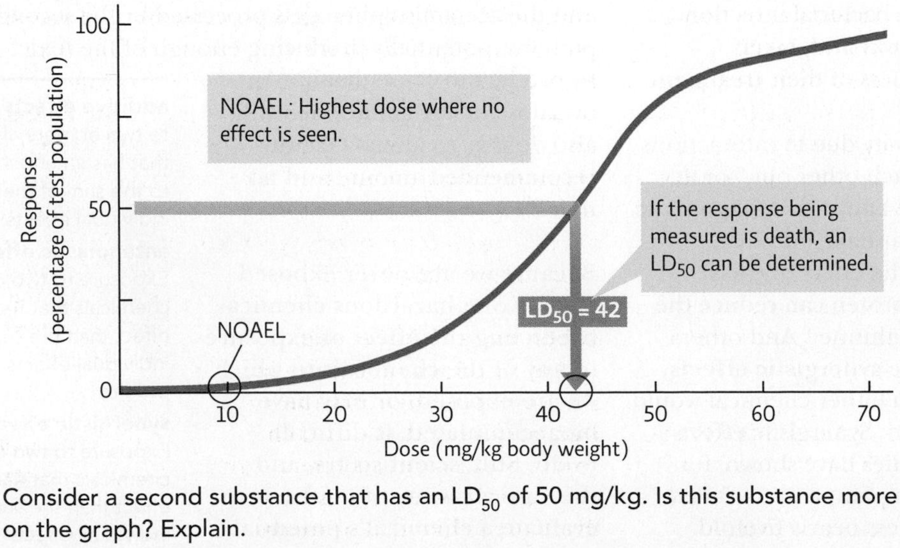

"SAFE DOSE"
Because there is uncertainty in the determination of a "safe dose," regulatory agencies factor in a safety margin of 100 or even 1,000 to be on the safe side. Example: If we have evidence indicating that a chemical is safe at 100 mg/kg, we might set the environmental standard (what is allowed) at 1 mg/kg (or even 0.1 mg/kg if children are at risk).

? Consider a second substance that has an LD_{50} of 50 mg/kg. Is this substance more or less toxic than the one shown on the graph? Explain.

To understand whether it is worthwhile to use a particular chemical in a particular situation, or whether to remove it if it's been found in an environment, public health scientists conduct a **risk assessment**, which involves weighing the risks and benefits of taking such an action. Not all chemicals will cause problems; determining which ones will, and in which situations, is a difficult challenge. Researchers have to consider the *likelihood* of a bad event occurring as well as the *seriousness* of the event if it were to happen. When one or both of these go up, it becomes more reasonable to take prompt measures to mitigate risks.

A risk assessment can help officials decide whether to invoke the **precautionary principle**—the idea that, when there is uncertainty about an outcome but the potential for serious problems, it makes sense to add in a margin of safety. Regulatory agencies use this "better safe than sorry" principle all the time—once safety testing is done, they leave a wide safety margin of 100 or even 1,000 when setting the acceptable exposure limits. Indeed, the WHO incorporated a safety margin into its risk assessment for Abate: Although the estimated NOAEL for humans is 2.3 mg/kg body weight per day, the WHO set the acceptable daily intake 100 times lower, at 0.023 mg/kg body weight per day—an amount far below the maximum an average person should consume. Still, the WHO recommends that individuals—children and bottle-fed babies especially—rely on other sources of drinking water for several weeks after a pond is treated with Abate, just to be extra cautious.

risk assessment The systematic process of weighing the risks and benefits of an undertaking to determine whether to take action.

precautionary principle Acting in a way that leaves a safety margin when the data are uncertain or severe consequences are possible.

7 ADDRESSING ENVIRONMENTALLY MEDIATED HEALTH PROBLEMS

Key Concept 7: Policies and actions that improve air and water quality, regulate dangerous chemicals or actions, and empower communities to meet their own needs will reduce environmental health hazards.

Governments often pass laws to protect their citizens from environmental hazards. In the United States, the **Toxic Substances Control Act (TSCA)** is the primary federal law governing chemical safety. It was first passed in 1976, but at that time, some 65,000 existing chemicals were "grandfathered in," meaning that they were not subject to the law's safety testing provisions. Of the roughly 20,000 chemicals that have been introduced to the marketplace since then, only about 200 have been rigorously tested.

The original TSCA had some major flaws that seriously restricted the ability of the U.S. Environmental Protection Agency (EPA) to identify and regulate toxic substances. For example, the EPA was required to show that the benefit of regulation outweighed the costs of determining if it was hazardous—an almost impossible task: How could they show benefits outweighed costs before they determined its safety? In addition, trade secrets regarding chemical composition had strong protections and the manufacturer did not have to show that a chemical was safe; the EPA had to prove it was unsafe. The TSCA was in need of major reform.

That reform came in 2016 with the passage of the Frank R. Lautenberg Chemical Safety for the 21st Century Act (a law passed with substantial bipartisan support), which shifted the law into a more precautionary stance. A manufacturer must now show a chemical is safe before it receives approval to go on the market. On its end, the EPA must complete a risk assessment in a timely fashion to determine what type of testing is required. And what constitutes a trade secret is more strongly regulated, requiring much more transparency from chemical manufacturers.

Implementing the new law, however, may be difficult with recent changes at the EPA. Under the Trump administration, the EPA has been accused of undermining the law by changing the risk assessment process in ways that could underestimate risk. For example, when evaluating risk, the EPA no longer considers the effects of long-term exposure to chemicals via air, water, or soil—it only considers the effects of direct-contact exposure—and critics have raised concerns that the studies the EPA considers during risk assessments favor industry reports over independent research.

Other major U.S. laws that regulate toxic substances include the **Resource Conservation and Recovery Act (RCRA)**, which regulates the management of solid and hazardous waste, ensuring that waste is handled properly.

Toxic Substances Control Act (TSCA) The primary federal law governing chemical safety.

Resource Conservation and Recovery Act (RCRA) The federal law that regulates the management of solid and hazardous waste.

(See Module 5.3 for more details.) The Clean Water Act, passed in 1972, regulates how pollutants are released into lakes, rivers, streams, wetlands, and coastal areas, while the Clean Air Act regulates air pollution. (See Modules 6.2 and 10.1 for more on these acts.) To reduce our exposure to physical threats, other regulations focus on safety in the workplace, in buildings (e.g., building codes), and on roadways.

Effective policies that address environmentally mediated diseases often focus on ways to reduce the factors that increase the occurrence or transmission of the disease. This includes taking steps such as protecting the food supply as well as keeping water supplies safe through sanitation and purification—the exact approach Ruiz-Tiben hoped to use when he set out to control the spread of GWD in Nigeria.

Ruiz-Tiben has worked tirelessly to convince villagers to support his efforts to combat GWD, and those efforts are paying off. Villagers guard treated ponds to ensure that nobody with emerging worms comes into contact with the water. The sick are quarantined and cared for during the month or more that it takes the worms to emerge. Young boys—travelers and hunters—carry whistle-shaped cylinders tied to strings around their necks—portable filters that allow the boys to safely drink directly from an environmental water source. And women are teaching their protégés the importance of filtering water before giving it to their families.

It's the picture of success that Ruiz-Tiben and his colleagues have envisioned for decades. "It just shows what you can accomplish when the support is there," Ruiz-Tiben says. His work ties in closely to the United Nations' Sustainable Development Goals (SDGs): SDG 3 is Good Health and Well-Being. A close look at the 17 goals shows a clear link between improving environmental health and reaching these goals. For example, access to clean water and sanitation will improve health, helping families move out of poverty. (Healthy individuals are more likely to be able to work and support their families.) Likewise, providing education and job opportunities will provide the funds a family needs to acquire clean water and pay for sanitation programs. (See Module 1.1 for more on the UN Sustainable Development Goals.) **INFOGRAPHIC 7**

As with most other environmental problems, there are no easy solutions. To conquer GWD, environmental health workers like Ruiz-Tiben have had to battle indifference, poverty, human stubbornness, and now a new complication: It appears that dogs can act as a host for the species of Guinea worm that infects humans. Cases of dog GWD have increased dramatically in the

INFOGRAPHIC 7 REDUCING ENVIRONMENTAL HEALTH HAZARDS MOVES US CLOSER TO SUSTAINABILITY

The 17 UN Sustainability Development Goals (SDGs) are closely tied to environmental health, especially Goal 3: Good Health and Well-Being. Steps that improve air and water quality and reduce exposure to disease-carrying vectors can reduce environmental health hazards but require education and effective public policy for success.

ACTIONS THAT REDUCE ENVIRONMENTAL HEALTH HAZARDS

STEP	EXAMPLES
Provide access to clean water	Dig wells and filter surface water, including using personal filters such as the LifeStraw; provide financial assistance to low-income areas for these technologies.
Improve sanitation and hygiene	Keep sewage out of surface waters by building latrines, planting streamside vegetation to reduce runoff, and keeping animals out of water sources.
Reduce vector exposure	Remove vector habitat (like standing water for mosquitoes), provide barriers like mosquito netting, and apply pesticides to kill vectors; vaccinate pets.
Reduce air pollution	Use cleaner-burning fuels, better-ventilated indoor stoves, and solar ovens to reduce indoor air pollution; adopt and enforce air quality standards to reduce outdoor air pollution.
Provide education	Teach individuals how to avoid exposure to pathogens and how to protect themselves from infection and hazardous chemicals.
Establish effective public policy and/or funding	Pass laws and regulations aimed at reducing environmental hazards and improving health care. In areas without government backing or funding, seek funding from nonprofits.

? Which do you think will be the biggest impediment to the eradication of GWD in Africa—the implementation of technical solutions to provide safe water or the education of people in local communities on ways to reduce their exposure to contaminated water? Explain.

past few years, most notably in Chad where, in the first half of 2019, more than 1,500 dogs were infected with Guinea worm. Dogs cannot transmit the disease to people, but they can introduce Guinea worm larvae into water sources and thus act as an avenue of transmission to copepods, which could go on to infect humans. Efforts now include steps to eradicate the infection from dogs as well as humans, including things like tying up infected dogs to prevent their access to water sources and experimental trials to test the effectiveness of medications to kill the worms in infected dogs.

The battles Ruiz-Tiben and others have been fighting to reduce human cases of GWD have been paying off: In 2009, Nigeria became the fifteenth African country to rid itself of the ancient worm. At that time, it was estimated that just 3,500 or so cases remained throughout the entire continent, and those numbers were dwindling rapidly. By 2012, only 542 human cases were reported in all of Africa; in 2014, that number had dropped to 126. In 2018, there were only 28. As Ruiz-Tiben said in 2018, "President Carter has said he wants the worm gone before he is. We may not be able to fulfill his wish, but we'll do the best we can."

Select References:

Callaway, E. (2016). Dogs thwart end to Guinea worm. *Nature, 529*(7584), 10–11.

Carter Center. (N.d.). *Health programs.* Retrieved from https://www.cartercenter.org/health/index.html

LoGiudice, K., et al. (2003). The ecology of infectious disease: Effects of host diversity and community composition on Lyme disease risk. *Proceedings of the National Academy of Sciences, 100*(2), 567–571.

Pruss-Ustun, A., Wolf, J., Corvalan, C., Bos, R., & Neira, M. (2016). *Preventing disease through healthy environments: A global assessment of the burden of disease from environmental risks.* Geneva, Switzerland: World Health Organization.

World Health Organization. (2018). *Air pollution and child health: Prescribing clean air. Summary* (WHO/CED/PHE/18.01). Geneva, Switzerland: Author. (License: CC BY-NC-SA 3.0 IGO.)

Guinea worm disease is a major impediment to a farmer's ability to work. A health volunteer in Ghana educates children on how to use pipe filters when they go to the fields with their families. Pipe filters, individual filtration devices worn around the neck, work similarly to a straw, allowing people to filter their water to avoid contracting Guinea worm disease while away from home.

Courtesy Louise Gubb/The Carter Center

GLOBAL CASE STUDIES ENVIRONMENTAL HEALTH CONCERNS

Achieve

A wide variety of infectious diseases have a strong environmental component. Some of these are on the rise due to a confluence of conditions that favor their occurrence or spread. Though many of these diseases are predominately found in less developed countries, others are also found in the more developed countries. Several are highlighted on the map below.

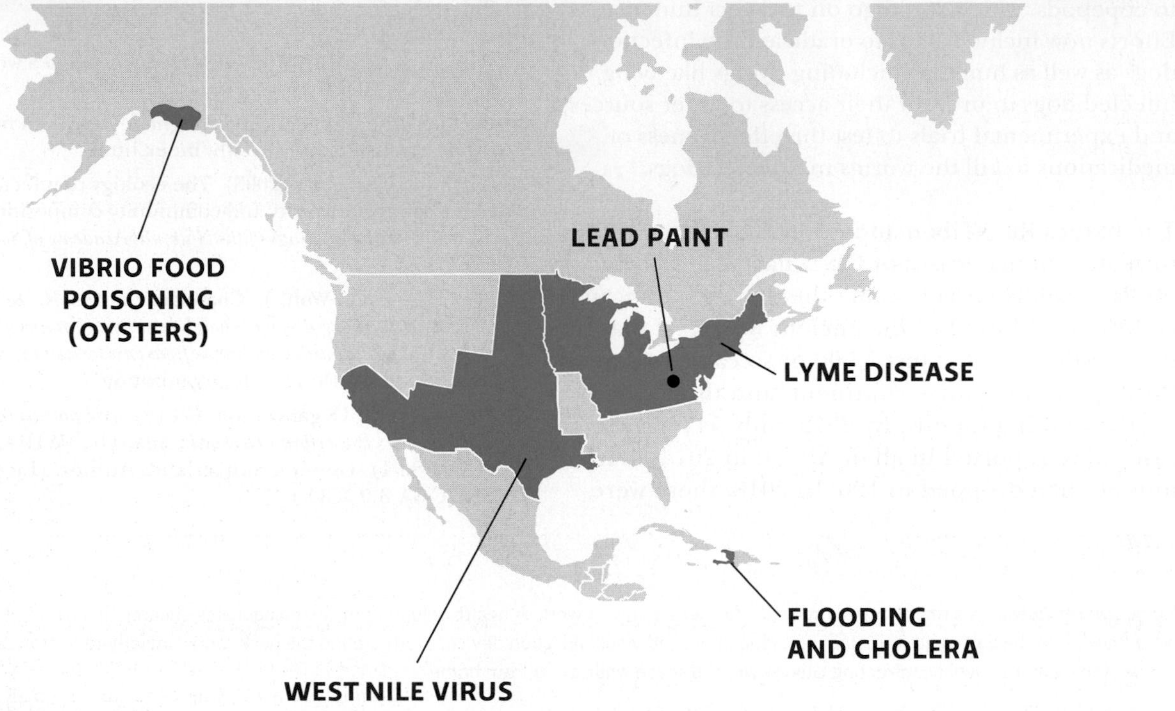

BRING IT HOME

PERSONAL CHOICES THAT HELP

People in developed countries typically do not experience the same prevalence of infectious disease as do those in the developing world. However, outbreaks of illnesses like whooping cough, West Nile virus, bacterial food poisoning, and antibiotic-resistant bacterial infections do occur in the developed world and are largely preventable.

Individual Steps

- Many diseases are spread by contaminated hands. The most effective way to remove infectious bacteria and viruses is with 20 seconds or more of hand washing with soap and water. This is even more effective than using hand sanitizer.
- Follow guidelines by local public health departments when outbreaks occur (e.g., wear face masks and practice social distancing) to limit disease spread.
- Reduce the likelihood that antibiotic-resistant bacteria will emerge by taking the entire prescription of any antibiotic you are prescribed.
- The Safe Drinking Water Act does not require the same level of testing for contaminants in bottled water as it does for tap water. Ask your legislators what steps could be taken to hold bottled water to similar standards.

Group Action

- Mosquitoes are responsible for spreading many diseases. Organize your neighbors to take preventive steps to reduce mosquito breeding, including removing containers that might trap rainwater, draining areas of standing water, and cleaning out rain gutters.
- Organize a fundraising campaign to help purchase pipe filters such as the LifeStraw or finance well digging in areas that need access to clean water.

REVISIT THE CONCEPTS

- A variety of environmental factors negatively affect human health. While air and water pollution are the leading environmental health threats globally, the spread of infectious diseases can also be environmentally mediated.

- Infectious diseases — including zoonotic diseases, which are spread between animals and people — threaten many human populations. Human actions that put people in close contact with one another or that increase habitat for pathogens or their vectors can facilitate the spread of these diseases.

- Unlike an individual's personal physician (whose task is improving the health of their patients), public health officials work to improve the health of a population as a whole by identifying risks to the people in a population and recommending steps to reduce that risk.

- Harnessing the power of chemicals, both natural and synthetic, has greatly improved the quality of life, but our actions have also resulted in the release of many hazardous chemicals that pollute our environment and threaten human and environmental health.

- Characteristics (e.g., potency, persistence, and solubility) of a chemical will affect its toxicity, but the route of exposure, personal traits of individuals (e.g., age, health status, and genetics), and interactions with other chemicals also influence its potential for harm.

- Both experimental and observational studies are needed to determine toxicity and safe doses for chemical exposure. Risk assessments allow us to weigh the pros and cons of a chemical. When there is uncertainty about how dangerous a chemical might be but there is the possibility that it might cause significant harm, we should apply the precautionary principle by factoring in a margin of safety when setting a "safe dose" standard.

- Policies and actions that reduce environmental hazards will also improve the quality of life and are in keeping with the United Nations' Sustainable Development Goals. Several U.S. laws target environmental pollution, and others focus on keeping workplaces, homes, and public places safe. Any actions that improve air and water quality and empower communities to meet their own needs will reduce environmental health hazards.

ENVIRONMENTAL LITERACY Understanding the Issue

1 What environmental hazards do humans face?

1. Give an example of a physical hazard, a chemical hazard, and a biological hazard. Describe a scenario in which exposure to one of the hazards could make a person more vulnerable to another type of hazard.
2. Globally, what is the leading environmental threat to human health and what health problems can it cause?
3. How do factors that affect human health differ between more and less developed nations?

2 How do environmental factors and human actions contribute to the spread of infectious diseases?

4. Explain why GWD is both a waterborne and a vector-borne disease.
5. Identify the risk factors that increase *your* chance of acquiring a zoonotic disease.
6. Give some examples of human modifications of the environment that increase the risk of contracting an infectious disease.

3 What is the field of public health and how does it work to improve the health of human populations?

7. What is the focus of environmental health programs?
8. Some parents choose not to vaccinate their child for childhood illnesses such as measles, even though their child is old enough or healthy enough to be given the vaccination. How can this personal health decision negatively impact public health?

4 What health, societal, and environmental problems result from exposure to chemical hazards?

9. Distinguish between chronic and acute effects caused by exposure to a toxic substance.
10. Why are toxic substances so pervasive in our environment today?
11. Why was lead poisoning in Flint, Michigan, considered an environmental health emergency and a violation of environmental justice?

5 What factors influence the toxicity of a substance?

12. Distinguish between bioaccumulation and biomagnification.
13. Explain the phrase "The dose makes the poison."
14. Explain the factors that could make a given dose of a toxic substance more dangerous for one individual than for another.

6 How is toxicity determined and what role does risk assessment play in determining the "safe dose" of a toxic substance?

15. How is a dose-response study done and what can it tell us?
16. What does LD_{50} measure and why is this measurement useful?
17. Why is it so hard to determine a safe exposure for a given chemical? How do regulators setting safe exposure standards deal with the uncertainty associated with these factors?

7 What actions and regulations reduce the risk of environmentally mediated health problems?

18. Distinguish between the regulatory approach used by the Toxic Substances Control Act of 1976 and the precautionary approach of the Lautenberg Act of 2016. What are the advantages and disadvantages of each?
19. How will reducing poverty help to improve environments and reduce environmental health risks?

SCIENCE LITERACY Working with Data

Cases of Guinea worm disease (GWD) have dropped dramatically in humans but are increasing in dogs. Researchers note that these numbers tend to increase when GWD eradication efforts in humans make progress. Why this happens is not understood.

GUINEA WORM CASES IN CHAD, SHOWN BY MONTH OF WORM EMERGENCE

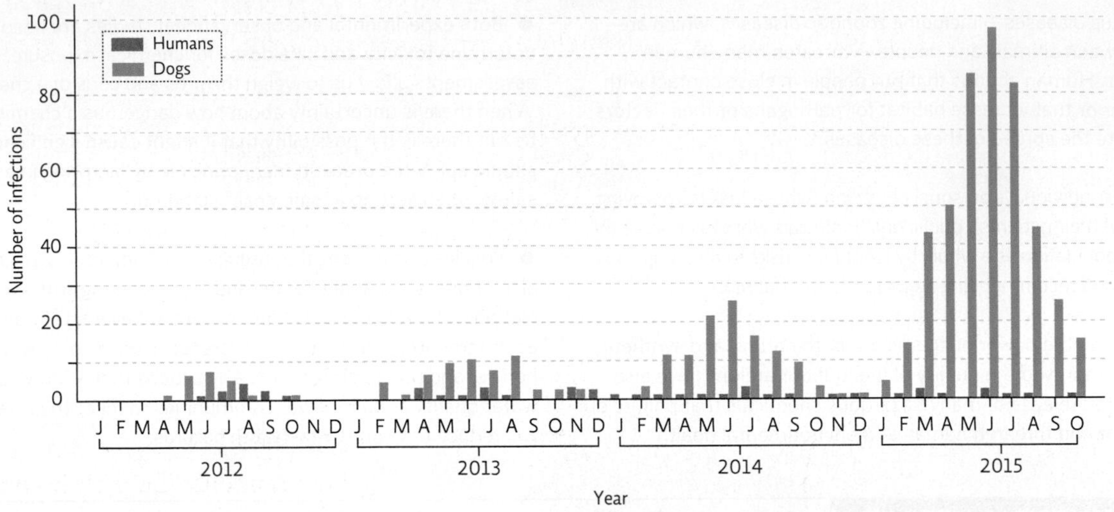

Interpretation

1. How many dogs were found with GWD in 2012, 2013, 2014, and 2015? (Note: 2015 data only goes to October.)
2. What is the general trend regarding GWD cases in humans between 2012 and 2015? How does this compare to the trend of GWD cases in dogs?
3. In what season are the greatest number of Guinea cases (emergence of worms) seen in dogs? In humans?

Advance Your Thinking

4. Worms begin to emerge a year after infection. What season poses the greatest risk of infection? Why do you think this is so?
5. In 2015, Chad began offering a reward for the reporting and containment of dogs with emerging Guinea worms. How might this have affected the totals shown here, or conclusions drawn from these data?

INFORMATION LITERACY Evaluating Information

Humanitarian organizations are increasingly interested in ensuring that their time and energy are invested in the most efficient and effective manner possible. The Bill & Melinda Gates Foundation has led the way with this approach and has donated a large amount of money to the Institute for Health Metrics and Evaluation (IHME) with this goal in mind.

Go to the IHME website (www.healthdata.org). Evaluate the website and work with the information to answer the following questions:

1. Review the topics listed under the "About" tab. What is the overall mission of IHME? Do you believe this mission is reasonable? Explain.
2. Select the "News" link under the "News & Events" tab, and look over the titles shown. Choose two of the articles and read each. Identify each article by name and complete the following evaluation for each:
   a. Is this a primary, secondary, or tertiary information source? Justify your answer.
   b. Identify the authors of each article. Do their credentials qualify them as suitable information sources?
   c. Identify a claim made in each article. Does the article give supporting evidence for this claim? If so, identify the evidence.
   d. Based on your evaluation of the two articles you examined, do you feel that the news coverage is consistent with the IHME's stated goals? Explain.
3. Select the "Research Articles" link under the "Results" tab. Look over the titles and choose two articles to evaluate. For each article, answer a–e:
   a. What type of article is this—primary research, a review article, or an opinion piece (editorial or blog)? How do you know?
   b. How does this differ from the news article you read?
   c. Try to access the actual article: Click on the "Read the article" link above the list of authors. (You may then have to click on another link to access the "Full Text" or a PDF of the article.) Is the full article available for you to read?
   d. Explain what steps you would take to access the article if a full copy is not available from this website. (Do this even if the full article is available for the article you chose.)
   e. How useful are these research articles? Do they help fulfill the mission of the IHME? Explain.

Additional study questions are available at Achieve.macmillanlearning.com. Achieve

CHAPTER 5

Managing Resources: Environmental Economics and Policy

How we choose to manage resources greatly affects our environmental impact. These choices are influenced by economics and bound by policies. As an example of how to turn an environmental problem into a resource, Module 5.3 will look at our options for managing solid waste.

Module 5.1

Environmental Economics and Consumption

An examination of the way economic choices affect human environmental impact and a look at new economic models that may be able to reduce that impact

Module 5.2

Environmental Policy

A look at the importance of environmental policies and how they are made in the United States and globally

Module 5.3

Managing Solid Waste

An evaluation of the problems created by solid waste and ways to reduce the production of waste and its impact

Romolo Tavani/iStock/Getty Images

Module 5.1 Environmental Economics and Consumption

WALL TO WALL, CRADLE TO CRADLE

A leading carpet company takes a chance on going green

Carpet tiles are paving the way for a more sustainable way to produce and market carpet. *Sophie James/Alamy*

After reading this module, you should be able to answer these GUIDING QUESTIONS

1. What are ecosystem services, and why are they important to ecosystems and human populations?
2. What is an ecological footprint?
3. What does the IPAT equation tell us about our impact and the potential to reduce that impact?
4. What are natural capital and natural interest, and how can they be sustainably used?
5. What is true cost accounting, and why should we employ it?
6. What are the main differences between the assumptions and economic models of environmental versus mainstream economics?
7. How can consumers make better choices that reduce their overall environmental impact?

It was the summer of 1994, and Ray Anderson was feeling pretty good about things. His Atlanta-based company, Interface Carpet, was the world's leading seller of carpet tiles—small, square pieces of carpet that are easier to install and replace than rolled carpet—and it was raking in more than $1 billion per year. One day, though, an associate from Anderson's research division approached him with a question. Some customers apparently wanted to know what Interface was doing for the environment. One potential customer had told Interface's West Coast sales manager that, environmentally speaking, Interface "just didn't get it."

Anderson was dumbfounded. The carpet industry was not generally an eco-conscious industry; after all, synthetic carpet is made from petroleum in a toxic process that releases significant amounts of air and water pollution, along with solid waste. Indeed, Interface used more than 1 billion pounds of oil-derived raw materials each year, and its plant in LaGrange, Georgia, released about 5.5 tons of carpet trimming waste to landfills each day. "I could not think of what to say, other than 'we obey the law; we comply,'" he recalled—in other words, his company did things by the book, in terms of the environment. Wasn't that enough? His research associate suggested that the company launch a task force to create a companywide environmental vision. Anderson agreed, albeit reluctantly.

Desperate for inspiration, Anderson began leafing through *The Ecology of Commerce*, a book by environmental activist, entrepreneur, and writer Paul Hawken, which one of his sales managers had lent him. The book told the story of a small island in Alaska, on which the U.S. Fish and Wildlife Service had introduced a population of reindeer during World War II. Although the reindeer thrived for a time on the available plants, eventually the population exploded beyond what the environment could support. The reindeer ultimately died out because, as Anderson explained, "you can't go on consuming more than your environment is able to renew." Yet that, he suddenly realized, was precisely what Interface was doing—using more resources than it could possibly renew.

◉ WHERE IS LAGRANGE, GEORGIA?

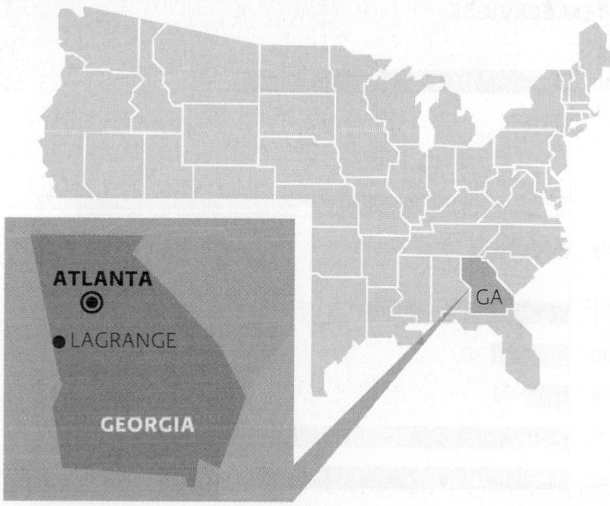

Ray Anderson, founder of Interface. The background displays sample pieces of his carpet tiles.

1 ECONOMICS AND THE ENVIRONMENT

Key Concept 1: Life on Earth depends on ecosystem services provided by nature. Recognizing the value of these services may motivate us to protect them.

Anderson took his quest for sustainability to heart. Hawken's book was a turning point. "As I read the book, it became clear that, God almighty, we're on the wrong side of history, and we've got to do something."

The choices businesses (and, by extension, consumers) make have tremendous impacts on the environment. The amount and type of energy and water they use, the way they handle the waste they produce, the raw materials they use—these decisions affect not only business operations themselves but also Earth as a whole, especially considering the magnitude of the resources and waste that some large businesses use and produce. Anderson realized that he had to make changes to Interface; he needed to build it into a **sustainable**, environmentally sound business. "I didn't know what it would cost, and I didn't know what our customers would pay, so it was a leap of faith," he recalled.

sustainable Capable of being continued indefinitely.

economics The social science that deals with the production, distribution, and consumption of goods and services.

ecosystem services Essential ecological processes that make life on Earth possible.

Businesses that are environmentally mindful aren't limited to simply trying to minimize their impact on nature; they can look to nature as an economic system from which to learn and model their choices. After all, **economics**—the social science that deals with how we allocate scarce resources—is not just about money.

All the resources we depend on come from the environment. Ecological processes like water purification and pollination are essential and economically valuable **ecosystem services** that help produce the resources we depend on, such as timber, food, and water. We take many of these services and resources for granted, but some are priceless because there are no substitutes—such as the oxygen produced by green plants, which we need to survive.

Economist Robert Costanza was the first to try to quantify the economic value of these resources, estimating their value in 1997 to be more than the global GDP (gross domestic product)—that is, more than the dollar value of all the goods and services to change hands in that year. By assigning a dollar value to these valuable, sometimes irreplaceable, services, even though his figures are probably underestimates, we are acknowledging that they have value and this, in turn, might make us more likely to protect the environment that supplies those services. Costanza's most recent analysis reveals that ecosystem services have steadily declined, and he predicts this decline will continue unless policy changes aimed at restraint and wise use of resources are pursued. **INFOGRAPHIC 1**

INFOGRAPHIC 1 VALUE OF ECOSYSTEM SERVICES

Robert Costanza and his colleagues evaluated Earth's ecosystem services and quantified their 2011 values to be around $125 trillion (in 2007 U.S. dollars). When we degrade ecosystems, we reduce their ability to provide these services. He estimates that we have lost $20 trillion in ecosystem services since 1997 due to ecosystem degradation.

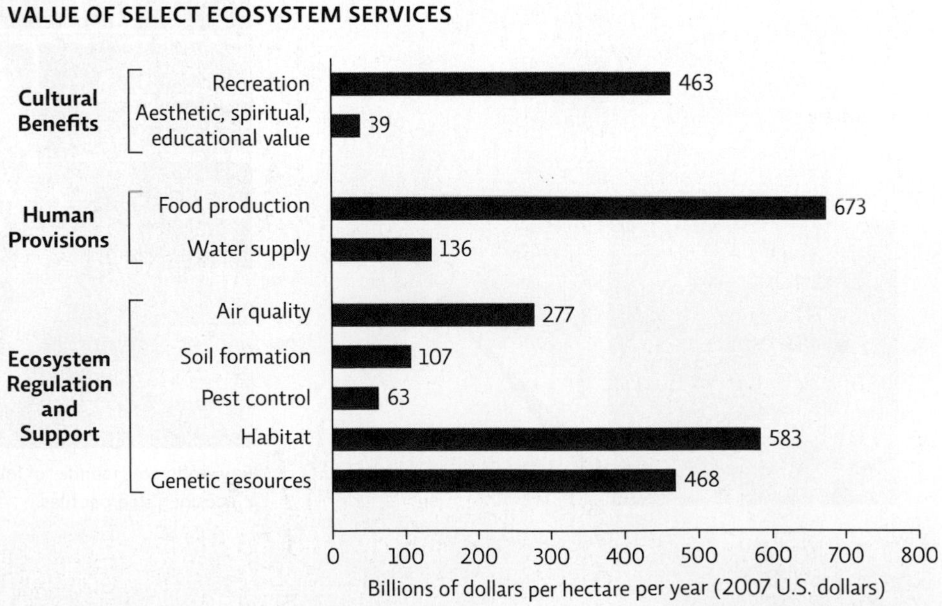

? How would the price of food be affected if we incorporated the ecosystem services of pest control, soil formation, and water supply into that price?

2 MEASURING OUR IMPACT: THE ECOLOGICAL FOOTPRINT

Key Concept 2: Human impact can be measured in terms of our ecological footprint—the amount of land needed to support our lifestyle.

When ecosystems are intact, they are naturally sustainable: They rely on renewable resources and provide services that help replenish and recycle these resources. But ecosystems will only be able to provide us with their valuable goods and services as long as we let them. As Anderson came to realize, when we degrade ecosystems by using more from them than can be replenished, we threaten our planet's ability to provide the services we need, and this ultimately threatens our own future.

Like many other businesses, Interface Carpet uses a lot of resources and generates a lot of waste, giving it a large **ecological footprint**. Businesses, individuals, and nations can calculate their ecological footprint as a way to quantify their impact on the environment as well as identify actions that can decrease that impact. **INFOGRAPHIC 2**

The United States, for instance, has a particularly high per capita (per person) footprint, in that it requires much more land area to support each person than it actually possesses. The country is forced to import resources from other countries and even to export some waste. (See Module 5.3 for more on solid waste issues.) In fact, if the more than 7 billion people who populate the planet all lived like the average person in the United States, we would need the landmass of more than five Earths to sustain everyone.

Like the United States, most developed nations have a high per capita footprint, but even developing nations with smaller per capita footprints contribute to overall impact, some a great deal due to the sheer size of their populations. According to the Global Footprint Network's 2018 assessment, humanity's combined footprint exceeds what is ultimately sustainable—and has done so since 1970. (See Infographic 7 in Module 4.1.)

ecological footprint The land area needed to provide the resources for, and to assimilate the waste of, a person or population.

INFOGRAPHIC 2 ECOLOGICAL FOOTPRINT

The ecological footprint is the land area needed to provide the resources for, and assimilate the waste of, a person or population and may extend far beyond the actual land occupied by the person or population; it is usually expressed as a per capita value (hectares or acres per person). The current world footprint would require about 1.5 Earths to maintain, but obviously we just have one to work with.

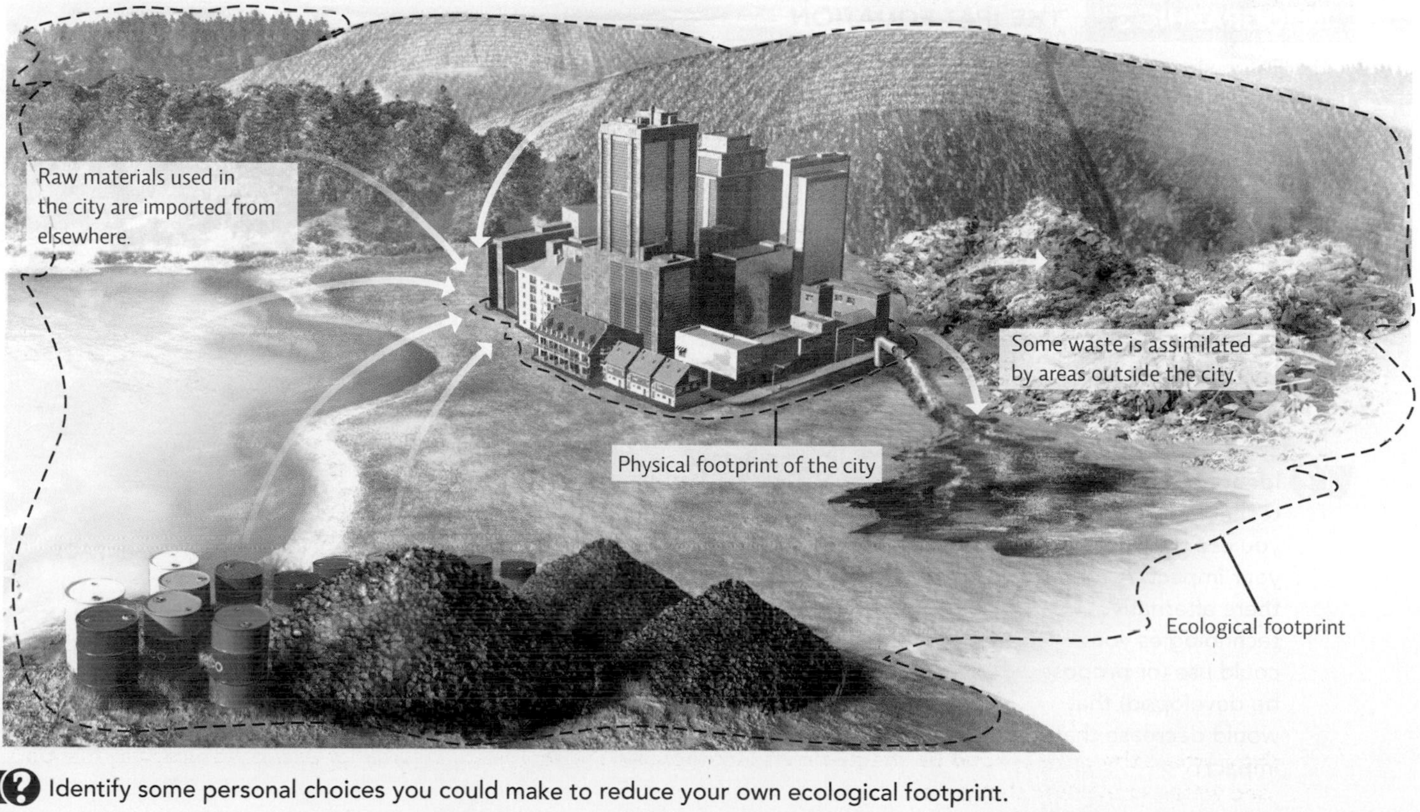

Raw materials used in the city are imported from elsewhere.
Some waste is assimilated by areas outside the city.
Physical footprint of the city
Ecological footprint

Identify some personal choices you could make to reduce your own ecological footprint.

3 FACTORS THAT AFFECT OUR ECOLOGICAL FOOTPRINT: THE IPAT EQUATION

Key Concept 3: The impact of a population generally increases as its size, affluence, and use of technology increase. However, the right technology can reduce resource use and pollution generation, thus helping to decrease impact.

Researchers can use the **IPAT model** to estimate the size of a population's ecological footprint, or impact (I), based on three factors: population (P), affluence (A), and technology (T). The premise is that as population size increases, so does impact. More affluent and technology-dependent populations use more resources and generate more waste than do less affluent and technology-dependent populations; technology allows us to build more things, dig deeper, and fly higher, all of which drain the environment. **INFOGRAPHIC 3**

One caveat with regard to this model is that technology can have the opposite effect: Some technologies can decrease rather than increase, environmental impact. In 2006, for instance, Interface invented a new technology called TacTiles: 2.5 × 2.5-inch squares of adhesive tape that join carpet tiles together. In contrast with traditional "spread on the floor" adhesives, Interface's new tape is safer to use and makes it possible for customers to replace single carpet tiles easily—no need to replace the entire carpet when only a portion is worn or damaged. Interface also developed a way to recycle discarded carpets and fishing nets into new carpet. Their ReEntry 2.0 program has diverted more than 100,000 tons of material from landfills.

IPAT model An equation (I = P × A × T) that measures human impact (I) based on three factors: population (P), affluence (A), and technology (T).

The company is close to attaining its "Mission Zero" goal to eliminate any negative impact on the environment, and the next step, according to CEO Jay Gould, who is carrying on Anderson's legacy, is to go from having no negative impact to having a positive impact—to be what Anderson called a "restorative enterprise." To do that, they are tackling climate change with their "Climate Take Back Plan." As part of that plan, they have developed a carpet backing made from recycled material and plant-based oils that they calculate stores more carbon than it emits. What Hawken realized and Anderson put into action was that if we continue to use resources faster than they are produced, we are going to run out. But it's not just about using up some valuable resources—it's also about not damaging the very ecological processes and species that provide them.

INFOGRAPHIC 3 THE IPAT EQUATION

The IPAT model suggests that the environmental *impact* (I) of a society is based on the size of its *population* (P), its *affluence* (A), and its use of *technology* (T). As any or each of these factors increase, so does the population's overall impact. The right kind of technology, however, can lower overall impact.

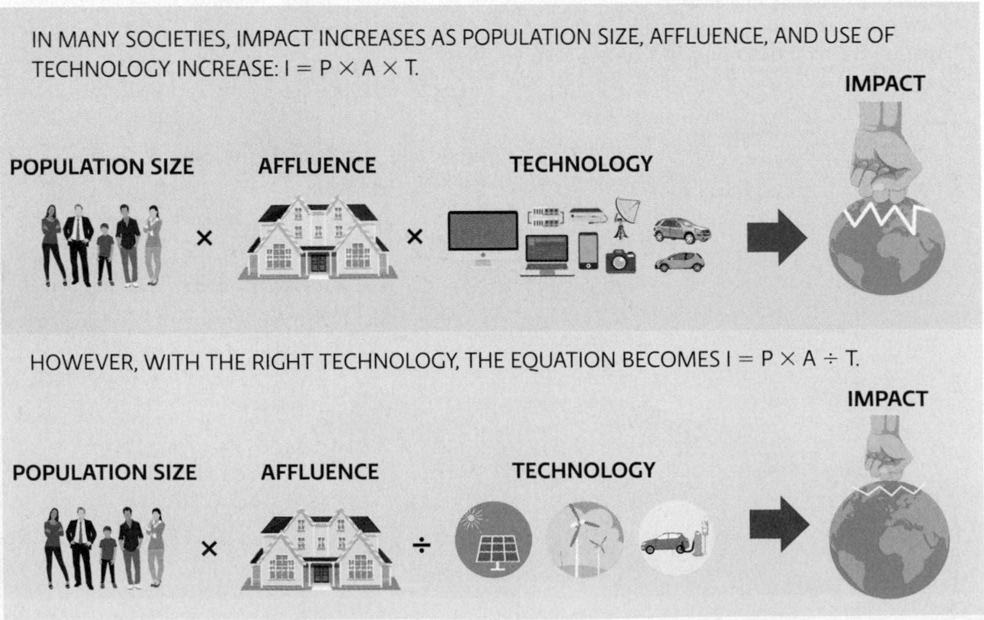
IN MANY SOCIETIES, IMPACT INCREASES AS POPULATION SIZE, AFFLUENCE, AND USE OF TECHNOLOGY INCREASE: I = P × A × T.
IMPACT
POPULATION SIZE
AFFLUENCE
TECHNOLOGY
×
×
HOWEVER, WITH THE RIGHT TECHNOLOGY, THE EQUATION BECOMES I = P × A ÷ T.
IMPACT
POPULATION SIZE
AFFLUENCE
TECHNOLOGY
×
÷

? Identify some technologies that you use that increase your impact. Are there alternative technologies you could use (or propose be developed) that would decrease that impact?

4 NATURAL RESOURCES AS CAPITAL AND INTEREST

Key Concept 4: If we harvest resources at or below the rate at which they are produced—that is, take only the natural interest—we will leave behind enough natural capital to replace what we took.

What kinds of essential resources does Earth provide us? Considered in financial terms, **natural capital** includes the natural resources we consume, like oxygen, trees, and fish as well as the natural systems—forests, wetlands, and oceans—that produce these resources. **Natural interest** is what is produced from this capital, over time—more trees and oxygen, for example—much like the interest you earn with a bank account. Natural interest represents the amount of readily produced resources that we *could* use and still leave enough natural capital behind to, in time, replace what we took. Natural interest might be represented by an increase in a fish population, for instance, or new growth in a forest—basically, the extra that is added in a given time frame. **INFOGRAPHIC 4**

natural capital The wealth of resources on Earth.

natural interest Readily produced resources that we could use and still leave enough natural capital behind to replace what we took.

INFOGRAPHIC 4 CAPITAL AND INTEREST

Natural resources can be compared to the financial concepts of capital and interest. Natural capital is the wealth of resources on Earth and includes all the natural resources we use as well as the natural systems that produce some of those resources (forests, wetlands, oceans, etc.). Natural interest is the amount produced regularly that we could use and still leave enough natural capital behind to replace what we took.

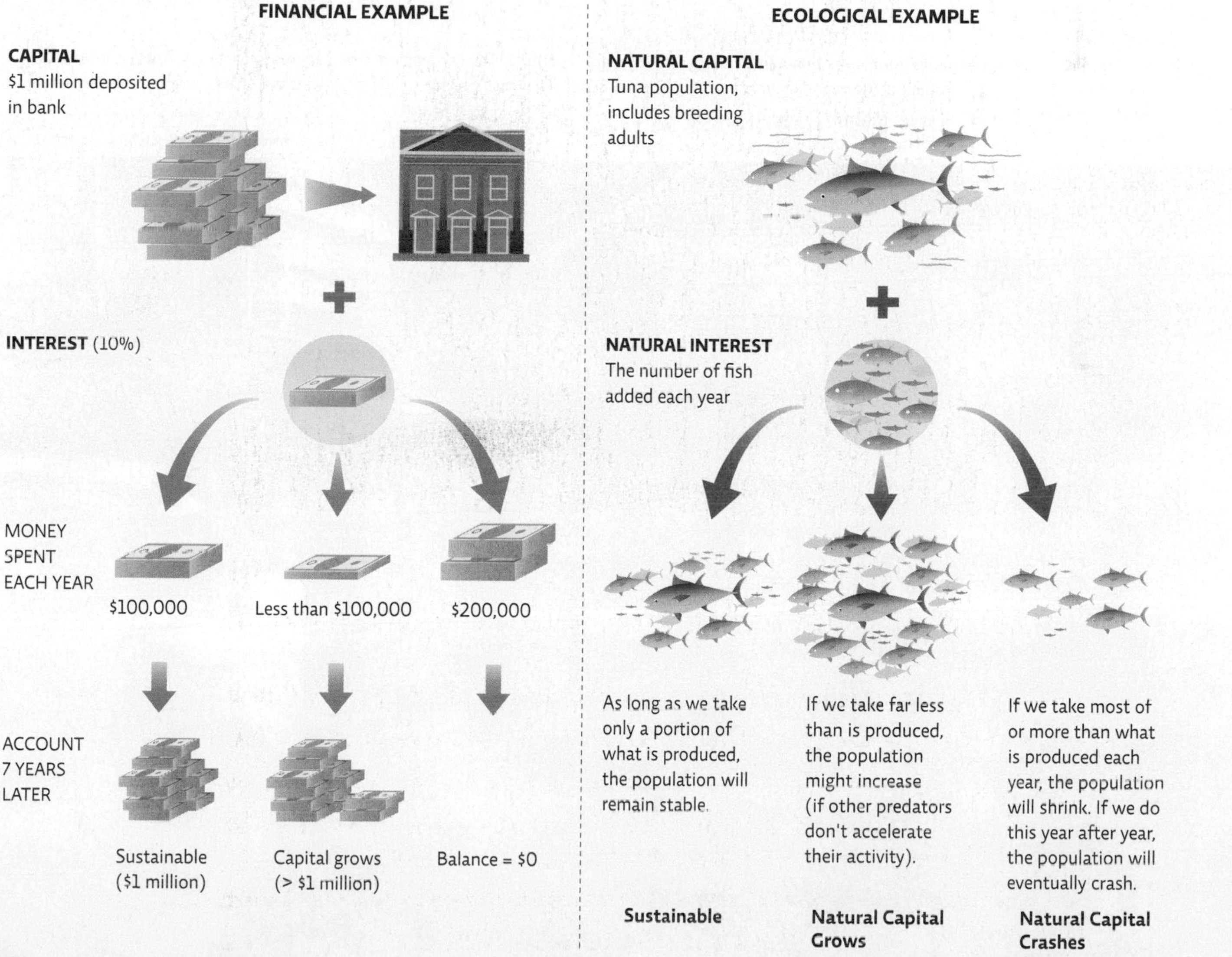
FINANCIAL EXAMPLE
CAPITAL
$1 million deposited in bank
INTEREST (10%)
MONEY SPENT EACH YEAR
$100,000
Less than $100,000
$200,000
ACCOUNT 7 YEARS LATER
Sustainable ($1 million)
Capital grows (> $1 million)
Balance = $0
ECOLOGICAL EXAMPLE
NATURAL CAPITAL
Tuna population, includes breeding adults
NATURAL INTEREST
The number of fish added each year
As long as we take only a portion of what is produced, the population will remain stable.
If we take far less than is produced, the population might increase (if other predators don't accelerate their activity).
If we take most of or more than what is produced each year, the population will shrink. If we do this year after year, the population will eventually crash.
Sustainable
Natural Capital Grows
Natural Capital Crashes

? Why might it be difficult to harvest a natural resource like a tuna population sustainably, even if we set that as our goal?

When Anderson spoke to his employees in the summer of 1994 about his new plan for sustainability, he stressed that his goal was to begin putting back more than the company took from the planet.

Up to that point, the company was using up far more natural capital in the form of resources like petroleum and water than was ultimately sustainable. Anderson realized that if we take more than is replaced, capital will shrink and therefore produce less the next year. Essentially, by taking 50% more resources than is sustainable, we are taking resources away from the future, in what eco-architect Bill McDonough calls *intergenerational tyranny*. When we liquidate our natural capital more quickly than it can be replaced and call that "income," the question becomes this: Where will future income come from?

This can be an especially big problem with commonly held resources like water: Once we remove it from wells or rivers, we have to wait for the next rainfall to replenish it. When many users are accessing the resource, it can quickly become depleted or degraded if they do not work together to manage it—a tragedy of the commons (see Module 1.1).

Starting in 1994, Interface made major changes in the pursuit of its new goals. By 2017, the company had cut the amount of energy it derived from fossil fuels by 88% and reduced its total energy use by 45%. It did this in part by maximizing energy efficiency in its facilities; installing skylights to replace artificial, electricity-dependent lighting; and installing more energy-efficient heating, ventilation, and air conditioning systems. In one of its factories, Interface also installed a real-time energy tracker that displays energy use prominently for its employees to see, inspiring them to think of new ways to conserve energy. Although Interface declined to reveal how much money it invested in such improvements and technology, the company has ultimately recouped its costs in energy savings, according to a company spokesperson.

Interface produces carpet made from yarns that contain up to 100% recycled content. The company has procedures in place to retrieve thread and carpet trimming from the production floor and recycle those into new product. They also reclaim used fishing nets and used carpet to recycle the components into new carpet.

Michiel Wijnbergh/Hollandse Hoogte/Redux

5 TRUE COST ACCOUNTING

Key Concept 5: When the price of a product does not reflect the social and environmental external costs, those costs are paid by others rather than being passed on to the consumer. Internalizing these external costs better reflects the true cost of a product.

Current-day, or "mainstream," economics allows managers to evaluate possible resource-use choices and make the most profitable decisions, often by seeking to *maximize value*— that is, achieving the greatest benefit at the lowest cost. However, one of the complaints against mainstream economics is that it doesn't consider *all* potential costs when trying to maximize value. For instance, a carpet tile might require a certain amount of material that has a recognized monetary cost; but what about the environmental costs associated with drilling enough oil to make that material in the first place or the costs associated with cleaning up the pollution it creates?

To determine how much to charge for its goods or services (in its quest to earn a profit), a business, such as a company that produces paper, must account for all its expenses. Wages, fees, insurance, building maintenance, and other expenses are all part of the **internal costs** of doing business. But nature also provides some ecosystem services that might be important to the business, such as nutrient cycles that support tree growth or the water cycle that provides water for the trees. Because these services are supplied by nature "free of charge" and are not part of the cost of doing business, they are classified as **external costs**.

However, the price of a good or service that is only based on internal costs is often incomplete because there can also be *negative* external costs—problems that result from doing business that are not accounted for in an internal cost assessment, such as the health costs associated with the waste produced by making a product or the environmental damage caused by pollution generated by the company. Historically, economists have regarded these, too, as external to the business (the business doesn't pay for them), and they aren't reflected in the price the consumer pays for the good or service. But if the business doesn't pay for the costs or pass those costs on to the consumer, who does pay? Other people, present and future, and other species do. They pay in the form of degraded health, ecosystems, and opportunities.

An assessment of the cost of a good or service (or any of our choices) should include more than just the economic costs; it should also include the social and environmental costs—the **triple bottom line.** (See Module 1.1 for an introduction to the concept of the triple bottom

internal cost A cost—such as for raw materials, manufacturing costs, labor, taxes, utilities, insurance, or rent—that is accounted for when a product or service is evaluated for pricing.

external cost A cost associated with a product or service that is not taken into account when a price is assigned to that product or service but rather is passed on to a third party who does not benefit from the transaction.

triple bottom line The combination of the environmental, social, and economic impacts of our choices.

East Valley Tribune, Tim Hacker/AP Photo

Brandon Sargent co-founded EcoScraps with two college friends in 2010. The company collects food waste from local restaurants and grocery stores, which it composts into potting soil.

line.) By ignoring external costs, economies create a false idea of the true and complete costs of particular choices. A customer may pay $8 for every carpet tile, but the **true cost** for that piece of carpet would be much higher if it included costs such as greenhouse gas emissions and the cost of treating people for asthma who have fallen ill as a result of breathing the particulate matter released during the carpet's production. The inadequate valuation of a product could eventually lead to the exploitation or overuse of resources needed to produce it—an example of market failure. When external costs are internalized, on the other hand, people (or species) who don't benefit from the transaction do not pay for it. In this case, the product or service can be more appropriately priced or valued; this new price more accurately reflects the true cost of the product or service.

true cost The sum of both external and internal costs of a good or service.

Because we are unaccustomed to paying true costs, we would most likely be surprised at how much some goods and services would really cost if all externalities were internalized. Although it sounds discouraging, any time we purchase products that were made in a more environmentally sound or socially ethical manner, we come a little closer to bearing the responsibility for our choices. We also create a demand for these products in the marketplace. And if businesses are forced to internalize external costs, it then becomes profitable for them to take steps to lower those costs—for example, by installing pollution prevention technologies—a benefit that could lower the environmental and societal costs overall. **INFOGRAPHIC 5**

INFOGRAPHIC 5 TRUE COST ACCOUNTING

Many environmental and health costs of our goods and services are externalized (not included in the price the consumer pays). But if consumers don't pay all the costs to produce a product, such as paper, who does?

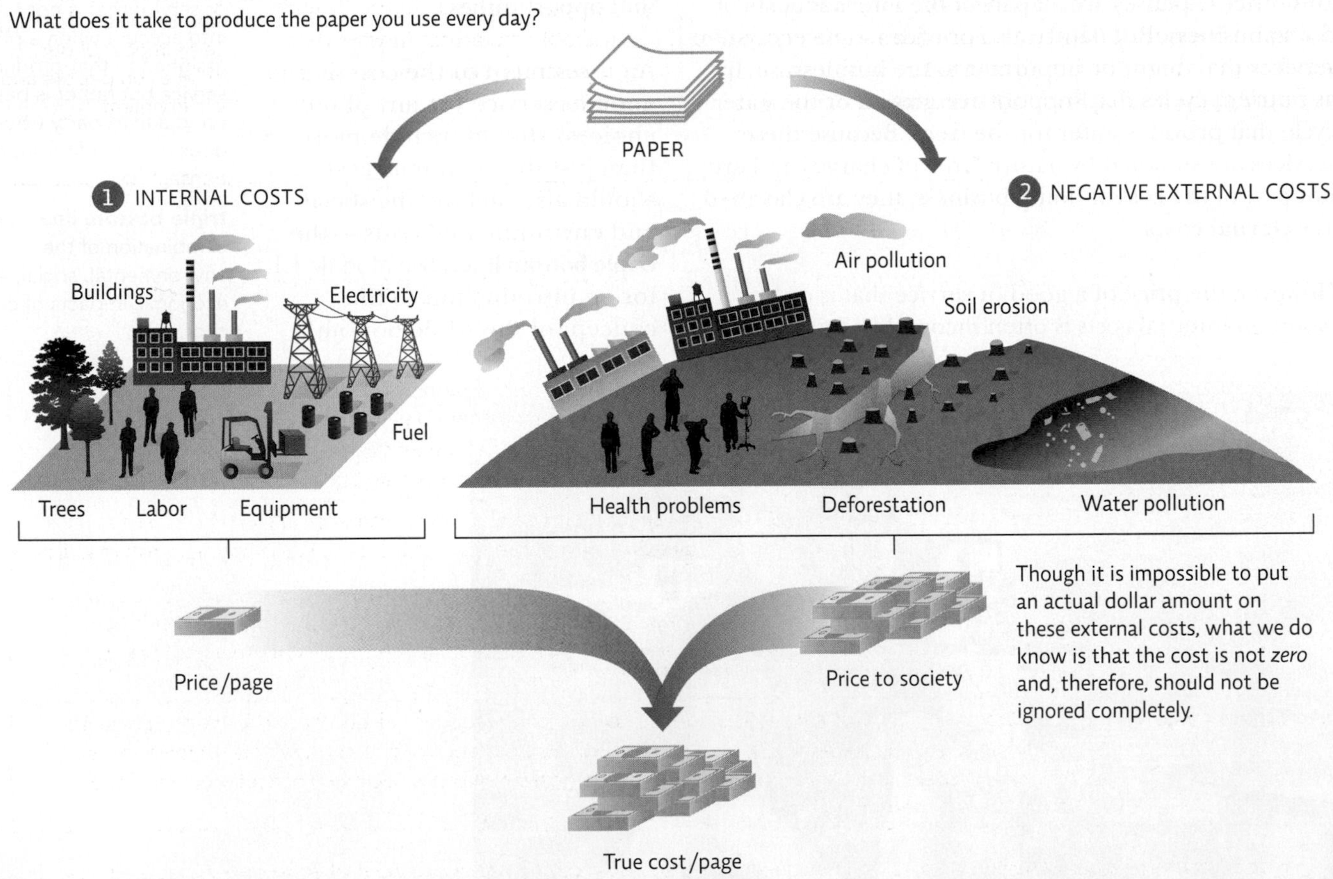

What does it take to produce the paper you use every day?
PAPER
1 INTERNAL COSTS
2 NEGATIVE EXTERNAL COSTS
Buildings
Electricity
Fuel
Trees
Labor
Equipment
Air pollution
Soil erosion
Health problems
Deforestation
Water pollution
Though it is impossible to put an actual dollar amount on these external costs, what we do know is that the cost is not *zero* and, therefore, should not be ignored completely.
Price/page
Price to society
True cost/page

What could a paper company do to reduce the external costs of harvesting trees to make paper?

6 ENVIRONMENTAL ECONOMICS VERSUS MAINSTREAM ECONOMICS

Key Concept 6: Mainstream economics is criticized because it makes some assumptions about resource availability and value that are inconsistent with the way nature operates. Transitioning from linear economic models to circular systems that depend on renewable resources and see waste as a useful input can help address these shortcomings.

Ecologically minded economists advocate incorporating environmental considerations into economic decisions. **Environmental economics** is a discipline that considers the long-term impact of our choices on human society and the environment. (There are actually two camps—environmental economics and ecological economics—but their distinctions are not important for this discussion. For simplicity, in this text, we will combine them and use the term *environmental economics.*) Environmental economists argue that without including environmental considerations, mainstream economic theory will fail because it is based on several erroneous assumptions. **INFOGRAPHIC 6A**

One inaccurate assumption of mainstream economics is that natural and human resources are either infinite or substitutes can be found if needed. This is true for some but not all resources. For instance, though we don't worry about running out of solar energy, fossil fuels are finite and will run out even with technological advances that allow us to access more of the fuel that is left. Crop productivity has limits—soil can be amended with fertilizer additives, but it cannot be replaced for most crops—and there is no substitute for water. In addition, our actions can degrade the quality of air, water, and soil resources faster than nature can restore them.

The fact that some resources are limited (finite or replaced over a defined time frame) leads to other erroneous assumptions. For example, mainstream economics also assumes that economic growth will go on forever. Since there are inherent limits to what Earth can provide, unlimited economic growth (at least that which depends on finite resources and the rate at which renewable resources are replaced) is not, in fact, possible. We have to work within the limits of available resources in ways that allow essential ecosystem services to continue.

environmental economics New theory of economics that considers the long-term impact of our choices on people and the environment.

INFOGRAPHIC 6A ASSUMPTIONS OF MAINSTREAM AND ENVIRONMENTAL ECONOMICS: A COMPARISON

Environmental economists criticize several assumptions of mainstream economics that they say will fail in the long run.

FACTOR	MAINSTREAM ECONOMICS	ENVIRONMENTAL ECONOMICS
Resource availability	Natural resources are either infinite or substitutable.	Some natural resources are finite, and there are not suitable substitutes for all. Therefore, when using finite resources, we should use those that are renewable resources and use them at a rate that is equal to or less than the rate of replacement.
Economic growth	Economic growth will go on forever.	Because it is based on finite resources or the rate of their replacement, economic growth based on consumption of resources has limits.
Resource value	The future value of a resource (its value if we save it) is worth less than its value if used today.	Recognizing that short-term exploitation of a resource may diminish its potential to be replaced, more weight is given to long-term benefits and costs than to short-term ones.
Production models	Models of production typically follow a linear sequence: Raw materials come in (inputs), products are made, and waste is discarded (outputs).	Production models should be closed loops like nature, focusing on renewable resources used at or below their rate of replacement and treating waste so that it becomes an input that creates new resources.

? Not all economic growth requires the consumption of finite resources. Give an example of an economic pursuit that does have the capacity to keep growing every year without depending on higher use of natural resources.

Another problem with mainstream economics is that it **discounts the future**: It tends to give more weight to short-term benefits and costs than it does to long-term ones. In other words, mainstream economics considers something that benefits or harms us today more important than something that might do so tomorrow. For instance, we value the tuna we can harvest today more highly than tuna we might harvest 10 years from now, so the value of taking a large harvest of tuna today outweighs the benefits of taking less now to ensure that there is still some later. If the money we could earn by using the resource now is higher than sustainable harvesting yields, modern economics tells us it is more profitable to use it now and invest the resulting money in another venture. But this investment approach doesn't take into account where those other ventures might come from or whether they are in any way diminished by the elimination of the first resource—we are back to the erroneous assumption that any resource can be replaced by another. How might the loss of tuna affect the ecosystem and other populations? Will there always be another fish population to harvest?

discounting the future
Giving more weight to short-term benefits and costs than to long-term ones.

linear economic system
A production model that is one-way: Inputs are used to manufacture a product, and waste is discarded.

circular economic system
A production system in which the product is returned to the resource stream when consumers are finished with it or is disposed of in such a way that nature can decompose it.

These assumptions lead to yet another misconception: that models of production typically follow a linear sequence—a "take, make, dispose" approach (see Module 5.3). In this **linear economic system**, raw materials come in, humans transform those materials into a product, and then they discard the waste generated in the process. But because some resources are finite, if the waste created from the use of these resources is stored away permanently as "waste," we will eventually run out of a needed input as supplies dwindled. In addition, generating waste in the form of pollution that damages natural capital such as air, water, and soil also reduces the amount of matter available to us. It is this unsustainable use of resources that will eventually cause linear models of production to fail. For instance, most traditionally produced carpets are made from fossil fuels, a practice that is not sustainable. Old, unwanted carpets are then discarded, and some are eventually burned, releasing toxic pollutants and greenhouse gases.

A sustainable approach would be more cyclical, where "waste" becomes the raw material once again and can be used to make new products. Interface's ReEntry 2.0 program uses old carpet tiles to make new ones, and it uses old carpet backings to make new carpet backings. This is an example of a **circular economic system**, where the product is folded back into the resource stream when consumers are finished with it or is disposed of in such a way that nature can decompose it and use it once again. **INFOGRAPHIC 6B**

More and more industries are investing in wind turbines and solar panels as a way to decrease the ecological footprint of their operations.

lumen-digital/Shutterstock

INFOGRAPHIC 6B ECONOMIC MODELS

Environmental economics recognizes that natural ecosystems provide our resources and assimilate our wastes. If companies could fold "waste" back into production or make sure it can be decomposed by nature, we could reduce our extraction costs and operate in a sustainable circular system.

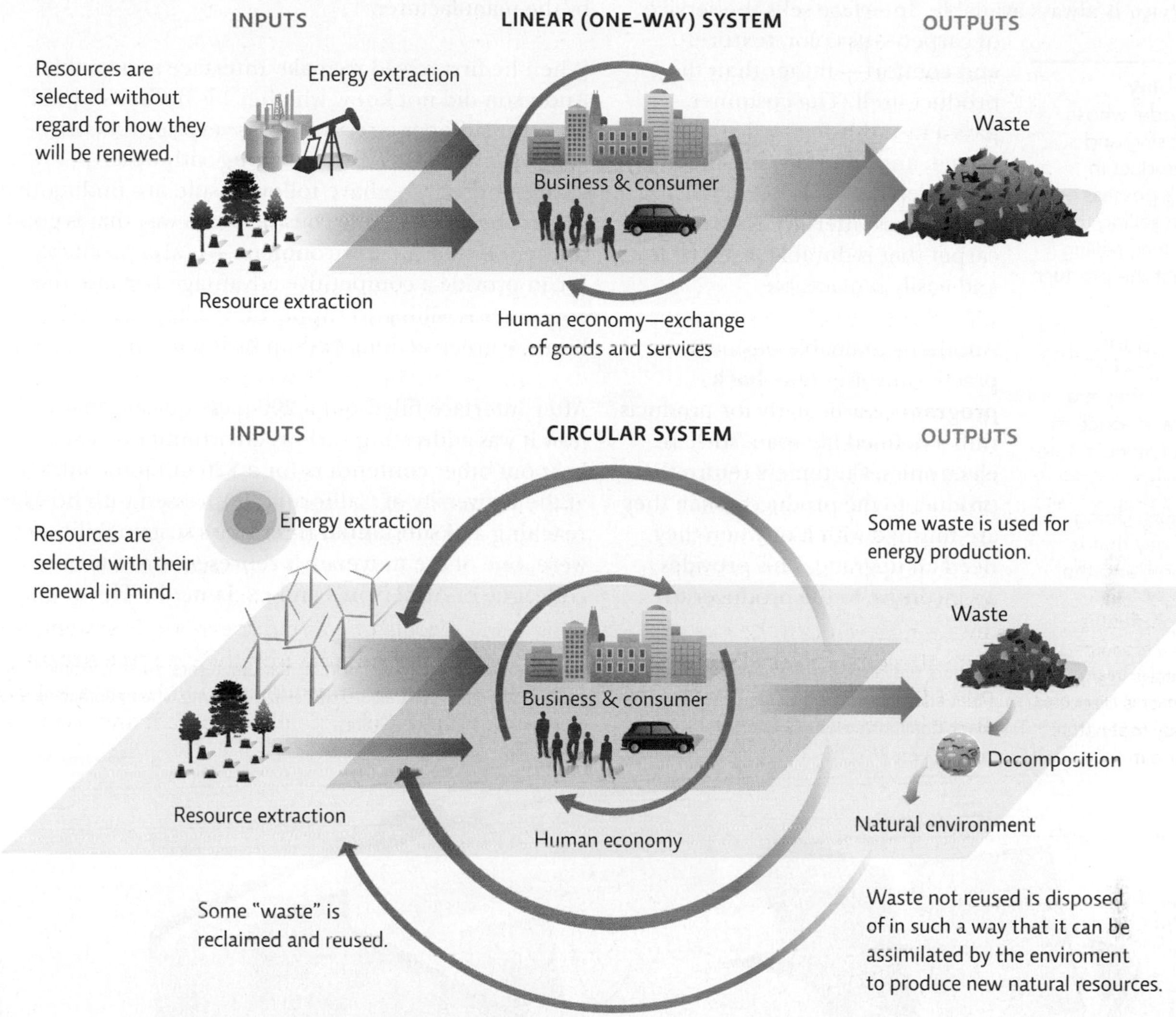
INPUTS
LINEAR (ONE-WAY) SYSTEM
OUTPUTS
Resources are selected without regard for how they will be renewed.
Energy extraction
Resource extraction
Business & consumer
Human economy—exchange of goods and services
Waste
INPUTS
CIRCULAR SYSTEM
OUTPUTS
Resources are selected with their renewal in mind.
Energy extraction
Resource extraction
Business & consumer
Human economy
Some waste is used for energy production.
Waste
Decomposition
Natural environment
Some "waste" is reclaimed and reused.
Waste not reused is disposed of in such a way that it can be assimilated by the enviroment to produce new natural resources.

Look at the size of the arrows in these two diagrams. Which ones increase and which ones decrease in size? Why?

To combat the erroneous assumptions of mainstream economics, environmental economists support actions such as improving technology to increase production efficiency and reduce waste, valuing resources (including waste) as realistically as possible, moving away from dependence on nonrenewable resources, and shifting away from a product-oriented economy. They look to natural ecosystems as models for how to efficiently use resources and live within the limits of nature.

Many businesses today account for all the steps needed to produce a product and deal with any waste produced—a *cradle-to-grave* approach. But that approach is incomplete. Interface seeks to operate as a circular system by managing its product in a **cradle-to-cradle** fashion: It considers the entire *life cycle* of the product, from the beginning (acquisition of raw materials) to the end of its useful life (disposal), and is responsible for the impact of its use at every stage of the process. This can lead to better material choices (less toxic, more sustainable) and better process choices (reusable materials, less waste and pollution).

Interface also revolutionized its operations by considering itself

cradle-to-cradle Refers to management of a resource that considers the impact of its use at every stage, from raw material extraction to final disposal or recycling.

part of a **service economy**; it focuses on selling a *service* rather than a *product*. The idea is simple: A customer pays for the service, such as the ability to photocopy or use a car when needed, and the vendor makes sure that the service is always available. Interface sells the service of carpet—its color, texture, and comfort—rather than the product itself. The customer pays a monthly fee to "lease" the carpet, and Interface maintains it and replaces it as needed. This encourages Interface to produce carpet that is durable, recyclable, and easily replaceable.

service economy A business model whose focus is on leasing and caring for a product in the customer's possession rather than on selling the product itself (i.e., selling the service that the product provides).

take-back program Program that allows a consumer, once they are finished with a product, to return it to the manufacturer that made it.

green business Doing business in a way that is good for people and the environment.

Another sustainable business practice involves **take-back programs**, particularly for products with a defined life span, such as electronics: Customers return the product to the producer when they are finished with it or when they need an upgrade. This provides an incentive to the producer to make a durable, high-quality product that can be reused or recycled and ensures proper disposal of hazardous materials. Some of these programs are mandated by law (take-back legislation); others are voluntarily put in place by the manufacturer.

When he first vowed to make Interface sustainable, Anderson did not know whether his business would thrive or suffer as a result. "I was very apprehensive about it," he recalls. But he, along with other entrepreneurs who have followed suit, are finding that **green business**—doing business in a way that is good for people and the environment—is also profitable. It can provide a competitive advantage because the consumer is willing to support the company's efforts or because green actions end up saving money.

After Interface filled out a 200-page questionnaire about how it was addressing various environmental issues, it beat out other contenders for a $20 million contract at the University of California. Impressed with how far-reaching and substantial Interface's sustainability actions were, one of the university's representatives turned to a colleague of Anderson's and exclaimed, "This is *real*."

Though take-back legislation is more prevalent in the European Union than the United States, many electronics manufacturers, such as Apple, Dell, and Samsung, have programs that allow consumers to mail back used devices free of charge or that award trade-in discounts when purchasing a new product. The Electronics Takeback Coalition's website offers details on these programs in the United States.

maryolyna/Shutterstock

7 SUSTAINABLE PRACTICES: THE ROLE OF THE CONSUMER

Key Concept 7: Transparency in how a business operates will allow consumers to make better choices and level the playing field for businesses that are trying to operate more sustainably. Finding ways to consume less is also an important green choice.

What about consumers? We can all decrease our impact by making more sustainable choices and by consuming less. This doesn't necessarily mean "doing without," but it does mean being mindful of our choices and opting for sustainable or low-impact choices whenever possible. However, this requires transparency from the industries that produce and sell us goods and services. That is hard to come by with current business models, often because the businesses themselves don't know all the external costs associated with their products. For its new plan to be successful, Interface was counting on its customers to make more sustainable choices as well. And make them they did. ReEntry 2.0 drew many new customers to Interface, including the Georgia state legislature, which purchased 13,000 square yards of carpet.

Changing the way we do business to give consumers the ability to choose sustainably produced goods and services is not going to be easy. Even Interface has progressed slowly, despite its strong desire to become sustainable. Start-up or upgrade costs can be substantial, and even though improvements may pay for themselves in the long run, many businesses simply do not have the funds to pay for them. In addition, they may find themselves at a competitive disadvantage with businesses that are not trying to internalize costs. Consumers also have a role to play.

Cal Cam/Alamy Stock Photo

Many retailers are taking steps to reduce their impact. For example, REI pursues a variety of actions that address its energy and materials footprint, including constructing energy-efficient buildings, installing solar panels, reducing product packaging, and sourcing recycled materials.

For example, recycled paper's higher cost may more closely reflect the true cost of paper, but if consumers are not willing to put their buying dollars behind their environmental ideals, businesses that make and sell paper from trees will still be more successful. In other words, it will take changes from both consumers and producers to put business and industry on the path to sustainability. But there are things that can be done to level the playing field.

Governments can encourage sustainability by providing incentives for businesses to account for true costs rather than just internal costs. This could be accomplished by taxing companies based on how much pollution they generate, subsidizing environmentally friendly processes, or giving out pollution "permits" that companies could sell if they release less pollution than they are allowed (*cap-and-trade*; see Module 10.1). For instance, if there is a pollution tax, it will be passed on to the consumer, who then decides whether to buy the product. The manufacturer that minimizes waste production and relies on lower fossil fuel inputs than its competitors would be able to meet the regulations at the lowest cost; offer lower prices; and, as a result, have a major market advantage. While promising, these policies can be complex to implement because external costs are hard to quantify (how is a pollution tax fairly assessed?) and because they may put a burden on smaller manufacturers that have less ability to absorb the cost of upgrades. Passing these costs on to the consumer also stresses low-income households.

According to Anderson, the consumer needs to know how a good or service is made—what the overall environmental impact of that product is. "You lay out for the consumer everything that goes into that product, and you lay it out for your competitors too—it's a totally transparent revelation of how you made that product, and what that footprint is at every step," he explained.

One way to communicate this information is through **ecolabeling**. But consumers have to be wary of labels because as

ecolabeling Providing information about how a product is made and where it comes from. Allows consumers to make more sustainable choices and support sustainable products and the businesses that produce them.

INFOGRAPHIC 7 GREEN CONSUMERISM

The choices we make as consumers are extremely important in any quest to pursue environmental sustainability. Supporting green products supports those companies; we vote with our dollars. Responsible consumerism also means consuming less and looking for ways to meet a need other than buying new products. Green consumer practices include:

- **Ecolabels:** Look for transparency in labeling to identify sustainable options, such as certified organic or 100% postconsumer recycled content.
- **Greenwashing:** Beware of claims with no verification of accuracy or environmental relevance.
- **Fair Trade:** Support fair trade products made with methods that support sustainable practices and social justice.
- **Share Programs:** Don't buy when you can rent or borrow a product.
- **Green Businesses:** Support businesses that use sustainable methods and resources. Be willing to pay more for goods and services whose price represents their true cost.
- **Consumption:** Lower your consumption of goods and services by purchasing high-quality products that are durable and will last a long time. Try to limit your overall purchases.

What do you feel are the biggest impediments to becoming a green consumer? Why?

"green" products become more attractive to consumers, more companies engage in **greenwashing**—claiming environmental benefits for a product when they are minor or nonexistent. **Fair trade** items are, however, more likely to be sustainably produced. For a product to be certified as fair trade, workers must be paid a fair wage, they must work in reasonable conditions to produce the goods or services, and the production process must have a low ecological impact. In other words, they have been produced in an environmentally and socially responsible way.

Consuming less is also an important consideration. *Share programs* are another useful option for items that people need infrequently, such as a car for those who live in a large city. Rather than buying, owning, and then storing the product for a large part of the time, consumers share ownership and use the product only when they need it. INFOGRAPHIC 7

greenwashing Claiming environmental benefits about a product when the benefits are actually minor or nonexistent.

fair trade A certification program whose products are made in ways that are environmentally sustainable and socially beneficial (e.g., fair wages, good working conditions).

Although Interface has come a long way since 1994, it is still working hard to achieve its sustainability goals. In June 2011, the company began producing its first 100% nonvirgin fiber carpet tiles, made from reclaimed material: fiber derived from salvaged commercial fishnets and postindustrial waste. By 2018, it was using 100% plant-based or recycled material as the raw materials for all its products. Practices like intercepting industrial waste destined for landfills have a positive effect, while other efforts lessen the company's overall negative impact: less toxic glues, less carpet waste. All the while, Interface is still the world's leading manufacturer of commercial carpet tiles and is growing (they saw 18% growth from 2017 to 2018). "I think we're on the right track, and we'll keep on going," said Anderson, who stepped down as the company's CEO in 2001 but still played the role of the company's conscience until his death in 2011 at the age of 77. "We'll get to the top of that mountain."

Select References:

Anderson, R. (1998). *Mid-course correction: Toward a sustainable enterprise: The Interface model.* Atlanta, GA: Peregrinzilla Press.

Anderson, R. (2009). The business logic of sustainability. http://www.ted.com/talks/ray_anderson_on_the_business_logic_of_sustainability.

Costanza, R., et al. (2014). Changes in the global value of ecosystem services. *Global Environmental Change, 26,* 152–158.

Kubiszewski, I., Costanza, R., Anderson, S., & Sutton, P. (2017). The future value of ecosystem services: Global scenarios and national implications. *Ecosystem Services, 26,* 289–301.

McDonough, W. (2000). A boat for Thoreau: A discourse on ecology, ethics, and the making of things. *The Ruffin Series of the Society for Business Ethics, 2,* 115–133.

GLOBAL CASE STUDIES GREEN BUSINESSES

Achieve

A wide variety of businesses are taking major strides to reduce their ecological footprint. Some of these actions profit the business by lowering production or operating costs, while others reach beyond business operations and contribute to environmental protection or provide societal benefits. The profiles presented here are not meant to be an endorsement of the businesses, but simply highlight some players that are taking extra steps to decrease their impact or even to help restore the environment their industry affects.

BRING IT HOME

PERSONAL CHOICES THAT HELP

You have an impact on creating a sustainable society. Every time you buy a product or service, you are telling the manufacturer that you agree with the principles behind the product. You can use your purchasing power to show companies that people are interested in good-quality products that support environmental and social values.

Individual Steps

- Reduce the amount of stuff you accumulate by buying fewer items and by choosing products that are well made and last longer.
- Ask your local food store or pharmacy to stock fair trade–certified products if it doesn't already.
- Start a blog or Facebook page to chronicle the changes you make in your buying habits and encourage others to do the same. Discuss the companies whose environmental policies you agree with.

Group Action

- Instead of buying a new or used car, join a car-share program like Zipcar.
- Get together with family and friends and write a letter to your favorite companies, asking them to reduce their ecological footprint. You can ask the company to become more transparent by publishing how their business practices impact the environment. For ideas, visit the *Letters to the Planet* website at https://letterstotheplanet.weebly.com for links to letter-writing campaigns.
- Investigate the investments of your university and push for more socially and environmentally responsible investing, such as ESG investments (a sustainability measure that considers the environmental, social, and governance impact of a company).

Tara Walton/The Toronto Star/ZUMAPRESS.com/Newscom

REVISIT THE CONCEPTS

- How best to allocate scarce resources is the focus of economics, but many economists are applying ecological concepts to facilitate the wise use of resources for long-term sustainability, not just for short-term profitability. Life on Earth, and the goods and services we enjoy daily, depend on ecosystem services provided by nature. Recognizing the value of these services may motivate us to protect them.

- Human impact can be measured in terms of our ecological footprint — the amount of land needed to provide the resources we use and assimilate the waste we produce. By identifying how we impact our environment, we can take steps to reduce that impact.

- The environmental impact of a population generally goes up as its size, affluence, and use of technology increase. However, some technologies allow us to use resources more efficiently and/or generate less pollution, thus helping to decrease impact.

- We can harvest natural resources in a sustainable fashion if we use them at or below the rate at which they are produced — that is, take only the natural interest. This will leave behind enough natural capital to replace what we took.

- To fairly set a price for an economic good or service, we should consider the internal and external costs of that product — its true cost. Ignoring external costs passes on those costs (environmental damage or health problems) to others who did not benefit from the transaction — someone other than the buyer or seller.

- Environmental economists argue that mainstream economics makes some assumptions that are inconsistent with the way nature operates. Linear economic production models use inputs and produce waste without regard to sustainability. In circular (cradle-to-cradle) management systems, waste is seen as a resource, which encourages manufacturers to make durable, recyclable, and nonhazardous products and product components.

- Consumers can decrease their impact by consuming less and by choosing products with lower environmental and societal impacts. Demanding transparency in how a product was made will allow consumers to make better choices and help businesses operating sustainably to compete with those that prefer to hide the impact of their operation.

ENVIRONMENTAL LITERACY Understanding the Issue

1 What are ecosystem services, and why are they important to ecosystems and human populations?

1. Give three examples of ecosystem services and for each, identify one or more human endeavors that would be affected if nature's ability to provide these services declined.
2. How can it be useful to place a monetary value on ecosystem services even if we know it will not be accurate?

2 What is an ecological footprint?

3. What is the unit of measure for determining an ecological footprint, and what kind of actions affect the size of a footprint?
4. Why is the per capita ecological footprint of the United States so much higher than that of most other nations?

3 What does the IPAT equation tell us about our impact and the potential to reduce that impact?

5. What is the IPAT model? How have P, A, and T changed here in the United States in recent decades, and how has that affected our impact?
6. Which parameter — P, A, or T — has the potential to decrease impact as it grows rather than increase it? Explain.

4 What are natural capital and natural interest, and how can they be sustainably used?

7. Explain the economic concepts of capital and interest in terms of a bank savings account and then apply the same concepts to natural resources.
8. The sap of maple trees (sap is "food" for the tree) can be tapped to make maple syrup, but taking too much will kill the tree. In this example, what would constitute the natural capital and what would be the natural interest?

5 What is true cost accounting, and why should we employ it?

9. Give some examples of the internal and external costs of coal mining.
10. What does it mean to "internalize all external costs," and why would it be good for the environment if businesses did this?

6 What are the main differences between the assumptions and economic models of environmental versus mainstream economics

11. Identify and explain four erroneous assumptions that mainstream economics makes with regard to the environment.
12. Why do environmental economists say that linear economic models of production will ultimately fail?
13. What does the term *cradle-to-cradle* mean in a discussion about product management?
14. Using the example of a lawncare company, explain how this is part of a service economy and how this could reduce the environmental impact of lawn care for its customers.

7 How can consumers make better choices that reduce their overall environmental impact?

15. Explain why purchasing a green product is not always the best green consumer choice.
16. What are fair trade products?
17. What is greenwashing, and how can a consumer avoid it?

SCIENCE LITERACY Working with Data

Robert Costanza and colleagues estimated the value of ecosystem services of many of Earth's major ecosystems as well as the area each one covered on the planet for the year 2011. Look at the following graph and data table to answer the questions that follow. (Land area is given in hectares, a unit of measure equal to 2.5 acres.)

ANNUAL VALUE PER HECTARE

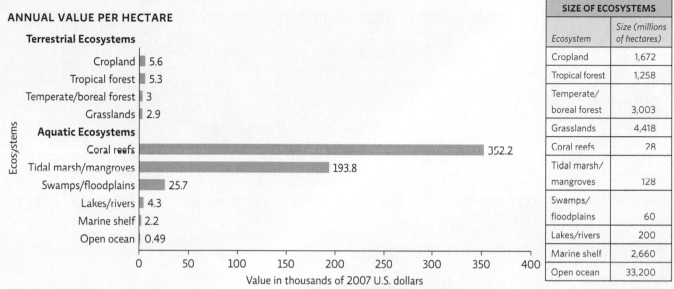

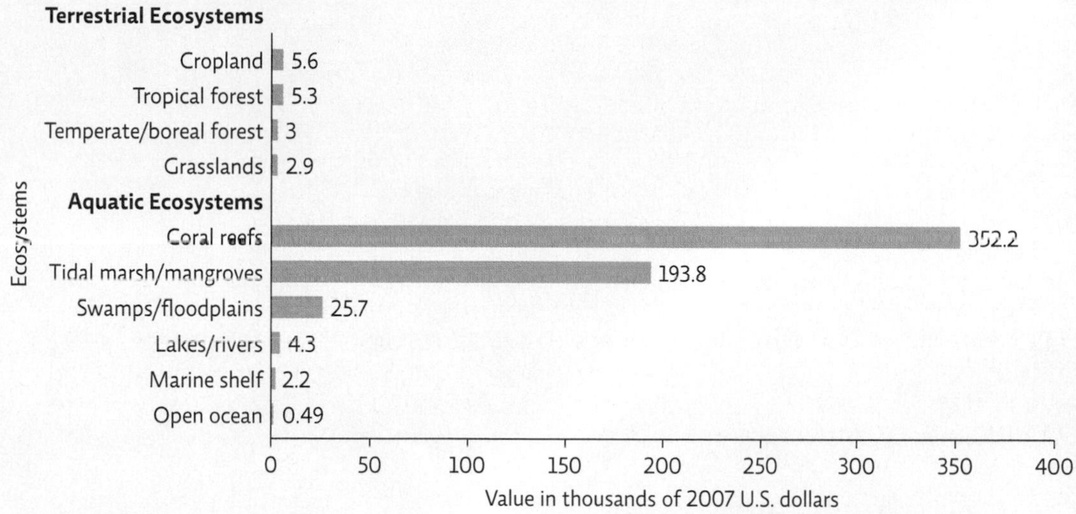

SIZE OF ECOSYSTEMS

Ecosystem	*Size (millions of hectares)*
Cropland	1,672
Tropical forest	1,258
Temperate/ boreal forest	3,003
Grasslands	4,418
Coral reefs	28
Tidal marsh/ mangroves	128
Swamps/ floodplains	60
Lakes/rivers	200
Marine shelf	2,660
Open ocean	33,200

Interpretation

1. Based on the graph, which ecosystem has the highest value per hectare per year? The lowest?
2. In total dollars, what is the value of Lakes/Rivers per hectare?
3. Based solely on the graph, how do terrestrial ecosystems compare to aquatic ones in terms of their annual contribution/hectare to ecosystem services? Give data to support your answer.

Advance Your Thinking

4. Look at the data table that shows the size of each ecosystem in millions of hectares. Using these data, calculate the total annual value in millions of 2007 U.S. dollars for each ecosystem and create a bar graph similar to the one above that shows the total annual value for each ecosystem.
5. Reevaluate the contribution of terrestrial ecosystems versus aquatic ones based on the total annual value of each. How does this compare to the answer you gave for question 3?
6. Explain the value of each graph and why it is useful to show the data both ways.

INFORMATION LITERACY Evaluating Information

Being an informed consumer begins with understanding what the labels on the products you buy really mean. For explanations of food labels, explore the Greener Choices website at greenerchoices.org and answer the questions below.

Evaluate the website and work with the information to answer the following questions:

1. Determine if this is a reliable information source with a clear and transparent agenda:
   a. Who runs this website? Do the credentials of the organization make the information presented reliable/unreliable? Explain.
   b. What is the mission of this website? What are its underlying values? How do you know this?
2. Under the "Labels" tab, select at least three labels and for each, answer the following questions:
   a. What does the label mean, and why is it important? Is it a reliable indicator of an environmentally sound consumer choice?
   b. What data sources were used in the label assessment? Are its sources reliable? Why or why not?
3. Select one of the "Reports" produced by this organization.
   a. Evaluate the information in terms of authorship (credentials).
   b. What data sources were used in the report? Are its sources reliable? Why or why not?
   c. Does the report provide useful information for a consumer who wants to buy this food product? Explain.
4. Overall, is the Greener Choices website a reliable source of information? Explain.

Additional study questions are available at Achieve.macmillanlearning.com. Achieve

Module 5.2 Environmental Policy

THE WORLD TACKLES OZONE DEPLETION

Dealing with ozone depletion taught nations how to address global environmental issues

Earth's thin atmosphere, visible here just below a crescent moon, offers protection from dangerous UV radiation from the Sun. Elen11/Getty Images

After reading this module, you should be able to answer these GUIDING QUESTIONS

1. What causes stratospheric ozone depletion, and why should it be addressed?
2. What is the purpose of environmental policies, and why are they needed?
3. How has environmental policy evolved in the United States?
4. How are policies developed and administered in the United States?
5. What factors influence policy decisions?
6. What policy tools can be used to implement and enforce environmental policy?
7. How are international policies established and enforced?
8. How has the international community responded to ozone depletion?

There's a point of no return halfway into the 9-hour flight from New Zealand to the Antarctic. Once that 4.5-hour mark passes, if something goes wrong with the plane, there's nowhere to stop in the Southern Ocean for repairs.

Susan Solomon is very familiar with that trip's all-important midway point. During her very first flight to the southernmost continent, the pilot told the passengers that their plane was not working properly: The front ski at the nose of their C-130 was frozen and couldn't be lowered into position, so landing on the packed snow at the Antarctic research station would be impossible. They had to turn around.

It was August 1986—late winter in the Antarctic—and Susan Solomon, an atmospheric chemist, was on her way to investigate a mystery. She and a team of scientists from various institutions were gathering at McMurdo Station for a highly coordinated research expedition to tackle a question of global significance: Why was the ozone layer above the South Pole disappearing?

Suddenly, the remoteness of where she was going hit home. "I remember, as we were flying back to New Zealand, thinking, 'Wow, I really am going to the Antarctic,'" Solomon says.

The next night the atmospheric scientist and her team managed to make it to Antarctica, landing at McMurdo Station. On this, her very first excursion, Solomon and her colleagues collected the data that would eventually grab the world's attention and resolve an ongoing, hard-fought scientific debate that was taking place on an international stage. That debate was not only over what was causing the ozone layer to thin but also over what exactly humans should do about it.

WHERE IS MCMURDO STATION?

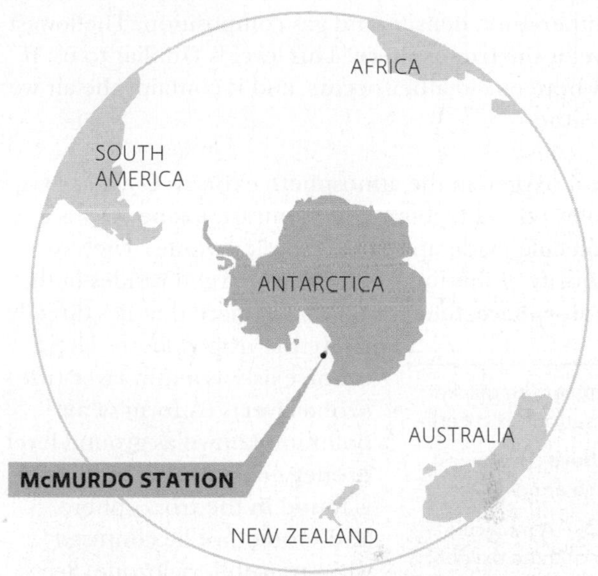

Courtesy Dr. Susan Solomon

Susan Solomon at McMurdo Station in 1987.

1 THE ISSUE: STRATOSPHERIC OZONE DEPLETION

Key Concept 1: Stratospheric ozone depletion, caused by CFCs (chlorofluorocarbons) and other anthropogenic chemical pollutants, increases our exposure to dangerous UV radiation.

Earth's **atmosphere**, the blanket of gases surrounding our planet, is made up of discernable layers that differ in temperature, density, and gas composition. The lowest level is the **troposphere**. This level is familiar to us: It is where our weather occurs, and it contains the air we breathe.

Most oxygen in the atmosphere exists as O_2 (two oxygen atoms bound together). In contrast, **ozone (O_3)** is a molecule made up of three oxygen atoms. The vast majority of this unusual form of oxygen resides in the **stratosphere**, the layer of atmosphere that lies directly above the troposphere. There, ozone exists as a thin layer (the ozone layer). (A form of air pollution known as ground-level ozone, a component of smog, is found in the troposphere and should not be confused with stratospheric ozone. See Module 10.1 for more on ground-level ozone pollution.)

atmosphere The blanket of gases surrounding Earth.

troposphere The lowest level of the atmosphere.

ozone (O_3) A molecule made up of three oxygen atoms.

stratosphere A layer of atmosphere that lies directly above the troposphere.

ultraviolet (UV) radiation High-energy radiation that is harmful to living things.

ozone-depleting substances (ODS) Chemicals that break down stratospheric ozone.

Earth's stratospheric ozone layer protects the planet from much of the Sun's **ultraviolet (UV) radiation**. Excessive exposure—especially of the most dangerous type to reach the surface, UV-B radiation—can cause human health problems such as cataracts and DNA mutations that can lead to skin cancer. It also hampers plant growth, interferes with embryonic and larval development, and hinders reproduction in many species—impacts that could disrupt community connections and impair ecosystem functions.

Normally, stratospheric ozone is continuously created and destroyed (alternating between O, O_2, and O_3) in a natural cycle powered by UV radiation that keeps overall O_3 levels constant. In this region of the stratosphere, UV-C is absorbed by O_2, breaking it into two oxygen atoms; each of these can bind to another O_2 molecule to form O_3. Ozone molecules absorb UV-B radiation, breaking them back apart into O_2 and O to continue the cycle. This absorption prevents most UV-B and all UV-C radiation from reaching Earth's surface.

However, O_3 molecules can also be broken down by some industrial chemicals, including chlorine-based refrigerants such as chlorofluorocarbons (CFCs) and hydrochlorofluorocarbons (HCFCs) as well as bromine compounds used in fire extinguishers and pesticides. These are known as **ozone-depleting substances (ODS)** because they break down ozone faster than it is formed, which in turn causes the ozone layer in Earth's atmosphere to become less concentrated or too "thin," forming what became known as the "ozone hole."

This breakdown is more likely to happen when two requirements are met: a great deal of solar radiation and ice crystals that can act as a platform on which the CFC molecules can be broken down. Early spring provides

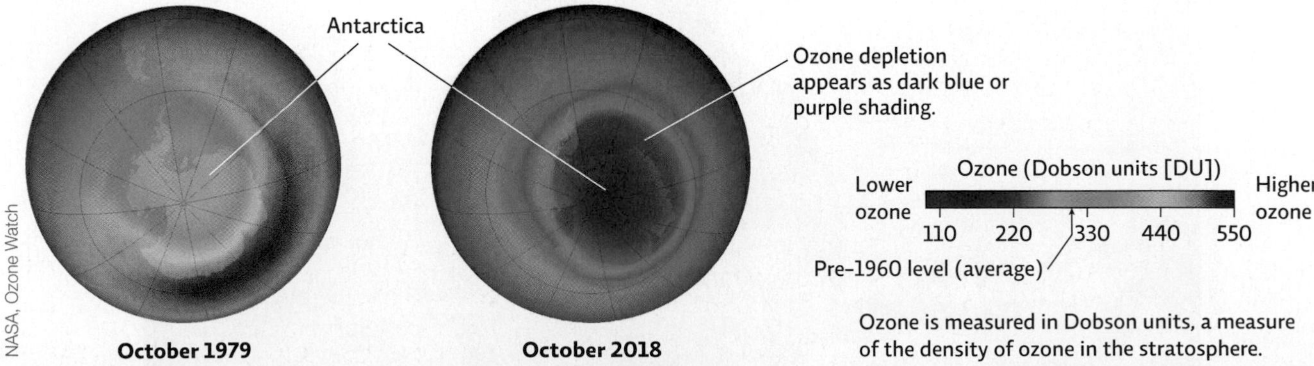

These images show the amount of ozone over Antarctica in 1979 (the first year that ozone satellite images were available) and a more recent year, 2018. The 2018 image shows an ozone "hole" (levels less than 220 DU) covering much of the continent. Prior to 1979, ozone levels below 220 DU were not observed.

these conditions—the Sun has returned to the area, but it's still cold enough to provide the ice crystals that facilitate the breakdown of CFCs. The hole in the stratospheric ozone layer over the Antarctic is therefore most pronounced during September and October (springtime in the Southern Hemisphere). A similar but less severe hole forms above the Arctic (which is not as cold as Antarctica) during the Northern Hemisphere's springtime.

Because stratospheric ozone is so important to life on Earth, uncovering the cause of ozone depletion so it could be addressed quickly became a quest of international urgency. Even though CFC molecules are very stable and do not break down in the troposphere, when they eventually reach the stratosphere, exposure to UV radiation breaks them apart, releasing a chlorine atom. This chlorine atom can then break apart ozone, forming chlorine monoxide (ClO) and O_2. This reaction happens again and again because chlorine is easily released from ClO, making it available to attack another O_3. Solomon's team found the evidence they were searching for: ClO in the stratosphere. (There is no natural source of this molecule; it could only come from the breakdown of CFCs.) In 1988, they published the studies that would provide the definitive link between CFCs and ozone depletion. **INFOGRAPHIC 1**

In a way, this was good news—since CFCs were human creations, we could simply stop creating and using them. Fortunately, by the time that research was made public, the wheels of policy were already in motion.

INFOGRAPHIC 1 THE ATMOSPHERE AND OZONE DEPLETION

Achieve

Earth's atmosphere is composed of layers that differ in chemical composition. The stratosphere contains a region with more ozone (O_3) than other parts of the atmosphere—the ozone layer. Ozone is important because it reduces the amount of harmful UV radiation that reaches Earth's surface. A variety of chemicals, such as CF_2Cl_2 (a CFC), destroy stratospheric ozone.

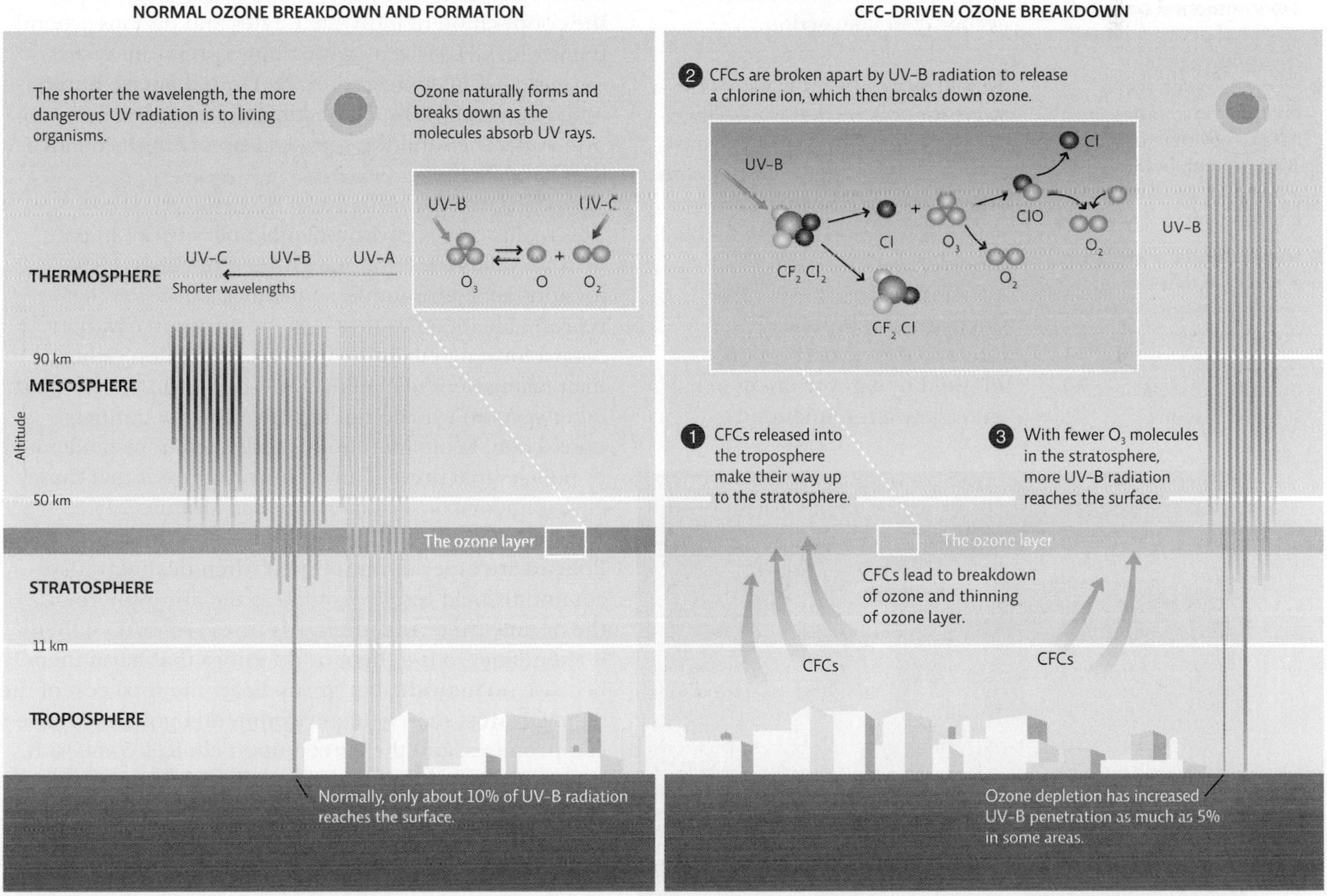

Looking at the chemical pathway that illustrates how CFCs break down ozone, which product of the reaction would you look for in the stratosphere as evidence that this reaction was occurring: CF_2Cl, O_2, Cl, or ClO? Explain.

2 ENVIRONMENTAL POLICY: PURPOSE AND SCOPE

Key Concept 2: Environmental policies protect the environment and human wellbeing. When environmental problems transcend national borders, international policies are needed to address them.

The development of CFCs back in the 1930s had been a cause for celebration among scientists. These new industrial coolants were far less toxic than the ammonia and sulfur dioxides that were being used at the time, and there seemed to be no limit to their possible uses not only as a coolant in refrigerators and air conditioners but also as a key ingredient in aerosol sprays and as a stabilizing substance in Styrofoam food containers. Like most American children of her generation, Solomon spent her childhood surrounded by CFCs. By the time she reached graduate school at the University of California at Berkeley in the late 1970s, they were truly ubiquitous. But at that point, scientists were beginning to raise alarm bells about the link between CFCs and ozone depletion. They didn't yet have definitive evidence of the connection, but they also felt they couldn't wait for certainty to take action.

environmental policy A course of action adopted by a government or an organization that is intended to improve the natural environment and public health or reduce human impact on the environment.

transboundary problem A problem that extends across state and national boundaries; pollution that is produced in one area but falls in or reaches other states or nations.

So they began pressing legislators to create policies that would limit or eliminate the use of suspicious chemicals (CFCs, HCFCs, and some others), just in case they were—as scientists suspected—responsible for ozone depletion.

Environmental policy refers to a course of action adopted by a government or an organization intended to either restore or protect the natural environment and resources—sometimes by repairing damaged ecosystems, other times by reducing or mitigating the impact we humans have on our planet. In many cases, the protection of human health and welfare is at the heart of environmental policies—clean air and water aren't just good for the environment; they are also necessary for human wellbeing.

Dorling Kindersley/Getty Images

In 1978, CFC propellants used in aerosol cans like this were banned and replaced with chemicals that don't harm the ozone layer, such as carbon dioxide and nitrous oxide.

As discussed in Module 1.1, environmental problems are often *wicked problems*, meaning that they tend to be very complex; they have multiple causes and consequences and multiple stakeholders. Most of the biggest environmental issues that we now face—like pollution, species endangerment, and climate change—are also **transboundary problems**; they occur across state and national boundaries. Therefore, solving them requires the cooperation of individual states and nations around the world. CFC-driven ozone depletion is one such example—CFCs released in the United States, Europe, and other regions were affecting stratospheric ozone over Antarctica and the southern tip of South America. **INFOGRAPHIC 2**

This reality makes environmental policy tricky, in part because what is best for one group is not always best for another. For example, sometimes a decision that is profitable for a business—to use a resource without regard for renewability or to release pollution rather than taking steps to eliminate it—can harm people (and other species) who do not benefit from the business transaction. Economic choices may need to be moderated by policies that prevent abuse of resources or that cause environmental or societal harm (see Module 5.1).

Policies are especially important when dealing with commonly held resources such as the atmosphere or the oceans. Since no one "owns" these resources, there is a tendency to use them or do things that harm them because no individual or group bears the total cost of the damage. Most successful environmental policies involve compromises, and the agreed-upon choices come with trade-offs. Still, despite these hurdles, there are many successful environmental policies in place—including those with such far-reaching goals as limiting pollution or habitat destruction, restoring damaged environments, and managing commonly held resources such as rivers or public resources such as national forests in a way that avoids the *tragedy of the commons* (see Module 1.1).

Other factors also make formulating and following through with environmental policies difficult. Those who suffer the

INFOGRAPHIC 2 ENVIRONMENTAL POLICY GOALS

Public policy in general is a course of action adopted by a government or an organization intended to enhance society as a whole.

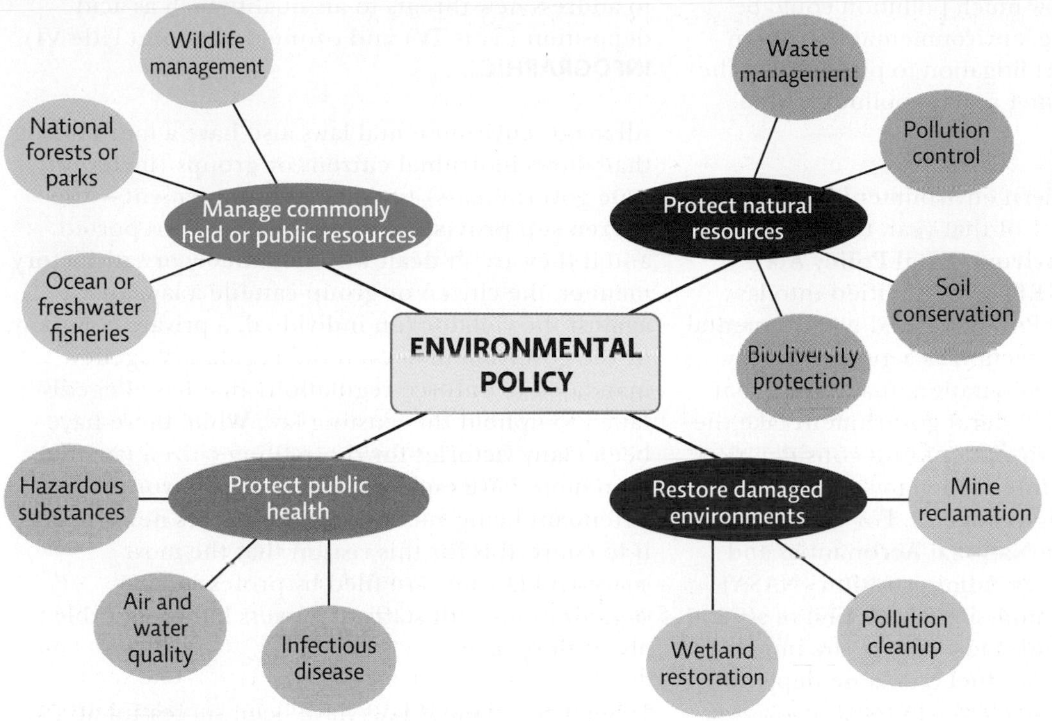

What other policy goals or examples could you add to this concept map?

burden caused by environmental degradation are often not the ones who caused the problem in the first place—we are disconnected from the outcome of our actions and may not know that what we do is harming someone else. There is often a significant lag time between the effect of an action and that action—it takes a while for enough pollution to build up in the air or water to cause health effects, for example. And most impacts have multiple causes, so it is difficult to identify the role of each cause (e.g., many things cause cancer, including environmental exposures).

As early as the 1970s, scientists involved in laboratory studies on CFCs and ozone began calling publicly for an end to CFC production. Their argument—that society should not wait until the link was definitive before taking some precautions—was a long shot. At the time, there were very few laws on the books designed exclusively to protect the environment and certainly none based on an unconfirmed hypothesis. But it turned out that right at that moment, the U.S. federal government's approach to environmental protection was changing dramatically.

3 U.S. ENVIRONMENTAL POLICY

Key Concept 3: Many iconic U.S. environmental laws originated in the 1960s and 1970s and became models for similar laws worldwide. Some of these laws allow citizens to sue the violator or government to ensure that the laws are properly enforced.

Before the 1960s, environmental policies mostly dealt with how best to use resources. Addressing pollution or environmental damage was not a key objective. And environmental problems were primarily handled at the state level because federal environmental regulations were considered an intrusion on state sovereignty. In fact, most environmental problems were addressed only after the fact, through litigation—an arrangement that too often favored the polluters. It was even more difficult back then than it is today to prove that toxins from a factory or dump that had seeped into the water or permeated the air were killing livestock or causing human illnesses.

Eventually, though, things began to change. Industry grew, and so did pollution. And as it did, environmental problems began slipping across state lines so that water and air pollution from one state affected another. In the 1960s and 1970s, a massive national outcry forced legislators to acknowledge that cohesive federal policies were needed. New laws were passed, and existing environmental legislation was updated. Regulators began to set environmental **performance standards** (also known as

performance standards The levels of pollutants allowed to be present in the environment or released over a certain time period.

pollution standards) by identifying the levels of pollutants allowed to be present in the environment or released over a certain time period—at the federal level. By determining how much pollution could be released in the first place, environmental regulation shifted from after-the-fact litigation to prevention. The shift proved effective: Water and air pollution from industry began to drop.

By 1970, the era of modern environmental policy had begun. On January 1 of that year, the **National Environmental Policy Act (NEPA)** was codified into law. NEPA established environmental protection as a guiding policy for the nation, mandating that the federal government take the environment into consideration before taking any action that might affect it. For example, the National Aeronautics and Space Administration (NASA) commissioned the first of several studies to evaluate the impact of rocket fuel on ozone depletion.

National Environmental Policy Act (NEPA) A 1969 U.S. law that established environmental protection as a guiding policy for the nation and required that the federal government take the environment into consideration before taking action that might affect it.

citizen suit provision A provision that allows a private citizen to sue, in federal court, a perceived violator of certain U.S. environmental laws, such as the Clean Air Act, in order to force compliance.

In NEPA's wake came a wave of iconic legislation—including the Clean Water Act, the Clean Air Act, and the Endangered Species Act—all passed with overwhelming bipartisan support. Environmental laws are amended as needed. For example, in 1990, the Clean Air Act (CAA) was amended by Congress to address new threats to air quality such as acid deposition (Title IV) and ozone depletion (Title VI). **INFOGRAPHIC 3**

All major environmental laws also have a mechanism that allows individual citizens or groups (including state governments) to demand enforcement—the **citizen suit provision**. Violations can be reported, and if they aren't dealt with in a timely or satisfactory manner, the citizen or group can file a lawsuit against the violator (an individual, a private company, or the government—even the regulatory agency mandated to enforce regulation) that has allegedly failed to uphold the existing law. While there have been many victories for those filing citizen lawsuits, even more have ended in rulings that favor the defendant being sued, and many others never make it to court. It is for this reason that the most successful lawsuits are filed by professional organizations with staffs of lawyers knowledgeable about the process.

U.S. environmental laws have been successful at improving the environment—our air is cleaner, our water is safer, and our wild areas and wildlife are better protected than before these laws were passed. In fact, these laws became models for environmental legislation in countries around the world.

INFOGRAPHIC 3 NOTABLE U.S. ENVIRONMENTAL LAWS

Many landmark U.S. environmental laws were passed, beginning in the 1960s, during a period of tremendous bipartisan cooperation and support for taking steps to ensure a clean and healthy environment.

LAW	DESCRIPTION
National Environmental Policy Act (1969)	Mandates that the federal government take the environment into consideration before pursuing any federal action that might have an environmental impact.
Clean Air Act (1970)	Regulates the amount of hazardous pollutants that can be present or released into the air. *See Module 10.1.*
Clean Water Act (1972)	Regulates water quality by setting standards for the release or presence of specified toxic or hazardous water pollutants. *See Module 6.2.*
Endangered Species Act (1973)	Protects and aids in the recovery of endangered and threatened species of fish, wildlife, and plants in the United States. *See Online Module 3.3.*
Safe Drinking Water Act (1974)	Protects public drinking water supplies through water quality standards set and enforced by the Environmental Protection Agency. *See Module 6.1.*
Toxic Substances Control Act (1976)	Protects consumers from toxic chemicals in the products they buy by regulating and monitoring industrial chemicals. *See Module 4.3.*
Nuclear Waste Policy Act (1982)	Mandates that the federal government provide for permanent disposal of high-level nuclear waste. *See Module 11.1.*

? Do you feel that these U.S. environmental laws should be strengthened, be weakened, or remain as they are?

4 THE POLICY CYCLE: DEVELOPMENT AND ADMINISTRATION OF POLICIES

Key Concept 4: Policy making includes systematically considering all options before setting policy and periodically evaluating policy after it is implemented.

NEPA did more than commit the federal government to environmental protection—it established a process for generating environmental policy that includes a scientific evaluation of the environmental problem and the consideration of various approaches to address that problem, including the economic and technical feasibility of each option. A decision is then made as to the best course of action. The process itself is responsive and allows for policy revision as new or changing information comes to light.

NEPA's signature feature has been the **environmental impact statement (EIS)**—a report that details the likely effect of a proposed federal action, such as building a road or upgrading a nuclear facility. The goal of an EIS is to identify problems before they occur so that stakeholders can choose the most acceptable course of action. To keep the process transparent, the findings are made public, and everyone is given a chance to respond (through letters and public hearings). It was an EIS, released in 1988, that first evaluated the health, societal, and environmental effects of potential regulatory actions to address ozone depletion. This document gave policy makers the information they needed to formulate useful and appropriate policies to protect atmospheric ozone. **INFOGRAPHIC 4**

NEPA has changed very little since its passage, but the Trump administration would like to significantly reform the law to reduce the time it takes to authorize infrastructure projects. It takes, on average, 4.5 years to complete an EIS; as part of the Trump administration's efforts to reform NEPA, President Trump issued an executive order in 2017 to streamline the process by mandating that one lead agency be assigned to oversee the production of the EIS (rather than multiple agencies) and setting a 2-year time limit (on average) on EIS completion. In that same year the Interior Department issued a rule that restricted most EISs to 150 pages and set the target for completion at 1 year, a move criticized as an arbitrary revision of a long-standing process. To further streamline the process, a rule proposed in early 2020 would prohibit the consideration of climate change in EIS evaluations. The administration's NEPA reforms have been welcomed by some project developers but are raising alarm among those who might oppose a particular development or who serve as watchdogs for environmental protection.

In 1970, Congress established the U.S. **Environmental Protection Agency (EPA)** to implement and enforce all the new federal environmental laws that were then being passed as part of a greater

environmental impact statement (EIS) A document outlining the positive and negative impacts of any federal action that has the potential to cause environmental damage.

Environmental Protection Agency (EPA) The federal agency responsible for setting policy and enforcing U.S. environmental laws.

INFOGRAPHIC 4 POLICY DECISION MAKING—THE NEPA PROCESS

Policies are created and revised using some basic steps that allow policy makers to systematically evaluate the situation and possible responses. Regulations based on bills that are passed and signed into law are proposed and administered by regulatory agencies (like the EPA) in a similar manner. The process itself is responsive and allows for revision as new or changing information comes to light.

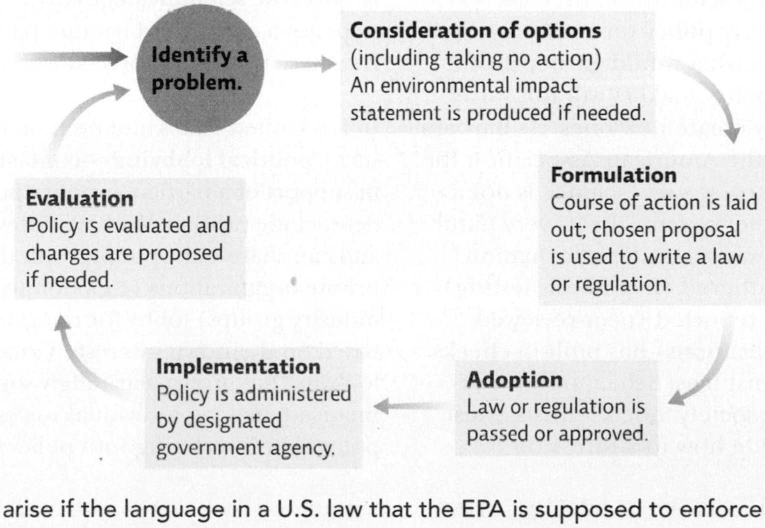
Identify a problem.
Consideration of options
(including taking no action)
An environmental impact statement is produced if needed.
Formulation
Course of action is laid out; chosen proposal is used to write a law or regulation.
Adoption
Law or regulation is passed or approved.
Implementation
Policy is administered by designated government agency.
Evaluation
Policy is evaluated and changes are proposed if needed.

? What problems can arise if the language in a U.S. law that the EPA is supposed to enforce is vague or ambiguous?

effort to coordinate environmental protection at the federal level. (The National Oceanic and Atmospheric Administration [NOAA] was also created at the same time as part of the national strategy to safeguard our environment.) The EPA and other agencies (such as the U.S. Department of Agriculture) follow the same NEPA process to establish rules and regulations in support of environmental laws. These agencies are administered by the executive branch, funded by the legislature, and subject to judicial oversight.

The EPA sets rules or standards (such as performance standards) that ensure that the goals of any given law are met. They are also tasked with holding individual states and corporations accountable. The EPA works with states and industry to help them achieve compliance, but if a given entity fails to comply with a given rule, the EPA has the authority to step in and mandate changes. It can, for example, force a power plant to make upgrades that decrease pollution, close a factory for repeated violations, or fine an individual state for failing to curb its vehicle-generated air pollution. It can also force entities to pay cleanup costs, revoke operating permits, or bring criminal charges against violators. For example, in 2019, Trident Seafoods Corporation, which had been in violation of Title VI of the CAA due to leaking refrigeration units on fishing vessels, agreed to spend $23 million to address the problem by fixing leaks and switching to non-ODS refrigerants. The company was also fined $900,000.

Of course, the EPA's reach extends only to U.S. borders. And as the case of CFCs shows, most environmental problems tend to stretch way beyond those.

5 FACTORS THAT INFLUENCE POLICY FORMULATION

Key Concept 5: Sound science is needed for effective policy decision making; political lobbies, public opinion, and the press also strongly influence the process.

Early on, the concerns raised by scientists over CFCs led to a U.S. ban on the chemicals in aerosol sprays. But when it came to further action—banning the chemicals completely—industry dug its heels in. The chemicals were in wide use by then in many different industries; they were useful and generated billions of dollars in profit for the chemical companies that made them and the manufacturers who used them. These business entities used their money and influence to argue against rushing to judgment. They insisted that there was far too little evidence to get rid of something so ubiquitous and important to the U.S. economy. The science community disagreed.

Depending on high-quality science—that is, peer-reviewed studies with strong scientific consensus—is crucial to good policy. Basing policy on what one would like to be true or on what would garner profit or political advantage to policy makers will not solve societal problems and may create new ones. As Barbara Schaal, then President of the American Association for the Advancement of Science, wrote, "science is not a political construct or a belief system." By its very nature, the way scientific information is gathered (hypothesis testing) and reported (peer-reviewed publications) has built-in checks against bias. Schaal points out that society, not scientists, must decide how to address the problems it faces, but science should be the basis for those policies.

political lobbying
Contacting elected officials in support of a particular position; some professional lobbyists are highly organized, with substantial financial backing.

Indeed, in an ideal world, all policies would be based on sound science and established need. Legislators would receive scientific information about a proposed policy from unbiased sources from the natural and social science community and use this to write bills. Once passed and signed into action, the judiciary would weigh in on the constitutionality of the new law, and rules to implement that law would be quickly set. But that's not quite how it works.

The science is often hard to distill into feasible actions, and getting timely policies in place can be challenging, with policy making often proceeding at a snail's pace. To be sure, the scientific, legislative, executive, and judicial steps are a crucial and routine part of any policy making. But there is another force at work, too: political lobbying.

In the United States and even on the international stage, **political lobbying**—contacting elected officials in support of a particular policy position—is part of the democratic process. We have access to our elected officials and can share our opinions with them. Citizens and private organizations (e.g., nonprofits, labor unions, and industry groups) lobby for or against specific proposals, based on their own interests. Critics say that professional lobbying has grown alarmingly sophisticated and well financed, making individual voices harder to hear and potentially interfering with policy makers' judgment.

Besides running ad campaigns promulgating ideas that serve their own best interests ("clean" coal, for example, is not as clean as it sounds; see Module 9.1), industries also contribute large sums of money to candidates for elected office, hoping to influence those candidates, if elected, to act in ways favorable to industry.

Nonprofit organizations like the Natural Resources Defense Council or the Sierra Club also have professional lobbying divisions that promote their positions to elected officials and the general public. Taken together, these environmental nonprofits spend tens of millions of dollars per year in federal lobbying efforts—but their investments are still dwarfed by the hundreds of millions of dollars spent by industry.

The regulation of toxic chemicals, for instance, is sometimes more heavily influenced by industry pressure than scientific evidence. The tobacco industry argued successfully for years that smoking was not linked to health problems, despite ample evidence to the contrary. And in a 2006 *New York Times* article, an EPA pesticide analyst complained (anonymously, citing fear of retribution) about industry's influence over policy, saying, "You go to a meeting, and word comes down that it is an important chemical, this is one we've got to save. It's all informal, of course. But it suggests that industry interests are governing the decisions of the EPA."

Of course, it doesn't necessarily take money to influence policy. Citizens and grassroots movements can influence Congress by calling their representatives and making their opinions known or by voting with their dollars (refusing to purchase products with impacts they don't want to support). For example, plastic beverage bottles used to contain a chemical known as bisphenol-A (BPA) that was known to be an endocrine disruptor (it acted like a hormone in the body)—the worry was that exposure to this substance could interfere with fetal or child development. Parents, physicians, scientists, and environmental groups called for a ban on BPA while industry touted its long history of safe use and lack of evidence of direct harm. The wheels of policy turned slowly, but when major retailers such as Whole Foods and Walmart announced they would no longer sell drink bottles, baby bottles, and sippy cups that contained BPA, industry responded. Bottle plastics were reformulated to exclude BPA (though its replacement doesn't appear to be much better) a full 4 years before a U.S. ban of the product was announced. This example illustrates that no matter how intense the professional lobbying—or, for that matter, how sound the science or how well established the need for a given policy—a strong public voice can influence policy. **INFOGRAPHIC 5**

INFOGRAPHIC 5 INFLUENCES ON U.S. ENVIRONMENTAL POLICY DECISION MAKING

Many organizations and individuals influence not only whether we institute a policy to deal with an environmental issue but also the design of that policy—what it covers and how it will be implemented and enforced. The wide variety of voices, many representing differing viewpoints, can make it difficult to create new policies. Though political ideologies might influence how one goes about addressing a problem, policy makers ideally look to the best available science when making decisions about whether a policy is needed to protect the health and wellbeing of the public and environment.

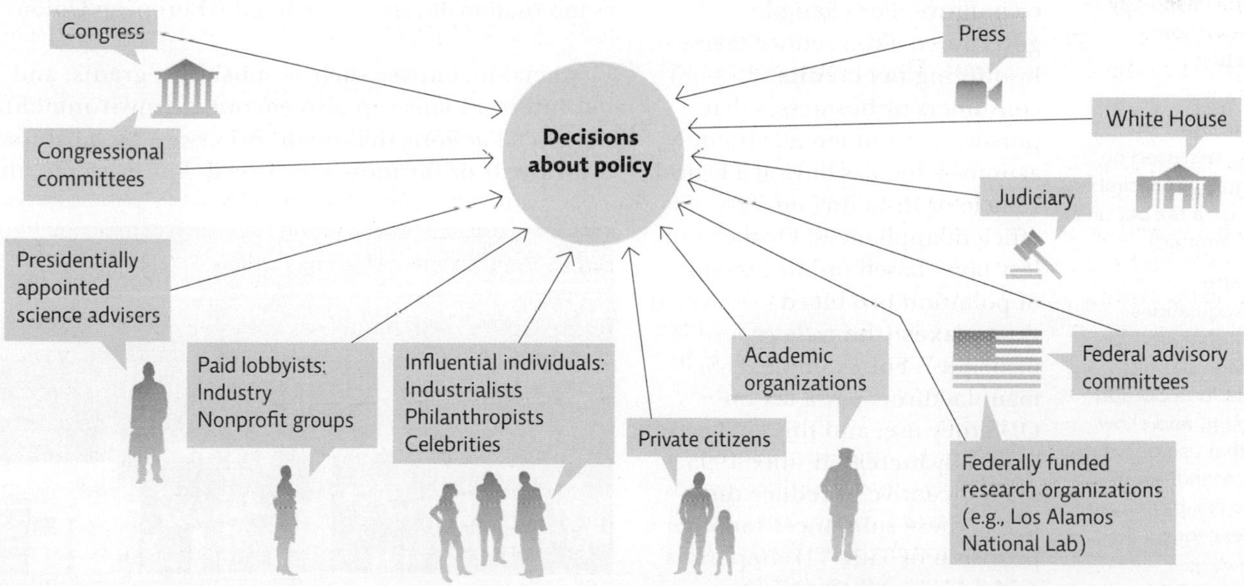

Based on how you see the process playing out in the United States today, rank the parties that influence U.S. policy making from "most influential" to "least influential." In your opinion, is this ranking as it should be? Explain.

6 IMPLEMENTATION AND ENFORCEMENT OF POLICIES

Key Concept 6: Policy can be enforced with command-and-control regulation or through economic incentives (market-driven approaches) that favor preferred responses.

A range of *policy tools* are used to enforce environmental laws. Governmental **command-and-control regulation** is one approach. This option relies on setting enforceable regulations, identifying performance that constitutes compliance, and setting penalties for noncompliance. This type of regulation can take several different forms: issuing permits to authorize operation of equipment or facilities that may pose risk to the environment, establishing performance or technology standards that regulate emissions or specify the practices that must be used to meet those standards, or simply banning the use of a given substance. For example, Title VI of the CAA gave the EPA the authority to regulate ODS. Along with other provisions, a phase-out program was established for ODS that were currently in use, with some exemptions for compounds without suitable replacements, and acceptable procedures were laid out for the destruction of ODS stockpiles.

Alternatively, *market-based approaches* create economic incentives for the private sector to reduce environmentally harmful actions, without dictating exactly how to reach a desired target. Economic incentives can be reward-based or punitive. For example, governments can reduce taxes by offering **tax credits** to consumers or businesses that pursue environmentally friendly actions—such as buying a hybrid vehicle or installing energy-efficient appliances. Or they can levy taxes based on the amount of pollution produced—so-called **green taxes** (the *polluter pays* principle). For example, U.S. manufacturers pay a tax on ODS they use, and this tax has gradually increased since 1995 as an incentive to reduce the use of these substances (another provision of Title VI). Importers to the United States of any product that may contain ODS must also pay a tax based on the weight of ODS in the product.

command-and-control regulation Legislative control of an activity or industry via rules that identify acceptable actions.

tax credit A reduction in the tax one has to pay in exchange for some desirable action.

green tax A tax (a fee paid to the government) assessed on environmentally undesirable activities (e.g., a tax per unit of pollution emitted).

cap-and-trade program Regulations that set an upper limit for pollution, issue permits to producers for a portion of that amount, and allow producers that use or release less than their allotment to sell permits to those who exceeded their allotment.

subsidies Financial assistance given by the government to promote desired activities.

Another market-based approach used to implement the initial phase-down of ODS that began in 1991 was **cap-and-trade** (also known as *permit trading*). This policy involves setting an upper limit (the cap) on the use or emission of certain pollutants, distributing permits for a portion of that limit to each user, and then letting individual businesses freely trade (or sell) those permits if they have some of their allotment left over because they used or emitted less of the regulated substance. Violators are fined: In the cap-and-trade program that addressed acid rain, that fine was set at $2,000 per ton of emissions that exceeded allowances. Caps can be lowered over time to bring down the level of pollution as needed. (See Infographic 7 in Module 10.1.) An advantage of this approach is the freedom for plant operators to choose which avenue to compliance works best for them—reducing actual emissions or purchasing additional permits.

One of the most successful cap-and-trade programs to date reduced sulfur pollution from coal-fired power plants, a major cause of acid deposition in the eastern United States. Targets were met ahead of time, and the program experienced almost 100% compliance (very few fines were assessed) and came in at a much lower price tag than predicted, demonstrating the power of market-driven approaches to pollution control. Carbon emissions are currently being reduced with cap-and-trade programs at the state and regional levels in the United States as well as internationally, such as within the European Union.

Financial incentives such as **subsidies**, grants, and low-interest loans can also encourage environmentally beneficial actions that might otherwise be hard for individuals or businesses to afford. For example, the

Daniel Schoenen/LOOK-foto/Getty Images

Houses with solar panels, like these in Baden-Wuerttemberg, are common in Germany, thanks to governmental policies that offered financial incentives such as fixed payments for power produced and subsidies to purchase the units.

INFOGRAPHIC 6 POLICY TOOLS

COMMAND-AND-CONTROL REGULATION		MARKET-BASED APPROACHES (ECONOMIC INCENTIVES)	
ADVANTAGES	**DISADVANTAGES**	**ADVANTAGES**	**DISADVANTAGES**
• Simple in concept and may achieve goals quickly. • Directly changes the behavior of the regulated industry. • Useful when the potential for severe environmental or health impact is high. • Useful when level of control is known and uniform across all regulated industries.	• Policy updates take time, which makes it hard to adapt to changes. • A one-size-fits-all approach may limit some industries' ability to use the most cost-effective methods to address their impact. • No incentive to reduce pollution below mandated limits. • Sufficient funding needed to enforce compliance.	• Fund actions too costly for individuals or business to afford. • Taxes generate revenue that could benefit environmental causes. • Provide incentives to reduce pollution below requirements. • Encourage innovation. • Stimulate the economy.	• Citizens may oppose tax dollars being used for endeavors they do not support. • A one-size-fits-all approach may limit some industries' ability to use the most cost-effective methods to address their impact. • The cost of green taxes may be passed on to the consumer. • May inadvertently support undesirable activities (perverse subsidies).

? Climate change is another serious global environmental problem. Which approach do you support to reduce our use of fossil fuels and other activities that contribute to climate change: command-and-control regulation or economic incentives? Explain.

United Nations and some individual nations gave out subsidies to businesses to assist their transition away from ODS to more ozone-friendly replacements (e.g., replacing CFC-based air conditioning with chlorine-free alternatives). Unfortunately, sometimes subsidies inadvertently support environmentally damaging actions (*perverse subsidies*) such as the health and environmental impacts caused by the heavily subsidized fossil fuel industry. **INFOGRAPHIC 6**

In deciding which of these tools to use for any given policy, policy makers must consider several factors, including whether the tool will actually help attain the desired environmental goal, who will bear the financial cost and other burdens of the policy if this tool or that one is used, and how adaptable the tool itself will be to changes in the underlying policy. They must also establish clear methods of enforcement: Someone has to oversee the implementation of a given policy tool and hold the relevant parties accountable (by assessing fines or penalties, for example, or by revoking privileges—or even in some cases by charging someone with a crime) in a predetermined and straightforward manner.

7 INTERNATIONAL ENVIRONMENTAL POLICY

Key Concept 7: International policy is established through treaties that range from a simple agreement that action is needed to protocols that specify procedures and targets for participants. Enforcing compliance for international treaties is difficult but may include incentives such as technological or financial assistance or disincentives such as economic sanctions.

Even as the tide started to turn against the use of CFCs by various industries, industry itself had one powerful argument against a full CFC ban. If U.S. companies were forced to curb their use of these chemicals while other countries still used them, besides the ozone layer not being saved, American industry would lose its competitive edge as it scrambled to come up with replacement chemicals (a tragedy of the commons dilemma). For the ban to achieve its desired goal of protecting the ozone layer, industries in other countries needed to be held to the same standard.

The efforts of individual states and countries are essential to environmental protection. But ultimately, global problems require international efforts. International agreements go by a litany of different names: treaties, conventions, accords, platforms, and so on. A **convention** is an international agreement that represents a position on an issue and identifies general goals that the signing countries agree to pursue, such as the protection of species or the wise use of wetlands. To lay

convention An international agreement that represents a position on an issue and identifies general goals that the signing countries agree to pursue.

out the specific actions that will be taken as well as the time frame over which these actions must occur, a **protocol** is often drafted to set precise goals and targets.

International treaties are drafted at meetings, often with intense negotiations, and signed by the attending representatives. The treaty must then be ratified by each nation; in the United States, new treaties are ratified by the U.S. Senate; amendments or subsequent agreements can be signed by the President (though the Senate sometimes disagrees with the President as to what constitutes a "new" treaty). For multilateral treaties involving many nations, the draft may identify a minimum number of nations that must ratify it before it "enters into force"; this could take one or more years. There are currently more than 2,500 international agreements that pertain to the environment: They cover things like whaling (where and how and how much it can be done), endangered species protection (marking certain habitats off-limits to loggers or banning the sale of ivory, for example), and the conservation of global resources like the atmosphere. **INFOGRAPHIC 7**

For any of these agreements to work—for the environment to be protected—signatory parties (countries that sign on) must keep their word and do what they say they will do. In other words, they must comply with the terms laid out in the agreement. Achieving compliance is even trickier at the international level than it is for individual countries or states. To be sure, international agreements often include mechanisms for tracking participants' performance around a certain goal or promise and to catch failures of compliance early on. And most of them also have dispute settlement mechanisms in place that may include conciliation (reducing targets or goals), negotiation to redefine responsibilities, or binding arbitration. Economic aid or technological assistance might also help a nation comply. But few agreements have a robust capacity to enforce the established standards or punish the participants who fall short of their promises. Incentives can help the participants reach their goals, and disincentives such as penalties, stricter surveillance and report requirements, trade sanctions, and even suspension of certain relevant privileges can help get participants back on track when they fall short. But those efforts may be ineffective if a country is not fully invested in the goals at hand. Including harsher penalties at the outset is difficult because it may mean that fewer countries ultimately sign on to the agreement at all.

protocol A document that sets precise goals and targets.

AP Images/Paul White

Swedish environmental activist Greta Thunberg speaks at the 2019 UN Climate Change Conference. Her activism has inspired young people around the world to speak up about climate change.

INFOGRAPHIC 7 INTERNATIONAL AGREEMENTS

More than 2,500 international environmental treaties are currently in place, covering a wide range of topics from protecting wetlands and species to the use or disposal of hazardous materials.

SIGNED/ENTERED INTO FORCE	TREATY	DESCRIPTION
1971/1975	Ramsar Convention on Wetlands of International Importance *(See Online Module 3.3.)*	Promotes the conservation and wise use of wetlands; protects vital biodiversity habitat and water resources
1973/1975	Convention on International Trade in Endangered Species of Wild Fauna and Flora (CITES) *(See Online Module 3.3.)*	Regulates the sale and trade of endangered or threatened species or products
1987/1989	Montréal Protocol	Oversees the phase-out of ozone-depleting substances to protect stratospheric ozone
1992/1994	UN Framework Convention on Climate Change *(See Module 10.2.)*	Established a framework for addressing climate change
1997/2005	Kyoto Protocol *(See Module 10.2.)*	Set binding greenhouse gas reduction targets for developed nations
2015/2016	Paris Agreement *(See Module 10.2.)*	Oversees greenhouse gas reduction targets for parties with a goal of keeping global temperature increase well below 2°C relative to preindustrial levels

? Why is it important that international treaties be revised or amended over time?

8 RESPONDING TO OZONE DEPLETION: THE MONTRÉAL PROTOCOL

Key Concept 8: The Montréal Protocol is successfully addressing ozone depletion thanks to a strong international commitment and the flexibility to update it as needed.

Attempts to address ozone depletion on the international stage began in the late 1970s. After much debate and political wrangling, the 1985 Vienna Convention, a treaty that established the need to respond, was passed. This was followed by the 1987 Montréal Protocol on Substances that Deplete the Ozone Layer, a protocol that established a planned phase-out of CFCs and other stratospheric ODS. By 2009, the protocol had been ratified by all nations of the world.

Interestingly, the 1987 Montréal Protocol and the international commitment to address CFCs and ozone depletion came before Susan Solomon's final study definitively linking CFCs to ozone depletion was published. We didn't know all the details in 1987, but because the potential loss of our protective ozone layer was such a serious problem, we chose to take action. This is an example of applying the *precautionary principle*—acting in the face of uncertainty when there is a chance that serious consequence might occur. (See Modules 1.2 and 4.3 for more on the precautionary principle.)

As more information poured in, it quickly became apparent that the Montréal Protocol targets would not be sufficient to stop ozone depletion in a timely fashion, so the phase-out dates were moved up for CFCs. Amendments to the Montréal Protocol are still negotiated in annual meetings that strengthen the response, such as adjusting the target dates to phase out harmful compounds or adding new chemicals to the list of regulated substances. This ongoing process is an example of **adaptive management**—allowing room for altering strategies as new information comes in or the situation itself changes. **INFOGRAPHIC 8**

Within a decade or so of the Montréal Protocol's acceptance, scientists began to see evidence that it was working. The rate of ozone thinning in polar regions slowed down, stabilized, and eventually began to reverse. The data now suggest that the ozone hole is finally beginning to "heal," defined as an observable increase in ozone. Based on the recovery to date, scientists are projecting that the Antarctic

adaptive management
A plan that allows room for altering strategies as new information becomes available or as the situation itself changes.

INFOGRAPHIC 8 MONTRÉAL PROTOCOL

Actual and projected change over time for total global emissions of ozone-depleting substances (ODS) with and without the Montréal Protocol and its amendments. Adjustments to the phase-out schedule of various ODSs in the form of amendments represent the success of adaptive management in dealing with complex environmental issues.

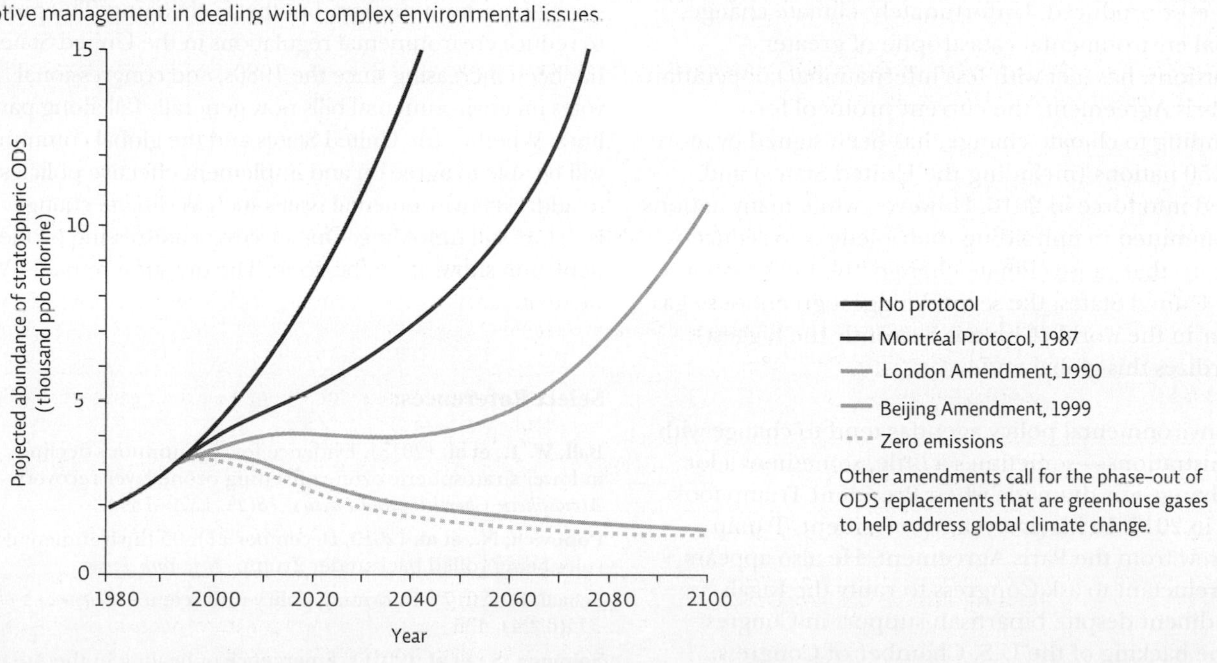

Looking at the graph that shows the projected decrease in chlorine in the stratosphere, would you say the Montréal Protocol is a success? What role did the amendments play in this success or lack of success? Explain.

ozone hole will have fully recovered by 2050. (See the *Science Literacy* activity at the end of this module.)

Unfortunately, the story is not over yet. A 2018 study published by William Ball and an international team of scientists found that stratospheric ozone appears to be declining in equatorial and mid-latitude regions. This is especially concerning since many more people live in these regions than in the Antarctic or Arctic and because these regions naturally receive more UV radiation. Scientists do not yet understand the cause of this decline, but it may be related to climate change.

You may be aware that our planet's climate is changing in the direction of warming—global warming. (See Module 10.2.) Warming may be changing stratospheric air circulation patterns and increasing the rate at which low-ozone air is delivered to these regions. Much more research is needed to unravel this new mystery.

There is another connection between ozone depletion and climate change. It turns out that CFCs are greenhouse gases (atmospheric molecules that trap heat). While phasing out CFCs would help reduce warming, unfortunately, some of the replacement chemicals (hydrochlorofluorocarbons [HCFCs] and hydrofluorocarbons [HFCs]) are even more potent greenhouse gases. In effect, some steps taken to address ozone depletion have accelerated climate change. The latest amendment to the Montréal Protocol, the Kigali Amendment, will phase out HCFCs and HFCs.

The Montréal Protocol represented the first and arguably most successful international environmental policy ever produced. Unfortunately, climate change, a global environmental catastrophe of greater proportions, has met with less international cooperation. The Paris Agreement, the current protocol for responding to climate change, has been signed by more than 150 nations (including the United States) and entered into force in 2016. However, while many nations are committed to upholding their pledges to reduce emissions that cause climate change, the weak response of the United States, the second biggest greenhouse gas emitter in the world (China is currently the biggest), jeopardizes this global endeavor.

U.S. environmental policy agendas tend to change with administrations—sometimes a little, sometimes a lot. The change was dramatic when President Trump took office in 2017. In his first year as president, Trump withdrew from the Paris Agreement. He also appears to be reluctant to ask Congress to ratify the Kigali Amendment despite bipartisan support in Congress and the backing of the U.S. Chamber of Congress. (The United States has signed all previous Montréal Protocol amendments.) President Trump sees many environmental regulations as excessive—holding back economic development—and his administration is taking steps to address that. In an analysis published by the *New York Times*, as of December 2019, the Trump administration had rolled back 58 environmental regulations with another 37 in the works. The EPA has been affected by budget cuts, gag orders, the loss of science advisers, and staff reductions.

In a 2019 House Energy and Commerce Committee hearing, former EPA head Christine Todd Whitman, who served under President George W. Bush, was one of four former EPA heads to voice concerns over actions by the Trump administration. Todd told the committee, "This unprecedented attack on science-based regulations designed to protect the environment and public health represents the gravest threat to the effectiveness of the EPA—and to the federal government's overall ability to do the same—in the nation's history." This is not to say that there is not room for regulatory reform. As Todd explained, "some regulations outlive their usefulness and regulatory reform can, if done properly, produce greater results than originally envisioned. But changes to, or the elimination of, existing regulations must be driven by careful scientific analysis and solid facts. Deregulation for its own sake is rarely prudent and often unwise."

The U.S. federal government brought environmental policy into focus in the 1960s; the world followed the nation's lead, and our environment is cleaner today than it was then. But that will only be true as long as we continue to uphold those laws and make new ones when needed. NEPA and the environmental laws that followed were passed with strong bipartisan support, but that is changing. The push to reduce environmental regulations in the United States has been increasing since the 1980s, and congressional votes on environmental bills now generally fall along party lines. Whether the United States and the global community will be able to agree on and implement effective policies to address environmental issues such as climate change is a story still unfolding. Our success at addressing ozone depletion shows it can be done. The question remains: Will we do it?

Select References:

Ball, W. T., et al. (2018). Evidence for a continuous decline in lower stratospheric ozone offsetting ozone layer recovery. *Atmospheric Chemistry and Physics, 18*(2), 1379–1394.

Popovich, N., et al. (2019, December 21). 95 Environmental rules being rolled back under Trump. *New York Times.*

Schaal, B. (2017). Informing policy with science. *Science, 355*(6324), 435.

Solomon, S., et al. (2016). Emergence of healing in the Antarctic ozone layer. *Science, 353*(6296), 269–274.

GLOBAL CASE STUDIES POLICY IN ACTION

Achie/e

Many U.S. environmental problems are being effectively addressed with national policies that got their start with grassroots action. See these online case studies for some notable U.S. examples.

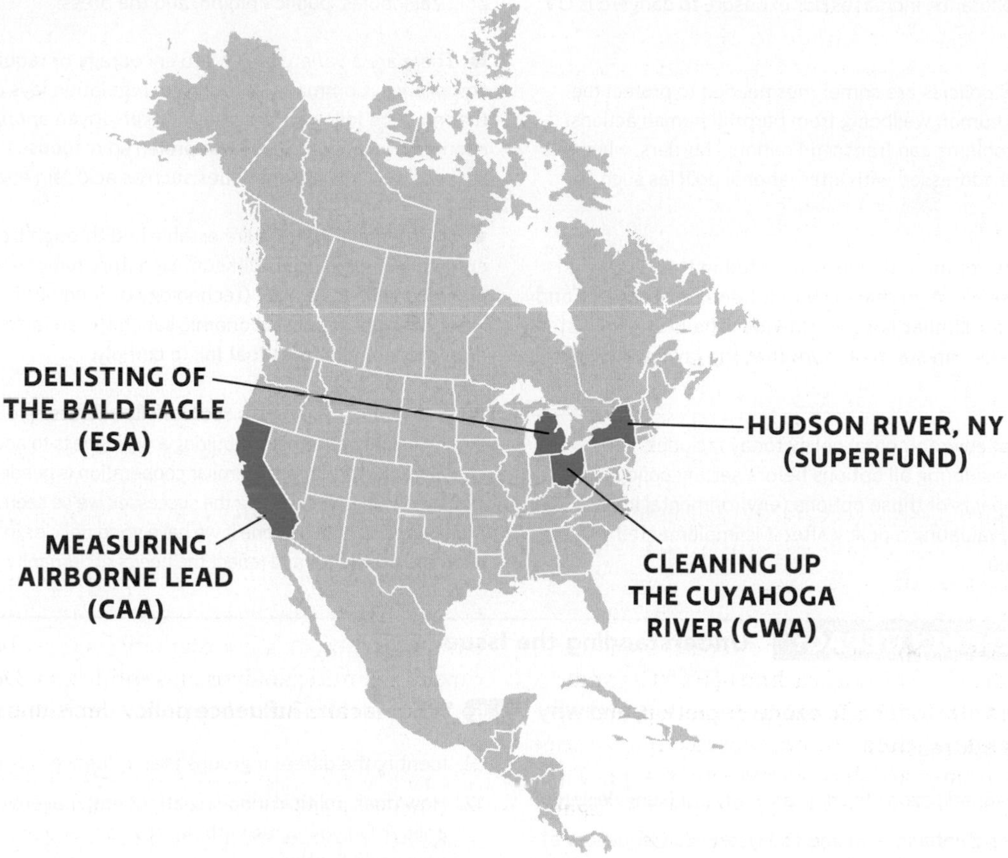

BRING IT HOME

PERSONAL CHOICES THAT HELP

The process of writing and revising policy, proposing it, voting on amendments, and finally enacting it as law is a complex and often messy one. The legislative process is often referred to as "sausage making" because of all the steps and input as well as the fact that the final product often looks much different than the original.

Individual Steps

- Call, email, write, or visit your elected officials to voice your opinion on policy issues of interest to you. Find your national congressional legislators at www.usa.gov/elected-officials; visit the websites for your state and local governments to do the same.
- Find out who and what is influencing your elected politicians. The website www.opensecrets.org allows you to look up the top individuals and industries that contribute to any candidate's campaign and the top recipients of those contributions.
- Actions speak louder than words. Visit www.votesmart.org and find the voting record of your representative. How does he or she vote on environmental issues like climate change? If you do not feel your representative's record is moving the country in the right direction, volunteer for another candidate whose policies you support during the next election cycle.

Group Action

- When an important issue is not adequately addressed, concerned citizens often form petition drives. Signatures are collected and delivered to politicians, who can propose new legislation. Form a group to petition for an important issue that you feel is being overlooked and present the collected signatures to any politician who can propose new policy. Organizations such as www.change.org can help get your petition out to a wider audience.

REVISIT THE CONCEPTS

- Environmental problems can be serious, and many demand our attention. For example, stratospheric ozone depletion, caused by anthropogenic pollutants, increases our exposure to dangerous UV radiation.
- Environmental policies are sometimes needed to protect the environment and human wellbeing from harmful human actions. Environmental problems can transcend national borders; when this happens, they are addressed with international policies such as treaties.
- Many U.S. environmental laws originated in the 1960s and 1970s in response to an increasingly degraded environment and became models for similar laws worldwide. The EPA administers these laws; citizens can sue to ensure that the laws are properly enforced.
- The creation of environmental policy today includes systematically considering all options before setting policy; evaluating the impacts of those options (environmental impact statement); and evaluating a policy after it is implemented, making changes as needed.
- Depending on high-quality science is crucial to developing sound policies, but other voices can also influence policy formation, such as political lobbies, public opinion, and the press.
- There are a variety of ways to encourage or require policy compliance. Command-and-control regulation lays out the response that must be followed, whereas market-driven approaches offer economic incentives that favor preferred responses. Both have been successful in addressing issues such as acid rain and water pollution.
- International policies are established through treaties, but enforcing compliance is difficult. Signatory nations might be provided with assistance (technology or funding) to help them meet their obligations. Economic sanctions are also sometimes employed for countries that fail to comply.
- The Montréal Protocol is successfully addressing ozone depletion, though unknowns remain, requiring a willingness to address new challenges as they appear. Similar cooperation is needed to address climate change. Maintaining the successes we've seen in environmental protection can only be done if we have good policies in place that are informed by science and reflect the needs of humanity.

ENVIRONMENTAL LITERACY Understanding the Issue

1 What causes stratospheric ozone depletion, and why should it be addressed?

1. How is stratospheric ozone formed, and why is it being depleted?
2. What are the potential health and ecosystem consequences of stratospheric ozone depletion?

2 What is the purpose of environmental policies, and why are they needed?

3. Why are national or international laws and policies necessary to address some environmental issues?
4. Why are environmental policies needed to address issues that are affected by the tragedy of the commons?

3 How has environmental policy evolved in the United States?

5. Identify some of the iconic U.S. environmental laws and describe their focus.
6. What is the citizen suit provision of environmental laws?
7. How effective have U.S. environmental laws been in addressing our environmental problems?

4 How are policies developed and administered in the United States?

8. Describe the NEPA process. Why is it considered a cycle rather than a linear process?
9. Explain the roles of Congress, the President, and the EPA with regard to environmental policy.
10. What is an environmental impact statement, and why is it produced?

5 What factors influence policy decisions?

11. Identify the different groups that influence U.S. public policy.
12. How does political lobbying affect national environmental policy? Do you agree with the critics that political lobbies are too powerful? Explain.

6 What policy tools can be used to implement and enforce environmental policy?

13. What is command-and-control regulation, and what are its pros and cons?
14. Give an example of a market-driven approach to solving environmental problems. How does this differ from command-and-control regulation of environmentally damaging behavior?

7 How are international policies established and enforced?

15. In terms of international treaties, distinguish between a convention and a protocol.
16. What limitations exist for the enforcement of treaties?

8 How has the international community responded to ozone depletion?

17. What is adaptive management?
18. When is it reasonable to invoke the precautionary principle when setting policy?
19. How are ozone depletion and climate change connected?

SCIENCE LITERACY Working with Data

The size of the ozone hole has been tracked since 1979 by a variety of methods. Two of those data sets are shown in the graph below: measurements taken by the Total Ozone Mapping Spectrophotometer/Ozone Monitoring Instrument (TOMS/OMI) and simulated data produced by the Chem-Dyn-Vol model that takes into account the chemical processes, dynamical processes, and volcanic eruptions believed to affect stratospheric ozone.

SIZE OF OZONE HOLE IN SEPTEMBER

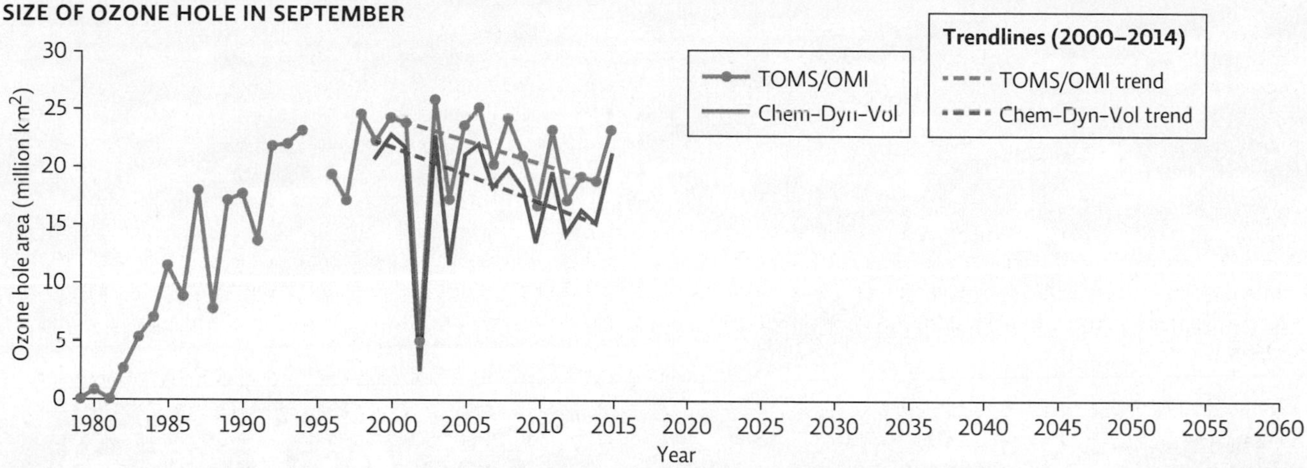

Interpretation

1. For what years are data for TOMS/OMI observations shown? For what years was the Chem-Dyn-Vol model simulation run?
2. Why is there no line connecting the TOMS/OMI data point for 1994 to 1996?
3. What trend was seen between 1989 and 2000 regarding the size of the ozone hole? What trend was seen after 2000?
4. Extend the trendlines out (following the same slope of the line or trend) and estimate the year that each line predicts the ozone hole will be gone.

Advance Your Thinking

5. Why are data shown for the month of September for each year?
6. How well does the model simulation (the red line) match the observed data (the blue line)? What does this tell us about the validity of the model — does it appear that researchers have identified the relevant parameters that affect ozone depletion?
7. What is the purpose of creating a computer model such as Chem-Dyn-Vol?

INFORMATION LITERACY Evaluating Information

The Montréal Protocol has been amended several times as new information comes to light regarding the proposed bans of ozone-depleting substances. The 2016 Kigali Amendment calls for the phase-out of hydrofluorocarbons (HFCs).

Answer the following questions after learning about the basics of the Kigali Amendment. Two good starting places are the EPA website at www.epa.gov/ozone-layer-protection/recent-international-developments-under-montreal-protocol (or search for Kigali Amendment on the EPA home page: www.epa.gov) and the *Nature* article "Nations Agree to Ban Refrigerants That Worsen Climate Change" at http://www.nature.com/news/nations-agree-to-ban-refrigerants-that-worsen-climate-change-1.20810. Consult other sources as needed.

1. Why were HFCs originally developed (what is their industrial use)?
2. Why are HFCs being targeted for phase-out? What is the expected advantage to global climate change if HFCs are phased out?
3. What is the timeline for HFC phase-out?

Next, conduct Internet searches to find articles about the proposed HFC phase-out from an industry point of view (e.g., the chemical industry or the heating and air conditioning [HVAC] industry) and an environmental point of view (e.g., nonprofit environmental groups such as the National Resources Defense Council or the Environmental Defense Fund). Evaluate the response from one industrial source and one nonprofit environmental source by answering the following questions for each:

4. Identify the information sources you visited for this information. What position does each take on the phase-out of HFCs? What inherent bias might each group have regarding HFCs?
5. Evaluate the quality of evidence (peer-reviewed scientific studies? newspaper articles? government websites? etc.) offered by each entity.

After learning about HFCs, the Kigali Amendment, and various viewpoints on this topic, draw your own conclusions about the soundness of the proposed phase-out of HFCs. Write a short paper (1–2 pages) that addresses this question: Do you support the Kigali Amendment? Provide evidence to back up your position. Provide in-text citations and a bibliography at the end of your paper.

Additional study questions are available at Achieve.macmillanlearning.com. Achieve

Module 5.3 Managing Solid Waste

MICROPLASTICS

Are tiny plastics a big problem?

Plastic trash washed up on a beach in Spain. *Image Broker/Media Bakery*

After reading this module, you should be able to answer these GUIDING QUESTIONS

1. What types of solid waste do we produce, and why do we say waste is a "human invention"?
2. What is municipal solid waste (MSW), and what are the types and proportion of waste in the U.S. MSW stream?
3. What problems are caused by the improper management of solid waste?
4. Compare the design, advantages, and disadvantages of open dumps and sanitary landfills.
5. What are the pros and cons of waste incinerators?
6. What are some common household hazardous wastes, and how should individuals deal with this waste?
7. What is composting, and how can it help us in our quest to deal with solid waste?
8. What are the four Rs of waste reduction?

When Sherri Mason pulled up the net after plunging it into Lake Superior, she couldn't believe her eyes. It was July 2012, and Mason, a chemist, was aboard the U.S. *Niagara*, a 198-foot-long teaching sailboat, conducting the world's first survey of plastic pollution in Lake Superior. "I'm looking down at the net, and I see this one little ball of Styrofoam, just sitting there in the middle of all this algae," Mason recalls. It was unreal: She'd found plastic in her very first sample.

On that trip, Mason and her students found several other big plastic pieces, including a wrapper from a cigarette box and parts of a bottle cap. But they didn't see the rest until they got back to the lab. When they used a microscope to count particles that were too small to see, they found that Lake Superior's water harbored some 30,000 plastic particles per square kilometer. In subsequent sampling trips, they found even more: Lake Ontario had 230,000 plastic particles per square kilometer, which meant that its waters harbored concentrations of plastic particles higher than reported for any other body of water on the planet.

WHERE ARE THE GREAT LAKES?

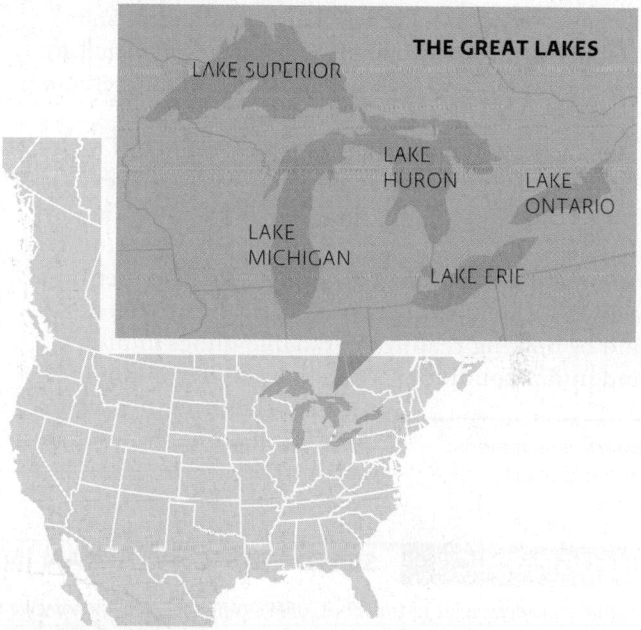

These plastic particles, known as "microplastics," that float through the Great Lakes are tiny—most are less than 1 millimeter in diameter (by definition, they are plastic less than 5 millimeters long)—but that doesn't mean they're harmless. To the contrary, many are embedded with hazardous chemicals—dyes, plasticizers, and flame retardants, among other things—that likely leak out from the plastics into the surrounding water. Microplastics can also absorb toxic substances from water. "The plastic becomes like little poison pills," says Mason, who is now a professor at Pennsylvania State University Behrend. "They're so incredibly tiny that they can be easily ingested by organisms at the very base of the food chain, like plankton, and they can be carried by these organisms up the food chain and ultimately into us, carrying those chemicals all along the way." A 2019 study by Phillip Schwabl of the Medical University of Vienna reported finding microplastics in the stool samples of human subjects, and it was not the first to do so.

Sherri A. "Sam" Mason, Ph.D.

Microplastics in a personal care product shown at 40× magnification. The blue spheres are microbeads, and the white flakes are additional microplastics included for their abrasive qualities.

As Mason and other scientists have shown, microplastics are virtually everywhere: in waterways, oceans, soil, and even the air. Mason discovered microbeads from personal care products—facial scrubs and toothpaste—and tiny fibers from polyester clothing, for example. Microplastics are turning up everywhere—in ocean water and sediments, in arctic snow, and in the bellies of fish living deep below the surface of the Gulf of Mexico waters. When Steve Allen's research team surveyed air in the pristine French Pyrenees mountains in a 2018–2019 study, they found 249 microplastic fragments floating around per square meter.

Considering that the world produces more than 380 tons of plastic each year—8,300 million tons since the production first began—is it any wonder that tiny plastic fragments are now scattered around the planet, covering nearly every surface? Scientists have many questions about how these microplastics are affecting the environment, marine and land organisms, and human health—and they're starting to get answers.

1 WASTE: A HUMAN INVENTION

Key Concept 1: Solid waste can be said to be unique to human societies because in nature there is no "waste": The discarded matter of one organism becomes a resource for another.

Solid waste—all manner of materials that we discard, including trash, sewage sludge, and agricultural waste, to name a few categories—is a uniquely human concept. In natural ecosystems, there is no such thing as *waste.* In a circular pathway, matter expelled by one organism is taken up by another organism and used again. Forms of matter that are dangerous to living things (think arsenic and mercury) tend to stay buried deep underground and are released only during extreme events like volcanic eruptions.

Human ecosystems are another story. We tend to handle matter in a linear manner—matter is extracted, transformed, used, and then discarded, often in ways that make it hard for us to reuse or for nature to reclaim. We do this by converting usable matter into synthetic chemicals that can't easily be broken down and by burying readily degradable things in places and under conditions where natural decomposition processes can't run their course. However, when we throw trash away, it isn't gone for good—it's still somewhere, in some form. This is consistent with the *Law of Conservation of Matter,* which tells us that matter can't be created or destroyed; it can only change forms. Whatever we dig up or create stays with us; whatever we throw away stays with us too. Plastic trash may break down into smaller and smaller pieces, but it doesn't cease to exist.

solid waste Any material that humans discard.

One solution to keeping plastics out of the environment is to change our focus away from *waste disposal* to *waste management*—preventing its generation in the first place. To do that, we must go back to the beginning, before the things we throw away are even made. By conducting a life cycle analysis to assess the environmental impact of every stage of a product's life—from production to use to disposal—companies are looking for ways to reduce the amount of waste generated by their products. More and more manufacturers are tackling the problem earlier still, considering reuse potential in the design stage, making it easier to turn *waste* back into *resource* (see Module 5.1). **INFOGRAPHIC 1**

INFOGRAPHIC 1 SOLID WASTE: A HUMAN INVENTION

Humans produce a lot of trash but, unlike other species, the way we discard it and the types we generate make it hard or impossible for nature to reuse or recycle. Our linear system of "*TAKE—MAKE—USE—DISCARD*" needs to be replaced with a circular system (like that of nature) of "*TAKE—MAKE—USE—PUT BACK—USE AGAIN*" (used by us or other species).

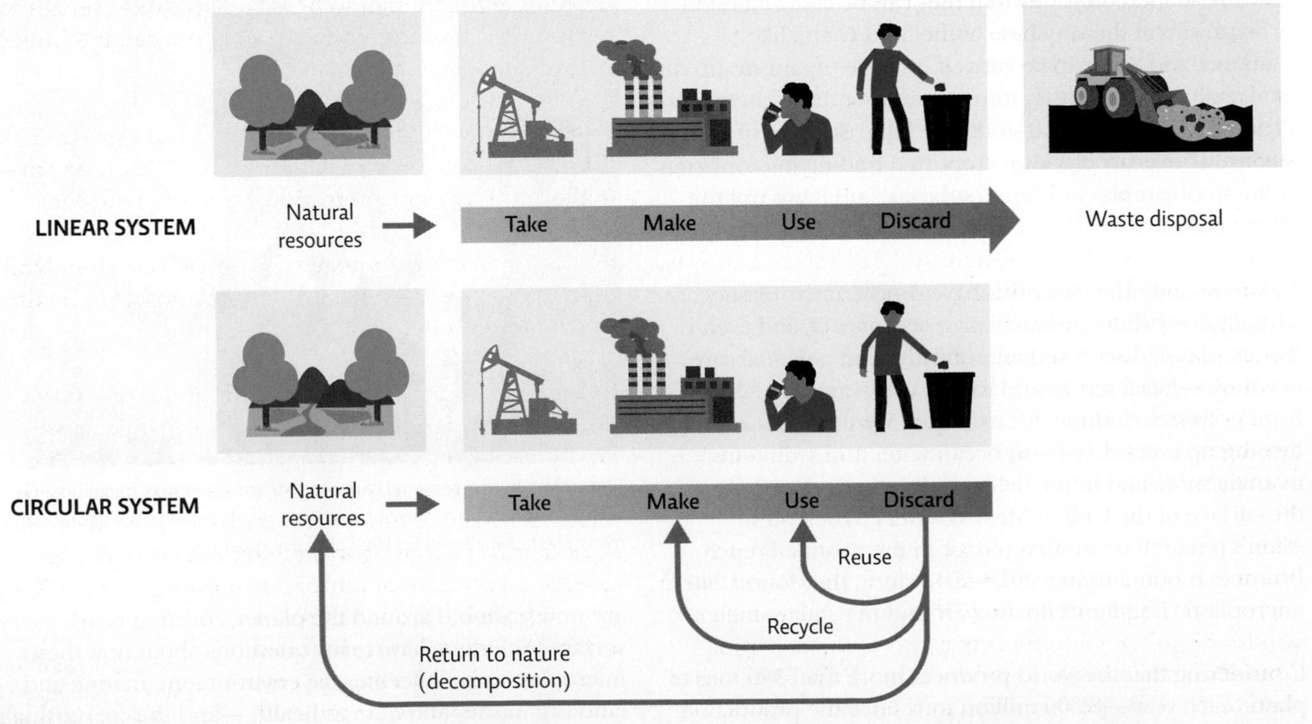

How does a circular system of production compare to natural matter cycles, such as the carbon cycle? (See Infographic 5 in Module 2.1.)

Not all waste is created equal. Waste that can be broken down by chemical and physical processes is considered *degradable* even if that degradation takes a long time. Waste that can be broken down by living organisms such as microbes is considered **biodegradable**. Some waste — mostly synthetic molecules like the pesticide DDT and the chlorofluorocarbons (CFCs) once found in aerosols — is considered **nondegradable**. These molecules are chemically stable and don't readily break down.

Different types of plastics have different degradation rates, but for all, it is generally quite slow. After all, one of the perks of plastics is their durability. Exposure to sunlight in an oxygen-rich environment causes plastics to chemically change and become brittle, breaking into smaller pieces. This breakdown can occur more quickly with the additional weathering action of wind and waves, which produces the tiny plastic bits found in waterbodies everywhere. But breaking up into pieces is not the same thing as biodegrading into simpler structures that the natural environment can use. "There are very few organisms that can use [plastic] as a food source, so it doesn't get turned back into soil the way natural materials would," Mason explains. "Even as these pieces get smaller and smaller and smaller, it's still plastic. It still isn't going away."

biodegradable Capable of being broken down by living organisms.

nondegradable Incapable of being broken down under normal conditions.

2 MUNICIPAL SOLID WASTE

Key Concept 2: Household trash contains a variety of items, much of it food or packaging waste. Though this municipal solid waste is a small part of the total waste produced, volumes are still significant and worth efforts aimed at reducing the trash we produce.

Almost any human activity you can think of generates some form of waste. Processes that produce food, consumer goods, and our roads and buildings generate agricultural, industrial, and construction waste. The harvesting of coal and valuable mineral resources like gold and copper generates mining waste, which can pollute air, water, and soil. Together, industrial, agricultural, construction, and mining waste make up most of U.S. solid waste (estimates range from 91% to 97%). The remainder is produced in the increasingly complicated act of living — in houses, apartments, dormitories, and small businesses around the world — and is referred to as **municipal solid waste (MSW)**.

Currently, urban dwellers produce most of the world's municipal trash — about 2 billion tons of solid waste each year — and this amount is expected to almost double by 2025. The per capita rate of trash generation is positively correlated with the income level of a country. (Curiously, plastic trash is the one type of waste that is not as closely tied to income — its production rates in countries around the world are about the same, regardless of income.)

Americans create more garbage per person than just about any other country in the world (only Canada has a higher per capita value). In 2015 alone, each American produced about 4.5 pounds of solid waste per day, up from about 2.7 pounds per day in the 1960s. With more than 300 million people living in the United States, this adds up to about 260 tons of household trash per year; that's twice the per capita amount produced by some other wealthy countries such as Belgium, Norway, or Japan, and as much as 10 times the amount produced by most less developed countries.

The vast majority of this garbage comes from a familiar array of goods: paper, wood, glass, rubber, leather, textiles, and of course plastic — cheap enough to have become a staple of both advanced and developing societies, light enough to float, and durable enough to persist for hundreds of years across thousands of miles of ocean. **INFOGRAPHIC 2**

Of course, there are good reasons to use plastic. Plastic helmets are strong but light. Plastic car parts make vehicles energy-efficient, since they reduce overall weight and are highly durable. Single-use medical plastics minimize the risk of infection, while plastic prosthetic limbs are lighter and more flexible than wooden ones.

But we also find plastic in parts and places that may not be needed: in packaging, straws, bags, and toys. Plastic packaging can help to extend food's shelf life, but amounts used may have more to do with marketing than safety. Plastic now makes up about 13% of our MSW stream, with packaging making up the largest single category. Very little of our plastic waste — about 10% — is recycled, while 67% is buried in a landfill and 14% is incinerated. Unfortunately, our love affair with plastic only seems to be growing: On average, plastic production doubles every 11 years.

municipal solid waste (MSW) Everyday garbage or trash (solid waste) produced by individuals or small businesses.

INFOGRAPHIC 2 **U.S. MUNICIPAL SOLID WASTE STREAM**

Municipal solid waste (MSW) is the trash produced by homes and businesses. Though the volume of industrial and agricultural wastes far exceeds the amount of MSW in the United States, we still produce millions of tons of MSW annually. This makes actions that reduce MSW worth the effort.

MSW GENERATION RATES, 1960 TO 2015

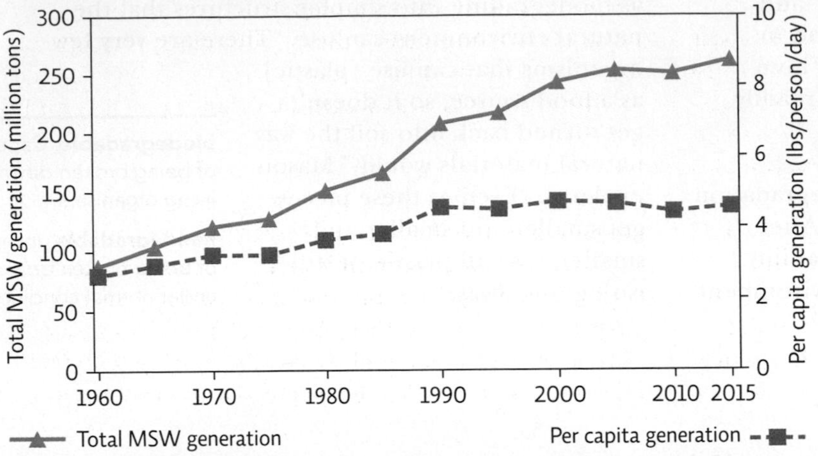

TOTAL U.S. MSW GENERATION BEFORE RECYCLING, 2015

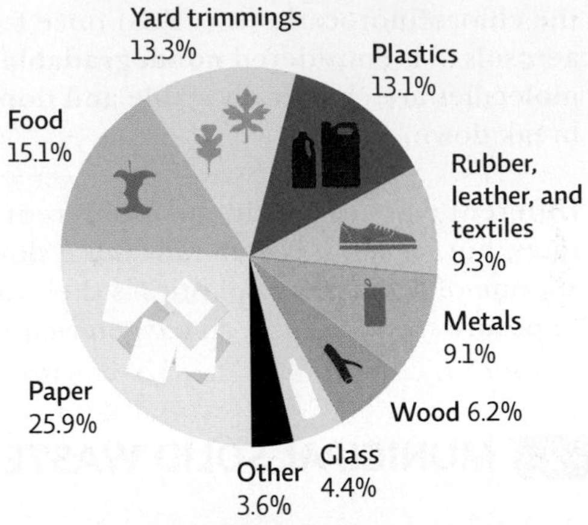

Which of the categories of waste shown in the pie chart are naturally degradable? Which could be recycled?

3 IMPACTS OF MISMANAGED SOLID WASTE

Key Concept 3: Uncollected and mismanaged solid waste contributes to a variety of environmental problems (e.g., flooding; air and water pollution), which can contribute to environmental damage and health problems for humans and other species.

In cities of the United States and other developed countries, MSW makes its way from garbage cans in our homes and places of work, to garbage trucks, and finally to one of several kinds of waste facilities. Unfortunately, close to half of the world's population—3 billion people—do not have access to any form of waste management. Their garbage is not picked up and hauled off; there are no municipal dumpsters where they can throw their trash. So they toss it where they can—on the ground or in the nearest river.

When no formal waste disposal system is in place (or it is ignored), trash often ends up on the street, where it can block storm drains, cause flooding, and create puddles that attract disease-carrying mosquitoes, flies, and rodents. Dust from open dumps pollutes the air along with smoke from burned trash. Plastic bags clog outboard boat motors and make fishing a daily struggle, and in many coastal areas, discarded plastics cover beaches, making sand barely visible. In addition, chemical waste tucked into this trash has the potential to wreak havoc on plant and animal life, including, of course, humans. (See Module 4.3 for more on chemical hazards and environmental health.) Even our modern disposal methods can contribute to health and environmental problems; incinerators create air pollution, and landfills produce methane.

Improperly discarded trash often ends up in bodies of water where record amounts are accumulating. In the ocean, plastic debris causes all sorts of problems. Many discarded plastics collect in what are called ocean "garbage patches"—regions of the ocean where strong currents encircle central regions with weaker currents (called *gyres*), making it difficult for waste to drift out once encircled. More than 100,000 plastic pieces per square kilometer have been found in the North Atlantic Gyre. And surrounding or just below the surface of all of these visible plastics are the invisible ones—the microplastics that Mason and others know are virtually everywhere.

Plastic debris in the ocean can harm marine organisms in various ways. First, plastics can transport organisms to places they don't belong and where they can't survive. Floating plastics can become covered in fish eggs, barnacles, and other young organisms, which then travel to new locales where they can become invasive species. Sea life can also get fatally tangled in discarded shipping nets, shopping bags, and plastic six-pack rings. Organisms eat the plastics, too. Many seabirds are particularly vulnerable, ingesting large amounts of plastic—plastic bits they pick up, mistaking them for food, or plastic that is in the prey they capture. To a sea turtle or whale, plastic bags look much like their jellyfish prey. A gut that fills with plastic may no longer be able to hold enough food to keep the animal alive. When an emaciated whale found in the waters off the Philippines was necropsied after its death, researchers found its stomach packed with 88 pounds of plastic bags, nylon ropes, and other debris; it likely died of starvation.

Organisms can also be harmed by the toxic chemicals the plastics harbor. Many plastics also release hormone-like compounds. A 2011 study by Chun Yang found that 70% of the plastics his team tested released chemicals that mimic the hormone estrogen. These compounds may interfere with reproduction or development of exposed organisms. UK researcher Matthew Cole and his colleagues found that marine copepods (tiny crustaceans) exposed to low levels of polystyrene microbeads do not feed, grow, or reproduce normally—among other things, they produce smaller eggs that are less likely to hatch. (See the *Science Literacy* activity at the end of this module for more on this study.) **INFOGRAPHIC 3**

And if marine organisms are eating plastic, then we are eating it, too. For example, shellfish (e.g., mussels and oysters), an important human food source, can accumulate large amounts of microplastics. A recent study by University of Victoria researcher Kieran Cox estimated that the average American consumes 50,000 microplastic particles a year via food and drinks and

INFOGRAPHIC 3 **NEGATIVE IMPACTS OF SOLID WASTE**

The negative impacts of improperly managed solid waste are varied and serious, threatening the lives and well-being of humans and other species worldwide.

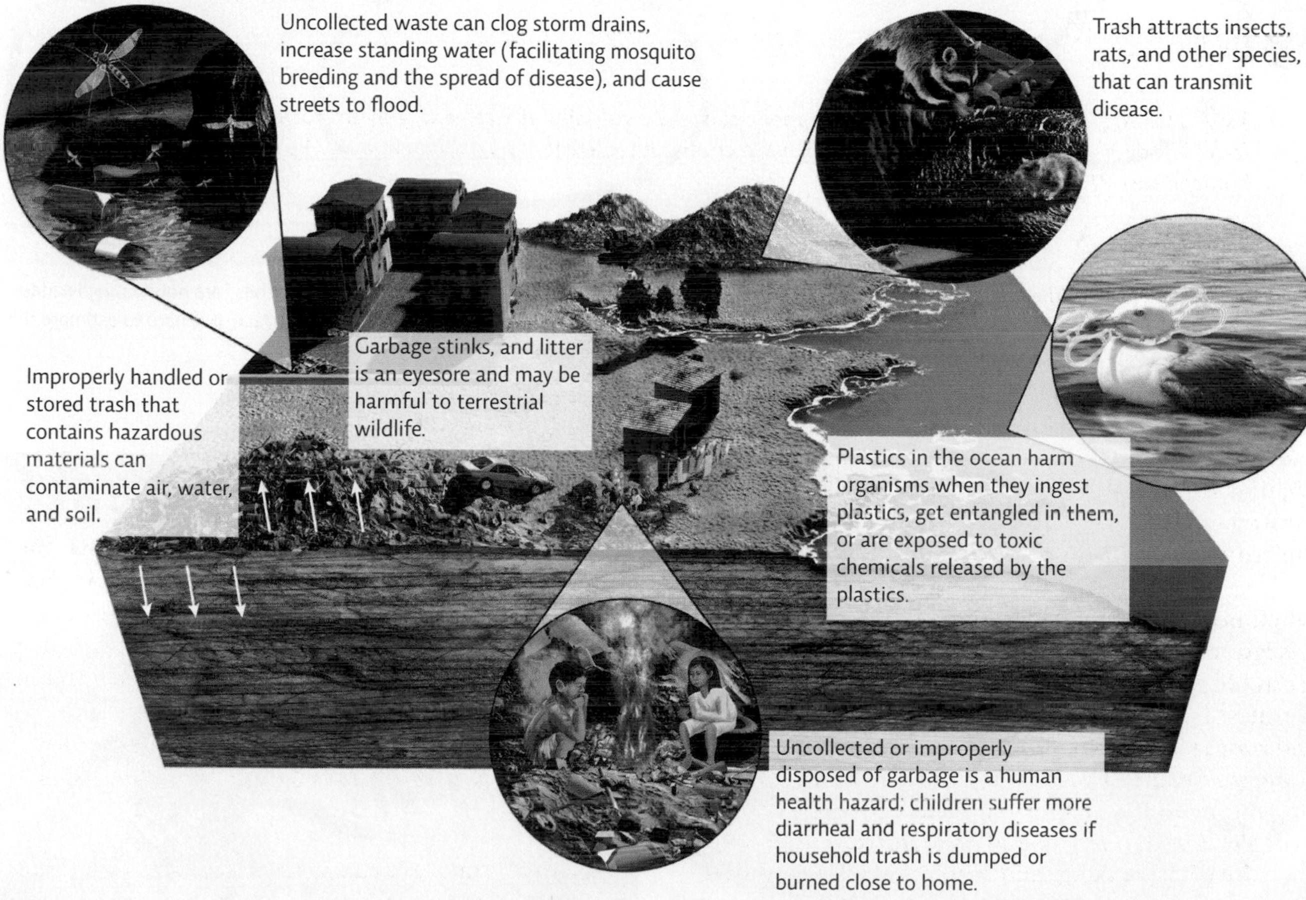

What are some of the negative impacts of improperly managed solid waste in your area?

inhales another 70,000 particles from the air. Mason's own research has found microplastics in tap water and even higher amounts in bottled water and beer. It's unclear what health effects, if any, these exposures may have on humans, but the varied effects seen in animal studies are cause for alarm.

To be sure, not all of the trash accumulating in the ocean gyres is plastic, but other types of waste—textiles, glass, wood, and rubber—sink or degrade relatively quickly. Plastic just floats along. Eventually, time, saltwater, and sunlight break it down from its recognizable everyday forms—combs, candy bar wrappers, plastic straws—into fragments so tiny that even thousands of them together can't be seen by a naked eye trained on a calm sea. According to a report from the MacArthur Foundation, the amount of trash entering the ocean will double by 2030 and quadruple by 2050. By that point, our oceans may contain more plastic by weight than fish.

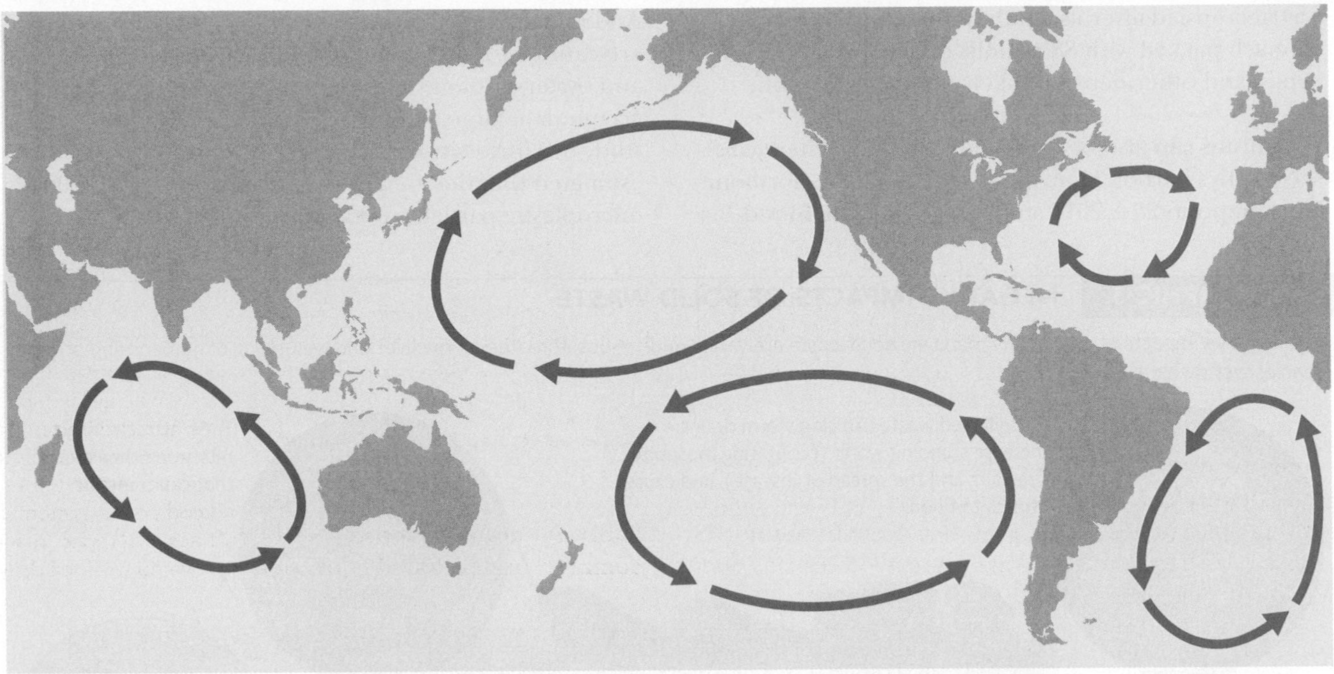

Five major gyres are found in the world's oceans, and there are floating bits of plastic in all of them. These "garbage patches" are not floating islands and aren't even necessarily visible when gazing at the water. Much of the debris is very small and lies just below the surface. It is hard to estimate the size of any garbage patch since the material is so spread out and may reach down to depths of 60 feet.

Courtesy Giora Proskurowski, Sea Education Association

This bucket holds the tiny bits of visible plastics, and microplastics you can't see, from a sample taken from the North Atlantic Garbage Patch.

Paulo Oliveira/Alamy

Many seabirds ingest plastics, mistaking them for food. The animal can starve to death because there is not enough room in the gut for food.

4 DISPOSAL METHODS: OPEN DUMPS AND SANITARY LANDFILLS

Key Concept 4: While better than open dumps because they bury trash in a way that reduces air and water pollution, sanitary landfills reduce decomposition rates of trash and produce methane, a powerful greenhouse gas.

Proper solid waste management is crucial to keeping it out of lakes, rivers, and oceans; out of forests and grasslands; out of gutters and backyards; and away from disease-carrying insects and rodents. Here in the United States, we have a long history of dealing with waste and a variety of ways to deal with our trash. While all the methods have drawbacks, some are more environmentally damaging than others.

All solid waste in the United States is regulated by the 1976 **Resource Conservation and Recovery Act (RCRA)**. As with many federal environmental laws, the Environmental Protection Agency (EPA) is tasked with providing the rules and regulations to administer the law; states are in charge of implementing it within their jurisdiction.

Through RCRA, the United States has long banned the use of **open dumps** for waste disposal, as have many other developed countries, but because dumps are one of the cheapest ways to get rid of human trash, they are common in less developed countries. Indeed, in many areas, entire communities spring up around the dumps and people survive on what they scavenge from the waste piles. However, there are some good reasons open dumps are banned in many countries—reasons linked to health and environmental problems.

Open dumps attract disease-carrying pests such as flies and rats, a human health hazard. Open dumps also contribute to water pollution: Rain either washes chemical pollutants away from the dump to surrounding areas or pulls it along as it soaks into the ground. If this contaminated water, called **leachate**, continues to travel downward, it can contaminate the soil and groundwater. These open dumps are also key contributors of the plastic pollution reaching the sea, beginning as larger pieces of plastic—bottles, food packages, clothing, straws, bags—that break down into smaller and smaller fragments and fibers.

Sanitary landfills, more common in developed countries,

Resource Conservation and Recovery Act (RCRA) The federal law that regulates the management of solid and hazardous waste.

open dumps Places where trash, both hazardous and nonhazardous, is simply piled up.

leachate Water that carries dissolved substances (often contaminated) that can percolate through soil.

sanitary landfills Disposal sites that seal in trash at the top and bottom to prevent its release into the atmosphere; the sites are lined on the bottom, and trash is dumped in and covered with soil daily.

Jens Grossmann/laif/Redux

At Phnom Penh, Cambodia's municipal garbage dump, people gather plastic, metals, wood, cloth, and paper to sell to recyclers, exposing themselves to dangerous chemicals, air pollution, and disease.

seal in trash at the top and bottom in an attempt to prevent its release into the environment. Several protective layers of gravel, soil, and thick plastic prevent leachate from delivering toxic substances to groundwater below the landfill. The trash is covered daily with a layer of soil that reduces unpleasant odors, thus attracting fewer pests.

But there are downsides to landfills. The compacting of trash under layers of soil excludes oxygen and water so well that the aerobic bacteria (those that require oxygen to live) and other organisms that normally decompose the biodegradable waste can't survive. Newspaper that would degrade in a matter of weeks is preserved in landfills for decades. Anaerobic bacteria (those that live in oxygen-poor environments) pick up some of the slack. But they break down the trash much more slowly and produce a great deal of methane in the process, a combustible greenhouse gas that contributes to climate change (see Module 10.2). A way to deal with this methane is to capture it to make electricity, turning trash into an energy resource. (This process does release CO_2 when the methane is burned, but since methane is a much more potent greenhouse gas than CO_2, this is preferable to letting the methane escape unimpeded to the atmosphere.) **INFOGRAPHIC 4**

Landfills also take up a large amount of space. This means communities where land is at a premium may have a hard time finding land to site a landfill. And then there's the NIMBY problem—*not in my backyard.* Environmentally undesirable projects like landfills, incinerators, or other objectionable land uses are often met with strong and vocal opposition. Since few people want waste disposal sites in their community, it is often the disenfranchised—minority or poorer communities—that find themselves saddled with these installations. The inequitable exposure of minority or low-income populations to actions that degrade their environment is considered an *environmental justice* issue. (See Module 4.2 for an introduction to environmental justice.)

Each method of dealing with trash comes with its own trade-offs, and the EPA has identified a hierarchy of preferred MSW disposal methods. At the top of that list is source reduction—ways to prevent materials from ever entering the waste stream—followed by recycling and composting and then disposal methods that allow for energy capture. Landfills are the last choice. Sadly, we are not meeting those goals: Landfills handle not the smallest percentage of our trash, but the greatest percentage—slightly more than half of our waste.

INFOGRAPHIC 4 **HOW IT WORKS: SANITARY LANDFILLS** Achieve

In a sanitary landfill, an area is dug out and lined to prevent groundwater contamination from leachate; trash is dumped and covered with soil frequently. (This soil may take up to 20% of the landfill area.) Newer landfills have a leachate-collection system built in; older landfills can be retrofitted to collect leachate. Leachate from holding ponds is treated before being released into the environment.

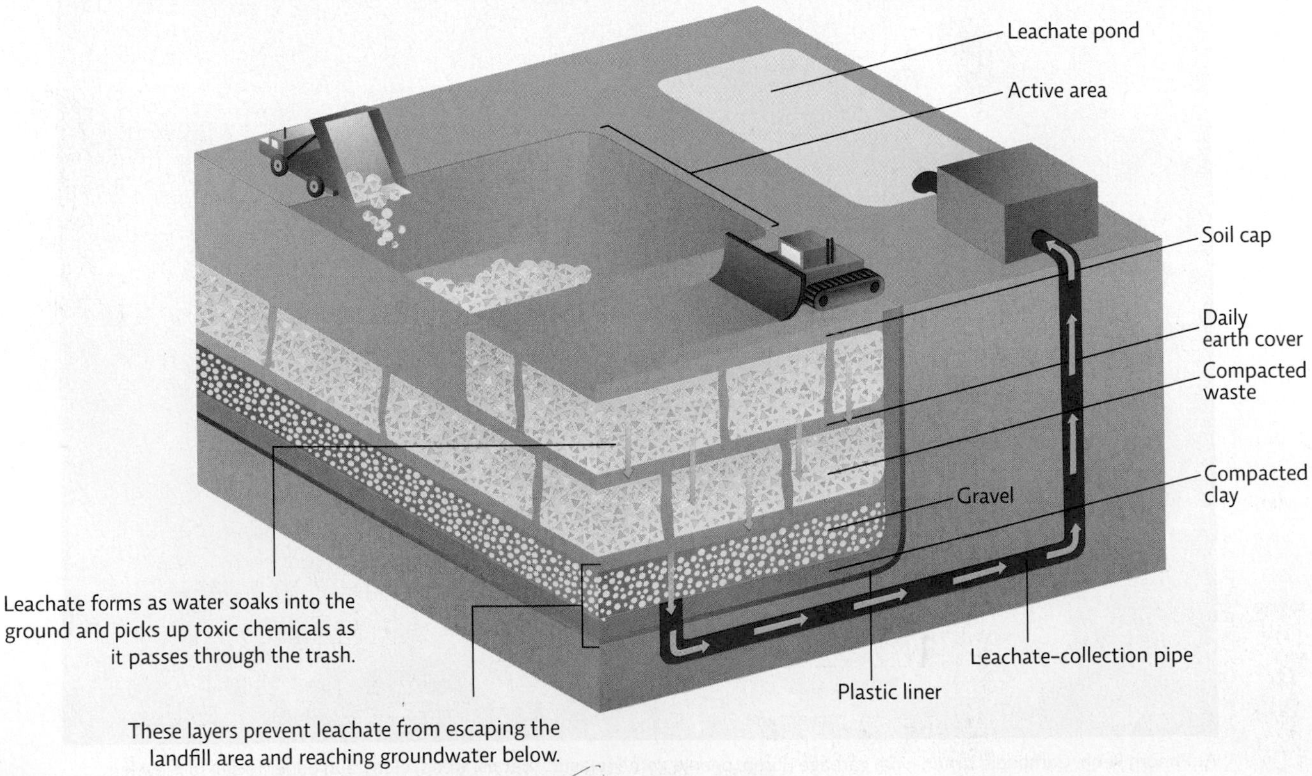
Leachate pond
Active area
Soil cap
Daily earth cover
Compacted waste
Compacted clay
Gravel
Leachate forms as water soaks into the ground and picks up toxic chemicals as it passes through the trash.
Leachate-collection pipe
Plastic liner
These layers prevent leachate from escaping the landfill area and reaching groundwater below.

Why is the rate of decomposition so slow in a landfill?

5 DISPOSAL METHODS: WASTE INCINERATORS

Key Concept 5: Incinerators can reduce the volume of trash tremendously and generate electricity in the process, but they produce air pollution, and the ash is hazardous due to the presence of chemicals in the trash.

Joining landfilling at the bottom of the EPA's list of disposal methods is incineration—burning the trash. A lot of garbage—thousands of tons per day—winds up in specially designed **incinerators**. In the United States, about 12.5% of our trash is incinerated; Europe incinerates a much higher proportion at 42%. Burning waste in this way reduces its volume dramatically—by about 80% to 90%. In addition to decreasing the volume of trash that must be disposed of, incinerators offer another benefit: converting garbage into usable energy. The heat produced during incineration can be converted into steam energy or used to produce electricity. For this reason, the EPA considers incineration with energy capture as a better waste disposal option than landfilling (or incineration without energy capture).

Despite these advantages, incinerator projects are often met with NIMBY protests. Burning waste that contains plastics and other chemicals releases toxic substances into the air, polluting local air and water. Burning also produces toxic ash, which must be disposed of in a separate, specially designed landfill. Incinerators are also extraordinarily expensive to build, and tipping fees (fees charged to drop off trash) are usually much higher at an incinerator than at a landfill. Some communities even face the problem of needing to import trash to keep their incinerators running most efficiently.
INFOGRAPHIC 5

incinerators Facilities that burn trash at high temperatures.

INFOGRAPHIC 5 **HOW IT WORKS: AN INCINERATOR** Achieve

Trash can be burned at very high temperatures in incinerators (some of which are designed to also generate electricity); fuel oil must sometimes be added for more complete combustion. In modern facilities, cleaning systems remove particulates, sulfur, and nitrogen pollutants as well as toxic pollutants like mercury and dioxins. The ash is considered toxic waste and must be buried in the hazardous waste landfills. Municipal solid waste, medical waste, and some hazardous waste are incinerated in the United States.

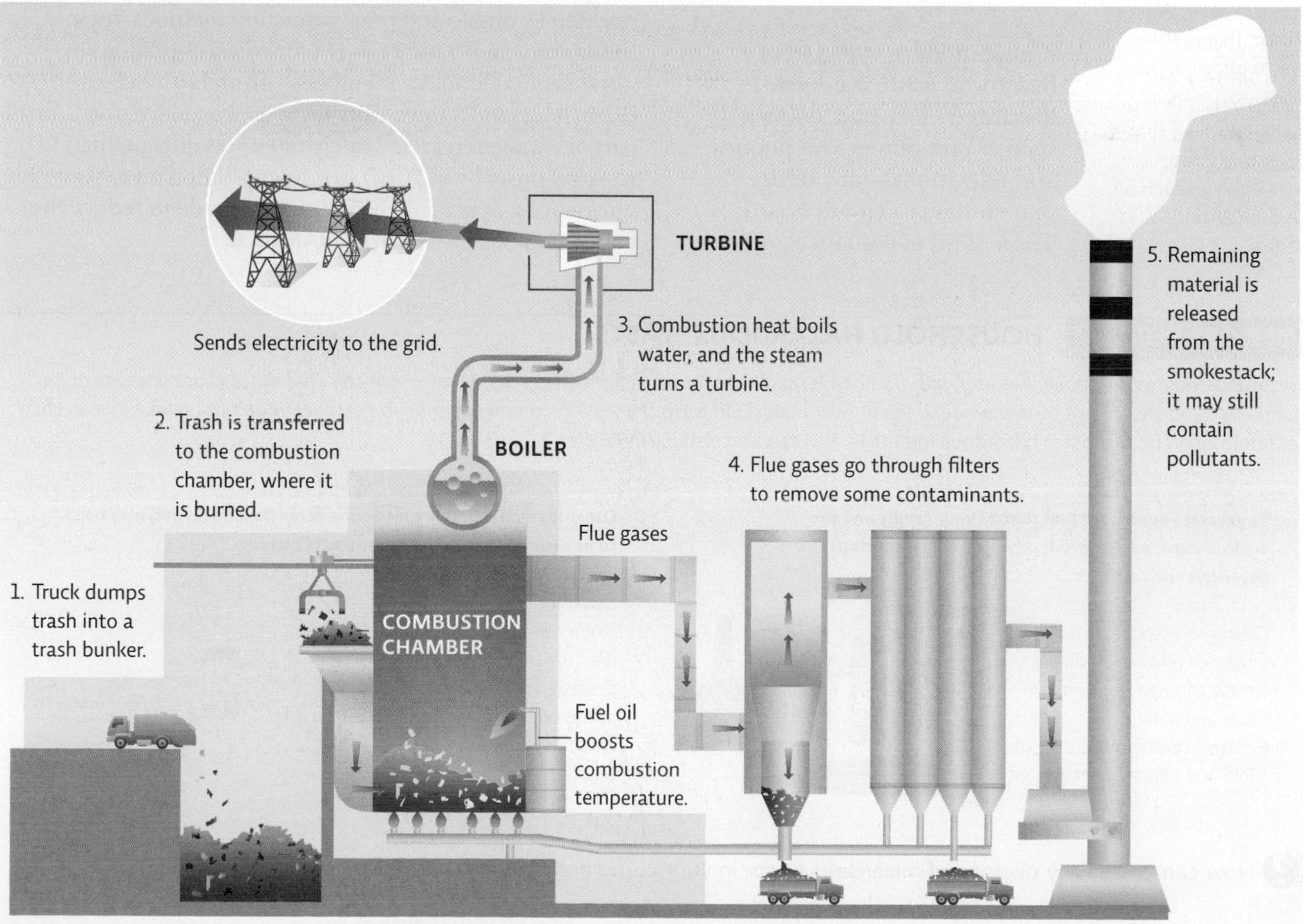

TURBINE
Sends electricity to the grid.
3. Combustion heat boils water, and the steam turns a turbine.
2. Trash is transferred to the combustion chamber, where it is burned.
BOILER
4. Flue gases go through filters to remove some contaminants.
5. Remaining material is released from the smokestack; it may still contain pollutants.
Flue gases
1. Truck dumps trash into a trash bunker.
COMBUSTION CHAMBER
Fuel oil boosts combustion temperature.

What could be done to reduce the toxicity of incinerator ash?

6 HOUSEHOLD HAZARDOUS WASTE

Key Concept 6: Some municipal solid waste is considered hazardous and must be properly managed (recycled or disposed of) to reduce health risks and environmental contamination.

A portion of our solid waste stream presents a health risk of some kind, making it **hazardous waste**. Modern households often contain a bevy of dangerous chemicals and products, so much so that most of us are exposed to more hazardous chemicals indoors than we are outdoors. Household cleaners, paints, automotive supplies, fluorescent light bulbs, and batteries—all of these common household items (and many others) are hazardous and should not be disposed of with the regular trash.

Unfortunately, many citizens do not know how to properly dispose of hazardous waste (or even which items are hazardous) and simply toss those dead batteries or half-full cans of paint into the trash where they could end up in a landfill or an incinerator and possibly contaminate the environment. Avoiding this contamination requires some investigative research on the part of every citizen. Look into the local plan for your community or university so that you can safely dispose of used batteries and chemicals—the Solid Waste Department for your city or county is a good place to start.

hazardous waste Waste that is toxic, flammable, corrosive, explosive, or radioactive.

e-waste Unwanted computers and other electronic devices such as discarded televisions and cell phones.

A new, modern category of hazardous waste is **e-waste**, discarded electronic devices such as computers, cell phones, and televisions—any device that contains a circuit board. These devices contain significant amounts of precious elements (see Module 7.1); reclaiming these materials from e-waste can reduce our need to extract them from mines, a process with a huge environmental impact. Organizers of the 2021 Olympics plan to use gold, silver, and copper retrieved from e-waste to make the medals for the Tokyo Olympics, just as Vancouver did for the 2010 Winter Olympic games.

Because these electronic devices contain hazardous materials, they should not be thrown out with the household trash. Currently, 20 U.S. states prohibit the landfilling of e-waste, but because there is no federal e-waste policy, much of our country's e-waste still winds up in landfills or is exported in convoluted e-waste streams that most often send it to countries in Asia or Africa. Many of these devices are harvested for usable parts or repaired; the rest is broken down to retrieve useful minerals or discarded (often burned). Unfortunately, despite international scrutiny, this dangerous process is still often done manually by workers using rudimentary tools and no protective equipment. When hazardous chemical components are released by unsafe and poor extraction methods, these substances cause a wide range of medical conditions in workers or community members—from birth defects to brain, lung, and kidney damage to cancer. Since only a small part of e-waste is recycled safely (20% was documented to be recycled properly in 2016), consider holding on to your old ones as long as possible before that upgrade to reduce the generation of e-waste. **INFOGRAPHIC 6**

INFOGRAPHIC 6 HOUSEHOLD HAZARDOUS WASTE

Hazardous wastes are those that are toxic, flammable, explosive, or corrosive (like acids). Many chemicals that enter your home are quite dangerous and should not be discarded in the regular household trash. The EPA recommends that you contact your local solid waste agency for information on disposing hazardous materials. You can also visit Earth911.com.

To protect your health and that of your family and the environment, avoid or reduce your use of hazardous chemicals such as:

Drain and oven cleaners
Engine oil and fuel additives
Grease and rust removers
Glues and paints
Pesticides and insecticides
Mold and mildew removers

Other materials that are also considered hazardous and may need to be disposed of as hazardous waste include:

Batteries
Fluorescent lightbulbs
Mercury thermometers
E-waste (cell phones, computers, printers, televisions, video game consoles, etc.)

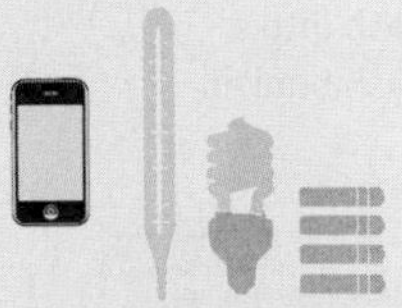

How can you safely dispose of hazardous waste in your community?

7 REDUCING SOLID WASTE: COMPOSTING

Key Concept 7: Composting organic trash is a waste disposal method that mimics nature—it allows waste to decompose quickly and produces a mulchlike product that can be used to return nutrients to soil.

There are many ways to reduce the solid waste we send to landfills or incinerators. While nondegradable trash like plastics presents a challenge for disposal, much of what we throw away is biodegradable, and we have a disposal method at hand that emulates nature to deal with this part of our waste stream. **Composting** can turn organic material into a soil-like mulch that nourishes the soil and can be used for gardening and landscaping.

Composting can be done on a small scale (in homes, schools, and small offices) or on a large one (in municipal "digesters"). Because they expose biodegradable waste to excellent conditions for decomposition, both small- and large-scale composting can break down organic waste, such as paper, kitchen scraps, and yard debris, very quickly. In a home compost pile or bin, kitchen waste might be turned into a soil-like product in just a few weeks or months, depending on the temperature, the amount of moisture, and the number and variety of decomposing organisms at work. Stirring frequently (to keep it well aerated) and adding worms (to break down the material and begin the transformation process) to a home composting bin can speed up the process.

On an industrial level, a municipal composter can take what was once household trash—smelly, full of food waste, packaging, and all the other biodegradable things we throw away—and in just a few days, turn it into compost. In one such facility located in Sevier County, Tennessee, this accelerated rate of decomposition is accomplished in a huge, cylindrical, rotating digester that keeps the waste moving (to prevent compaction) and exposes the material to ideal temperature, air, and moisture conditions. Sewage sludge is often added to provide a bacterial boost. The compost is then removed from the digester and piled up to allow it to continue to decompose in a sheltered place. In just a few weeks, it is fully cured and ready to be used on gardens or in landscaping. **INFOGRAPHIC 7**

composting Allowing waste to biologically decompose in the presence of oxygen and water, producing a soil-like mulch.

INFOGRAPHIC 7 HOW IT WORKS: COMPOSTING

Achieve

Composting can reduce the amount of a household's trash tremendously. A simple compost pile can be started in the backyard, or a compost bin can be built or purchased. Municipal composting facilities process household waste after recyclables and nonbiodegradable materials have been removed.

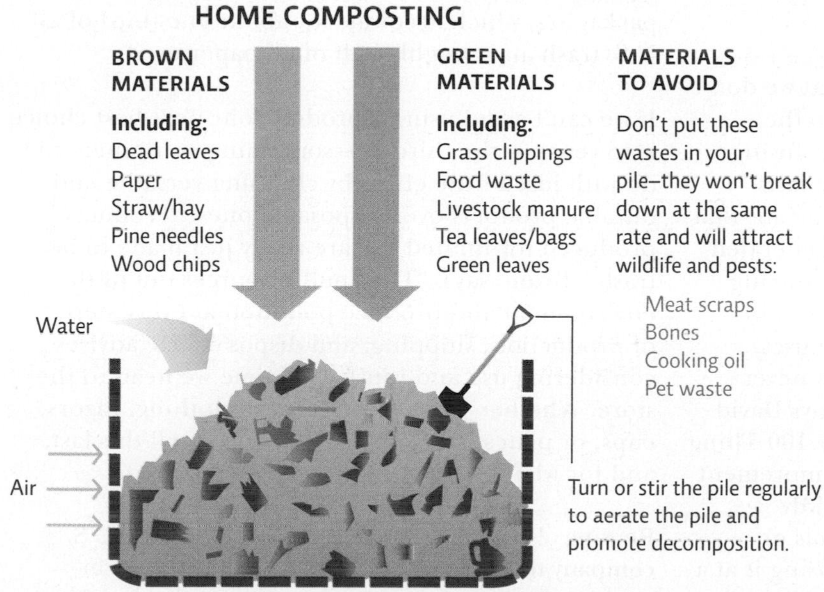

Alan Marsh/Wave/Corbis

A variety of compost bins can be used. Small tubs and countertop green bins can be used indoors. Outdoor compost piles can be set up anywhere it is convenient.

Africa Studio/Shutterstock

The end product is a rich, soil-like mulch that can be used in gardens.

? Does your school's dining services program have a composting program? If not, what would it take to implement one?

Composting is a natural way to deal with biodegradable trash, reducing the volume of trash destined for landfills or incinerators, but ecosystems offer another insight into the problem of solid waste. Just as diversity in an ecosystem enhances its functionality (the fourth characteristic of a sustainable ecosystem; see Infographic 6 in Module 1.1), we stand a better chance of solving our solid waste problems if we pursue a variety of solutions. One of the most meaningful steps we can take is to produce less waste in the first place. Manufacturers can take steps to reduce the waste they produce by recycling product components, increasing the efficiency with which they use resources, and working with other industries—the waste of one industry could be the raw material needed for the production processes of another. But it doesn't stop there—consumers also have an important role to play.

8 REDUCING SOLID WASTE: CONSUMERS AND THE 4 Rs

Key Concept 8: Consumers can reduce their generation of waste by using less, reusing or recycling what is used, and "closing the loop" by purchasing recycled products.

Advertising—a virtual staple of advanced societies—bombards us from every corner of modern life, not just on the televisions in our living rooms but on taxicab computer screens, billboard-laden subway cars, and the pop-up ads that invade our laptops. The message is surprisingly uniform: To live a happy, more fulfilling life, we simply must have more "stuff." But this stuff requires resources to produce and package, and waste is generated every step of the way—not just when we discard it. One solution to the problem of waste is using less stuff so there is less waste to begin with. It is for this reason that the EPA identifies *source reduction* as the number one choice for addressing our solid waste problems.

So how do we start making different choices? As any good environmentalist will tell you, it comes down to the "four Rs": Refuse, reduce, reuse, and recycle.

First, we can simply **refuse** to use things that we don't really need, especially if they are harmful to the environment. When we save a resource by refusing to use it, that resource lasts longer, which in turn means that less pollution will be generated disposing of it and producing replacements, and more will be available for future uses. "Refusal doesn't mean never using the resource," says David Bruno, founder of the 100 Thing Challenge, a popular movement to pare down our worldly possessions to 100 items or fewer. "It just means using it at a more sustainable rate." Avoiding disposables, choosing goods with no packaging, or declining a bag at checkout are all examples that help us refuse. Mason, for instance, has made a commitment to never use plastic bags, water bottles, or straws. "I have metal straws that I keep in my purse and in my backpack, but I also order all of my beverages, 'no straws please,'" she says.

refuse The first of the four Rs of waste reduction: Choose not to use or buy a product if you can do without it.

reduce The second of the four Rs of waste reduction: Make choices that allow you to use less of a resource by, for instance, purchasing durable goods that will last or can be repaired.

reuse The third of the four Rs of waste reduction: Use a product more than once for its original purpose or for another purpose.

If we can't completely refuse a given commodity, we can still try to **reduce** our consumption of it or minimize our overall ecological footprint by making careful purchases. Choose durable repairable goods that will last. If you find that you need to purchase a bottle of water, buy the largest one you can use rather than several smaller bottles—this reduces the amount of plastic you will be using. And all consumers can greatly reduce the amount of waste they generate by paying special attention to packaging, which accounts for about one-third of all U.S. trash and roughly half of all paper used.

If we can't avoid using a product, our next best choice is to **reuse**—the third *R*—something consumers can do with just a little effort by choosing versatile and durable products over disposable ones. "Products produced for limited use are really just made to be trash," Bruno says. "They pull resources out of the environment and produce pollution at every step of production, shipping, and disposal." He advises considering use and reuse each time we head to the store. Whether you are purchasing clothing, razors, cups, or plates, ask yourself: How long will this last, and for what other purpose might it be used?

Reusing also applies to industry. TerraCycle is a U.S. company that began its life packaging liquid worm-compost fertilizer in used soda and water bottles. The company has expanded to offer a number of services aimed at reducing waste, such as an enterprise known as Loop. Though currently available in only certain parts of the country, with expansion on the horizon, Loop offers name-brand food and household products packaged in reusable containers, shipped directly to the customer's door. Consumers use the products, and when finished, the empties are picked back up to be returned to the company where the containers are cleaned and readied for reuse. This is reminiscent of "bottle deposit" programs of the past in which consumers paid a small fee when they bought a drink and were reimbursed when the bottle was returned for cleaning and reuse.

Once we've refused, reduced, and/or reused a given commodity as much as possible, we are left with the final R, **recycling**, the reprocessing of waste into new products. Recycling reduces the trash we generate, but it does more than that. By reclaiming raw materials from an item that we can no longer use, we limit the amount of raw materials that must be harvested, mined, or cut down to make new items. In most cases, this helps conserve limited resources—not only trees and precious metals but also energy and water. However, notice that recycling is our fourth R—our last choice. While better than discarding the item to be landfilled or incinerated, recycling requires more energy, water, and other resources than the first three Rs. **INFOGRAPHIC 8**

recycle The fourth of the four Rs of waste reduction: Return items for reprocessing into new products.

INFOGRAPHIC 8 REDUCING WASTE WITH THE 4 Rs

There are ways to reduce the amount of waste we generate, but for the waste we do have, there are better options than simply throwing away many products. An item like a plastic bottle can be recovered and reused, or the bottle may be recycled into another product.

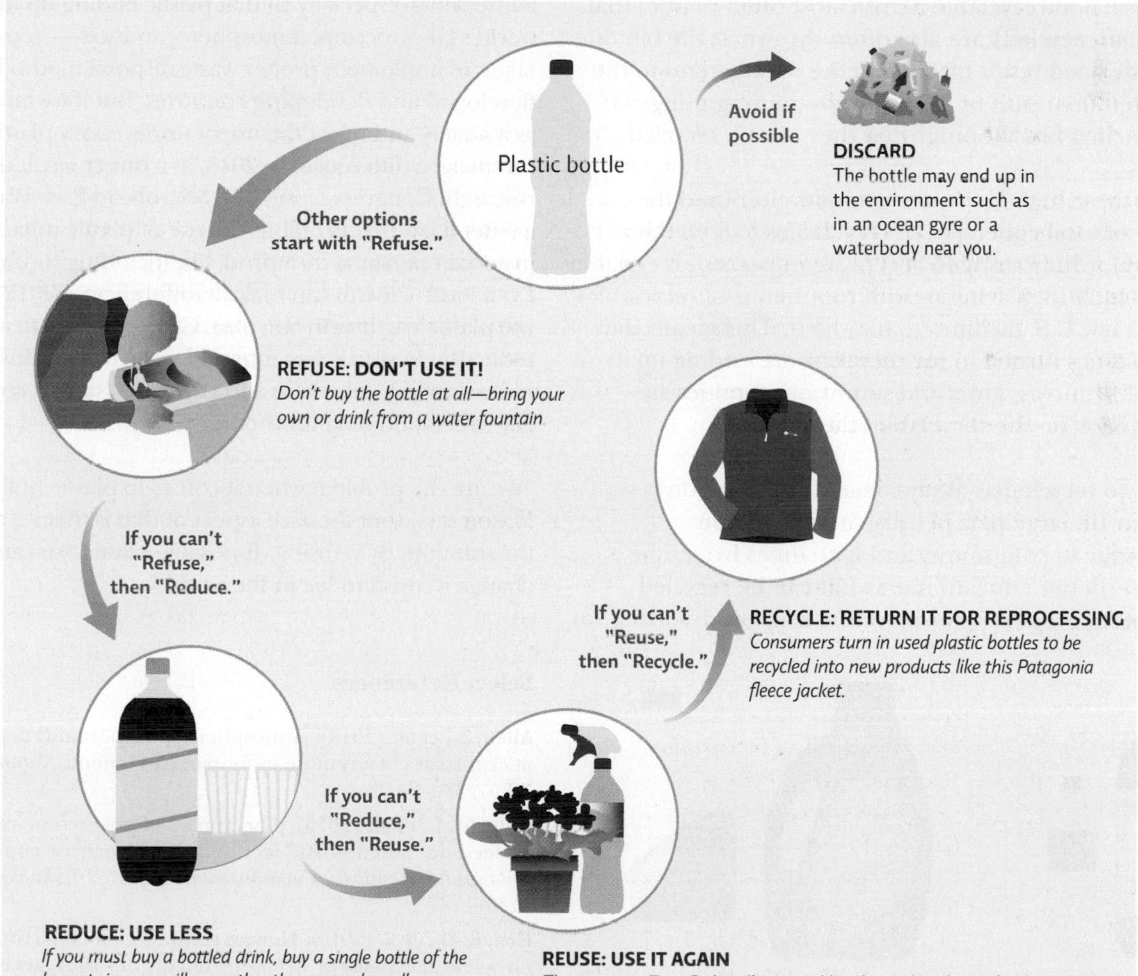

Plastic bottle
Avoid if possible
DISCARD
The bottle may end up in the environment such as in an ocean gyre or a waterbody near you.
Other options start with "Refuse."
REFUSE: DON'T USE IT!
Don't buy the bottle at all—bring your own or drink from a water fountain
If you can't "Refuse," then "Reduce."
REDUCE: USE LESS
If you must buy a bottled drink, buy a single bottle of the largest size you will use rather than several small ones.
If you can't "Reduce," then "Reuse."
REUSE: USE IT AGAIN
The company TerraCycle collects used bottles and packages fertilizer in them.
If you can't "Reuse," then "Recycle."
RECYCLE: RETURN IT FOR REPROCESSING
Consumers turn in used plastic bottles to be recycled into new products like this Patagonia fleece jacket.

How does the prevalence of "disposable/single-use" products and those with short life spans (planned obsolescence) affect our ability to implement the 4Rs?

Recycling of plastic and aluminum containers is being encouraged in some European countries and U.S. states with container deposit programs that repay the deposit when the bottle or can is turned back in, often in a *reverse vending machine*—rather than putting in money to receive a drink, you put in the bottle or can and receive money.

Unfortunately, recycling programs have taken a hit in recent years thanks to policy changes in China. The United States and most other developed countries used to send the bulk of their recycling to China, whose rapidly growing economy thrived on the recyclable material they collected. This even influenced the way recyclables were collected in the United States, changing in many places from collection sites that asked consumers to sort their recyclables to single-stream recycling bins into which all our recyclables would be placed. But single-stream recycling leads to high rates of contamination. In this method, many recyclable plastic or paper items become contaminated with food or grease; non-recyclable items (most often plastics that cannot be recycled) are also often thrown in the bin due to uninformed residents adding the wrong items to the bin, often the result of *wish-cycling*—putting things in the recycling bin "hoping" that they can be recycled.

Due to these high levels of contamination (and the growth of local collection of recyclables to feed their industry), China stopped accepting imports of recyclable plastics in 2018, leaving us with mountains of recyclables and too few U.S. facilities to handle it. This means that some plastics turned in for recycling are ending up in landfills or incinerators and some communities are cutting back on the recyclables they accept.

"What" to recycle has always been an "education" problem (in large part because it differs from community to community and it changes from time to time). Better education on what can be recycled and how to prepare it for recycling is needed; sorting recyclables at drop-off centers or changing back to multi-stream recycling (different bins for collection of different items) tends to lead to better net recovery rates because there is less waste that must be removed and less chance recyclables will become contaminated and rendered unusable.

But properly turning in recyclables is only part of the answer. To execute this step properly, we must first have purchased items that can be recycled. We must also *close the loop* by purchasing items that are made of recycled materials to encourage manufacturers to make those products. This last step is key. Reconfiguring the plastics industry from a linear production model (take, make, discard) to a circular one that produces items that are recyclable and then recovers and recycles plastics is seen as a key part of any solution that effectively addresses plastic pollution. (See Module 5.1 for more on economic models of production.)

For the world's solid waste problem to properly be addressed—especially all that plastic ending up in the world's lakes, oceans, atmosphere, and soil—steps must be taken to implement proper waste disposal methods in both developed and developing countries. But if we take a stand as a society and reject the use of unnecessary plastics, we can make a difference. In 2015, as a direct result of Mason's research, Congress passed the Microbead-Free Waters Act, a federal law that prohibits the use of plastic microbeads in certain personal care products, including toothpaste. Even local activism can make a difference: A 2012 single-use plastic bag ban in San Jose, California, led to an 89% reduction in plastic bag litter in local storm drains, a 60% reduction in local creeks and rivers, and a 59% reduction in city streets and neighborhoods.

"We are the problem when it comes to plastic pollution," Mason says, "but the nice aspect of that is that we are also the solution. By refusing disposable plastic, we can be the change we wish to see in the world."

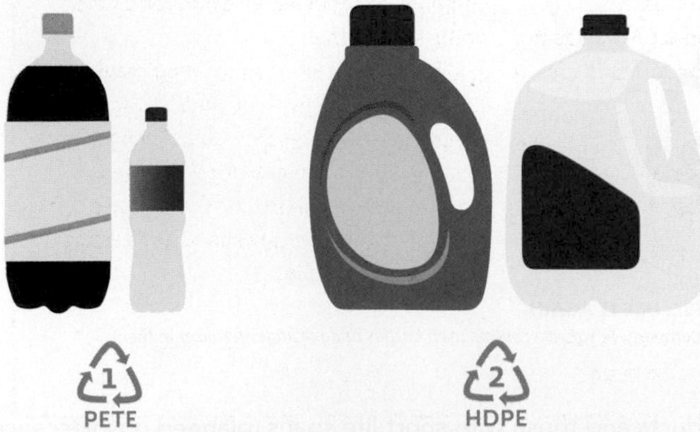

The number code on a plastic item indicates the type of resin with which it is made. The most commonly accepted items in most communities are #1 plastics (e.g., water and soda bottles) and #2 plastics (e.g., milk jugs). Check with your local recycling service to see what is accepted in your area.

Select References:

Allen, S., et al. (2019). Atmospheric transport and deposition of microplastics in a remote mountain catchment. *Nature Geoscience, 12*(5), 339–344.

Cole, M. J., et al. (2015). The impact of polystyrene microplastics on feeding, function and fecundity in the marine copepod *Calanus helgolandicus. Environmental Science & Technology, 49*(2), 1130–1137.

Cox, K. D., et al. (2019). Human consumption of microplastics. *Environmental Science & Technology.* doi.org/10.1021/acs.est.9b01517

Mason, S. A. (2019). Plastics, plastics everywhere: Studies in the Great Lakes and beyond highlight the ubiquity of microplastics in our rivers and drinking water. *American Scientist, 107*(5), 284–288.

Parker, L. (2019, May 1). Sea of plastic. *National Geographic,* 41–55.

Schwabl, P., et al. (2019). Detection of various microplastics in human stool: A prospective case series. *Annals of Internal Medicine, 171*(7), 453–457. doi:10.7326/M19-0618

GLOBAL CASE STUDIES GARBAGE: PROBLEMS AND SOLUTIONS

Achieve

There's no debating the fact that humans produce a lot of trash. Addressing the problem will involve a combination of efforts to reduce the production of trash and to properly deal with the garbage that is created. Read about some research and efforts underway to learn about and deal with this issue.

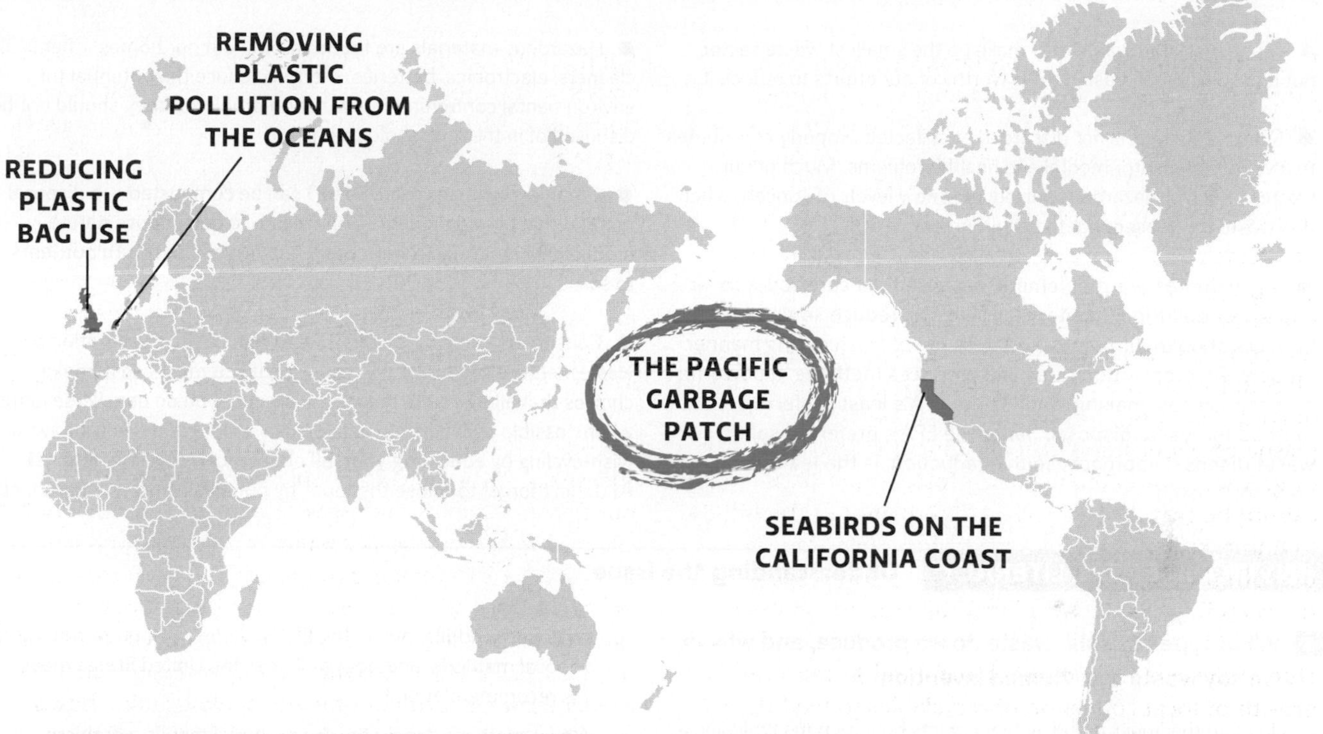

BRING IT HOME

PERSONAL CHOICES THAT HELP

How much solid waste you produce is under your control. By reducing the amount of waste you produce, you reduce how much money we spend on waste disposal as a whole and at the same time place less pressure on the resources used to produce consumer goods. Reducing your solid waste is very easy and can save you money in the process.

Individual Steps

- Track your trash. Record what you throw out for a week by category and volume. How could you reduce your total trash weight by one-quarter? By half?
- Use the information in Infographic 8 (strategies to refuse, reduce, reuse, and recycle) to identify five changes you can make to reduce your solid waste.
- Talk to community leaders to discuss the possibility of starting a communitywide composting program.
- Research rates of recycling participation in your community. Advocate for recycling education and curbside recycling programs.

Group Action

- Start recycling unusual items in your community. TerraCycle is a company that offers services for the collection of hard-to-recycle items; school or other community groups can partner with the company to reduce waste and raise funds at the same time.
- Talk to friends and family about having "no gift" or "low gift" celebrations. Instead of buying many presents, treat friends to a dinner or a fun activity. For large families, use a grab bag or draw names and buy or upcycle (give used goods) for only specific people.

REVISIT THE CONCEPTS

- Human handling of resources tends to be linear — a *take-make-discard* pathway that views waste as useless. This differs from the matter cycles of natural ecosystems in which the discarded matter of one organism becomes a resource for another.

- Municipal solid waste may make up the smallest waste sector, but it is still a lot of trash, so it is worthy of our efforts to reduce it.

- Solid waste that is not discarded or collected properly contributes to a variety of environmental and health problems. Much of our waste is toxic or hazardous, introducing new levels of concern when this waste is released into the environment.

- Open dumps are problematic because they contribute to air and water pollution. Sanitary landfills can reduce some of the impacts of open dumping, but the burial of trash in this manner reduces decomposition rates and produces methane, a powerful greenhouse gas, making landfills the EPA's least preferred method for waste disposal. Sadly, the EPA's preferred solid waste disposal approach, source reduction, is the least commonly used.

- Incinerators can address two problems at once — they dispose of our trash and generate electricity in the process. However, they produce dangerous air pollution and the ash is hazardous. Like landfills, they may also spark NIMBY protests.

- Hazardous materials are found throughout our homes (chemical cleaners, electronics, batteries, etc.). To reduce the potential for environmental contamination or health problems, they should not be disposed of in the household trash.

- Organic trash (e.g., food waste) can be composted — a disposal method that allows biodegradable waste to decompose naturally, producing a mulchlike product that can be used to return nutrients to soil.

- Consumers can employ the 4 Rs to reduce their generation of waste: *Refuse* a product if you can — or, if you must use it, make choices that allow you to *reduce* the amount you do use. *Reuse* items when possible and *recycle* what you can no longer reuse (but avoid wish-cycling by educating yourself about local recycling options). And don't forget to "close the loop" by purchasing recycled products.

ENVIRONMENTAL LITERACY Understanding the Issue

1 What types of solid waste do we produce, and why do we say waste is a "human invention"?

1. Explain the linear pathway that depicts how humans typically handle resources, the problems this causes, and how it could be transformed to a circular pathway.
2. Distinguish among degradable, biodegradable, and nondegradable waste. How well do plastics degrade?

2 What is municipal solid waste (MSW), and what are the types and proportion of waste in the U.S. MSW stream?

3. Why might the U.S. per capita rate of MSW production be so much higher than in less developed countries or even in other wealthy countries?
4. Why are plastics so abundant and what is the fate of plastic waste in the United States?

3 What problems are caused by the improper management of solid waste?

5. What are the ocean "garbage patches," and in what ways does the plastic trash in oceans harm ocean life?
6. Why might the fact that aquatic organisms like oysters or fish consume microplastics be a concern for humans?

4 Compare the design, advantages, and disadvantages of open dumps and sanitary landfills.

7. Why are landfills considered a better waste management option than open dumps?
8. Where do landfills rank in the EPA's hierarchy of preferred MSW disposal methods, and how well does the United States meet this recommendation?
9. Why is methane production by sanitary landfills a problem, and how can it be turned into an advantage?

5 What are the pros and cons of waste incinerators?

10. Why is the ash produced by an incinerator a hazardous waste?
11. Why don't more U.S. communities have waste incinerators?

6 What are some common household hazardous wastes, and how should individuals deal with this waste?

12. What is e-waste, and why is it a concern?
13. Identify some hazardous substances in your home or dorm room.

7 What is composting, and how can it help us in our quest to deal with solid waste?

14. Describe how one would set up and implement a home composting system.
15. Compare and contrast landfilling and composting. What are the trade-offs for each option?

8 What are the four Rs of waste reduction?

16. Give an example of each of the *four Rs* that you could use in your own day-to-day life.
17. What does it mean to "close the loop," and why is it important?
18. Why is "recycling" considered the fourth choice of the four Rs?

SCIENCE LITERACY Working with Data

To determine whether microplastics are harmful to small marine organisms, researchers exposed marine copepods to either their normal diet of algae or to algae plus microplastic beads (20 μm polystyrene beads at a concentration of 75 beads per mL of water). As mentioned in the text, copepods exposed to these microplastic beads produced smaller eggs that had a lower hatching rate. Researchers also examined a variety of dietary parameters including the size, total number, and weight of algal cells ingested. Look at their data below and answer the questions that follow. (Bars with an asterisk above them are significantly different from the control at $p < 0.05$. See online Appendix 2 for more on *p*-values and statistical significance.)

DIET AND MICROPLASTIC EXPOSURE

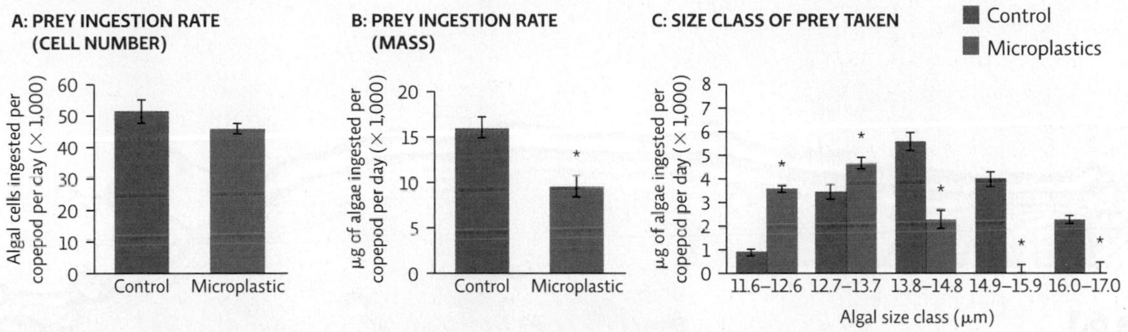

Interpretation

1. In one or two sentences, describe the result depicted in Graph A. (Be sure to reference statistical significance in your answer.)
2. Estimate how much total food (algae) was eaten by each group (Graph B). Can you conclude that there is a difference between the two groups for this parameter?
3. In one sentence describe the result obtained in Graph C (i.e., explain what differences are seen in the size of algal cells taken by the control versus the microplastic group).

Advance Your Thinking

4. What overall conclusion can you draw from the three graphs regarding the ability of copepods to acquire food?
5. Offer a hypothesis to explain the feeding difference seen between copepods in the two groups.
6. Other data gathered showed that copepods in the microplastic group produced the same number of eggs but that these eggs were smaller and had a lower hatching success rate. Based on the data presented here, why might this be true?

INFORMATION LITERACY Evaluating Information

Curbside recycling programs are a very convenient system for recycling metals, paper, glass, and plastic. However, they are available to only about half the population in the United States today, and what can be recycled is limited. So where do you go if you do not have curbside recycling or if you want to recycle or safely dispose of such things as electronics, batteries, or books? One source that may be helpful is the Earth911 website.

Go to the Earth911 website (www.earth911.com).

Evaluate the website and work with the information to answer the following questions:

1. Determine if this is a reliable information source with a clear and transparent agenda. Begin your search using the key words "About Us."
   a. Who runs this website? Do the organization's credentials make the information presented reliable or unreliable? Explain.
   b. What is the mission of this website? What are its underlying values?
2. Click on any of the tabs at the top of the home page and read at least three articles. Identify which tab you chose and answer the following questions:
   a. Briefly describe the content of each and identify the information source(s) used. Are these reliable information sources? Explain.
   b. Does the content seem useful and credible? Explain your response.
3. Choose the "Where to Recycle" link at the top of the webpage.
   a. Choose BATTERIES from the list provided. Select three different kinds of batteries, one at a time. How close is the closest recycling center near you for each? Is this information dated to let you know how current it is?
   b. Choose ELECTRONICS. Where is the nearest location where you could recycle a desktop computer? A cell phone? A video game console?
4. Choose the "How to Recycle" link at the top of the webpage. Investigate at least three of the links that appear. Is this information valuable? Accurate? Explain.
5. How useful is a website like Earth911? How can a website like this influence societal understanding of waste issues and facilitate a change in behavior on the part of individuals and businesses? What is missing from this website that would make it more useful?

CHAPTER 6

Water Resources

Water is vital to all life on Earth. Human impact is affecting both water quantity (freshwater supplies) and water quality (water pollution). Our actions are also harming ocean ecosystems, ecosystems on which we and countless other species depend.

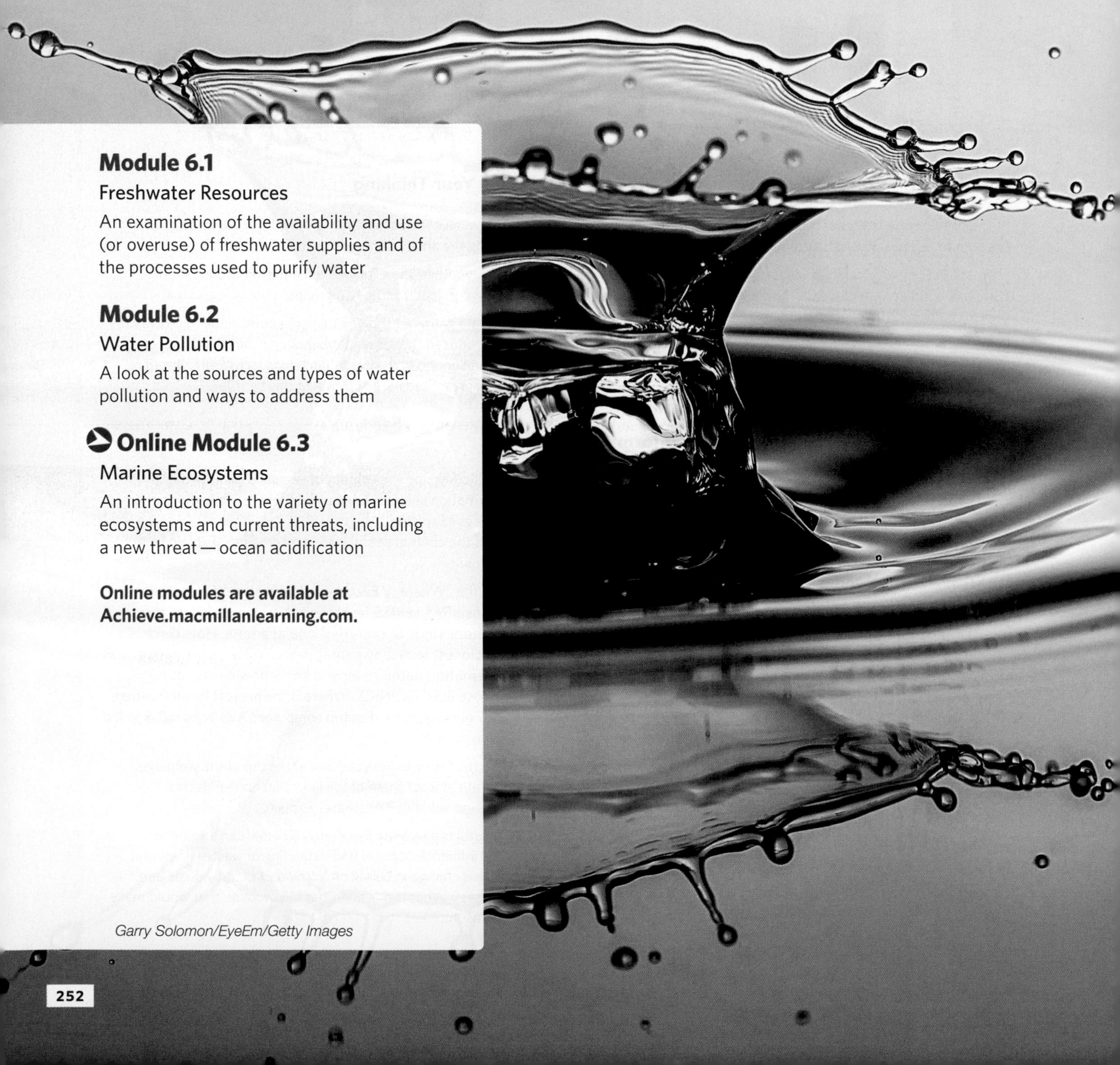

Module 6.1

Freshwater Resources

An examination of the availability and use (or overuse) of freshwater supplies and of the processes used to purify water

Module 6.2

Water Pollution

A look at the sources and types of water pollution and ways to address them

Online Module 6.3

Marine Ecosystems

An introduction to the variety of marine ecosystems and current threats, including a new threat — ocean acidification

Online modules are available at Achieve.macmillanlearning.com.

Garry Solomon/EyeEm/Getty Images

Module 6.1 Freshwater Resources

WATER WARS

Fighting over water in the American Southwest

The Colorado River passes through the Grand Canyon on its journey from its headwaters in Colorado to the Gulf of California. *Mlenny/E+/Getty Images*

After reading this module, you should be able to answer these GUIDING QUESTIONS

1. How is water distributed on Earth, and what are the sources of freshwater?
2. How does water cycle through the environment?
3. How are surface water and groundwater supplies accessed, and what problems do these water sources face?
4. How does water use differ between sectors and income levels, and what problems does water scarcity cause?
5. How is domestic water regulated in the United States, and what problems does our water supply face?
6. How can wastewater be treated to make it safe to release into the environment?
7. What technologies can address water scarcity and what are the trade-offs of each?
8. How can conservation help address water scarcity issues?

The Colorado River is emblematic of the breathtaking beauty of the American Southwest: It even helped carve, over the course of millions of years, the iconic Grand Canyon in Arizona. The Colorado, flowing through a region known as the Colorado River Basin (the area that feeds the Colorado and through which the river flows—also known as its *watershed*; see Module 6.2), also provides life-sustaining freshwater to some 36 million people as well as countless animals, plants, and other organisms in Arizona, California, Colorado, New Mexico, Nevada, Utah, and Mexico. Wyoming, too, is part of this river system; its Green River is the largest tributary that feeds into the Colorado. Over the course of its long life, the Colorado River has had a rich and dramatic history, having sparked lies and theft and even war—yet its future, today, is uncertain.

The modern saga of the Colorado River—the backstory that explains why its fate is now so precarious—begins in 1922. That's when officials from the seven U.S. states that lie in the Colorado Basin signed an agreement that evenly divided its flow between the four upper states (the Upper Basin) and the lower three states (the Lower Basin). Unfortunately, the agreement didn't determine how the water would be divvied up between the lower three states, and Arizona and California immediately began bickering over their share. The two states even briefly went to war over these water rights in 1934.

Arguments over the Colorado's water have once again reached fever pitch, and the cause, yet again, can be traced back to that century-old agreement. Because water flows were unusually high in 1922, officials vastly overestimated the river's annual flow, and states began using more water than was sustainable. Now, 100 years later, their continued overuse of the Colorado's water—combined with decades-long droughts and the effects of climate change—have put the Southwest in a precarious situation. In 2019, Lake Mead, a reservoir that provides water to Las Vegas and parts of Arizona and California, was only 40% full, with water at 1,088 feet above sea level; if it drops below 950 feet, the water will no longer turn the dam's turbines. At 895 feet, water will stop flowing out at all. Water from the Colorado rarely reaches its outlet at the ocean in the Gulf of California, the narrow body of water that separates the Mexican state of Baja California from the rest of Mexico; the river's southernmost reaches in Arizona could soon begin to see a similar drawdown and loss of flow.

In response, officials from the seven U.S. states signed the Colorado River Basin Drought Contingency Plan in May 2019, an important step to managing the ongoing Colorado water crisis—a crisis expected to worsen. In testimony given to the House of Representatives' Natural Resources Committee in February 2019, Brad Udall, a senior water and climate research scientist at the Colorado Water Institute, said that states need to be doing a lot more. "To minimize this threat," he said, "we must act now."

◉ WHERE IS THE COLORADO RIVER BASIN?

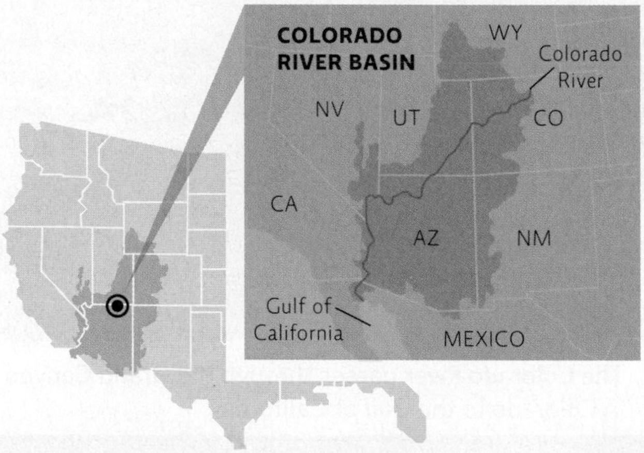

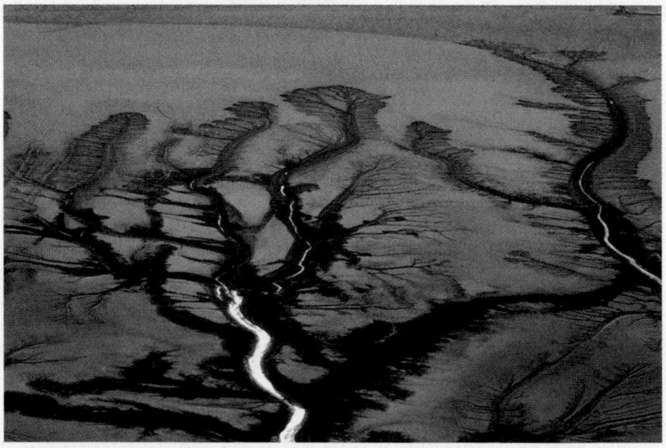

Pete Mcbride/National Geographic Image Collection/ Getty Images

The lower Colorado River as it passes through the Sonoran Desert. This dry delta is evidence that its flow is severely reduced.

1 FRESHWATER DISTRIBUTION AND SOURCES

Key Concept 1: Only 2.5% of water on Earth is freshwater, and very little of that is accessible to humans. Fortunately, even this small percentage represents a large amount of water.

Water is vital to life on Earth. It connects ecosystems and provides many important ecosystem services that animals and plants require to live. Water is habitat to countless species, and it is the major component of most cells. Up to 75% of the human body, for instance, consists of water. Our cells must have water for needed chemical reactions; water also facilitates the transport of nutrients and the regulation of body temperature, just to name a few vital functions.

But our bodies need liquid *freshwater*; ocean water is too salty for human consumption and is toxic in large doses. However, the vast majority of water on Earth is saltwater—around 97.5%. Freshwater makes up barely 2.5% of the total, and nearly 70% of that is trapped in glaciers and ice caps at the poles. To meet our needs, we depend on the freshwater trapped underground (**groundwater**) and in **surface waters** (lakes, streams, and rivers)—that is slightly less than 1% of all the water on Earth. But there is a lot of water on the planet—about 75% of its surface is covered with water—totaling around 370,000,000 trillion gallons of water. Even 1% of such a large number is still a lot of water! **INFOGRAPHIC 1**

Like many bodies of water, the Colorado is a lifeline for the human communities that depend on it as their primary water source. But that can also lead to water shortages if more is removed than replaced. The Colorado River and the tributaries that empty into it are largely fed by Rocky Mountain snowmelt—but as Earth's climate changes, this mountain snowpack may lessen, worsening potential water shortages in the Southwest.

In 2018, very little snow fell in the Rockies, intensifying fears over whether there would be enough water in the Colorado for all the states relying on it. Thankfully, the winter of 2019 had above-average snowfalls, and residents breathed a collective sigh of relief. But who knows how long the respite will last: Climate scientists say that droughts will likely recur as a result of climate change, so freshwater in these states may become even more scarce in the near future. (See Module 10.2 for more on climate change.)

groundwater Water found underground trapped in soil or porous rock.

surface water Any body of water found above ground, such as oceans, rivers, and lakes.

INFOGRAPHIC 1 **DISTRIBUTION OF WATER ON EARTH**

Most of the water on Earth is found in the oceans, and most of the freshwater is tied up in ice and snow. A small fraction of all of Earth's water is available for us to use, but with more than 370,000,000 trillion gallons of water on the planet, that is still a lot of water.

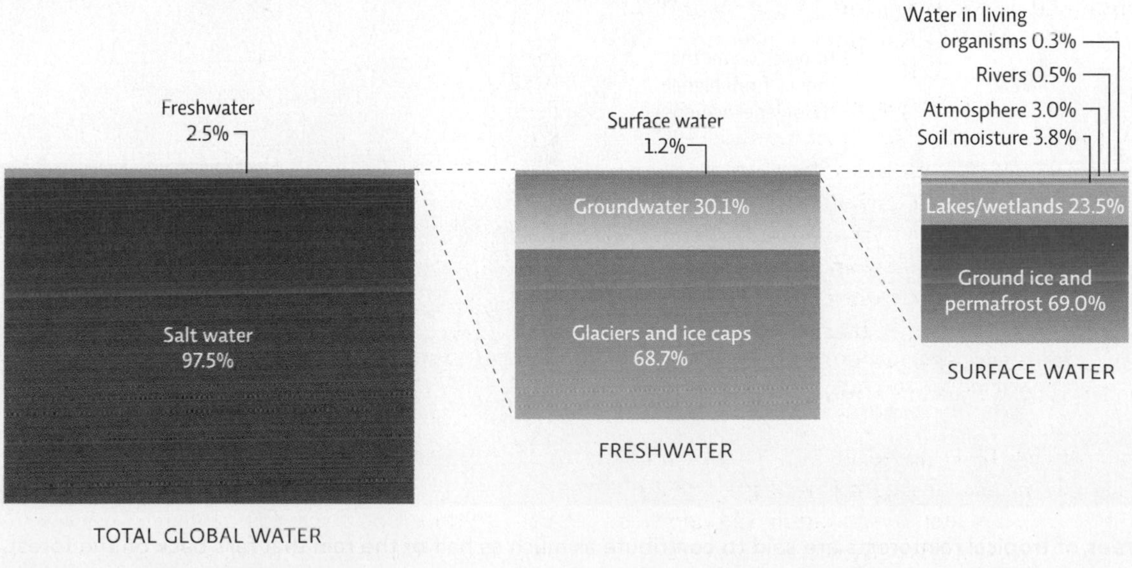

? Based on this diagram, what percentage of the total water supply on Earth is found in groundwater? What percentage is found in rivers?

2 THE WATER CYCLE

Key Concept 2: Water moves through the environment via the water cycle, a process that constantly recycles water on Earth through surface and groundwater, plants, and the atmosphere.

Wherever there is water, it is constantly moving through the environment via the **water cycle** (hydrologic cycle). Heat from the Sun causes water to evaporate from surface waters and land surfaces (evaporation). At the same time, plant roots pull up water from the soil and then release some into the atmosphere in a process called **transpiration**. Plants with deep roots, like trees, may bring up thousands of gallons of water a year, releasing much of this to the atmosphere. Altogether, the combination of evaporation and transpiration—*evapotranspiration*—sends more than 17,000 trillion gallons of water vapor into the atmosphere every year. Once aloft, that water condenses into clouds (condensation) and may fall back to Earth as precipitation (rain, snow, sleet, etc.). **INFOGRAPHIC 2**

water cycle The movement of water through various water compartments such as surface waters, atmosphere, soil, and living organisms.

transpiration The loss of water vapor from plants.

Almost all precipitation ends up falling on the oceans, and a tiny remainder falls on land. This latter portion is the part humans can harvest for their own use. We access freshwater from lakes and rivers and from groundwater. Water from both surface and groundwater sources eventually makes its way back to the oceans.

INFOGRAPHIC 2 THE WATER CYCLE Achieve

Water cycles between liquid and gaseous forms as it moves through space and time. Ocean water (which we cannot use) is converted to freshwater when it evaporates and falls back to Earth as precipitation, refilling surface and underground freshwater supplies. Liquid freshwater is a renewable resource as long as we don't use it faster than it is naturally replenished.

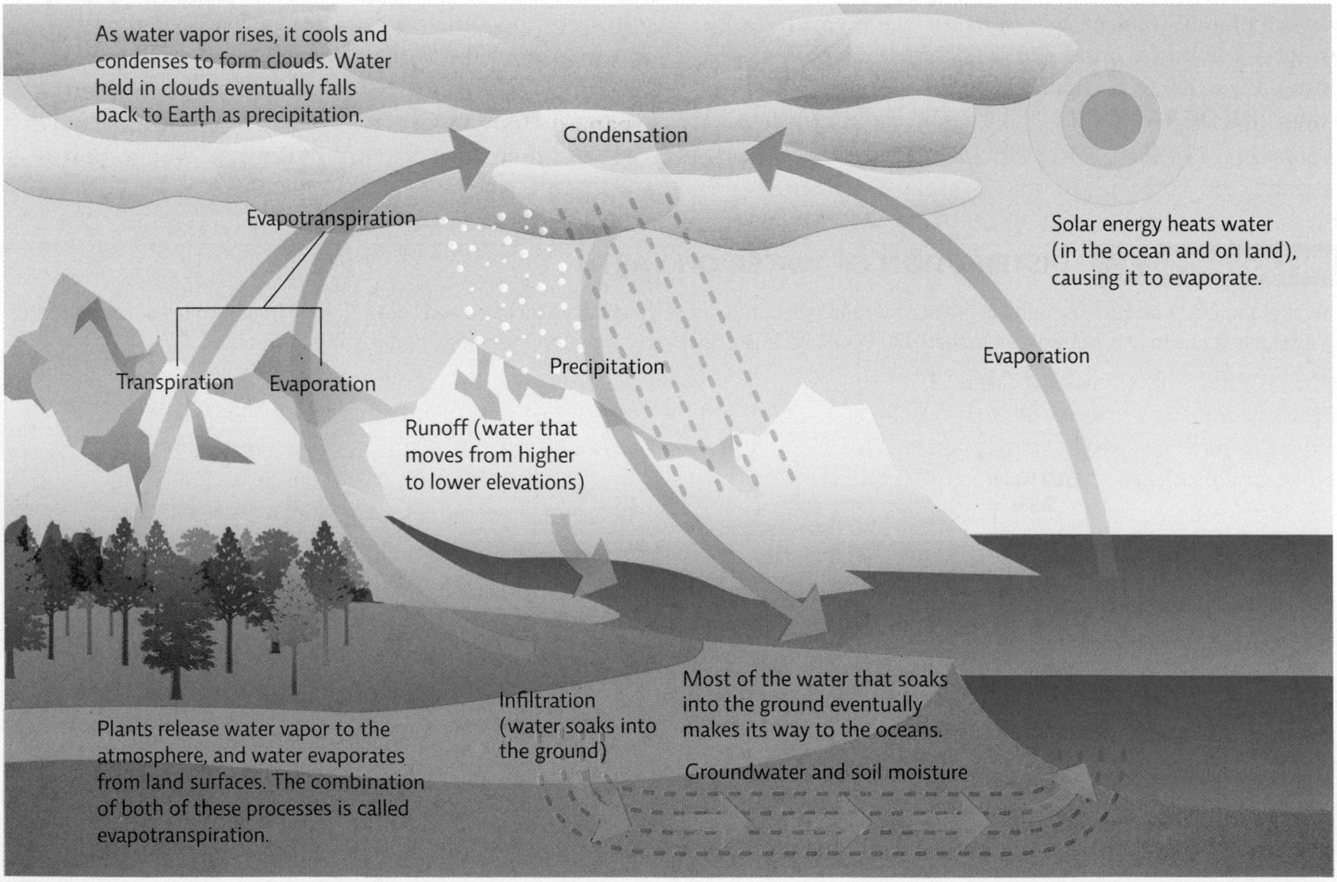

As water vapor rises, it cools and condenses to form clouds. Water held in clouds eventually falls back to Earth as precipitation.
Condensation
Evapotranspiration
Solar energy heats water (in the ocean and on land), causing it to evaporate.
Transpiration
Evaporation
Precipitation
Evaporation
Runoff (water that moves from higher to lower elevations)
Most of the water that soaks into the ground eventually makes its way to the oceans.
Infiltration (water soaks into the ground)
Plants release water vapor to the atmosphere, and water evaporates from land surfaces. The combination of both of these processes is called evapotranspiration.
Groundwater and soil moisture

? The trees of tropical rainforests are said to contribute as much as half of the rain that falls back on the forest. Explain how this occurs.

3 FRESHWATER SUPPLIES: SURFACE AND GROUNDWATER

Key Concept 3: Surface waters provide freshwater to human communities, but if we remove water faster than it is replaced the sources can run dry; water wars can even erupt. Groundwater supplies can shrink due to over-withdrawal and to decreased infiltration, causing wells to run dry or become contaminated with saltwater (in coastal regions).

Of all the water we use in the United States, more than 70% comes from surface water sources, though groundwater, too, is an important source of freshwater in many areas. Surface water can be accessed on a large scale with pumping stations and on a small scale by hand. It can also be diverted via canals and pipes to deliver water to places that need it, such as farms and power plants or even distant cities.

Surface water sources can be threatened by problems such as climate change, pollution (see Module 6.2), and excessive water withdrawal. Over-withdrawal is a problem not just for the Colorado River in the arid Southwest—disputes over water in the Chattahoochee River, which runs through Georgia and Florida, have reached the U.S. Supreme Court. Much of the city of Atlanta, Georgia, relies on the Chattahoochee River for its drinking water, and as its population has grown, so has the city's withdrawals from the river. Georgia farmers farther south also divert water from the Flint River, which joins the Chattahoochee before both of them drain into the Apalachicola River in the Florida Panhandle. Paired with a regional drought, all this upstream withdrawal means that less water reaches Florida and the coastal regions there that depend on it.

Water scarcity threatens not just the people who need the water but all the other organisms and plants that need it, too. In 2012, the once-thriving Apalachicola Bay oyster industry, which provides 90% of Florida's oysters, collapsed as oyster numbers plummeted. Although the causes of the sudden shortage were likely multifaceted, in a 2013 report, the Florida Fish and Wildlife Commission highlighted a lack of freshwater flow into rivers and estuaries as a major culprit. Oysters thrive in regions where freshwater meets the sea, but as freshwater flows diminished, the salinity of the water increased to levels too high for the oysters to tolerate. The collapse intensified an already existing Chattahoochee **water war**, a political conflict over the allocation of water resources, similar to the water war ongoing in the Southwest.

Groundwater is another significant source of freshwater for the United States. Many people draw their water from an underground region of permeable soil or porous rock saturated with water, called an **aquifer**. These groundwater stores receive water from rainfall and snowmelt that soaks into the ground through **infiltration**. Plant roots take up some of the water along the way, but much of the water continues to move downward, filling every available space in the aquifer. As the water trickles down, it becomes naturally filtered by rocks and soil, which trap bacteria and other contaminants as the water passes by. The top of this water-saturated region, referred to as the **water table**, rises and falls due to seasonal weather changes or human withdrawals.

In addition to withdrawals, anything that reduces infiltration will also decrease the amount of water we can sustainably remove by reducing the rate at which the aquifer refills. Infiltration is hampered in urban and suburban settings because of all the hard surfaces, such as roads and buildings; even a typical suburban lawn is so compacted from the home construction process that very little water infiltrates the ground. Urban and suburban designs that provide ways for water to soak into the ground—such as *permeable pavement* and *rain gardens*—can help refill aquifers as well as help prevent flooding events after heavy rainfalls. (For more ways to increase infiltration, see Module 6.2.)

Overdrawing water from aquifers in coastal areas can result in **saltwater intrusion**—the movement of saltwater into aquifer spaces left by missing freshwater. Groundwater levels are typically

water wars Political conflicts over the allocation of water sources.

aquifer An underground, permeable region of soil or rock that is saturated with water.

infiltration The process of water soaking into the ground.

water table The uppermost water level of the saturated zone of an aquifer.

saltwater intrusion The inflow of ocean (salt) water into a freshwater aquifer that happens when an aquifer has lost some of its freshwater stores.

INFOGRAPHIC 3 **GROUNDWATER IS FOUND IN AQUIFERS OF POROUS ROCK OR PERMEABLE SOIL**

Groundwater in aquifers is naturally replenished as water soaks into the ground. Humans can access this groundwater through wells, but we can pull out water faster than it is naturally replaced. This can lead to saltwater intrusion in coastal areas or dry wells in inland areas. Surface pollution can also seep into the ground and contaminate groundwater.

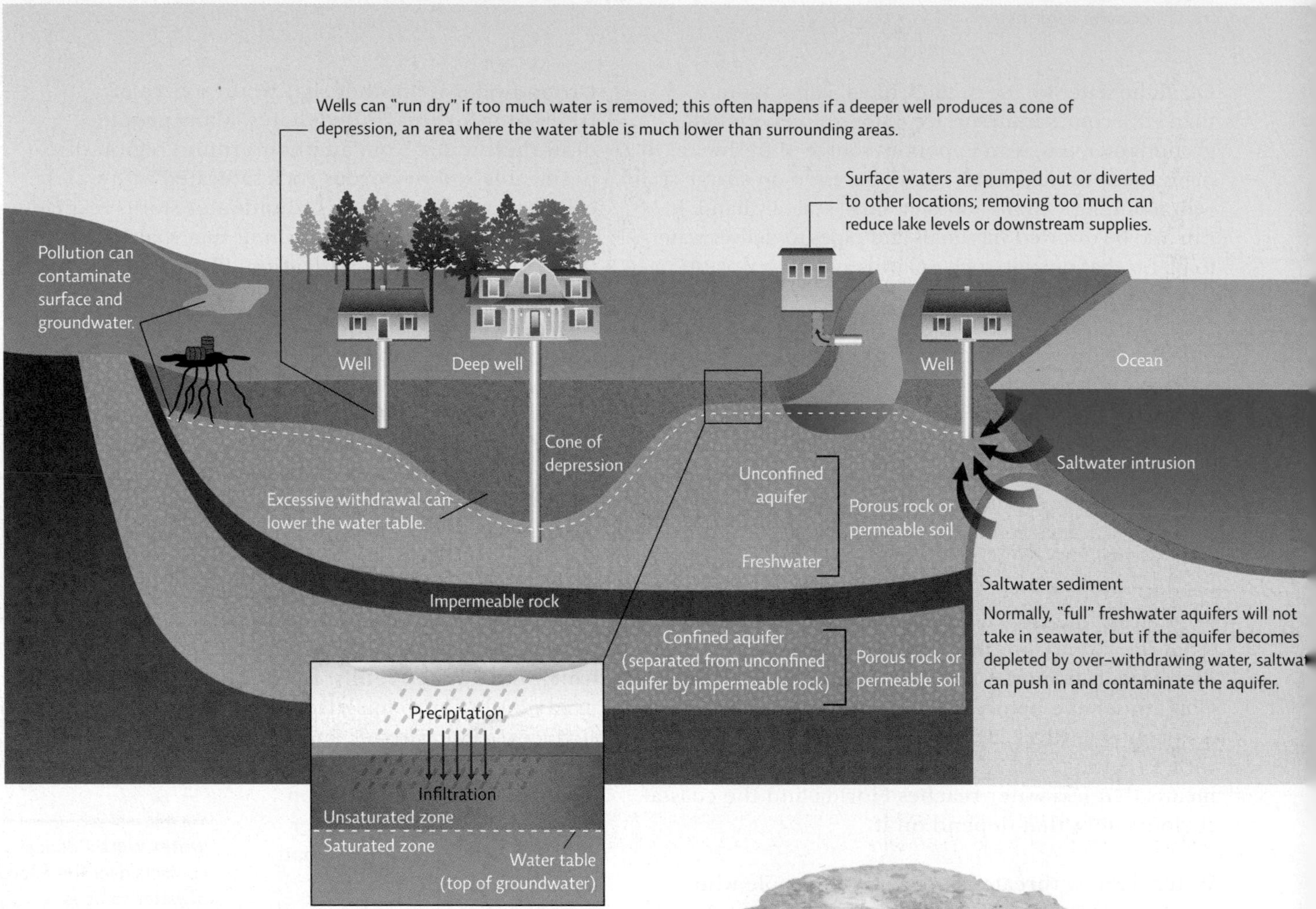

Why might deep cones of depression and dry wells (formed from overdrawing well water) be more common in inland areas than in coastal ones?

higher than sea level, so saltwater doesn't infiltrate saturated aquifer regions. But when too much water is being removed, saltwater can seep into empty spaces, potentially rendering an aquifer unusable. To prevent this from happening, some coastal communities pump water into their aquifers: For instance, Orange County, California, which gets about half of its water from the Colorado River, treats its sewage water to the point that it is drinkable and then pumps it into the aquifer to prevent saltwater intrusion. **INFOGRAPHIC 3**

Courtesy of the St. John's River Water Management District, Palatka, Fla.

An aquifer core sample shows the porous nature of a limestone rock formation.

Zbigniew Bzdak/Chicago Tribune/MCT via Getty Images

A sinkhole swallows three cars during a heavy rainstorm in Chicago, Illinois, in 2013. Aquifers can be weakened when they lose water, and then a heavy rain event may add enough weight to collapse the aquifer, opening up a sinkhole.

4 GLOBAL WATER USE AND SCARCITY CONCERNS

Key Concept 4: Agriculture is the biggest user of water, followed by industrial use, and then domestic use. Not surprisingly, domestic use goes up as income goes up. Water shortages can be physical (lack of water) or economic (the inability to pay for water).

Agriculture is, by far, the sector that uses the most freshwater. To increase yields or grow high-value crops in areas that are naturally too dry for those crops, we add water. Crop irrigation in the United States consumes 80% of our water (it's even higher in some western states); in the countries of South Asia, closer to 90% of total water usage goes to irrigation. And in the United States, irrigation water comes, about equally, from surface and groundwater sources, depleting both in some areas.

The Ogallala aquifer, the largest groundwater system in North America, underlies eight states in the Great Plains of the American Midwest, from Nebraska to Texas, supplying about 30% of all U.S. irrigation water to these farm belt states. But withdrawals far exceed recharge rates, which are only replacing about 15% of what is removed. This has lowered the water table an average of 15 feet—a loss of 30% of its historic capacity—with most of this drawdown coming since 1960, according to an analysis by Kansas State University researcher David Steward. The southern reaches of the aquifer are most severely impacted (from Kansas to Texas); wells have run dry in some areas, and drilling deeper won't help—the groundwater there is gone. Steward predicts that the aquifer will be 69% depleted by 2060 (a number that could increase if climate change reduces rainfall in the area, which it is projected to do). Local water supplies and future agricultural productivity depend on this aquifer; steps taken today to improve irrigation efficiencies or reduce the irrigation water needed for a crop (e.g., plant a crop that needs less water) are crucial for the future of this region.

In the Colorado Basin states, the river is the biggest source of water and an estimated 78% is used for agriculture—the lion's share of that is used by farmers in the Imperial and Yuma valleys of California and Arizona where it supports the growth of vegetable crops. A smaller proportion is used by farmers farther north, often to grow crops like alfalfa and corn, which are then used to feed cattle. The rest of the withdrawn water flows to homes and businesses—some diverted and sent to distant thirsty cities such as San Diego and Los Angeles.

Industrial use of water varies from country to country. It is understandably higher in more developed nations than less developed ones. Thermoelectric power plants such as coal or nuclear facilities use large amounts of water for cooling, and in many developed nations, this represents the biggest portion of industrial water use; in the United States, it accounts for around 85% of the industrial sector's use of water.

domestic water Indoor and outdoor water used by households and small businesses.

water scarcity Not having access to enough clean water.

Domestic water use—that of homes and businesses—is the sector that uses the least amount of water, but there are big differences between nations. In general, water use goes up as income levels go up, but some wealthy nations use much more water than others. For example, on average, a citizen in the United States uses more than four times what a citizen of Denmark uses and about twice as much as the world average. **INFOGRAPHIC 4**

People don't always live near abundant sources of freshwater, making access a vital issue. Around the world, many areas suffer from **water scarcity**—not having sufficient access to clean water supplies. In some dry regions, there is simply not enough to meet needs; many arid nations like those of the Middle East, parts of Africa, and much of Australia face water shortages as a way of life. The Middle Eastern countries of Bahrain, Qatar, Kuwait, and Saudi Arabia have the lowest per capita water availability in the world, but these oil-rich nations can afford to invest in costly technology to access water (like facilities to remove salt from seawater). In other areas, particularly in sub-Saharan Africa, there may be enough water, but people do not have the money to purchase it or dig wells to access it. People in these areas may be getting by on just a few gallons of water a day—and that water may not even be safe to use.

Ensuring that all people have access to clean water and sanitation is Goal Six of the United Nation's (UN) Sustainable Development Goals, a goal, the UN points out, that affects the successful achievement of every other goal. Currently, more than half the world's population, around 4 billion people, faces physical water scarcity at least part of the year. By 2050 as many as 5.7 billion could

As many as 4 billion people lack access to dependable supplies of clean water. Water scarcity may be physical (not enough water is present) or economic (cannot afford to buy or access water).

Likhitha/Getty Images

INFOGRAPHIC 4 GLOBAL WATER USE

Agriculture is by far the biggest water user globally. By 2050, it is projected that we will need 20% to 30% more water: much of this increased demand will come from industry and domestic use.

Individuals in more developed countries use far more water per person than those in less developed countries. In some areas, individuals must make do with only a few gallons a day.

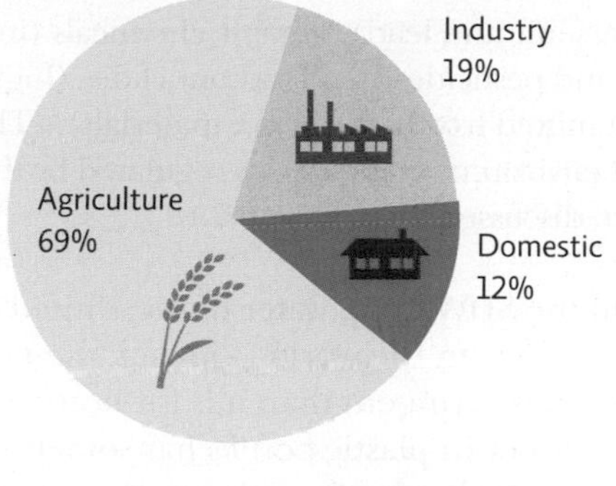

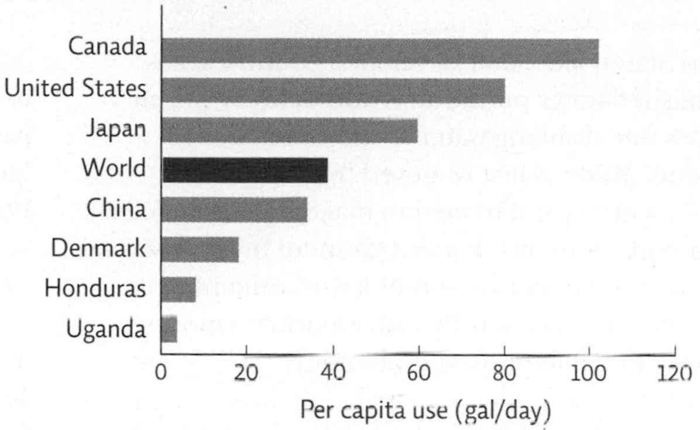

The United Nations recognizes meeting water needs as key to sustainability, but presently almost 600 million people lack physical access to clean water; another 1.5 million must travel at least 30 minutes to acquire clean water. More than 500 million people live in areas with severe water scarcity year-round, and as many as 4 billion face water shortages at least 1 month of the year.

Goal 6 of the UN's Sustainable Development Goals: Ensure access to water and sanitation for all.

Targets to be reached by 2030 include:

- Access to safe and affordable drinking water for all.
- Access to adequate sanitation for all and end open defecation.
- Reduce pollution; halve the proportion of untreated wastewater.
- Ensure sustainable withdrawals of freshwater.
- Protect and restore water-related ecosystems.
- Support local communities in improving water and sanitation management.

The United States and Denmark are both developed countries with high standards of living. What might account for the large difference in domestic water use per person in these two countries?

face water shortages. (See Infographic 4 in Module 1.1 for the UN's Sustainable Development Goals and see this module's Infographic 4 for more on Goal Six.)

Much of the water that is available in regions where scarcity is a problem is contaminated with fecal material (human or animal waste), often the result of inadequate sanitation. According to the World Health Organization (WHO), around 300,000 gallons of raw sewage enter the Ganges River of India every minute. These waters can be contaminated with raw sewage that contains dangerous pathogens, increasing the chance for disease transmission. (See Module 4.3 for more on environmental health and waterborne diseases.) All told, as many as 4.5 billion people lack access to sufficient sanitation facilities, with serious consequences. The UN estimated that in 2018, 5.6 times more people died from contaminated drinking water than from flood, drought, earthquakes, epidemics, and conflict combined.

Water wars like the ones over the Colorado and Chattahoochee rivers take place in developing countries, too—and in economically disadvantaged regions, their effects can be devastating. Syria, for instance, has always been an arid country, but its most recent drought, from 2006 to 2011, has been called the worst since farmers first began tilling the Fertile Crescent many millennia ago. The drought has affected millions of Syrian farmers, causing so many crop failures that more than 1.5 million farmers have migrated from rural lands into cities, trying to find new ways to make a living. This influx led to social unrest and urban unemployment and is cited as a contributing factor to the ongoing civil war that began in 2011.

More water wars may be just around the corner. As glaciers that provide water for billions of people in Asia and South America melt at accelerated rates, for example, concerns over water are setting the stage for water wars to come. (See the *Global Case Studies* map at the end of this module for a closer look at some additional water wars.)

5 U.S. WATER SUPPLIES

Key Concept 5: The United States has one of the world's most dependable purification and delivery systems for tap water thanks to the Safe Drinking Water Act; however, thousands of violations occur each year and infrastructure upgrades are needed to continue providing safe drinking water.

The United States, like other developed countries, has a well-established water purification and delivery system that provides safe drinking water to the vast majority of its residents. Water is first removed from surface or groundwater sources and treated to make it **potable** (clean enough for consumption). Water treatment involves several steps to remove sediment and other large contaminants, followed by disinfection, usually with chlorine. Fluoride might also be added to prevent tooth decay.

In the United States, public drinking water is protected under the **Safe Drinking Water Act (SDWA)**. The Environmental Protection Agency (EPA) administers the act by setting water quality standards and works with state and local governments and other federal agencies to see that those standards are met. For example, the EPA sets standards for the way water is treated and sets limits for six categories of contaminants: microorganisms (pathogens such as bacteria and protozoa), disinfectants (such as chlorine), by-products of disinfectants, inorganic chemicals (such as arsenic and lead), organic chemicals (including benzene and pesticides), and radionuclides (high-energy particles emitted from radioactive materials). (The water quality of environmental waters is regulated by the Clean Water Act, discussed in Module 6.2.)

potable Water that is clean enough for consumption.

Safe Drinking Water Act (SDWA) Federal law that protects public drinking water supplies in the United States.

Thanks to the SDWA, tap water in the United States is among the safest in the world—in fact, the purification process is more stringent than it is for bottled water, and as a result, water in plastic bottles has sometimes been found to have higher levels of chemicals (see Module 4.3) and microplastics (see Module 5.3) than tap water does, although it's not clear whether these higher levels pose a health risk.

That said, the U.S. system isn't perfect, and municipal water supplies sometimes get contaminated. The most common contaminant is fecal material, which can cause gastrointestinal illnesses. Water polluted with nitrogen fertilizers can also cause serious problems, especially for infants and young children (see Module 6.2). Long-term exposure to disinfection by-products can cause kidney and liver damage, while heavy metals such as arsenic and lead can cause serious harm, too.

Jim West/Alamy

Workers replacing lead pipes in Flint, Michigan. Old lead water pipes need to be replaced in communities across the country.

Many U.S. water treatment facilities are old or too small to handle growing local populations, and much of our water infrastructure (e.g., pipes) is old, leading to lost water (leaks) and contamination. When the city of Flint, Michigan, switched to the nearby Flint River as its new drinking water source in 2015, the water was so corrosive that it caused lead to leach out of old pipes. Before long, Flint's drinking water became contaminated with lead at levels that were up to seven times higher than EPA safety limits. **INFOGRAPHIC 5**

The EPA estimates it will cost billions just to fix existing problems with drinking water infrastructure, such as replacing old pipes. In many communities, water and sewage treatment plants need upgrading too.The EPA's 2016 Action Plan lays out a framework for addressing these and other issues.

INFOGRAPHIC 5 VIOLATIONS OF THE SAFE DRINKING WATER ACT

In most places, tap water in the United States is very safe thanks to the Safe Drinking Water Act (SDWA) but there are still thousands of violations each year. Fecal contamination from sewage and livestock facilities is the most common contaminant. The presence of disinfection by-products can also exceed allowable limits; nitrate violations are most common in agricultural areas. The presence of toxic heavy metals (e.g., arsenic, lead, copper) occurs less frequently, but when it is present, it presents a major health risk, as in the case of lead contamination in Flint, Michigan.

NUMBER OF HEALTH VIOLATIONS OF THE SDWA (1982–2015)

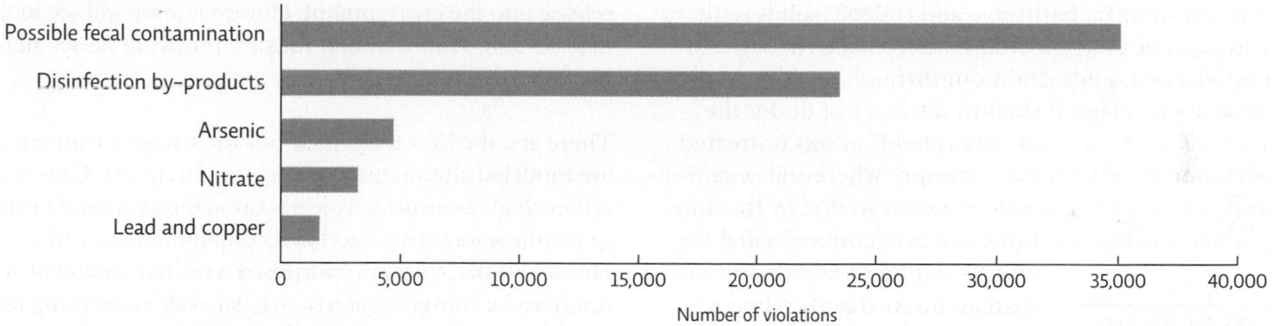

LEAD CONTAMINATION IN THE TAP WATER OF FLINT, MICHIGAN

Moussa81/iStock/Getty Images

In 2015, residents of Flint, Michigan, noticed strange colors and odors in their tap water, and many suffered from unexplainable health problems such as rashes, headaches, and hair loss. They would later discover that their water was contaminated with lead, a potent neurotoxin. Flint had switched its water source from Lake Huron to the Flint River, a river with heavy industrial pollution. The Flint River water was so corrosive, it leached lead out of the old pipes that delivered water to much of the city — 40% of the homes had tap water lead levels that were as much as seven times higher than EPA allowable limits. Children are especially vulnerable to lead exposure, and in some neighborhoods, 15% of the children had blood-lead levels that were dangerously high.

The lead contamination in Flint's water didn't have to occur — safety measures were already in place to prevent this problem, but Flint officials didn't implement them and took steps to hide the seriousness and extent of the contamination. Before all was said and done, some 100,000 residents, including tens of thousands of children, were potentially exposed to elevated levels of lead in their drinking water for nearly 2 years. The hardest-hit areas were in low-income, minority communities; Paul Mohai of the University of Michigan's School for Environment and Sustainability calls the lead crisis in Flint one of the biggest environmental justice disasters in U.S. history.

Flint's pipes are being upgraded to get rid of the old lead pipes at tremendous cost. Similar upgrades are needed nationwide. The EPA estimates upgrades needed across the country will cost more than $650 billion. They have developed a plan — the 2016 Action Plan — that lays out a framework for addressing these and other water issues. This plan just needs funding to carry it out completely.

Why might low-income, minority communities suffer from SDWA violations more often than other parts of a city? What could be done to address this?

6 WASTEWATER TREATMENT

Key Concept 6: Wastewater can be decontaminated using high-tech methods that use advanced filtration and chemicals or low-tech methods that mimic the way wetlands purify water.

Wastewater is any kind of used or contaminated water produced by industry, agriculture, homes, and businesses. The water that arrives at municipal wastewater treatment facilities contains used water washed down drains and sewage flushed down toilets; in many communities, it also contains stormwater delivered to the facility by storm drains that collect runoff from city streets.

Rural homes in the United States that are not connected to a wastewater treatment facility (a city sewage system) use septic systems. A waterproof (e.g., concrete-lined) pit buried in the yard receives the wastewater from the home (from sinks, bathtubs, and toilets); solids settle to the bottom of this septic tank where bacteria begin to digest it. The liquids drain out through perforated pipes buried in gravel-lined shallow ditches just under the surface of the yard. This "drain field" allows untreated wastewater to soak into the ground, where soil organisms digest much of the organic material so that by the time the water reaches the aquifer, it is decontaminated (as long as the water table is not too close to the surface). When it fills up, the septic tank must be pumped out and the contents disposed of as sewage sludge.

wastewater Used and contaminated water that is released after use by households, businesses, industry, or agriculture.

wastewater treatment The process of removing contaminants from wastewater to make it safe enough to release into the environment.

Municipalities offer a more technologically advanced system for treating sewage wastewater. In a traditional "high-tech" facility, **wastewater treatment** includes initial steps that filter the water and then send it to settling tanks where much of the remaining suspended solids sink to the bottom. The water continues on to other tanks where bacteria digest much of the remaining organic matter. Final treatment includes chemical treatment (such as exposure to chlorine to kill some pathogens or the addition of other chemicals to remove phosphates or other substances that should not be released). Proper treatment is critical. Sewage can carry pathogenic viruses and bacteria; swimmers and surfers get exposed to these if poorly treated wastewater is discharged into recreational waters. The goal is not to purify the wastewater for consumption, but rather to make it safe to release into the environment. However, as we will see in the next section, some communities are purifying wastewater to the point that it is potable.

There are also "low-tech" methods for sewage treatment that are modeled after natural ecosystems. In Arcata, California, rather than construct a typical wastewater treatment facility to handle sewage that had been contaminating nearby Humboldt Bay, the city repurposed a retired landfill near the coast by converting it to a *wetland*—an ecosystem that is permanently or seasonally flooded. A slow river meanders through the wetland, where organisms there purify it. To them, it's not "sewage"; it's food. The Arcata facility depends on nature to perform the job of water purification—no toxic chemicals are used. The water discharged into the ocean is very clean, and the health of the bay ecosystem has improved. Similar facilities have been constructed on a smaller scale, providing wastewater treatment for a single building. **INFOGRAPHIC 6**

This wetland marsh in Arcata, California, is actually part of a wastewater treatment system that uses nature to help purify sewage. The wetland is now an Audubon bird sanctuary.

David Howell

INFOGRAPHIC 6 HOW IT WORKS: WASTEWATER TREATMENT

Sewage must be treated before it can be safely released to the environment. Most communities use chemical- and energy-intensive high-tech methods, but systems that mimic nature can also effectively purify water.

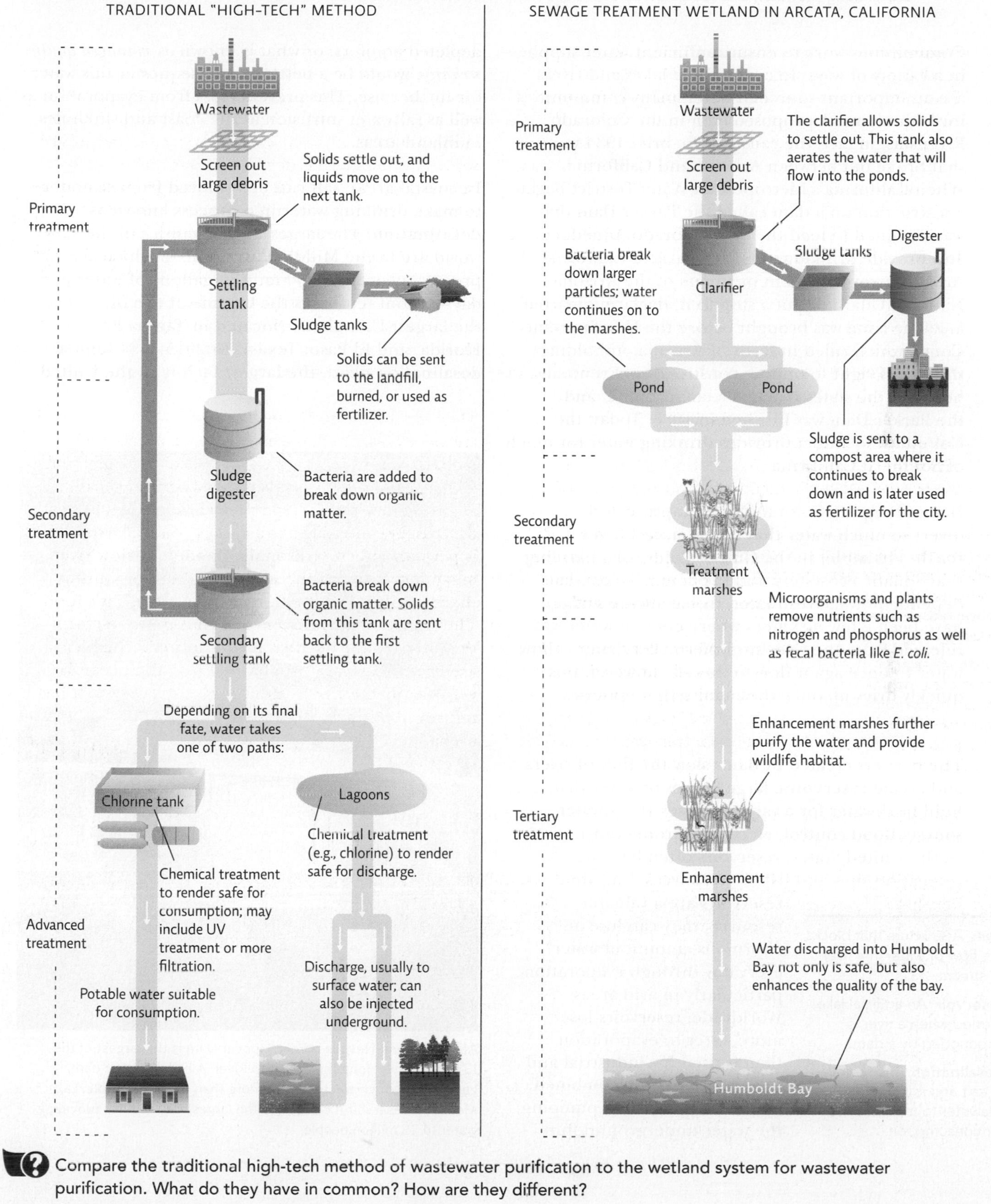

TRADITIONAL "HIGH-TECH" METHOD
Wastewater
Screen out large debris
Solids settle out, and liquids move on to the next tank.
Primary treatment
Settling tank
Sludge tanks
Solids can be sent to the landfill, burned, or used as fertilizer.
Sludge digester
Bacteria are added to break down organic matter.
Secondary treatment
Bacteria break down organic matter. Solids from this tank are sent back to the first settling tank.
Secondary settling tank
Depending on its final fate, water takes one of two paths:
Chlorine tank
Lagoons
Chemical treatment (e.g., chlorine) to render safe for discharge.
Chemical treatment to render safe for consumption; may include UV treatment or more filtration.
Advanced treatment
Potable water suitable for consumption.
Discharge, usually to surface water; can also be injected underground.
SEWAGE TREATMENT WETLAND IN ARCATA, CALIFORNIA
Wastewater
Primary treatment
The clarifier allows solids to settle out. This tank also aerates the water that will flow into the ponds.
Screen out large debris
Digester
Sludge tanks
Bacteria break down larger particles; water continues on to the marshes.
Clarifier
Pond
Pond
Sludge is sent to a compost area where it continues to break down and is later used as fertilizer for the city.
Secondary treatment
Treatment marshes
Microorganisms and plants remove nutrients such as nitrogen and phosphorus as well as fecal bacteria like E. coli.
Enhancement marshes further purify the water and provide wildlife habitat.
Tertiary treatment
Enhancement marshes
Water discharged into Humboldt Bay not only is safe, but also enhances the quality of the bay.
Humboldt Bay

Compare the traditional high-tech method of wastewater purification to the wetland system for wastewater purification. What do they have in common? How are they different?

7 ADDRESSING WATER SHORTAGES WITH TECHNOLOGY

Key Concept 7: Water scarcity can be addressed by storing water (using dams or underground storage), desalinating seawater, and using new solutions like purified wastewater. All these solutions come with trade-offs.

Communities work to ensure sufficient water supplies in a variety of ways. In areas where lakes and rivers are an important source of water, many communities invest in **dams**. A proposed dam in the Colorado River was, in fact, the cause of the brief 1934 war that broke out between Arizona and California. When California's Metropolitan Water District began construction on a dam called the Parker Dam that was designed to feed the new Colorado Aqueduct and provide drinking water for parts of California, Arizona's governor sent members of the Arizona National Guard to put a stop to it, declaring martial law. The issue was brought before the U.S. Supreme Court, which ruled in favor of Arizona, upholding the state's right to oppose construction. Eventually, however, the states reached a compromise, and the Parker Dam was finished in 1938. Today, the Colorado Aqueduct provides drinking water for much of Southern California.

Dams along the Colorado River capture and divert so much water that the river no longer reaches its outlet in the Gulf of California, harming that aquatic ecosystem and the communities that depended on its freshwater. In recent years, there have been a few instances where enough water was released from the river's southernmost dam to allow water to once again flow to the sea; however, this quickly dries up once these pulse-flow episodes are over.

The barriers created by dams slow the flow of rivers and create **reservoirs**, large bodies of water that hold freshwater for a variety of uses (freshwater source, flood control, electricity production). In the United States, reservoirs often become recreation sites and fishing resources. But while reservoirs are a valuable resource, they can lose an enormous amount of water every day through evaporation, particularly in arid areas. Worldwide, reservoirs lose more water to evaporation than is used for industrial and domestic purposes combined. Critics point out that pumping the water underground into depleted aquifers, or what is known as *managed aquifer recharge*, would be a better way to sequester this water for further use. This prevents loss from evaporation as well as saltwater intrusion at the coast and sinkholes in inland areas.

dam A structure that blocks the flow of water in a river or stream.

reservoir An artificial lake formed when a river is impounded by a dam.

desalination The removal of salt and minerals from seawater to make it suitable for consumption.

In coastal areas, salt can be removed from seawater to make drinking water in a process known as **desalination.** The largest such facilities in the world are in the Middle East, some of which are processing around 200 million gallons of water per day—about ten times the volume of two of the largest U.S. plants (located in Tampa Bay, Florida, and El Paso, Texas). In 2015, a $1 billion desalination plant, the largest facility in the United

Robert Fulton/Shutterstock

At the Arizona-Nevada border, Hoover Dam is the largest of the 15 dams found along the Colorado River. A hydroelectric dam, it generates enough electricity for more than 1 million people. Lake Mead, the reservoir it created, is the largest on the river, providing water to 25 million people.

States, came online in Carlsbad, California, and began producing potable water for San Diego; it is expected to meet about 7% of the city's water needs (about 50 million gallons a day). But the briny water that results from the desalination process is toxic and must be dealt with safely. Furthermore, removing salt and other minerals from seawater uses a large amount of energy and is very expensive. Still, thousands of desalination plants worldwide operate today to meet some of the water needs of their regions.

As is the case in Orange County, California, communities in the Colorado River Basin recycle their wastewater—purifying it to the point that it is safe to consume. Las Vegas, dependent on Lake Mead for its water, sends roughly 90 million gallons of highly purified wastewater per day back to the reservoir, replacing much of what it took. The idea of drinking treated wastewater might not sound appealing, but keep in mind that if your community withdraws water from a surface water source to purify for drinking water, it is likely that an upstream community has discharged its treated wastewater into that same waterbody. In other words, most of us are already drinking purified wastewater. **INFOGRAPHIC 7**

Mary Knox Merrill/Getty Images

The Groundwater Replenishment System in Fountain Valley, California, solves two problems at once: addressing water scarcity and dealing with wastewater. This $480 million water treatment system converts the sewage water of Orange County into drinking water, producing more than 100 million gallons of drinking water every day.

INFOGRAPHIC 7 TECHNOLOGIES THAT ADDRESS WATER SCARCITY

Ensuring ample water supplies is vital for society, and we have some high-tech methods to increase our water supply or to make it more predictably available. They all come with trade-offs that must be evaluated to determine if they are suitable for a given location.

DAMS

Pros
- Dependable water source
- Electricity generation
- Flood control
- Recreation

Cons
- Habitat destruction
- Water loss from evaporation
- "Water wars" between upstream and downstream areas

DESALINATION

Pros
- Ample supply

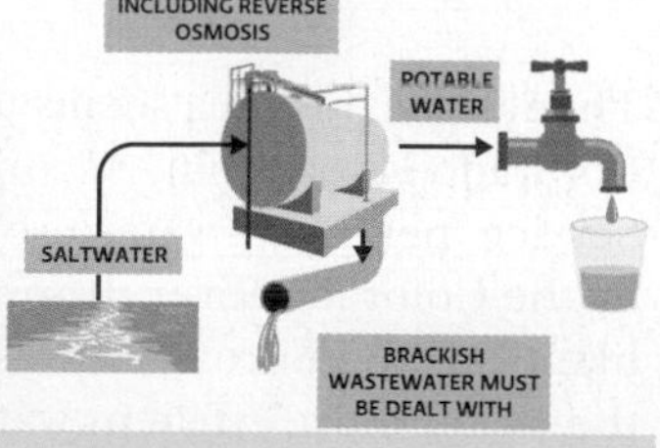

Cons
- Expensive, energy-intensive process
- Toxic wastewater

RECYCLING WASTEWATER

Pros
- Addresses water scarcity and wastewater

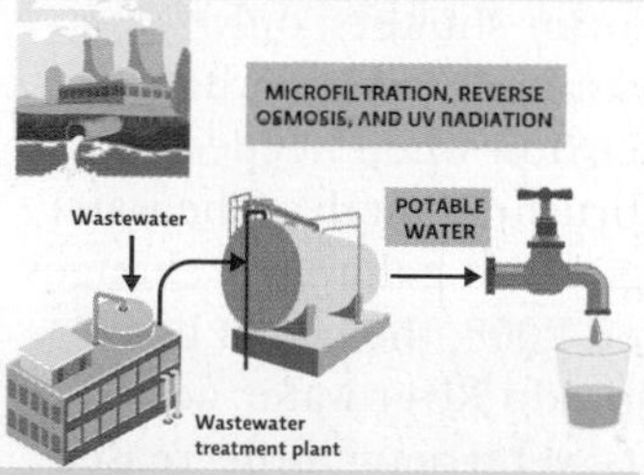

Cons
- Expensive
- Negative public perceptions

UNDERGROUND STORAGE

Pros
- Prevents loss from evaporation
- Prevents saltwater intrusion and sinkhole formation
- Especially useful in areas with variable seasonal rainfall

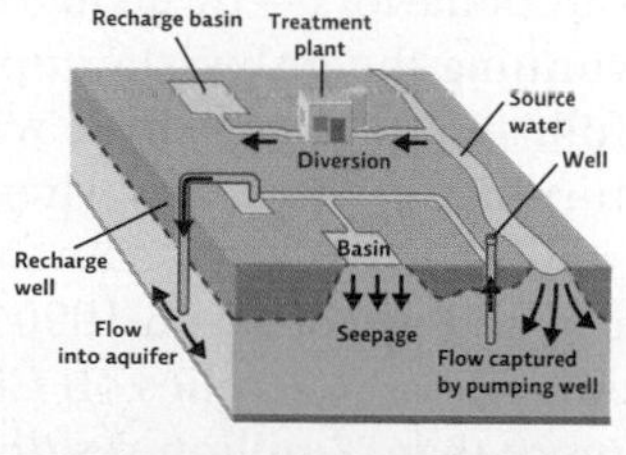

Cons
- Too much river water diversion could reduce flow to downstream areas
- A high water table can harm plants that cannot tolerate waterlogging

Which technology would you recommend for use in your area if there was a need to increase water supplies?

8 ADDRESSING WATER SHORTAGES WITH CONSERVATION

Key Concept 8: Conservation can effectively address scarcity and includes the use of water-saving technologies, behavioral changes that decrease water use, and consumer choices that minimize our water footprint.

While we have many technologies to address water shortages, an easier and cheaper way to maintain water supplies is simply not to waste so much. For example, water-saving irrigation methods limit loss to evaporation and runoff, thus significantly reducing the water that is used—by as much as half, according to Tess Russo of the Columbia Water Center at Columbia University. This has the added advantages of protecting surface waters and preventing soil salinization (the buildup of salt as water evaporates), a common problem in dry climates. Choosing to plant crops more suited to the environment and water availability will decrease agricultural water use. And many industrial processes are now designed to reuse water rather than discharge it into the environment.

The average U.S. citizen uses about 140 gallons of water per day in the home. This is considered *direct use*—water used to meet personal needs of cooking, washing, toilet flushing, and so on. Small individual changes in the household can save a lot of that water. In a typical home, toilets are the biggest water users, but low-flow models can save a significant amount. Toilets with two flush buttons save even more water—they minimize water used to flush liquids, reserving the stronger water flow for solids. New dishwasher models also conserve water compared with older models, so if you have an old dishwasher, consider upgrading. Low-flow shower heads and faucets can make a difference, too.

Still, we can make an even bigger impact by changing our behaviors—by taking shorter showers, only running the dishwasher or washing machine when it is full, and only letting the water flow when needed (e.g., turning off the faucet when brushing teeth). The water savings from these technologies and behavioral changes can add up. Between 1990 and 2008, the city of Los Angeles, which relies on Colorado River water, added more than 3 million residents, yet it actually decreased its overall water use by 4%, in part due to drops in everyday household use.

water footprint The water consumed by a given group (that is, person or population) or appropriated and/or polluted by industry to produce products or energy.

It is also important to remember that our personal use of water is not limited to direct use of water from the tap. Water is also used on our behalf by industry to produce the products and energy that we consume (*indirect use*). When the **water footprint** of the goods we consume is added in, the average daily use of water goes up to 1,100 gallons per person in the United States!

For example, blue jeans and other cotton clothing items have a particularly large water footprint because cotton, typically grown in arid climates, is a water-thirsty crop. It requires large amounts of irrigation water, and if improperly applied, much of this irrigation water is lost to evaporation. It is also heavily treated with pesticides, which can pollute local water supplies and raise its water footprint. Further processing (making the fabric, dying the fabric, etc.) also uses water. Choosing clothing made from organically grown cotton (cotton grown in a way that uses water sustainably), buying used clothing, or buying fewer clothing items will all reduce the impact of cotton. Levi Strauss & Co. is working to source only sustainably produced cotton for its jeans—cotton that is grown in a manner that improves the livelihoods and working conditions of the farmers and reduces the environmental impact of growing the cotton with changes such as more efficient irrigation and fewer pesticides. **INFOGRAPHIC 8**

Freshwater is a precious but limited resource—it always has been and always will be. Access to freshwater is unevenly distributed across the globe, and limited access leads to conflict. In any given locale facing a water shortage, we need to use strategies and technologies that make the most sense then and there; what works well in one location or at one point in time may not be ideal in others. But water conservation is needed everywhere—because by conserving this precious resource, we will make the water we do have go further and reduce the tensions that make water wars a reality.

Thankfully, this idea seems to be catching on in the Colorado River Basin. According to the U.S. Geological Service, per-capita water use in the seven states served by the Colorado River dropped between 1985 and 2010, likely due to improved infrastructure, conservation, and the use of more efficient water-using appliances. This will help, but additional reductions may be needed.

It was in anticipation of further reduced flows that the seven states that share the Colorado River came together in 2019 to draft the Colorado River Basin Drought Contingency Plan. This agreement contains provisions that call for voluntary reductions in water withdrawals if water levels in Lake Mead and Lake Powell (the two largest reservoirs on the Colorado River) fall below a specified level. This is a short-term plan—a seven-year agreement that goes out to 2026. After that, more

INFOGRAPHIC 8 REDUCING OUR WATER FOOTPRINT

Understanding how we use water helps us make water-saving decisions, which may include behavioral changes or the use of water-efficient technologies. Much water usage can be reduced by buying less stuff and, for those things we do buy, choosing products with lower water footprints. Because water is needed in thermoelectrical production of electricity, energy conservation or use of sustainable energy sources will also reduce our water footprint.

U.S. HOUSEHOLD WATER USE

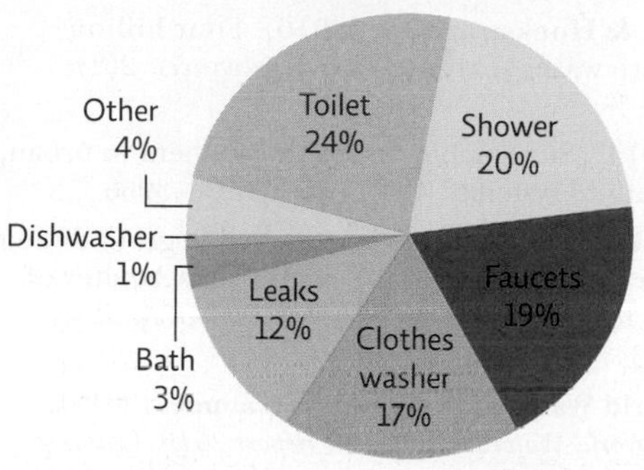

Johnrob/iStock/Getty Images

On average in the United States, 13,500 gallons of water are used to produce 900 kilowatt-hours of electricity (the average monthly electricity consumption). Any actions that save energy will also save water.

GALLONS OF WATER NEEDED TO PRODUCE 1 POUND OF FOOD

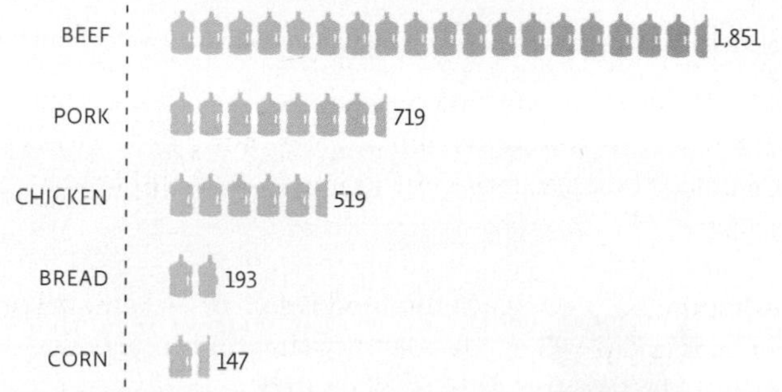

GALLONS OF WATER PER PRODUCT

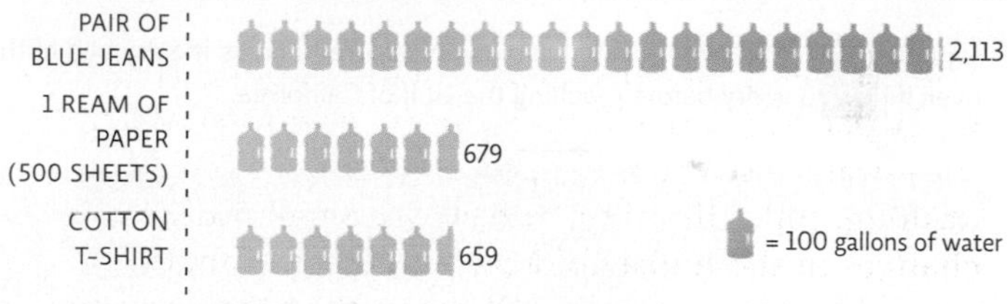

WATER-SAVING TECHNOLOGIES AND ACTIONS

Our water usage can be reduced by using new water-efficient technologies and by making behavioral changes that don't waste water.

OLD TECHNOLOGY	NEW TECHNOLOGY	BEHAVIORAL CHANGE
Toilet 6 gallons/flush	**Low-Flow Toilet** 1.3 gallons/flush	Don't flush tissues—use the trash. Flush liquid waste less frequently.
Shower 3.8 gallons/minute **Bath** 35 gallons	**Low-Flow Shower Head** 2.3 gallons/minute	Take a shorter shower or a "Navy" shower: Turn off the shower head except to rinse (some shower heads come with a convenient valve that allows you to switch off the water without turning it off at the source).
Faucet 5 gallons/minute	**Low-Flow Faucet** 1.5 gallons/minute	Don't leave the faucet running while brushing your teeth, shaving, or washing your face.
Washing Machine 40 gallons/load	**Washing Machine (Energy Star)** 22 gallons/load	Don't wash a clothing item unless it needs it (those jeans can probably be worn several times before washing) and run the washer only when it is full.
Dishwasher 9 gallons/load	**Dishwasher (Energy Star)** 4 gallons/load	Run the dishwasher only when it is full and limit the amount of rinsing you do before loading dishes into the dishwasher; if you have a new dishwasher, rinsing isn't needed.

? Estimate how long your typical shower lasts, and then calculate how much water you would use over the course of a year if you used a traditional 3.8-gallon/minute shower head. Do the same calculation for the low-flow shower head. How many gallons per year would you save by switching to a low-flow shower head? Compared to the water used in a typical shower using a traditional shower head, how much would you save in a year if you reduced the duration of your shower by half and used a low-flow shower head?

aggressive steps may be needed, reigniting tensions between the North Basin, where most of the water enters the river, and the thirsty South Basin that uses most of the water. Mexico, too, is still part of this equation with hopes that future withdrawal reductions will restore flow all the way to the Gulf of California.

Ultimately the problem is one of more people and less water. Whether the players in this drama will be able to make the changes needed to save the iconic river is a story still unfolding.

Select References:

Environmental Protection Agency Office of Water. (2016). *Drinking Water Action Plan.* https://www.epa.gov/sites/production/files/2016-11/documents/508.final_.usepa_.drinking.water_.action.plan_11.30.16.v0.pdf

Gleick, P. H. (2014). Water, drought, climate change, and conflict in Syria. *Weather, Climate, and Society, 6*(3), 331–340.

Heffernan, O. (2014). Bottoms up. *Scientific American, 311*(1), 69–75.

Hoekstra, A., & Chapagain, A. (2007). Water footprints of nations: Water use by people as a function of their consumption pattern. *Water Resources Management, 21*(1), 35–48.

Maupin, M. A., et al. (2018). *Estimates of water use and trends in the Colorado River Basin, southwestern United States, 1985–2010* (U.S. Geological Survey Scientific Investigations Report 2018-5049). https://pubs.er.usgs.gov/publication/sir20185049

Mekonnen, M. M., & Hoekstra, A. Y. (2016). Four billion people facing severe water scarcity. *Science Advances, 2*(2), e1500323.

Russo, T., et al. (2014). Sustainable water management in urban, agricultural, and natural systems. *Water, 6*(12), 3934–3956.

Steward, D. R., et al. (2013). Tapping unsustainable groundwater stores for agricultural production in the High Plains Aquifer of Kansas, projections to 2110. *Proceedings of the National Academy of Sciences, 110*(37), E3477–E3486.

United Nations World Water Assessment Programme. (2019). *The United Nations World Water Development Report 2019: Leaving No One Behind.* Paris: UNESCO. https://unesdoc.unesco.org/ark:/48223/pf0000367306

Jonathan Waterman walks with his blowup pack raft across the dry bed of the Colorado River. Because so many communities withdraw water from the river, it now runs dry before reaching the Gulf of California.

PETE MCBRIDE/National Geographic Creative

GLOBAL CASE STUDIES WATER WARS

Achieve

Water is a precious resource — important enough to fight over. Some fights are in court, and others are on the battlefield. Growing populations and climate change will further stress our water supplies and increase the lengths many populations will go to attain access to water. Here are a few examples of recent "water wars."

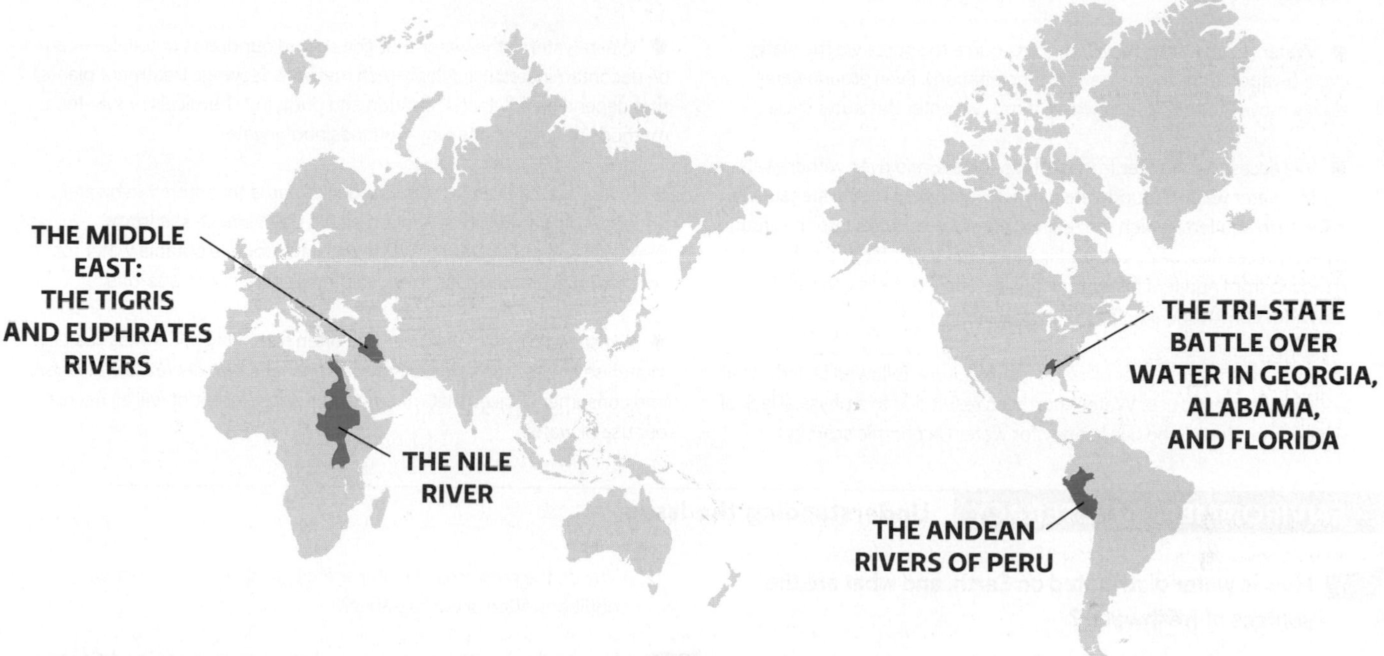

BRING IT HOME

PERSONAL CHOICES THAT HELP

Regardless of whether our water comes from an aquifer or a local reservoir, we can make those water sources last longer by taking steps to use our water as efficiently as possible.

Individual Steps

- If you have a smartphone, download a water usage tracking app. Once you have a baseline, try to reduce it by 10%.
- Time your shower and try to reduce it by 1 to 2 minutes.
- Have a container by the sink or shower to catch water while it warms up; make sure not to get soap in it. Use this water for watering plants both inside and out.
- Install a rain barrel at home. Rain barrels allow people to use the rain that falls on the roof of a building to water plants, as opposed to letting it run off into the storm drain. If you live in a dorm or an apartment, see if you can get permission to have a rain barrel installed.

Group Action

- Do you know where your water comes from? Talk to a city representative to find out where your water comes from and what steps are being taken to make sure it lasts as long as possible.
- Encourage local policy makers to ban the watering of lawns or restrict the use of water for landscaping to certain days of the week.

kislev/iStock/Getty Images

REVISIT THE CONCEPTS

- Water is vital to life, but less than 3% of all water on Earth is freshwater, and most of that is frozen in glaciers or polar ice. Fortunately, because there is so much water on the planet, even this small percentage represents a large amount of water.
- Water is constantly recycled from source to source via the water cycle (evaporation, condensation, precipitation). Even groundwater slowly moves to the oceans where it too can enter the water cycle.
- We access freshwater from surface waters, and over-withdrawals can lead to water wars. Groundwater, also an important freshwater supply, is found in aquifers, which are refilled when water soaks into the ground. Land surfaces like roads and parking lots limit this infiltration. If water is removed from aquifers faster than it is resupplied, wells can run dry or saltwater can invade aquifers in coastal areas.
- By far, the biggest user of water is agriculture, followed by industrial and then domestic use. Water shortages result due to a physical lack of water or for those who cannot pay for water (economic scarcity).
- The Safe Drinking Water Act helps ensure that tap water in the United States is safe to consume; however, thousands of violations occur each year across the country. Many municipalities need infrastructure upgrades to continue providing safe drinking water.
- Wastewater—the water that goes down our drains or toilets—can be decontaminated using high-tech methods (sewage treatment plants) that depend on advanced filtration and purifying chemicals or low-tech methods that mimic the way wetlands purify water.
- Water scarcity can be addressed with dams that store freshwater, underground storage, desalination plants that remove salt from seawater, and purification of wastewater to produce potable water for consumption. However, all these solutions come with trade-offs.
- Conservation is an effective (and often free) way to address scarcity. Water-saving technologies, behavioral changes that decrease water use, and consumer choices that minimize our water footprint will all decrease our use of water.

ENVIRONMENTAL LITERACY Understanding the Issue

1 How is water distributed on Earth, and what are the sources of freshwater?

1. About how much of Earth's surface is covered with water, and how much of that water is freshwater that is accessible to humans?
2. Many island nations face freshwater shortages. How can this be when they are surrounded by water?

2 How does water cycle through the environment?

3. Draw a flowchart of the water cycle. (Don't copy from the book; create your own small drawing.) Follow a single water molecule from a cloud through some portion of the cycle, including a living organism, and back to a cloud.
4. What is evapotranspiration?
5. Why is it appropriate to say that the water cycle is solar-powered?

3 How are surface water and groundwater supplies accessed, and what problems do these water sources face?

6. Distinguish between surface and groundwater freshwater sources. Which is most abundant (see Infographic 1)? Here in the United States, which accounts for most of the U.S. water supply?
7. What is an aquifer, and how does it get its water?
8. What can happen if too much water is removed by a well in coastal areas?
9. What is a water war, and why are they "fought"?

4 How does water use differ between sectors and income levels, and what problems does water scarcity cause?

10. From high to low, rank these sectors (agriculture, domestic, industry) by water use.
11. How is domestic water use affected by income?
12. Why do the problems of water scarcity and unsanitary water conditions often occur together?

5 How is domestic water regulated in the United States, and what problems does our water supply face?

13. What is the purpose of the Safe Water Drinking Act, and what role does the EPA play in administering this law?
14. What water quality problem are residents of Flint, Michigan, dealing with? What would it take to address this issue in Flint and other U.S. communities?

6 How can wastewater be treated to make it safe to release into the environment?

15. What is the typical goal in terms of water purity for wastewater treatment—how pure does it need to be?
16. How do septic systems work to handle wastewater?

7 What technologies can address water scarcity, and what are the trade-offs of each?

17. Which technology would you recommend using to address water scarcity in an arid area far from the coast—a dam that impounds a river to form a reservoir or the capture and injection of river water into depleted aquifers? Explain.
18. Most water on Earth is saltwater. Why, then, is desalination not the answer to water shortages everywhere?

8 How can conservation help address water scarcity issues?

19. Identify at least four ways you can reduce your direct use of water and four ways you can reduce your indirect use of water.
20. Identify one behavioral action that could reduce one's use of water for clothes washing and one water-efficient technology that could do so.

SCIENCE LITERACY Working with Data

Water enters the Colorado River via snowmelt and precipitation that run off the land and wash into the river (runoff). A warmer climate would decrease runoff due to more evaporation. Below are annual data from the Upper Basin region for average regional temperature and runoff efficiency (the amount of runoff per precipitation amount — the lower the value, the less rainwater reached in the stream). These graphs show Z-score values (the departure from average value from 1906 to 2012; a Z-score of 0 equals the average for the period).

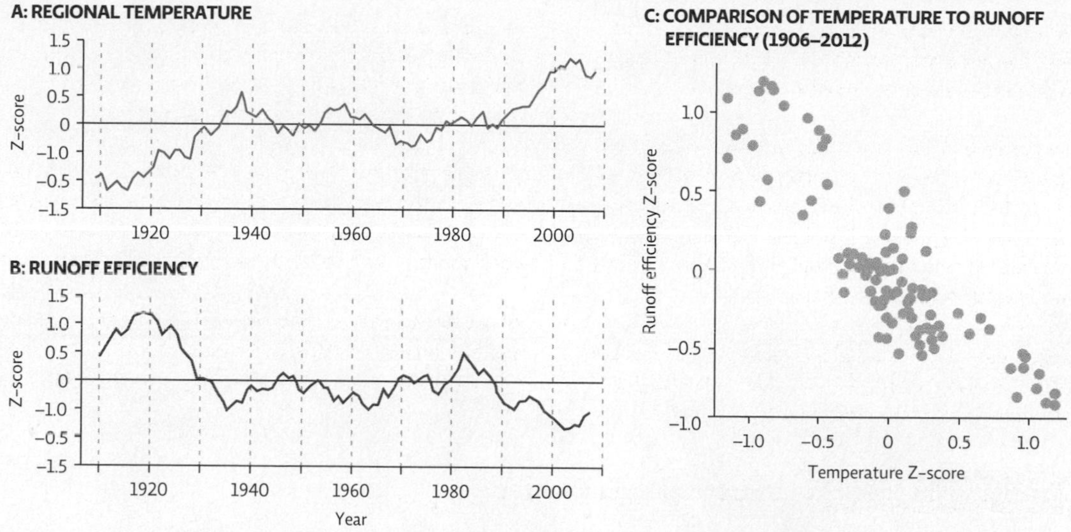

Interpretation

1. How did temperature in the region change between 1906 and 2012? How did runoff efficiency change for the same period?
2. Which time periods were cooler than normal? Warmer than normal? Which time periods saw less runoff efficiency than normal? Greater than normal?
3. Look at the scatterplot (Graph C). Describe the trend seen between temperature and runoff efficiency.

Advance Your Thinking

4. For Graph C, why might this relationship between temperature and runoff efficiency be seen?
5. If a cause and effect relationship exists here, which of these factors is likely the cause and which is the effect? Explain.
6. Do these data offer evidence that climate change is affecting the Colorado River?

INFORMATION LITERACY Evaluating Information

Clean water is vital to life, but as many as 2.8 billion people do not have access to enough clean water, exposing them to parasites and other waterborne pathogens. According to Water.org, a child dies every 20 seconds from a water-related illness, and women in some water-stressed areas spend several hours every day collecting water for their families' basic needs.

Modern technology could vastly improve access to clean water in regions that suffer from water scarcity or contaminated water. But the lack of access in these areas is often coupled with a lack of the electricity, developed roads, machines, and equipment necessary to be able to support digging municipal wells; providing pumping stations, reservoirs, and pipelines; and installing adequate sanitation.

Go to the Global Water website (www.globalwater.org) and explore the links under "Projects." Then go to the Water.org website (www.water.org) and look at some of the featured projects.

Evaluate the websites and work with the information to answer the following questions:

1. Determine if these authors/sponsoring groups are reliable information sources:
   a. Do they give supporting evidence for their claims?
   b. Do they give sources for their evidence as well as clear explanations?
   c. What is the mission of the organization? How do you know this?
   d. Does the organization appear to have a workable solution or solutions?

Now search the Internet for information about two low-tech filtration devices: fog harvesters and the LifeStraw.

2. Evaluate the proposed solutions for price, ease of use, scale of use (individual or village), and limitations.
3. Does either of the proposed solutions stand out as a good option for remote or undeveloped areas, such as the ones featured in the Global Water website or the Water.org website? Explain your answer.

Additional study questions are available at Achieve.macmillanlearning.com. Achieve

Module 6.2 Water Pollution

SUFFOCATING THE GULF

Researchers try to pin down the cause of hypoxia in the Gulf of Mexico

An algal bloom colors Gulf Coast waters off Fort Walton Beach, Florida. Algae population explosions can occur when warm summer waters receive an influx of fertilizer. This fertilizer can wash into the ocean from surrounding land areas or be delivered by rivers that carry fertilizers and sediments picked up as they flow toward the Gulf.

AP Images/Nick Tomecek

After reading this module, you should be able to answer these GUIDING QUESTIONS

1. What is water pollution, and how is it classified by source?
2. What are the causes and consequences of eutrophication?
3. What is a watershed, and how does it affect surface water and groundwater?
4. How is water quality assessed?
5. What public policies are in place to protect water quality?
6. What role does watershed management play in preventing water pollution?
7. How can nonpoint source water pollution be reduced?
8. What strategies can be used to restore damaged aquatic ecosystems?

Back in 1974, when he had just begun his career at Louisiana State University, biologist Eugene Turner took a 15-foot skiff out along the Gulf Coast to survey the water. He brought a handheld oxygen meter along with him. Other researchers had measured oxygen levels in the same waters and had come up with some disturbingly low numbers—levels low enough to essentially "suffocate" any aquatic organism that couldn't relocate to more oxygen-rich waters. But none of them had followed up, and Turner was curious. Were those earlier measurements wrong? Flukes?

Sure enough, Turner's own readings came up low as well—much lower than expected. His curiosity deepened: What would cause low oxygen levels in these waters? He suspected the myriad oil rigs in nearby waters might have something to do with it. But due to the complicated nature of water pollution, he also knew that the true culprit could be hiding hundreds, or even thousands, of miles away.

Thus began a study of what would become known as the Gulf of Mexico dead zone by Turner and his colleague, Nancy Rabalais, that is still going strong more than four decades later. Month after month, year after year, in bigger and better equipped boats, Turner and Rabalais have surveyed the water, mapping out a hypoxic zone that grew from 15 square miles in 1988 to a peak (so far) of 8,776 square miles in 2017 (that's about the size of New Hampshire). As they quickly learned, the oil rigs were not the main problem.

VICTORIA LOE/Tribune News Service/XSEA-Newscom

Louisiana State University biologist Nancy Rabalais taking water samples on a research vessel. She has studied hypoxia in the Gulf of Mexico since 1985.

WHERE IS THE GULF OF MEXICO WATERSHED?

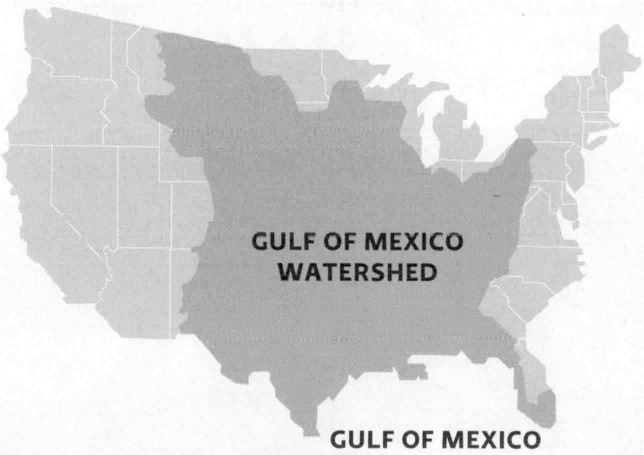

1 WATER POLLUTION: TYPES AND CAUSES

Key Concept 1: Water pollution may come from readily identifiable sources such as discharge pipes (point sources) or from more dispersed sources such as stormwater runoff or atmospheric fallout (nonpoint sources).

Water pollution is the addition of any substance to a body of water that might degrade its quality. The list of such substances, or pollutants, is depressingly long: Raw sewage and industrial chemicals like petrochemicals get dumped directly into bodies of water. Meanwhile, contaminants like mercury and acid-forming precursors, along with other air pollutants from fossil fuel combustion or industry, fall back to Earth with the rain and flow as **stormwater runoff** into rivers, streams, lakes, and seas. Nutrients (from fertilizers and animal waste) and pesticides also enter from farm and lawn runoff. Sediments from soil erosion can flow into surface waters from farms, construction sites, or heavily eroded stream banks that are no longer shored up by a well-rooted plant community.

Water pollution is not limited to the introduction of chemicals or sediment; municipal trash often finds its way to rivers, streams, and oceans (see Module 5.3). The heated water released into surface waters near power plants causes *thermal pollution*, raising the temperature of the water enough to impact many of the organisms that live in that stretch of river, lake, or ocean. Even groundwater sources can become polluted from underground chemical storage tanks or from the movement of surface pollutants down through the soil.

There are two classes of pollution, defined by how they are delivered to the water. **Point source pollution** is water pollution whose discharge source can be clearly identified (that is, one can "point" to the "source"). This includes pollution from large discharge pipes of wastewater treatment plants or industrial sites. The discharge itself is known as **effluent**.

Some of this effluent has the capacity to be quite dangerous. In most cities, sewer pipes run alongside streams because engineers assume that piped sewage will flow in the same direction and therefore reach treatment plants via gravity. But in many urban areas, those pipes are old and beginning to leak. With such a leak, raw sewage flows directly into the stream. Contamination is particularly bad in cities of the developing world, where raw sewage is often directly released, untreated, into bodies of water. In many such places, waterborne *pathogens* (disease-causing organisms) in raw or partially treated human and animal waste represent a leading cause of sickness and death. (See Module 4.3 for more on waterborne illnesses.)

But the good news is that because point sources can be easily identified, they can also—at least hypothetically—be remedied.

As researchers would soon discover, the low oxygen levels in the Gulf of Mexico were due to a different and even more challenging problem: **nonpoint source pollution**. The origins of nonpoint source pollutants are not easily identifiable. Though some arrive by air, most enter the water from overland flow (stormwater runoff); this means they can come from any part of the land that drains into a given body of water. Heavy rains that produce large amounts of runoff or even flood conditions can also release toxins trapped in soil, reintroducing these to water or air. **INFOGRAPHIC 1**

Nonpoint source pollutants can include everything from human-made toxic substances to natural substances such as silt, sand, and clay, which can enter the water as *sediment pollution* (eroded soil that is washed into the water through runoff). To be sure, these substances deliver valuable nutrients to aquatic ecosystems. But excessive amounts of them can cloud the water, making it hard for sunlight to penetrate and thus disrupting photosynthesis. Sediment pollution can also harm organisms directly by clogging gills. And when it covers the sea or river bottom, sediment can smother the nooks and crannies that serve as habitat or spawning areas.

Robert Brook/Science Source

Discharges from a discrete point of entry, such as a pipe, are known as point source pollution.

water pollution The addition of any substance to a body of water that might degrade its quality.

stormwater runoff Water from precipitation that flows over the surface of the land.

point source pollution Pollution from wastewater treatment plants or industrial sites, such as that from discharge pipes or smokestacks.

effluent Wastewater discharged into the environment.

nonpoint source pollution Runoff that enters the water from overland flow.

INFOGRAPHIC 1 MAJOR CAUSES OF WATER POLLUTION

Water pollution comes from a variety of point and nonpoint sources. Although the main threat to the Gulf of Mexico is excess nutrient and sediment runoff, the EPA identifies pathogens as the leading cause of impaired waters in the United States as a whole. Found in sewage or animal waste, pathogens can come from both point and nonpoint sources.

POINT SOURCES

Some industrial and agricultural sources discharge pollutants directly into a body of water.

NONPOINT SOURCES

A variety of sources contribute pollutants that can run off the surface of the land during rainfall and enter the water; air pollutants can fall directly with the rain.

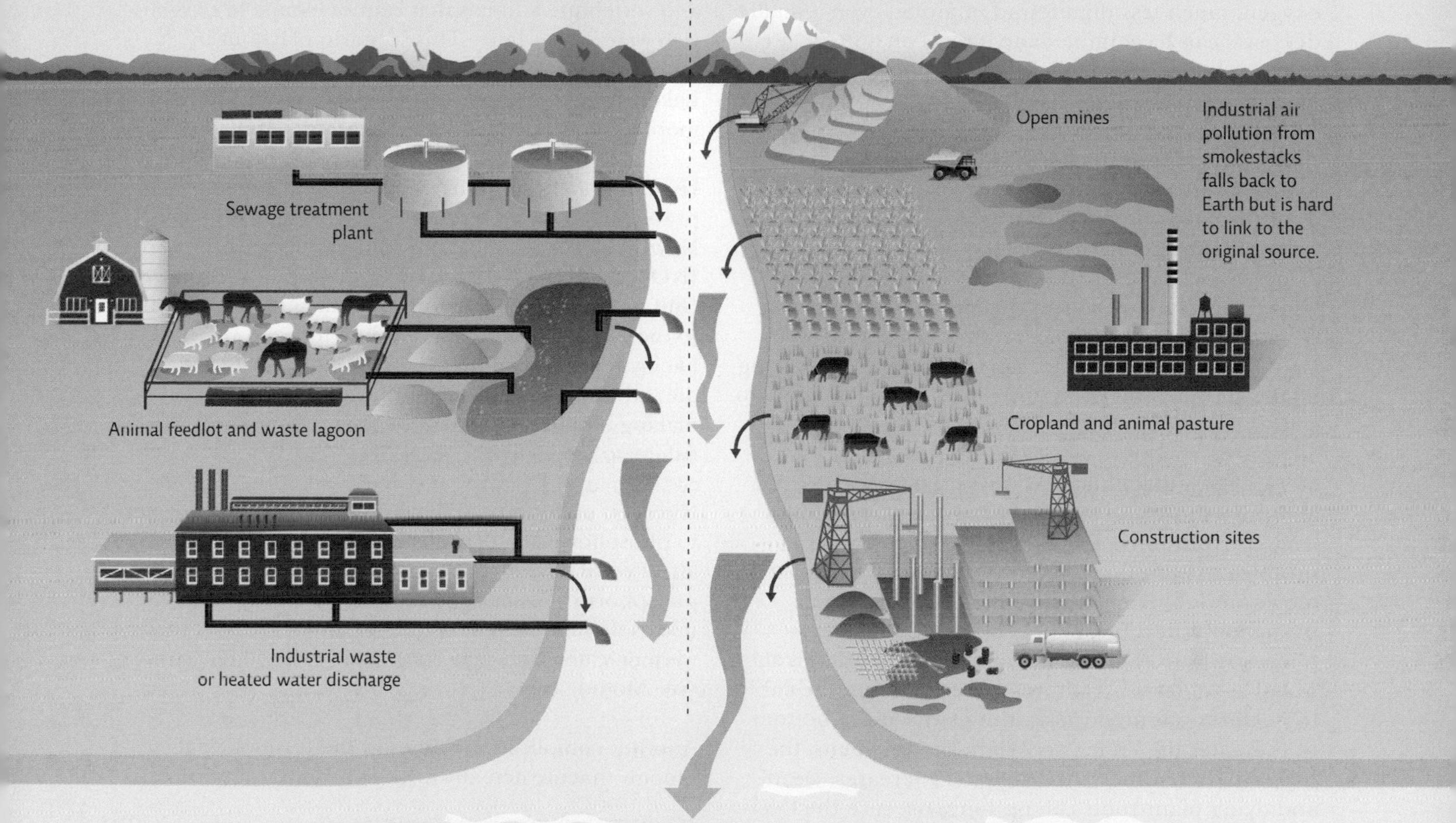

Various pollutants contribute to the many miles of U.S. surface waters that are "impaired"—waters that are too polluted to meet state or local water quality standards.

LEADING CAUSES OF IMPAIRED SURFACE WATERS (2016)

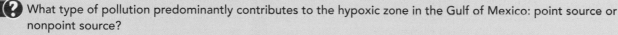

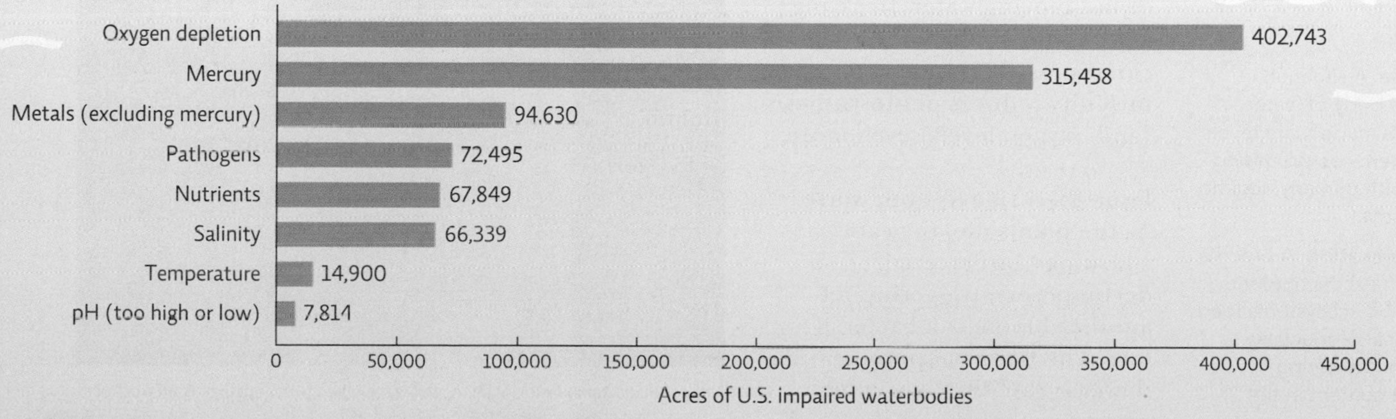

What type of pollution predominantly contributes to the hypoxic zone in the Gulf of Mexico: point source or nonpoint source?

2 EUTROPHICATION

Key Concept 2: The influx of excess nutrients into a body of water may spur algal growth and bacterial population explosions, which may ultimately result in hypoxia severe enough to harm aquatic life.

Something in the Gulf waters—some pollutant—was causing the levels of **dissolved oxygen (DO)** to plummet—a condition known as **hypoxia**. Turner and Rabalais knew that hypoxia was a serious problem. Water can hold only a limited amount of oxygen, much less than found in air. So even a small decrease can have immediate effects on aquatic life. Even underwater organisms need oxygen to survive. Like terrestrial beings, they use it in the process of cellular respiration. Waters so depleted of oxygen that aquatic life suffers are known as *dead zones*. It turns out an excess of two nutrients, nitrogen and phosphorus, were triggering a process known as **eutrophication.**

"Imagine stretching a giant sheet of plastic wrap from the Mississippi River's mouth, straight across to Galveston [Texas]," Turner says. "Now imagine sucking all the air out and leaving the whole ecosystem there to suffocate."

Here's how eutrophication works: Because nitrogen and phosphorus fuel plant growth, extra amounts trigger explosions of algal growth. Though algae produce oxygen through photosynthesis, they also block sunlight from reaching underwater plants in shallow waters, ultimately blocking much more photosynthesis than they conduct. Oxygen levels start to fall as underwater photosynthesis declines. Unable to perform photosynthesis, the plants at the bottom of the water die en masse. When that happens, the turbidity (cloudiness) of the water increases: Dead and dying plant roots can no longer secure the river or seabed in shallow areas, and bottom sediments can easily enter the water column if disturbed by waves or fast-moving water. Decaying matter also contributes to the cloudiness. This increased turbidity reduces photosynthesis (and oxygen levels) even more.

dissolved oxygen (DO) The amount of oxygen in the water.

hypoxia A situation in which a body of water contains inadequate levels of oxygen, compromising the health of many aquatic organisms.

eutrophication A process in which excess nutrients in aquatic ecosystems feed biological productivity, ultimately lowering the oxygen content in the water.

From there, it gets even worse. As the plants die, they are consumed by bacterial decomposers, triggering yet another bloom—a bacterial one. The bacterial populations increase rapidly, consuming oxygen as they digest the dead plants, quickly depleting any remaining oxygen in the water. This high rate of decomposition in the bottom layer is the major cause of hypoxia in deeper water, such as those further offshore in the Gulf of Mexico. Oxygen levels can drop low enough to kill invertebrate and vertebrate animals that cannot escape to oxygen-rich waters elsewhere. This scenario plays out in waterbodies everywhere—in small ponds, the Great Lakes, and huge expanses of coastal waters; there are more than 500 dead zones worldwide. **INFOGRAPHIC 2**

Hypoxic waters are more than an environmental problem—they are also an economic concern. The National Oceanic and Atmospheric Administration (NOAA) estimates that the oxygen-poor waters of the Gulf of Mexico cost millions of dollars in lost revenue to the seafood and tourism industries each year. Species like oysters that can't escape the oxygen-poor waters die; others move away from the coastal dead zones, making it harder (and more expensive) for commercial fishers to catch them.

Nitrogen pollution can also upset the ratio of nitrogen to phosphorous in ocean water, which can negatively affect coral, making it harder for them to take up phosphorus—essentially starving the coral. This makes them more susceptible to other stressors such as warmer waters and may contribute to coral bleaching (see Module 6.3).

Another complication comes in the form of algal blooms that are actually toxic. Nutrient pollution

Lynn Betts/NRSC/USDA

Unprotected farm fields lose topsoil as well as farm fertilizers and other potential pollutants when heavy rains occur.

INFOGRAPHIC 2 **EUTROPHICATION CAN CREATE DEAD ZONES**

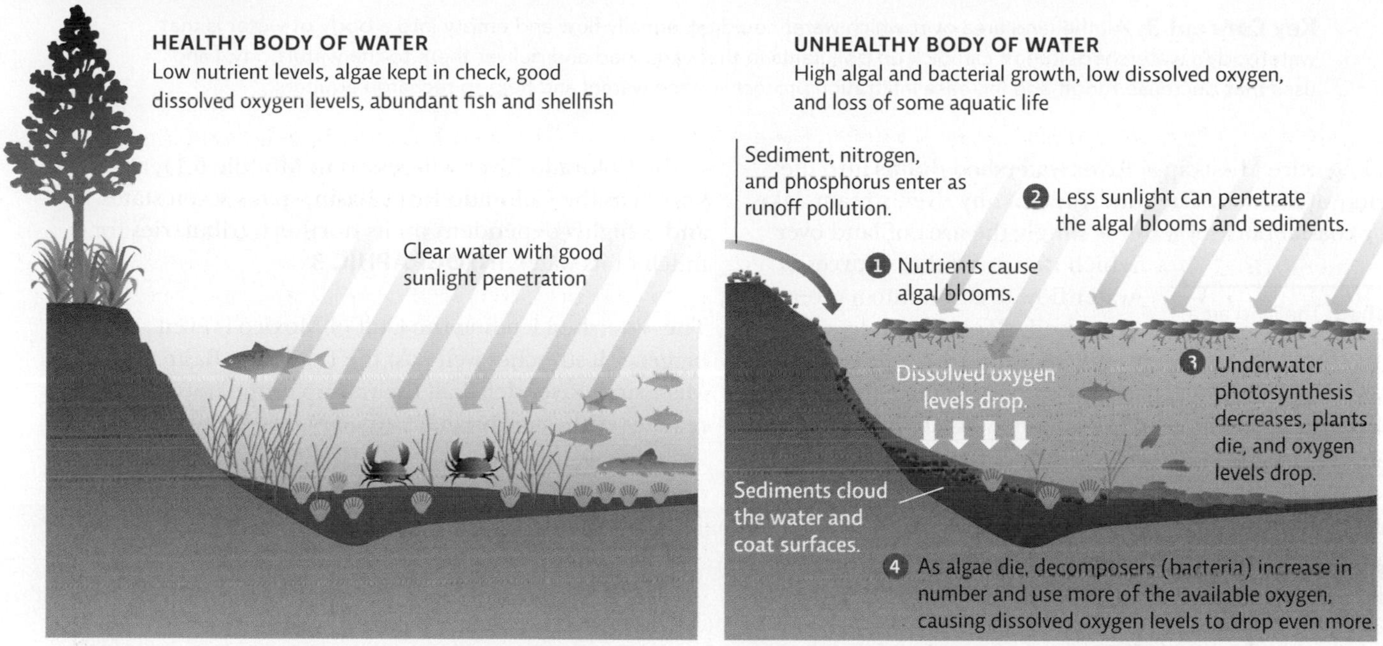
HEALTHY BODY OF WATER
Low nutrient levels, algae kept in check, good dissolved oxygen levels, abundant fish and shellfish
Clear water with good sunlight penetration
UNHEALTHY BODY OF WATER
High algal and bacterial growth, low dissolved oxygen, and loss of some aquatic life
Sediment, nitrogen, and phosphorus enter as runoff pollution.
2 Less sunlight can penetrate the algal blooms and sediments.
1 Nutrients cause algal blooms.
Dissolved oxygen levels drop.
3 Underwater photosynthesis decreases, plants die, and oxygen levels drop.
Sediments cloud the water and coat surfaces.
4 As algae die, decomposers (bacteria) increase in number and use more of the available oxygen, causing dissolved oxygen levels to drop even more.

Healthy Gulf water is relatively clear with abundant sea life. Oysters and forage fish filter out particles, keeping water clean, while submerged vegetation produces oxygen to support healthy fish and shellfish populations.

Algal blooms, sediments that block sunlight and prevent photosynthesis, or an influx of bacteria that deplete the water's oxygen can each produce hypoxic regions.

Why is it that in eutrophic waters, hypoxia can develop from both a decline in oxygen production and an increase in oxygen consumption?

may trigger the growth of photosynthetic bacteria (cyanobacteria) that produce deadly toxins—these bacteria can grow in fresh or saltwater. The U.S. Environmental Protection Agency (EPA) recommends (but does not require) that public water departments test drinking water for these toxins if their presence is suspected and take steps to purify the water or issue a "Do Not Drink" advisory if levels exceed recommended amounts.

Even swimming in or coming into contact with water contaminated with cyanobacterial toxins can sicken individuals; children and the elderly are most at risk for serious health problems. In the summer of 2019, the waters along the entire coastline of Mississippi were closed to swimming due to a cyanobacterial outbreak fueled by extremely heavy spring rains in the Midwest that washed fertilizer and sewage into the Gulf of Mexico. Though most people will avoid exposure simply because they are unwilling to swim or play in water covered with green scum, our pets may not be so picky. A Labrador will readily chase a ball thrown into the water. In the process of retrieving that ball several times or licking itself clean, the dog may consume enough toxin to be killed. Fish, too, succumb. The sight of dead fish washing ashore after a eutrophic event is a clear sign of cyanobacterial toxins.

But where was all that extra nitrogen and phosphorus coming from in the first place?

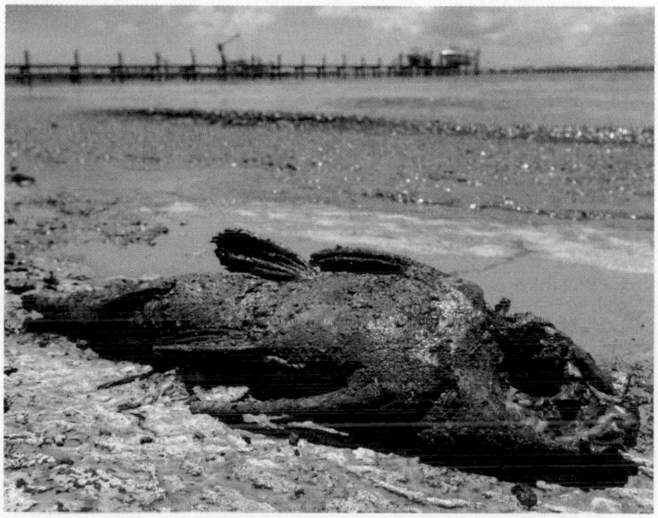

CRISTOBAL HERRERA/European Pressphoto Agency/Newscom

Cyanobacterial blooms, like this 2016 bloom in Stuart, Florida, plague much of Florida in the summer months. Caused by fertilizer and sewage runoff pollution, these blooms kill fish, close beaches, and may even be linked to the deaths of several manatees.

THE WATERSHED CONCEPT

Key Concept 3: All the land area over which water could potentially flow and empty into a body of water is that waterbody's watershed. Runoff can pick up pollutants in the watershed and deliver them to the waterbody. Land uses that decrease runoff and increase infiltration protect surface waters and help to recharge aquifers.

The entire Mississippi River **watershed** drains into the Gulf of Mexico. The watershed of any river, stream, lake, or coastal body of water is simply the area of land over which rain and other sources of water flow to drain into a given waterbody. It also includes all the smaller streams (*tributaries*) that empty into the waterbody and their watersheds. The watershed of the Colorado River (discussed in Module 6.1), also known as the Colorado River Basin, spans seven states and is highly dependent on its northern tributaries for much of its water. **INFOGRAPHIC 3**

watershed The land area surrounding a body of water over which water such as rain can flow and potentially enter that body of water.

The watershed feeding the Gulf of Mexico is even bigger—it stretches well past the Louisiana Basin and the Mississippi River delta, all the way up to the northernmost reaches of the continental United States. Indeed, to

INFOGRAPHIC 3 WATERSHEDS

Anything that happens in the watershed can potentially affect the quality of a body of water as well as the quantity of groundwater. This is especially important in terms of nonpoint source pollution that originates on land. Mapping the watershed is an important tool in watershed management. Here, the watershed of this river and coastal area is outlined as a black dashed line.

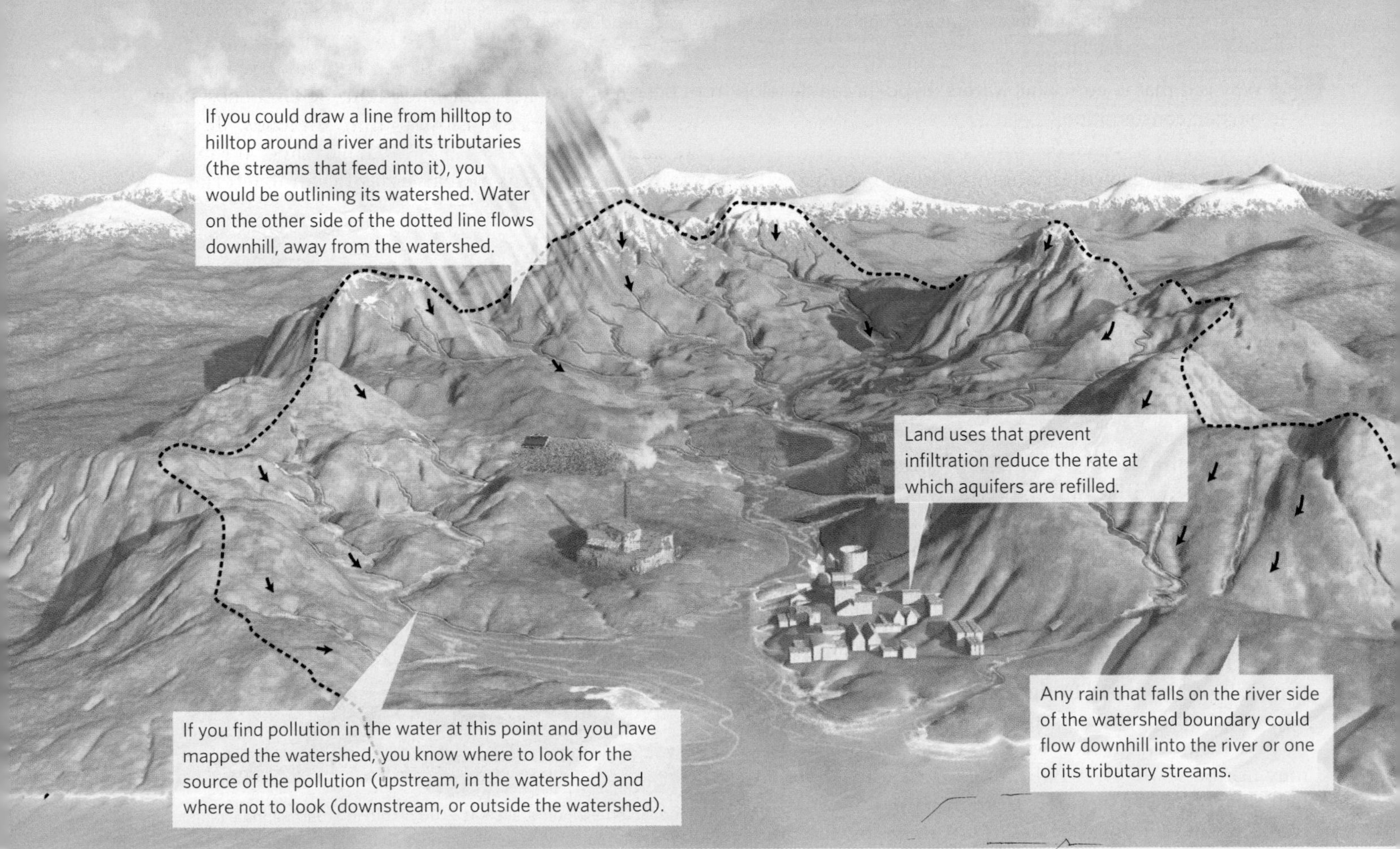
If you could draw a line from hilltop to hilltop around a river and its tributaries (the streams that feed into it), you would be outlining its watershed. Water on the other side of the dotted line flows downhill, away from the watershed.
Land uses that prevent infiltration reduce the rate at which aquifers are refilled.
If you find pollution in the water at this point and you have mapped the watershed, you know where to look for the source of the pollution (upstream, in the watershed) and where not to look (downstream, or outside the watershed).
Any rain that falls on the river side of the watershed boundary could flow downhill into the river or one of its tributary streams.

Suppose you discover that the stream by your home appears to be polluted (i.e., unusual smell, many dead fish, noticeable pollution, etc.). Why might it be helpful to have mapped the watershed of this stream?

glimpse the causes of—and possible solutions to—the hypoxia uncovered by Turner and Rabalais, we must travel to Minnesota, where the Mississippi River begins.

It's tough to imagine that the homesteads of the far north—a place of corn fields and dairy farms—have anything to do with the rollicking backwater swamps of Louisiana. But it is here, along the tributaries that feed the nation's farmlands, that our story begins. "The nitrogen that's killing the Gulf starts here, gets added to the ecosystem here, in these farms and fields," says Alex Echols, former head of the U.S. Fish and Wildlife Service and a consultant to the Minnesota-based Sand County Foundation (an environmental group). "This is where it's all coming from."

The flow of that nitrogen (and other fertilizer components) over a watershed is well mapped and shows that runoff pollution can directly affect surface water *quality* by delivering pollutants. But that's not all. This runoff also impacts groundwater. When it rains, some rainfall soaks into the soil, infiltrating the ground below; some eventually reaches groundwater in the aquifer, which people can tap into to create a well. If the groundwater is deep enough, infiltration can act as a filtering system that purifies the water. If, however, the water table (upper level of the groundwater) is close to the surface, pollutants can make it to the groundwater; for example, nitrate pollution from fertilizer runoff can contaminate well water enough to be life-threatening, especially to young children. (See Module 6.1 for more on aquifers.)

But groundwater *quantity* can also be affected. Human land uses have altered the way water drains through watersheds and waterways. Pavement and even suburban lawns enhance runoff by preventing water from soaking into the ground, reducing the rate at which aquifers are refilled; they can be rapidly depleted if we remove well water faster than it is replaced. Meandering streams have been channelized and their water diverted for other uses. Wetlands, too, have been completely drained to allow urban or suburban development or for agriculture. These activities reduce the infiltration of rainwater into the ground by speeding the flow of water across and out of a watershed, sending with the water whatever industrial, agricultural, or municipal pollutants it encounters on its trip downhill. In the Mississippi River watershed, this includes runoff pollution from roadways, urban areas, and industries throughout the region, as well as animal waste and excess fertilizer from hundreds of thousands of acres of farmland. In all, some 1.5 million tons of nitrogen and phosphorus now flow through the Mississippi River watershed and into the Gulf of Mexico every year.

In 2008, the newly formed Gulf of Mexico Watershed Nutrient Task Force set a goal: to reduce the hypoxic zone down to 1,930 square miles. That same year, Rabalais and Turner concluded that the Gulf ecosystem was becoming more sensitive to nutrient loads. "It's not the same system as the 1960s," Rabalais says. "It's taking less nutrients now to fuel the hypoxia."

Oscar Garcia Bayerri/AGE Fotostock

All the land uphill of a body of water is part of its watershed and can potentially send stormwater runoff into the water. Evaluating the land uses in the watershed allows one to better anticipate and address potential sources of water pollution.

4 ASSESSING WATER QUALITY

Key Concept 4: Chemical and physical characteristics of water can be monitored to assess water quality. A biological assessment (determining the type and number of species living in the water) is also used to evaluate water quality.

Scientists continually assess the impact of nutrient runoff on the Mississippi River watershed's various ecosystems. Chemical parameters of water quality such as pH and dissolved oxygen (DO) content are often monitored to see if these values are within acceptable limits for the organisms who live there (e.g., a pH around neutral, high DO). If a particular pollutant is suspected, such as nitrates from fertilizer pollution, that too can be checked. Physical parameters that are assessed include the water's turbidity (cloudiness) and temperature (higher temperatures mean lower DO levels).

biological assessment The process of sampling an area to see what lives there as a tool to determine how healthy the area is.

benthic macroinvertebrates Easy-to-see (not microscopic) organisms such as insects that live on the stream bottom.

But not everyone has the equipment or supplies to measure these chemical and physical parameters; for those who do, chemical and physical measurements that are within acceptable limits can sometimes be deceiving. Water might look crystal clear because it supports little life; an undetected chemical might have killed everything in the water. Another way to assess water quality is a process known as **biological assessment**. This can be as simple as netting, identifying, and counting **benthic macroinvertebrates**, such as insects and crayfish that inhabit the bottoms of streams, such as the ones that empty into the Mississippi River. If a stream is unhealthy, there won't be many organisms present that are sensitive to pollutants. The abundance and diversity of pollution-tolerant and pollution-sensitive species in the sample can be used to "rate" the stream quality. Poor stream quality can sometimes indicate that nutrient runoff or toxic substances are polluting the water or that sediments are smothering needed stream-bottom habitat. Similar protocols are in place to assess other populations that can serve as indicators of water quality, such as fish or algae. **INFOGRAPHIC 4**

INFOGRAPHIC 4 BIOLOGICAL ASSESSMENT

A simple way to investigate the water quality of a stream is a biological assessment—using a net to collect a sample of aquatic organisms to see what is living in the water. If there is good diversity and abundant aquatic life, especially of those organisms that are sensitive to pollution, then it is reasonable to conclude the water is clean. A poor sample suggests there is a problem, prompting one to look more closely at water quality to determine the issue.

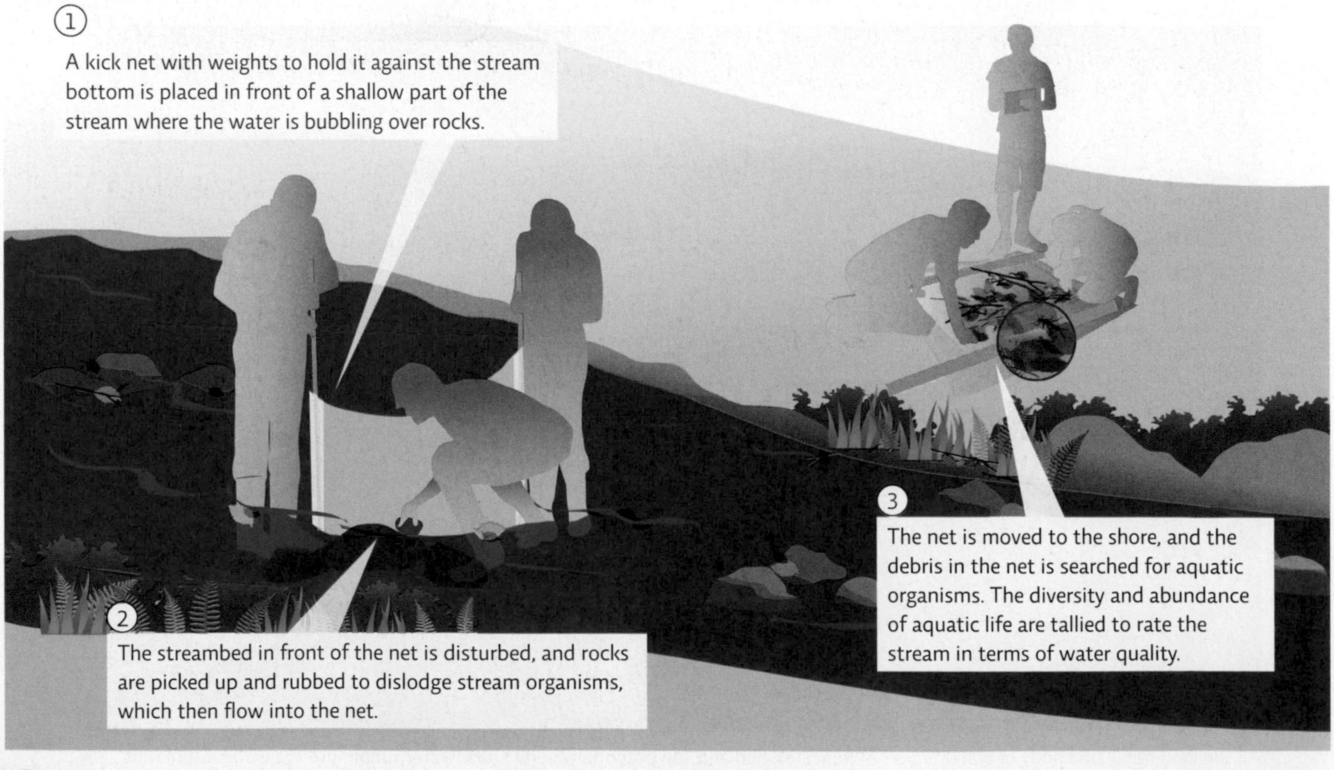

①
A kick net with weights to hold it against the stream bottom is placed in front of a shallow part of the stream where the water is bubbling over rocks.
②
The streambed in front of the net is disturbed, and rocks are picked up and rubbed to dislodge stream organisms, which then flow into the net.
③
The net is moved to the shore, and the debris in the net is searched for aquatic organisms. The diversity and abundance of aquatic life are tallied to rate the stream in terms of water quality.

? If you were conducting a biological assessment and were hoping for a healthy stream, which would you rather have in your net: pollution-tolerant organisms, pollution-intolerant organisms, or both? Explain.

5 LEGAL PROTECTION: THE CLEAN WATER ACT

Key Concept 5: The Clean Water Act sets pollution standards to limit the release of pollution from point sources and has been effective at reducing this form of pollution. It has been less effective at controlling nonpoint source water pollution.

Water quality has always been a concern but perhaps never more so than in the late 1960s. At the time, there were no restrictions on the release of industrial chemical pollutants into bodies of water. From the sewage-filled Potomac River flowing through Washington, D.C.—declared a "national disgrace" by President Johnson in 1965—to Lake Erie's massive algal blooms that made it one of the nation's early dead zones, U.S. environmental waters were in trouble. In 1969, the Cuyahoga River in Cleveland, Ohio, made headlines when it caught fire (and not for the first time) because so much oil and other flammable industrial pollutants floated on the surface of the water. This dramatic event helped spur the passage of the 1972 **Clean Water Act (CWA)**, which regulates industrial and municipal (such as sewage treatment plants) point source pollution, with the goal of making all environmental waters "fishable and swimmable." As with all environmental laws, the EPA administers the CWA by writing guidelines (known as "rules") that comply with the scope and intent of the law. The EPA accomplishes the fishable and swimmable goal by setting **performance standards**—allowable levels of a pollutant that can be present in environmental waters or released over a certain time period (see Module 5.2).

Clean Water Act (CWA) U.S. federal legislation that regulates the release of point source pollution into surface waters and sets water quality standards for those waters. It also supports best management practices to reduce nonpoint source pollution.

performance standards The levels of pollutants allowed to be present in the environment or released over a certain time period.

Firefighters stand on a bridge over the Cuyahoga River in Cleveland, Ohio, to spray water on a tugboat as a fire—started in an oil slick on the river—moves toward the docks at the Great Lakes Towing Company site. This 1952 blaze, one of 13 fires on the river since the late 1800s, was the costliest, destroying three tugboats, three buildings, and the ship repair yards.

Bettmann/Getty Images

For example, the CWA protects the public against pathogens in *recreational* waters (rivers, lakes, and coastal areas) — a leading cause of impaired waters in the United States and a major problem worldwide. Globally, exposure to pathogen-polluted water is one of the leading causes of infection; close to 1 million people die each year of diarrheal diseases linked to unsafe water. The United Nations cites the provision of clean water and sanitation as one of its 2030 Sustainable Development Goals (see Infographic 4 of Module 6.1). In the United States, the EPA allows only very small amounts of fecal bacteria such as *E. coli* in recreational waters; the presence of fecal bacteria indicates sewage or animal waste contamination. (The Safe Drinking Water Act mandates that absolutely no pathogens are allowed in drinking water.) Violations are investigated and addressed at either the federal or state level. Today, about 47% of U.S. streams and rivers meet the fishable/swimmable goal; other bodies of water don't fare as well — only 28% of assessed coastal waters meet this criterion, and less than 3% of Great Lakes waters do so. Though still far from 100% compliance with CWA guidelines, this is higher than the number that met that goal prior to the passage of the CWA(~40%).

The CWA, like many environmental protection laws, has a *citizen suit provision* that allows citizens or groups to sue a violator of the act, or the state or federal government if they feel the statutes of the CWA are not being upheld. This empowers citizens to work to ensure the protection of the surface waters in their own communities. (See Module 5.2 for more on the citizen suit provision of environmental laws.)
INFOGRAPHIC 5

In the original rules approved by the EPA for CWA enforcement, only navigable waters and those of immediate municipal or recreational value were slated for protection. But that left many upstream areas unprotected, making them potential sources of pollution for the downstream waters the CWA was mandated to protect. The more recent EPA Clean Water Rule (2015), developed over several years with bipartisan support under the Obama administration, added to that any upstream waters that have been shown to impact the health of downstream waters. However, before it could be implemented, newly elected President Trump took steps to rescind the rule. Early efforts to scrap the rule failed legal challenges, so the administration has revised the rule to reduce its reach.

Another Obama-era rule recently eliminated is the Stream Water Protection Rule, a regulation aimed at preventing coal-mining operations from dumping mining waste in ways that would result in the pollution of nearby streams (especially relevant for mountaintop removal mining — see Module 9.1).

INFOGRAPHIC 5 THE CLEAN WATER ACT

The Clean Water Act is the landmark environmental federal law that established the modern framework for the regulation of water pollution in U.S. surface waters. Many other nations have copied this framework in drafting their own water protection legislation.

GOAL: Make all of our environmental waters safe for fishing and swimming.

IMPLEMENTATION
- Pollution standards
- Permits issued to limit polluted industrial discharge
- Best management practices recommended for nonpoint source pollution

ENFORCEMENT
- Penalties: fines, revoked permits, incarceration
- Citizen suit provision

EFFECTIVENESS
- Good control of point source pollution
- Less effective for nonpoint source pollution

The idea of pollution standards implies that there is some level of a pollutant that is acceptable to have in our waters. Do you agree with this or should we strive for "zero pollution"? Explain your answer.

6 WATERSHED MANAGEMENT

Key Concept 6: Good watershed management can reduce nonpoint source pollution. Limits can be set for the maximum amount of a pollutant that can enter the water, and steps can be taken to meet those targets. For example, well-vegetated riparian areas reduce runoff and act as nutrient sinks.

"We did a good job on point source [pollution]," Echols says. "Back in the late 1960s, we said 'thou shall not pollute,' and we made rules and we made people follow them."Nonpoint source pollution is a bit trickier. The CWA does not specify, for example, how much nitrogen fertilizer a farmer can apply. And this is not feasible because recommended amounts vary from farm to farm, and runoff potential varies according to rainfall, terrain, and even the crop that is planted. But scientists, farmers, and homeowners are working to fix it, just the same. Although they differ in their preferred approaches, most scientists agree that **watershed management**—management of what goes on in an area around streams and rivers—will be the key to saving the Gulf. They've identified *best management practices* (agreed-upon actions that minimize pollution problems caused by human actions) that reduce the amount of pollution being delivered to the Mississippi River. In agriculture, many of these practices focus on ways to decrease the amount of chemicals applied to land areas in the first place and to reduce the potential for soil erosion and runoff.

Recent CWA amendments advocate the use of these best management practices and provide funding for their implementation. However, watershed management is not easy to do; after all, many individuals and groups are engaged in a wide variety of different land uses within any watershed. Identifying how each should address its contribution to nonpoint source water pollution—much less enforcing compliance—is no easy task.

One approach being used is to set **total maximum daily loads (TMDLs)** for a specific pollutant or pollutant group (e.g., nutrient pollution)—this is the maximum amount of the pollutant that can enter the water on a daily basis. The state agencies responsible for meeting the goals of the CWA often set TMDLs. Targets can be adjusted downward to improve water quality over time.

For point source pollution, sources of a pollutant (e.g., industries, sewage treatment plants) are allotted a portion of that TMDL via a permit system. But administering a TMDL system for nonpoint source pollution is harder. For example, it is not feasible to assign each farmer or suburban homeowner a portion of the nitrogen/phosphorus TMDL for their nearby body of water.

One approach used to address eutrophication in the Chesapeake Bay has been to establish TMDLs for the Bay and then to allocate a portion of that to different regions within the watershed (which spans six states and the District of Columbia). Actions to help the region meet reduction targets include banning the use of fertilizer in some areas and applying more stringent standards of waste handling to large animal operations. In addition, farmers and urban planners are assisted (financially and with technical guidance) in implementing best management practices that reduce runoff. In its midpoint assessment, seven years after the Chesapeake Bay TMDL plan was implemented, the EPA reported that the area was on track to meet targets for phosphorus TMDL; it is still short of its targets for nitrogen pollution in some areas, but progress has been made.

One of the key steps in preventing runoff pollution from reaching a body of water is to restore natural areas within the watershed's **riparian areas**, the land areas close to the water. This is done by maintaining or planting vegetated buffer zones that slow runoff and give the rainwater time to soak into the ground before it reaches a nearby body of water. "Most of our riparian areas—in the Mississippi River watershed, anyway—are not functioning as riparian areas anymore," Echols says. "We can't really fix hypoxia until we fix that."

Healthy (well-vegetated) riparian areas are widely recognized as critical to maintaining good water quality, and projects are underway across the United States to restore and revegetate the riparian areas and watersheds of rivers and streams. In the late 1990s, New York City invested billions of dollars to restore and protect areas in the Catskills and the Delaware watershed supplying its water; thanks to this investment, the city avoided the need to construct high-tech and costly filtration systems. Wetlands, too, can be targeted for restoration, such as the multibillion-dollar restoration project underway to restore some of the natural water flow in the Florida Everglades, a major water source for South Florida (see Module 2.3). **INFOGRAPHIC 6**

watershed management Management of what goes on in an area around streams and rivers.

total maximum daily loads (TMDLs) The maximum amount of a pollutant allowed to enter a waterbody so that the waterbody will meet water quality standards.

riparian areas The land areas close enough to a body of water to be affected by the water's presence (e.g., areas where water-tolerant plants grow) and that affect the water itself (e.g., provide shade).

INFOGRAPHIC 6 HEALTHY RIPARIAN AREAS PROVIDE MANY BENEFITS

The area next to a body of water that impacts that water (provides shade and nutrients) and is itself impacted (water-tolerant species live here) is the riparian area. A well-vegetated riparian area reduces the runoff that reaches a body of water by slowing the water's movement across the land so that it soaks into the ground rather than flows into the stream.

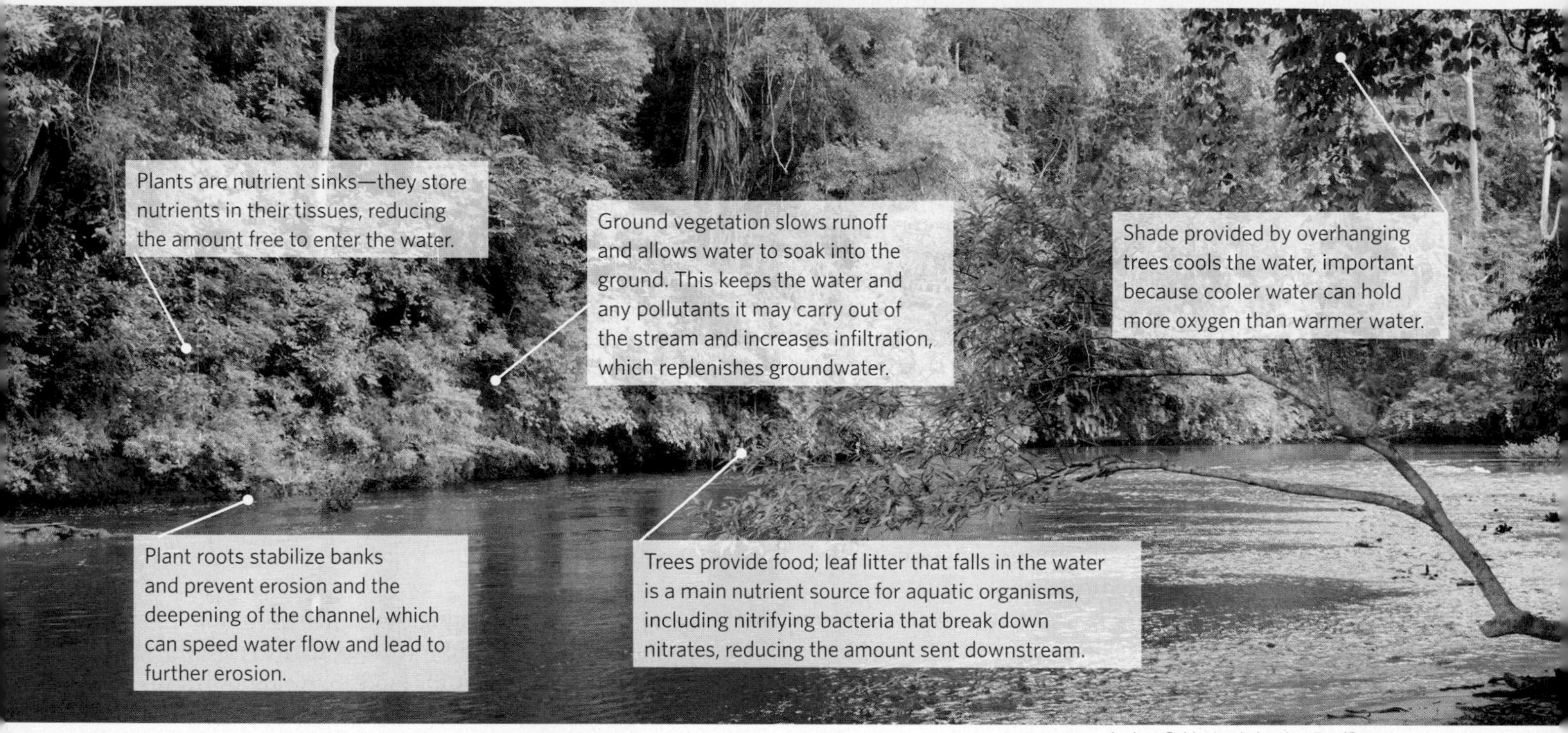

Plants are nutrient sinks—they store nutrients in their tissues, reducing the amount free to enter the water.
Ground vegetation slows runoff and allows water to soak into the ground. This keeps the water and any pollutants it may carry out of the stream and increases infiltration, which replenishes groundwater.
Shade provided by overhanging trees cools the water, important because cooler water can hold more oxygen than warmer water.
Plant roots stabilize banks and prevent erosion and the deepening of the channel, which can speed water flow and lead to further erosion.
Trees provide food; leaf litter that falls in the water is a main nutrient source for aquatic organisms, including nitrifying bacteria that break down nitrates, reducing the amount sent downstream.

Jochen Schlenker/robertharding/Getty Images

Shown here are the U.S. Department of Agriculture's recommendations for land use in the riparian area. This includes setting aside at least 75 feet of land in managed and undisturbed forest. More may be required for suitable protection in areas with steeper terrain—that is, where runoff flow would be faster.

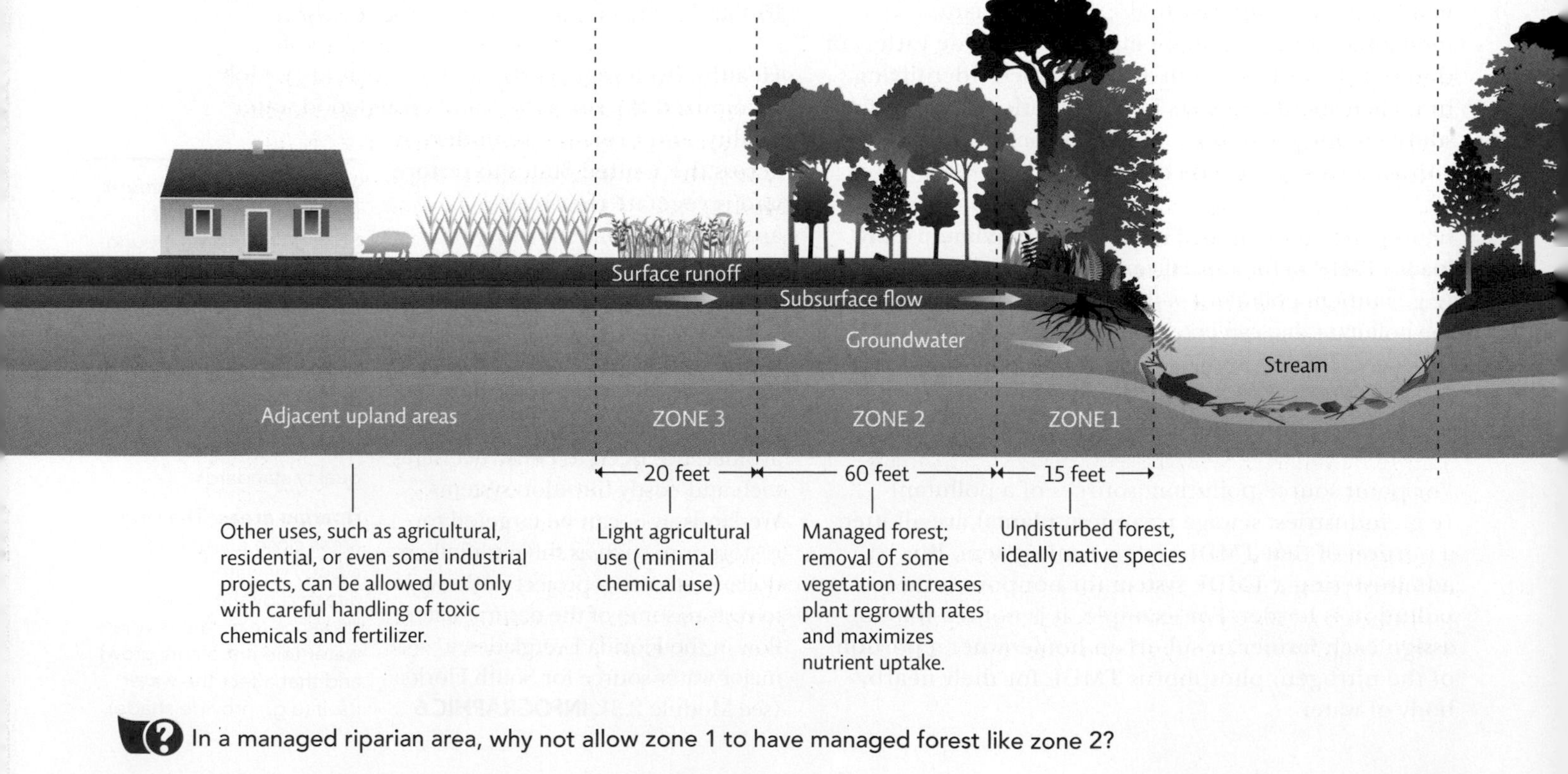

In a managed riparian area, why not allow zone 1 to have managed forest like zone 2?

7 ADDRESSING NONPOINT SOURCE POLLUTION

Key Concept 7: Decreasing nonpoint source pollution can be accomplished by minimizing the use of chemicals on lawns and farms and by redirecting or capturing stormwater to prevent runoff from reaching bodies of water.

Protecting or restoring riparian areas can decrease the amount of runoff from farm fields that reaches nearby waters, but farmers can also take steps to reduce the runoff that leaves their fields in the first place. It is common for farmers to install systems to drain the subsurface water out of wet soils — this allows them to plant earlier in the spring, and it protects crops from heavy rains that might flood the field. "But at the same time you've just established a very efficient mechanism for moving water off the field, and keep in mind, nitrogen is water soluble. Water is its ticket down to the party in Louisiana," Echols says. Those drainage pipes (commonly called *tile lines* because they were originally made of short, perforated clay pipes known as *tiles*) end up being its passageway.

Echols and others say that tile lines may also be *our* passageway — to addressing the problem of nutrient runoff. "It's going to take huge, tectonic shifts — in federal policies, and in the global food economy, and even in the culture — to move farmers off corn and soybean [crops that tend to "leak" nitrogen back into the soil], to move them off synthetic fertilizers, to move them off fossil fuels," Echols says. "But managing the tile lines? We can do that. We can do that farm by farm."

Jeffrey Strock, a soil scientist at the University of Minnesota, is investigating ways to do just that. In 2003, he teamed up with Minnesota farmer Brian Hicks to study new methods to reduce runoff and fertilizer contamination of nearby streams and rivers. Hicks's land was perfect: a snowbound expanse of 1,500 acres that straddles the Cottonwood River, which feeds the Minnesota River, which in turn feeds the Mississippi River. And because Hicks and his father used to run cattle and had yet to convert some of their pastureland to corn and soybean, there was still virgin prairie left. That meant Strock and his team could use untouched land to install and compare two different drainage treatments.

On the east side of their experimental plot, Strock and Hicks installed a conventional drainage system with buried pipes that drained away excess water from the soil at a steady rate. On the west side, they installed a controlled drainage system that could alter the depth of the water (via a small wooden box that would serve as a dam) to meet the needs of the crop. The hope was that the controlled drainage site would store more water — and nutrients — in the top layer of soil, which would then be available during crop production.

They discovered that during normal rainfall years, there was a 50% reduction in nutrient loss and a slight increase in yield for the controlled drainage zone compared with the conventional drainage zone. "We don't always get the agronomic benefit [of higher crop yields]," Strock says. "But we have consistently reduced the amount of nutrients running off the field. So, that environmental benefit is definitely there."

Other agricultural best management practices will also reduce the amount of nitrogen and phosphorus available to runoff. One recommendation is to frequently test the soil to ensure that only the correct type and amount of fertilizer are added as needed and to avoid adding fertilizer right before a rain (to prevent it from washing away). Tilling fertilizer directly into the soil (rather than spreading it on the surface) also reduces the potential for loss due to runoff or wind. In addition, a variety of erosion prevention methods are available to farmers, such as planting winter crops in the off-season to hold soil in place or planting trees as windbreaks to reduce soil erosion from wind. (For more on sustainable farming techniques, see Module 8.2.) Farmers are also going high-tech, pursuing *precision agriculture* by using GPS technology to guide farm equipment and reduce overlap when working a field, thus reducing disturbance of soil that might lead to erosion, and to guide site-specific applications of chemicals within a single field to minimize the amount used.

Of course, not all of hypoxia's solutions will come from the farm belt. In suburban areas, lawns can be major nonpoint sources of nitrogen pollution if homeowners apply too much fertilizer. However, grass also sequesters nitrogen because of its long growing season, preventing it from flowing to the nearest stream. Echols encourages homeowners to limit fertilizer use on lawns and to plant native plants and grasses that do not need fertilizer. Planting a *rain garden* of water-tolerant plants in low-lying areas that tend to flood in rain events can also help capture water and reduce runoff.

In urban and suburban areas, replacing some hard surfaces with porous surfaces such as green space or permeable pavers reduces runoff by allowing water to seep through. For this reason, Chicago has changed a large part of its alley pavement to porous concrete. Green roofs can also help capture rainwater and slow or prevent its release to the environment. Installing curb cutouts in roadways can direct stormwater flow onto natural areas where the water has a chance to soak into the ground rather than flowing directly into a storm drain, which often takes it to a nearby waterbody. **INFOGRAPHIC 7**

Peter Wynn Thompson/The New York Times/Redux

"Green alleys" like this one in Chicago are those that have been resurfaced with permeable pavement to allow rainwater or snowmelt to pass through the pavement and soak into the ground rather than run off into storm drains that would eventually empty into nearby Lake Michigan.

INFOGRAPHIC 7 INCREASING INFILTRATION OF STORMWATER

Stormwater that doesn't soak into the ground can enter storm drains that flow directly to rivers and streams or can cause floods, especially in heavily built-up urban settings. Anything that increases infiltration can help avoid these stormwater problems.

Trees slow and allow infiltration of runoff.
A green roof has vegetation over a waterproof layer, which can trap some water and reduce roof runoff.
Rain barrel captures runoff from roof.
Redirected downspout
Rain gardens capture gutter and lawn runoff.
Curb cutouts reduce street runoff by diverting it to the ground, where it can infiltrate.
Storm drain
All these efforts help reduce the amount of polluted stormwater entering storm drains that lead directly to local rivers, streams, and lakes.

What are some other building or infrastructure methods that could reduce the flow of stormwater? (Consider things like roof size, road width, other pavement options, etc.)

8 HOLISTIC STRATEGIES TO PROTECT AND RESTORE AQUATIC HABITATS

Key Concept 8: Addressing water pollution includes identifying where problems exist, reducing pollution at the source, employing watershed management to reduce runoff potential, and restoring wetlands.

Water pollution is a wicked problem with many causes and consequences, as well as multiple stakeholders who often have different opinions about what actions to take. But the success of the CWA in controlling point source water pollution in the United States shows that progress can be made. Likewise, steps to address watershed health and reduce the potential for runoff are reducing nonpoint source pollution.

To address hypoxia in the Gulf of Mexico, the Gulf Task Force has identified a multipronged, holistic approach that considers all the causes and consequences of the problem in a plan that addresses the *triple bottom line* (see Module 1.1). The *environmental* needs of the coastal and offshore ecosystems are addressed by pursuing best management practices that reduce the occurrence or impact of water pollution. For example, steps that restore and conserve riverside or coastal habitats in the Gulf watershed will reduce the amount of pollution that reaches the Gulf and protect the species that live there. This, in turn, will protect the ecosystem services these species provide, such as water purification. Taking steps to reduce pollution at the source (e.g., reducing industrial point source and agricultural nonpoint source water pollution) will make these restoration efforts even more effective.

These efforts and others will also address the *social* impacts on human communities that are affected throughout the Gulf region as important fisheries are replenished and protected. Coastal communities also benefit from improved protection against storm surges or coastal erosion.

Finally, the plan accounts for the *economic* realities that require us to prioritize our actions, first pursuing either those actions that are less expensive and are easy to implement or those that are more costly with high payoffs to the environment or community. **INFOGRAPHIC 8**

INFOGRAPHIC 8 GULF OF MEXICO REGIONAL ECOSYSTEM RESTORATION

GOAL	STRATEGY
Restore and conserve habitat.	• Restore natural flow of rivers and streams to wetland and delta regions. • Restore riparian areas and wetland habitats. • Implement measures to restore seagrass beds and protect seagrass from boat propeller damage.
Replenish and protect marine resources.	• Restore and manage coral reefs and oyster beds; sustainably harvest fish and shellfish. • Track indicator species, such as sea oats, pelicans, bluefin tuna, and oysters, to monitor progress or problems. • Minimize or eliminate invasive species, such as lionfish and nutria.
Restore water quality.	• Implement better regulation of industrial point source pollution. • Reduce runoff from animal operations by fencing off streams and construct wetlands to capture runoff. • Implement precision fertilizer application to reduce overuse.
Enhance community resilience.	• Improve coastal protections against storm damage. • Promote low-impact community growth plans. • Enhance education and outreach.

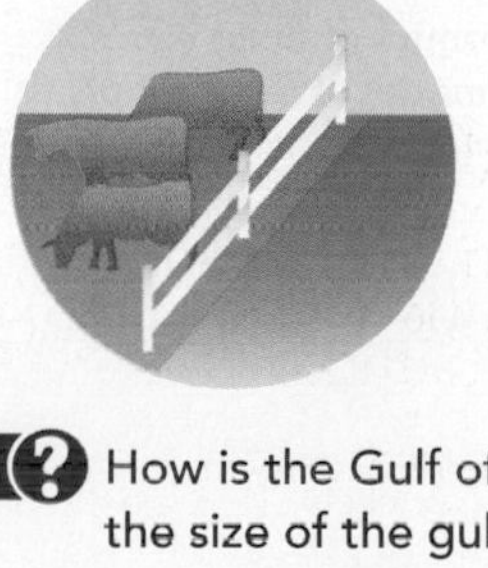

How is the Gulf of Mexico Hypoxia Task Force considering the triple bottom line with its recommendations to reduce the size of the gulf hypoxic zone?

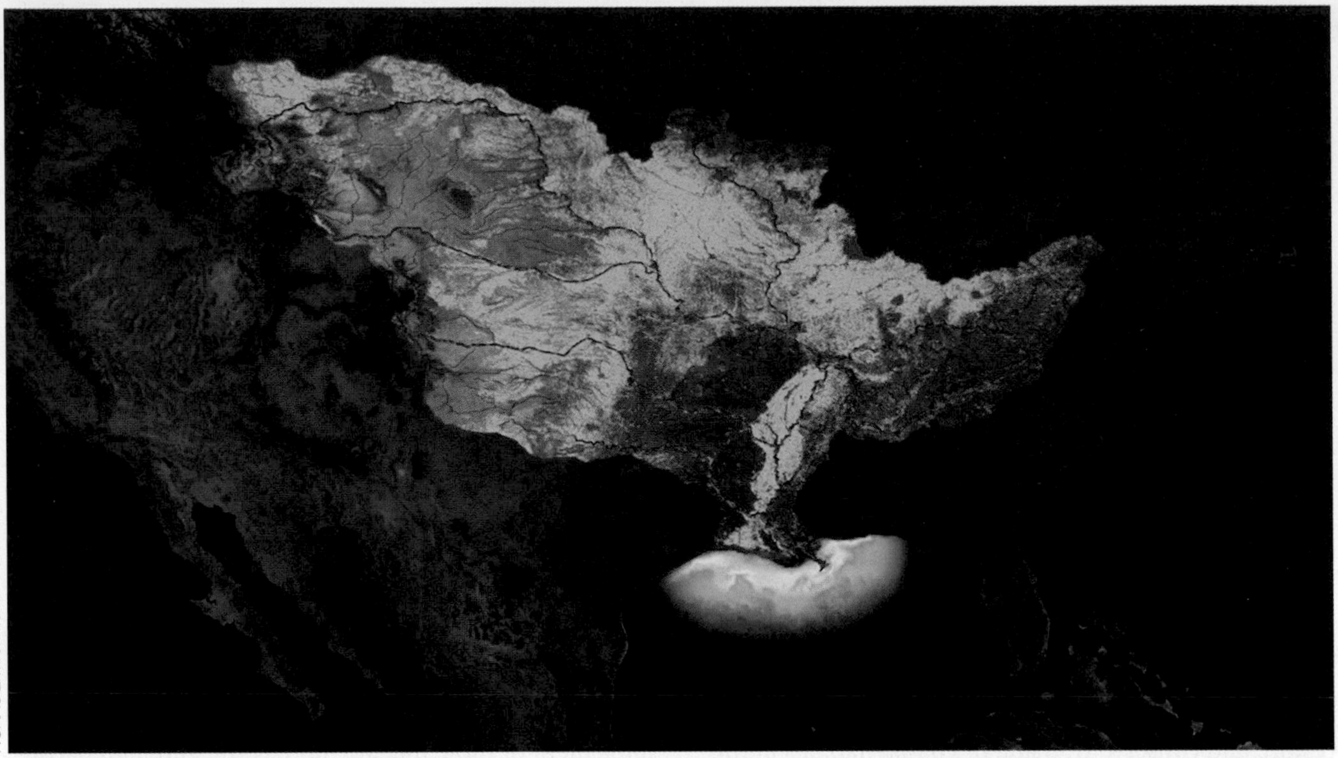

Cities (shown in pink) and farmland (shown in green) within the watershed of the Mississippi River fuel the Gulf of Mexico's hypoxic zone in late summer. The severity of the hypoxic zone in the Gulf of Mexico is indicated by color ranging from "lowest oxygen levels" (red) to "highest oxygen levels" (blue).

These types of control efforts are gaining increasing significance in a world where the human population is still growing and the need for higher agricultural productivity grows with it. Meeting these needs and protecting our water at the same time may get even more difficult due to climate change that is reducing agricultural productivity in many areas worldwide, a trend that is expected to get worse in the future. The warmer temperatures that come with recent climate change are expected to accelerate the formation of dead zones and/or extend their reach due to an increased rate of bacterial decomposition of matter, the fact that water's ability to hold oxygen goes down as temperature goes up, and the reduced ability of aquatic species under stress to deal with hypoxic conditions.

Climate change may also be contributing to dead zone formation via weather changes—exceptionally high rainfall amounts in 2017 sent high loads of nutrients into the Gulf, creating a record-setting dead zone. The summer of 2018 saw a smaller-than-average dead zone (less than 3,000 square miles), but 2019 saw it tick back up to rank as the eighth largest on record at 6,952 square miles.

The almost 7,000-square-mile hypoxic zone in 2019 was substantially greater than the five-year average of 5,400 square miles, and more than 3.5 times the size of the Gulf of Mexico Hypoxia Task Force goal, a goal it had hoped to reach by 2015. That target date has been extended to 2035.

Select References:

Altieri, A. H., & Gedan, K. B. (2015). Climate change and dead zones. *Global Change Biology, 21*(4), 1395–1406.

Diaz, R. J., & Rosenberg, R. (2008). Spreading dead zones and consequences for marine ecosystems. *Science, 321*(5891), 926–929.

Echols, A., & Vitousek, P. (2012). Managing water, harvesting results. *Frontiers in Ecology and the Environment, 10*(1), 3.

Environmental Protection Agency. (2011). *Gulf of Mexico Regional Ecosystem Restoration Strategy.* https://archive.epa.gov/gulfcoasttaskforce/web/pdf/gulfcoastreport_full_12-04_508-1.pdf

Rabotyagov, S. S., et al. (2014). The economics of dead zones: Causes, impacts, policy challenges, and a model of the Gulf of Mexico hypoxic zone. *Review of Environmental Economics and Policy, 8*(1), 58–79.

Turner, R. E., & Rabalais, N. N. (2019). The Gulf of Mexico. In *World seas:An environmental evaluation* (pp. 445–464). New York: Academic Press.

GLOBAL CASE STUDIES WATER POLLUTION

Achieve

Water pollution comes in many forms. Effective programs to reduce it address pollution at the source. Here are four examples in the United States of communities addressing water pollution in their area.

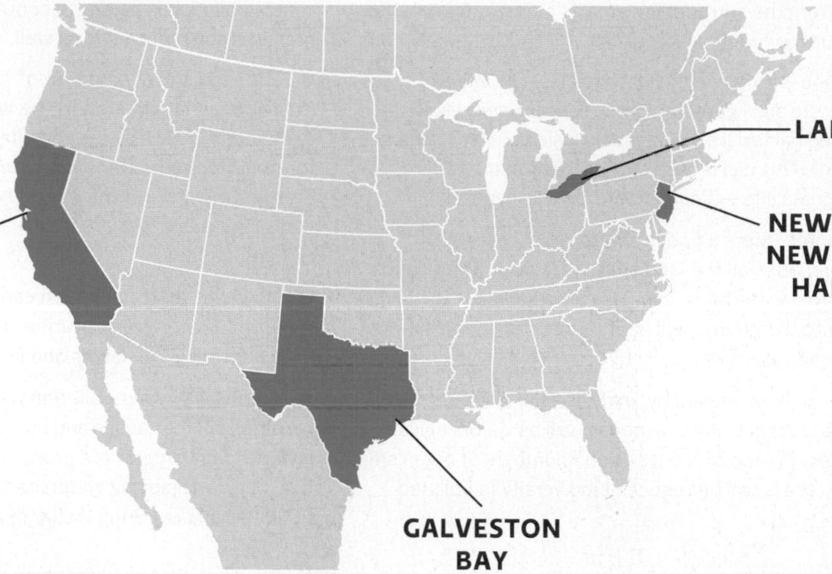

BRING IT HOME

PERSONAL CHOICES THAT HELP

We are facing not only shrinking supplies of easily accessible water but also the potential degradation of this resource due to pollution. By changing products and modifying common practices, we can improve our water quality for years to come.

Individual Steps

- Read your city's water quality report to see which pollutants are prevalent in your area.
- Decrease your use of chemicals (fertilizers, pesticides, harsh cleaners, etc.) that will end up in the water supply. For safer alternatives to lawn care, see www.epa.gov/safepestcontrol/lawn-and-garden.
- Always dispose of pet waste properly. In high quantities, it acts as an oxygen-demanding waste and can also spread disease.
- August is National Water Quality Month. Take steps every day to reduce your water pollution and help raise awareness in August by writing a letter to your local newspaper outlining simple steps people can take to improve water quality.

Group Action

- Marking storm drains with "Don't Dump" symbols can remind people not to dump waste liquids down sewer drains. If the drains in your area are not marked, talk to city officials to see if you and other volunteers can mark them.
- Contact the Izaac Walton League of America at www.iwla.org/sos to receive tools and training to conduct a biological assessment of a stream through its *Save Our Streams* program. Work with other volunteers to regularly monitor a stream in your area.

Mario Tama/Getty Images

REVISIT THE CONCEPTS

- The source of water pollution may be easy to identify, such as discharge pipes (point sources), but it can also be more difficult to pinpoint, coming from the surrounding area as stormwater runoff or atmospheric fallout (nonpoint sources).
- Nutrient pollution is damaging to waterbodies. The influx of excess nutrients can promote algal growth (which reduces underwater photosynthesis) and bacterial population explosions (which consume underwater oxygen). This can reduce oxygen levels in the water enough to harm aquatic life—a condition known as hypoxia.
- Stormwater runoff enters a body of water from within that waterbody's watershed—all the land area over which water could flow and empty into a waterbody. Land uses that increase the ability of water to soak into the ground will limit runoff, protecting surface waters and recharging aquifers.
- Water quality can be assessed by analyzing the chemical and physical characteristics of the water as well as via biological assessment—an evaluation of the types and numbers of organisms living in the water. There will be reduced biodiversity in polluted waters.
- U.S. surface waters are protected by the Clean Water Act. It has been effective at reducing point source pollution. Nonpoint source pollution is harder to control, but following best management practices can reduce this as well.
- The key to the reduction of nonpoint source pollution is managing what goes on in the watershed. Setting total maximum daily loads (TMDLs) for certain pollutants can provide a target for pollution reductions, and steps that decrease the likelihood that stormwater runoff will deliver it to waterbodies (e.g., well-vegetated riparian areas) are the focus of water pollution reduction plans.
- Nonpoint source pollution can also be decreased by preventing runoff (redirecting or capturing stormwater) and by minimizing the use of chemicals on lawns and farms.
- Addressing water pollution requires a holistic approach that considers all the causes and consequences of the pollution. Steps include identifying where problems exist, reducing pollution at the source, employing watershed management to reduce runoff potential, and restoring wetlands.

ENVIRONMENTAL LITERACY Understanding the Issue

1 What is water pollution, and how is it classified by source?

1. Distinguish between point source and nonpoint source water pollution.
2. Compare three typical point source pollutants and three nonpoint source pollutants from the area where you live.

2 What are the causes and consequences of eutrophication?

3. What is hypoxia, and why is it dangerous to aquatic organisms?
4. Many cattle pastures have ponds to provide water to the livestock, but often by summer's end, these ponds are covered in a thick green scum and fish in the pond die. What is causing the green scum to form, and what might be causing the fish kills?

3 What is a watershed, and how does it affect surface water and groundwater?

5. In general, which land areas and bodies of water would be part of the watershed for a river?
6. Where does the nitrogen and phosphorus pollution that reaches the Gulf of Mexico come from?

4 How is water quality assessed?

7. Distinguish between chemical, physical, and biological assessments of water quality.
8. Why is it incorrect to assume that a stream that looks crystal clear is environmentally sound?

5 What public policies are in place to protect water quality?

9. What is the main goal of the Clean Water Act (CWA), and how is this law implemented and enforced?
10. For what type of water pollution is the CWA most effective?
11. What is the citizen suit provision of the CWA?

6 What role does watershed management play in preventing water pollution?

12. The CWA regulates point source pollution with pollution standards. What approach does it take to address nonpoint source water pollution?
13. What are the benefits of a healthy riparian area?

7 How can nonpoint source water pollution be reduced?

14. Identify actions farmers can take to reduce agricultural nonpoint source water pollution.
15. Identify actions that will reduce nonpoint source water pollution from urban and suburban areas.
16. What is a rain garden, and how can it reduce stormwater runoff?

8 What strategies can be used to restore damaged aquatic ecosystems?

17. Why are holistic strategies needed to address water pollution issues in the Gulf of Mexico and elsewhere?
18. Identify the four goals of the Gulf of Mexico Regional Ecosystem Restoration and some of the strategies proposed to meet each of those goals.

SCIENCE LITERACY Working with Data

The application of road salt in cold climates can result in runoff pollution that can damage aquatic ecosystems. Researchers at Saint Mary's University in Halifax, Nova Scotia, surveyed local roadside ponds (some of which were occupied by amphibians and some of which were not) to determine chloride concentration and species richness of amphibians. They also investigated the vulnerability of five amphibian species to sodium chloride (NaCl) by determining the LD_{50}, the dose or concentration of chloride that kills 50% of the population.

A: EFFECTS OF CHLORIDE ON AMPHIBIAN SPECIES RICHNESS

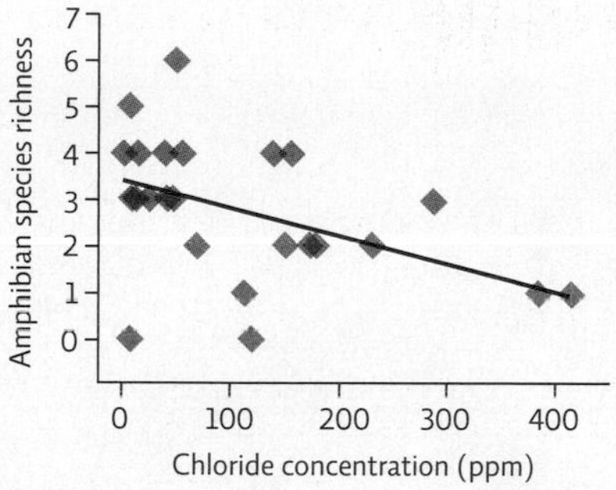

B: CHLORIDE CONCENTRATION IN OCCUPIED AND UNOCCUPIED PONDS FOR 5 AMPHIBIAN SPECIES

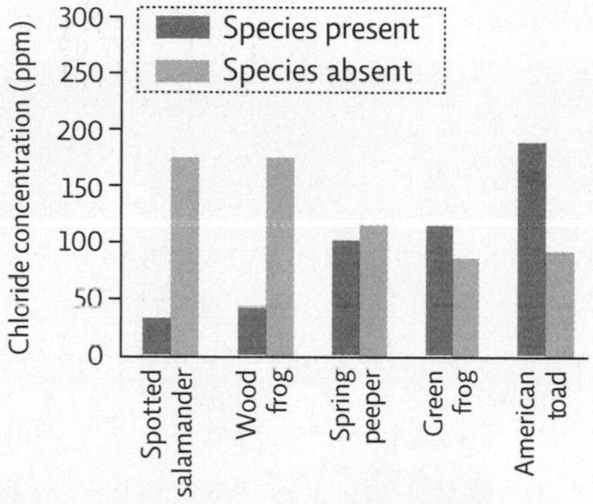

C: COMPARISON OF MEDIAN LD_{50} VALUES IN AMPHIBIANS

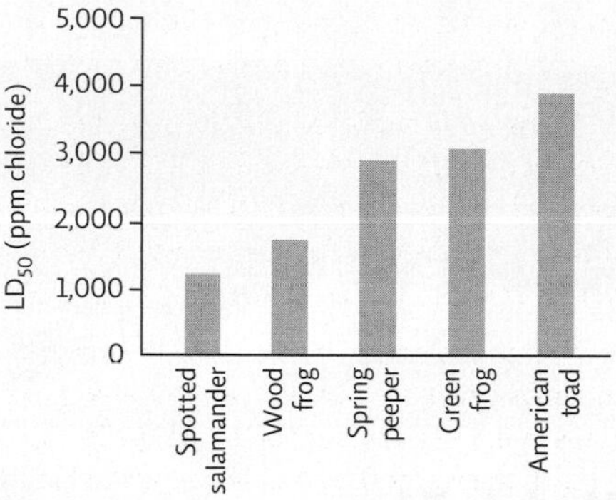

Interpretation

1. Look at Graph A. In a single sentence, explain the relationship between chloride concentration and species richness.
2. Look at Graph B. Which species of amphibian is least affected by the presence of chloride in the water? Explain your reasoning. (*Note*: The *y* axis shows chloride concentration, not population size.)
3. Look at Graph C. Which species is the most sensitive to chloride? Which is the least sensitive?

Advance Your Thinking

4. Look at Graph B and Graph C. Does the LD_{50} for each species shown in Graph C correlate with the data (presence or absence of each species according to chloride concentration) shown in Graph B? Explain.
5. Suggest two actions that could be taken to reduce the chance that road salt would enter ponds in runoff.

INFORMATION LITERACY Evaluating Information

A wide variety of industrial chemicals pollute waterbodies and those deemed dangerous are regulated by the Clean Water Act (CWA). But there is another type of chemical pollution of increasing concern that the CWA does not address — pharmaceuticals: medicines or their metabolized by-products.

Investigate this problem

1. Do an Internet search using the key words *pharmaceutical water problems.* From your search results, select three websites that you think are reliable sources of information on this topic and create a citation for each (identify the article or website by title, author and date [if provided], and URL). Then answer questions a, b, and c below. For each answer, cite the source(s) you used from your citation list.
   a. How do medicines enter our water system? How widespread is this problem?
   b. What are the environmental and health concerns of this type of water pollution?
   c. What can be done to address this type of water pollution?

Compare the Three Articles/Websites

2. Did any of the websites you visited disagree with, or contradict, each other for any of the three questions above? Explain.
3. Why did you choose each website that you used for this exercise — what made you think it was reliable? After reading each article or examining each website, do you still feel each was a reliable information source? Explain.

Dig Deeper

4. Choose one of the first three questions above and dig deeper to find more information on that question. Summarize your search (What key words did you use? What search engine(s) did you use?) and the information you found to answer that question, citing your source(s).

Draw Conclusions

5. Based on your research, do you feel pharmaceutical water pollution is something that should be addressed? What do you feel is the most appropriate response to this water pollution issue? Support your answer.

Additional study questions are available at Achieve.macmillanlearning.com. Achieve

CHAPTER 7

Geological and Land Resources

Terrestrial ecosystems provide many ecosystem services including the provision of many vital biological and mineral resources. Our actions can overexploit these resources or damage the ecosystems that provide them, but we can make choices to protect the natural ecosystem and the resources they provide.

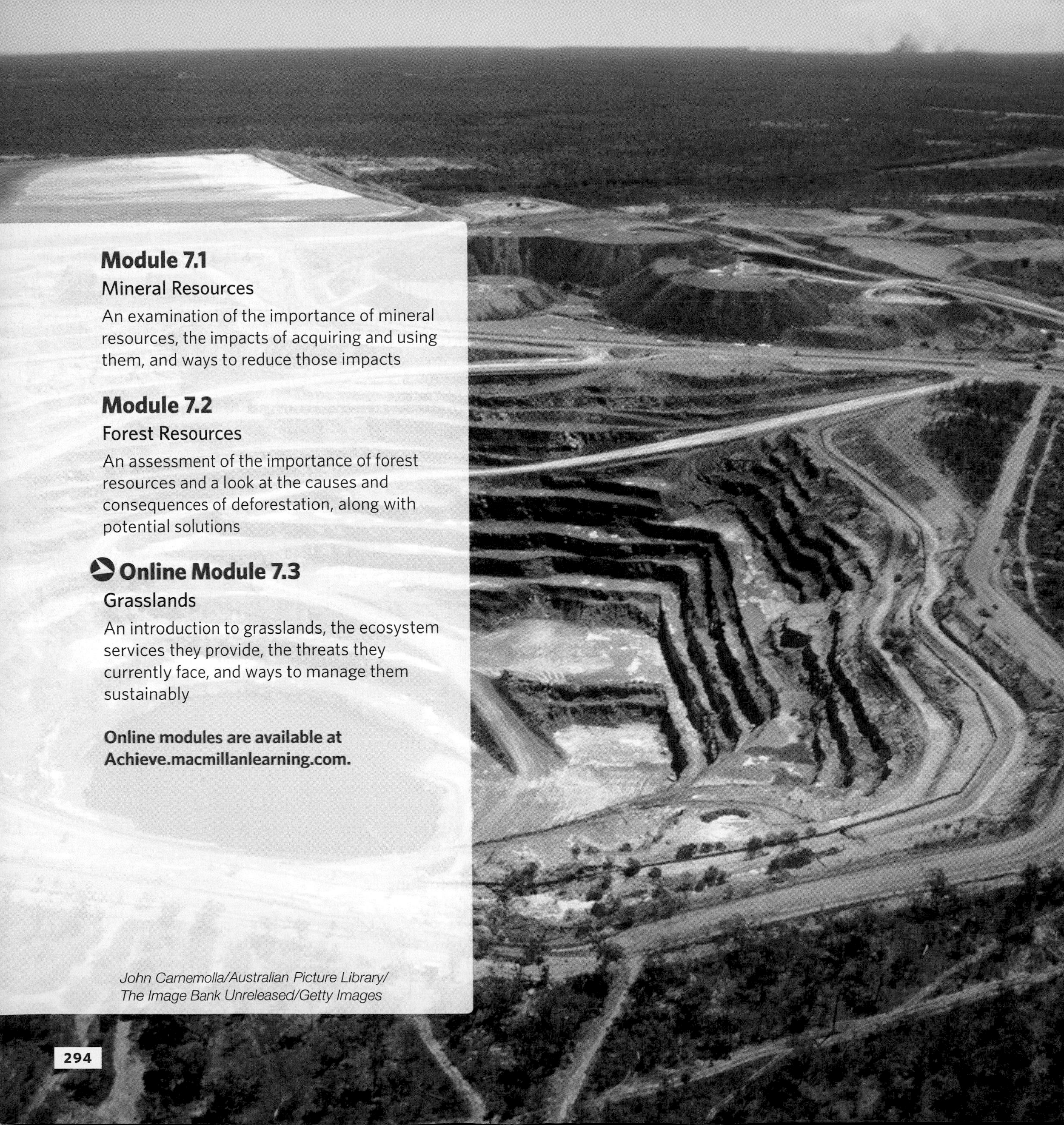

Online modules are available at Achieve.macmillanlearning.com.

John Carnemolla/Australian Picture Library/ The Image Bank Unreleased/Getty Images

Module 7.1 Mineral Resources

LITHIUM: THE NEW PETROLEUM?

Green technology threatens a Chilean desert

Lithium carbonate dries in the sun next to brine pools at a lithium mine located in Chile's Salar de Atacama. *Bloomberg/Getty Images*

After reading this module, you should be able to answer these GUIDING QUESTIONS

1. What are mineral resources, and how do we use them in modern society?
2. What geological processes affect Earth's outer layer (lithosphere)?
3. What hazards result from Earth's strong geological forces?
4. How are rocks formed via the rock cycle?
5. How are mineral resources mined and processed?
6. What environmental and societal problems arise from mineral mining and processing?
7. How can the impact of using mineral resources be decreased?

In October 2019, protesters from a handful of indigenous Chilean communities created a roadblock to hobble operations for two of the world's top producers of lithium. The lightest metal on Earth, lithium is a coveted commodity because of its use in the manufacturing of glass and ceramics and, more recently, for the production of rechargeable lithium-ion batteries, which are used to power, among other things, electric cars. Soon, researchers hope mega-batteries that store excess energy from wind and solar power installations will be available. The promise of rechargeable lithium-ion batteries helping the world transition from fossil fuels to cleaner sources of energy has fueled a rapid increase in lithium mining, and demand is expected to increase tenfold by 2030. So important was the development of this high-capacity battery, its inventors won the Nobel Prize in 2019.

Chile possesses one of the world's largest lithium deposits, found in brine (saltwater) deposits just beneath the Salar de Atacama, a 1,200-square-mile expanse in northern Chile's Atacama Desert. Other commercially valuable materials such as copper and potash (for fertilizer) are also found there. Lithium mining began in Chile in the 1970s and picked up in the 1990s, shortly after lithium-ion batteries hit the market. The mining operation extracts vast quantities of freshwater (in addition to the saltwater withdrawn from the salar), and the area is already among the driest places on Earth. As such, many scientists, as well as local residents, worry that mining is harming the Chilean environment beyond its ability to recover and worsening social inequality in a region that is already economically struggling.

"I am very concerned about the destiny of the Salar de Atacama," says Cristina Dorador, a microbial ecologist at the Universidad de Antofagasta in Chile. As water is depleted and salinity levels increase in the soil, Dorador worries that precious biodiversity could be lost, including microbes that live in the salty water found in the Atacama—life-forms so unique they give scientists a window into what microbial life, if it exists, might be like on Mars. Villagers also rely on the region's freshwater to graze their animals and grow crops like maize and alfalfa; without enough, farming could become impossible. As local farmer Jorge Cruz told the BBC in August 2019, "The birds have gone, we can't keep animals anymore. It's getting harder and harder to grow crops. If it gets any worse . . . we will have to emigrate."

WHERE IS THE SALAR DE ATACAMA?

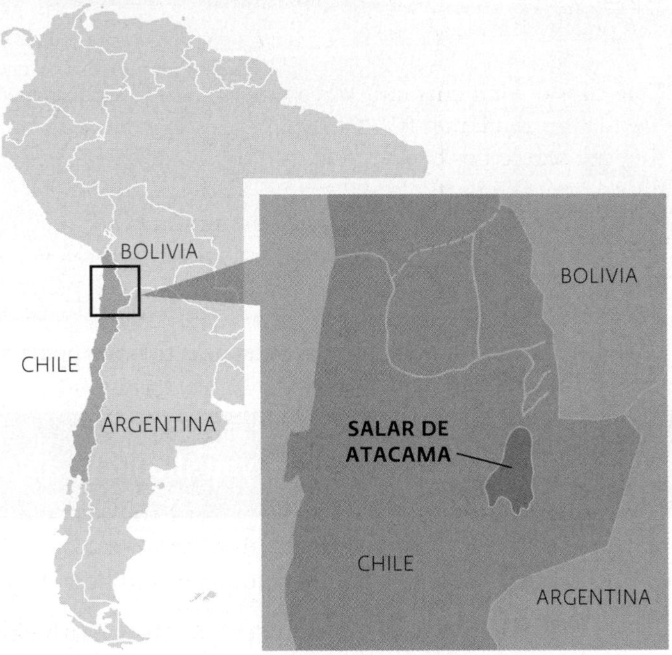

So the Chileans are doing what they can to fight for their land. During the October 2019 roadblock, locals waved colorful Wiphala flags representing the native people of the Andes, and they prevented mining trucks from transporting lithium along the usual routes. "We hope to continue protesting until the state hears us and attends our legitimate demands," said Sergio Cubillos, president of the Atacama People's Council, in an interview with Reuters. In the meantime, global demand for lithium continues to soar, and questions abound as to how we can meet our need for this metal without also destroying the lands that supply them.

1 MINERAL RESOURCES IN MODERN SOCIETY

Key Concept 1: Ore minerals are finite resources that provide a wide variety of materials essential to the production of construction materials, electronics, machinery of all kinds, and an array of household products.

Chances are, wherever you're reading this right now, you are surrounded by a variety of different materials—aluminum, copper, lithium, zinc, granite, marble, and so on—derived from mineral resources. **Minerals** are naturally occurring chemical compounds that exist as solids with predictable, three-dimensional, repeating structures (crystals). Their chemical composition and crystalline structure give each mineral distinct physical properties. Most **metals** and rare elements are bound up with more common elements like oxygen or sulfur in mineral deposits; when these exist in formations from which they can be profitably extracted, they are called **ore minerals** or **ores**. In addition to its presence in the Atacama, dissolved in the brines found there, lithium is found in hardrock ore deposits, where it is bound up in minerals such as spodumene [$LiAl(SiO_3)_2$] and petalite ($LiAl(Si_2O_5)_2$)].

One class of elements mined from mineral deposits is the rare earth elements (REEs), so named not because they are scarce but because they are not commonly found in concentrated ore deposits. REEs are a group of chemically similar elements that are finding their way into all kinds of modern products, used to, for instance, tint sunglasses, power laptops and cell phones, and put the color into color televisions. Precious metals like gold and silver used in jewelry are also mineral resources. In fact, metals are an essential component of virtually all electronic devices, thanks largely to their malleability and their ability to conduct electricity.

Mineral deposits are almost always found as a component of **rock**—aggregates of one or more minerals. Rocks can be used as resources themselves, for instance as building materials like concrete or gravel, or they can be sources for specific minerals that are used once these minerals are extracted. Often, rocks contain common nonmetallic minerals like quartz and feldspar, the most abundant minerals on Earth. Minerals are almost everywhere in almost everything. "They are a deeply ingrained part of everything we know and love," says California Institute of Technology geologist George Rossman. "Human society as we know it would not exist without them."

Today our demand for minerals is higher than ever. Annual per capita use in the United States is slightly more than 40,000 pounds of minerals each year. Of this, nearly half goes toward the stone, sand, and gravel used to make roads, buildings, and bridges. Yet many minerals we use go toward things we can't readily see: the circuit boards and rechargeable batteries in electronic devices, computers, and cars. Lithium-ion batteries especially are crucial to these products, and the need for lithium, cobalt, nickel, and other components of these batteries is only going to increase as electric vehicles, seen as the up-and-coming replacement for fossil fuel–powered cars, become more mainstream. **INFOGRAPHIC 1**

mineral A naturally occurring chemical compound that exists as a solid with a predictable, three-dimensional, repeating structure.

metal A malleable substance that can conduct electricity; usually found in nature as part of a mineral compound.

ore mineral/ore A rock deposit that contains economically valuable amounts of metal-bearing minerals.

rock Solid aggregate of one or more minerals that occurs in a variety of configurations.

NurPhoto/NurPhoto/Getty Images

With demand for lithium set to double by 2025 as the need for lithium-ion batteries increases with more cell phones in people's hands and more electric vehicles on the road, investors and manufacturers are looking with increased interest to lithium-rich countries like Bolivia, Chile, and Argentina.

INFOGRAPHIC 1 **MINERAL RESOURCES ARE A PART OF OUR EVERYDAY LIVES**

We depend on many mineral resources to make the products we use every day. Common minerals you are familiar with, as well as rare ones you may never have heard of, can be found in our homes and household items. This diagram points out some of the minerals you might find in the average dorm room.

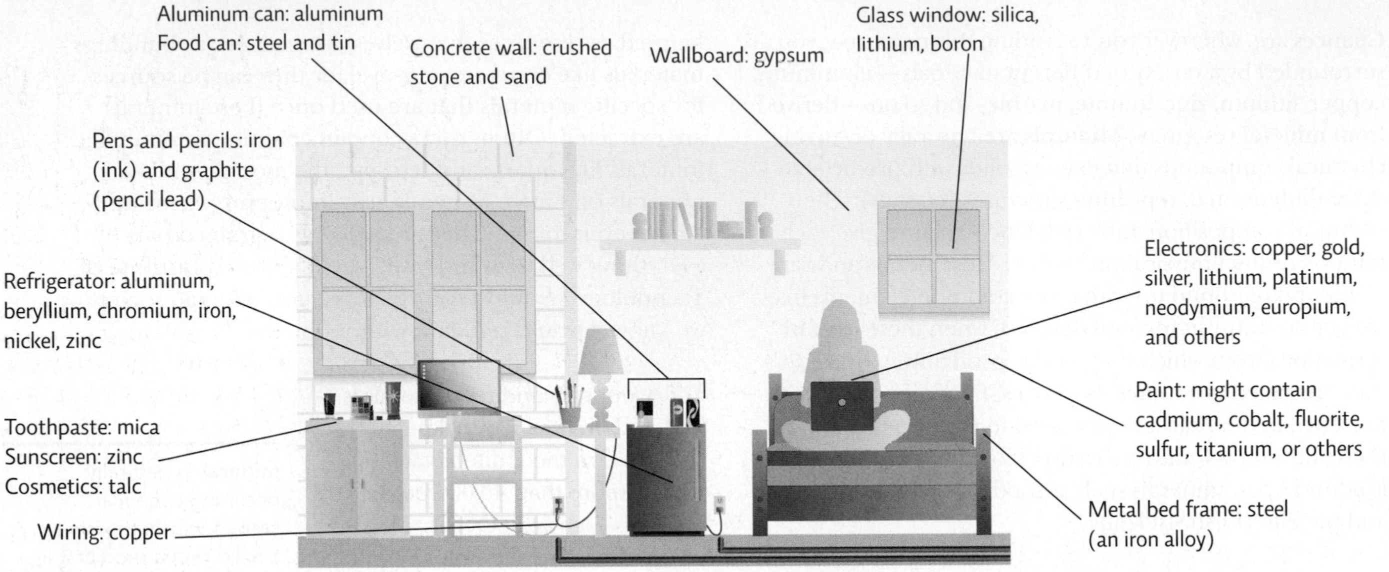

Lithium is used in the production of lithium-ion batteries. How many electronic devices and computers do you own that might contain these batteries?

But extracting minerals from Earth has brought an array of negative consequences—not just environmental (air and water pollution) but also societal (health hazards and human rights violations) consequences—felt by Atacama farmers and others around the globe. What's more, demand for all types of resources—from REEs to precious gems to familiar metals like nickel and copper—is rising slowly but steadily as population continues to climb, standards of living increase across the globe, and unprecedented technological advances create a litany of new uses for many metal and nonmetal mineral resources. The lengths we will go to in order to extract the minerals we need are remarkable and sometimes surprising. Who would have thought that we would need to mine salt flats that have been sitting undisturbed in ancient lakes in rural Chile in order to power our electric cars?

2 GEOLOGICAL FORCES SHAPE EARTH

Key Concept 2: Strong geological forces, such as those that power plate tectonics, slowly but constantly reshape Earth's surface and interior. New minerals are formed at plate boundaries, and the movement of tectonic plates redistributes minerals in Earth's crust.

Minerals include both metals (like gold, silver, lithium, and the REEs) and nonmetals (like quartz and rock salt) and are classified according to several physical properties, such as hardness and luster, that arise from their chemical composition and structure. To make use of minerals—to find large deposits that we can mine or to design synthetic compounds with similar properties—we must first understand the geological forces that create and concentrate them. The scientific field of geology examines the physical and chemical makeup of Earth and the processes that shaped, and continue to shape, our planet.

Earth is composed of discrete layers of solid rock and liquid metal that differ according to chemical composition and temperature. Earth's layers get hotter at increasing depths, and at its center is the hottest, densest portion of Earth, the *core,* made predominantly of iron. There is a solid inner core as well as a liquid outer core. The uppermost portion is the *crust,* made up of a thin (relative to the size of the Earth) layer of rock—oceanic crust is primarily basalt, whereas the continental crust is largely granite, a rock that usually contains the minerals quartz and feldspar as well as others. The continental crust is thicker than the oceanic crust, but the oceanic crust is denser. Valuable mineral deposits and fossil fuels are found in the crust, and our mines and drilling operations tap these resources.

Between the core and the crust is the *mantle*—a region composed of solid rock. The uppermost mantle region is rigid, like the crust, but is composed of a different type of rock, primarily peridotite, so it is not considered part of the crust. Together the crust and this hard, upper region of the mantle are known as the **lithosphere**. (The term *lithosphere* is sometimes used to describe all the rock found on Earth, from crust to core.) The lithosphere does not exist as one continuous shell around Earth, but rather is broken into discrete **tectonic plates** that fit together like the pieces of a jigsaw puzzle. These plates can consist of continental crust, oceanic crust, or a combination of the two.

The mantle region directly below the lithosphere, the *asthenosphere,* is so hot that the peridotite rock begins to soften and become more malleable, like putty or molding clay, allowing it to flow (slowly)—this is called "plastic behavior," the ability to deform without breaking. Though softer than the peridotite of the upper mantle, rock in the asthenosphere is not molten (liquid) rock; it is still considered a solid. However, this layer is the source of magma (molten rock) that forms in locations where the temperature rises or pressure is lowered (or both)—physical conditions that begin to melt the rock.

The asthenosphere is in motion due to convection—rock deeper in this part of the mantle is hotter and less dense than the rock above, so it flows upward; cooler rock flows downward as it is displaced by the upward-flowing hotter rock, which cools as it rises and then sinks again. This creates a *convection cell* (the cyclical movement of a hot and cold substance) that keeps the asthenosphere moving. This convection is powered by heat produced deep within Earth from the decay of radioactive material and by residual heat remaining from the formation of the planet.

The tectonic plates that sit on the asthenosphere move very slowly (0.5 to 6 inches per year)—your fingernails grow at similar rates. The continents that ride on these plates have taken up different positions on the planet in the past and at least twice in Earth's history have come together to form a single supercontinent. Earthquakes and volcanoes can give scientists clues about the locations of plate boundaries (areas where two or more plates meet), because they occur in regions where plates move relative to one another. The energy released at plate boundaries fuels volcanic and hydrothermal activity that end up creating new minerals, and redistributing others, within Earth's crust.

Geologists recognize different types of plate boundaries. Plates approach each other at **convergent plate boundaries**. Where they meet, the denser (oceanic) plate will **subduct** (or slide under the other), forming an ocean trench. (If two oceanic plates converge, the older, denser plate will subduct.) It is this action that fuels the movement of tectonic plates—as gravity pulls the denser plate deeper into the mantle, the rest of the plate is dragged along behind it.

Volcanic activity is seen in subduction zones—magma forms in these areas and then rises to create volcanic mountains or island chains. It is in these areas that rich veins of metal can be found, delivered to the plate boundary by the rising magma. This is the case in Chile's Atacama region, where localized volcanic activity fuels the accumulation of lithium in its salars. (The Spanish noun *salar* is translated as "salt flat" or "salt lake.") Similar deposits are found in salt flats in neighboring Bolivia and Argentina, whose salars share a geological history with Atacama. The Salar de Uyuni in Bolivia is the largest salt flat in the world and may be the world's single largest source of lithium. The region, known as the "Lithium Triangle," holds an estimated 70% of the world's lithium.

When two continental plates collide, subduction does not occur because their density is less than that of the mantle below (so they will not sink into the mantle). Rather they crash together in slow-motion collisions that cause the lithosphere to bend and buckle, reforming the face of the Earth; this is the power that fuels *mountain building*. As with the other plate boundaries, earthquake activity is seen at convergent plate boundaries.

Plates move apart at **divergent plate boundaries**. At these boundaries, magma travels upward from the asthenosphere, through the fissure formed as the plates diverge, and cools, creating new crust. This new crust takes the form of underwater ridges at oceanic divergent plate boundaries (e.g., the mid-Atlantic ridge) or becomes the floor of a rift valley if divergence occurs on a continent. Many important metallic minerals are formed at these ridges and can be mined.

A third type of plate boundary exists, too: a **transform plate boundary** where two adjacent plates slide past each other laterally, parallel to the boundary—one in one direction and the other in the opposite direction. These regions are often the sites of earthquakes, created when plates slide to release built-up tension. The crust may be cracked or broken apart after this movement, but no new crust is created and no crust is destroyed. Most transform plate boundaries are in the oceans, but the San Andreas Fault that runs through California is an exception. **INFOGRAPHIC 2**

lithosphere The rigid outer layer of Earth made up of the crust and the hard uppermost layer of mantle.

tectonic plates Rigid pieces of Earth's lithosphere that move above the asthenosphere.

convergent plate boundary A place where tectonic plates are moving toward each other.

subduct The movement of one tectonic plate below another at a convergent plate boundary.

divergent plate boundary A place where tectonic plates are moving away from each other.

transform plate boundary A place where two tectonic plates slide side to side relative to each other.

INFOGRAPHIC 2 EARTH—A DYNAMIC PLANET

Earth is composed of discrete layers. The crust, mantle, and core are distinguished by their distinct chemical compositions; regions within these layers that have different physical properties are also recognized. Mineral and fossil fuel deposits are found in Earth's crust.

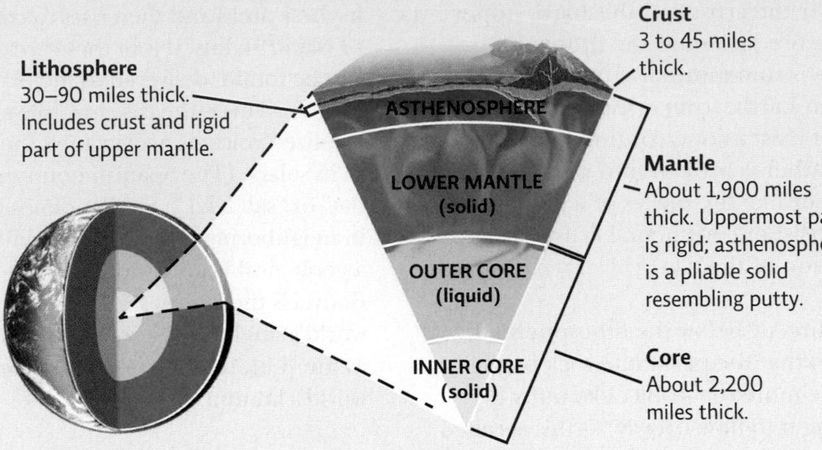

Lithosphere
30–90 miles thick. Includes crust and rigid part of upper mantle.
Crust
3 to 45 miles thick.
ASTHENOSPHERE
LOWER MANTLE (solid)
OUTER CORE (liquid)
INNER CORE (solid)
Mantle
About 1,900 miles thick. Uppermost part is rigid; asthenosphere is a pliable solid resembling putty.
Core
About 2,200 miles thick.

Tectonic Forces

Powerful geological forces are constantly but slowly rearranging the face of Earth. The lithosphere is made up of tectonic plates that move slowly, powered by heat deep within Earth.

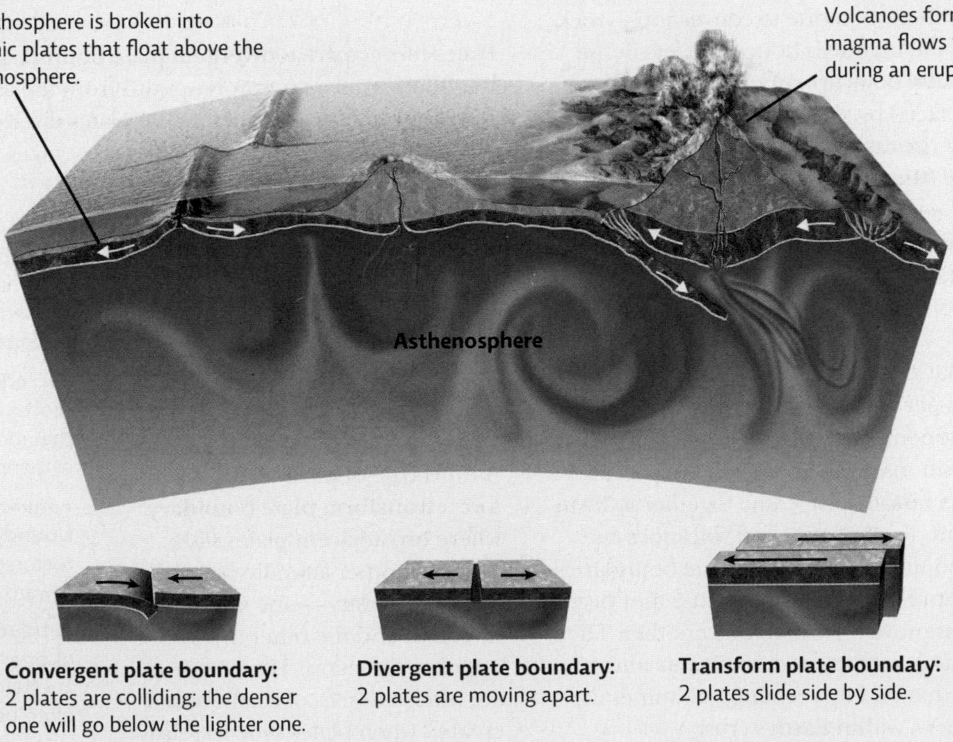

The lithosphere is broken into tectonic plates that float above the asthenosphere.
Volcanoes form where magma flows to the surface during an eruption.
Asthenosphere
Convergent plate boundary:
2 plates are colliding; the denser one will go below the lighter one.
Divergent plate boundary:
2 plates are moving apart.
Transform plate boundary:
2 plates slide side by side.

Why is the crust above a convergent or divergent plate boundary a good location to search for mineral deposits?

Just as the theory of evolution is a unifying theory of biology, *plate tectonics* is a powerful unifying theory of geology that explains the dynamic processes that shape and reshape the face of Earth. As plates move, they cause earthquakes, create mountains and volcanoes, and form trenches or depressions at the plate boundaries. Powerful forces form and move mineral resources around, often concentrating them at plate boundaries. We mine these regions of the lithosphere (both on land and at sea) to retrieve the valuable minerals and ores they contain. But we also must endure the hazards created by these forces, destructive events that can overwhelm even our best attempts to be prepared.

3 GEOLOGICAL HAZARDS

Key Concept 3: Geological forces contribute to destructive events such as earthquakes, landslides, tsunamis, and volcanic eruptions.

On May 22, 1960, Chile was devastated by a magnitude 9.5 **earthquake**—a sudden shaking of the ground most commonly caused by tectonic plates moving relative to each other or volcanic activity. It was the strongest earthquake ever measured. In 1994, Bolivia was shaken by a massive magnitude 8.2 earthquake that was felt as far away as Toronto, Canada. A magnitude 6.1 earthquake that hit Argentina on January 4, 2020, killed a child when a wall collapsed. The fact that such lithium-rich countries are also at risk for earthquakes is no coincidence. The regions of Earth that are richest in minerals tend to be most at risk for geological hazards, because both are created by the same strong geological forces.

Earthquakes occur at *faults*, underground regions where cracks are found in the lithosphere. Plate boundaries are faults but so are cracks occurring within a single plate. Rocks on either side of these faults typically move slowly, but if friction keeps them static for a period of time, stress builds up from accumulating pressure. When the energy from this stress is suddenly released, it moves through the ground in the form of seismic waves, causing the ground to shake.

This energy disperses from what is called the earthquake's *focus*—the underground location where the earthquake originated—to the *epicenter*, the point directly above the focus on Earth's surface. The epicenter is typically where most damage from an earthquake occurs, but damage can occur for miles around the epicenter, too, depending on the earthquake's strength. (That said, most earthquakes are hardly felt and cause little to no damage.) The damage done by an earthquake is shaped by its depth, by the nature of the geological materials through which the seismic waves travel (shaking is greater in less consolidated material such as silt or sand compared to waves that pass through rock), and by what has been built above ground near the epicenter. If buildings and other structures are constructed to withstand earthquakes, damage is often minimized.

The deadliest earthquake in recorded history happened in 1556 in Shaanxi, China. It was a magnitude 8.0 earthquake—less than the one that hit Bolivia in 1994—yet it killed approximately 830,000 people, many of whom lived in cave settlements that collapsed. The second most fatal earthquake was the magnitude 7.0 earthquake that struck Haiti in 2010. This earthquake was so deadly (perhaps as many as 316,000 deaths by one estimate) because it occurred close to a populated area with buildings poorly equipped to handle the shaking.

Earthquakes can also be caused by human activities, such as deep drilling for geothermal power plants (see Module 11.2), the use of injection wells to dispose of fracking fluid (see Module 9.2), and even damming up rivers and flooding river valleys—the extra weight of the water in a reservoir behind a dam can trigger earthquakes in seismically active areas.

Volcanoes are another geological hazard. They form when magma works its way up from the asthenosphere, exiting the volcano as *lava*. When lava and erupted ash cool, they harden into rock that forms the conical volcano shape. Like earthquakes, most volcanoes form near plate boundaries. They're found at divergent boundaries, where plates are spreading apart, as well as near subduction zones at convergent boundaries. The region where the Pacific Plate meets its adjacent plates is known as the *Ring of Fire*—a seismically active area encircling much of the Pacific Ocean where 90% of earthquake activity occurs and 75% of volcanoes are found. Argentina and Bolivia are close to the Ring of Fire, which extends up the western coast of South America. Chile sits directly along it and has as many as 2,000 active and inactive volcanoes. As an active earthquake zone, it is not unusual for Chile to experience more than 100 earthquakes in a month's time. **INFOGRAPHIC 3**

earthquake A sudden shaking of the ground caused by movement of tectonic plates at a plate boundary or an intraplate fault, as well as volcanic activity

volcano An opening (vent) through which lava, gases, and other material escapes from beneath Earth's crust, often accumulating to form a mountain or hill.

INFOGRAPHIC 3 THE RING OF FIRE

Earthquakes and volcanic activity are most often found above divergent and convergent plate boundaries. The Ring of Fire represents the area where the Pacific Plate meets surrounding tectonic plates, forming an almost continuous series of trenches and volcanoes. This seismically active area accounts for the majority of earthquake and volcanic activity on Earth.

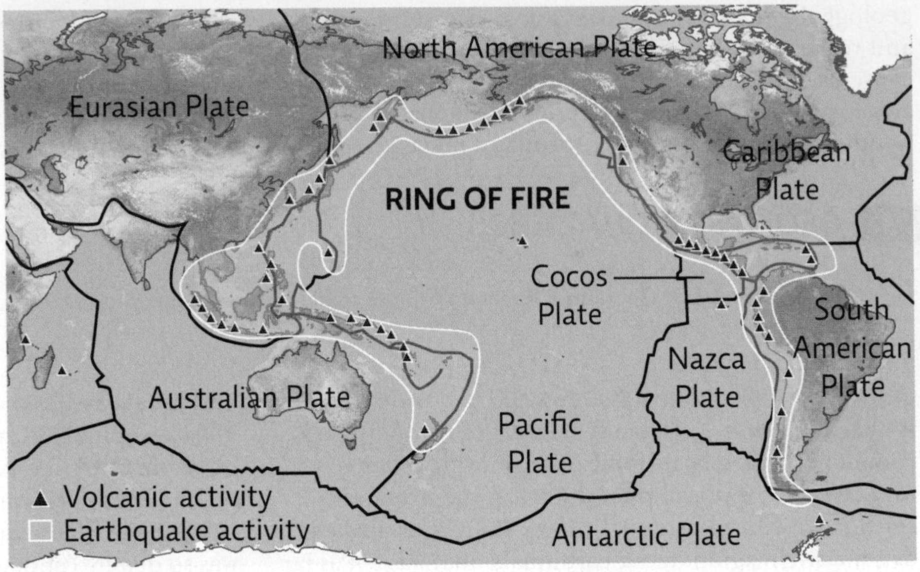

Based on the location of the Ring of Fire, which U.S. states do you predict have the most volcanoes and earthquakes?

Volcanoes can also form away from plate boundaries where "hotspots" bring high heat close to the crust. The Hawaiian Islands, in the middle of the Pacific Plate, formed one after the other as the plate drifted over just such a hotspot. Because hotspots remain stationary while the plate slides past, an evaluation of hotspot island chains helps geologists reconstruct past rates and direction of tectonic plate movement.

Volcanic eruptions can be explosive and extremely dangerous. The eruption of Mt. Vesuvius in Italy in 79 A.D. buried the Italian town of Pompeii, killing more than 20,000 people. Sometimes, earthquakes and volcanic eruptions happen at the same time or one after the other. In March 1980, scientists noticed a sudden increase in earthquake activity in the Cascade mountain range around Mount St. Helens in southern Washington state, with as many as three earthquakes happening each hour. Shortly thereafter, Mount St. Helens erupted, forming a crater on top. It soon became clear that magma was rising within the volcano and pressure was building up inside—a visible bulge appeared near the top of the mountain and grew outward as much as 6 feet per day. On May 18, 1980, the volcano erupted again, this time with massive force, blowing off the side of the mountain. Superheated gas and rock blew out of the volcano in a blast that created the largest landslide in recorded history. Though residents had been evacuated, 57 people were killed, including David Johnson, a volcanologist 6 miles away who sent word via radio of the eruption as it started. "Vancouver! Vancouver! This is it!" was his last transmission.

landslide The sudden movement of unstable rock or soil material down a slope due to the force of gravity, often triggered by heavy rain or an earthquake.

tsunami A series of high, long, fast-moving water waves caused by the displacement of a large volume of water by an underwater earthquake, landslide, or volcanic eruption.

When unstable soil suddenly gives way on a slope, **landslides**, another geological hazard, can occur. They often happen after earthquakes, when material is washed away during rainstorms, or when soil becomes heavily saturated with water, such as after a heavy rainfall. Landslides can also develop due to an excess weight of snow as well as after eroded riverbanks, dikes, or coastal areas collapse. The 2014 Oso landslide near Oso, Washington, that killed 43 people happened after 2 months of unexpectedly heavy rainfall caused slope failure, sending mud and debris traveling at speeds of up to 40 miles per hour. Worldwide, landslides cause about $2 billion in damage each year.

When earthquakes or landslides occur under the ocean, they can create **tsunamis**. (Sometimes they are mistakenly called tidal waves, although they are not caused by tides.) When the sea floor suddenly moves during an earthquake, the energy is transferred to the water, creating waves that ripple out and can travel for miles. In the open ocean, these waves look like large swells (not towering waves), but as they approach land, water piles up behind the waves, making them extremely tall. After an underwater earthquake shook Japan in 2011, a powerful tsunami came ashore. It killed close to 20,000 people and inundated the nuclear power facility at Fukushima, causing a meltdown that will cost trillions of dollars to deal with. (See Module 11.1 for more on this disaster.)

4 THE ROCK CYCLE

Key Concept 4: Rocks are formed and transformed by melting, cooling, erosion and deposition, and pressure.

It is through the internal processes described in the earlier sections—through the movement of tectonic plates and the resulting release of energy—that many minerals are created. Under extraordinary heat and pressure, atoms of naturally occurring elements combine to form new minerals, many of which are stable enough to endure for long periods of time. The relative abundance of elements in any given location determines which chemical reactions will be favored and, thus, which minerals will form. Because of this, minerals are not distributed evenly throughout the planet's lithosphere; different minerals are abundant or uncommon in different parts of Earth.

But it is not just internal forces that influence Earth's geology. While internal forces can create mountain ranges, external forces like wind and rain wear them back down. **Weathering**, the breakdown of rock by physical (wind, ice, etc.) or chemical (acid, water, etc.) means, and **erosion**, the movement of this broken-down material to another location, reshape the Earth's surface, form soil (in conjunction with the process of biological decomposition), and redistribute, bury, or uncover mineral deposits.

As mentioned earlier, most minerals are found, alone or in combination, as a component of rock—an aggregate of one or more minerals. Rocks, the most common material found on Earth, occur in a variety of configurations, categorized by how they form. In fact, rocks are constantly formed and transformed from one type to another via the **rock cycle**.

Igneous rocks form where magma rises and cools at plate boundaries; they are made of silicate minerals (those that contain silicon, oxygen, and other elements). The magma that forms them can cool slowly and crystallize while still underground, forming *intrusive* igneous rocks such as granite. Pegmatite, a type of rock mined for lithium, is an intrusive igneous rock. When magma exits the ground in a volcanic eruption, it cools and crystallizes quickly, forming *extrusive* igneous rocks above ground such as basalt or obsidian.

All types of rock can be weathered by physical or chemical means to produce **sediments**. Erosion transports these rock particles, such as sand and silt (as well as other materials such as shells and organic detritus), to locations where they pile on top of each other in rivers, streams, and deep basins under the sea. When these sediments accumulate and become compacted, **sedimentary rocks** are created.

Sedimentary rocks can form through one of three processes. In *clastic sedimentation*, eroded rock sediments accumulate, get compacted by pressure from above, and are cemented together by dissolved minerals in between the clasts. (A clast is a fragment or chunk of rock broken down by weathering.) This is how shale is made from tiny clay particles and how sandstone is made from sand. Conglomerates and breccias are also clastic sedimentary rocks, even though the sediments they contain can vary from small pebbles to large boulders. Clastic rock formations may also contain fossil fuel deposits (see Chapter 9 for more on fossil fuel formation).

In *biogenic sedimentation*, the sediments are biological in origin. Living organisms extract materials from water to form structures such as shells and bones, which, after death, accumulate on lake or sea floors and get compressed over time. Clastic and biogenic sedimentation happen in similar ways; the difference lies in the source material. Limestone and chalk are biogenic sedimentary rocks: They are formed from the accretion and compaction of the shells or exoskeletons of marine organisms over time. These rocks can be a window into the geological and biological history of an area, containing fossils that provide clues about the climatic conditions of the area at the time they were formed.

The third way to form sedimentary rock is through *chemical sedimentation*, when rock forms from concentrated solutes (material dissolved in a liquid) that precipitate out of solution. Some limestones form by direct precipitation of calcium carbonate on the sea floor. Salt flats, such as the Salar de Atacama, are *evaporite deposits* that form via chemical sedimentation in areas where no outlet for water exists—water that collects from precipitation does not flow away; it evaporates and leaves behind dried salt deposits or concentrated solutions of dissolved minerals.

weathering The breakdown of rock by physical or chemical forces.

erosion The movement of broken-down rock, soil, and other materials from one location to another.

rock cycle The process in which rock is constantly made and destroyed.

igneous rock Rock that forms when molten rock cools and solidifies.

sediments Fragments of mineral, rock, or organic material.

sedimentary rock Rock that forms when fragments of mineral or biological origin are deposited, accumulate, and are compacted and cemented.

INFOGRAPHIC 4 THE ROCK CYCLE

Rocks form and are transformed when they are subjected to high heat and pressure underground and when they are exposed to wind and water on Earth's surface.

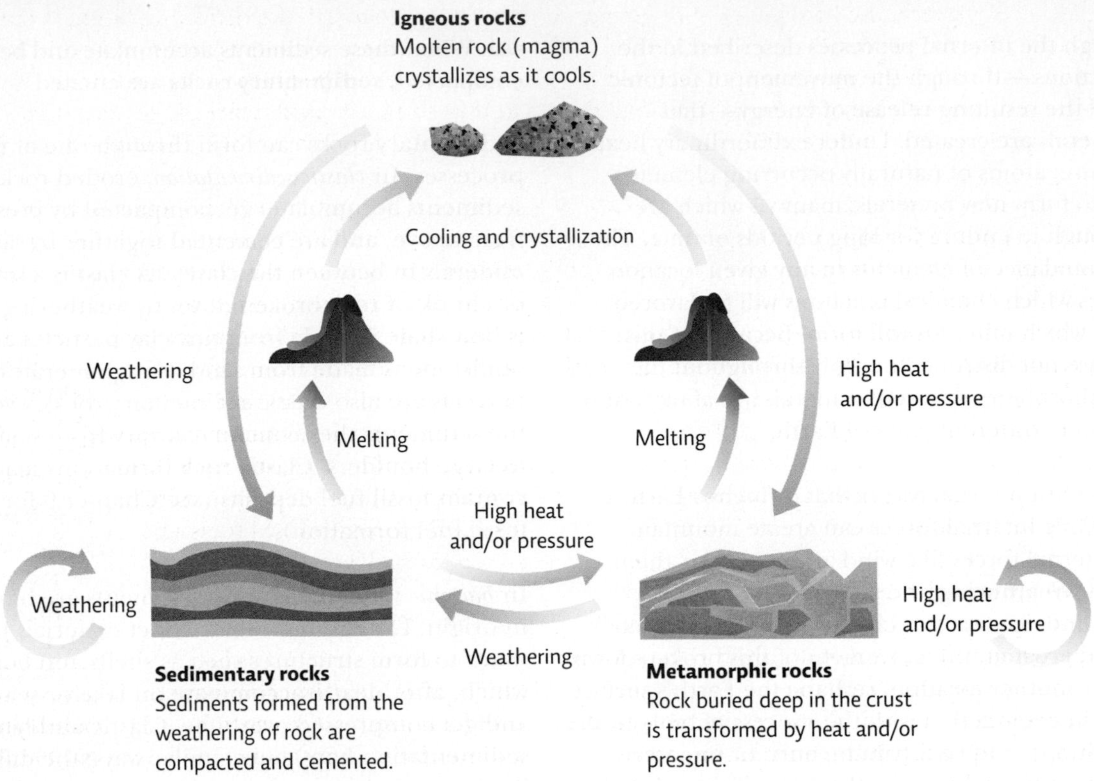

Which type of rock is likely to contain fossils? Why?

Metamorphic rocks are those that have been altered from their original solid rock state by the pressure exerted by moving plates and from deep burial, which heats them enough to cause a physical change (but not enough to melt them and form igneous rock). Being buried deep, these rocks are put under extraordinary amounts of pressure that can slowly transform them. The mineral crystals in the rock may change their orientation or recrystallize to form larger crystals. For example, the metamorphic rock marble is formed when limestone is exposed to sufficient heat to cause the calcium carbonate molecules to recrystallize. Metamorphic rock forms distinctive combinations of minerals under different conditions; by analyzing these rocks, geologists can tell us much about local conditions at the time of formation, allowing them to determine the geological past of the area. **INFOGRAPHIC 4**

metamorphic rock Rock that forms when existing rock in Earth's crust is transformed by high heat and pressure.

5 MINERAL RESOURCE MINING AND PROCESSING

Key Concept 5: Minerals are mined in both high-tech and low-tech ways, depending on the location, depth, and type of deposit. After mineral resources are extracted, additional processing steps are often needed.

Minerals are the building blocks of rocks, but not all rocks contain economically accessible deposits of useful minerals. As mentioned earlier, ores are rock deposits that contain economically valuable amounts of metal-bearing minerals; this means that the minerals within not only have practical applications but also are concentrated enough to be worth the effort of **mining** and extracting them.

mining The extraction of natural resources from the ground.

Minerals are considered nonrenewable resources. Yes, they form naturally through the aforementioned

Earth processes. But these processes are extremely slow relative to human lifespans, and for most minerals, we are using up existing deposits faster than new ones are being created. As with many of Earth's resources, the vast majority of mineral use has occurred in recent decades. In fact, of all of the minerals used in the 20th century, more than half were used between 1975 and 2000; according to one recent study, only 2.5% of all the copper ever mined was used before 1900.

Mining begins with geologists exploring various rock formations to pinpoint mining-worthy deposits. A *mineral resource* is a deposit with reasonable prospects for eventual economic extraction, but mining efforts are generally restricted to *mineral reserves,* those deposits that are economically and technically feasible to tap. Mineral exploration is a science in itself and one that requires remote sensing, exploratory digging or drilling, and chemical testing to evaluate samples. Once a suitable deposit is located, minerals are extracted using one of several mining techniques.

In **subsurface mining**, the deposits are found deep underground rather than close to the Earth's surface. Zinc, silver, gold, uranium, lead, diamonds, salt, and coal (a fossil fuel) are mined this way. To reach these deep deposits, workers create a vertical mineshaft down to the ore deposit. Once the deposit is reached, they create horizontal tunnels that follow the ore **vein**, the distinct region within the rock that contains a mineral ore or a coal seam. Elevators and vehicles transport workers deep into the mine. In the room and pillar mining method, enough rock is left behind as pillars to support the roof of the mine, and the ore is then removed. After the ore has been taken, the furthermost pillars can be removed, allowing the tunnel to collapse a little at a time behind the retreating miners. In mountainous regions, a sloping shaft rather than a vertical shaft is sometimes made to access the veins.

In developed countries like the United States, or in mines that have a lot of financial backing in less developed countries, modern mechanized equipment is used to extract the ore. Subsurface mining is typically more expensive than surface mining, in part because the deep mines have to be ventilated to bring oxygen to the workers and because the mines sometimes get so hot that air conditioning is necessary. In less developed countries, miners still use hand tools such as picks and shovels. In many instances, mining operations in these locations are extremely hazardous, and resulting accidents have led to human fatalities in the thousands.

In **solution mining**, the desired mineral is found in a natural liquid deposit—like the lithium salts found in Chile's salars, which are essentially ancient underground lakes. When a solid deposit contains material that is easily dissolved in a liquid, such as water or acid, a borehole is drilled into the deposit to create an injection well through which the dissolving liquid is pumped. Then the mineral-laden liquid is pumped out through a recovery well drilled into the same formation. When the underground deposit is already dissolved in water, as is the case for lithium, it is simply pumped out.

Surface mining is used when deposits are found close to the surface. If the deposit is on fairly level ground, **strip mining** can be done, which involves using explosives and heavy machinery to remove soil and rock above the deposit (called the **overburden**) in strips, and then removing the mineral resource. In each subsequent strip, the overburden goes on top of the previously mined strip. Strip mines are used to extract fossil fuels such as coal as well as the oil found in tar sands.

A variation of surface mining known as **mountaintop removal (MTR)** is used in the central and southern Appalachian Mountains, where coal deposits are found close to the surface in mountainous areas. The overlying soil and rock are removed to reach the coal seams below; but because the area is mountainous and not flat, there is no place to store the overburden, so it is dumped down the side of the mountain into the valley below, which can be extremely environmentally damaging (see Module 9.1 for more on strip and MTR coal mining).

When the top of a fairly deep mineral deposit is found close to the surface and the mineral is evenly dispersed throughout the deposit, **open-pit mining** is the method of choice. In this process, heavy machinery and explosives are used to remove the overburden and then the mineral ore. The pit is excavated in terraces or "benches" to reduce the distance that falling rocks can travel (for safety reasons). These pits can be

subsurface mines Sites where tunnels are used to access underground fossil fuel or mineral resources.

vein Distinct region within a rock that contains a mineral ore.

solution mining Extraction method where desired minerals are dissolved by a liquid injected into the deposit, then pumped back out and purified.

surface mining A form of mining that involves removing soil and rock that overlay a mineral deposit close to the surface in order to access that deposit.

strip mining A surface mining method that accesses fossil fuel or mineral resources from deposits close to the surface on level ground, one section at a time.

overburden The rock and soil removed to uncover a mineral deposit during surface mining.

mountaintop removal (MTR) A surface mining technique that involves using explosives to blast away the top of a mountain to expose the coal seam underneath; the waste rock and rubble are deposited in a nearby valley.

open-pit mining A surface mining method that extracts rock or mineral from a pit excavated for that purpose.

huge; rain and groundwater can collect in the crater, so this water must be removed by drilling drainage holes or pumping it out. Copper, iron, gold, uranium, aluminum, diamonds, and stones like granite, quartzite, and marble are mined this way, as are lithium-containing deposits of pegmatite. *Quarrying* is open-pit mining for materials such as sand, gravel, chalk, and slate.

When minerals are dispersed in sediments (such as riverbeds) that contain heavy minerals or metals, **placer mining** is used. The material is suspended in water, and the heavy mineral or metal falls to the bottom while the lighter sediments can be washed away. A *sluice*—a trough that allows the lighter sediments to flow away while the heavier ones fall—is often used for separation. Panning for gold is an example of placer mining that is still used today, but more modern methods are available. These include dredging—when a machine positioned in water excavates sediment from the riverbed via suction or mechanical dragging and then sends it through a sluice—and hydraulic blasting, in which a strong jet of water is fired at sediments and the slurry is then sent through a sluice. **INFOGRAPHIC 5**

placer mining The mining of underwater sediments (e.g., streambeds) for minerals.

smelting Mineral processing step in which the material is melted at high temperatures and mixed with chemicals that separate the mineral from the rock.

Digging out the ore is just the first step—the ore must be processed to yield a useful form of the mineral. At large-scale ore-processing facilities, this includes heavy machinery, some that crush the ore and others that grind it to a fine powder. The material may also need to be heated to high temperatures and mixed with other chemicals that bind to the mineral (**smelting**), freeing it up from the crushed rock; hazardous substances found in the ore can be released in the process, contributing to acid rain or toxic air pollution. After smelting, the material is then typically sent through additional purification steps to extract the metal of interest.

INFOGRAPHIC 5 MINING TECHNIQUES

A variety of mining techniques are available to access mineral resources. The nature of the deposit (depth, geometry, concentration, type of ore) influences the type of mining that will be done. Modern mining depends on heavy equipment and can move tremendous amounts of material in search of the desired mineral.

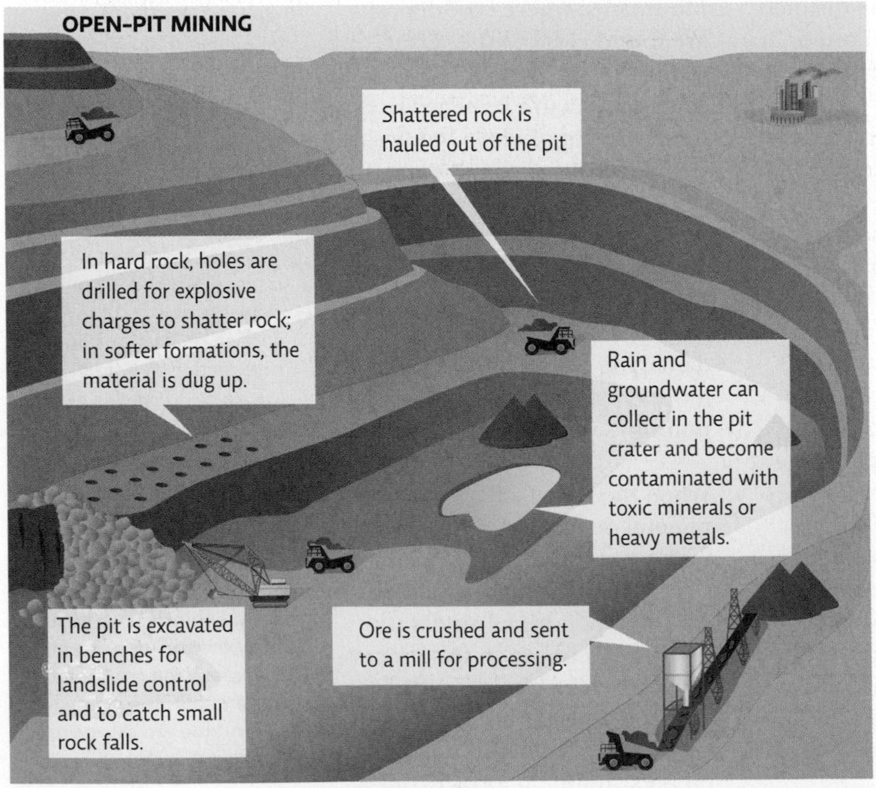

OPEN-PIT MINING
Shattered rock is hauled out of the pit
In hard rock, holes are drilled for explosive charges to shatter rock; in softer formations, the material is dug up.
Rain and groundwater can collect in the pit crater and become contaminated with toxic minerals or heavy metals.
The pit is excavated in benches for landslide control and to catch small rock falls.
Ore is crushed and sent to a mill for processing.

OPEN-PIT MINING is used when the top of a deep ore deposit is found close to the surface. A wide variety of minerals are mined this way, including aluminum, copper, and diamonds. Hardrock lithium found deposits are also mined this way.

LOES SCHALEKAMP/National Geographic Creative

PLACER MINING is used when sediments contain heavy minerals or metals that will easily settle out when mixed with water. Some operations, like this Mali gold mine, still depend on traditional methods.

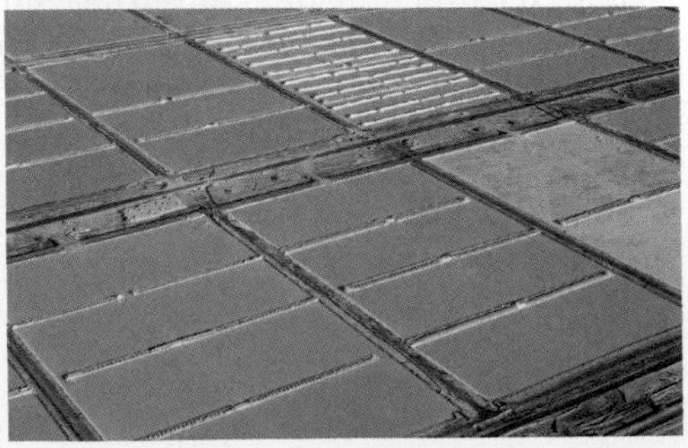

COLIN Matthieu/hemis.fr/Alamy

SOLUTION MINING extracts mineral-laden liquid from which the mineral is extracted.

Why are a variety of mining techniques used to access mineral deposits?

Electrolysis is sometimes used to extract metals put in solution during the refining process; an electric current is passed through aluminum oxide, for instance, to extract pure aluminum. Smelting is also used to combine metals to form *alloys* like steel, which is formed when iron is combined with small amounts of carbon.

The processing of lithium brines is done with lower-tech methods—after being pumped to the surface, the briny water is kept in holding ponds to allow the Sun's rays to evaporate the water, leaving behind a concentrated solution of lithium and other salts, a process that can take 1 to 2 years. After this, the concentrate is delivered to another facility where it is further processed to produce lithium carbonate or lithium hydroxide, the compounds used by industry in the manufacturing of lithium-ion batteries.

electrolysis Processing step that uses an electric current to separate metal from ore.

6 IMPACTS OF ACQUIRING MINERAL RESOURCES

Key Concept 6: Mining and processing of mineral resources can require high energy and water inputs. Additional environmental impacts of mining include the production of hazardous wastes; habitat destruction; and air, water, and soil pollution. Societal impacts include health problems for miners and community members, the use of child labor, and outright violence.

The same sequence used at large mining operations—search, discover, dig, extract, process—plays out at smaller-scale mines around the world where impoverished individuals use rudimentary tools and simpler strategies to achieve the same ends. (These are known as "artisanal" mines to distinguish them from industrial mines.) Similar processing steps are used by artisanal miners, but often without safety protocols to protect themselves or their surroundings from the hazardous materials used or created.

As demand for mineral resources has grown and prices have increased in tandem, large-scale operations have come to dominate the landscape. Such mechanization has increased mine productivity by an order of magnitude—mostly by enabling us to access previously unreachable deposits or to reopen shuttered mines. But it has also increased the carbon footprint of mining and the amount of waste produced—and in so doing has amplified the environmental and health consequences that mineral mining and processing bring with them. In addition, large industrial mines often require that vast swaths of land be stripped bare and require the construction of an extensive web of roads for access and transport, both of which can cause serious habitat damage. These combined forces—deforestation and road building—contribute to both soil erosion and water pollution as disturbed topsoil washes into nearby streams.

Mining the salt flats of South America, too, destroys habitat. By pumping out both the lithium-rich saltwater (around 320 million gallons per week) and removing freshwater from even deeper regions (to support the mining operation), mining efforts in the salars of the Atacama Desert of Chile have altered habitat and are adversely affecting nearby residents in this place known as the driest nonpolar desert on Earth. Water extraction has been linked to a decline in vegetation and soil moisture, higher daytime temperatures, and an increase in drought conditions.

The negative effects of mineral extraction aren't restricted to mining. Each processing step produces finely ground rock waste known as **mine tailings**, some of which may be hazardous. These tailings can contain the hazardous chemicals used or produced during processing and may be radioactive (if the rock formation contained radioactive material). Water traveling through underground mines or above-ground piles of waste rock and mine tailings can collect all sorts of contaminants such as heavy metals like mercury and arsenic that are detrimental to human health. In addition, when sulfur-containing rocks at mining sites are exposed to water and oxygen, sulfuric acid forms. This acid lowers the water's pH and produces **acid mine drainage** that in turn pollutes surface water and groundwater and harms aquatic ecosystems. The now-closed Richmond mine in northern California still releases acidic water into local bodies of water, including the Sacramento River and Keswick Reservoir; the water draining out of the mine has a pH similar to stomach acid (around 1.5)—acidic enough to kill local fish and other wildlife. (See Module 6.3 for more on the pH scale.)

There are two basic approaches to address acid mine drainage—prevent its formation or reduce its impact after formation. Preventing the formation of sulfuric acid in waste rock piles can be accomplished by burying mine tailings or immersing them in water (to exclude the oxygen needed for the chemical reaction that forms the acid). If acid mine drainage does occur, there are treatment protocols

mine tailings The waste of mining operations; includes mill tailings, the finely ground rock left over from processing mineral ores.

acid mine drainage Water that flows past exposed rock in mines and leaches out sulfates. These sulfates react with the water and oxygen to form acids (low-pH solutions).

that aim to neutralize acids through the addition of buffering compounds such as calcium carbonate—crushed limestone is commonly used.

Indeed, it is nearly impossible to dig so much, and so deeply, into Earth without affecting the environment. Blasting, digging, crushing, and grinding all create dust, which pollutes the air with small particles and toxic substances and in some cases contains radioactive material. Large mining operations have developed a range of strategies to reduce this dust; they use huge amounts of water when crushing and grinding ore, and they store lightweight waste in tailings ponds (adding to the high water footprint of the operation) to keep it from floating through the air. But these solutions bring their own set of environmental problems—namely, they generate thousands of gallons of hazardous waste for every ton of minerals mined. And because that waste is so concentrated, the threat posed by a single spill or leak can be significant.

In the largest tailings dam failure ever recorded, a section of an earthen dam near Minas Gerais, Brazil, collapsed in 2015, releasing more than 60 million tons of sediment and water that destroyed entire villages, contaminated water supplies, and killed 19 people. The discharge contaminated more than 400 miles of nearby rivers, eventually reaching the Atlantic Ocean, causing significant environmental damage all along the way.

When a Canadian-based mining giant arrived in Cabanas, El Salvador, wanting to build a large open-pit gold mine, area residents were wary. They knew from the experience of neighbors that such an operation could divert water from their farms, pollute their air and streams, and threaten their crops. So they formed an environmental group and began protesting. Their own government listened—denying permits to the company in response to public pressure. But that success was costly, resulting in violence that caused four deaths.

The thread between mining and violence spans the globe and involves a range of minerals—from gold and silver to diamonds and emeralds. And it's not just protesters who suffer. Independent diamond miners in Zimbabwe have been assaulted and threatened by security guards working for larger-scale mines whose operators want to maintain a monopoly on the region's treasures. Elsewhere in Africa, minerals and precious gems such as diamonds harvested from areas controlled by warlords—known as "conflict minerals" or "blood diamonds"—have long been used to support armed conflicts. Such violence underscores a point that has long been central to any discussion of minerals' *true costs*: The environment is not the only casualty. Human health and human rights costs need to be considered as well.

In Chile, indigenous protestors want a say in lithium mining that takes place in their ancestral homes.

Child labor is common in above- and below-ground mines. Here, children gold miners are at work in the village of Gam in the Central African Republic.

ISSOUF SANOGO/AFP/Getty Images

Though Chile ratified International Labor Organization Convention No. 169, a treaty that requires governments to consult indigenous peoples who would be affected by projects such as mining, lithium mining began before the treaty was signed. Recent protests are centered on a new deal between the country and a Chilean mining company that would triple production—a move human rights advocates say does fall under the treaty. So far, the Chilean courts have rejected formal requests and appeals by the Atacama People's Council to be given a chance to voice their concerns.

Among the most disturbing human rights violations is the fact that some mining companies routinely violate child labor laws. For example, the International Labor Organization reports that tens of thousands of children work in gold mines. Other children do backbreaking work in artisanal cobalt mines in the Democratic Republic of Congo. (Cobalt makes up the cathode in many rechargeable lithium-ion batteries, and demand for it has tripled over the past 5 years.) In December 2019, Congolese families launched a landmark legal class action suit against Apple, Dell, Microsoft, and Tesla, claiming that their children were killed or hurt while mining for the cobalt used to make these tech companies' products.

Even for adults the work is dangerous. Underground mines can collapse or develop pockets of methane that then cause explosions or gather high concentrations of other hazardous inhalants, leading to lung disorders and other illnesses such as black lung disease, an ailment suffered by coal miners (see Module 9.1). And the hazards don't end when the ore is brought to the surface. Toxic chemicals used in processing can pollute the environment even when protective regulations are followed, but this pollution can reach extreme levels when artisanal miners employ the same range of toxic chemicals used in industrial mining—mercury and cyanide chief among them—in their own crude smelting processes but without safeguards employed at industrial sites. Artisanal gold miners may use their bare hands to mix crushed ore with mercury; mercury concentrations in the air inside such artisanal shops are known to reach 1,000 times the World Health Organization (WHO) limit for exposure. **INFOGRAPHIC 6**

INFOGRAPHIC 6 THE IMPACTS OF MINING

Because so much earth and rock are dug up and moved, mining produces unavoidable environmental impacts. Though they may not be the norm, human rights violations such as displacement, violence, and child labor also occur at some mines.

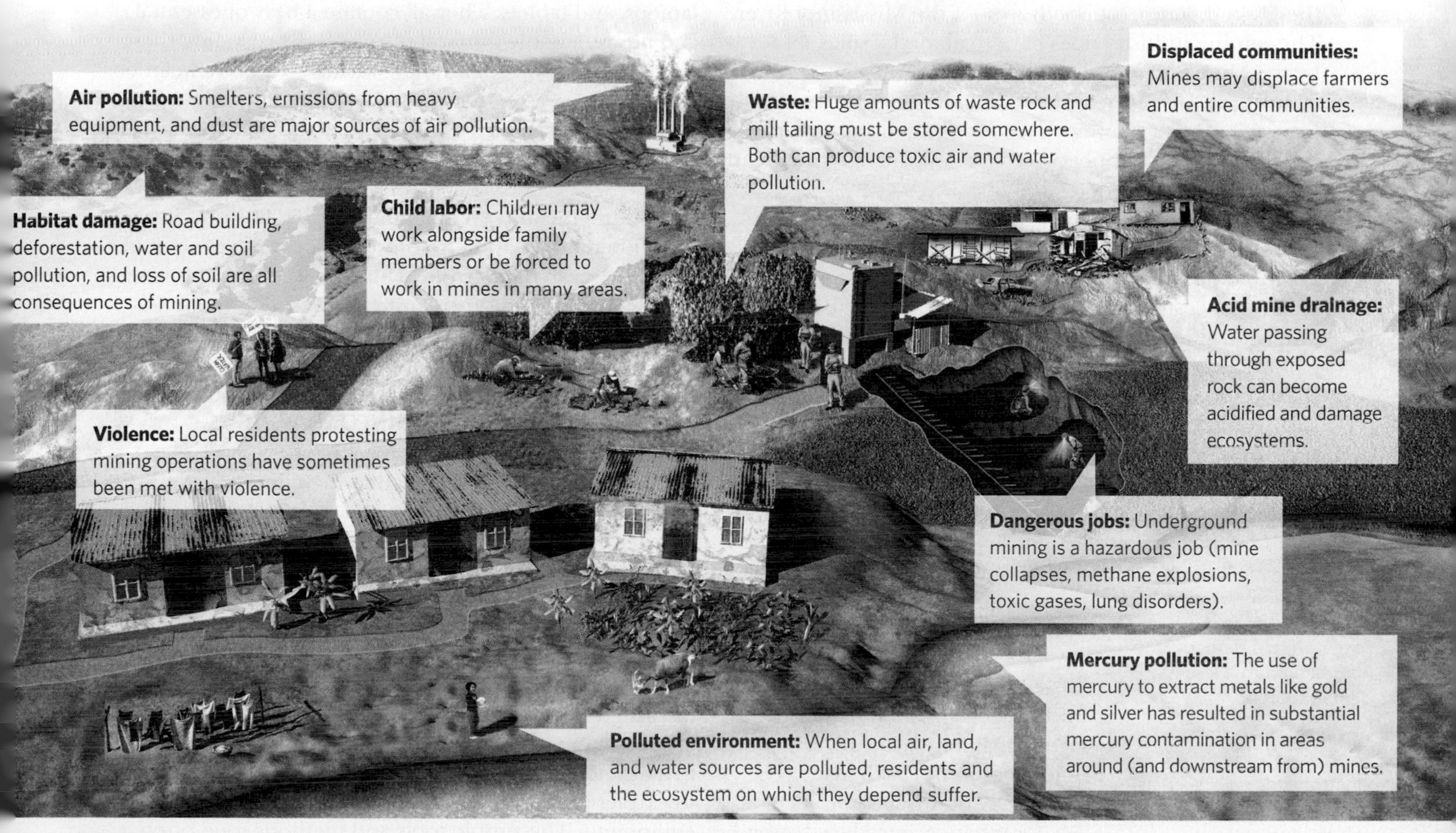
Air pollution: Smelters, emissions from heavy equipment, and dust are major sources of air pollution.
Waste: Huge amounts of waste rock and mill tailing must be stored somewhere. Both can produce toxic air and water pollution.
Displaced communities: Mines may displace farmers and entire communities.
Habitat damage: Road building, deforestation, water and soil pollution, and loss of soil are all consequences of mining.
Child labor: Children may work alongside family members or be forced to work in mines in many areas.
Acid mine drainage: Water passing through exposed rock can become acidified and damage ecosystems.
Violence: Local residents protesting mining operations have sometimes been met with violence.
Dangerous jobs: Underground mining is a hazardous job (mine collapses, methane explosions, toxic gases, lung disorders).
Mercury pollution: The use of mercury to extract metals like gold and silver has resulted in substantial mercury contamination in areas around (and downstream from) mines.
Polluted environment: When local air, land, and water sources are polluted, residents and the ecosystem on which they depend suffer.

? Why do you think human rights violations in the mining industry continue to be a problem in the modern world?

7 REDUCING THE IMPACT OF MINERAL RESOURCE USE

Key Concept 7: The impacts associated with mineral use can be reduced by using safer mining and processing techniques, implementing more efficient manufacturing processes (to reduce waste or the volume of material needed), recycling minerals, and extending the life of mineral-containing products.

We are always going to need mining in order to meet the growing needs of our modern society. Unfortunately, in nations across the globe, regulations often fall short of protecting the environment and surrounding communities from the negative impacts of mining. In the United States, mineral mining on federal land is still largely governed by the *General Mining Act of 1872*, which was put into place to protect miners and their claims and to encourage western settlement. (More recent mining laws uphold the goal of fostering the domestic mineral industry, too.) This law allowed a miner or a group to stake a claim on any public land open to mining and to purchase the mineral rights for as little as $2.50 per acre. Though claim holders must pay a nominal annual maintenance fee ($155 to $165, depending on the type of claim), they do not have to pay royalties on any mining income to the government.

Today much oversight of mining occurs at the state level. Federal mining laws still tend to focus on protecting the rights of miners rather than environmental protection, which has resulted in serious environmental damage in the United States, especially west of the Mississippi River. The mining industry does have to abide by rules set forth by the Mine Safety and Health Administration to protect its workers. However, at the federal level, environmental protection from mining operations largely falls under other laws that might become relevant, such as the Clean Water Act or the Endangered Species Act.

Federal law does mandate, however, that some closed and abandoned mines undergo **reclamation**—the process of restoring a damaged natural area to a less damaged state. In many cases, the responsible parties can't be found or can't pay for reclamation, but when they are identified and they can, they are expected to cover these costs. Closed coal mines, for instance, have to be reclaimed according to the *Surface Mining Control and Reclamation Act of 1977* (see Module 9.1), but hardrock mines, which are mined for mineral ores and metals in hardrock deposits (as opposed to soft rock deposits, such as those that contain coal), do not have such a mandate.

reclamation The process of restoring a damaged natural area to a less damaged state.

e-waste Unwanted computers and other electronic devices such as discarded televisions and cell phones.

On several occasions, Congress has introduced legislation to update the General Mining Act of 1872, but little progress has been made. The Hardrock Mining and Reclamation Act, for instance, introduced several times since 2007, would require that hardrock mining operations pay the government royalties on the minerals they extract, and also pay a fee based on the amount of material they mine toward a reclamation fund, but this law has yet to be passed.

If there is a silver lining to the story of mineral resources—resources that fuel modern life, resources that we dig into Earth to access and fight and even die over—it is this: Many are eminently reusable. In fact, the greatest reserves yet of gold, silver, and lithium may not be hidden in some desert canyon but sitting in the backs of our own closets and garages or scattered throughout our landfills as **e-waste** (electronic waste). An average gold mine produces a mere 5 grams of gold per metric ton of rock—and sometimes less, depending on the ore being mined. A metric ton of cell phones, on the other hand, might contain nearly 200 grams of gold—plus over 100 kg of copper, 3 kg of silver, 1.2 kg of lithium, and a smattering of neodymium and other rare earth elements. Likewise for our flat-screen televisions, laptops, and tablets: They all contain a bevy of essential and nonrenewable mineral resources. By recycling these products, we can reclaim these metals so that they can be used again; e-waste, in fact, may become the main source of many of the REEs used to manufacture electronics as ground reserves become depleted.

To be sure, metal recycling is already big business in the United States and elsewhere. Indeed, the process of recycling scrap metal from vehicles and home appliances has long been an industry unto itself. And aluminum, the most recycled metal on Earth, is so effectively recycled in the United States that, on average, the aluminum in any given beverage can is back on the shelf in 2 months, in another can. **INFOGRAPHIC 7**

It's clear from such successes that recycling can supply some of society's needed minerals and, in doing so, reduce the amount of ore minerals that must be mined. But for that to work, recycling must be done properly. And when it comes to e-waste, we don't yet have widely available recycling programs. Most of our discarded cell phones, computers, and flat-screen televisions collect dust in the crevices of our homes and offices or are discarded with the trash. Even when sent to recyclers, electronics may end up in poorly managed operations in a developing country where low-paid workers labor to extract the various metals contained within the equipment. The simple tools and methods used often

INFOGRAPHIC 7 ALUMINUM RECYCLING: A SUCCESS STORY

Recycling metal products to recover the metals for reuse extends their useful life and reduces the need to obtain the metals from mined ores, decreasing the environmental and health impacts associated with mining and processing the ore. A comparison of the two methods for obtaining aluminum to make a new aluminum can—production from the raw material (bauxite ore) or from recycled aluminum cans—highlights the differences between the two processes, especially with regard to energy savings.

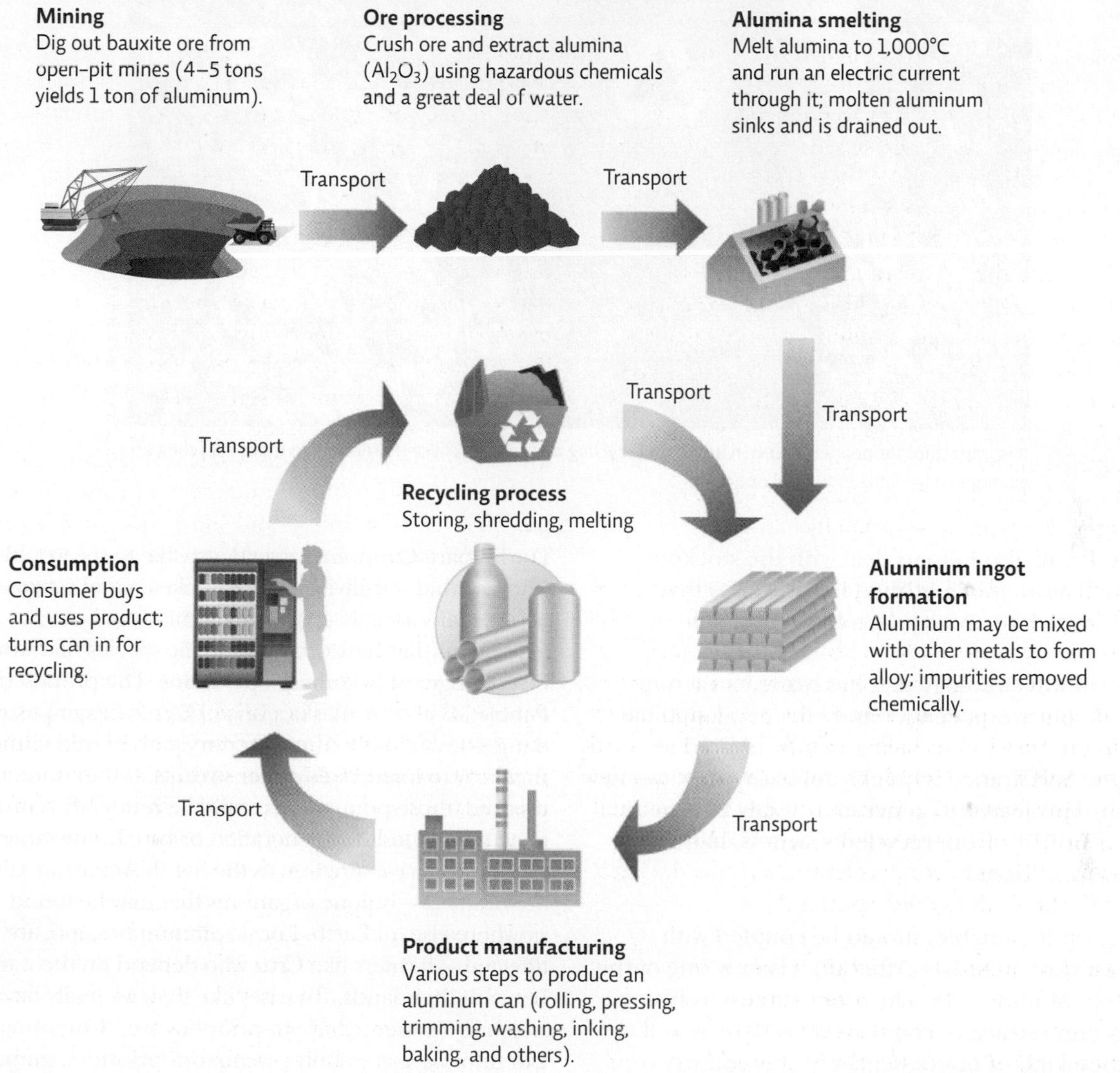
Mining
Dig out bauxite ore from open-pit mines (4–5 tons yields 1 ton of aluminum).
Ore processing
Crush ore and extract alumina (Al_2O_3) using hazardous chemicals and a great deal of water.
Alumina smelting
Melt alumina to 1,000°C and run an electric current through it; molten aluminum sinks and is drained out.
Transport
Transport
Transport
Transport
Recycling process
Storing, shredding, melting
Transport
Consumption
Consumer buys and uses product; turns can in for recycling.
Aluminum ingot formation
Aluminum may be mixed with other metals to form alloy; impurities removed chemically.
Transport
Transport
Product manufacturing
Various steps to produce an aluminum can (rolling, pressing, trimming, washing, inking, baking, and others).

? Why is the energy needed and air/water pollution generated so much less for a recycled aluminum can compared to a can made from aluminum that comes from virgin ore?

expose the workers to significant health hazards and contaminate the local environment. So far, the world has generated more than 45 million tons of e-waste, but only 20% is being safely recycled.

There is, of course, a better way. "Programs similar to the ones we have for plastic recycling, but focused on e-waste, could make a huge difference," says retired mineralogist Robert Housley. "We need a system whereby discarded electronics are picked up from people's homes and delivered to facilities specializing in e-recycling—facilities where lawmakers can mandate—and regulators can enforce—proper health and safety standards."

In other parts of the developed world, an entire industry is being born around this very idea. The growing demand and rising price for elements like gold and neodymium have made recovering them from hybrid vehicles, electronics, and other products an attractive endeavor—one that several big-name producers such as Honda, Toyota, and Hitachi are pursuing. With the ever-increasing growth in the cell phone market and the push toward electric vehicles—both of which rely on lithium-ion batteries—the need to reclaim and recycle lithium, cobalt, and other valuable materials from e-waste rises every day. This will not only help meet lithium and cobalt needs but also reduce the environmental damage that comes from mining and processing these mineral

Erik Isakson/Getty Images

It is important for people to turn in items for recycling and to buy recycled products to make recycling an economically viable industrial option.

resources. It will also help us deal with the stockpile of discarded automobile lithium-ion batteries that is expected to grow to around 11 million by 2030.

Thanks to new technology that has made extracting metals from electronics easier and cheaper, Japan has already opened several recycling plants devoted to electronics. And France is quickly following suit; two new factories are projected to generate roughly 200 metric tons a year of REEs from recycled magnets, batteries, and fluorescent lamps.

Recycling, while valuable, should be coupled with other conservation efforts. After all, it is only one of the four *R*s (see Module 5.3). The other three—refusing, reducing, and reusing—could serve us here as well. "We've gone kind of upgrade mad in this country," Brendan Cummings, an environmental attorney with the Center for Biological Diversity, says. "If we used our cell phones and laptops until they naturally expired . . . we could really make a difference."

Most experts agree that neither recycling nor scientific advances will replace all that we dig from Earth. Indeed, mining will be necessary for many lifetimes to come. This is why we need to do it more responsibly, says Cummings. "The key is to minimize the damage as much as possible." That means employing *best practices*—methods recognized as the most efficient and safest available at the time. More sensitive sensing equipment, for example, reduces the need for exploratory drilling or digging; more energy-efficient and cleaner-running mine equipment can dramatically reduce the carbon footprint.

For his part, Cummings would also like to see would-be miners avoid sensitive ecosystems like coastal estuaries where many aquatic organisms come to spawn or ecosystems that house rare, endemic species that would be endangered by mining operations. The proposed Pebble Mine near Alaska's Bristol Bay, for example, cuts dangerously close to the migratory path of wild salmon on their way to inland freshwater streams. If their migration is blocked, those populations could be rendered reproductive dead ends—lost in a generation or two. In the same way, organisms face extinction as the South American salars are drained—unique organisms that may be found nowhere else on Earth. Local communities, too, are affected—farmers like Cruz who depend on the water beneath their lands. "In cases like that, we really have to weigh as a society what our priorities are," Cummings says. But until we successfully revamp our priorities, mining operations in Chile, the Democratic Republic of Congo, and elsewhere will continue to extract and sell minerals to the electronics industry to feed a hungry public anxious for the latest technology.

Select References:

Cubillos, C. F., et al. (2019). Insights into the microbiology of the chaotropic brines of Salar de Atacama, Chile. *Frontiers in Microbiology, 10,* 1611.

Hudson-Edwards, K. (2016). Tackling mine wastes. *Science, 352*(6283), 288–290.

Jacoby, M. (2019). It's time to get serious about recycling lithium-ion batteries. *Chemical and Engineering News, 97*(28), 1–3.

Sherwood, D. (2019). *Chile protesters block access to lithium operations: Local leader.* Retrieved from https://www.reuters.com/article/us-chile-protests-lithium/chile-protesters-block-access-to-lithium-operations-local-leader-idUSKBN1X42B9

GLOBAL CASE STUDIES MINING "SUPERLATIVES"

Humans have long dug into the ground to harvest valuable mineral resources. In the process, we've created mining sites large enough to be visible from space and we have experienced some tragic mining disasters. Here are some notable examples.

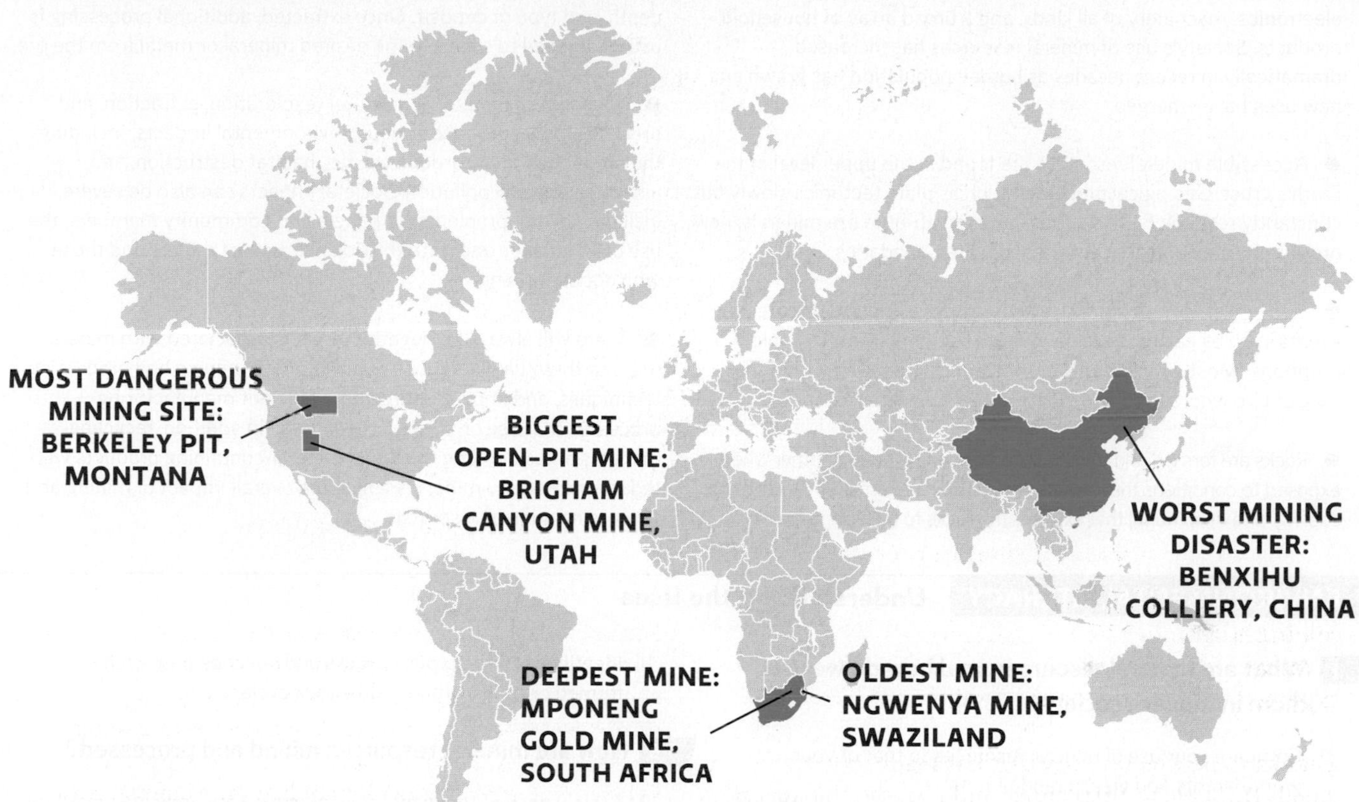

BRING IT HOME

PERSONAL CHOICES THAT HELP

While mining is necessary to extract many of the minerals and metals needed for the technology that runs our society, we can take action to minimize the negative environmental and health impacts.

Individual Steps

- Reduce your use of new consumer goods (e.g., buy used goods or use less) and recycle (e.g., aluminum and steel food and beverage cans; appliances) or reuse (e.g., building materials, automotive parts) when you can. (See Module 5.3 for more suggestions.)
- Hold onto your cell phones, computers, and other electronic devices as long as possible to reduce the need to acquire more minerals for the production of new ones. When you purchase a replacement, recycle the old item. Find a local vendor or electronics recycling event at https://earth911.com/recycling-center-search-guides/.
- Before your next electronic purchase, use the EPA's online Electronic Product Environmental Assessment Tool at https://www.epeat.net/ to evaluate the environmental attributes of computers, TVs, cell phones, and other electronics.

Group Action

- Organize a campus screening and discussion of a documentary on the social cost of mineral mining such as *When Elephants Fight*, a film that examines the toll of conflict minerals in the Democratic Republic of Congo.
- Donate your old computers to a charity such as InterConnection, a Seattle-based organization that refurbishes old computers and donates them to nonprofit groups.
- Sponsor a collection event or permanent site on campus or in your community to collect used cell phones for Cell Phones for Soldiers (www.cellphonesforsoldiers.com), a nonprofit organization that provides free phones to veterans and active-duty military members.

REVISIT THE CONCEPTS

- Ore minerals are used to produce a wide variety of materials we use or depend on every day, such as construction materials, electronics, machinery of all kinds, and a broad array of household products. Society's use of mineral resources has increased dramatically in recent decades as human population has grown and new uses have emerged.

- Accessible mineral resources are found in the upper level of the Earth's crust. Geological processes such as plate tectonics slowly but constantly reshape Earth's surface and redistribute ore minerals; new ores are formed and concentrated at plate boundaries.

- Geological forces contribute to destructive and life-threatening events such as earthquakes, landslides, tsunamis, and volcanic eruptions. Most earthquakes and volcanoes are found within the Ring of Fire where the Pacific Plate meets adjacent plates.

- Rocks are formed and transformed from one type to another when exposed to conditions that include weathering and erosion, melting and cooling, and deep burial that exposes the rocks to high pressure.

- Minerals are extracted from surface and subsurface mines using both high-tech and low-tech methods, depending on the location, depth, and type of deposit. Once extracted, additional processing is usually needed to separate the desired mineral or metal from the ore.

- All stages of mineral acquisition (exploration, extraction, and processing) can produce serious environmental impacts, including the production of hazardous wastes, habitat destruction, and air, water, and soil pollution. Societal impacts can also be severe, including health problems for miners and community members, the use of child labor, and outright violence against miners and those who oppose mining.

- There will always be negative impacts associated with mineral use, but these can be reduced by using safer mining and processing techniques, and implementing more efficient manufacturing processes that use or waste less material. In addition, recycling minerals and extending the life of mineral-containing products will reduce the need to mine, lessening the overall impact of mining and processing mineral resources.

ENVIRONMENTAL LITERACY Understanding the Issue

1 What are mineral resources, and how do we use them in modern society?

1. Compare your use of mineral resources to that of your grandparents and great-grandparents.
2. Distinguish between "minerals" and "ore minerals."
3. List three ways in which you personally use minerals. What is something that you could do to decrease your mineral "footprint"?

2 What geological processes affect Earth's outer layer (lithosphere)?

4. Identify Earth's layers. Include in your answer a definition of the lithosphere and tectonic plates.
5. What happens when an oceanic and a continental tectonic plate converge and when two continental plates converge? Why is the outcome different?
6. Describe the three types of plate boundaries that are seen between tectonic plates. Why is an understanding of plate tectonics useful for mineral mining?

3 What hazards result from Earth's strong geological forces?

7. What can cause an earthquake, and what are its focus and epicenter?
8. Why do volcanoes and earthquakes often occur in similar locations or even occur together?
9. What is the relationship among earthquakes, landslides, and tsunamis?

4 How are rocks formed via the rock cycle?

10. Distinguish between weathering and erosion.
11. Identify the three types of rocks and describe how each is formed or transformed via the rock cycle.

5 How are mineral resources mined and processed?

12. Explain each of these methods of mining and indicate what types of mineral deposits are extracted with each method: subsurface, open-pit, strip, and placer mining.
13. What type of mineral deposits are extracted with solution mining, and how is it done?
14. Looking at Infographic 2, what might be an advantage and a disadvantage of mining at the mid-Atlantic ridge?

6 What environmental and societal problems arise from mineral mining and processing?

15. Why are our methods of mineral ore processing hazardous? Are the processing steps used by small-scale (artisanal) miners safer than those used at large-scale industrial mines? Explain.
16. What is acid mine drainage, and how can it be addressed?
17. What are some of the negative societal impacts that mining may bring about?

7 How can the impact of using mineral resources be decreased?

18. Why is it important for consumers to turn in materials for recycling and to buy recycled products?
19. Identify ways that the environmental impacts of mining can be reduced.
20. Give an example of how each of these can reduce the amount of mining needed to meet society's need for minerals: electronics manufacturers, e-waste recyclers, and consumers.

SCIENCE LITERACY Working with Data

Lithium mining began in the late 1800s in the pegmatite mines of the United States. Today the major producers of lithium are Australia (pegmatite) and Chile (brine salars). The U.S. Geological Survey (USGS) estimates global lithium resources (those deposits with reasonable prospects for eventual economic extraction) to be more than 39 million metric tons, enough, they predict, to meet global demand to the year 2100. Refer to this area graph with data from 1935 to 2007 to answer the following questions. (Consult online Appendix 2 for help reading an area graph.)

WORLD LITHIUM PRODUCTION BY DEPOSIT

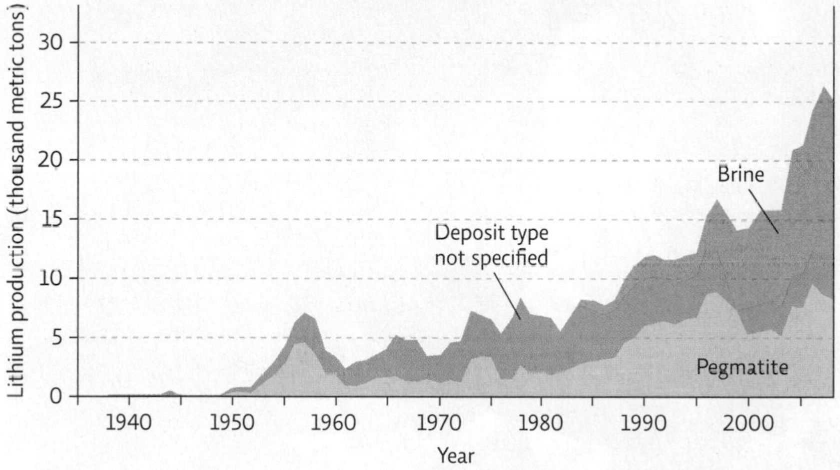

Interpretation

1. In metric tons, about how much lithium was harvested from each source in the year 2000?
2. In which year did brine production of lithium first exceed that of pegmatite sources? Did pegmatite production ever exceed that of brine production after that date?
3. Estimate the total increase in lithium production between 1975 and 2007.

Advance Your Thinking

4. What likely accounted for the rapid increase in lithium production that began in the late 20th century?
5. If global lithium resources total 39 million metric tons, about how much of that total was harvested in 2007?
6. If we exhaust world lithium supplies by 2100, as the USGS projects, what do you predict will happen to the future annual rate of usage as compared to 2007 — will it increase, decrease, or stay the same? Explain.

INFORMATION LITERACY Evaluating Information

As cell phone use has skyrocketed, more energy and mineral resources are needed to produce the phones and more waste is generated when they are produced and when they are discarded. Recycling our cell phones can decrease the environmental impact of our increasing use of electronic devices. How are we doing?

Investigate the Topic

Look online for information to answer this question: What percentage of cell phones are recycled in the United States? Using the search engine of your choice, search the Internet for information using key words such as *cell phone recycling statistics*. Select three different information sources and answer the following questions about each source and the information you uncovered:

1. What was the title of the webpage or article you visited?
2. What made you choose that website or article?
3. Summarize the information presented. Did it provide the statistics you were looking for?

Evaluate the Information Sources

4. Is the information current? What was the latest year for which statistics on cell phone recycling were provided by this information source?
5. Evaluate the quality of the information source:
   a. Who provided the information? Give the author of the article and the author's credentials, or the organization responsible for the website and their credentials.
   b. Is this author/organization qualified to present information on this topic? Explain.
   c. Is supporting evidence given for claims made in the article or on the website?
   d. Does the author or organization appear to be biased in its presentation of information? Explain.
6. Of the three information sources you visited, which do you feel was the most reliable and useful information source? Explain.
7. Did you find the answer to the question posed? If not, how would you modify your search to find that information?

Additional study questions are available at Achieve.macmillanlearning.com. Achieve

Module 7.2 Forest Resources

RETURNING TREES TO HAITI

Repairing a forest ecosystem one tree at a time

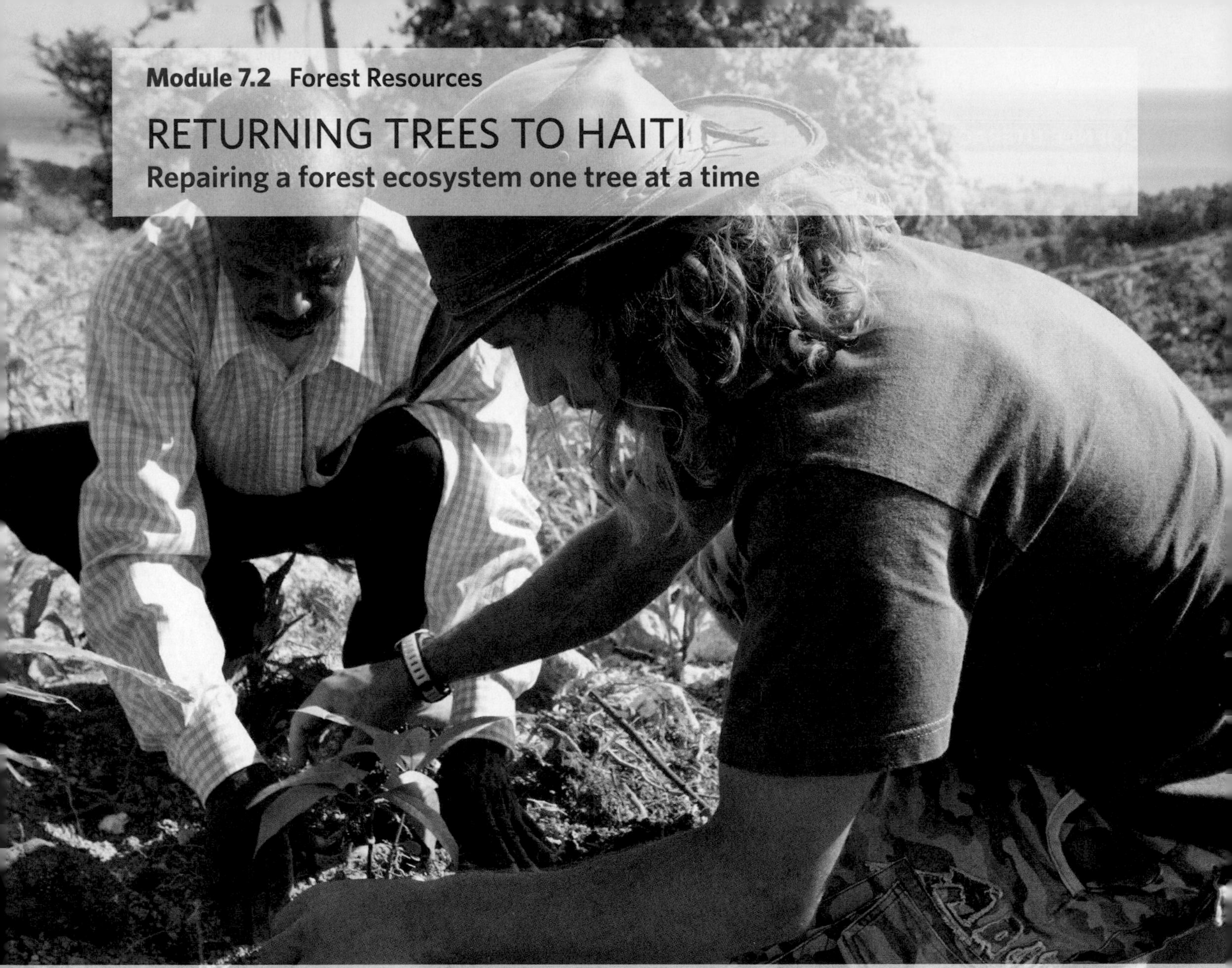

Volunteers plant trees near Port Salut, Haiti. *JS Callahan/tropicalpix/Alamy*

After reading this module, you should be able to answer these GUIDING QUESTIONS

1. What is a forest, and what influences which forest type is found in a given area?
2. What is the three-dimensional structure of a forest, and how are the plant species found there adapted to their level of the forest?
3. What ecosystem services do forests provide?
4. What threats do forests face?
5. What timber-harvesting techniques are available, and how do they compare in terms of method and sustainability?
6. How can we protect and sustainably manage forest resources?

When Jean Robert was a young boy, the mountains surrounding the Haitian city of Gonaïves (pronounced go-nah-EEV) were still lush and green with trees and his family's hillside farm produced a variety of crops. But today the forests are gone, and the mountains are bare. Besides crop yields having shrunk drastically, the mountain homes and the city below have been left defenseless against the tropical storms that pound Haiti every summer. Unencumbered by trunks or roots or shrubs, water rushes freely downward, gathering into apocalyptic mudslides that destroy homes, crops, and livelihoods. In 2004, a single storm claimed more than 2,000 lives from this one city.

Now Robert and his neighbors are trying to bring back trees. Working as a team, the community work group, or *kombit,* plants a carefully planned mix of fruit and timber trees. By the end of the month, they say, each member will have his or her own saplings to tend. If the saplings survive, their efforts will give rise to a new era of sustainable forestry, which strikes a balance between what people need from the forest and what the forest itself needs to survive.

The kombit will plant three main types of trees. They start with fast-growing, multipurpose trees like the moringa. Because it is a nitrogen-fixing plant, the moringa helps fertilize the soil. And because it grows quickly, it can provide a sustainable source of food (the leaves are edible) and fuel wood. Next, they plant fruit trees — mangoes, avocados, and citrus. These trees take longer — 3 to 5 years — before their fruit is ready to harvest, but they put down roots and thus stabilize the soil in just a few months' time. "Farmers are less likely to cut down fruit trees for charcoal, because they know it will provide the kind of food they can both eat and sell for profit," says Haitian ecologist Timote Georges, who is working with the U.S.-based nonprofit Trees for the Future to reforest the mountains around Gonaïves. The last thing the kombit plants is the slow-growing timber trees, which may be sustainably and profitably harvested in the future but don't provide any immediate benefits to the farmers. "The goal is to mix it up," says Georges. "You can create a whole stable system that's going to provide money and food throughout the seasons and across the years."

WHERE ARE HAITI AND GONAÏVES?

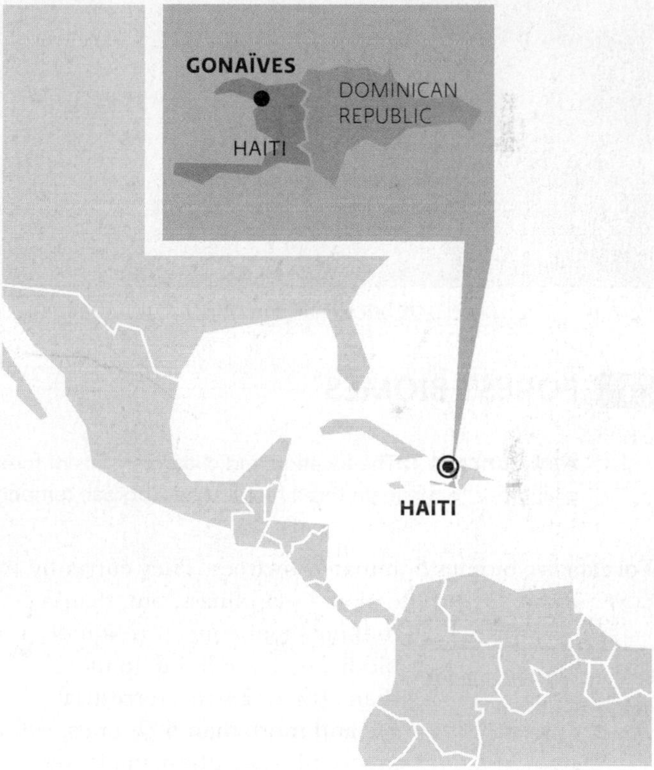

But to really understand how trees might help alleviate poverty and protect communities, we must first consider just what forests are, what they do for us, and why so many of them are being chopped down in the first place.

AP Photo/Ariana Cubillos

Volunteers begin planting 25,000 donated trees in Mahotiere, Haiti, to combat soil erosion.

1 FOREST BIOMES

Key Concept 1: The location and characteristics of forest biomes are influenced by temperature and precipitation, giving rise to the main three forest types (boreal, temperate, and tropical) and a variety of subtypes of each.

forest An ecosystem made up primarily of trees and other woody vegetation.

boreal forest Coniferous forest found at high latitudes and altitudes characterized by low temperatures and low annual precipitation.

temperate forest Forest found in areas that have four seasons and a moderate climate, receive 30 to 60 inches of precipitation per year, and may include evergreen and deciduous conifers and broadleaf trees.

Forests are biomes dominated by trees. They currently cover about 25% of the planet's landmass, but, thanks to their sheer concentration of biodiversity, are home to more than 50% of Earth's terrestrial life and more than 60% of its green, photosynthesizing leaves. There are many types of forest biomes around the globe, each determined by the temperature and amount of precipitation the area receives.

Boreal forests, characterized by evergreen coniferous (cone-bearing trees with needles or scale-like leaves) species like pine, spruce, and fir, cover vast tracts of land in the higher latitudes and altitudes and are characterized by low temperatures and precipitation levels. They represent some of the most expansive forests left on Earth — some of these stands are still *primary* or *old-growth* forests; other areas that are harvested for timber have regrown as *secondary* forests.

Temperate forests, which contain deciduous (trees that shed leaves seasonally) broadleaf trees like oak, hickory, and maple that lose their leaves in winter, are found in midlatitudes. (Evergreen broadleaf trees like the southern magnolia are found in warmer regions of these latitudes.) These forests are not as expansive as the boreal forests because they are found in latitudes with high human populations. Most temperate forests today represent secondary growth rather than primary stands due to extensive timber harvesting. However, some areas have seen remarkable regrowth in the 20th century.

The original forests of Haiti would have looked like this one on the Barahona Coast in neighboring Dominican Republic.

For example, there is now more forest cover in the eastern United States than there was in the 19th century, following heavy clearing. (This regrowth can be sustained because forests elsewhere are harvested.)

Tropical forests contain a diverse mix of tree and undergrowth species and are found in tropical latitudes where temperatures do not vary much throughout the year. Though vast tracts of primary tropical forest still remain, these forests are experiencing the highest losses of any forest biome, especially in Southeast Asia and South America. **INFOGRAPHIC 1**

When Europeans first arrived in Haiti some 500 years ago, two-thirds of the land was covered in forests so majestic that explorers dubbed this region "the Pearl of the Antilles." But as centuries passed, the forests were cleared — sometimes with amazing breadth to make way for coffee and sugar plantations, and at other times they were cleared in discrete chunks that Haitians would then convert into subsistence farms. Many of the trees — along with the coffee and sugar crops that displaced them — were sold overseas. Most of the rest provided fuel for cooking Haitian meals.

In 2018, Temple University researcher Blair Hedges estimated that less than 1% of that original forest remains; 6% of the land has no soil left at all. This isn't to say that Haiti has few trees. As much as 30% of the land area contains trees, but almost all of this is secondary growth existing in small patches, and many of these forest fragments are too small to offer suitable forest habitat for local species. If deforestation continues along its current path, Hedges cautions that Haiti could lose all its primary forest — and as many as 83% of the species that depend on it — in the near future.

tropical forest Forest found in equatorial areas with warm temperatures year-round and high rainfall; some have distinct wet and dry seasons, but none has a winter season.

INFOGRAPHIC 1 FORESTS OF THE WORLD

Forests are biomes primarily made up of trees and other woody vegetation. There are three main types of forests, classified according to climate, and many subdivisions within these three types.

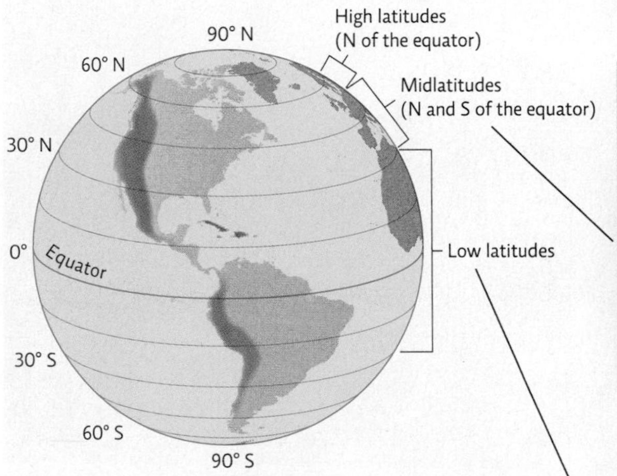

Boreal forests (taiga) represent the largest terrestrial biome; they stretch from Canada to Siberia and are found at higher elevations in lower latitudes. They have a short growing season with little precipitation, most of it snow. Soils here are thin and acidic, and the major tree species are evergreen conifers with needlelike leaves.

Temperate forests have distinct seasons. The soil is fertile, with a thick layer of decomposing leaf litter that supports the plant life. Depending on how much precipitation an area gets, the forest may be predominantly coniferous or broadleaf.

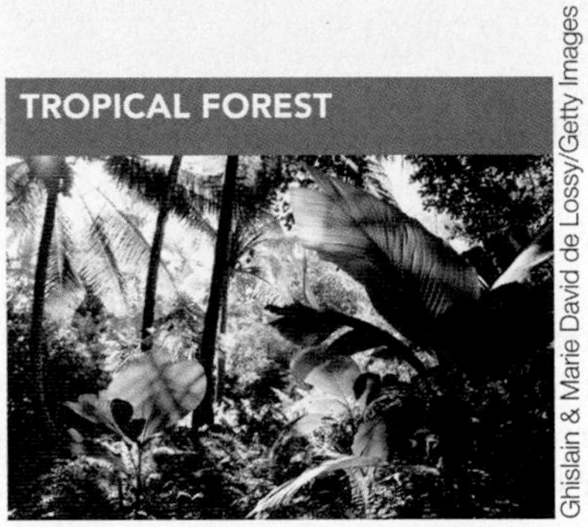

Tropical forests have similar temperatures year-round. Dry tropical forests have distinct wet and dry seasons, whereas tropical rainforests receive rain year-round. The soils are thin, acidic, and low in nutrients. Rapid decomposition by fungi and bacteria supports the dense vegetation found in these soil-poor areas.

? Which forest type is the most expansive? Why do you think this is true?

2 FOREST STRUCTURE

Key Concept 2: Forests are stratified, having distinct vertically stacked layers, each of which contains species adapted to the level of sunlight and moisture available.

canopy The upper layer of a forest, formed where the crowns (tops) of the majority of the tallest trees meet.

emergent layer The region where a tree that is taller than the canopy trees rises above the canopy layer.

Most forests consist of four distinct layers. The **canopy**, formed by the overlapping crowns of the tallest trees, makes up the ceiling of the forest. In some cases, an even taller tree may reach above the canopy as part of an **emergent layer**. Beneath the canopy is the **understory** layer, where shade-tolerant shrubs, smaller trees, or the saplings of larger trees grow. Sometimes these trees are dense enough to form a lower canopy. The lowest level is the **forest floor**, which is typically made up of seedlings, herbs, wildflowers, and ferns. The forest floor also contains soil, which is composed of leaf litter and other debris—branches, logs, and stumps—that decompose over time.

INFOGRAPHIC 2 **CROSS SECTION OF A FOREST**

A look at the three-dimensional structure of a forest reveals layers that house distinct species adapted to the environmental conditions in each layer.

Emergent layer: A few trees grow above the general level of the forest canopy.

Canopy: The crowns of the dominant trees shade the layers below.

Understory: Trees and shrubs here are adapted to shade; saplings will grow rapidly when a spot in the canopy opens up.

Forest floor: This lowest level contains leaf litter, decomposing plant and animal material, herbs, flowers, and seedlings.

What kind of plants might be found in a region of the understory that has been opened up to sunlight by a fallen tree?

Within each forest layer is a range of species uniquely suited to the temperature, humidity, and amount of sunlight reaching that layer—species well adapted to their particular neighbors. For instance, in a temperate deciduous forest, wildflowers on the forest floor will bloom early in the spring, before the bigger trees "leaf out" and block the Sun. Sunlight is a precious commodity on the forest floor, and wildflowers compete for it. One wildflower species may bloom one week and another the next week. Disruption to one part of the forest (for instance, cutting down a tree that opens the canopy) can have a trickle-down effect that impacts each subsequent layer. **INFOGRAPHIC 2**

understory The smaller trees, shrubs, and saplings that live in the shade of the forest canopy.

forest floor The lowest level of the forest, containing herbaceous plants, fungi, leaf litter, and soil.

3 ECOSYSTEM SERVICES OF FORESTS

Key Concept 3: Forests provide a wide variety of critical ecological services to water, soil, air, and wildlife; have great economic value (e.g., jobs, consumer products); and provide sociocultural benefits (e.g., cultural, spiritual, and recreational value).

Due in part to unchecked deforestation, Haiti has become both the poorest and the most environmentally degraded country in the Western Hemisphere. With few trees and little soil, drinking water has grown polluted, crops have dwindled, and the people have suffered.

Farmers like Jean Robert say that forests are the key to changing all of that. They have planted thousands of trees in these mountains in recent years and hope to plant thousands more. "Almost all of the country's problems—natural disasters, food shortages, poverty—can be traced back to rampant **deforestation**," says Georges. "So if we want to fix the country, we have to put the forests back and then find a way to manage them better."

deforestation Net loss of trees in a forested area.

That's no small feat. Forests, after all, are one of our most contentious resources. Such is the range of **ecosystem services** they provide that any one use must be weighed against a host of others and that immediate human needs are frequently pitted against long-term conservation goals. In Haiti, where most people live on less than $2 a day, trees provide food, energy, building material, and desperately needed income. To stop them from being chopped down, Georges and his fellow Haitians will have to find alternative sources of each, not to mention a farming method that doesn't require clearing more forestland.

Food, fuel, and building materials are perhaps the most obvious ecosystem services provided by forests. However, even though trees are the largest and most notable life-forms present, a forest is much more than just its trees. Together, the species inhabiting the different layers participate in a delicate symphony of chemical and physical cycles that produce an invaluable range of ecosystem services for the planet.

It starts with the soil, which the forest itself helps to form and maintain: Leaves and branches fall to the ground and decay, forming a brown layer of nutrients in which all future generations of plant life will take root. (For more on soil, see Module 8.2.)

While dead and decaying plants help form the soil, living ones hold it in place. During rainstorms, roots—especially tree roots—anchor soil in place so that it can't flow as easily down hillsides into nearby surface waters (lakes, streams, oceans) with the rain. Soil and roots also slow the flow of rainwater across the ground's surface (called **stormwater runoff**), preventing potentially polluted water from reaching surface waters. And by slowing runoff, a well-vegetated area allows more water to soak into the ground, recharging the groundwater supplies that provide area residents with their major source of drinking water. The soil also traps chemicals that might otherwise contaminate that drinking water. (See Module 6.2 for more on stormwater runoff.)

ecosystem services Essential ecological processes that make life on Earth possible.

stormwater runoff Water from precipitation that flows over the surface of the land.

carbon sink An area such as a forest, ocean sediment, or soil, where accumulated carbon does not readily reenter the carbon cycle.

While plants, soil, and water are playing off one another in this manner, the forest is also conducting another important cycle: pulling carbon dioxide (CO_2) out of the atmosphere and replacing it with oxygen (through the process of photosynthesis)—forest leaves produce so much oxygen that they are commonly referred to as "the lungs of the planet." But it's the ability of forests to capture atmospheric CO_2 that is garnering more and more attention in a world struggling to address climate change. Forests as a whole store more carbon in their biomass, litter, and soil than all the carbon in the atmosphere, making this biome the world's largest terrestrial **carbon sink**—an area that stores more carbon than it releases, such as the standing timber in a forest or organic matter in soil. In fact, reforestation on a grand scale is seen as a vital tool in the fight against climate change—by one calculation, the planting of 1.2 trillion trees (there are about 3 trillion on the planet today) would offset 10 years of human CO_2 emissions. (The oceans hold more carbon, but their capacity to absorb more may be diminishing; see Online Module 6.3.)

Last but not least, virtually every layer of forest provides food and habitat for a bevy of animals (vertebrates and invertebrates), fungi, and microbes; these creatures all do their part to contribute to the functioning of the forest ecosystem as a whole. It is difficult to overstate the importance of forests to the biosphere.

The sociocultural benefits are also far-reaching. Some forest stands are considered sacred by indigenous people, connecting them to their past and representing their future. Many people view a forest as the ideal place to reconnect with nature, become rejuvenated, or find

JAMES P. BLAIR/National Geographic Creative

South of Dajabón, Haiti's brown landscape on the left contrasts sharply with the rich forests of its neighbor, the Dominican Republic, on the right.

artistic inspiration—benefits that, for some, can't be replicated in our modern, built-up world.

In addition to all these ecosystem services and sociocultural benefits, forests also provide a range of economic benefits. In fact, humans have relied on forests for millennia for a multitude of consumer goods. Wood products, including lumber, firewood, charcoal, paper pulp, and some medicines, account for around $100 billion in global trade every year. On top of that, the wildlife supported by forests provides humans with food and recreational hunting opportunities, both of which have economic value. **INFOGRAPHIC 3**

However, taking advantage of the economic benefits of a forest can lead to its loss and the loss of the ecosystem services it provides. In Haiti, the cost of deforestation spun out over decades and proved catastrophic. Soil eroded into streams, rivers, and gullies, clogging them with sediment and disrupting aquatic ecosystems. With nothing to absorb the water, floods became more severe, and groundwater sources were quickly depleted as the water flowed away rather than soaking into the ground. Slowly, crop yields shrank. As the forest habitat was fragmented, biodiversity dwindled. And as the trees vanished, the people of Haiti suffered. Unlike wealthier countries, Haiti could not afford expensive water purification systems—a service the forests once provided for free. "It took some years before we could feel the other effects of deforestation," says Georges. "The floods got worse, and we lost drinking water to runoff and pollution. And once those problems started, there was no easy way to fix them."

As time passed, more and more people abandoned the countryside for the capital city. Even that did not stop the tide of deforestation. As the population swelled, so did the demand for fuel, which in Haiti still comes almost exclusively from trees.

INFOGRAPHIC 3 ECOSYSTEM SERVICES OF FORESTS

ECOLOGICAL SERVICES	ECONOMIC VALUE	SOCIOCULTURAL BENEFITS
Watershed services: Water purification and provision. **Climate regulation:** A major sink for CO_2; increases rainfall in some areas; biggest contributor of oxygen to the atmosphere. **Soil maintenance and protection:** Soil production and recycling of nutrients; reduction of soil erosion. **Disturbance regulation:** Protection from storm damage, especially in coastal areas. **Habitat and genetic resources:** Food and habitat for biodiversity; a rich storehouse of genes that might prove useful to improve our crops or provide as-yet-undiscovered medicines.	**Goods:** Provide many of the basic goods we depend on, including: • Food • Fuel • Building materials • Other products, such as rubber and cork • Raw material for paper and other industrial products • Medicines **Recreation and ecotourism opportunities** **Jobs:** More than 10 million people make their living in and from forests.	The beauty of forests provides a place for spiritual renewal, artistic inspiration, and stress reduction. Ancient stands of trees provide a connection to the past; many indigenous people are an important part of their forest ecosystem, possessing ancestral knowledge of the forest and its inhabitants.

Robert Costanza and his colleagues evaluated ecosystem services of forests and quantified their value (in 2007 U.S. dollars) to be $16 trillion. The values for some of those services are shown below.

ESTIMATED ANNUAL VALUE OF THE ECOSYSTEM SERVICES OF FORESTS

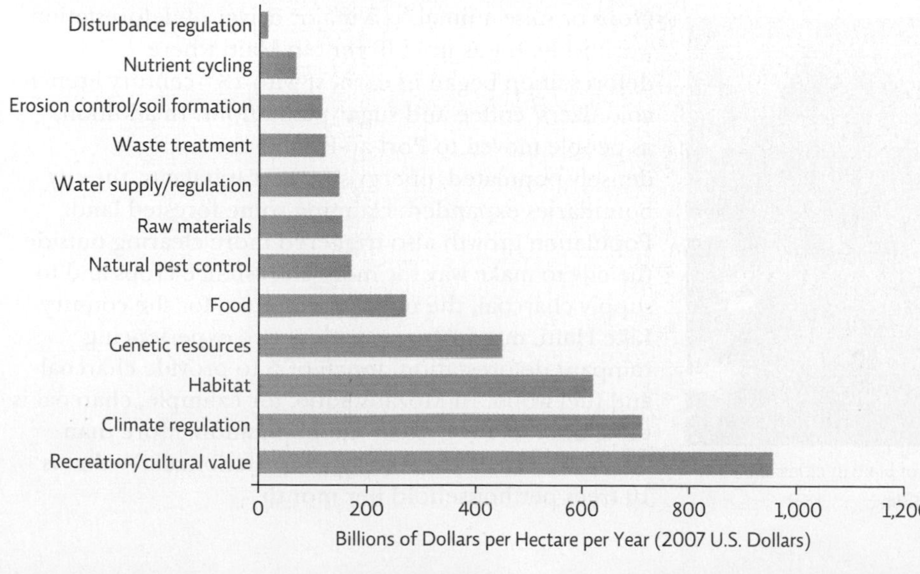

How might the local climate of an area that undergoes heavy deforestation change?

4 THREATS TO FORESTS

Key Concept 4: Globally, forest cover is shrinking due largely to harvesting for lumber and fuel, clearing for agriculture or mining, pest infestations, and fire suppression.

Over the last 300 years, at least 35% of the world's forests have been lost—many of them cut down for the timber they hold or the land they occupy. Of those forests that remain, more than 80% are degraded. According to the United Nations (UN) Food and Agriculture Organization (FAO), today global deforestation has slowed, from 20 million acres (ac) per year (in the 1990s) to 19 million ac each year between 2010 and 2015. Forest gains between 2010 and 2015 helped offset this loss, resulting in a net decrease of 8.2 million ac.

While global deforestation rates have declined in most areas, they are increasing in some tropical areas, such as Indonesia and Malaysia. (See Module 3.2 for more on deforestation for oil palm plantations in these regions.) Brazil, where rates had been declining, began experiencing a sharp uptick in deforestation in 2019. This increase is linked to policy changes put in place by the newly elected president, Jair Bolsonaro, that according to Minister of the Environment Ricardo Salle are "dismantling the country's environmental agencies and effectively halting fines for illegal clearing."

Some of our actions intentionally lead to the deforestation of tropical (and other) forests. The most common culprits are the harvesting of forests for wood and wood products and the removal of forests to use the land for other purposes such as agriculture (see Chapter 8) and mining (see Module 7.1). Growing populations are also spreading out as urban and suburban areas claim forested land (see Module 4.2).

The way we manage fire is also a factor in forest destruction, especially temperate forests. Frequent fires remove deadwood and other flammable material. If we suppress these fires, the deadwood builds up so that when a fire does come through, it can burn so hot it catches the entire forest on fire. Fire is actually needed to maintain some forests whose trees are fire-adapted, with seeds that germinate only when exposed to the heat of a fire.

Climate change, too, is impacting forests. Warmer winters allow pest populations that would normally die back during cold months to attack trees year-round, taking a toll on temperate and boreal forests. Though extra atmospheric CO_2 (a major cause of climate change) and warmer temperatures may extend the growing season in some areas, precipitation declines will negate this benefit in many places. The incidence of fire has also increased, especially in drought-stricken areas; this directly destroys forests, and some may not be able to recover as easily as in years past due to an altered climate. In general, range shifts for many tree species will occur (and are already occurring) as young trees take root and grow with more success at the highest latitudes and altitudes of their historic range; trees growing at the lowest latitudes or altitudes in their current range may die out, and the forest may be replaced with a different ecosystem (different type of forest) or even a different biome (shrubland or grassland). Steps to address climate change and restrict global warming to no more than 3.6° F are needed to protect the forests that remain. (See Module 10.2 for more on climate change.)

The nature and degree of each of these threats vary by country because forests are often used and managed differently in developing countries than they are in developed ones. **INFOGRAPHIC 4**

As mentioned earlier, the clearing of land to grow crops or raise animals is a major driver of deforestation worldwide. It was no different in Haiti where deforestation began in earnest with 18th-century French colonizers' coffee and sugar plantations. In addition, as people moved to Port-au-Prince, the nation's densely populated, energy-starved capital city, the city boundaries expanded, claiming some forested land. Population growth also triggered more clearing outside the city to make way for more subsistence crops and to supply charcoal, the major fuel source for the country. Like Haiti, many African nations are experiencing rampant deforestation, much of it to provide charcoal and fuel wood. In Mozambique, for example, charcoal is the main fuel for 80% of the population. More than 2.5 million trees are cut each year in Somalia—that is 10 trees per household per month.

Abbie Trayler-Smith/Panos

Charcoal is the main fuel source for the people of Haiti in cities like Port-au-Prince, shown here, and in the countryside.

INFOGRAPHIC 4 THREATS TO FORESTS

A variety of "drivers" are responsible for deforestation around the world.

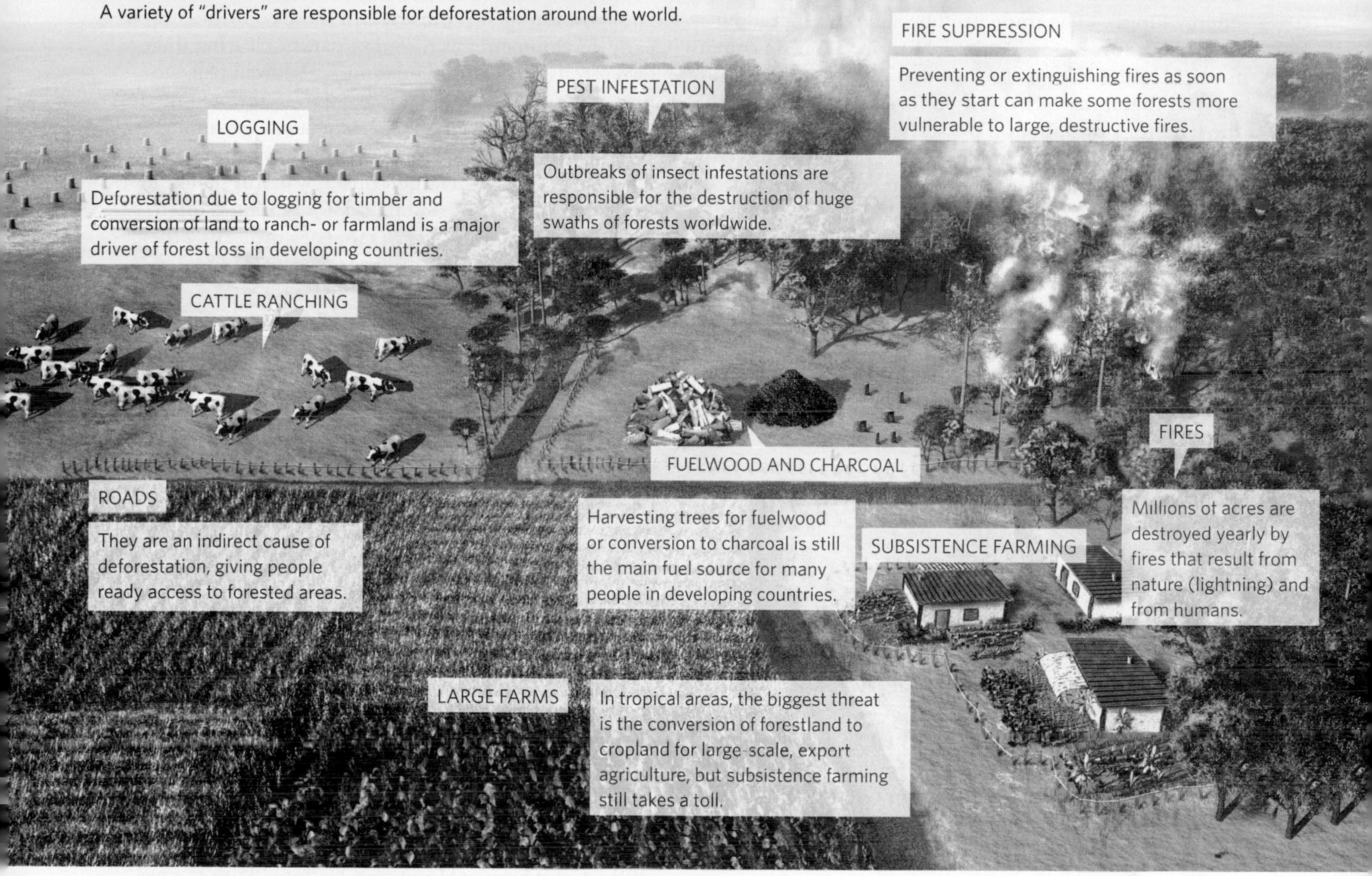

FIRE SUPPRESSION
Preventing or extinguishing fires as soon as they start can make some forests more vulnerable to large, destructive fires.
PEST INFESTATION
Outbreaks of insect infestations are responsible for the destruction of huge swaths of forests worldwide.
LOGGING
Deforestation due to logging for timber and conversion of land to ranch- or farmland is a major driver of forest loss in developing countries.
CATTLE RANCHING
FIRES
Millions of acres are destroyed yearly by fires that result from nature (lightning) and from humans.
FUELWOOD AND CHARCOAL
Harvesting trees for fuelwood or conversion to charcoal is still the main fuel source for many people in developing countries.
ROADS
They are an indirect cause of deforestation, giving people ready access to forested areas.
SUBSISTENCE FARMING
LARGE FARMS
In tropical areas, the biggest threat is the conversion of forestland to cropland for large-scale, export agriculture, but subsistence farming still takes a toll.

NET CHANGE IN FOREST AREA BY COUNTRY, 1990–2015

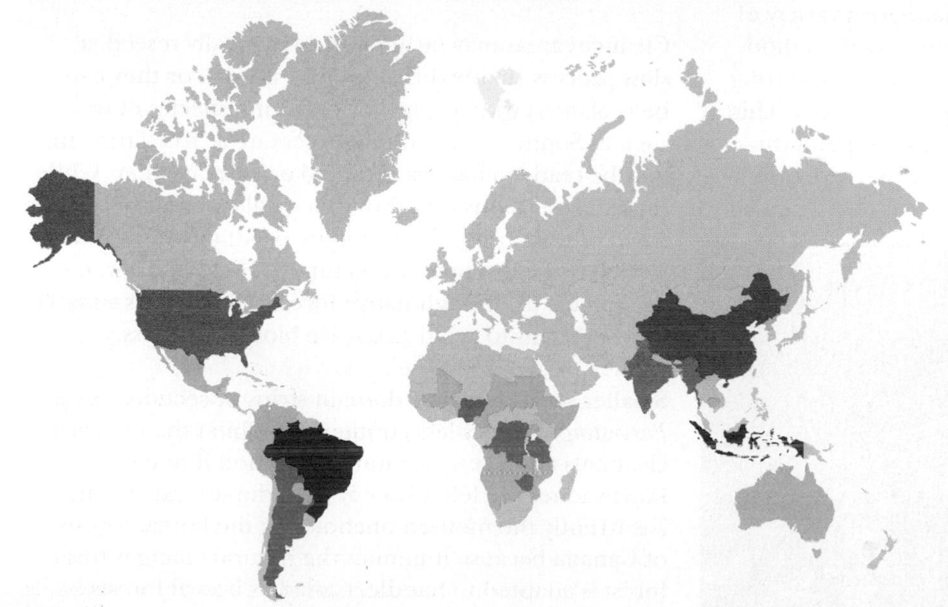

? How does the presence of roads lead to loss of forests?

In time, Haiti's charcoal trade grew to account for 20% of the rural economy and 80% of the country's energy supply. "The trade itself became this incredibly destructive force," says Andrew Morton, a forest ecologist for the United Nations Environment Programme. "And the fact that it was based not on foreigners exploiting the land for profit, but on poor Haitians trying to earn money and feed their families . . . made it impossible to stop. There really was no other source of energy."

In most cases, developing countries have far greater remaining forest stands than developed countries, but many of these stands are falling fast in places like Haiti, where dire poverty and a lack of alternatives force people to harvest their forests or remove them for other land uses. They need wood and charcoal for fuel and housing, they need more space for agriculture, and they need commodities to sell in the marketplace.

Deforestation rates are lower in many, if not most, developed countries, including the United States, Canada, and most European countries. In these countries, whose early development was fueled in part by clearing their own forests, stringent regulations are now in place to protect those forests that are left, and reforestation efforts have increased forest cover in some areas, including the eastern United States. Unfortunately, any forest gains in these areas are still largely offset by deforestation in developing countries, and that deforestation is often linked to developed countries that are protecting their own forests. "Industrial countries may be leading the way in conserving their own forests," says Morton. "But their demand for wood drives much of the deforestation elsewhere." Multinational corporations have simply moved deforestation operations to developing nations, where regulation and enforcement are often lacking and people are desperate for income.

5 TIMBER HARVESTING: SUSTAINABLE OPTIONS

Key Concept 5: Selective and shelterwood timber harvesting maximize long-term profits and protect biodiversity; clear-cutting methods maximize short-term profits at the expense of the local ecosystem. Tree farms are a highly productive alternative that may allow managers to leave natural stands alone.

Harvesting forests for timber or other forest products is not something that is going to stop—and there is no reason to expect that it must. There are ways to sustainably use forests. In Haiti, for example, much of the wood now harvested for charcoal production is provided by trimming branches rather than cutting down the entire tree—a method that appears to be sustainable.

Of course, there are times when larger-scale timber harvesting is needed and for that there are a variety of methods that can be used. A commonly used method is known as a **clear-cut**—cutting all the trees in a given area. This method maximizes short-term profit but does so at the expense of the local ecosystem—the forest habitat is gone, at least for a while, and the risk of soil erosion is high especially if the harvest was done on a slope. Destructive mudslides that invaded Asheville, North Carolina, in the early 1900s were the devastating consequence of extensive clear-cutting on the surrounding mountains. (This rampant deforestation spurred the establishment of the Great Smoky Mountains National Park to save those mountains from the same fate.)

clear-cut Timber-harvesting technique that cuts all trees in an area.

robert cicchetti/Shutterstock

Clear-cutting on a slope is never advised due to the tremendous potential for soil erosion. Without trees to hold the soil in place, it washes down the hill with every rainfall.

Clear-cut areas may be allowed to naturally reseed, a slow process (made difficult if soil is lost), or they can be replanted with a single fast-growing species of tree such as Southern yellow pine to create a "tree farm" that may be ready to harvest again 50 or 60 years later. While these forest stands lack the ecological diversity of a native stand of trees, they can be highly productive and reduce pressure to cut remaining natural forests. (However, there must be enough native forests left in areas suitable for tree farms to avoid extensive biodiversity loss.)

Smaller scale clear-cuts done in strips or sections (*strip harvesting*) destroy less contiguous habitat than massive clear-cuts and cause less water pollution if uncut buffer zones are left adjacent to bodies of water. This is currently the method of choice in the boreal forests of Canada because it mimics the natural changes this forest is adapted to handle. Canada's boreal forests are dynamic and ever changing largely due to natural disasters (e.g., fire or insect infestations) that open up sections of forest. These forest openings will be colonized by seedlings from neighboring trees, creating an even-aged stand that grows until the next disturbance. Trees and wildlife in these areas are adapted to tolerate this type of forest loss as long as the area cut is not too large, and nearby strips or "islands" of standing trees are left untouched.

Still, clear-cutting is often ill-advised as it is very damaging to the integrity of most forests. Luckily, it is not the only profitable method available for harvesting timber. **Selective harvesting** and **shelterwood harvesting** reduce disruption to the ecosystem while still providing economically valuable forest products. In natural forests, these methods have lower environmental impacts and better long-term yields than clear-cuts. In each method, only some trees are harvested (the choice of tree differs between the two methods), leaving behind an area that still contains trees. Foresters can return to the area to harvest additional trees much more quickly than an area that has been clear-cut, providing steady income in a given forest.

These appear to be the most environmentally sound methods for harvesting mixed-age stands of temperate forests. Though many foresters recommend selective harvesting in tropical forests, debates are ongoing about whether commercially viable amounts of timber can be harvested sustainably from these forests. Felling one tree in a tropical forest typically causes others to fall—the forest is densely populated with trees and with vines that connect neighboring trees to one another. Simple steps that reduce damage include cutting vines before felling a tree and cutting the tree in such a way that it hits the fewest neighboring trees when it falls. Strip harvesting, in very narrow strips, may be a reasonable alternative in some tropical forests. And because soils are thin and fragile in tropical forests, efforts to reduce damage when trees are removed, such as using draft animals instead of trucks, are advised.

Ideally, the stand of trees—its health, age, and species composition—and the slope of the land determine the harvesting method. Consideration is also given to the other species that reside there. When trees are harvested properly, they can provide immediate and long-term economic benefits without serious environmental damage. Even clear-cutting can be appropriate when it increases the overall health and viability of a future forest by removing invasive, unhealthy, or genetically inferior trees and replacing them with better stock. **INFOGRAPHIC 5**

Forest management is not without its critics. Conflicting interests make it difficult to achieve a balance between multiple forest uses and ecosystem protection. For example, harvesting trees in the Pacific Northwest provides many jobs, good profits for timber companies, and useful products for homes and businesses, but it can reduce biodiversity and harm salmon runs, which are vital for ecosystem health, native cultures, and the tourism industry.

selective harvesting Timber-harvesting technique that cuts only the highest-value trees; the remaining trees reseed the plot.

shelterwood harvesting Timber-harvesting technique that cuts all but the best trees, which reseed the plot and are then harvested.

Foresters evaluate individual trees in a forest to determine which to cut and which to leave behind prior to harvesting with shelterwood or selective harvesting methods.

sdigital/iStock/Getty Images

INFOGRAPHIC 5 TIMBER-HARVESTING TECHNIQUES

Achieve

There are many ways to harvest trees from a forest, each with its own economic and ecological trade-offs. Here are three techniques. To evaluate the impact of a particular method, consider what the area will look like 50 years after a harvest.

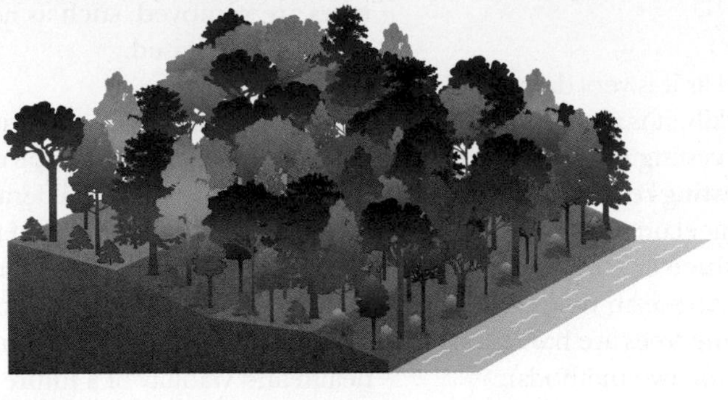

ORIGINAL FOREST

CLEAR-CUTTING

All trees are cut; usually replanted with a fast-growing species.

IMMEDIATELY AFTER HARVEST

- High profits at harvest, then no profits until forest regrows.
- Water is polluted by heavy erosion, especially on steep slopes.
- Biodiversity is very low after the cut.

50 YEARS AFTER HARVEST

Even-aged, single-species stand.

- Tree farms produce harvestable timber in a short time span.
- Water may be polluted by runoff from the open stand, but to a lesser degree than immediately after clear-cut.
- Tree farm has less biodiversity than the original forest.

SELECTIVE HARVESTING

High-value trees are cut, leaving others to reseed the plot.

IMMEDIATELY AFTER HARVEST

- Short-term profits lower than clear-cut.
- Land can be harvested again in less than 50 years.
- Biodiversity declines after harvest but less than in clear-cut lands.

50 YEARS AFTER HARVEST

Mixed-age, multispecies stand; remaining large trees are harvested.

- Remaining large trees can be harvested; since the poorer-quality trees were left to reseed the area, the forest quality may decline.
- Biodiversity is higher than after the harvest but lower than the original forest.

SHELTERWOOD HARVESTING

The best trees are left behind to reseed the plot while others are harvested.

IMMEDIATELY AFTER HARVEST

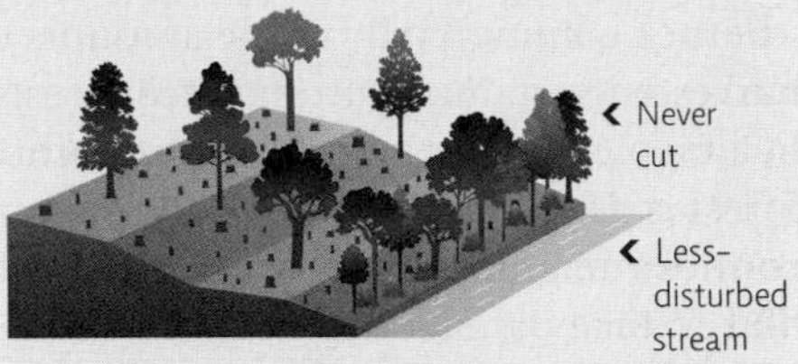

- Profits similar to selective harvest initially but improve on later cuts since high-quality trees remain.
- Biodiversity declines after each harvest but may be higher than selective harvest.

50 YEARS AFTER HARVEST

Mixed-age, multispecies stand; new growth is progeny of the best trees; remaining large trees are harvested.

- Harvesting the large trees left behind provides additional income.
- Biodiversity increases as the forest grows and may be higher than in any of the other harvested stands.

? Clear-cutting radically changes the ecosystem and can cause serious ecological damage. When then might it be ecologically sound to clear-cut, and what restrictions would you place on such an action?

6 SUSTAINABLE FOREST MANAGEMENT

Key Concept 6: Forests can be protected by encouraging the sustainable harvesting of forest products, finding alternatives to replace wood as a raw material, and promoting the value of intact forests through the realistic valuation of ecosystem services.

Funding and technical expertise can facilitate more effective forest management. For example, in 1905, the United States established the National Forest Service. Its first director, Gifford Pinchot, challenged the prevailing notion that U.S. forests were inexhaustible and introduced the idea of sustainable forestry—taking only what the forests could sustainably produce or replace. (See Module 5.1 for more on environmental economics.) The focus in those early years was **maximum sustainable yield (MSY)**: harvesting as much as sustainably possible (but no more) for the greatest economic benefit. In 1960, the United States expanded on Pinchot's ideas and enacted the **Multiple-Use Sustained-Yield Act**, which mandated that national forests be managed in a way that balances a variety of uses—outdoor recreation and timber interests as well as the health of watersheds, fish, and wildlife. No single use could predominate.

Today the U.S. National Forest Service, administered by the U.S. Department of Agriculture, oversees 155 national forests (and 20 national grasslands) and provides guidance for the management of private forests and grasslands, both nationally and internationally. Its overriding objective is the management of these areas in an ecologically sustainable manner, and it promotes **forest ecosystem management (FEM)** as the best way to meet this mandate. Rather than focus exclusively on timber harvests and maximum sustainable yields, FEM aims to manage the forest ecosystem as a whole. This includes a variety of techniques for timber harvesting, vegetation removal, and controlled burns to remove deadwood and stimulate seed germination in fire-adapted species, as well as restoration of forested areas and forest research.

The trade-offs associated with the use of forest resources are the subject of much debate and conflict because each decision impacts both human livelihood and the health of the environment. To be sure, the economic value of wood may be dwarfed by the ecosystem services lost if the area is overharvested, and protecting the ecosystem may actually prove more profitable, even in the short term. But people still need fuel, building material, and income. "Of course, trees and forests are important for the environment," says Georges, "We know that they protect us from floods and help keep drinking water clean and plentiful. But for many Haitians, selling those same trees is the only way to feed a family. So how can you ask them not to?"

One solution, according to a growing number of experts, is to price the ecosystem services themselves. In Costa Rica, for example, higher water bills offset the costs of maintaining rainforests that purify and replenish the water people use every day. This money is given directly to landowners who would otherwise have to chop down the trees to sell as fuel or timber or to convert the land to agriculture.

Other options include the promotion and increased availability of sustainable wood products. The Forest Stewardship Council (FSC) certifies lumber and other timber products through a process that evaluates the timber-harvesting techniques and the forest itself in terms of wildlife, water, and soil quality. Worldwide, more than 500 million acres of forest are certified by the FSC as sustainably managed (about 5% of the world total).

Consideration can also be given to resources, such as latex and tree nuts, that can be sustainably harvested from standing forests. Alternatives for wood products also exist and can reduce the pressure on forests. For example, lumber from old buildings can be salvaged to provide quality building materials. Paper can be made from old paper (recycled) or from fast-growing crops such as kenaf, jute, flax, and hemp.

Alternate energy sources are also needed in charcoal-dependent places like Haiti. One potential alternative is jatropha, a fast-growing plant whose oil-rich seeds have been hailed as a promising biofuel source. Even the material left after the oil is pressed out can be digested by bacteria to produce biogas, a fuel similar to natural gas. Another option is the production of composite briquettes from a variety of flammable materials, such as grass, paper, or sawdust. Shredded plastic can be added to make it more combustible. (See Module 11.2 and Online Module 11.3 for more on sustainable energy.)

Reforestation projects are increasing in number around the world. For example, the Kenyan Green Belt Movement, begun by Nobel Laureate Wangari Maathai,

maximum sustainable yield (MSY) The amount that can be harvested without decreasing the yield in future years.

Multiple-Use Sustained-Yield Act U.S. legislation (1960) mandating that national forests be managed in a way that balances a variety of uses.

forest ecosystem management (FEM) A system that focuses on managing the forest as a whole rather than on maximizing yields of a specific product.

has planted and protected more than 50 million trees and has sparked similar movements worldwide. The UN's Trillion Tree Campaign has planted almost 14 billion trees so far. A unique method of tree planting employed in Myanmar to restore coast mangrove forests deploys seeds from drones—in a year's time, the mangrove seedings that germinated from these "seed missiles" grew into 20-inch saplings. This method holds promise with the potential to plant tree seeds much more quickly than can be done by hand and in locations difficult to reach by foot. But reforestation efforts take commitment, money, and expertise. In Haiti, the success of reforestation efforts is strongly tied to the availability of water, especially on steep slopes; engineering terraced water catchment systems to catch and retain rainwater enhances growth and survival of the trees.

For some, saving the forests may come down to recognizing their intrinsic value: their natural beauty, their inherent sacredness, and their right to exist as living things, regardless of what we humans might extract from them. Naturalists from John Muir to Wallace Stegner have argued for the protection and preservation of forests on these grounds alone. "We simply need that wild country available to us, even if we never do more than drive to its edge and look in," 20th-century U.S. naturalist Wallace Stegner wrote in his famous *Wilderness Letter*. "For it can be a means of reassuring ourselves of our sanity as creatures, a part of the geography of hope."

This appreciation for the intrinsic value of a forest has been successfully translated into the economic enterprise known as **ecotourism**. Ecotourism is a viable option for many areas, especially less developed countries. These areas often possess high biodiversity precisely because they are less developed, and they are often found in tropical or subtropical regions with naturally high biodiversity. In the quest to develop economically, these countries may find that the highest economic value for their resources lies in keeping them intact. Ecotourism allows a way for funds to enter the country while protecting the natural areas at the same time. **INFOGRAPHIC 6**

There is a Haiti that people like Georges and Robert talk about in the quiet moments after a day's planting. It's a Haiti lush and green with trees, a country where families earn their living selling mangoes and moringa leaves instead of charcoal and firewood. Whether this country resembles its past as much as its future will depend on an infinite number of variables—its ability not only to manage forests and reforestation efforts but also to find alternative building materials or establish a reliable energy sector based on something other than wood.

ecotourism Low-impact travel to natural areas that contributes to the protection of the environment and respects the local people.

Micheline Pelletier/Corbis/Getty Images

Wangari Maathai's Green Belt Movement of Kenya has spread to more than 30 other African nations. The first Kenyan woman to earn a PhD, Maathai received several environmental awards, including the Nobel Peace Prize. She passed away in 2011, but the Green Belt Movement lives on.

Three months after the kombit visited his hillside farm, Jean Robert is harvesting moringa leaves. They're tasty and packed with protein, and when harvested properly, they regrow rather quickly. If everything goes as expected, he will eventually have a bevy of crops to see him through the year: mangoes

INFOGRAPHIC 6 PROTECTING FORESTS

STRATEGY	EXAMPLE
Assign a monetary value to forest ecosystem services.	A cost analysis of various ways to provide water supplies for New York City revealed that protecting the forests of the nearby Catskills Mountains so that they could continue to provide their water capture and purification services was cheaper than building a state-of-the-art water filtration facility.
Sustainable harvest of forest products.	In the Brazilian state of Acre, local management of tropical forests for the harvest of tree nuts, latex (from tree sap), and sustainably harvested timber produces more than $90 million per year.
Use alternatives to wood for products and fuel.	Paper can be made from fast-growing nontree plants like bamboo and kenaf or can be made by recycling used paper; jatropha is replacing charcoal in Haiti and other regions of the world.
Designate forests as protected areas.	U.S. national forests are managed for multiple uses, including recreation, wildlife, and forest resource harvesting; resource harvesting is prohibited in wilderness areas and national parks. (See Online Module 3.3 for more on protected areas.)
Promote ecotourism.	With 36% of the country under protected status, Belize has a thriving ecotourism industry that brings in more than $100 million annually.

? Explain how the sustainable harvesting of forest products, which provides less annual income than harvesting the trees themselves for timber, can be more economically valuable than the wholesale removal of the valuable timber in that forest.

in the summertime; coffee in the fall; and, if he's lucky, oak and mahogany stands that will yield high prices in the timber market down the road. In the meantime, the moringa trees provide his family with a sustainable supply of protein and fuel wood.

Select References:

Brandt, J. P., et al. (2013). An introduction to Canada's boreal zone: Ecosystem processes, health, sustainability, and environmental issues 1. *Environmental Reviews, 21*(4), 207–226.

Costanza, R., et al. (2014). Changes in the global value of ecosystem services. *Global Environmental Change, 26*, 152–158.

Food and Agriculture Organization (FAO). (2015). *Global forest resources assessment 2015: How have the world's forests changed?* Rome, Italy: Author.

Hedges, S. B., et al. (2018). Haiti's biodiversity threatened by nearly complete loss of primary forest. *Proceedings of the National Academy of Sciences, 115*(46), 11850–11855.

Kubiszewski, I., Costanza, R., et al. (2017). The future value of ecosystem services: Global scenarios and national implications. *Ecosystem Services, 26*, 289–301.

Sprenkle-Hyppolite, S. D., et al. (2016). Landscape factors and restoration practices associated with initial reforestation success in Haiti. *Ecological Restoration, 34*(4), 306–316.

Watson, J. E., et al. (2018). The exceptional value of intact forest ecosystems. *Nature Ecology & Evolution, 2*(4), 599.

Hataigan Doungbal/AGE Fotostock

The nutritious leaves of the moringa tree are high in protein, vitamins, and minerals and contain enough iron to treat mild anemia; they can be harvested without killing the tree, and they can be eaten fresh or dried for later use. The seeds and roots are also edible, and cuttings can be planted to start new trees.

GLOBAL CASE STUDIES FORESTS OF THE WORLD

Achie/e

The type of forest found in a given area depends on its climate, which is determined in large part by latitude and altitude. Read about three examples to learn more about each of the three major forest biomes.

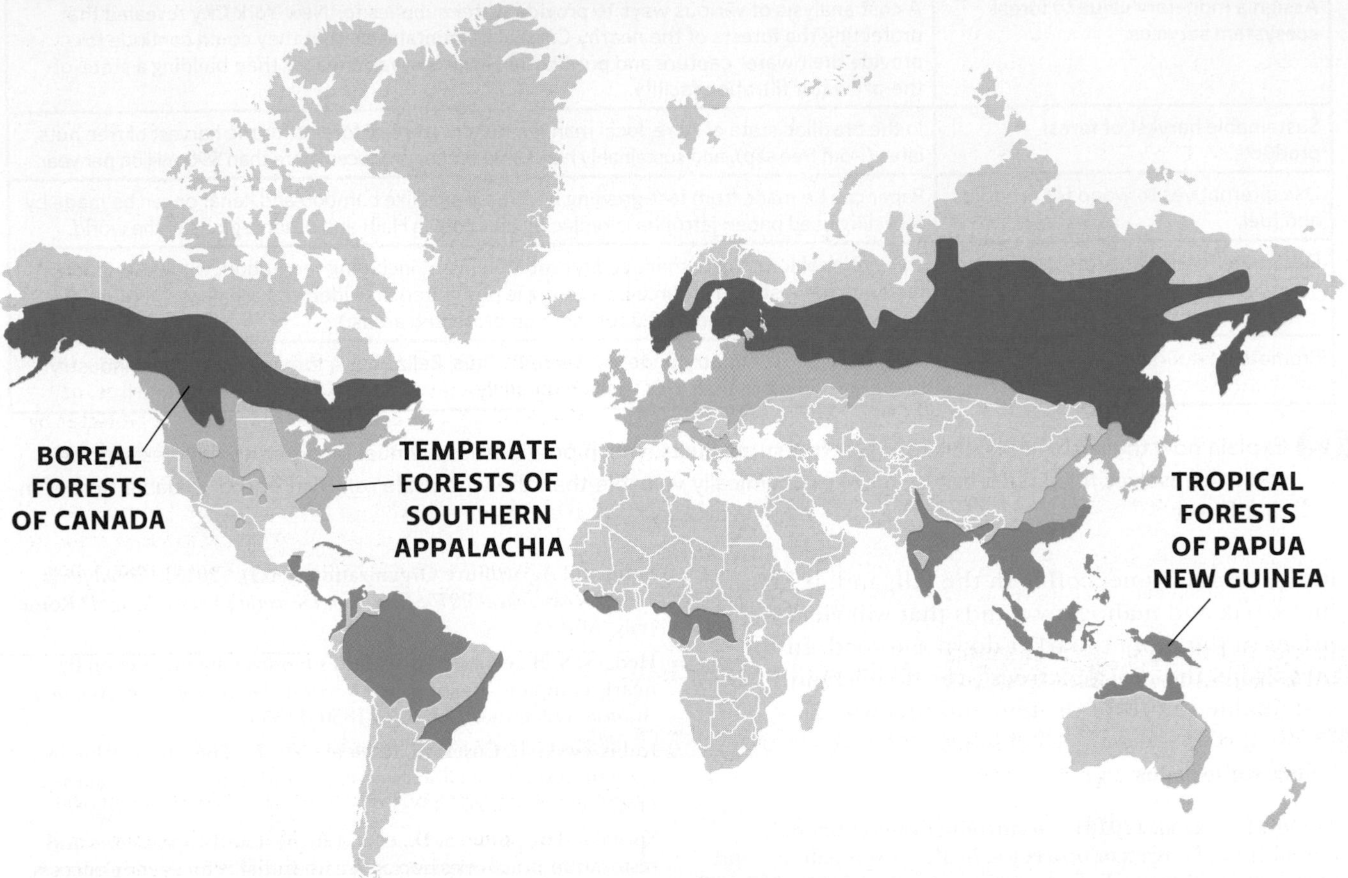

BRING IT HOME

PERSONAL CHOICES THAT HELP

Forests are a renewable resource that can be used sustainably for many years under proper management conditions. Using harvesting systems such as selective cutting or strip cutting can allow economic use of forests without eliminating the ecological functions that forests provide.

Individual Steps

- Buy paper products (toilet paper, facial tissue, and notebook paper) made of recycled content to decrease the unnecessary cutting of trees and to encourage recycling.
- When purchasing lumber for projects, look for wood that has been certified as sustainably managed by the Forest Stewardship Council.
- Avoid buying noncertified furniture made from tropical wood such as rosewood, teak, and ebony; harvesting these woods contributes to tropical deforestation.
- Support legislation that protects roadless areas and old-growth forests.
- Ask your local grocery stores, restaurants, and college dining facilities to offer shade-grown coffee and chocolate, which can help encourage preservation of tropical forests.

Group Action

- Like these volunteers removing garlic mustard plants that have invaded a Detroit park, organize or volunteer for a workday to clear invasive species such as common buckthorn, English ivy, and Japanese honeysuckle to ensure that our remaining forests provide high-quality habitat.
- Participate in local tree-planting events or donate to forest conservation and reforestation efforts.

Jim West/AGE Fotostock/Media Bakery

REVISIT THE CONCEPTS

- Forests are found in areas where precipitation is sufficient to support trees, but the amount of precipitation and its seasonal variability, as well as temperature, influence the type of forest biome that is found in an area. There are three main forest types (boreal, temperate, and tropical) and a variety of subtypes of each.

- Due to the height of trees and the variety of understory plants that live beneath them, forests are stratified — they have a three-dimensional structure of more or less distinct layers from forest floor to canopy. This stratification creates different habitats, each of which contains species adapted to the level of sunlight and moisture available.

- Forests provide a wide variety of ecosystem services. They contribute critical ecological services such as water provision and purification, air purification, and wildlife habitat. We also derive economic benefits from forests in the form of jobs and consumer products. In addition, forests provide sociocultural benefits (e.g., cultural, spiritual, and recreational value).

- Globally, forest cover is shrinking due to human actions such as harvesting for lumber and fuel, clearing for agriculture or mining, and fire suppression. Pest infestations (often made worse by human actions) are also damaging forests. So far, 35% of forests have been lost; most of the remaining forests are degraded, and many are fragmented, decreasing their ability to provide ecological services on which we and other species depend.

- Timber-harvesting techniques such as selective and shelterwood timber harvesting maximize long-term profits and protect biodiversity; clear-cutting methods maximize short-term profits, but these profits often come at the expense of the local ecosystem. Replanting a harvested area with one type of fast-growing, commercially valuable tree (a tree farm) may allow managers to leave remaining natural stands alone.

- U.S national forests are managed in a way that balances a variety of uses — outdoor recreation and timber interests as well as the health of watersheds, fish, and wildlife. Forests can be protected by encouraging the sustainable harvesting of forest products, finding alternatives to replace wood as a raw material, and promoting the value of intact forests through the realistic valuation of ecosystem services. Reforestation efforts are underway worldwide; ecotourism may provide another incentive for protecting the forests and their biodiversity that remains.

ENVIRONMENTAL LITERACY Understanding the Issue

1 What is a forest, and what influences which forest type is found in a given area?

1. Describe the climate and general characteristics of tropical, temperate, and boreal forests.
2. Tropical rainforests have thin, acidic soils, yet they contain dense vegetation and high biodiversity. How can these tropical forests have poor soil but support such diverse arrays of plant and animal life?

2 What is the three-dimensional structure of a forest, and how are the plant species found there adapted to their level of the forest?

3. Identify and describe the four layers of a forest.
4. Explain why many forest wildflowers emerge and bloom in early spring, before the trees "leaf out."

3 What ecosystem services do forests provide?

5. What is an "ecosystem service"? Describe three such services provided by forests.
6. Why might it be useful to put a monetary value on the ecosystem services of forests even if we know it is not exactly right?
7. How does deforestation contribute to loss of drinking water in Haiti?

4 What threats do forests face?

8. What are the causes (past and present) of deforestation in Haiti?
9. How is the current status of forests different in developing versus developed countries? What factors account for these differences?
10. How can fire suppression actually harm a forest?
11. How does climate change impact forests? How can forests help address climate change?

5 What timber-harvesting techniques are available, and how do they compare in terms of method and sustainability?

12. Identify the economic and ecological pros and cons of clear-cutting.
13. Compare and contrast selective and shelterwood harvesting.

6 How can we protect and sustainably manage forest resources?

14. Discuss how U.S. forest policy has changed over time.
15. What are some ways a forest can be used for economic gain without cutting down the entire forest?
16. How does ecotourism help protect forest resources?
17. What is the reforestation strategy employed by the Haitian kombit? Explain the rationale behind the approach and discuss whether it is likely to be successful.

SCIENCE LITERACY Working with Data

The Hubbard Brook Experimental Forest research center in the White Mountains of New Hampshire consists of several valleys and streams that can be manipulated as test and control plots for experimental purposes. Small dams allow researchers to monitor the volume of water that flows through the forest. The data below result from a study that compared the volume of stream-water flow from streams in similar valleys close to one another that either were left forested or were completely deforested.

STREAM-WATER FLOW COMPARISON OF FORESTED AND DEFORESTED AREAS

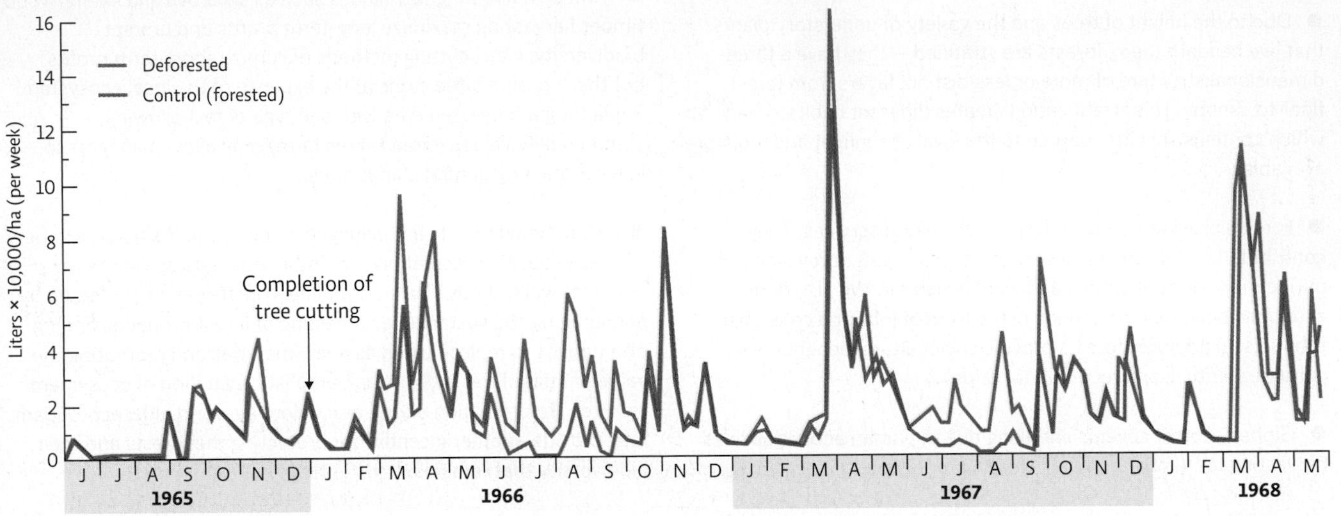

Interpretation

1. Which months typically have the greatest flow of water through streams in the evaluated area?
2. How did stream flow in the two forests compare before tree cutting in the test forest?
3. How did stream flow in the two forests compare after tree cutting in the test forest?

Advance Your Thinking

4. Why did deforestation impact the volume of water flowing in the stream in the way observed?
5. What do you think the water quality was like in the stream in terms of sediment loads? Explain.
6. Which downstream area would be more likely to experience a flood during a heavy rain event? Explain.

INFORMATION LITERACY Evaluating Information

Many of the world's forests are severely degraded. Yet forests produce many consumer products that we depend on, such as food, medicine, building materials, and raw materials for industrial products like paper. So what should a conscientious consumer do?

Explore the Forest Stewardship Council (FSC) website (https://us.fsc.org).

Evaluate the website and work with the information to answer the following questions:

1. Determine if this is a reliable information source with a clear and transparent agenda:
   a. Who makes up this organization? Does its membership make the FSC reliable or unreliable? Explain.
   b. What is this organization's mission? What are its underlying values? How do you know this?
   c. Does the FSC give supporting evidence for its claims about forest resources and its vision to address the problem? Does the website give sources for its evidence?
   d. Identify a claim the FSC makes and the evidence it gives in support of this claim. Is it sufficient? Explain.
2. Understand products and certification:
   a. What kind of products can be FSC certified?
   b. Read about the certification process. Do you feel this process is sufficient? Explain.
   c. How can a consumer select FSC-certified products, according to this website? Is it easy for consumers to know if the wood products they are purchasing are FSC certified? Explain.
3. How might FSC certification help forests?

Additional study questions are available at Achieve.macmillanlearning.com. Achieve

CHAPTER 8

Food Resources

Feeding a world of more than 7 billion people is a daunting task. Productive and sustainable approaches to growing crops, rearing livestock, and obtaining fish will be needed if we are to meet humanity's food needs now and in the future.

Online Modules are available at Achieve.macmillanlearning.com.

Fotokostic/Shutterstock

Module 8.1 Feeding the World

BANKING ON SEEDS

The role of seed banks in addressing hunger

Maintaining the genetic and species diversity of our crops is vital to our ability to find crop varieties suitable for changing environmental conditions. *Helen Camacaro/Getty Images*

After reading this module, you should be able to answer these GUIDING QUESTIONS

1. How prevalent is hunger, and what are its causes?
2. What is malnutrition, and what problems can it cause?
3. What were the intent, scope, and outcome of the Green Revolution?
4. What is industrial agriculture, and what are the pros and cons of farming this way?
5. What are the trade-offs of fertilizer and pesticide use in agriculture?
6. How can genetic engineering be used in agriculture, and what are its trade-offs?
7. What are food self-sufficiency and food sovereignty, and why are they important?

Every winter, Arizona horticulturalist Sheila Murray receives a master list of 20 plants that will determine her fate over the upcoming spring and summer months. Murray collects seeds for England's Millennium Seed Bank, a global network that safekeeps seeds from plants around the world, including the wild relatives of modern food crops. According to an extensive report on the state of the world's plants prepared by the United Kingdom's Royal Botanical Gardens, one in every five plants is now threatened with extinction, and the Millennium Seed Bank is working to protect their future.

As soon as Murray gets her hands on the list, "I go out and see what I can find," she explains. She'll start with online research to predict where the coveted plants might grow, and then she'll head out to hunt for them, sometimes setting up camp for a few days to make sure she has enough time to scour a promising area. Murray closely monitors each plant she finds, visiting it every couple of weeks, like a doctor tending an important patient. She doesn't want the plants to die, and she doesn't want them to go to seed when she's not there or she'll miss the opportunity to collect the all-important seeds.

When the plants do start to seed, it's collection time. Murray and her intern meticulously bag seeds for hours, even days, in order to gather at least 20,000 of each species for the bank, making sure not to collect too many seeds from any one plant so as not to hamper its ability to successfully reproduce—a rule of thumb is to take no more than 10% of the seeds from any single plant or population. Many of the seeds she collects are stored in the secure, temperature-controlled National Center for Genetic Resources Preservation in Fort Collins, Colorado, which collaborates with the Millennium Seed Bank. There, the seeds are kept safe and secure, a back-up plan for the world in the event that these plants go extinct in the wild. Scientists can also study her banked seeds to find ways to grow better, more resilient plants in the future. "It feels good," Murray says about her job and her mission. "This is something for future generations."

WHERE IS THE MILLENNIUM SEED BANK?

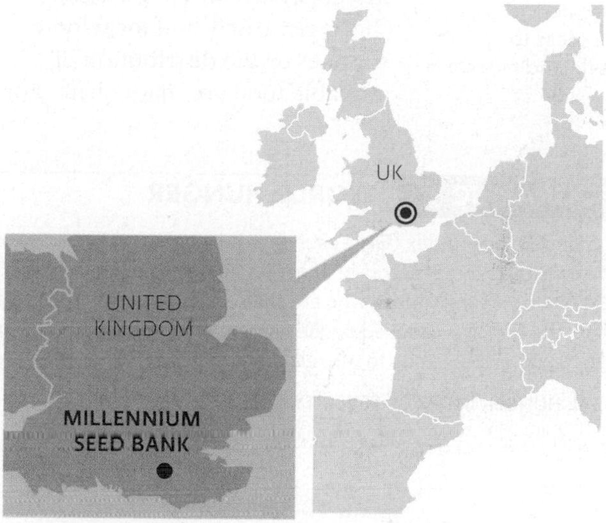

The Millennium Seed Bank, part of Britain's Royal Botanic Gardens of Kew, holds thousands of seed varieties in cold storage.

Peter Nicholls/Reuters/Newscom

B WORLD HUNGER

Key Concept 1: Although we currently grow enough food to feed everyone, nearly a billion people are undernourished due to poverty, war, environmental degradation, and inadequate food distribution or preservation.

By most accounts, we currently grow enough food to feed the almost 8 billion people who inhabit Earth, yet according to the Food and Agriculture Organization (FAO), an estimated 820 million (slightly more than 1 in 10 people on Earth) lack **food security**—access to sufficient amounts of safe and nutritious food.

food security Having physical, social, and economic access to sufficient safe and nutritious food.

The reasons for this food insecurity are varied. Physical shortages arise when local food supplies or the distribution of available food are inadequate. For example, natural disasters or environmental degradation by humans can seriously hinder the ability to grow or acquire food. Insufficient distribution may be logistical (it may be difficult to deliver food to remote areas) or arise from political roadblocks or corruption that hamper distribution. Social obstacles to food security also exist for some populations, such as discrimination that limits the ability of marginalized groups to grow their own food or acquire culturally appropriate food or that keeps food from reaching them. However, in many cases, food insecurity is driven by poverty—the lack of sufficient funds to buy food. **INFOGRAPHIC 1**

INFOGRAPHIC 1 WORLD HUNGER

The Global Hunger Index (GHI) estimates hunger for each country by considering the percentage of the population with insufficient caloric intake as well as three metrics that assess hunger in children under 5 years of age: the proportion of children who have a low weight for their height (child wasting), those who are stunted (low height for their age), and the child mortality rate (deaths per 1,000 children). Worldwide in 2019, the GHI score ranked on the border between serious and moderate, an improvement over recent years, though some countries saw hunger levels increase due to war, economic slowdowns, or environmental challenges related to climate change.

GLOBAL HUNGER INDEX BY SEVERITY (2019)

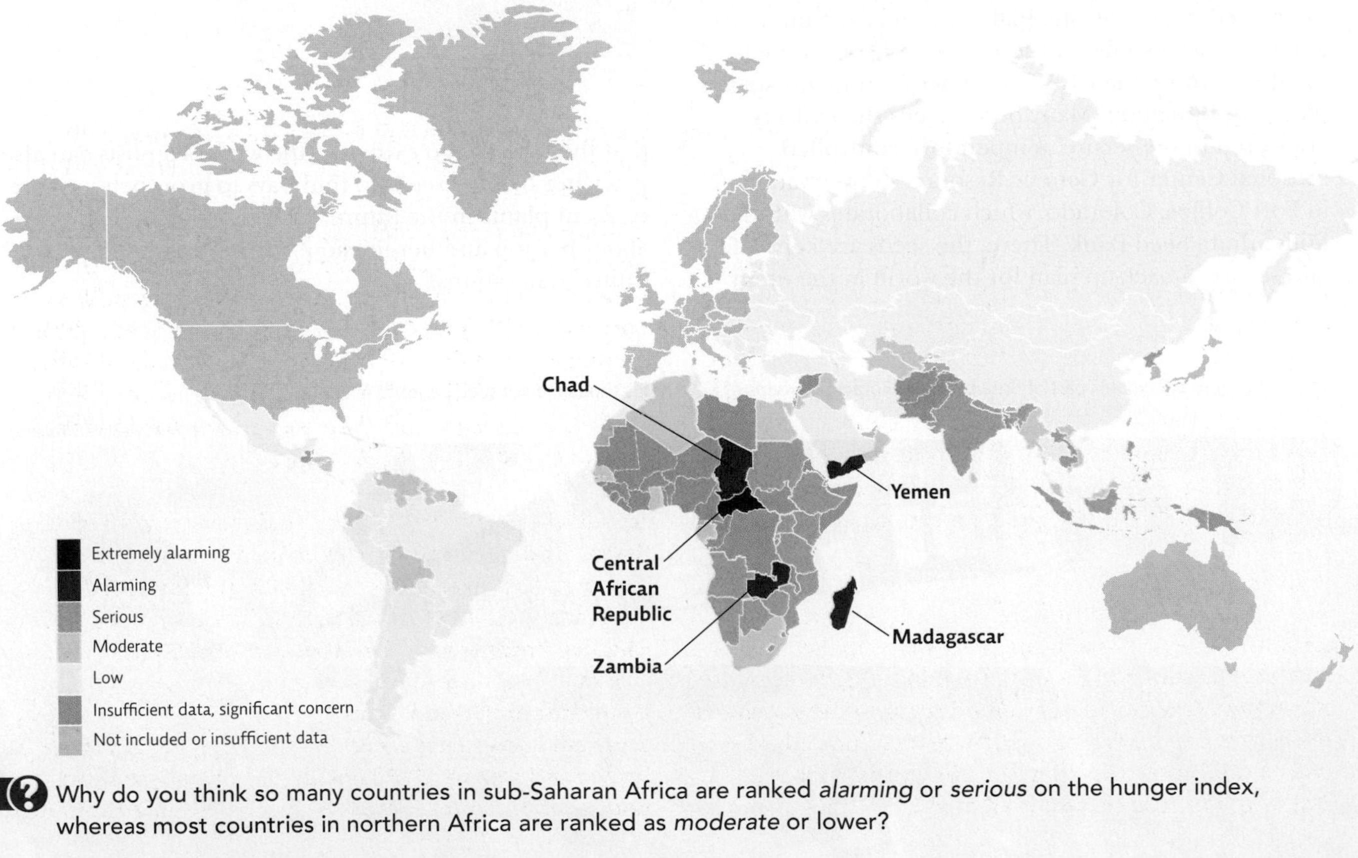

Why do you think so many countries in sub-Saharan Africa are ranked *alarming* or *serious* on the hunger index, whereas most countries in northern Africa are ranked as *moderate* or lower?

"Ugly" produce is often discarded by wholesalers or grocery stores and never hits the shelves, but some companies and nonprofits are stepping in to make this food available to consumers at reduced or no cost.

Food waste also eats into food supplies. This includes food losses that occur before it reaches retailers as well as the remains of uneaten food and fresh or prepared food that spoils before it can be sold or used. The FAO estimates that 1.3 billion tons—as much as one-third of food grown for human consumption—are lost annually to waste. In low-income countries, this loss most often occurs during early stages of production (e.g., harvest, transport, storage), whereas in higher-income countries the majority is wasted later in the production chain, discarded by grocery stores, consumers, and restaurants—perfectly good food that ends up in the trash.

Even in wealthy countries like the United States, food insecurity exists. Though the percentage of Americans who are food-insecure is declining, in 2018, an estimated 37 million Americans lacked food security. This includes college students, who, as a group, experience a higher rate of food insecurity than the general population. A number of studies have evaluated U.S. colleges and universities, and though it varies from institution to institution, food insecurity among students was found to be as high as 67%, due in many cases to the economic hurdles that often come with the pursuit of a college degree: a lack of income and the cost of attending college. This is not a trivial matter. Food insecurity correlates with lower academic performance and poorer mental and physical health overall, potentially affecting the student's ability to succeed. In recognition of this, many U.S. colleges and universities stock food pantries available to students free of charge.

As is the case with most college students, in the United States, food insecurity is often economic—households simply lack enough money to reliably put food on the table. However, in many cases, it is also due to living in what has become known as a **food desert**. Found in most, if not all, large U.S. cities (and in large cities around the world), these are areas where access to fresh, nutritious foods is limited, nonexistent, or prohibitively

food desert A locale where access to affordable, fresh, and nutritious food is limited or nonexistent.

expensive. Most often these food deserts are in low-income areas that lack traditional grocery stores. Even in urban areas with public transportation, the ride to and from the nearest grocery may take an hour or more, leaving residents no option but to buy most of their food from convenience stores and fast-food restaurants. Low-income residents in rural areas may face the same food desert obstacles with little access to nutritious, affordable food.

The United Nations (UN) has set itself the task of eradicating hunger (and extreme poverty along with it) with its Sustainable Development Agenda, identifying 17 Sustainable Development Goals (SDGs) that it hopes to reach in an effort to "ensure prosperity for all people." (See Infographic 4 in Module 1.1 for more on the UN's SDGs.) SDG 2 is Zero Hunger, a goal the UN hopes to achieve by 2030. The more than 800 million who were hungry in 2018 is unacceptably high, but this does represent significant success—from 1992 to 2015, world hunger was cut almost in half, even with an extra 1.9 billion mouths to feed. But more work needs to be done to eliminate hunger—we have to improve access to food for the hundreds of millions who are currently underfed, and we must be able to feed an additional 2 billion or so people by 2050.

To end hunger and achieve global food security, the UN is also calling for a doubling of the agricultural productivity of smallholder family farms in developing countries by 2030 (as a source of food and income in these poverty-stricken areas) as well as programs to address inequities in food distribution and access. Ending war and protracted conflict, and finding ways to get food to people in affected areas, are also hurdles that must be addressed. Social protections to assist those in need and bring them out of poverty are seen as vital to breaking the cycle of poverty and hunger. Indeed, Eliminating Poverty is SDG 1. The UN estimates that an investment of $267 billion over the next 15 years would eliminate hunger and extreme poverty. That is slightly less than half of the U.S. military budget for a single year—globally, a very accessible target.

In order to feed all these hungry mouths, we have to think not only about how to grow and distribute food, but also how to preserve its source—the seeds that grow crops in the first place. With many plants on the brink of extinction, governments have begun to realize that they need to take responsibility for saving the world's seeds, which is why seed banks like Millennium have come to be.

Many inner-city areas contain food deserts where the nearest grocery store or farmers' market is a mile or more away, meaning access to affordable, nutritious food is lacking for many residents.

David Grossman/Alamy

2 MALNUTRITION

Key Concept 2: Malnutrition can result from undernutrition or overnutrition, both of which can lead to serious health problems.

The UN's mission is not just a matter of food quantity but also of improving food quality. A healthy diet contains a variety of foods that provide the proteins, carbohydrates, and fats needed for good health, as well as enough calories to meet daily energy needs. The food that supplies these calories must also supply micronutrients such as vitamins and minerals. When a person's diet falls short of these basics—when he or she does not consume enough calories, protein, vitamins, or minerals—that person is undernourished. *Undernutrition*, a form of **malnutrition**, can disrupt growth and development in the young and can serve as a prelude to a whole host of diseases, from blindness (a result of vitamin A deficiency) to impaired immunity (a zinc deficiency). Undernutrition is often caused by food insecurity, but it can also be caused by illness (which may interfere with nutrient absorption or a willingness or ability to eat) or poor eating choices.

Malnutrition can start in the womb: Malnourished women give birth to malnourished children who fail to grow and develop normally; worldwide one in four children experiences stunted growth due to undernutrition. The UN estimates that the cost of treating malnourishment in children under age 2 is double what it would cost to prevent malnourishment in the first place.

Overnutrition, or the consumption of too many calories or too much of a specific nutrient (e.g., too much fat or sugar), is also considered malnutrition. Overdosing of vitamins or minerals can also occur by taking higher than recommended doses of dietary supplements. By some estimates, nearly 2 billion people around the world—more than twice the number that are undernourished—consume excess calories and are thus vulnerable to another set of nutrition-related conditions such as heart disease and type 2 diabetes. Contrary to intuition, this is not just a problem of the wealthy—low-income populations also suffer from overnutrition. Cheaper foods may be calorie-laden (from high sugar or fat content), but they are often low in essential nutrients. **INFOGRAPHIC 2**

Of course, to conquer global hunger, we must grapple with

malnutrition A state of poor health that results from inappropriate caloric intake (too many or too few calories) or deficiency in one or more nutrients.

INFOGRAPHIC 2   MALNUTRITION

Malnutrition is defined as a state of poor health that results from inadequate or unbalanced food intake. This includes diets that don't provide enough calories (and thus are nutrient-deficient) as well as those that may provide enough calories but are deficient in one or more nutrients. Consuming too many calories also leads to a variety of malnutrition illnesses.

Undernutrition			Overnutrition
Protein deficiency	Calorie & protein deficiency	Vitamin/mineral deficiency (some examples)	Overweight and obesity
↓	↓	↓	↓
Kwashiorkor	Marasmus (wasting disease)	• **Vitamin A:** blindness • **Vitamin D:** impaired bone health • **B Vitamins:** impaired nervous system • **Iron:** anemia • **Zinc:** stunting and impaired immunity • **Iodine:** impaired mental development and thyroid function • **Calcium:** weak bones	Increases the risk of acquiring a variety of diseases such as: • Diabetes • Heart disease • Stroke • Alzheimer's disease • Cancers of the esophagus, pancreas, colon and rectum, breast, endometrium, kidney, and possibly others

How might a plan to combat undernutrition be similar to one designed to combat overnutrition?

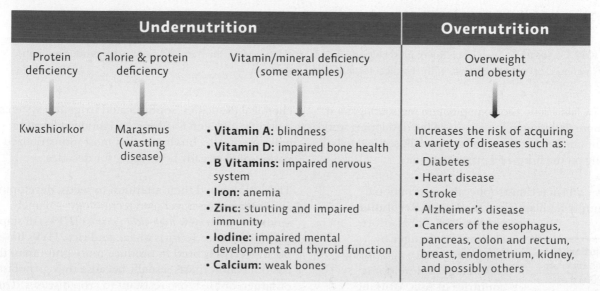

Per-Anders Pettersson/The Image Bank/Getty Images

Food insecurity plagues millions of people, especially in drought and war-torn regions of the world. In rural Africa, women do much of the farming with rudimentary tools. Here, a young mother with her baby on her back works with other women of her Rwandan village to prepare their field for planting.

more than just nutrients and calories. Indeed, we must tackle a vast array of problems that lie at the root of the brewing food crisis. Political instability and ecological degradation play significant roles, to be sure. According to the FAO, undernourishment is more than two times greater in developing nations experiencing a prolonged armed conflict, drought, or natural disaster than it is in countries not in such crises. But, as mentioned earlier, there are other contributors, too—namely, social disempowerment and poverty. In every country, there are groups of people who have access to food and groups of people who do not. To reach the UN's lofty goals, we must first understand why and how that came to be.

And to do that, we must travel back through recent history, to the last time global hunger was the stuff of headlines.

3 THE GREEN REVOLUTION

Key Concept 3: The Green Revolution helped address world hunger in the mid-20th century by producing high-yielding crop varieties grown with chemical fertilizers and pesticides.

It was the late 1960s. Global population was soaring; food crops were flatlining in some parts of the developing world and plummeting in others. India especially seemed to be hovering on the brink of a massive famine.

To stave off such catastrophe, the international community had launched the **Green Revolution**—a coordinated global effort to eliminate hunger by bringing modern agricultural technology to developing countries in Asia. Working across the globe, scientists, farmers, and world leaders introduced India and China to chemical pesticides, sophisticated irrigation systems, synthetic nitrogen fertilizer, and modern farming equipment—technologies that most industrialized nations had already been using for decades.

Green Revolution A plant-breeding program in the mid-1900s that dramatically increased crop yields and paved the way for mechanized, large-scale agriculture.

They also turned their attention to seeds, developing and introducing novel seed technology—namely, seeds that grew new *high-yield varieties (HYVs)* of staple crops like maize (corn), wheat, and rice. HYVs have been selectively bred to produce more grain than their natural counterparts, usually because they grow faster or larger, or are more resistant to crop diseases. These HYVs were most often hybrid seeds, developed from crossing two varieties that produced a high-yielding crop but whose seeds cannot be harvested for next year's

INFOGRAPHIC 3 THE GREEN REVOLUTION INCREASED AGRICULTURAL PRODUCTIVITY

The Green Revolution transformed agriculture into the industrial model we see today. Through the selective breeding of the most productive plants, high-yield varieties of crops were developed that more than doubled production per acre (when grown with inputs like fertilizer and pesticides).

AVERAGE CROP PRODUCTIVITY (WHEAT, MAIZE, AND RICE)

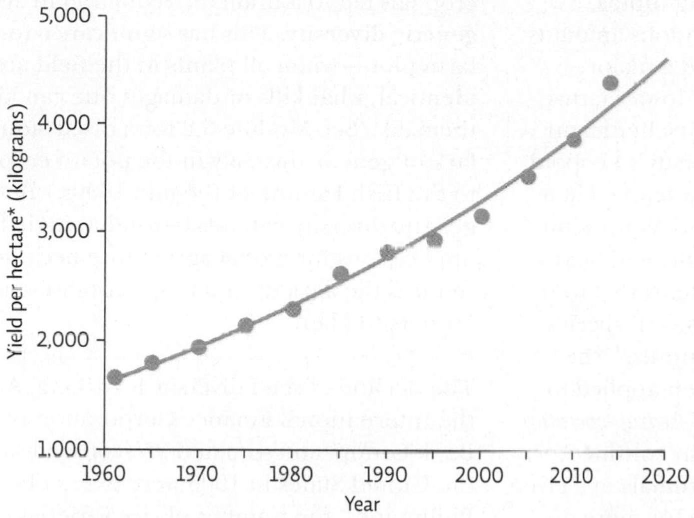

* One hectare equals 2.47 acres.

The trendline for this data shows an upward slope. Even if higher-yielding plants are developed, what might prevent productivity per hectare from continuing to rise at this rate?

planting—those seeds are not viable. Rather than save harvested seeds for next year's planting as could be done with traditional varieties, farmers simply bought new hybrid seeds at the start of each planting season.

In order to develop these extremely productive crops, scientists needed to have an arsenal of different crops to evaluate in their search for the traits they needed for these new high-yielding strains. Luckily, at that time, seeds of multiple varieties of our staple food crops were plentiful, so researchers had more than enough raw material to find the traits they wanted. Productivity per acre increased dramatically.

If all that were not enough, developed countries like the United States also began implementing a litany of agricultural policies—tax breaks, government subsidies, and insurance plans—that encouraged their own farmers to plant crops "fencerow to fencerow" and kept the price of food low, thus adding substantially to the world's food supply.

To be sure, the adoption of modern, large-scale farming methods in India and China was expensive (paying for costly inputs like fertilizers and pesticides and the necessary equipment), but in the short term, the Green Revolution accomplished its goal of addressing world hunger. The combined force of HYVs, existing technology, and new food policies resulted in a 100% increase in global food production and a 20% reduction in famine between 1960 and 1990. Today most experts credit the initiative with the fact that we are now producing enough food to feed every one of more than 7.5 billion people on Earth. **INFOGRAPHIC 3**

4 INDUSTRIAL AGRICULTURE

Key Concept 4: Modern industrial agriculture produces huge amounts of food but leads to ecological and social problems and a reduction in crop genetic diversity.

As it turns out, the Green Revolution owes its success to modern **industrial agriculture**. Industrial-scale farming is dependent on machinery as well as synthetic fertilizers (to boost plant growth) and pesticides (to protect crops from pests)—methods that allow the farmer to maximize productivity and profits. Typically, a single, genetically uniform crop, called a **monoculture**, is planted in rows over huge swaths of freshly plowed land, making it possible for fewer farmers to farm larger tracts of land using heavy equipment

industrial agriculture Farming methods that rely on technology, synthetic chemical inputs, and economies of scale to increase productivity and profits.

such as tractors and combine harvesters. In many cases, mechanized irrigation, drawn from surface water or groundwater supplies, is also used and allows farmers to grow crops that could not be sustained by local rainfall.

Monoculture industrial farming, paired with mechanization and these modern farming inputs, increases efficiency and produces tremendous amounts of food on a large scale. It has also caused a major shift in agriculture. "In the 1920s, half of Iowa's farms produced 10 commodities each," Fred Kirschenmann, a Distinguished Fellow at Iowa State University's Leopold Center for Sustainable Agriculture, told attendees at a 2003 Biodynamic Farming and Gardening Association conference. "Today 92% of the state's cultivated land is exclusively corn or soybean. Farming systems that were once supported by complexity and diversity of species have now been replaced by reliance on inputs." The same monoculture approach has also been applied to rearing livestock. In a *concentrated animal feeding operation (CAFO)*, livestock and poultry are raised in confined spaces, with a focus on raising as many animals in a given area as possible. (See Online Module 8.3 for more on CAFOs.) Worldwide, 90% of our food comes from just 15 crop species and 8 species of livestock.

Although industrial agriculture can be very productive, it has unintended consequences. As discussed further in the next section, the chemical inputs used in large-scale monoculture operations can degrade soil and pollute water supplies. Large-scale irrigation can deplete local water supplies, and in hot, arid areas, high rates of evaporation can leave a salt residue on soils that impedes plant growth or renders the soils unusable.

In addition, farm equipment runs on fossil fuels. Even the synthetic fertilizers and pesticides are derived from fossil fuels, made from natural gas or petroleum. This requires more fossil fuel extraction, with all its accompanying environmental issues (see Module 9.2). There are also local social and economic impacts: In developed countries like the United States, fewer farmers on larger farms threaten to make the family farm a thing of the past, impacting entire communities. In low-income developing countries, smallholder farmers may find the financial burden of the industrial farming model too much to handle—in debt for all the inputs, a bad harvest can be devastating. Such losses have been linked to high rates of farmer suicide in India and elsewhere—even in the United States, the suicide rate among farmers is higher than the national average.

monoculture A farming method in which a single variety of one crop is planted, typically in rows over huge swaths of land, with large inputs of fertilizer, pesticides, and water.

genetic diversity The heritable variation among individuals of a single population or within the species as a whole.

Monocultures, too, bring their own set of problems. The reliance on just a few high-yield varieties of each crop has led to a dramatic reduction in agricultural **genetic diversity**. This has significance for a given farm plot—when all plants in the field are genetically identical, what kills or damages one can kill or damage them all. (See Module 3.2 for a discussion of how the lack of genetic diversity in the potato crop contributed to the Irish Famine of the mid-1800s.) But this erosion of genetic diversity extends beyond a single field. It also has implications for global agriculture because it drastically reduces the agricultural crop varieties available for farmers to plant.

The decline of seed diversity is striking. According to the International Finance Corporation of the World Bank Group, an estimated 93% of seed varieties sold in the United States in 1900 were extinct by 1983; in the Philippines, the number of rice varieties once numbered in the thousands but has dwindled to only a few hundred today. This decline continues—by their calculations, one seed variety goes extinct every day.

This loss of crop varieties is one reason that seed collections such as the Millennium Seed Bank are so crucial for our future food supply: We need to retain as much crop diversity as possible to increase the chance that we'll be able to find varieties that will grow well in the face of changing environmental conditions. And

JIM RICHARDSON/National Geographic Image Collection

A toxic crust of salt and other minerals builds up in heavily irrigated fields in Colorado.

these banks need to be protected well enough that their seeds can't be accidentally lost due to a warming climate or natural disasters. One well-known international bank is the Svalbard Global Seed Vault, the world's largest collection of crop diversity. It is dug into the permafrost of a mountain on an island more than 600 miles north of Norway's mainland. The hope is even if its cooling units stop working, the seeds it houses—right now, more than 986,000—will be kept safe in cold storage because of the surrounding permafrost. It is "the best insulated freezer in the world," says Cary Fowler, the former executive director of the Global Crop Diversity Trust (though climate change may put even this facility at risk).

Without seed banks, as old crop varieties fall out of use—as farmers stop producing and conserving the seeds that have been bred through traditional agriculture over thousands of years—many valuable genetic traits face the possibility of being lost forever. Many of these traits are still found in *heirloom* crop varieties (traditional crops that existed before the hybrid varieties of the Green Revolution were developed), but because farmers are not planting many heirloom seeds anymore, many of these varieties are disappearing. In addition, because we are destroying habitats around the world, it is getting harder and harder to find the wild ancestors (the original source of all that genetic diversity) of key crops like wheat, corn, rice, and potatoes as well as those of other important plants—work that keeps Murray busy searching the terrain of the desert Southwest. This erosion of genetic diversity translates into a loss of the genetic raw material that allows crops to respond to changes, such as the arrival of a new fungal pest or a changing climate. **INFOGRAPHIC 4**

In fact, this new threat to agriculture—climate change—may be its greatest challenge. Unseasonable floods or extended droughts are destroying crops and diminishing yields worldwide. Though in some locales, a warmer or wetter environment or a longer growing season might enhance crop growth, in many areas, new climatic conditions are affecting temperature extremes and water availability, taxing crops that are adapted to an area's "former climate." Take for example coffee, the second-most traded commodity product in the world after petroleum. Coffee productivity is declining—50% lower in some areas—as the highlands where the trees are grown get warmer and drier. It is estimated that the area suitable to grow the two most popular varieties of coffee, Arabica and Robusta, will be cut in half by 2050. Research is underway to identify climate-resilient varieties that might become the coffee crops of the future.

INFOGRAPHIC 4 **THE TRADE-OFFS OF INDUSTRIAL AGRICULTURE**

Neil W Zoglauer/Getty Images

Mark Miller Photos/Getty Images

Noel Hendrickson/Getty Images

ADVANTAGES

- Large-scale farming
- Higher yields per acre
- Ability to grow crops in nutrient-poor soil (fertilizers)
- Fewer blemishes on crops (pesticides)
- Ability to grow crops in water-poor areas (irrigation)
- Higher efficiency (planting and harvesting) of monocultures
- Less labor-intensive

DISADVANTAGES

- Expense of mechanization and chemical inputs
- Fossil fuel dependence (for fuel and as raw material to make pesticides and fertilizers)
- Degraded soil quality
- Water scarcity and pollution
- Monoculture vulnerability to pests
- Loss of crop genetic diversity
- Pesticide resistance in pests
- Toxic residue on food
- Fewer jobs and family farms

? Which of the advantages of industrial agriculture do you feel is its greatest strength? Why? Which of the disadvantages of industrial agriculture worries you the most? Why?

5 SYNTHETIC FERTILIZERS AND PESTICIDES

Key Concept 5: The application of synthetic fertilizers and pesticides contributes to the high productivity of industrial agriculture, but their use comes with trade-offs.

The application of **fertilizer** to the soil can boost plant growth. Industrial agriculture typically relies on synthetic fertilizers, many of which are made from fossil fuels rather than natural products derived from plant or animal sources (which tend to contain less concentrated nutrient formulations). These potent synthetic fertilizers help industrial farms maximize their productivity.

Fertilizers boost growth because they provide nutrients that plants need, such as nitrogen, phosphorus, and potassium. (See Module 2.1 for more on nutrient cycles.) These nutrients are often in limited supply in soil, and plant growth slows when one nutrient starts to run out. Fertilizers overcome this deficiency, ensuring that plants can keep growing. They help farmers grow crops in marginal soils that wouldn't otherwise be able to accommodate crop growth, thereby helping to feed the world.

But while fertilizer can indeed boost crop production, its use has downsides. For one thing, fertilizer is expensive. And the excess nutrients (whatever the plants don't use) can wash off fields and contaminate waterways, potentially creating hypoxic (oxygen-poor) conditions that threaten aquatic life. This process, known as *eutrophication*, is a significant problem worldwide and the United States is no exception. For example, nutrient runoff from farms in states within the Mississippi River watershed has created a summertime dead zone in the Gulf of Mexico (see Module 6.2).

The application of fertilizer can also create a dependence on future fertilizer use. Many industrial farmers view the soil as merely a medium for plant growth—structural support for roots and a repository for nutrients and water their crops need. However, the provision of these soil services arises from a complex biological community of soil organisms that cooperate with each other and the plants growing in the soil to enhance soil fertility, to make these nutrients available to the plants, and to improve the soil's ability to trap water. Excessive fertilizer use disrupts these mutualistic relationships. The end result is a soil that is less fertile and in need of even more fertilizers, which only exacerbates the problem and reduces the fertility-enhancing processes even further. (See Module 8.2 for more on soil.)

fertilizer A natural or synthetic mixture of nutrients that is added to soil to boost plant growth.

pesticide A natural or synthetic chemical that kills or repels plant or animal pests.

In addition, synthetic fertilizers require a lot of energy to produce—one study estimated that fertilizers account for 30% of the energy that goes into growing a corn crop—and many are made from fossil fuels (natural gas and, to a lesser extent, petroleum), which are in limited supply and are environmentally damaging to extract, process, and ship.

As mentioned earlier, another problem facing industrial farms is that their monoculture crops are quite vulnerable to pests or disease: Pest populations can explode when they encounter acres and acres of a suitable crop on which to feed, and a single infestation can wipe out the entire crop because what kills one plant will likely kill them all. To deal with this, farmers have turned to **pesticides**, chemicals that reduce pest numbers by killing or repelling pests or reducing their ability to successfully reproduce. Pesticides are very useful for farmers because they increase productivity (by decreasing loss due to pests), improve the quality of the crop because there are fewer blemishes on them, and are less labor-intensive than non-chemical pest removal methods. (See Module 8.2 for more on non-chemical ways to control pests.)

Orest Iyzhechka/Shutterstock

Pests, such as these Colorado potato beetle larvae, cost farmers millions of dollars in lost productivity. Livestock too are attacked by pests. Pesticides can reduce infestations but come with many trade-offs such as cost, environmental damage, and the emergence of pesticide-resistant pests.

Yet the use of pesticides has also proven problematic. Because they are toxic, pesticides also pose a threat to human and ecosystem health. Toxic residues can be found on many foods, and applying pesticides is dangerous work for the farmer. Pesticides can also harm the soil organisms that enrich the soil, reducing fertility, and these poisons can contaminate surface and groundwater sources on which humans and other organisms depend.

Here's another problem: The very use of pesticides sets into motion events that render them ineffective, causing pest populations to develop **pesticide resistance**, an unintentional example of artificial selection (see Module 3.1). Herbicide-resistant weeds and insecticide-resistant insects are cropping up all around the world. Such resistance encourages us to employ more drastic measures in the form of higher doses or more toxic chemicals; as resistance to that next pesticide develops, the cycle repeats itself, becoming what some have called a "pesticide treadmill." It's like an arms race between humans and pests, with the deck stacked in favor of the pests. **INFOGRAPHIC 5**

In the ongoing fight between crops and their pests, it becomes more and more important to preserve a genetically diverse arsenal of seeds that can be used to breed new, more pest-resilient crops in the future.

pesticide resistance The ability of a pest to withstand exposure to a given pesticide; the result of natural selection favoring the survivors of an original population that was exposed to the pesticide.

INFOGRAPHIC 5 THE EMERGENCE OF PESTICIDE-RESISTANT PESTS Achieve

Exposure to a pesticide will not make an individual pest resistant; it will likely kill it. However, if a few pests survive because they happen to be naturally resistant, they will breed and their offspring (most of which are also pesticide-resistant) will make up the next generation. Over time, the original pesticide will no longer be effective and will have to be applied at a higher dose or a different pesticide will have to be used.

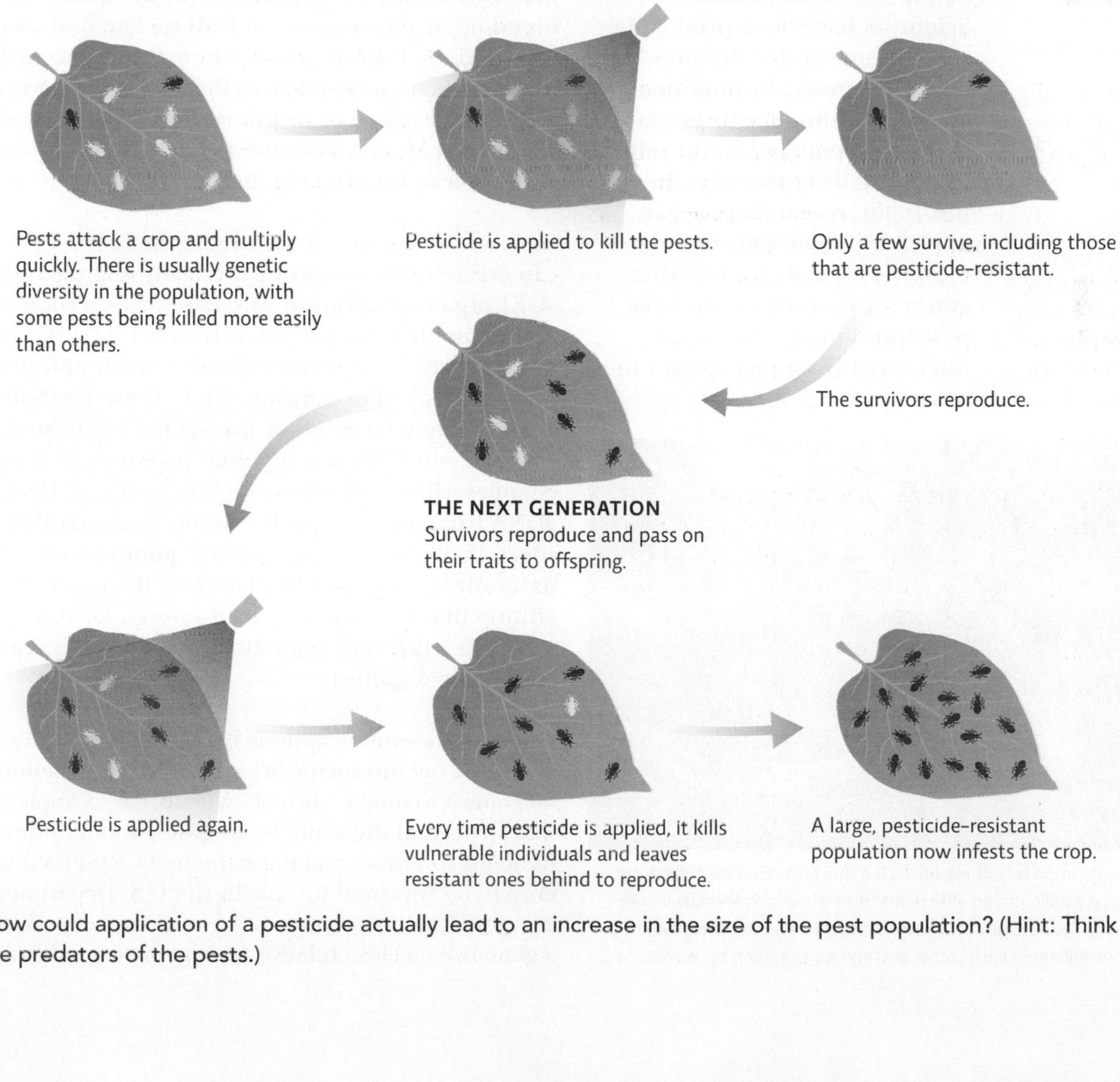

Pests attack a crop and multiply quickly. There is usually genetic diversity in the population, with some pests being killed more easily than others.
Pesticide is applied to kill the pests.
Only a few survive, including those that are pesticide-resistant.
The survivors reproduce.
THE NEXT GENERATION
Survivors reproduce and pass on their traits to offspring.
Pesticide is applied again.
Every time pesticide is applied, it kills vulnerable individuals and leaves resistant ones behind to reproduce.
A large, pesticide-resistant population now infests the crop.

? How could application of a pesticide actually lead to an increase in the size of the pest population? (Hint: Think about the predators of the pests.)

6 THE GENE REVOLUTION: GENETICALLY ENGINEERED CROPS

Key Concept 6: Genetically engineering crops (or animals) to contain useful traits such as faster growth or an expanded growing range may increase food supplies but comes with environmental, economic, and ethical concerns.

Humans have been altering the genetic makeup of crops ever since the first budding farmers started cross-breeding plants to produce new varieties, and it was with these same artificial selection techniques that plant scientists created the high-yielding crops of the Green Revolution. But scientists now have new methods to directly alter the genetic makeup of an organism, and these techniques form the basis of what some farmers and scientists like to think of as Green Revolution 2.0, or the *Gene Revolution*—the next battle against hunger.

genetically modified organism (GMO) An organism that has had its genetic information modified to give it desirable characteristics such as pest or drought resistance.

transgenic organism An organism that contains genes from another species.

intragenic organism An organism whose own DNA has been edited.

cisgenic organism An organism that received DNA from a close relative, DNA that could have been acquired via traditional breeding.

Genetically modified organisms (GMOs) are defined as those organisms that have had their genetic information altered in a way that would not be possible by natural means. Scientists have been producing such organisms for decades, coaxing genetically modified bacteria to produce important medicines such as insulin—like tiny living drug factories. In the 1990s, researchers began applying the same technology to food crops. By transferring genes for desirable traits (like pest resistance or herbicide tolerance) from one species to another, they have created a new suite of genetically modified food crops—plants that can grow more plentifully and thrive in a wider range of habitats than they ordinarily would. Organisms that receive DNA (genetic material) from another species with which they could not naturally breed (a sexually incompatible species) are known as **transgenic organisms**.

Edward Parker/Alamy

Genetic modification is nothing new. Like this farmer cross-pollinating the bloom of a vanilla orchid (the plant from which commercial vanilla flavoring is derived) by hand, humans have been altering the genetics of organisms for millennia with plant- and animal-breeding programs.

As powerful as genetic engineering might be, it cannot create the traits we would like our crops to have—it can only transfer them. Seed banks are an important resource not only because they preserve seeds that might otherwise go extinct, but also because they are a repository for diverse genetic material. When scientists analyze the genes of seeds kept in banks, they can identify useful genes for plant breeding or the creation of GMOs. The best place to find these traits is among the populations with the most genetic diversity—those plants growing wild in their place of origin (wild-type plants)—the very plants Murray and others like her search out to collect seeds for safekeeping.

In recent years, new technologies have emerged that can create other types of genetically engineered (GE) organisms. Known as *genome editing*, such as the technique that goes by the acronym CRISPR, these methods allow scientists to more quickly and precisely edit the DNA of organisms. While these methods could be used to produce transgenic organisms, they can also edit DNA in a way that produces desired changes without introducing any "foreign" DNA (DNA from an unrelated, sexually incompatible species). Because "same-species" genes could naturally be acquired by plant breeding, genome editing that doesn't introduce foreign DNA may avoid the ethical or regulatory concerns of creating a transgenic organism.

One genome-editing approach simply alters a cell's own DNA (an **intragenic organism**); the most common outcome is to simply "turn off" a gene. For example, a gene was silenced in the white button mushroom that reduces its ability to bruise, making it the first CRISPR-edited crop to be approved for sale by the U.S. Department of Agriculture. Genome editing can also be used to add a gene from a close relative (a **cisgenic organism**).

For example, in dairy cows, the gene for producing horns was removed and replaced with a gene for "hornlessness" from Angus cattle, creating two hornless dairy bulls. This useful trait would eliminate the need to dehorn calves, a painful process that is done for safety reasons.

A cisgenic change is one that could have been acquired naturally by cross-breeding (though it would take much, much longer), so it is not considered an "unnatural" change. This also means cisgenic organisms do not fit the U.S. regulatory definition of a GMO—they are not transgenic. (They *are* considered GMOs by the European Union.) Once the "poster child" of cisgenic engineering, the company that genetically engineered the two hornless bulls, Recombinetics, met an unexpected obstacle when it was discovered that a bacterial gene was inadvertently inserted during the process, technically making the bulls transgenic organisms.

CRISPR and other new genome-editing techniques are fast becoming the editing methods of choice in research, in part because they are fast, cheap, and easy to do. However, few intragenic or cisgenic foods have hit the market. Much work remains to be done to convince regulators and the public that these are safe foods to grow and consume.

Even before the development of CRISPR and other genome-editing techniques, a large percentage of U.S. grown staple food crops were GMOs—most of them the corn and soybeans that feed livestock. (Only a small percentage of the corn we eat is genetically modified.) But much of these crops end up in the processed food that lines our grocery store shelves—more than 75% of U.S.-processed food contains GMOs; for corn and soybeans, that number is even higher, around 85% to 90%. Most cotton grown in the United States is also genetically modified.

Currently, most GM crops are *herbicide-tolerant (HT)* in that they have a gene added that enables them to withstand huge doses of herbicides (chemicals that kill weeds). The most widespread HT crops are those with a gene that allows them to tolerate the herbicide glyphosate (Monsanto's Roundup®). This trait enables farmers to treat the field with Roundup and kill the weeds without threatening their harvest. Other crops, known as Bt crops, have been engineered to better resist pests; they contain a gene from *Bacillus thuringiensis*, a naturally occurring bacterium that produces a toxin that kills some insect pests. Some crop GMOs even contain both an HT gene and a Bt gene. **INFOGRAPHIC 6**

Proponents of the technology point to the possibility of not only increasing our food supply with genetically engineered foods but also improving it, such as the GMO known as "golden rice"—a rice modified to produce extra beta-carotene (a precursor to vitamin A), to help address vitamin A deficiency, a leading cause of blindness in children. Though it has been the focus of research since 1992, golden rice has not yet been approved for use, though Bangladesh may be the first country to do so.

To be sure, genetic engineering is a decades-old technology, and new technologies such as CRISPR have revolutionized the field, making it easier, cheaper, and faster to precisely edit the genetic makeup of crops and animals with less chance of introducing unwanted effects. But genetically modified food is unlike any of the other products that have been produced with genetic engineering. With the production of genetically engineered medications, the transgenic organism remains confined to a flask in a lab; GE crops, however, are out in the real world, growing and sharing genes with other organisms and being eaten by humans both directly and indirectly, through animals we are raising for food. Many of those humans are very uneasy about the prospect.

An early worry of those concerned about GE foods was the possibility that allergens or toxic substances would be present in GM foods. The evidence to date suggests that GM foods are safe to eat by humans and animals.

Ecologically, critics also worried that the genes introduced into genetically modified crops could be passed to wild plants through cross-pollination—that is, they could escape into the natural world and be incorporated into other plants for which they were not intended. This has occurred with the HT gene—it has been accidently transferred to at least 16 weed species, creating "superweeds" that can tolerate all the glyphosate herbicide a farmer can spray, reducing crop yields, often below what would have been realized with traditional (non-HT) crops. However, an extensive 2016 review of genetically engineered crops by the National Academy of Sciences concluded that while gene transfer between HT crops and wild plants does occur, it only appears to be a problem in fields where the herbicide was applied—the HT wild plants (weeds) had no competitive advantage in places where the herbicide was not used. This does, however, reduce the utility of the HT crops—if the herbicide cannot be used, the HT crop loses its value.

INFOGRAPHIC 6 GENETIC ENGINEERING

The genetic material of an organism can be modified in a variety of ways and with a variety of techniques to produce a crop or an animal with desired traits. Older methods used the bacterium *Argobacterium* to ferry new DNA into a target cell, but these protocols were time-consuming and outcomes were hard to control. Newer genome-editing methods can precisely edit genes or introduce new DNA in a much quicker and easier way, opening new opportunities to design crops, livestock, or even pets.

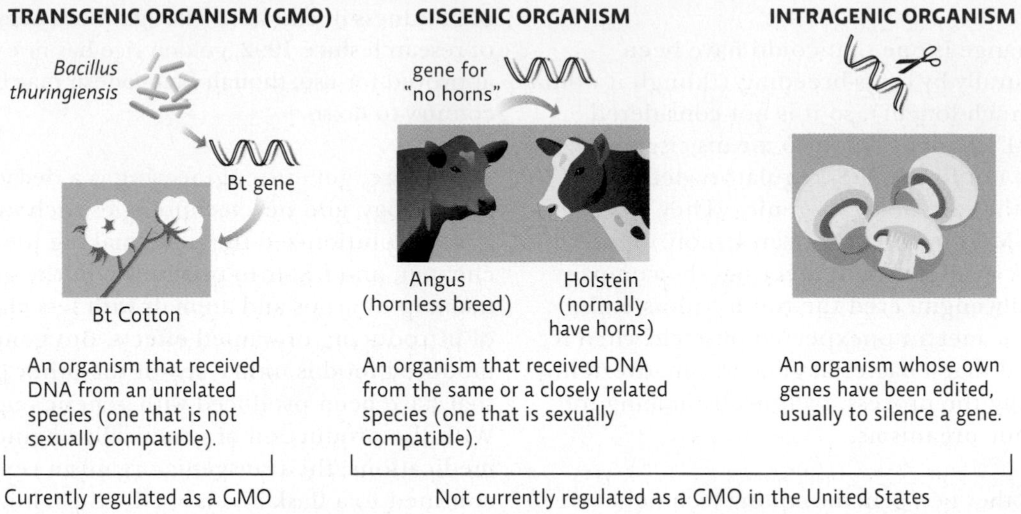
TRANSGENIC ORGANISM (GMO)
Bacillus thuringiensis
Bt gene
Bt Cotton
An organism that received DNA from an unrelated species (one that is not sexually compatible).
CISGENIC ORGANISM
gene for "no horns"
Angus (hornless breed)
Holstein (normally have horns)
An organism that received DNA from the same or a closely related species (one that is sexually compatible).
INTRAGENIC ORGANISM
An organism whose own genes have been edited, usually to silence a gene.
Currently regulated as a GMO
Not currently regulated as a GMO in the United States

Do you think that cisgenic and/or intragenic crops should be regulated the same way as transgenic crops (GMOs)? Explain.

Increased use of a pesticide is another worry—herbicide can be applied to HT crops while they are growing in the field (something that can't always be done with non-HT crops since the herbicide would harm their growth), extending the time frame for herbicide use and increasing the amount that can be applied. Another fear is that crops engineered to resist pests, such as Bt crops, could allow nontargeted (secondary) pests to increase, thereby requiring more pesticide application, or that they might give rise to a new population of Bt-resistant pests much more quickly than would traditional pesticide spraying. (Bt has long been used as a pesticide spray.) Recent Bt crop studies suggest that though its use did sometimes lead to an increase of secondary pests, overall the use of pesticide decreased.

One of the biggest criticisms is the prospect of putting even more of our food supply under the control of a few multinational corporations. In the United States, companies like Monsanto have been known to tightly guard their GM seeds. Because seeds are patented, farmers are not permitted to save seeds from one year to the next but instead must repurchase them year after year. Environmental activist Vandana Shiva calls this "seed slavery," arguing that farmers need control over seeds for food security. "After all," she said in an interview for *The Guardian*, "seeds are the first link in the food chain."

In 2001, the International Treaty on Plant Genetic Resources for Food and Agriculture was adopted to promote the conservation of plant genetic resources, but it also calls for the protection of farmers' rights to those resources—farmers today who are often the descendants of the farmers who developed these varieties over generations. Many argue that we not only need to save the seeds that might feed us in the future but also need to protect the rights of farmers to use and trade those seeds.

The current consensus presented in the National Academy report is that, in general, the use of GM crops has increased crop productivity (slightly), that it has not produced serious ecological problems (e.g., pest resistance, escaping GM traits) that can't be handled with modified agricultural practices, and that GM food is safe to eat and feed to livestock. However, in a quest to increase food production, they point out that it is but one part of the solution, acknowledging that in some areas, GM crops may not be the best option and traditional breeding practices may work better.

7 FOOD SELF-SUFFICIENCY AND SOVEREIGNTY

Key Concept 7: Modern agricultural techniques can help fight hunger in affected nations or regions, but solutions need to include food self-sufficiency and food sovereignty.

No one can dispute the fact that the Green Revolution fed, and continues to feed, a lot of people. Unfortunately, not everyone has benefited from this agricultural transformation. Africa, a continent plagued by hunger and largely bypassed by the Green Revolution, has nonetheless fallen prey to a different set of the Green Revolution's unintended consequences: a lack of **food self-sufficiency** (the ability of an individual nation to grow enough food to feed its people), a lack of **food sovereignty** (the ability of an individual nation to control its own food system), and ultimately a lack of food security.

Here's why: As industrialization and farm subsidies enabled (mostly U.S.) farmers to produce vast surpluses of wheat, corn, and soybeans, the global marketplace was flooded with cheap food from the developed world. Farmers in many African countries could not compete with such cheap and plentiful food imports—plagued as they were by land degradation (drought, soil erosion, water shortages) and armed conflict (violent clashes destroy existing crops and prevent new ones from being planted).

So they converted much of their farmable land to **cash crops** like cotton, coffee, and cocoa. (*Cannabis*, the opium poppy, and coca [for cocaine], grown in countries around the world, are the most valuable cash crops per acre.) Rather than feed local populations, these food and fiber crops are usually exported for profit. Though some income remains in the local community, much of the profit goes to middle managers and those in power. Thus, a system where grain was locally produced and supplied gave way to a system where it was imported from thousands of miles away. And as they became dependent on food imports, developing countries found themselves at the mercy of forces far beyond their own borders.

NIC BOTH-MA/European Pressphoto Agency/CAPE TOWN/South Africa/Newscom

Seed exchanges, like this one sponsored by SEED, a nonprofit community group in South Africa, helps farmers use and preserve a wide variety of locally adapted seeds.

More than once, this dependency has spurred violence as protesters in African nations have taken to the streets. In 2008, riots broke out in the tiny west African country of Burkina Faso. Angry protesters clogged the streets of the country's two major cities—shouting, throwing rocks, and flipping cars—rioting over the cost of food, which had doubled in the span of just 7 months, an unbearable increase after prices had already more than doubled in the 2 years prior. Soldiers mobilized to restore order, but the chaos only spread from there. In neighboring Côte D'Ivoire, tear gas was employed, and dozens were injured; in Cameroon, some two dozen people were killed. Before long, the violence spilled across Africa's borders. Protesters in Yemen torched police stations and blocked roads. In Bangladesh, they smashed cars and buses, vandalized factories, and ultimately injured dozens of bystanders. Such dramatic and rapid cost increases had pushed staple foods like bread and rice out of their reach and, in so doing, had nudged too many people toward the brink of starvation. "For countries where food comprises from half to three quarters of [income] consumption," explains World Bank President Robert Zoellick, "there is no margin for survival."

We may indeed be growing enough food to feed the world, but if this food is not available to the people who need it most, hunger and food insecurity will remain. In addition to producing enough food to feed everyone and producing more of that food locally, many advocates stress the importance of working toward food sovereignty. Social programs that provide food to those in need may help stave off hunger in the short term, but if programs are not also in place that address the underlying causes of food insecurity, food sovereignty will not be achieved. As Rajeev Patel explains in an essay published in *PLoS Medicine*, "It is possible, after all, to be food secure in prison where one might continually access safe and nutritious food yet remain fundamentally disempowered over the process and politics of the food's production, consumption, and distribution."

food self-sufficiency The ability of an individual nation to grow enough food to feed its people.

food sovereignty The ability of an individual nation to control its own food system.

cash crops Food and fiber crops grown to sell for profit rather than for use by local families or communities.

INFOGRAPHIC 7 PRINCIPLES OF FOOD SOVEREIGNTY

For many regions, achieving food security may hinge on attaining food sovereignty—putting control of the food supply in the hands of that region and its farmers. These six principles aim to protect local populations and their environment so they can grow their own food.

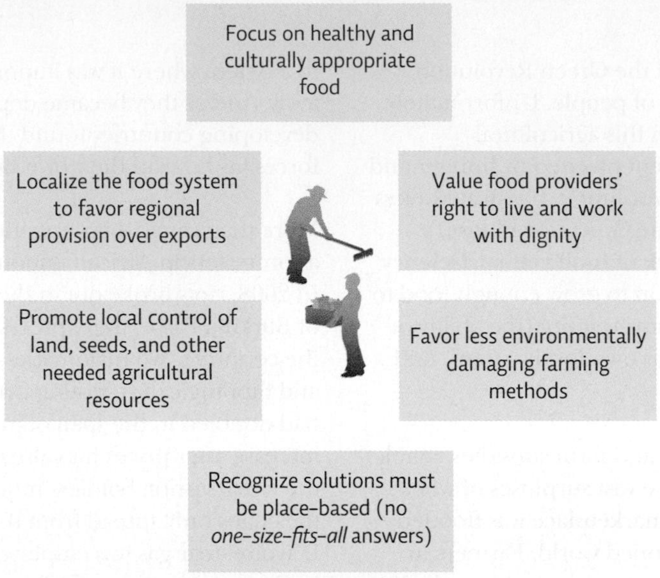
Focus on healthy and culturally appropriate food
Localize the food system to favor regional provision over exports
Value food providers' right to live and work with dignity
Promote local control of land, seeds, and other needed agricultural resources
Favor less environmentally damaging farming methods
Recognize solutions must be place-based (no *one-size-fits-all* answers)

Explain the connection between food sovereignty and social justice.

The international grassroots organization of small farmers, migrant workers, and indigenous people known as La Via Campesina pushes for changes that will empower communities to achieve food security through food sovereignty. This includes producing local food to enhance food self-sufficiency but also empowering people to determine what their own food-provisioning systems will look like—what types of food will be produced, how it will be produced, and how it will be distributed. Taking the power out of the hands of a few multinational agricultural and food corporations and giving it to the people themselves, La Via Campesina argues, is key to addressing food insecurity. That production will likely depend on a variety of farming methods—the industrial methods and GM crops described in this module—and a variety of low-tech and traditional techniques that may be better suited to some locales and local populations (see Module 8.2). The key to food security may be allowing each village, community, or region to choose for themselves which path they will follow. **INFOGRAPHIC 7**

Into this morass, an additional 2 billion people will soon be born. Global population is expected to approach 10 billion by 2050; experts say that to feed that many mouths, we will need to produce significantly more food than we are now producing and we will need to address social justice issues, such as poverty and discrimination, that contribute to food shortages and food deserts. And of course, we will need to do everything we can to ensure that the plants that feed us do not disappear forever. By banking the world's seeds, we make it more likely that we will have the plant varieties we need—both to feed local communities and to grow the cash crops providing communities with essential income.

Our futures may depend on our ability to safeguard seeds, which is why the work of seed warriors like Sheila Murray is so important. Every seed she collects—every seed that gets safely locked away and protected against environmental disaster, war, and climate change—is one more reason to have faith in humanity's future.

Select References:

Coleman-Jensen, A., et al. (2019). *Household Food Security in the United States in 2018*, ERR-270. Washington, DC: U.S. Department of Agriculture, Economic Research Service.

Confino, J. (2012, October 8). *Vandana Shiva: Corporate monopoly of seeds must end.* Retrieved from https://www.theguardian.com/sustainable-business/vandana-shiva-corporate-monopoly-seeds

Food and Agriculture Organization of the United Nations (FAO). (2018). *The State of Food Security and Nutrition in the World 2018: Building Climate Resilience for Food Security and Nutrition.* Rome: FAO.

National Academies of Sciences, Engineering, and Medicine. (2016). *Genetically Engineered Crops: Experiences and Prospects.* Washington, DC: The National Academies Press. doi:10.17226/23395.

Patel, R. C. (2012). Food sovereignty: Power, gender, and the right to food. *PLoS Medicine, 96,* e1001223.

RBG Kew. (2016). *The State of the World's Plants Report—2016.* Kew: Royal Botanic Gardens, Kew.

Wooten, R., et al. (2019). Assessing food insecurity prevalence and associated factors among college students enrolled in a university in the Southeast USA. *Public Health Nutrition, 22*(3), 383–390.

GLOBAL CASE STUDIES FEEDING THE WORLD

Achieve

Around the world, different approaches are being used to improve agriculture and crop yield. Each come with their own trade-offs. Check out these four examples.

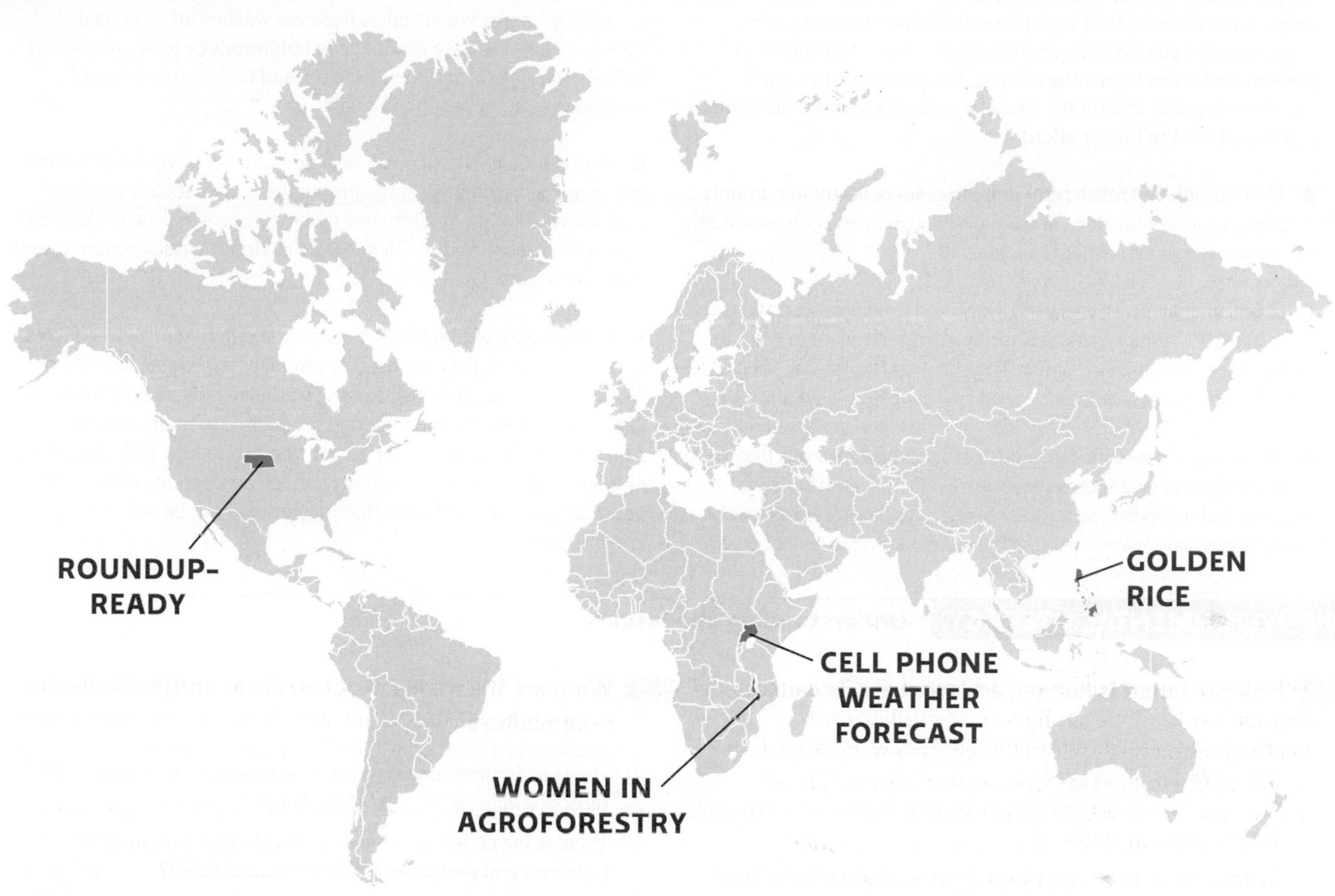

BRING IT HOME

PERSONAL CHOICES THAT HELP

TODD ANDERSON/The New York Times/Redux

The challenge of meeting the nutritional needs of the entire population seems overwhelming, but there are plenty of opportunities to make an impact within your local community. Two of the best ways to have an impact on those facing food insecurity are through education and financial support.

Individual Steps

- There are thousands of food banks all over the country that collect and distribute food to those who need it the most. Donate your time or nonperishable food items. Find the closest food bank at www.feedingamerica.org.
- Plant your own vegetable garden and donate part of the harvest to your local food bank. Some areas have community gardens that rent space during the growing season.
- Make a monetary donation to an international nonprofit organization that provides food aid or agricultural assistance, such as UNICEF (www.unicef.org) or Heifer International (www.heifer.org).
- GMOs may be part of the solution because they can increase crop yields around the world, but they may cause some unintended consequences. Contact your federal senators and representative: Express your opinions and discuss your representative's position on the regulation, testing, and labeling of GMOs.

Group Action

- Work with parents, community leaders, and educators to develop a summer food service program for school-age children on free or reduced-price lunch programs. Find contacts and resources at https://www.fns.usda.gov/sfsp/summer-food-service-program.
- Volunteer at or start a food pantry at your school to help address food insecurity on campus.

REVISIT THE CONCEPTS

- The United Nations has set a 2030 goal for all people to have food security — access to sufficient safe and nutritious food. Today nearly a billion people are undernourished due to poverty, war, environmental degradation, and inadequate food distribution or preservation. Even in wealthy nations, food deserts can exist in low-income urban and rural areas where access to fresh, nutritious foods is difficult to find or afford.

- Malnutrition can result from diets that are deficient in nutrients or calories (undernutrition) or from overnutrition, the consumption of excess calories or nutrients such as fats.

- Scientists engineered the Green Revolution to address world hunger in the mid-20th century by using plant-breeding techniques to produce high-yielding crop varieties. Grown with chemical fertilizers and pesticides in monocultures, these crops increased food supplies substantially.

- The widely adopted modern industrial agriculture methods that made the Green Revolution a success have also led to ecological problems (e.g., reduced soil fertility and water pollution), social problems (fewer family farmers and lower profits), and a reduction in crop genetic diversity.

- The application of synthetic fertilizers helps increase crop yields but damages the soil ecosystem that produces fertile soil and can pollute nearby waterbodies if excess washes off farm fields. Pesticides also increase productivity but introduce toxic substances to the environment, and their use can lead to the emergence of pesticide-resistant pest populations.

- A modern technique to increase productivity beyond what plant breeding can accomplish is the production of genetically modified organisms (GMOs). Though new techniques are making it easier to make these genetic edits, GMOs raise environmental, economic, and ethical concerns.

- Modern agricultural techniques can help fight hunger in affected nations, but when they are used to produce cash crops, local food supplies decrease. The environment may also be damaged from industrial farming. By keeping food production local (food self-sufficiency) and letting local populations decide how and what kinds of foods to grow (food sovereignty), those populations will be less dependent on outside forces that may or may not be able to help provide food.

ENVIRONMENTAL LITERACY Understanding the Issue

1 How prevalent is hunger, and what are its causes?

1. Define food security and identify at least five causes of food insecurity.
2. How does food waste contribute to food insecurity? What could be done to address it?
3. What is a food desert, and where are they most likely to be found?

2 What is malnutrition, and what problems can it cause?

4. Distinguish between undernutrition and overnutrition in terms of causes and consequences.
5. What is the relationship between malnourishment in a pregnant mother and her child?

3 What were the intent, scope, and outcome of the Green Revolution?

6. What did the Green Revolution accomplish, and how did it do so?
7. Explain the importance of genetic diversity of crops to the plant breeders of the Green Revolution.

4 What is industrial agriculture, and what are the pros and cons of farming this way?

8. Explain the advantages and disadvantages of monoculture use in industrial agriculture.
9. What are the advantages and disadvantages of irrigating crops?
10. How can seed banks help agriculture now and in the future?

5 What are the trade-offs of fertilizer and pesticide use in agriculture?

11. How can fertilizer use increase crop productivity? Is this a short-term or a long-term benefit? Explain.
12. What is the connection between fossil fuels and synthetic fertilizers and pesticides? Why is this a concern?
13. Explain how the use of chemical pesticides can lead to pesticide-resistant populations.

6 How can genetic engineering be used in agriculture, and what are its trade-offs?

14. Compare the types of genetically engineered organisms that are made today — transgenic, cisgenic, and intragenic organisms — in terms of the origin of the genetic material added and regulatory status.
15. Outline some of the concerns of growing or eating genetically modified crops and indicate what the evidence reveals regarding the validity of those concerns.

7 What are food self-sufficiency and food sovereignty, and why are they important?

16. What is the definition of a cash crop, and what food security problems can cash crops cause?
17. Outline and briefly explain the six principles of food sovereignty presented in Infographic 7.
18. Why does La Via Campesina argue that taking the power out of the hands of a few multinational agricultural and food corporations and giving it to the people themselves is key to addressing food insecurity?

SCIENCE LITERACY Working with Data

The Food and Agriculture Organization (FAO) of the United Nations monitors world hunger. The UN set a goal of reducing the proportion of undernourished people to half of the 1990 level by 2015 and to eradicate hunger by 2030. Evaluate the data shown in this graph to assess progress on addressing world hunger.

NUMBER AND PROPORTION OF UNDERNOURISHED PEOPLE IN THE WORLD

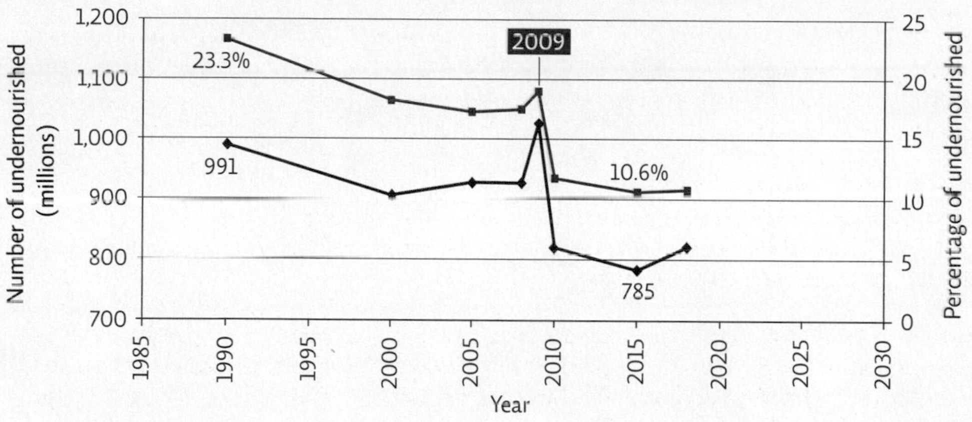

Interpretation

1. Describe in one sentence what the graph shows about the change in undernourishment over time. Why are both lines shown here?
2. In what year were the most people undernourished? In what year was the highest proportion of people undernourished? Why do you think these occurred in different years?
3. What happened to the number and percentage of undernourished people between 2008 and 2009? What do you think led to this change?
4. In general, what might have led to the decrease in undernourishment between 1990 and 2018?

Advance Your Thinking

5. By 2015, the FAO had hoped to cut the number and proportion of undernourished people relative to the 1990 levels in half. How did it do?
6. In looking at both the number and percentage of undernourished people, do you think there has been a complete recovery from the 2008 world food crisis? Why or why not?
7. Following what appears to be the general trend of the data from 1990 to 2018, extrapolate these lines out to 2030 by drawing a straight trendline. Do you predict that the UN will meet its target of eradicating world hunger by 2030? If not, what do you predict the number and percentage of undernourishment to be in 2030? (*Hint*: Be sure to consult the correct *y* axis when estimating the 2030 values.)

INFORMATION LITERACY Evaluating Information

La Via Campesina is an international movement that fights for farmers' rights and food sovereignty. Learn more about this organization at https://viacampesina.org/en/. (The site is also available in Spanish and French.)

Evaluate the website and work with the information to answer the following questions:

1. Why was La Via Campesina founded, and what is its mission today?
2. Explore the website and read two articles posted under the "Food Sovereignty" link on the "What Are We Fighting For?" page. Determine if the website is a reliable information source by answering these questions about the articles you read. For each article:
   a. Give the name of the article and briefly summarize its main point.
   b. Identify a claim made in the article and the evidence given in support of this claim. Is it sufficient? Explain.
   c. Are sources for its evidence provided? Are these reliable sources? If so, how do you know?
   d. Where would you attempt to find more evidence to support or refute the claim?
3. Explore the website further by visiting the page entitled "What Are We Fighting Against?" Choose two articles from this page and answer the same questions you answered in Question 2 above.
4. Overall, what is your impression of this organization? Would you support it with donations or through social media? Explain.

Additional study questions are available at Achieve.macmillanlearning.com. Achieve

Module 8.2 Sustainable Agriculture: Raising Crops

FARMING LIKE AN ECOSYSTEM

Learning to farm from nature

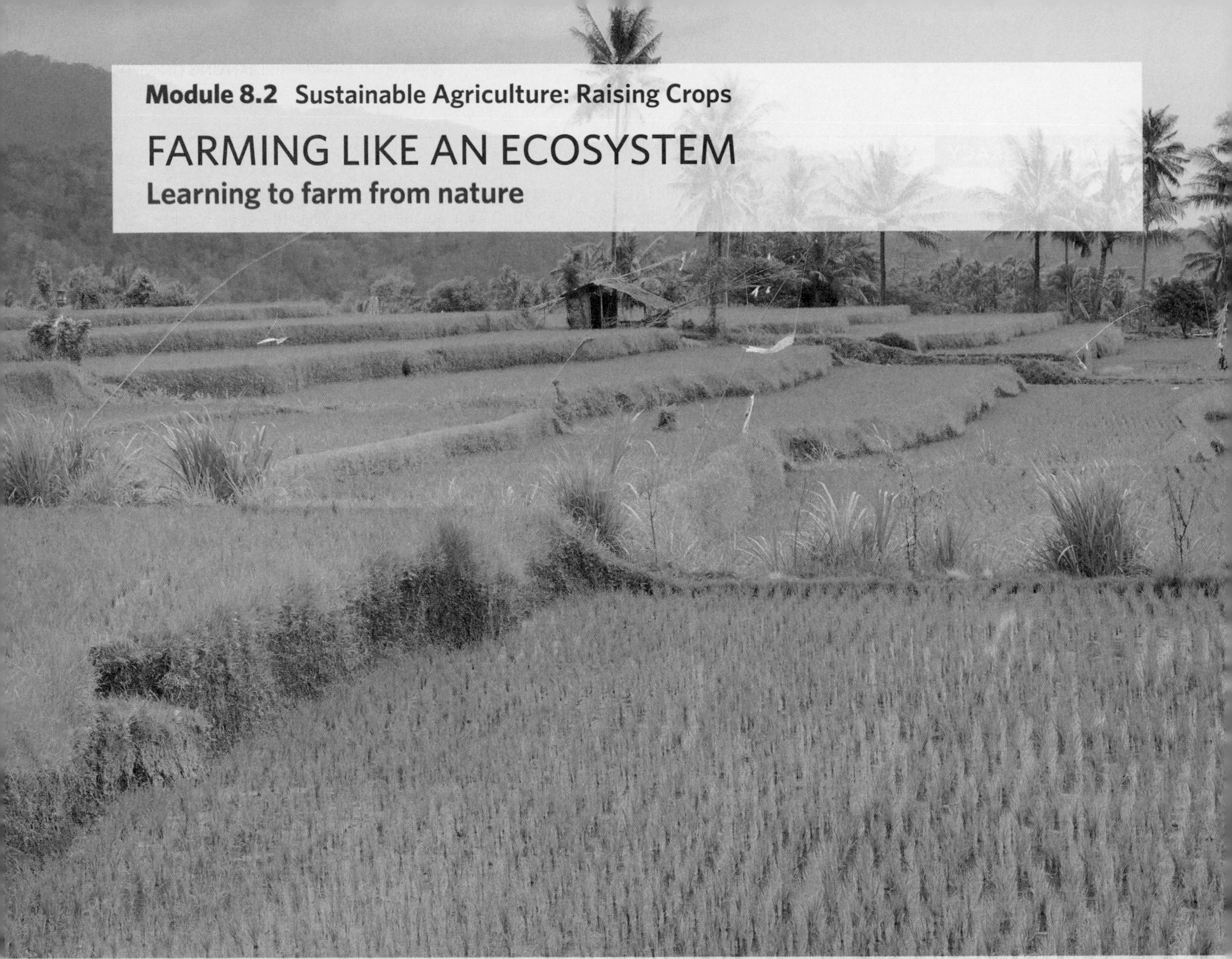

Rice is one of the most important crops in the world. An ancient Japanese rice-farming practice offers a more sustainable approach to growing the crop. *Image Broker/Media Bakery*

After reading this module, you should be able to answer these GUIDING QUESTIONS

1. What is sustainable agriculture?
2. What is the importance of fertile soil, and how does land use affect it?
3. What is agroecology, and what are its benefits?
4. How can pests be managed in a way that minimizes environmental damage?
5. How can traditional farming methods contribute to sustainable agriculture?
6. What role does the consumer play in helping build a sustainable food system?
7. What are the trade-offs of sustainable agriculture, and how might they be addressed?

When Erik Andrus bought his Vermont farm in 2005, he dreamed of growing wheat and barley, because he loved bread and beer. But the land and climate, he discovered, just wouldn't cooperate. His farm, which he named Boundbrook, was nestled within the valley around Lake Champlain in western Vermont, where it is swampy and wet—far too wet for growing cereal crops. Sometimes Andrus would overhear farmers in the local diner, joking, "We ought to be growing rice!" One day, it dawned on him that they were right.

So now Andrus grows rice. In fact, he is one of only a handful of American farmers using an ancient integrated farming method (one that uses multiple species) known in Japan as *aigamo*, or "integrated rice and duck farming." After growing rice seedlings in a nursery, Andrus and his farmers transfer them to a field in early spring. Then, their baby ducklings arrive, ready to be released into the rice field to do a very important job. Because rice grows so slowly, it can quickly be crowded out and damaged by quick-growing weeds. Farmers can pull these weeds by hand, but this is difficult and tedious; they can use pesticides, but these contaminate the paddy water. Ducklings, on the other hand, do a marvelous and efficient job of controlling the weeds without harming anything: They eat and trample the weeds, avoiding the rice (whose leaves contain abrasive silica), and their small bodies don't crush or knock over the delicate plants. By the time the ducks have grown big enough to damage the rice, the rice, too, has grown big enough that weeds no longer grow and the ducks are no longer needed. The method works well: Andrus's farm is, right now, the largest rice farm in the northeastern United States.

WHERE IS BOUNDBROOK FARM?

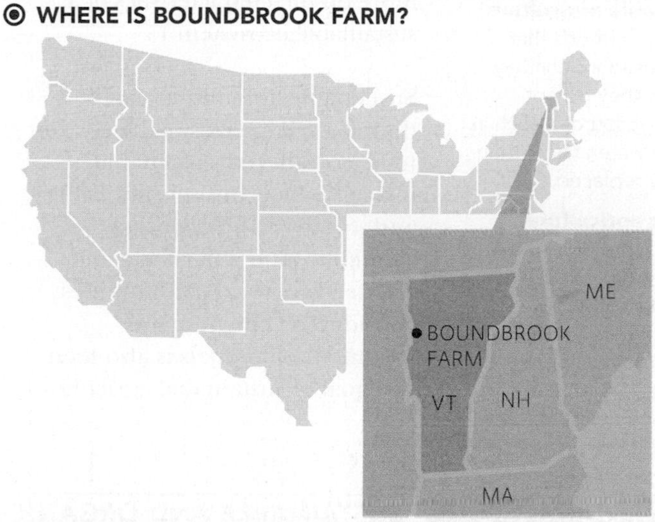

At a time when our world population has surpassed a staggering 7 billion and is growing toward 10 or 11 billion, the need to produce even more food has never been so urgent, and many farmers are looking for ways to grow food without damaging the environment that produces it. Andrus's farming method not only produces a profitable crop but also provides valuable ecosystem services to the area's human and ecological communities. "A paddy is like a wetland—it slows down the path of water, and the various ponds and reservoirs and canals provide resources for firefighting and erosion control," he says. The soil remains nutrient-stable, and the growing environment creates new habitats for wildlife. The rice, too, is delicious. "It is a special honor for a farmer to be able to open this world of experience to neighbors, and to begin to make a place for this ancient and venerable food in our landscape," Andrus says.

1 SUSTAINABLE AGRICULTURE

Key Concept 1: The goal of sustainable agriculture is to raise food without damaging the environment or future productivity while operating ethically with regard to animals, workers, and local communities.

Andrus wanted to find an alternative to industrial agriculture as it is commonly practiced because the methods employed have many downsides. Most industrial farms are monocultures of genetically identical crops, and they are raised with synthetic fertilizers and pesticides. Although they do produce a lot of food, they threaten long-term production by degrading soil, depleting or polluting water, and shrinking the genetic storehouses of crop and animal varieties. (See Module 8.1 for more on industrial agriculture.)

Sustainable agriculture, on the other hand, is farming that can be done indefinitely—without compromising the environment or future productivity—all while meeting the needs of the farmer and society as a whole. The techniques used will maintain or even enhance the environment, in many cases using farming methods that help build soil rather than deplete it. Farms often achieve this by mimicking the traits of a sustainable ecosystem: They rely on renewable energy and local resources for inputs, and they depend on biodiversity to trap energy, deal with waste, and control pest populations. (See Infographic 6 in Module 1.1 for more on the characteristics of a sustainable ecosystem.)

sustainable agriculture
Farming methods that can be used indefinitely because they do not deplete resources, such as soil and water, faster than they are replaced.

organic agriculture
Farming that does not use synthetic fertilizer, pesticides, GMOs, or other chemical additives like hormones (for animal rearing).

Sustainable agriculture can be done on a large or small scale, but to be considered sustainable, it must also be economically viable and socially ethical; workers should be paid a fair wage and the needs of local communities considered. The humane treatment of animals is also seen as a goal of sustainable agriculture.

Like many other farmers pursuing sustainable agriculture, Andrus cultivates his rice according to the methods of **organic agriculture**. Instead of using synthetic fertilizers and pesticides, organic farmers employ more natural, or "organic," techniques in the growing of crops—such as fertilizing with manure and luring in natural predators to control pests. In addition, genetically modified organisms (GMOs) cannot be grown on organic farms (see Module 8.1). Organic food is not "chemical-free" or "pesticide-free"—it is just largely "synthetic chemical-free." There are even some exceptions to the "all-natural" rule, since vaccinations, synthetic pheromones that repel pests, and some synthetic cleaners can be used, while some toxic natural substances, such as arsenic, cannot. Organic foods that are certified by the U.S. Department of Agriculture (USDA) are labeled as *USDA Organic*. **INFOGRAPHIC 1**

While there are certainly environmental benefits to growing foods without synthetic pesticides or fertilizers, there are health benefits as well. A 2014 meta-analysis of 343 studies found that, on average, conventionally grown produce has four times more synthetic pesticide residue than organic produce—pesticides that can be ingested when the food is eaten. (See the *Science Literacy* activity

INFOGRAPHIC 1 SUSTAINABLE AND ORGANIC AGRICULTURE

According to the U.S. Department of Agriculture (USDA), sustainable agriculture is farming that uses only limited amounts of nonrenewable resources (e.g., fossil fuels) and does not degrade the environment or the well-being of people or society as a whole. Organic agriculture is a subset of sustainable agriculture that specifies some requirements for raising the crops or animals. Organic farms that have their production methods verified by the USDA are certified organic.

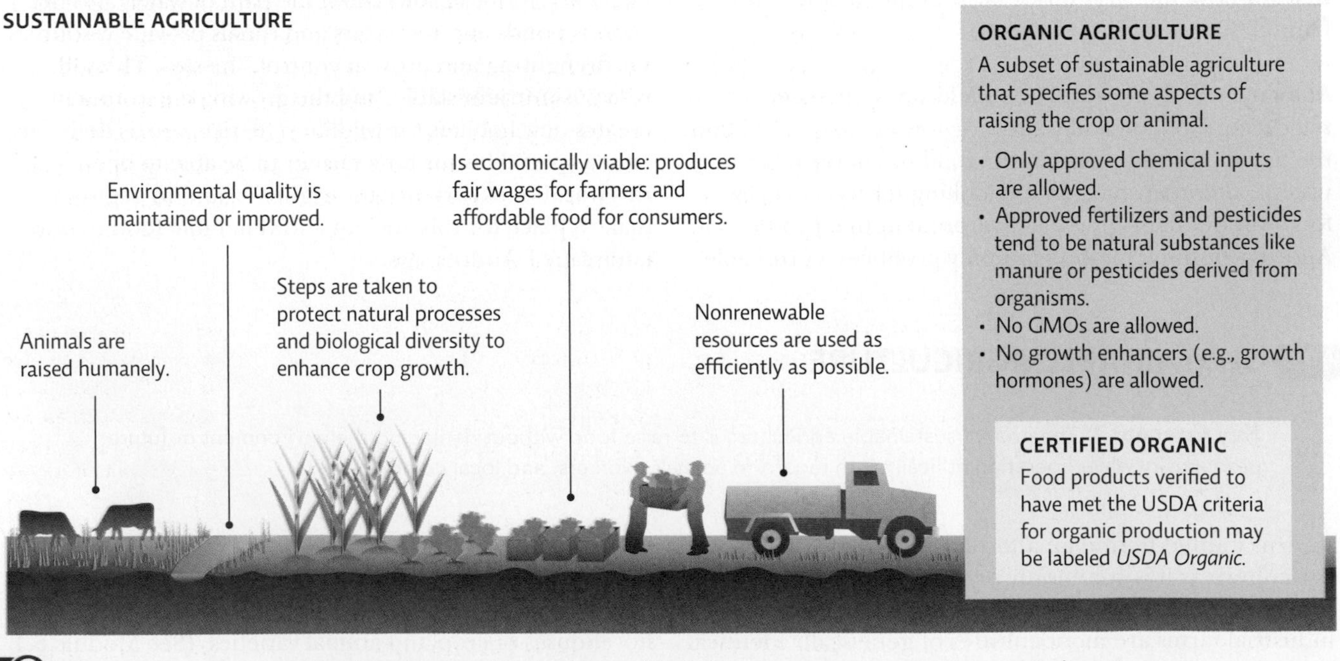
SUSTAINABLE AGRICULTURE
Is economically viable: produces fair wages for farmers and affordable food for consumers.
Environmental quality is maintained or improved.
Steps are taken to protect natural processes and biological diversity to enhance crop growth.
Animals are raised humanely.
Nonrenewable resources are used as efficiently as possible.
ORGANIC AGRICULTURE
A subset of sustainable agriculture that specifies some aspects of raising the crop or animal.
• Only approved chemical inputs are allowed.
• Approved fertilizers and pesticides tend to be natural substances like manure or pesticides derived from organisms.
• No GMOs are allowed.
• No growth enhancers (e.g., growth hormones) are allowed.
CERTIFIED ORGANIC
Food products verified to have met the USDA criteria for organic production may be labeled *USDA Organic*.

? Why are ethical concerns of workers and animals considered to be part of sustainable agriculture?

at the end of this module for more on this topic.) The meta-analysis also showed that some organically grown foods have higher levels of nutrients than conventionally grown crops; however, it is not known whether these differences translate into meaningful health effects.

Organic farming can be difficult, especially in the short term—it often has lower short-term yields and higher labor costs—but its environmental benefits keep production going for longer (so long-term productivity is higher) and keep the soil and surrounding ecosystems healthier. The profit margin for organic food is often larger than it is for conventionally grown food, too. Consumers are willing to pay more for it, both because of its consumer and environmental benefits and because, sometimes, it just tastes better. As Andrus says about his rice, "It has different culinary characteristics and a different story than anything else you can sell."

2 THE IMPORTANCE OF SOIL

Key Concept 2: Fertile soil is a diverse ecosystem that is dependent on a wide variety of biotic and abiotic factors. It provides many important ecosystem services; unfortunately, in many places, soil is being lost much more quickly than it forms.

The productivity of farmland is governed by the health of its **soil**: Without good soil, there is little, if any, terrestrial plant life, and without plant life, there is little life of any kind. Soil is not just a solid mineral matrix in which plant roots are anchored. It is a complex, living ecosystem full of a diverse assortment of bacteria, fungi, and invertebrate species that recycle matter and enrich the soil with nutrients. Soil contains open spaces that hold air and water needed by these soil organisms and the plants that grow in the soil. And while the plants rely on the soil that supports them, so, too, does the soil community depend on the plants growing in it.

Some species of bacteria and fungi in soil form mutualistic relationships with plant roots. Bacteria and fungi need energy, and they get it from sugars and other carbon molecules exuded from plant roots. (These secretions are called *exudates.*) In return, the soil organisms provide fixed nitrogen and soluble phosphorus to the plants. (See Module 2.1 for more on the nitrogen and phosphorus cycles.) And fungi and invertebrates break down biological matter, forming the organic material that supports both plants and soil microbes. The higher the diversity of plants above ground and soil organisms below ground, the better these processes will proceed.

The structure of the soil is an important part of this relationship. Consider the plant's need for nitrogen. While some plants (known as *legumes*) form symbiotic relationships with nitrogen-fixing bacteria that live in nodules on the plant roots, nonleguminous plants depend on free-living bacteria that live in the tiny pores that form between the clumps and lumps of soil for the nitrogen they need. This environment needs to be just right—pore sizes that aren't too big or too small, low oxygen levels, just the right amount of moisture—conditions that develop as organic material is decomposed. Known as *soil aggregates,* these clumps are formed and held in place by a kind of glue produced by the bacteria themselves.

Unfortunately, industrial farming methods that depend on heavy soil tilling result in soil compaction and an increased rate of **soil erosion**—a process in which soil is swept away by wind and rain into streams, rivers, and gullies faster than it can possibly be replenished. That's no small matter. Soil formation is a slow process that requires the weathering of rock and decomposition of organic material. Though it varies greatly from place to place, it has been estimated that in most places, it takes 500 years to more than 1,000 years to generate 1 inch of the precious brown "gold" in which our food grows. Any action that breaks down the architecture of the soil (destroys those pores and the glues that maintain soil's three-dimensional structure) and exposes soil to the elements can worsen soil erosion. When farmers ready their fields for planting by plowing to turn the soil and remove weeds or last year's crop residue, they increase the soil's vulnerability to erosion. The loss of these plants—even dead or decaying plant matter—means fewer roots to hold soil in place and fewer nutrients returning to the soil. Erosion can remove in days what took decades to form.

Soil also plays a role in our groundwater quality. As water soaks into the ground, the part that passes through soil is "filtered"—enough to be drinkable without further treatment from wells that are deep enough. Soils are also one of the most important carbon sinks, holding more than the atmosphere or the world's vegetation—making soil a key player in mitigating climate change (see Module 10.2).

soil A complex ecosystem of mineral and organic material, including living organisms such as bacteria, invertebrates, and fungi, that supports the growth of plants and is, in turn, affected by those plants.

soil erosion The process in which soil is moved from one location to another most often by wind or water.

JJ Gouin/Getty Images

Plowing farm fields disturbs the soil and increases the chance for soil erosion. Even after a crop starts to grow, the potential for erosion is great while the plants are still small, without deep roots to hold the soil in place.

Soils vary in quality, as anyone who has tried to grow a garden knows. The texture of soil is important and is determined by the size of the mineral particles that make it up and the amount of organic material mixed in. Sandy soils predominately have large particles (diameter 0.05 mm to 2 mm); roots can pass through these loose soils easily, but they do not offer good structural support to plants. In addition, water drains through these soils easily, leaving them dry much of the time. The smallest particles, which Andrus knows well, are clay (diameter less than 0.002 mm). The soil surrounding his farm is composed largely of clay, a heavily weathered soil that will hold more water than other types of soil. But the small particles can pack very tightly, making the soil so dense that roots have a hard time penetrating. Therefore, clay makes great bricks but poor soil for plant growth. But with the right techniques, this kind of soil works well for growing water-loving rice.

Silt particles are smaller than sand but larger than clay (diameter 0.002 mm to 0.05 mm). Silty soils are a nice compromise between sand and clay, offering structural support for root systems but also allowing water to percolate through at a rate that is more suitable for plants. Still, a soil made up entirely of silt would not be highly productive. In general, the best soil for plant growth is one with even proportions of large, medium, and small particles. These soils have enough clay to bind and keep water and nutrients in the soil, enough sand to help some water drain away, and enough silt to strike a balance between the other two.

A closer look at a soil's profile sometimes reveals distinct layers (called *horizons*) that reflect the formation process. Soil layers represent the dual nature of soil formation: the weathering of rock beneath the soil (the *C horizon*) and the deposition of organic material, such as leaf litter, at the top (in the *O horizon*). This surface litter is broken down by soil organisms (microorganisms and invertebrates such as worms, insects, and fungi) in the process of decomposition, releasing nutrients into the *A horizon*, the soil level where plant roots are most densely packed. The upper level of the A horizon is the most fertile region of soil; it is known as *topsoil*.

The loss of the O and A horizons due to soil erosion (or intentional removal) impoverishes the soil by removing not only the nutrient-rich soil but also the living organisms that help produce it. Fires that burn hot enough to kill soil organisms greatly impede ecosystem recovery. Transplanting a thin layer of soil from healthy grasslands to barren land (known as "soil inoculation") helps replace the missing microbes and invertebrates that form the soil community, substantially speeding the recovery of the area. **INFOGRAPHIC 2**

Protecting or repairing damaged farmland ultimately, then, comes down to restoring soil and soil fertility. The use of industrial fertilizers is a temporary fix that can eventually reduce soil fertility even more (see Section 5 of Module 8.1). Turning to nature—emulating how natural ecosystems recycle nutrients and keep soil fertile—is an option gaining traction on farm fields and pastures everywhere.

INFOGRAPHIC 2 **SOIL FORMATION**

Soil is produced by the decay of organic material and the weathering of rock. Distinct layers are seen in healthy soils, with the topsoil (A horizon) being the most fertile for plant growth. Soil erosion will reduce or remove the O and A horizons and produce drier B and C horizons.

JIM RICHARDSON/National Geographic Creative

Native prairie grasses have deep roots (up to 16 feet), which allow them to access deepwater supplies and to weather droughts. The native grass roots also hold the soil in place much better than do shallow-rooted annual crops like wheat.

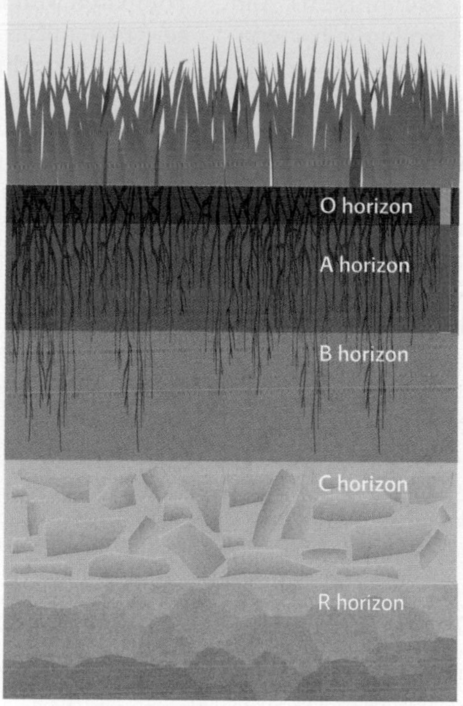

O horizon: Surface litter

A horizon: Topsoil: Contains decaying organic matter (humus) and living soil organisms

B horizon: Subsoil: Denser than A horizon, higher mineral content, lower fertility

C horizon: Contains rock in the process of being broken down (weathering) to produce new soil

R horizon: Solid rock that has not been broken down

? Why are perennial grasses (those that live for several years) less vulnerable to drought than annual grasses (those that must be planted every year)?

3 AGROECOLOGY

Key Concept 3: Modeling a farm after an ecosystem (agroecology) to include a variety of plants and animals can boost productivity and protect or even enhance the local environment.

Long before Andrus bought Boundbrook Farm, another rice farmer was uncovering the ancient secrets of rice and duck farming in Japan. Takao Furuno had been, by most standards, a very successful industrial rice farmer, with annual yields among the highest in southern Japan. But it was a tough grind. Each year he was forced to put much of his earnings back into the next year's crop—insecticides, herbicides, irrigation, and fertilizer—so that despite his success, he and his family were left with very little for themselves at the end of each season. In the late 1970s, he transitioned to organic farming but seemed to be fighting a losing battle against the pests that ate his rice and the weeds that outcompeted it in the paddies.

In searching for a better way, he turned, as he often did, to his forebears to see what he could learn from their knowledge. That was when he first heard about rice and duck farming, and initially, he was surprised: Like most other rice farmers, Furuno considered ducks a pest. Adult ducks eat rice seeds before they have a chance to grow, and as the ducks forage, they trample young seedlings into the mud. This disturbance creates open patches of water, which in turn invites more ducks.

But ducklings, Furuno soon realized, were too small to do such damage; for one thing, their bills were not big or strong enough to extract rice seeds from mud. Instead, they ate bugs and snails (pests that eat rice) and weeds (other plants that compete with rice for space and nutrients).

Furuno's forebears also grew a fish called the pond loach in their paddies. The loaches' waste provides nutrients rice can use, and the fish can be harvested and sold as food.

Together the ducklings and loaches would keep the weeds from strangling the rice crop, but Furuno noticed that if the animals did their job well, they soon ran out of food. To combat this, he began adding azolla to the rice paddy, a small fernlike plant that grows on the surface of the water. In early stages of rice farming, azolla is considered a harmful weed—one that grows quickly and crowds out the rice. But when introduced after the rice plants are well established, azolla provides food for the ducks without competing with the rice for soil nutrients. And because it is a legume that contains symbiotic bacteria that produce a usable form of nitrogen, it helps fertilize the rice. In fact, between the nitrogen from the azolla and the duck and fish droppings, Furuno found that he no longer needed to spend money on fertilizer. And in an unexpected benefit, Furuno found that the action of ducklings swimming around the rice plants strengthened the root crowns (where the root meets the stem), allowing each plant to produce more rice.

agroecology A scientific field that considers the area's ecology and indigenous knowledge and favors agricultural methods that protect the environment and meet the needs of local people.

Furuno's operation is an elegant example of *biomimicry*—a farm operating like a natural ecosystem. This type of farming is known as **agroecology**—a holistic approach that considers the local ecology of the area, the value of traditional farming methods, and the socioeconomic needs of the local community. Agroecology comes with many benefits compared with industrial farming. For one thing, it uses fewer chemicals (fertilizers and pesticides). This saves money and reduces environmental damage. Biodiversity is often enhanced, and the system self-regulates so the farmer doesn't have to micromanage the operations. And because nature performs some of the tasks that would otherwise require machinery powered by fossil fuels, these systems also tend to have lower greenhouse gas emissions.

Furuno found that there were financial gains, too. Duck eggs, duck meat, and fish all fetched a good price in the market. And because he was no longer using pesticides, Furuno could also grow fruit on the edges of his rice field. (He opted for fig trees since he could harvest the figs annually without having to replant.) **INFOGRAPHIC 3**

INFOGRAPHIC 3 AGROECOLOGY: THE INTEGRATED RICE AND DUCK FARM

Takao Furuno's farm is a self-regulating, multiple-species system that naturally meets the needs of the farm ecosystem. All of the species play a role in the system, helping each other and boosting overall production.

THE METHOD

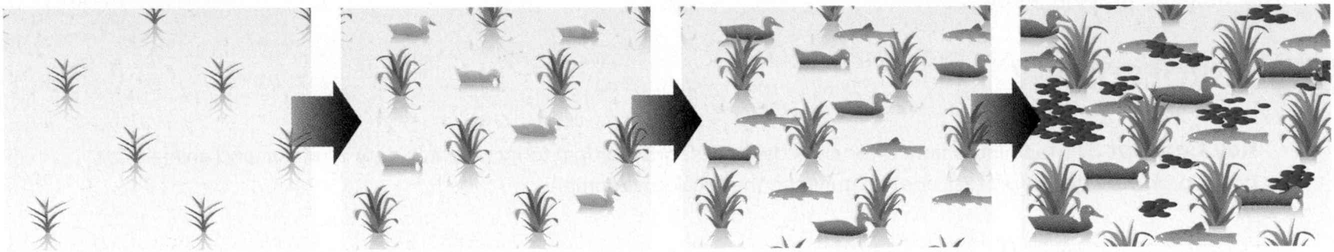

Rice seedlings are planted in flooded rice paddies.

Ducklings are introduced to eat weeds and provide "fertilizer."

Fish are introduced to eat weeds and provide "fertilizer."

Azolla is introduced to add nitrogen; ducklings and fish keep the azolla from growing too much.

THE FINAL PRODUCT: AN INTEGRATED SYSTEM **THE HARVEST**

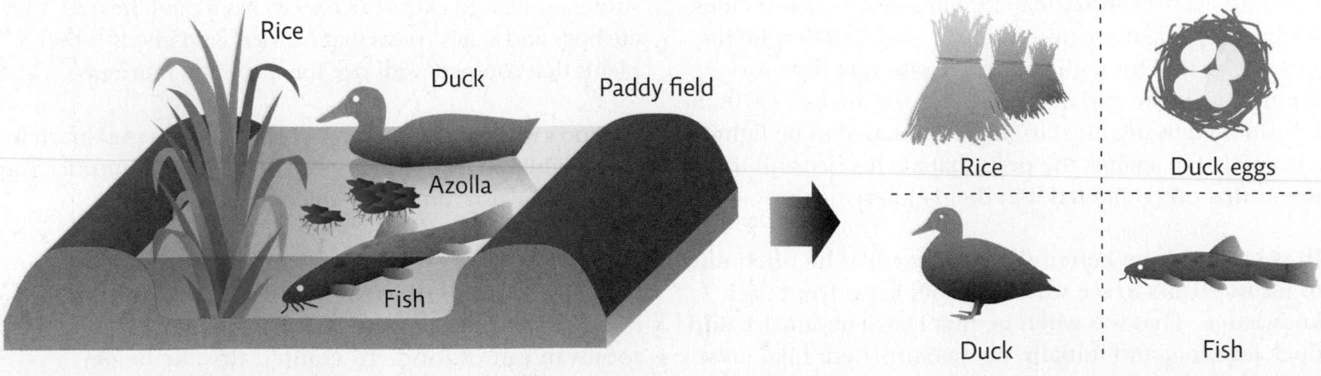

What normal industrial inputs can be averted by growing rice using the duck/rice farm model?

Boundbrook Farm

Ducks help control pests and their waste provides nutrients to the rice at Boundbrook Farm.

Furuno's integrated farm is an example of **polyculture** — intentionally raising more than one species on a given plot of land. In the decades since Furuno began rice and duck farming, his rice yields have increased by as much as 50%, making his among the most productive farms in the world. This kind of success is especially important for farmers in many developing countries, who struggle to produce enough food for current populations. Increasing production with lesser dependence on expensive inputs and using methods that enhance rather than diminish environmental quality can help communities become more self-sufficient and help them achieve *food security* (see Module 8.1).

Another example of an integrated farm gaining in popularity is an approach known as *aquaponics*. Here, crops and fish are raised together in a system that cycles water between fish tanks and crops. Plants are grown in hydroponic systems, meaning that they are grown in water rather than soil; the fish waste provides nutrients for the plants, and the plants filter the water so that it can be returned to the fish tanks. Aquaponics can be done on large scales in warehouses or outdoor areas or on small scales as "bench-top hobbies." Such operations are found in almost every state: in urban centers such as Chicago and Detroit and in small towns across rural America.

polyculture A farming method in which a mix of different species are grown together in one area.

In this aquaponics facility, water from fish tanks provides nutrients to plants; the plants then filter the water, which returns to the fish tank.

Mmdi/Digital Vision/Getty Images

4 MANAGING PESTS

Key Concept 4: Integrated pest management techniques can often effectively control pests while minimizing or eliminating the use of chemical pesticides.

In sustainable agriculture, farmers need to control pests without causing additional harm. They might try to do so by reducing or forgoing the use of pesticides, which minimizes the load of toxic substances on our food and in the environment and reduces the likelihood that pesticide resistance will develop. Pesticides are problematic for another reason, too: They often kill "nontarget" species such as honeybees or beetles that play a role in their ecosystem and may even be helpful to the farmer; honeybees pollinate, while beetles can be natural predators of crop pests. (See Module 8.1 for more on pesticide resistance and other issues related to pesticide use.)

Andrus's weed-control methods of employing ducks as natural predators and removing weeds by hand are examples of **integrated pest management (IPM)**, or the use of a variety of methods to help reduce a pest population. The goal of IPM is to successfully control pests while minimizing or eliminating the use of chemical toxins. First, the farmer must examine the life cycle of the pest and the pest's interactions with the environment to identify the best way to deal with the pest. In general, IPM techniques fall into four categories: biological control, cultural control, mechanical control, and chemical control.

integrated pest management (IPM) The use of a variety of methods to control a pest population, with the goal of minimizing or eliminating the use of chemical toxins.

Biological control involves using other species to keep pest populations in check; this is Andrus's approach, with the ducks as natural predators of insect pests and weeds. When farmers use cultural control, they plant their crop in such a way as to limit pest infestations. This might involve planting a variety of crops together in a polyculture to reduce the pests' food supply. With mechanical control, farmers employ methods that either physically exclude pests (such as nets or row covers), or that remove or trap the pests. In 2015, Andrus, with support from the organization SARE (Sustainable Agriculture Research & Education), compared the efficacy of using either ducks or azolla (biological controls) or hand-weeding (a mechanical control) to minimize the growth of weeds in his rice paddies. He found that the ducks worked best: In some cases, they were 100% more effective than hand-weeding alone.

When devising an IPM plan, farmers use a combination of biological, cultural, and mechanical controls to deal with pest problems, and they resort to using chemicals only if these methods don't adequately deter the pests. If a pesticide is going to be used, the preference is for natural, biodegradable chemicals that are toxic only to a limited group of organisms. For example, pyrethrum, a compound naturally produced by a flower in the chrysanthemum family, is directly toxic to insects but not to mammals; it is certified for organic agriculture because it is naturally produced (not a synthetic human creation) and because it breaks down quickly and does not linger in the environment. However, while it kills insect pests, it is also toxic to *good* insects like honeybees, so its use is avoided during times of pollination. Synthetic pesticides, like those commonly used in industrial agriculture, are not acceptable for use on certified organic crops but may be part of an IPM plan for conventionally raised crops. **INFOGRAPHIC 4**

Sticky traps offer a mechanical control mechanism to keep pests off crops. Many insects are attracted to blue and yellow, making these popular colors for traps.

INFOGRAPHIC 4 INTEGRATED PEST MANAGEMENT (IPM)

Controlling pests is important for our agricultural yields as well as for our health and for the health of our pets. Rather than using harsh methods in an attempt to completely eliminate the pest (which rarely works anyway), a combination of less hazardous methods can often reduce pest numbers to manageable levels.

INTEGRATED PEST MANAGEMENT HAS SEVERAL STEPS

1. IDENTIFY TRUE PESTS
Not all "bugs" and "weeds" are pests — some plants and animals are innocuous or actually beneficial. By working to control actual pests only, we save time and money and help the environment by maintaining diversity and preventing toxic pollution.

2. SET AN ACTION THRESHOLD AND MONITOR PESTS
The pest population size that is unacceptable must be identified. We may have zero tolerance for some pests (e.g., fleas and ticks in our homes) but be able to tolerate small populations of crop pests; we will only act if the action threshold is reached.

3. DEVELOP AN ACTION PLAN
This may include a variety of methods, each of which aids in pest control by excluding, discouraging, or killing pests. The goal is to control the pest while avoiding or minimizing the use of chemical control agents, which may be toxic and have unwanted health and environmental effects.

PREVENTION AND CONTROL METHODS

CULTURAL Usually the first method chosen as part of the action plan; involves cultivation techniques that minimize the habitat or food source for the pest so that other control methods can then be used to adequately control the pests.

BanksPhotos/E+/Getty Images

Strip cropping minimizes potential food for pests that don't disperse over great distances, lessening the chance of an outbreak of pests.

MECHANICAL Relies on methods that physically exclude, trap, repel, or remove pests or weeds; can be labor-intensive but are often inexpensive and may be particularly useful in developing countries with plentiful labor but little cash.

Cosmo Condina/The Image Bank/Getty Images

Netting can keep out birds and rabbits that would eat the crops. Reducing water levels in rice paddies to kill azolla is an example of mechanical control.

BIOLOGICAL Introducing predators, sterile males, or plants that repel the pest; this technique works best if it follows cultural and mechanical steps.

Herrik_L/iStock/Getty Images

Ladybugs, predatory beetles that eat pests such as these aphids, can actually be purchased and released, or steps can be taken to attract them naturally to the area. The ideal control agent is a specialist that prefers the pest in question as food and does not attack nontarget species.

CHEMICAL Applying chemicals that kill or repel pests; a last resort that is used only if the other three methods cannot control the pests.

Abid Katib/Getty Images News/Getty Images

To minimize health and environmental concerns, the preferred chemical is the one that can do the job while being the least toxic and the most degradable.

? In what way is integrated pest management similar to agroecology?

5 TRADITIONAL FARMING METHODS

Key Concept 5: Sustainable agriculture draws on a variety of traditional farming methods that can protect or improve soil and reduce pest problems.

Farmers have a variety of traditional methods—those used long before industrial methods emerged—to choose from to meet the needs of their crops, many of which are responsive to the local environment. These methods help to prevent soil erosion, maintain or even improve soil fertility, and control pests.

As discussed in Section 2, healthy soil is vital to life on Earth; it supports the growth of plants as well as the animals that feed on those plants. However, its slow rate of formation cannot keep up with losses that result from modern farming. David Montgomery, researcher at the University of Washington, estimates that soil erosion from industrial agriculture exceeds soil formation by 100-fold. As much as 5 tons of topsoil are lost for every ton of industrially produced grain in the United States. On the other hand, some traditional methods help restore or protect soil. For example, methods such as *terracing*, *contour farming*, and *reduced-tillage cultivation* can decrease soil erosion. Even though Andrus's fields are exceptionally flat, he still uses terracing to get his paddies just right. "The creation and maintenance of the paddy field is really the most important part of the whole farming enterprise," he says. "It has to be remade each year."

Other traditional methods include the planting of a *cover crop* in the off-season, a practice that prevents soil erosion and, if the right crop is chosen, restores fertility. Soil fertility can also be enhanced and pests controlled using *crop rotation* and *strip cropping*. And systems like Furuno's that combine animal and plant rearing return animal waste—a natural fertilizer—to the soil. **INFOGRAPHIC 5**

In addition to these traditional farming methods, farmers have long added substances to soil to improve its fertility. Fertilizers (natural or synthetic), organic matter (to loosen compacted soil), and substances such as lime (to alter the pH of the soil) are all soil amendments that can be added to address fertility problems.

annual crops Crops that grow, produce seeds, and die in a single year and must be replanted each season.

perennial crops Crops that do not die at the end of the growing season but live for several years, which means they can be harvested annually without replanting.

Synthetic fertilizers can increase levels of a particular nutrient beyond what might be found in a natural fertilizer, though this can upset the mutualistic relationship between the growing plant and the soil microbes that normally provide nutrients to the plant. (See Module 8.1 for more on synthetic fertilizers.) Natural fertilizers like manure or biochar (a form of charcoal produced when organic matter is partially burned) rarely upset the soil community in this way since they do not contain excessive amounts of nutrients. Farmers can minimize the use of fertilizers altogether through a technique known as *microfertilization*, which involves adding just the amount required, and no more, to the specific plants that need it.

Rather than working on ways to improve the soil, Wes Jackson and a team of dedicated scientists at the Land Institute in Kansas are pursuing an ambitious plan that he says will correct a mistake made well before the rise of industrial farming. He wants to replace virtually all of our existing grain crops—which are **annual crops**—with **perennial crop** varieties. Early

JIM RICHARDSON/National Geographic Image Collection

An agroecologist holds a perennial wheat plant to showcase its expansive root system. Compare that to the much shallower root system of the annual wheat plants to the right.

INFOGRAPHIC 5 TRADITIONAL FARMING METHODS

Many traditional farming methods are available that address problems with soil erosion and fertility, water use, and pest control and are useful for sustainable agriculture.

ClassicStock/Alamy

CONTOUR FARMING When farming on hilly land, rows are planted along the slope, following the lay of the land, rather than oriented downhill to reduce the loss of water and soil after a rainfall.

Bill Barksdale/Design Pics Inc/Alamy

REDUCED TILLAGE Planting crops into soil that is minimally tilled reduces soil erosion and water needs (it reduces water evaporation). It also requires less fuel because of less tractor use.

voraorn/iStock/Getty Images

TERRACE FARMING On steep slopes, the land can be leveled into steps. This reduces soil erosion and allows a crop like rice to stay flooded when needed.

Matt Meadows/Photolibrary/Getty Images

CROP ROTATION Planting different crops on a given plot of land every few years helps maintain soil fertility and reduces pest outbreaks since pests (or their offspring) from the year before will not find a suitable food when they emerge in the new season.

Andrew Holt/Photographer's Choice/Getty Images

STRIP CROPPING Alternating different crops in strips that are several rows wide keeps pest populations low; it is less likely the pests will travel beyond the edge of a strip, and they may not find another row of this crop.

Michael Melford/Getty Images

COVER CROPS During the off-season, rather than letting a field stand bare, a crop can be planted that will hold the soil in place. Nitrogen-fixing crops like alfalfa that improve the soil are often chosen.

Which of these traditional farming practices help reduce soil erosion, which help improve soil fertility, and which help reduce pest outbreaks?

farmers domesticated annual plants—which grow, produce seeds, and die in a single year—because they could manipulate them to produce higher yields from one year to the next (by crossbreeding plants with the best traits). But annuals must be replanted every season, which requires labor and fossil fuel energy, and necessitates tearing up the ground and disrupting the delicate balance of soil ecosystems.

Perennials, on the other hand, can be harvested year after year without disturbing the soil to replant—which means heavy equipment is used less often to manage perennial crops. Perennial plants hold a considerable amount of carbon in their roots, providing the valuable ecosystem service of carbon sequestration. And the deep roots that perennials develop not only hold soil in place but also tap much farther down into the soil than their annual counterparts, allowing them to access more of the soil's water, thus dramatically reducing the amount of irrigation needed. Less herbicide is needed for perennial plots as well; weeds do not readily sprout and grow among the established plants. This makes perennials especially attractive for regions with marginal land or an arid climate.

"We hope to advance and enlarge upon the idea that the ecosystem is the necessary conceptual tool for truly sustainable grain agriculture," Jackson told the *Atlantic* in a 2011 interview. "We believe we can have an agriculture where management by human intervention is greatly reduced."

6 THE ROLE OF CONSUMERS

Key Concept 6: Consumer choices can support sustainable agriculture. Buying locally grown, organic food is the best option for reducing the carbon footprint of that food.

The rice varieties that Andrus grows on his farm are Japanese short-grain varieties, delicious when served alongside other foods or when used to make sushi. Andrus wanted to create a farm that contributed to the formation of a local community food system, and he sells his rice direct from his farm to consumers who visit, as well as wholesale to distributors who sell his rice in specialty grocery stores. More and more consumers are buying food from local farmers; local agriculture supports local economies and provides fresher and thus healthier food to consumers. Because transportation depends on fossil fuels, the more **food miles** a product travels before reaching the consumer, the greater the **carbon footprint** of that food. And much of our food has traveled quite far—around 1,500 miles, on average.

However, transportation is not the main use of fossil fuels when crops are raised industrially. Research by Christopher Weber and Scott Matthews of Carnegie Mellon University determined that about 90% of the carbon footprint for food grown using conventional industrial methods is from the production of the crop (fuel for equipment, raw materials for pesticides and fertilizer production), not its transport. Therefore, buying organically grown produce—even from far away—may reduce the carbon footprint more so than buying locally grown industrial crops. Of course, the best option is to choose organic foods that are locally grown. Likewise, because beef raised in high-density industrial operations has one of the highest carbon footprints of all agricultural products (see Online Module 8.3), one of the best things you can do is to replace at least some of the beef you eat with chicken, pork, fish, or meatless dishes.

food miles The distance a food travels from its site of production to the consumer.

carbon footprint The amount of CO_2 (and other greenhouse gases that contribute to climate change) released to the atmosphere by a person, company, nation, or activity.

Consumers are becoming more aware that things like food miles and the way food is raised matter for the environment, their communities, and their own health. Organic foods and ethically raised animal products are claiming a bigger share of consumer dollars annually. But this has opened the door to *greenwashing*—making claims about the environmental benefits of sustainably raised or organic foods that are misleading (see Module 5.1). For example, organic cookies are probably not healthier than those made with conventional ingredients, and "cage-free" eggs may still come from chickens living in overcrowded conditions. Consumers need to be diligent about evaluating claims and make informed decisions about what to purchase. This is one reason why foods need meaningful, trustworthy labels—so consumers can know how the food was raised and whether it conforms to their personal preferences and values. **INFOGRAPHIC 6**

INFOGRAPHIC 6 CONSUMER CHOICES MATTER

Because the growing and transport of our food impacts the environment and our own health so much, choosing foods produced in a way that has a lower impact makes a difference. This also supports sustainable agriculture as an economic endeavor, helping the farmers and communities pursuing these methods.

CONSIDER HOW YOUR FOOD IS RAISED...

Richard Graulich/ZUMA Press/Newscom

Do some research and choose food grown in a sustainable manner or labeled *USDA Organic* when possible.

...AND HOW FAR IT IS SHIPPED

Mindy Schauer/ZUMA Press/Newscom

Buy locally grown food to reduce transportation costs and lower the carbon footprint of a food.

PERSONAL FOOD CHOICES

If you can't afford to buy all organic produce, consider steering your purchases away, when you are able, from the "dirty dozen" — the 12 fruits and vegetables most likely to be contaminated with pesticide residue. The "clean 15" are the products least likely to have pesticide residue, so if you can't afford to buy all organic produce, buy these from the regular produce shelf — but always wash all produce well before eating or cooking!

THE CLEAN 15
The fresh fruits and vegetables with the lowest pesticide residue

1. Avocados
2. Sweet corn
3. Pineapples
4. Sweet peas, frozen
5. Onions
6. Papayas
7. Eggplant
8. Asparagus
9. Kiwi
10. Cabbage
11. Cauliflower
12. Cantaloupe
13. Broccoli
14. Mushrooms
15. Honeydew melon

THE DIRTY DOZEN
(in order of pesticide exposure when eaten) Buy organic if possible.

1. Strawberries
2. Spinach
3. Kale
4. Nectarines
5. Apples
6. Grapes
7. Peaches
8. Cherries
9. Pears
10. Tomatoes
11. Celery
12. Potatoes

What information would you like to see on food labels that would allow you to make wise consumer choices when buying food? Does food labeling need to be regulated so that certain terms (e.g., organic, free-range) are specifically defined?

7 CAN SUSTAINABLE AGRICULTURE FEED THE WORLD?

Key Concept 7: Sustainable agriculture comes with trade-offs, but many feel that the disadvantages are less problematic than those of modern industrial agriculture, especially in the long term.

Andrus and Furuno are not the only ones experimenting with nature-based polyculture. Indeed, many other farmers and scientists in the United States and beyond are turning to agroecology—working to develop polyculture systems that use a mix of different species that better replicates the normal ecological community makeup of a given region. Evidence is mounting that such systems can increase a farm's productivity.

A 2010 report by the International Livestock Research Institute concluded that mixed polyculture farms—ones that, like Furuno's and Andrus's, grow both plants and livestock—hold the most promise for intensifying food production worldwide. "It is not big efficient farms on high potential lands but rather one billion small, mixed family farmers tending rice paddies or cultivating maize and beans while raising a few chickens and pigs, a herd of goats, or a cow or two . . . [who are] likely to play the biggest role in global food security over the next several decades," the institute's executive director, Knut Hove, wrote in the report. "These 'mixed extensive' farms make up the biggest . . . and most environmentally sustainable agricultural system in the world."

Feeding a world of 10 billion people will not be easy. It will require a combination of modern industrial and traditional techniques, native crop varieties, and perhaps genetically modified ones—the methods and crops chosen should fit the place and need of the farmers and their communities. But consideration needs to be given to the long-term viability of the farm, with preference given to techniques that maintain or improve the health of the ecosystem. As these many examples (agroecology, the resurrection of traditional farming techniques, IPM, and the development of perennial crops) illustrate, a multitude of methods are available to help facilitate a transition from the current industrial agriculture model to methods that are more sustainable. Of course, as with all environmental choices, it shouldn't be surprising that sustainable and organic agriculture have their own set of trade-offs. However, addressing these trade-offs may be more feasible than dealing with the disadvantages of large-scale industrial monocultures. **INFOGRAPHIC 7**

For Andrus, the decision to incorporate ducklings into his Vermont farm was part of an ongoing search for solutions to the challenges of modern rice farming. And he continues to experiment in order to find new and better approaches. This year, for instance, he plans to build a remote-controlled weeding drone to try out in shallow areas of his paddies, which his ducklings tend to avoid (and which then get overrun with weeds). "We're

Bill Barksdale/AGE Fotostock

Mature rice plants, ready to harvest.

INFOGRAPHIC 7 THE TRADE-OFFS OF SUSTAINABLE AGRICULTURE

FOR	ADVANTAGES	DISADVANTAGES
The Consumer	• Food may be fresher, tastier, and healthier. • Consumer gains satisfaction in making a more ethical and environmentally sound choice.	• Sustainably grown crops may be more expensive. • Greenwashing can mislead consumers. • Organic produce may have more blemishes. • Shelf life of organic produce is shorter (not waxed, not picked before ripe).
Farmer and Environment	• Using fewer inputs of water and fossil fuels saves money and causes less environmental damage. • Soil is not degraded and may be enhanced. • Less use of toxic chemicals benefits the environment and local communities. • More genetic diversity and species diversity makes it less likely that a pest outbreak or other problem will decimate the entire crop.	• May be more labor-intensive. • Crop productivity per acre may be lower than industrially farmed methods (in the short term). • Fewer government subsidies are available for sustainable agriculture compared to those for industrially grown crops. • The certification process for getting crops to be labeled as organic takes time and is costly to farmers.
Society	• Many sustainable methods are less expensive, so they are suitable for developing nations. • Methods are available that minimize water need — useful in arid areas. • Local production of food can increase food security (see Module 8.1).	• Research is needed to identify best methods and crops for a given area. • Farmers need training to implement these systems (though if indigenous methods are used, it may be the locals who educate the researchers).

? In your opinion, which advantages listed above are the most important? Which disadvantages are the most troublesome? Explain.

still doing R and D of various kinds," he says, with the goal of making his farm as efficient and sustainable as possible.

Furuno admits that managing this type of farm can sometimes be more time-consuming and requires more expertise, but when asked about the extra work this type of farming requires, he replied, "It gives you a sense of tranquility and peace. My role is to feed the ducks and later on, it's the ducks who feed me. All in all, it can be considered a rather fair exchange."

Select References:

Andrus, E. (2018). *Wet rice organic weed control trials: Final report for FNE15-816.* Retrieved from https://projects.sare.org/project-reports/fne15-816/

Baranski, M., et al. (2014). Higher antioxidant and lower cadmium concentrations and lower incidence of pesticide residues in organically grown crops: A systematic literature review and meta-analyses. *The British Journal of Nutrition, 112*(5), 794–811.

Bradman, A., et al. (2015). Effect of organic diet intervention on pesticide exposures in young children living in low-income urban and agricultural communities. *Environmental Health Perspectives, 123*(10), 1086–1093.

Crews, T. E., et al. (2018). Is the future of agriculture perennial? Imperatives and opportunities to reinvent agriculture by shifting from annual monocultures to perennial polycultures. *Global Sustainability, 1*, e11, 1–18. https://doi.org/10.1017/sus.2018.11

Furuno, T. (2001). *The power of duck: Integrated rice and duck farming.* Tasmania, Australia: Tagari Publications.

International Livestock Research Institute. (2010). *ILRI corporate report 2009–2010. Back to the Future: Revisiting mixed crop-livestock systems.* Nairobi, Kenya: ILRI.

Jackson, W. (2002). Natural systems agriculture: A truly radical alternative. *Agriculture, Ecosystems & Environment, 88*(2), 111–117.

Montgomery, D. R. (2007). Soil erosion and agricultural sustainability. *Proceedings of the National Academy of Sciences, 104*(33), 13268–13272.

Weber, C. L., & Matthews, H. S. (2008). Food-miles and the relative climate impacts of food choices in the United States. *Environmental Science and Technology, 42*(10), 3508–3513.

GLOBAL CASE STUDIES SUSTAINABLE AGRICULTURE

Achieve

Depending on the crop, high- and low-tech solutions can promote higher yields without jeopardizing the environment. Read these four success stories.

BRING IT HOME

PERSONAL CHOICES THAT HELP

While a typical supermarket may seem to present a dizzying array of food choices to the consumer, a look at the ingredient labels betrays our increasing reliance on growing monocultures of common strains of corn, soy, and wheat. These choices will change only if you, the consumer, demand it.

Individual Steps

- Carefully examine the labels on the food you buy. As your food budget allows, opt for food products that are organically grown and, if available, locally produced.
- Grow your own food if possible—anything from a small window herb garden to a large outdoor plot will reduce the need to buy food produced elsewhere.
- Support local farmers at farmers' markets or farm stands. Visit local farmers to get to know them and to see how their food is grown.

Group Action

- Organize a community garden that specializes in heirloom varieties of vegetables that might not be found in the local grocery stores. Start by requesting a seed catalog from www.seedsavers.org or www.rareseeds.com.
- Research specific farming practices that more closely mimic those found in natural ecosystems, such as Joel Salatin's Polyface Farm (www.polyfacefarms.com).
- Subscribe to a community-supported agriculture (CSA) farm and receive a weekly supply of sustainably grown produce. For a list of CSAs in your area, see www.localharvest.org.

Jean-Michel Groult/Superstock

REVISIT THE CONCEPTS

- The goal of sustainable agriculture is to raise enough food for the human population in a way that can be done indefinitely (i.e., without damaging the environment or future productivity). It also seeks to operate ethically with regard to animals, workers, and local communities. Organic agriculture is a subset of sustainable agriculture that avoids the use of synthetic fertilizers or pesticides or genetically modified organisms.

- The productivity of farmland is governed by the health of its soil. Fertile soil is a complex ecosystem made up of a diverse assortment of organisms, including bacteria and fungi that form mutualistic relationships with plant roots and support plant growth. Soil formation is a slow process, but in many places, soil is being lost much more quickly than it forms.

- One way to determine how to farm an area sustainably is to model the farm after an ecosystem (agroecology), including a variety of plants and animals rather than the monocultures typical in industrial agriculture. This can boost productivity and protect or even enhance the local environment.

- Integrated pest management techniques can often effectively control pests using a combination of biological, physical, and cultural pest control techniques. This often allows the farmer to minimize or even eliminate the use of chemical pesticides.

- Traditional farming methods from the past are useful options for sustainable agriculture that can be used with or without industrial techniques. These methods can protect or improve soil and reduce pest problems, and they can be used on large and small scales.

- Consumer choices can support sustainable agriculture. Buying locally grown, organic food is the best option for reducing the carbon footprint of that food, but to do this, consumers need access to meaningful labels that accurately depict the method used to grow the food and its place of origin.

- To meet the goal of feeding an ever-growing population sustainably, agriculture will need to diversify and use all the tools at its disposal — all while striving to farm in a way that is productive in the long term. Sustainable agriculture comes with trade-offs but many feel that the disadvantages are less problematic than are those of modern industrial agriculture, especially in the long term.

ENVIRONMENTAL LITERACY Understanding the Issue

1 What is sustainable agriculture?

1. Distinguish between sustainable and organic agriculture.
2. What are the criteria for certified organic food?
3. What are some of the benefits of sustainable agriculture to the human community? To the ecological community?

2 What is the importance of fertile soil, and how does land use affect it?

4. Why is soil formation such a slow process, and what does this mean for us in terms of the problem of soil erosion?
5. Distinguish among the three soil types in terms of relative particle size, structural support of plants, and water penetration and retention.
6. Explain the mutualistic relationship that exists between plant roots and soil bacteria and fungi.

3 What is agroecology, and what are its benefits?

7. Explain the integrated rice and duck farming method used by Takeo Furuno and point out the functional role of each component.
8. Look at the four characteristics of a sustainable ecosystem shown in Infographic 6 in Module 1.1. How does the ecosystem created by the ducks, rice, fish, and azolla fulfill these roles to make it sustainable?

4 How can pests be managed in a way that minimizes environmental damage?

9. Explain the goal and approach of integrated pest management and how it views the use of chemical pesticides.
10. Briefly define and give examples of biological, cultural, and mechanical pest control.
11. Why are multiple methods needed in integrated pest management?

5 How can traditional farming methods contribute to sustainable agriculture?

12. How might the combination of strip farming and crop rotation help combat pests better than either method alone?
13. Which of the traditional farming methods protect the soil aggregates found in healthy soil? Explain.
14. What are perennial crops, and what advantages do they offer over annual crops?

6 What role does the consumer play in helping build a sustainable food system?

15. To achieve the lowest carbon footprint as a consumer, should you opt for organic food grown far away or industrially grown food produced close to home? Explain.
16. What are the "dirty dozen" and the "clean 15"? Why might it be useful to know which fruits and vegetables belong to each group?

7 What are the trade-offs of sustainable agriculture, and how might they be addressed?

17. Why might mixed polyculture farms be very important for the future of sustainable agriculture?
18. Look over the disadvantages of sustainable agriculture in Infographic 7. Propose actions that could address each.

SCIENCE LITERACY Working with Data

A recent study compared pesticide uptake in children when they ate produce grown conventionally (grown with pesticides that are not allowed in organic farming) or organically. Uptake in the children was determined by the amount of pesticide breakdown products (metabolites) in their urine. During the study, the children ate conventionally grown produce for 4 days, switched to a diet with organically grown produce for 7 days, and then switched back to conventional produce. Data for three metabolites are shown below. (Bars marked with different letters denote statistical differences between groups for that metabolite.)

PESTICIDE METABOLITES IN URINE OF CHILDREN ON CONVENTIONAL OR ORGANIC DIETS

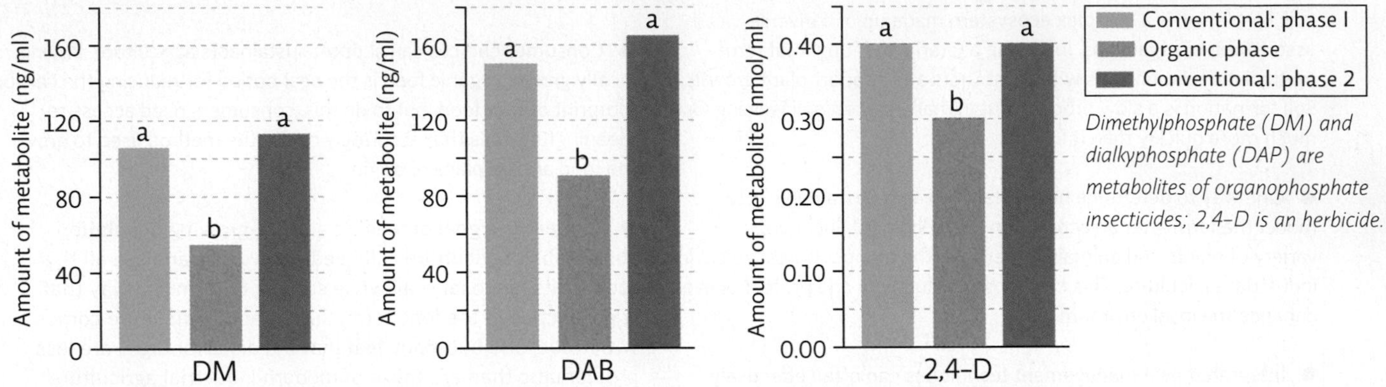

Dimethylphosphate (DM) and dialkyphosphate (DAP) are metabolites of organophosphate insecticides; 2,4–D is an herbicide.

Interpretation

1. How did the amount of pesticide metabolites in the urine of children change between the conventional and the organic diets for each of the substances shown here?
2. Look at the *y* axes. Which pesticide metabolite was present at the highest concentration in urine? The lowest?
3. Why was a different scale used on the *y* axis for the 2,4-D graph? Why is it important to always check the axes scales when comparing graphs?

Advance Your Thinking

4. Is there evidence that eating an organic diet reduces children's ingestion of pesticides compared to the consumption of conventionally grown foods? Explain.
5. Is there evidence that the amount of pesticide in the urine while on either diet causes a health problem? Propose an additional study that might clarify whether these levels of pesticide could be a health concern in children.

INFORMATION LITERACY Evaluating Information

Which seed varieties are best? There are groups, such as Renewing America's Food Traditions (RAFT) (www.albc-usa.org/RAFT/), that say we should preserve the original biodiversity and the heirloom varieties that our ancestors grew. Other groups believe that selective breeding and hybridization create superior crops, the downside being that farmers must purchase new seeds every year. Bayer is a huge conglomerate that sells a variety of hybrid seeds and chemicals to enhance growth (www.bayercropscience.us). Another group, AgBioWorld (www.agbioworld.org), says that it is neutral and reports the news on agricultural practices. Look at each website to help analyze the organizations.

Evaluate the websites and work with the information to answer the following questions:

1. Evaluate the agendas of the three organizations as well as the accuracy of the science behind their positions on heirloom varieties versus hybrid crops by answering questions a–d for each organization.
   a. Who runs each website? Do the person's/organization's credentials make the information presented on food and agriculture issues reliable or unreliable? Explain.
   b. What is the mission of each website? What are the underlying values? How do you know this?
   c. What claims does each website make about the current problems in food production and what the future of agriculture should be? Are their claims reasonable? Explain.
   d. How do the websites compare in providing scientific evidence in support of their assessment of agriculture and their position on the role of heirloom varieties versus hybrid crops? Is the evidence accurate and reliable? Explain.
2. How do the three organizations compare in engaging you, as a citizen, in agricultural policy? Do you think that citizen involvement in policy issues is necessary and effective? Explain your responses.

Additional study questions are available at Achieve.macmillanlearning.com. Achieve

CHAPTER 9

Conventional Energy: Fossil Fuels

Fossil fuels power modern society and are important raw materials for a wide variety of products. An understanding of the value and environmental/health costs of acquiring and using fossil fuels can guide our choices as we pursue more sustainable energy options.

Module 9.1

Coal

An evaluation of the pros and cons of coal as an energy source and ways to address coal's disadvantages to lessen its impact

Module 9.2

Oil and Natural Gas

An examination of the pros and cons of acquiring and using oil and natural gas, including a look at unconventional sources of these fossil fuels such as shale oil/gas and tar sands

ping han/Alamy

Module 9.1 Coal

BRINGING DOWN THE MOUNTAIN

In the rubble, the true costs of coal

Leveling a mountain for coal in Appalachia using a method known as mountaintop removal coal mining.

Alan Gignoux/Alamy

After reading this module, you should be able to answer these GUIDING QUESTIONS

1. What energy sources are used in modern society and for what purposes?
2. How important is coal as an energy source, and how is it used to generate electricity?
3. What is coal, how is it formed, and what regions of the world have coal deposits that are accessible?
4. What surface methods are used to mine coal, and what are their advantages and disadvantages?
5. How are deep deposits of coal mined, and what are the advantages and disadvantages of this method?
6. What are the advantages and disadvantages of using coal?
7. What new technologies allow us to burn coal with fewer environmental and health problems?
8. How can mining damage be repaired, and how effective is this restoration?

A thousand feet above the mountains of central Appalachia, near the Kentucky–West Virginia border, a four-seater plane ducks and sways like a tiny boat on an anxious sea. It's windier than expected, and Chuck Nelson, a retired coal miner seated next to the pilot, grips the door to steady his nerves. The passengers have come to survey the devastation wrought by mountaintop removal—a type of mining that involves blasting off several hundred feet of mountaintop, dumping the rubble into adjacent valleys, and harvesting the thin ribbons of coal beneath.

At first, the landscape looks mostly unbroken; mountains made soft and round by eons of erosion roll and dip and rise in every direction, carrying a dense hardwood forest with them to the horizon. But before long, a series of mountaintop mining sites come into view. Trucks and heavy equipment crawl like insects across what looks like an apocalyptic moonscape: decapitated peaks and acres of barren sandstone and shale. Smoke curls up from a brush fire as the side of an existing mountain is cleared for demolition. Orange and turquoise sediment ponds—designed to filter out heavy metal contaminants before they permeate the water downstream—dot the perimeter.

WHERE IS APPALACHIA?

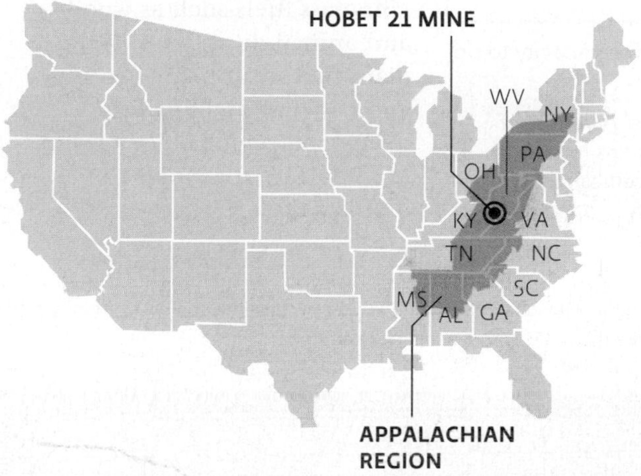

Here and there, tiny patches of forest cling to some improbably preserved ridge line. "That's where I live," Nelson says, forgetting his air sickness long enough to point out one such patch. "My God, you would never know it was this bad from the ground." The aerial tour has reached Hobet 21, which, at more than 20 square miles, is the region's largest mining operation. So far, sites like this one have claimed roughly 1 million acres of forested mountain, across just four states: Kentucky, West Virginia, Virginia, and Tennessee. But there is still more **coal** to mine. It could mean the destruction of hundreds of thousands of more acres in the coming years.

David Hosking/FLPA/Science Source

Seams of coal exposed at a mountaintop removal mining site in Welch, West Virginia.

To stem this tide of destruction, environmental activists have sued the coal industry, the state of West Virginia, and the federal government. They argue that mining coal in this manner destroys biodiversity, pollutes the water beyond recompense, and threatens the health and safety of area residents. And by obliterating the mountains, they say, it also obliterates the culture of Appalachia.

Coal industry representatives have countered by decrying the loss of jobs, tax revenue, and business the already impoverished region would suffer if the mines were to close under the weight of too much regulation. They also point out that the culture of Appalachia is as bound to coal mining as it is to the mountains. Both sides count area residents, including miners, among their ranks.

coal A fossil fuel that is formed when plant material is buried in oxygen-poor conditions and subjected to high heat and pressure over a long time.

1 ENERGY SOURCES

Key Concept 1: Modern society depends on stationary sources of energy to produce heat and electricity (e.g., coal, natural gas, nuclear energy) and mobile sources for transportation purposes (e.g., petroleum products).

Simply put, coal equals energy. **Energy** is defined as the capacity to do work; like all other living things, we humans need it, in a biological sense, to survive. But we also need it to run our societies: to heat and cool our homes; operate our cell phones, lamps, and laptops; fuel our cars; and power our industries. Most of modern society's energy comes from **fossil fuels**—nonrenewable carbon-based resources (e.g., coal, oil, and natural gas) that were formed over millions of years from the remains of dead organisms. (Biomass fuels such as wood and animal dung are still the main fuel sources for heating and cooking in many less developed regions. See Online Module 11.3 for more on biomass fuels.)

energy The capacity to do work.

fossil fuel A variety of hydrocarbons formed from the remains of dead organisms.

In general, modern society depends on two categories of energy: *stationary sources* that provide electricity for buildings and *mobile sources* that fuel our transportation fleets (cars, boats, planes, etc.). There are a variety of ways to produce electricity. The most common method comes in the form of thermoelectric power plants that burn a fuel like coal, natural gas, biomass, or even garbage (see Module 5.3, Infographic 5). Electricity can also be produced by harnessing nuclear energy or the geothermal heat of Earth or by using renewable energy sources such as the energy of moving water (hydroelectric), solar energy, or wind—these alternatives will be examined in Chapter 11.

Mobile energy sources primarily come from oil, a fossil fuel, which is refined to produce liquid fuels such as gasoline, diesel, and jet fuel; vehicles can also run on

An aerial view of a mountaintop removal coal mine in West Virginia, with several large dump trucks hauling rock that are barely visible in this photo, shows the scale of these mining operations.

MANDEL NGAN/AFP/Getty Images

INFOGRAPHIC 1 ENERGY SOURCES USED BY MODERN SOCIETY

A variety of energy sources are used today to supply stationary and mobile sources of energy. Nonrenewable sources still account for more than 80% of U.S. energy use.

TYPE	PREDOMINANT USE
NONRENEWABLE SOURCES	
Coal	Stationary source: electricity/heat
Oil/Petroleum	Mobile source: vehicles; also used for electricity or heat
Natural Gas	Stationary source: electricity/heat; also used in vehicles (compressed natural gas)
Nuclear	Stationary source: electricity; also used in nuclear submarines and ships
RENEWABLE SOURCES	
Solar	Stationary source: electricity/heat/light
Wind	Stationary source: electricity
Geothermal	Stationary source: electricity/heat
Hydroelectric	Stationary source: electricity
Biomass	Stationary or mobile source: electricity, heat, or vehicle fuel

Which of the nonrenewable and renewable energy sources listed above appear to be the most versatile (have the most uses)? Explain.

compressed natural gas. Alternatively, liquid fuels like biodiesel or ethanol can be made from biomass—an energy source that is renewable as long as we don't harvest it faster than it can regrow (see Online Module 11.3).

The line between stationary and mobile fuel sources is beginning to blur as electric vehicles, powered by rechargeable lithium-ion batteries (see Module 7.1), are more and more regarded as the future of light vehicle traffic—running on batteries that get their power from stationary sources. This will give electricity a larger share of the global energy market, increasing demand and scrutiny on how we generate that electricity.

All of these fuels have advantages and disadvantages. Fossil fuels are abundant and energy-rich. They powered the Industrial Revolution, and we built our societies, our machines, and our infrastructure (ways to distribute the fuel) around them. They still account for close to 80% of U.S. energy production. But like all energy sources, they come with negative trade-offs. Understanding where our energy sources come from (and the problems they cause) can allow us to make more informed decisions about how much and what types of energy sources to use in our everyday lives and which energy sources to support politically or economically. **INFOGRAPHIC 1**

2 GENERATING ELECTRICITY FROM COAL

Key Concept 2: Coal is mainly used for electricity production. It is burned to heat water to produce steam; the steam turns a turbine connected to a generator, producing electricity.

Worldwide, we used more than 8.5 billion tons of coal in 2018, the vast majority of it to generate electricity. **Electricity** is a natural form of energy (lightning and nerve impulses are electrical) that we have learned to create on demand; we produce it in a central location and send it out via transmission lines where we want it to go. In 2018, burning coal generated approximately 40% of electricity worldwide and just over 27% in the United States. Worldwide, the use of coal to produce electricity has almost tripled since 1985, rising steadily over most of that time period. Since around 2007, there have been slight declines in some years but the use of coal is still trending upward. Usage once again declined in 2019 compared to the previous year thanks in part to increases in alternative production methods such as nuclear, solar, and wind power.

In the United States, coal use has steadily decreased since

electricity The flow of electrons (negatively charged subatomic particles) through a conductive material (such as wire).

2007 due largely to an abundance of cheaper natural gas (acquired by fracking—see Module 9.2). More stringent air pollution standards for coal-fired power plants have also had a role in the downturn—natural gas releases less air pollution and greenhouse gas emissions than coal when burned. Some coal advocates hope that rolling back air and water quality standards will once again boost coal's use; however, as long as natural gas remains abundant and cheaper, most analysts say that is unlikely.

Coal-fired power plants work by feeding pulverized coal into a furnace where it is burned to generate heat. This heat is used to boil water, producing steam. The steam flows over a turbine, causing it to spin inside a generator (copper wiring positioned near magnets), a process that produces electricity. It takes roughly 1 pound of coal to generate 1 kilowatt-hour (kWh) of electricity; that's enough to run ten 100-watt incandescent lightbulbs for an hour, an energy-efficient refrigerator for 20 hours, or an older and less efficient refrigerator for 7 hours. The average U.S. family of four uses close to 11,000 kWh of electricity per year. If that electricity is produced by coal, that comes out to around 3,000 pounds of coal per person per year.

energy return on energy investment (EROEI) A measure of the net energy from an energy source (the energy in the source minus the energy required to get it, process it, ship it, and then use it).

So how does coal stack up against other fossil fuel energy sources? On one hand, it produces more air pollution than any other fossil fuel. On the other, it is safer to ship, cheaper to extract, and (in the United States at least) abundant.

The usefulness of an energy source is related to the *net energy* realized from its use—a metric known as the **energy return on energy investment (EROEI)**. An evaluation of the EROEI allows us to compare the amount of energy we get from any individual source to the amount we must expend to obtain, process, and ship it. In terms of electricity production, coal is neither the best nor the worst. It has an average EROEI of about 17:1 (17 units of energy produced for every 1 unit consumed). Steps that can be taken to reduce the impact of coal, such as taking steps to capture and sequester CO_2 emissions from coal burning (discussed later in this module), require extra energy and so reduce the EROEI of coal—by one estimate as low as 6.6:1.

There can be considerable variability in the EROEI estimate for a given energy source because many factors influence its net yield. This is especially true for wind or solar electricity production—"windy" or "sunny" areas will have a higher EROEI than areas with less dependable wind or less abundant solar radiation. As fossil fuel supplies have declined, the EROEI has also declined—more energy is required to extract the remaining supplies. On the other hand, improvements in technology, especially for renewable energy sources, are raising the EROEI of these methods. **INFOGRAPHIC 2**

There can be no denying the blessings of coal: This sticky black rock has powered several waves of carbon-based industrialization—first in Great Britain and the United States, now in China—and in so doing has shaped and reshaped the world as we know it. But as time marches on, the costs of those blessings have become all too apparent. They include an ever-growing list of health impacts—from birth defects to black lung disease—and an equally lengthy roster of environmental costs—not only habitat destruction but also the pollution of Earth's atmosphere with CO_2, a potent greenhouse gas. (See Module 10.2 for more on greenhouse gases and climate change.) Water quality, too, is impacted by both the mining and burning of coal.

"We're caught in a catch-22," says Scott Eggerud, a forest manager with West Virginia's Department of Environmental Protection. "On one hand, it's like we need the stuff to live; on the other hand, we see that it's kind of killing us." Nowhere is this paradox more pronounced than in the mountains of central and southern Appalachia.

Les Stone/The Image Works/TopFoto

A train carrying coal leaves a mountaintop removal mining site and travels through the backyards of homes in Welch, West Virginia.

INFOGRAPHIC 2 HOW IT WORKS: ELECTRICITY PRODUCTION FROM COAL

The most common way to generate electricity is to heat water to produce steam; the flow of steam turns a turbine inside a generator to produce electricity.

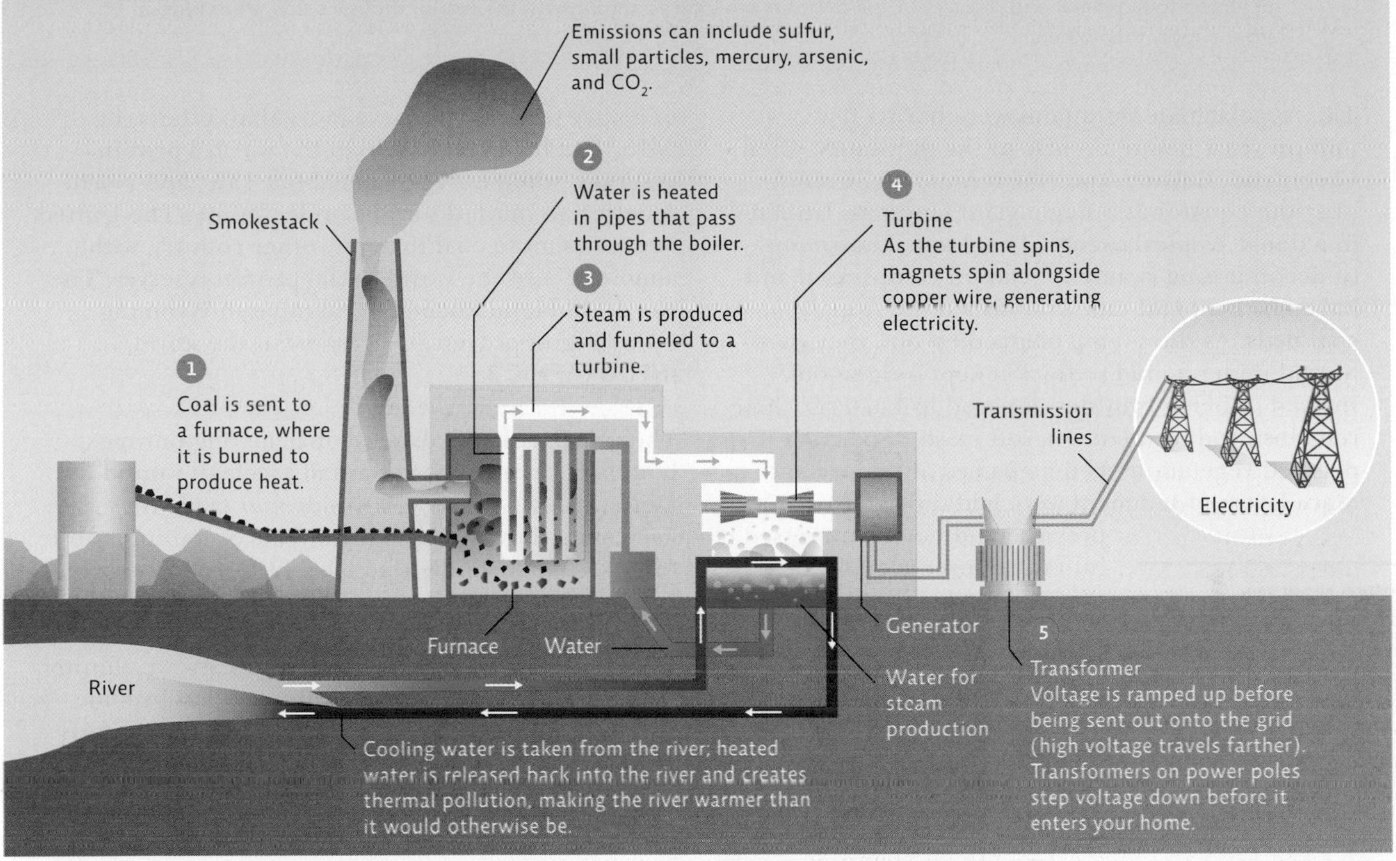

Emissions can include sulfur, small particles, mercury, arsenic, and CO_2.
Smokestack
2
Water is heated in pipes that pass through the boiler.
3
Steam is produced and funneled to a turbine.
4
Turbine
As the turbine spins, magnets spin alongside copper wire, generating electricity.
1
Coal is sent to a furnace, where it is burned to produce heat.
Transmission lines
Electricity
Furnace
Water
Generator
5
Transformer
Voltage is ramped up before being sent out onto the grid (high voltage travels farther). Transformers on power poles step voltage down before it enters your home.
River
Water for steam production
Cooling water is taken from the river; heated water is released back into the river and creates thermal pollution, making the river warmer than it would otherwise be.

In 2015, coal ceased being the leading energy source for U.S. electricity production. The falling price of natural gas has displaced coal in many areas.

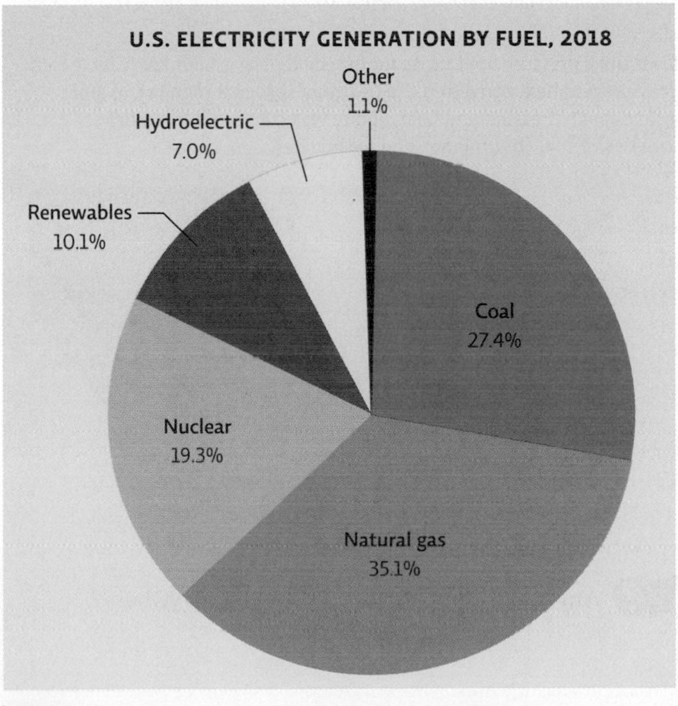

U.S. ELECTRICITY GENERATION BY FUEL, 2018
Other 1.1%
Hydroelectric 7.0%
Renewables 10.1%
Coal 27.4%
Nuclear 19.3%
Natural gas 35.1%

EROEI, an estimate of the net energy produced from an energy source, provides a useful means of comparing different energy sources.

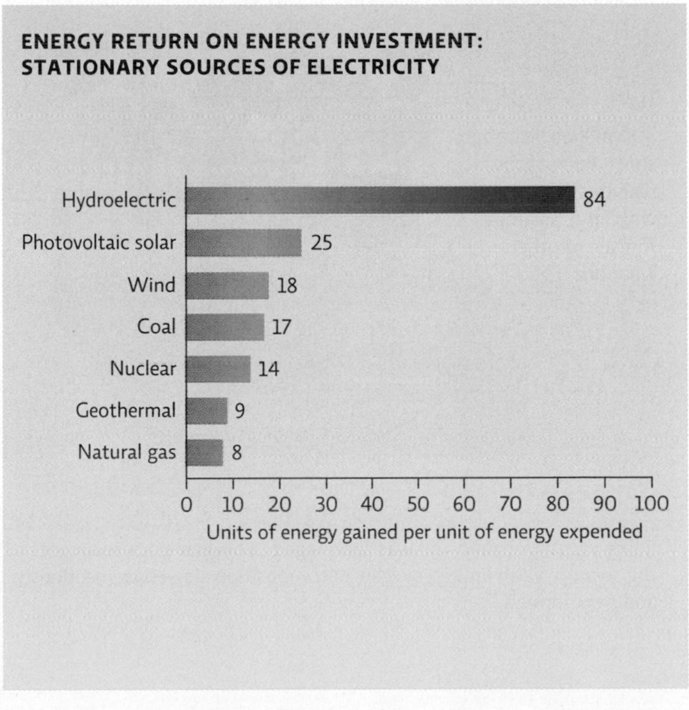

ENERGY RETURN ON ENERGY INVESTMENT: STATIONARY SOURCES OF ELECTRICITY
Hydroelectric 84
Photovoltaic solar 25
Wind 18
Coal 17
Nuclear 14
Geothermal 9
Natural gas 8
0 10 20 30 40 50 60 70 80 90 100
Units of energy gained per unit of energy expended

What other methods could be used to spin a turbine and generate electricity?

3 COAL DEPOSITS: FORMATION AND DISTRIBUTION

Key Concept 3: Coal formation occurred over long periods of time when dead plant material was buried and subjected to high heat and pressure. Coal deposits are located throughout the world, though some areas have a larger supply than others.

The Appalachian Mountains were born a few million years before the rise of the dinosaurs, when Greenland, Europe, and North America hovered near the equator as a single giant landmass bathed in a dense tropical swamp. The gradual accretion of decomposing swamp vegetation compressed and baked by heat and time established the Appalachian coal beds. As the swamp plants died out, they were buried under a mud so thick it kept oxygen out. Instead of being fully decomposed by bacteria, their remains produced *peat*—a soft mash of partially decayed vegetation. As time passed and more and more layers of sediment were laid down over the peat, pressure and heat compressed it into the denser rocklike material that we know as coal.

proven reserves A measure of the amount of a fossil fuel that is economically feasible to extract from a known deposit using current technology.

This same story has played out in numerous places around the globe. As a result, coal is found everywhere, though of course some places have more than others. In 2018, Asia held about 42% of the world's **proven reserves**, while Europe held about 13%, and North America accounted for just under 24.5%. The United States has more coal than any other country, with almost 24% of the world's total proven reserves. The North Antelope Rochelle coal mine in Wyoming is the largest proven coal reserve in the world. **INFOGRAPHIC 3**

The Appalachian beds were once among our most bountiful reserves; seams as tall as a man wound for miles through the mountainside and made for easy harvesting. But after 150 or so years of mining, those reserves have dwindled noticeably. At current rates of usage, proven coal reserves should last about 130 years—longer if deeper reserves can be accessed. And as the layers of minable coal have grown thinner and harder to reach, the coal industry has become both more sophisticated and more destructive in its approach to extracting the coal.

INFOGRAPHIC 3 COAL FORMATION

Coal is formed over long periods of time as plant matter is buried in an oxygen-poor environment and subjected to high heat and pressure. Places with substantial coal deposits that are retrievable with current technology are called coal reserves.

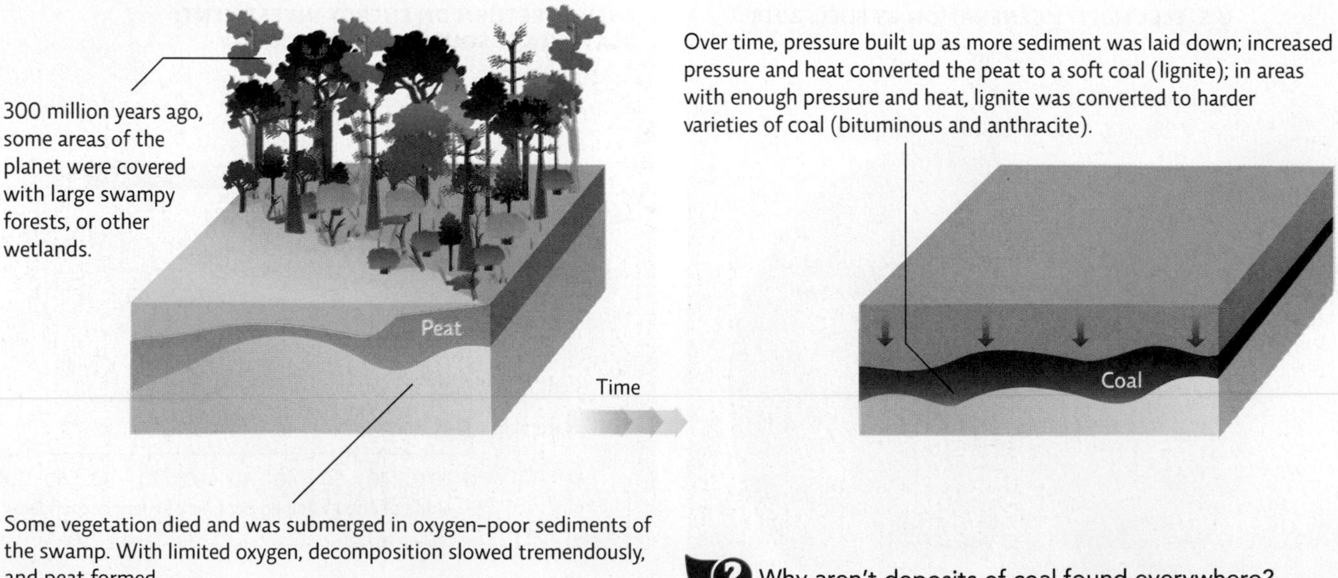
300 million years ago, some areas of the planet were covered with large swampy forests, or other wetlands.
Peat
Time
Over time, pressure built up as more sediment was laid down; increased pressure and heat converted the peat to a soft coal (lignite); in areas with enough pressure and heat, lignite was converted to harder varieties of coal (bituminous and anthracite).
Coal
Some vegetation died and was submerged in oxygen-poor sediments of the swamp. With limited oxygen, decomposition slowed tremendously, and peat formed.

Why aren't deposits of coal found everywhere?

4 COAL MINING: SURFACE TECHNIQUES

Key Concept 4: Coal close to the surface is accessed in flat areas via strip mining and in mountainous areas with mountaintop removal (MTR). Both are cost-effective mining techniques, but MTR creates significantly more environmental damage than traditional strip mining.

The method used to retrieve coal depends on the depth of the deposit and the thickness of the coal seam. **Surface mining** techniques tackle coal seams that are close to the surface of the ground by simply digging down to expose it. If the land is fairly level, such as in the rich deposits of Wyoming, **strip mining** is the method of choice: Vegetation, soil, and rock that lie above the coal seam are stripped away, and heavy equipment digs out the coal.

In a strip mine, the **overburden**, material that lies above the coal seam, can be systematically removed, typically in a narrow strip one layer at a time, and stored nearby. Coal is removed from that exposed strip, and once exhausted, the rock is returned to the pit and topped with the reserved soil. The area can be replanted with native grasses (or allowed to reseed) while the next strip is excavated and mining continues.

Hobet 21 is another matter. Located in the mountains of Appalachia, this method of surface mining is not an option. Though thick coal seams deep under a mountain are accessed with subsurface mining methods, in areas like Hobet 21 where coal seams are close to the surface or simply too thin to feasibly harvest with a traditional underground mine, **mountaintop removal (MTR)** is used—the entire mountaintop is removed to expose the coal seams below. This particular mountaintop removal site has claimed at least 12,000 acres of land and was once the site of several adjacent peaks.

To get at the coal beneath those peaks, miners began by clear-cutting the forest above. Next, they drilled holes deep into the side of the mountain, set dynamite in those holes, and blasted as much as 1,000 feet of mountain into a mass of rubble (the overburden). Then, using buckets big enough to hold 20 midsized cars, they scooped that rubble into waiting dump trucks that carried it away. The miners

surface mining A form of mining that involves removing soil and rock that overlay a mineral deposit close to the surface in order to access that deposit.

strip mining A surface mining method that accesses fossil fuel or mineral resources from deposits close to the surface on level ground one section at a time.

overburden The rock and soil removed to uncover a mineral deposit during surface mining.

mountaintop removal (MTR) A surface mining technique that involves using explosives to blast away the top of a mountain to expose the coal seam underneath; the waste rock and rubble are deposited in a nearby valley.

David R. Frazier/Science Source

Wyoming strip mine.

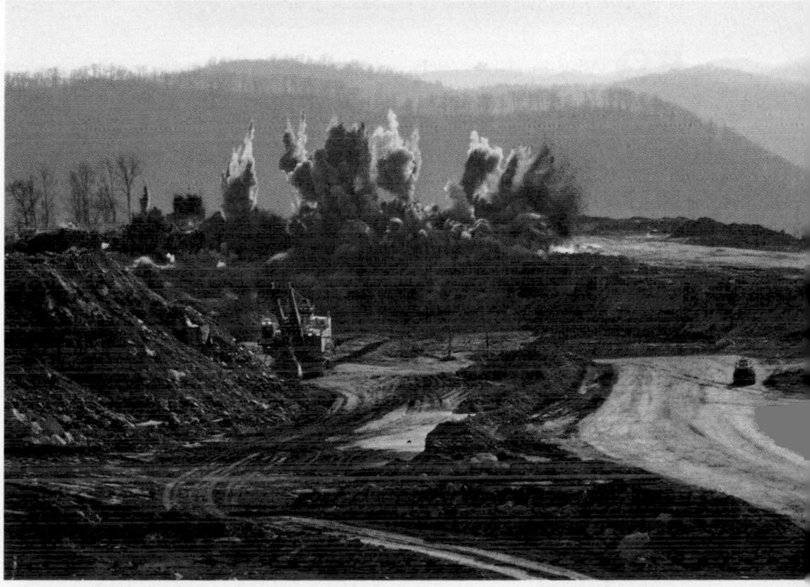

Antrim Caskey

Blasting at a mountaintop removal site.

INFOGRAPHIC 4 **SURFACE MINING TECHNIQUES**

STRIP MINING

Strip mines are used in areas where the ground is fairly level and coal seams are close to the surface. The overburden is removed and set aside. The exposed coal seam is harvested using heavy equipment, and once exhausted, the strip is filled back in with the overburden and vegetation replanted.

MOUNTAINTOP REMOVAL

In Appalachia, the forests are first clear-cut and then explosives are used to blast away part of the mountain. Heavy equipment then digs through debris, dumping the overburden (soil and rock) into the nearby valley, burying streams as the valley is filled in. The exposed coal is dug out and some processing is done on site. Coal sludge left over from processing is stored in ponds on the mining site.

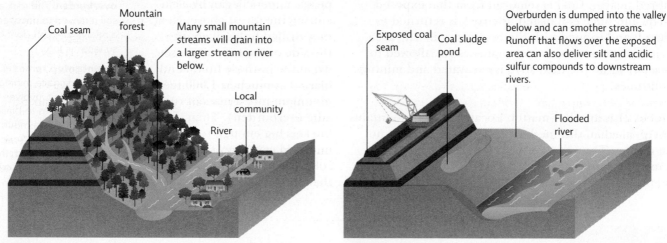

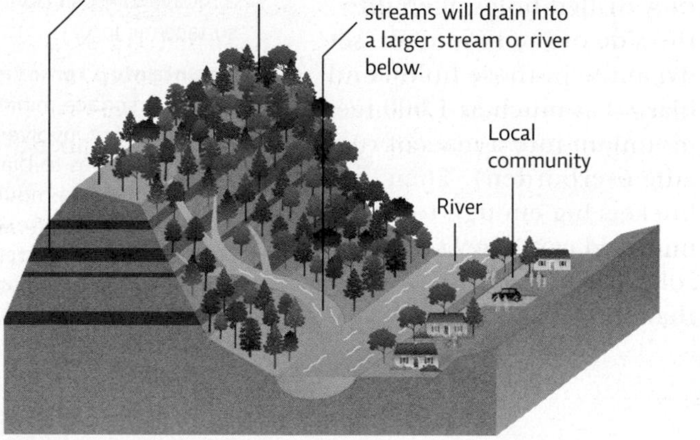

Coal seam
Mountain forest
Many small mountain streams will drain into a larger stream or river below.
Local community
River

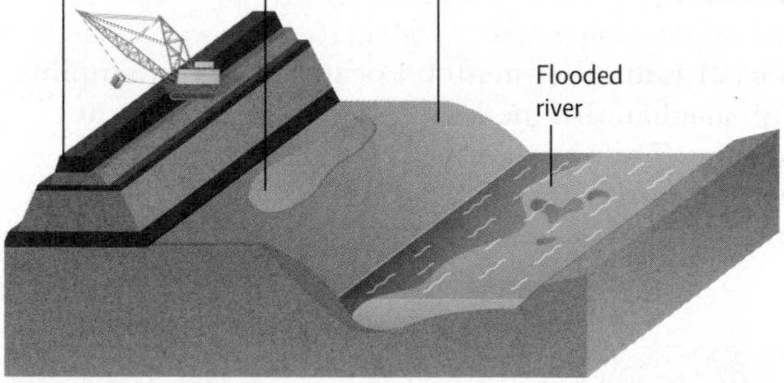

Exposed coal seam
Coal sludge pond
Overburden is dumped into the valley below and can smother streams. Runoff that flows over the exposed area can also deliver silt and acidic sulfur compounds to downstream rivers.
Flooded river

Why is the overburden dumped into the valley below when it is known that it damages streams and valley habitat?

repeated this process several times, until the layers of coal were exposed.

Unlike the flat strip mine areas of Wyoming, there is nowhere to store the overburden—MTR overburden is simply dumped down what is left of the mountainside, into the valleys (and streams) below. The process destroys the forest habitat, buries countless streams, and permanently reorders the land's natural contours. Acidic compounds can also be released from the rubble as rainwater reacts with iron sulfide components in the rock, converting it to sulfuric acid. Known as **acid mine drainage**, this acid goes on to contaminate soil and streams and has become a major problem in both active and closed mines. In 2011, Duke University ecologist Emily Bernhardt and her colleagues reported that acidic water not only is directly toxic to many aquatic plants and animals but also alters the nutrient cycle of streams in ways that reverberate all the way up the food chain. (See Module 7.1 for more on acid mine drainage and mining in general.) **INFOGRAPHIC 4**

acid mine drainage Water that flows past exposed rock in mines and leaches out sulfates. These sulfates react with the water and oxygen to form acids (low-pH solutions).

5 COAL MINING: SUBSURFACE TECHNIQUES

Key Concept 5: Subsurface mining is used to access deep, thick coal seams. It is less environmentally damaging and employs more workers than surface mining but is a hazardous job.

With their reliance on explosives and supersized heavy equipment, surface mines are a far cry from the underground mines used in **subsurface mining**—mines that sustained the Nelson family for so many generations. "When our daddies were mining, back in the '40s and '50s, the seams were as tall as full-grown men," says Nelson, who since retiring has become a spokesperson for the anti-mining Ohio Valley Environmental Coalition. "So you could get at them the old-fashioned way, with pickaxes and sledgehammers." Those days of plenty are gone, he says. Many of the coal seams that remain are too thin to be culled by human hands.

To be sure, subsurface mines (which make up about 50% of all U.S. coal mines) come with their own challenges. Water seeps easily into tunnels, and as it does, hazardous chemicals leach from the surrounding rocks into the gathering pools. Like surface mines, subsurface mines can also be a major source of acid mine drainage if water collects in the tunnels or flows through nearby overburden piles. Operators must continuously pump water out of mine tunnels into holding ponds or nearby ecosystems. Acidic water can also drain out of closed mines. This acidic water readily dissolves hazardous chemicals that might be in the rocks, such as heavy metals (e.g., mercury, arsenic, chromium), increasing the hazard it presents. **INFOGRAPHIC 5**

Subsurface mines are also dangerous places to work. Explosions and toxic fumes can be fatal to miners, and breathing in coal dust causes pneumoconiosis (black lung disease). In fact, more miners die from lung disease than from mining accidents. Fires, too, can start in a coal mine, and once they begin, it is hard to extinguish them. (See the *Global Case Studies* map at the end of this module for more on mine fires and other disasters.)

But subsurface mines also come with some advantages. Unlike surface mines, they don't disrupt or permanently alter large surface areas. And because subsurface mines require more workers, they employ more people. In Appalachia, 100,000 mining jobs were lost between 1980 and 1993 as underground mining gave way to mountaintop removal. More recently, Appalachian coal mining jobs have fallen around 40%, though numbers increased slightly in 2017 and 2018. "It's a double insult," says Tim Landry, a fourth-generation deep miner in West Virginia. "They're not only destroying the land that we love, but they're taking our jobs away, too."

But a loss of jobs is not the community's only—or even its most serious—concern.

subsurface mines Sites where tunnels are used to access underground fossil fuel or mineral resources.

INFOGRAPHIC 5 SUBSURFACE MINING

Subsurface mines are used to access large deposits or thick seams of many minerals and ores, including coal. Modern mining depends on powerful machinery to drill out tunnels and remove and transport the materials.

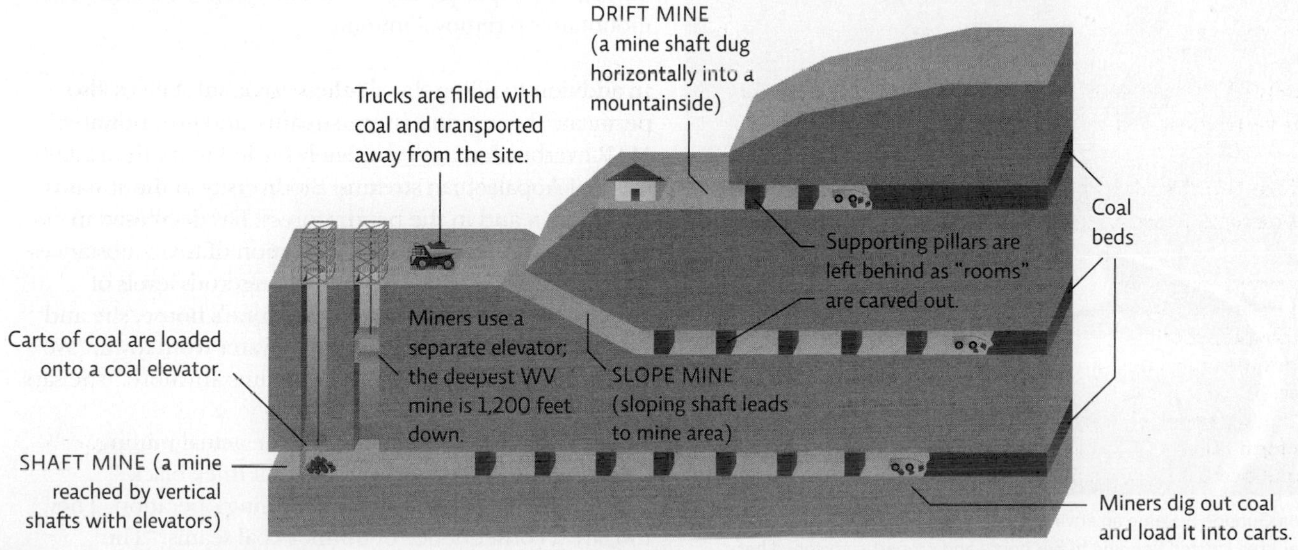
DRIFT MINE (a mine shaft dug horizontally into a mountainside)
Trucks are filled with coal and transported away from the site.
Coal beds
Supporting pillars are left behind as "rooms" are carved out.
Miners use a separate elevator; the deepest WV mine is 1,200 feet down.
Carts of coal are loaded onto a coal elevator.
SLOPE MINE (sloping shaft leads to mine area)
SHAFT MINE (a mine reached by vertical shafts with elevators)
Miners dig out coal and load it into carts.

? Why might a mining operation use mountaintop removal instead of subsurface mining?

6 THE ADVANTAGES AND DISADVANTAGES OF COAL

Key Concept 6: Coal is an abundant and high-energy fuel, but mining and burning it create significant environmental and health problems for human populations and ecosystems through habitat destruction and the pollution of air and water.

Coal is an energy-rich fossil fuel that has several advantages over other fossil fuels. Its wide distribution around the world means there is little, if any, geopolitical tension associated with coal supplies. Coal is also safer and easier to store and transport than liquid or gaseous fuels. And it is affordable.

But coal has many disadvantages, many of them felt by the residents of Appalachia, such as Maria Gunnoe—the daughter, sister, wife, niece, and aunt of coal miners who has lived on the same property in Bob White, West Virginia, all her life. She remembers having free run of the mountain as a child. "We had access to all the resources—food, medicine, water—that these mountains provided."

Things are different now. Gunnoe's children run into big yellow gates and No Trespassing signs wherever they go. The mountains, she says, have been closed off for blasting. And in the past decade, several million tons of overburden have been dumped into the valleys around Bob White. The upheaval has had a noticeable impact on area residents. For one thing, the loss of forest and the compaction of so much soil have increased both the frequency and severity of flooding; without trees and with the soil so compressed, the ground can't absorb water. "Floods are about three times more serious than they ever were before," she says. One 2003 flood nearly swallowed her entire valley.

Maria Gunnoe became an environmentalist after mountaintop mining in her area caused flooding to her home and property, poisoned her well water, and made her daughter sick. Gunnoe received the Goldman Environmental Prize for her organizing efforts in her southern West Virginia community.

Floods aren't the only problem. In fact, it's the blasting that most scares Gunnoe. The explosions that send rock and debris flying can damage homes and, in rare cases, trigger rockslides. In 2005, a 3-year-old child was killed when a boulder crashed through his house and landed on the bed where he was sleeping. Blasting also fills the air with tiny particles of coal dust—easily inhaled and full of toxic substances like mercury and arsenic. Studies show a higher-than-typical incidence of respiratory illnesses in mining communities. "When they were clearing the ridgeline right behind us, we'd get blasted as much as three times a day," Gunnoe says. "There were days when we'd have to just stay inside because you couldn't breathe out there. And now, my daughter and I both get nosebleeds all the time."

The health impacts are far-reaching—something that researchers Melissa Ahern and Michael Hendryx have been studying for more than a decade. Their 2011 study showed that children in MTR communities are more likely to suffer a range of serious birth defects, including heart, lung, and central nervous system disorders. And children aren't the only ones affected; another study they published in 2012 found increased rates of leukemia and lung, colon, kidney, and bladder cancer in adults living in Appalachian counties with mountaintop removal sites compared to non-mining counties. (See the *Science Literacy* activity at the end of this module for more on this study.) All told, Hendryx estimates that 1,200 people die each year due to the health impacts of mountaintop removal mining.

In addition to filling the air, these toxic substances also permeate the region's rivers, streams, and groundwater. MTR overburden has completely buried more than 2,000 miles of Appalachian streams. Biodiversity in the streams that remain and in the nearby forests has decreased in direct proportion to the concentration of toxic substances in the water. When tests revealed dangerous levels of selenium in the stream behind Gunnoe's house, she and her neighbors started getting their water from town. "We can't trust the water in our own streams anymore," she says.

But the damage doesn't stop with the actual mining. Slurry impoundments—reservoirs of thick black sludge—accompany each MTR mining operation. They, too, are a consequence of thinner coal seams. "The thinner seams are messier," says Randal Maggard, a mining supervisor at Argus Energy, a company that has several mining operations throughout Appalachia. "It's 6 inches of coal, 2 inches of rock, 8 inches of coal, 3 inches of rock, and so on. It takes a lot more work to process coal like that." To separate the coal from the rock, miners use a mix of water and magnetite powder known as *slurry* in which coal can float. Once the sulfur and other impurities have been washed out, the coal is sent for further processing, and the slurry by-product is pumped into artificial holding ponds.

Maggard insists that the impoundments are safe. "Before we fill it, we have to do all kinds of drilling to test the bedrock around it," he says. "And then a whole slew of chemical tests on top of that—all to make sure the barrier is impermeable." Still, area residents worry about a breach. With good reason.

In October 2000, the containment at a sludge pond holding coal slurry failed, pouring more than 300 million gallons (30 times the 1989 Exxon *Valdez* spill) of toxic sludge into the Big Sandy River of Martin County, Kentucky. The contamination killed all wildlife in streams and streamside areas in the immediate area and impacted waterways for more than 60 miles. The sludge was 5 feet deep in some places. The U.S. Environmental Protection Agency (EPA) would register it as one of the worst environmental disasters in U.S. history. Sludge can still be found a few inches under the river sediments.

Of course, it's not just the mining and processing of coal that pollute the environment. When coal is burned to produce heat energy, it releases a variety of substances that damage the environment and threaten human health: gases (sulfur dioxide, carbon monoxide, nitrogen oxide, and planet-warming carbon dioxide), radioactive material (uranium and thorium), and particulate matter (soot) that can irritate lung tissue or even enter the bloodstream if inhaled.

Heavy metals (such as mercury and arsenic) are also released. Worldwide, coal burning accounts for 24% of the total anthropogenic mercury released. Much of this mercury finds its way into the world's oceans and into fish. In fact, the main way humans are exposed to mercury is by eating contaminated fish, especially top predators like tuna that biomagnify mercury (see Module 4.3).

Coal-fired power plants also generate tons of toxic fly ash—fine ashen particles made up of silica and small amounts of toxic metal such as cadmium and lead. Some of that ash is diverted to industry, where it's used in concrete production. But most is buried in landfills or stored in open ponds like the ones used to contain the slurry waste from mining. In 2008, following a heavy rain event, a 50-foot-tall dike from a pond holding coal fly ash from the Kingston Fossil Plant in Tennessee failed, releasing 1.1 billion gallons of fly ash into nearby rivers and coating vast expanses of riverside land. The EPA spent 6 years and $1.1 billion in cleanup efforts, and in 2017 reported that the ecosystem was back to "pre-spill conditions." The cleanup has been costly in human health: A decade after the cleanup operation, 41 coal ash spill workers have died and more than 400 are ill—illnesses linked to their exposure to toxic chemicals in the coal ash. Lawsuits claim the workers were not properly

In 2008, more than 1 billion gallons of coal fly ash slurry surged into homes and waterways near Harriman, Tennessee, after breaching a coal ash containment pond. Nearby residents were permanently displaced, and many individuals who worked on the response team have suffered serious and sometimes fatal health problems linked to the cleanup operation.

Antrim Caskey

INFOGRAPHIC 6 **THE TRADE-OFFS OF COAL**

Though coal has many advantages as a fuel, mining, processing, and burning coal contribute to a variety of environmental problems, many of which lead to serious health issues for human populations.

ADVANTAGES	DISADVANTAGES
• Widely distributed and fairly abundant supply • Low geopolitical conflict over supplies • Energy-rich fuel • Safe to transport and store • Affordable	• Nonrenewable, finite resource • Hazardous air pollution from mining and burning coal • Most carbon-intensive fossil fuel to burn (high CO_2 emissions) • Water pollution (surface and groundwater) from mining and burning coal • Habitat/property destruction from surface mining, and ash and sludge pond failures • Biodiversity loss from mining (especially MTR mining) • Blasting hazards to nearby communities from MTR mining • Flooding risk increased due to loss of vegetation from surface mining • Health issues such as respiratory problems, cancers, birth defects, and cardiovascular diseases in miners and community members living near surface mines or coal power plants

? If we choose to continue to use coal, what might be done to address the disadvantages of mining, processing, or burning it?

protected during the cleanup operation—a claim that highlights the toxic nature of this ash.

In the wake of the Kingston spill (and others such as a 2014 spill that sent more than 36,000 tons of coal ash into the Dan River of North Carolina), arguments were made to classify coal fly ash as a hazardous material, but the EPA ultimately rejected these arguments and ruled to classify and regulate coal ash as a nonhazardous substance in December 2014. **INFOGRAPHIC 6**

In 2011—in an attempt to calculate the *true costs* of coal—Harvard University public health researcher Paul Epstein and colleagues tallied up the costs of mining and using coal. The analysis took into account many of the external costs from mining, shipping, burning, and waste production—that is, the costs that are not currently reflected in the market cost of coal. They estimated that coal costs the American public between \$300 billion and \$500 billion a year in externalized costs (health, environmental, and property costs). This amounts to an extra \$0.10 to \$0.26 per kWh in external costs—in some cases more than twice what consumers actually pay. (See Module 5.1 for an introduction to true cost accounting.)

7 REDUCING THE IMPACT OF BURNING COAL

Key Concept 7: The negative impact of burning coal can be addressed by capturing pollutants before they are released or converting coal to a liquid or gaseous fuel that burns more cleanly.

So what do we do? On one hand, we rely on coal to power our society. On the other, the processes of mining and then burning it are hurting us and our environment as much as they are sustaining our way of life.

One potential solution is "clean coal technology"—technology that minimizes the amount of pollution produced by coal. For example, scientists and engineers around the world are working on ways to capture the gases emitted from burning coal. We already have the capacity to capture some emissions like mercury, particulate matter, and sulfur (see Module 10.1). The next big challenge is **carbon capture and sequestration (CCS)**—the capture and storage of CO_2 in a way that prevents it from reentering the atmosphere and contributing to climate change. Several CCS facilities have begun operation, but most of these are industrial facilities (most commonly natural gas processing plants). As of 2019, only two post-combustion carbon capture systems from a coal-fired power plant were up and running.

carbon capture and sequestration (CCS) The process of removing carbon from fuel combustion emissions or other sources and storing it to prevent its release into the atmosphere.

The only operational U.S. CCS coal-fired power plant, Petra Nova, went online in early 2017, a \$1 billion retrofit of an existing power plant near Houston, Texas. (The only other CCS facility online is located in Canada—the Boundary Dam Power Station.) Petra Nova captures 90% of CO_2 from combustion, around 1.6 million tons a year; the CO_2 gas is injected into a nearby oil field to enhance oil recovery (see Module 9.2). Using the CO_2 to boost oil production gives the captured gas economic value and improves the bottom line, making Petra Nova a *CCUS* (carbon capture, utilization, and sequestration) *operation*. Industry analysts say that while the technology seems to work, the biggest obstacle is economic. Stronger incentives—a market for the captured product or policies that require or support carbon capture—may be the main hurdles to moving CCS and CCUS forward. **INFOGRAPHIC 7**

Another emerging technology creates what is known as **clean coal** in a process that chemically removes some of coal's contaminants before burning it. A liquid fuel can be produced in this process (liquefaction), but the most common clean coal technology converts coal to a gaseous fuel (syngas) that burns more efficiently than pulverized coal and releases fewer air pollutants. The gasification process is complex: It occurs at high pressure and temperature, and it requires chemical solvents to strip some of the impurities out of the coal. The process requires significant inputs of energy, thus lowering coal's EROEI despite its greater energy efficiency in electricity generation.

And it doesn't come cheap. While the $1 billion price tag for the CCS refit of Petra Nova is significant, it is much more affordable than the $7.5 billion that was spent on the Kemper County Energy Facility in Mississippi, a next-generation clean coal–powered power station with CCS that missed its start-up date (May 2014) by years and its projected price tag ($3 billion) by close to 300%. In 2017, the project was suspended; the Kemper facility now burns natural gas to produce electricity. Indeed, high costs and technical difficulties, compounded by low natural gas prices, have stalled the clean coal industry in the United

clean coal A liquid or gaseous product produced by removing some contaminant contained in coal so that the resulting fuel releases less pollution when burned.

INFOGRAPHIC 7 **HOW IT WORKS: CARBON CAPTURE AND SEQUESTRATION (CCS)** Achieve

The capture of CO_2 and subsequent storage that prevents it from reentering the atmosphere would greatly decrease coal's contribution to global climate change. A variety of CCS methods are currently in research and development or are being tested as pilot programs.

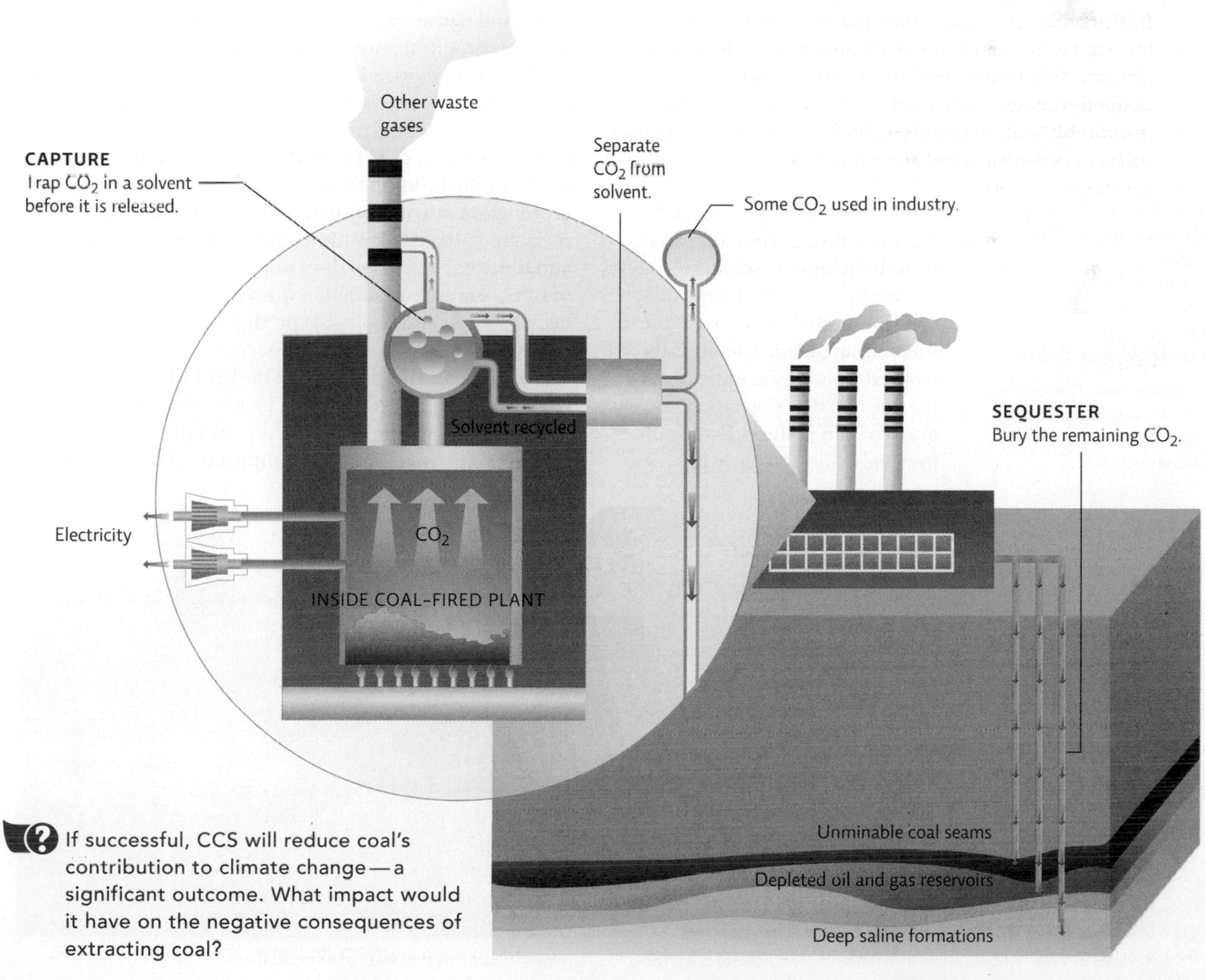

If successful, CCS will reduce coal's contribution to climate change—a significant outcome. What impact would it have on the negative consequences of extracting coal?

States. But in places without a domestic natural gas supply, coal gasification is still on the table with several projects in development or deployed in Asia.

These efforts, if implemented, will address some of the downsides to *burning* coal, allowing us to continue to use an abundant fossil fuel and, in doing so, save jobs associated with the coal industry at the same time. But as critics are quick to point out, even with clean coal technologies like CCS and cleaner coal-derived fuels, coal can never be truly clean. Coal liquefaction and gasification produce their own toxic by-products that must be dealt with. And CCS and clean coal fuels do nothing to address the negative impacts of *acquiring* that coal. For coal to truly be considered "clean," ecosystems and communities must also be protected during and after the mining process.

8 MINE RECLAMATION

Key Concept 8: Surface coal mines must undergo reclamation once they are closed; this reduces the environmental damage of mining but does not eliminate it.

In some ways, closed MTR mines—those where all the coal has been harvested—look even more alien than do active sites like Hobet 21. Instead of the natural sweep of rolling hills, staircase-shaped mounds are covered with grasses or stands of evenly spaced, young conifers. Such efforts represent the coal industry's attempt to honor the **U.S. Surface Mining Control and Reclamation Act**, which in 1977 mandated that areas that have been surface-mined for coal be "reclaimed" once the mine closes. **Reclamation** requires that the area be returned to its "approximate original contour" and to a state that benefits the local community (e.g., wildlands or lands suitable for public uses such as residential or industrial areas).

U.S. Surface Mining Control and Reclamation Act Federal law that mandates coal mines be restored to their approximate original condition after closure.

reclamation The process of restoring a damaged natural area to a less damaged state.

If the mined area was originally fairly flat, the reclamation process includes filling the site with the stockpiled overburden and contouring the site to match the surrounding land. The site is then covered with topsoil saved from the original dig. Vegetation, usually grass, is then planted, leaving other local vegetation to move in on its own. This method has successfully converted strip mines in western states back into grasslands that are once again home to native wildlife such as pronghorn antelope and grouse.

But returning MTR sites back to their approximate original contour is impossible—the mountain is gone. Even restoring the forest habitat is difficult due to the loss of soil and native species. MTR sites are excused from the original contour mandate, but they must be reclaimed to a usable state that provides benefits to the local community or economy equal to or better than the pre-mining state.

In the early days of MTR reclamation, the mined area was smoothed out, the soil packed down, and quick-growing grasses, usually non-native, were planted. These reclaimed areas look nothing like the original forest, and the compacted soil does not support the return of trees—areas reclaimed in this way remain treeless decades later, and ecologists predict that they will remain that way for decades more, if not centuries. Streams, too, are irrevocably damaged. The 1973 Clean Water Act requires that damaged streams be returned close enough to their original state such that the overall impact on the stream ecosystem is "nonsignificant." The method

Antrim Caskey

A valley in West Virginia after reclamation shows none of the original forest, ridges, or streams that were once found there.

INFOGRAPHIC 8 MINE SITE RECLAMATION

U.S. federal law requires that mine lands be restored close to their original state after surface coal mine operations cease. Though mountaintop removal mining sites can never be returned to their original state, the goal is to reclaim land so it is suitable for public land uses (e.g. residential, commercial, or industrial) or is returned to a natural state, preferably a forest similar to the original.

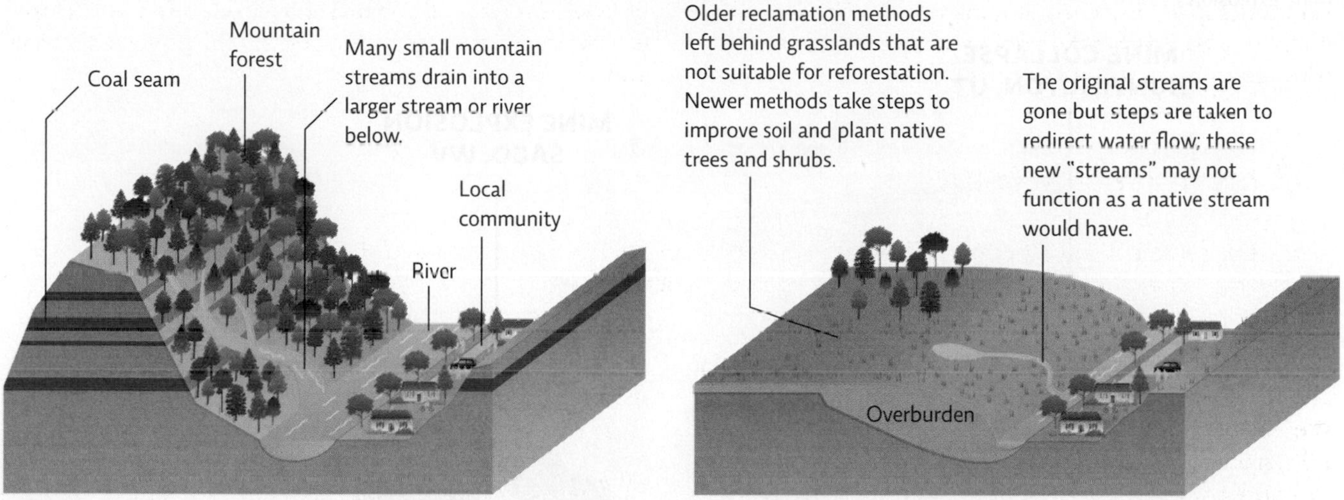

Mountain forest
Coal seam
Many small mountain streams drain into a larger stream or river below.
Local community
River
Older reclamation methods left behind grasslands that are not suitable for reforestation. Newer methods take steps to improve soil and plant native trees and shrubs.
The original streams are gone but steps are taken to redirect water flow; these new "streams" may not function as a native stream would have.
Overburden

If you lived near an area where mountaintop removal mining was being done, what type of reclamation would you like to see—reforestation to return the area to a wildland again or conversion to a public land use (residential, commercial, or industrial)? What are the advantages of each reclamation outcome?

of choice is to dig drainage ditches and line them with stones in a way that resembles a stream or river. But so far, research shows that such channels don't perform the ecological functions of a stream. "They may look like streams," says ecologist Margaret Palmer from the University of Maryland. "But form is not function. The channels don't hold water on the same seasonal cycle, or support the same aquatic life, or process contaminants out of the water—all things a natural stream does."

Recognizing that the grassy areas created by earlier MTR reclamation methods were falling short of restoration goals, especially with regard to water management and ecological health of the land, Dr. Chris Barton of the University of Kentucky developed the Forestry Reclamation Approach to bring forests back to these Appalachian mining sites. In this method, soil is "improved" by adding at least 4 feet of topsoil or other suitable growth medium. Quick-growing native tree and shrub species are planted to stabilize soil and benefit wildlife; commercially valuable trees are also planted for potential timber harvesting at a later date. Barton's approach is becoming the method of choice in Appalachia.

With this success in hand, Barton realized there was still more than a million acres that had already been reclaimed using the older "grade, compact, and plant grass" method. In 2009, he helped establish the nonprofit group Green Forests Work, which targets these sites for reforestation and gives jobs to Appalachian residents in the process. In its first ten years, Green Forests Work planted more than 2.8 million trees on 4,500 acres of previously reclaimed mine lands—areas that today are functioning forests. **INFOGRAPHIC 8**

In Appalachia, the arguments over when, where, and how to mine for coal are quickly boiling down to a single intractable question: Once it's all gone, how will we clean up the mess we've made? For a story that has played out over geologic time, the question is more immediate than one might think. In West Virginia, coal reserves are expected to last another 50 years, at best. That means no matter what regulations the government imposes, or what methods the coal companies resort to, the day of reckoning will soon be upon us.

Select References:

Ahern, M., et al. (2011). The association between mountaintop mining and birth defects among live births in central Appalachia, 1996–2003. *Environmental Research, 111*(6), 838–846.

Ahern, M., & Hendryx, M. (2012). Cancer mortality rates in Appalachian mountaintop coal mining areas. *Journal of Environmental and Occupational Science, 1*(2), 63–70.

Bernhardt, E. S., & Palmer, M. A. (2011). The environmental costs of mountaintop mining valley fill operations for aquatic ecosystems of the central Appalachians. *Annals of the New York Academy of Sciences, 1223*(1), 39–57.

Epstein, P. R., et al. (2011). Full cost accounting for the life cycle of coal. *Annals of the New York Academy of Sciences: Ecological Economics Review, 1219*(1), 73–98.

Green Forests Work. (2019). *Green Forests Work 2018 Annual Report.* Retrieved from https://www.greenforestswork.org/project-reports

Wagner, G. R. (2018). The public health conundrum of coal mining. *Annals of the American Thoracic Society, 15*(1), 11–13.

GLOBAL CASE STUDIES COAL MINING: A DANGEROUS OCCUPATION

Achie/e

While coal mining regulations have significantly reduced injuries and deaths in the United States, mining is still a dangerous job, especially in countries with lax or insufficient regulations in place to protect workers or the surrounding community. Here are accounts of five recent disasters related to coal. (See the *Global Case Studies* map in Module 7.1: Mineral Resources for the example of the Benxihu Colliery coal mine explosion.)

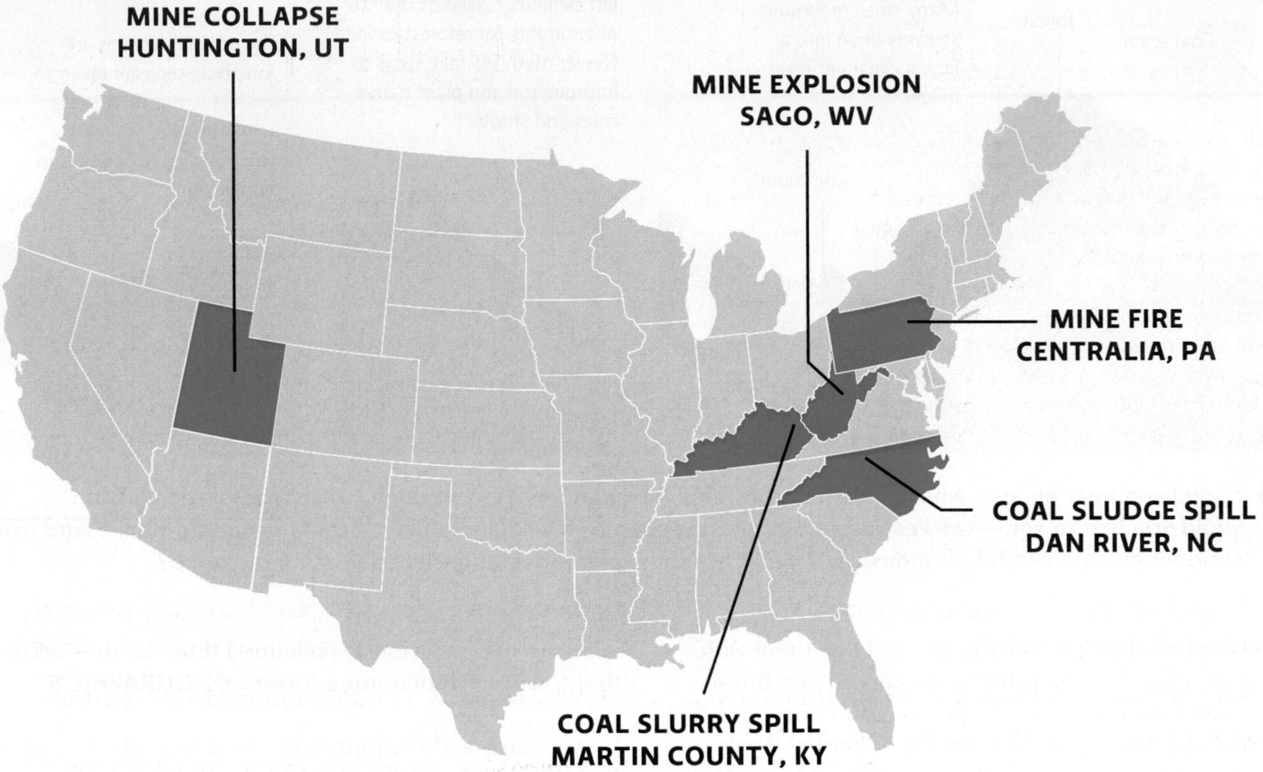

BRING IT HOME

PERSONAL CHOICES THAT HELP

Although coal is one of our most abundant fossil fuels, its drawbacks are significant. They include CO_2 emissions, the release of air pollutants that cause environmental problems such as acid rain; health problems such as asthma and bronchitis; and massive environmental damage from the mining process. One way to minimize the impact of coal is to reduce consumption of electricity.

Individual Steps

- Always conserve energy at home and in your workplace.
- Turn off or unplug electronics when not in use.
- Put outside lights on timers or motion detectors so that they come on only when needed.
- Dry clothes outdoors in the sunshine.
- Turn the thermostat up or down a couple of degrees in summer and winter to save energy and money.

Group Action

- Organize a movie screening of *Coal Country* or *Kilowatt Ours*, which present issues related to coal mining and mountaintop removal from many perspectives.
- The Appalachian Regional Reforestation Initiative is a great example of how groups, sometimes with very different objectives, can work toward a common goal. Go to http://arri.osmre.gov to see how this coalition of the coal industry, citizens, and government agencies is working to restore forest habitat on lands used for coal mining.

REVISIT THE CONCEPTS

- Modern society runs on energy — stationary sources of energy provide electricity and heat, and mobile fuels power our transportation fleets. Fossil fuels (coal, natural gas, and oil) provide most of this energy.

- The vast majority of coal is used for electricity production in coal-fired power plants that burn coal to boil water, producing steam that turns a turbine connected to a generator, producing electricity.

- The coal we use today formed millions of years ago when plant material was buried and subjected to high heat and pressure. Many areas have coal deposits that can be tapped, but because we are consuming it much faster than it is formed, coal is considered a nonrenewable resource.

- Coal located close to the surface is retrieved with surface mining techniques. In flat areas, it is accessed via strip mining, but in mountainous areas, a technique known as mountaintop removal (MTR) is used. MTR creates significantly more environmental damage and health problems for nearby communities than strip mining.

- When thick coal deposits are located deep underground, subsurface mining techniques are used. Though there is less damage to the environment and this method employs more workers than surface mining, the inherently hazardous working conditions make subsurface mining a dangerous job.

- While coal's advantages include the fact that it is an abundant and high-energy fuel, mining and burning it destroy habitat and create air and water pollution that is hazardous to human and ecological communities alike.

- The negative impact of burning coal can be addressed by capturing pollutants when the coal is burned or by converting it to a cleaner-burning fuel. Carbon capture and sequestration will decrease the contribution of coal combustion to climate change, but this industry is in its infancy.

- U.S. law requires that surface coal mines undergo reclamation once they are closed. When done according to best practices, reclamation can reduce the environmental damage of mining but does not eliminate it, especially for MTR mines since it is impossible to replace a mountain that is no longer there.

ENVIRONMENTAL LITERACY Understanding the Issue

1 What energy sources are used in modern society and for what purposes?

1. Distinguish between stationary and mobile sources of energy.
2. Explain the importance of fossil fuels to modern society.

2 How important is coal as an energy source, and how is it used to generate electricity?

3. Describe the process by which electricity is generated through the combustion of coal.
4. Explain the concept of energy return on energy investment (EROEI) and explain its significance.
5. What energy sources are used to produce electricity, and how do they compare in terms of EROEI?

3 What is coal, how is it formed, and what regions of the world have coal deposits that are accessible?

6. In your own words, describe the process of coal formation. Why is coal considered a finite resource?
7. Where are coal deposits found worldwide?

4 What surface methods are used to mine coal, and what are their advantages and disadvantages?

8. What types of coal deposits are accessed using the surface mining techniques of strip mining and mountaintop removal?
9. What are some of the community impacts, both good and bad, of mountaintop removal mining in Appalachia?
10. How does surface mining on flat ground compare to surface mining in mountainous areas in terms of environmental damage?

5 How are deep deposits of coal mined, and what are the advantages and disadvantages of this method?

11. What type of coal deposits are accessed with subsurface mining techniques?
12. How do subsurface and mountaintop removal mining in Appalachia compare in terms of jobs, health risks for workers, and environmental impacts on the nearby community?

6 What are the advantages and disadvantages of using coal?

13. Outline the pros and cons of mining, transporting, and burning coal — do the pros outweigh the cons?
14. If we paid the true costs of using fossil fuels like coal, how would this affect the price of our electricity? Why?

7 What new technologies allow us to burn coal with fewer environmental and health problems?

15. What are the pros and cons of converting coal to a cleaner fuel such as syngas?
16. What is carbon capture and sequestration (CCS), and why is it done? What are the pros and cons of the process?
17. Which problems associated with the traditional use of coal are not addressed by using clean coal technologies?

8 How can mining damage be repaired, and how effective is this restoration?

18. What is the goal of and general procedure for surface mine reclamation?
19. Why is coal mine restoration more difficult at an Appalachian MTR site than at a Wyoming strip mine?

SCIENCE LITERACY Working with Data

Cancer mortality rates were determined for Appalachian counties with MTR mines, counties with mines other than MTR mines, or counties with no mines. (*Note*: On the bar graph, the letter *a* beside a bar denotes significant difference from the "No mining" group; the letter *b* indicates it is significantly different from the "Other mining" group; $p < 0.05$ level.)

PROXIMITY TO COAL MINES AND CANCER MORTALITY RATES IN APPALACHIA

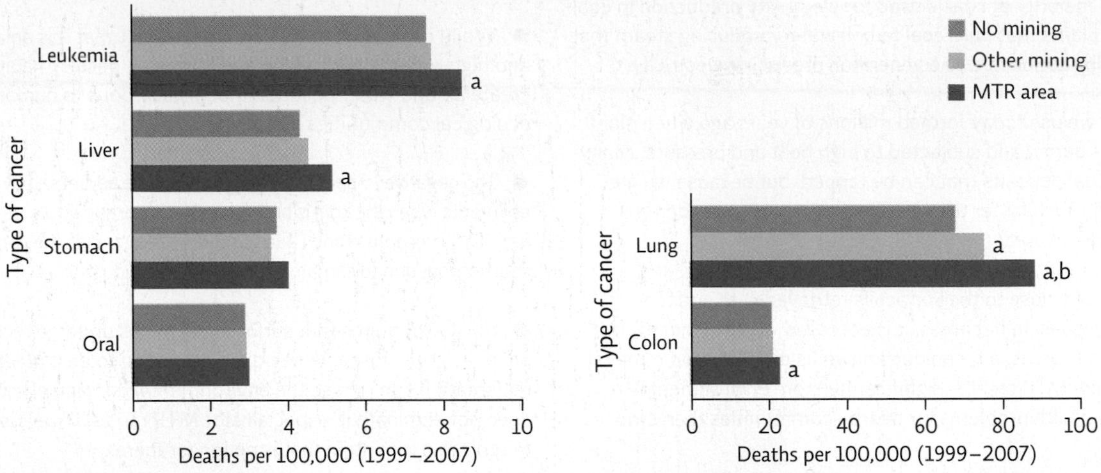

Interpretation

1. Which cancers show higher rates of mortality in MTR areas compared to areas with no mines? Compared to areas with other mines?
2. Why are the mortality rates for lung and colon cancer shown in a different graph?
3. Why is it incorrect to conclude that deaths due to stomach or oral cancer are more common in MTR counties even though the MTR bar for each indicates a higher number of deaths?

Advance Your Thinking

4. Why might lung cancer death rates be so high in Appalachian countries and highest in MTR areas?
5. Why did the researchers collect data on "Other mining" counties rather than simply comparing MTR counties to those with no mining?
6. In this study, researchers controlled for socioeconomic and health variables (such as smoking rates, obesity, and income). Why did they control for these variables?

INFORMATION LITERACY Evaluating Information

The Stream Protection Rule, part of the Clean Water Act, was signed by President Barack Obama on his last day in office in January 2017, after years of working to strengthen protections of streams against damage from surface coal mining. The rule was repealed soon after President Donald Trump took office that same year.

Read the following two articles about the rule and answer the questions that follow. (If these links are not active, search the Internet using the keywords "Stream Protection Rule Repeal" and read two articles—one for the rule and one against—and answer the questions below.)

1. Go to www.kentucky.com/opinion/op-ed/article133445464.html and read the article entitled "Stream Rule's Repeal First Step to Economic Relief for Coal Country."
   a. Who is the author of this article, and what web organization published the article? Are these reliable information sources? Why or why not?
   b. Is there any indication of bias that might influence the author's position? Explain.
   c. What position does the author take regarding the Stream Protection Rule? What evidence is offered for this position? Are reliable references provided to support this position or evidence? Explain.
   d. What information could have been provided to strengthen the position offered?
2. Now read the article "Trump's Repeal of Stream Rule Helps Coal at the Expense of Climate and Species" at www.insideclimatenews.org/news/16022017/coal-mining-environment-stream-rule-donald-trump-mussels-species and answer the same questions as above (a–d).
3. Compare the arguments presented by the two stories:
   a. Which argument prioritizes the economy and jobs? Which prioritizes the environment?
   b. Does each article address the main concern of the other (jobs, environment)? Explain.
   c. After reading these two articles, do you have the information you need to draw your own conclusion about the Stream Protection Rule? If so, why? If not, what additional information do you need and where would you find it?

Additional study questions are available at Achieve.macmillanlearning.com. Achieve

Module 9.2 Oil and Natural Gas

THE BAKKEN OIL BOOM

Is fracking the path to energy independence?

A fracking drill rig rises above the prairie in the western United States. grandriver/iStock/Getty Images

After reading this module, you should be able to answer the following GUIDING QUESTIONS

1. How are fossil fuels formed, and why are they considered nonrenewable resources?
2. How are oil and natural gas reserves classified, and where are they found?
3. How are conventional oil and natural gas extracted?
4. What are the trade-offs of acquiring and using conventional oil and natural gas sources?
5. What unconventional sources of oil and natural gas exist, and how are they extracted?
6. What are the trade-offs of pursuing unconventional oil and natural gas sources?
7. What obstacles stand in the way of the United States (or any nation) achieving energy independence or security?

"It is as you imagine it: Vast. Open. Windy. Stark. Mostly flat. All but treeless. Above all, profoundly underpopulated, so much so that you might, at times, suspect it is actually unpopulated. It is not. But it is heading there."

So went the opening paragraph of a 2006 feature article in *The New York Times Sunday Magazine* about the demise of the prairie states in general and of North Dakota in particular. The entire region, it seemed, was suffering a mass exodus: houses and churches and schools and stores, all completely abandoned.

That was 2006. Less than a decade later, the same sleepy patch of country was in the middle of a modern-day gold rush—an oil boom, to be precise. In 2012, journalists described local grocery stores barely keeping shelves stocked, town movie theaters being so crowded that people are sitting in the aisles, and housing costs approaching those of the most expensive cities. Towns once home to 1,200 residents claimed 10 times as many. "With the way it is now," Jeff Keller, a natural resource manager with the Army Corps of Engineers, told ProPublica in 2012, "you're getting to the crazy point."

Two short years later, the price of oil fell. Was the boom about to become a bust?

WHERE IS THE BAKKEN FORMATION?

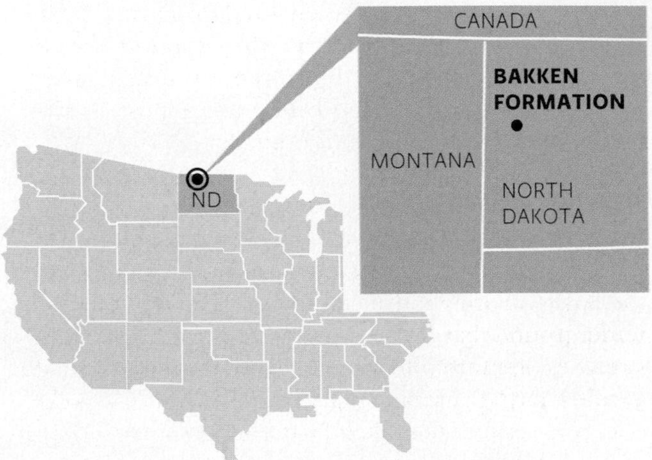

Rows of mobile homes house oil workers in Williston, North Dakota, where housing shortages drive up the cost of rent. In 2014, Williston was the most expensive place to rent a home in the United States.

Ben Garvin/Stringer/Reuters/Newscom

1 FOSSIL FUELS: OIL AND NATURAL GAS

Key Concept 1: Fossil fuels form when organic matter is buried and subjected to high temperature and pressure. Because we use them much faster than they form, they are considered nonrenewable resources.

Fossil fuels form over millions of years when organisms die and are buried in sediment under low-oxygen conditions that slow down decomposition tremendously. In time, as pressure and heat increase, the buried material goes through a chemical transformation to form **oil** (a liquid), **natural gas** (a gas), or coal (a solid). (See Module 9.1 for more on coal.) Oil and natural gas were formed from marine organisms that died and went through this process. (The starting material for coal formation was plant material.) Even though they are produced by natural processes, fossil fuels are considered a **nonrenewable resource** because we are using them up in a fraction of the time it takes for them to form. In other words, natural processes cannot possibly keep pace with human consumption. **INFOGRAPHIC 1**

Everywhere, countries are looking for new sources of oil and natural gas. The heart of the U.S. oil boom is the Bakken Formation, a vast shale formation deep underground that resides mostly in North Dakota but also stretches westward into Montana and north into Canada. (Shale is a type of sedimentary rock that is very dense and does not easily let the oil and natural gas trapped within escape.) It's the largest continuous oil accumulation the U.S. Geological Survey (USGS) has ever assessed.

The United States has other oil-rich shale formations, including the highly productive Eagle Ford Formation in Texas. Since 2014, Bakken has produced around 1 million barrels of crude oil each day—almost 14% of all oil produced in the United States (around 6% of the oil consumed per day). That number topped 1.3 billion/day in 2019. (A barrel of oil is equivalent to 42 gallons.)

So why didn't the oil industry go after the oil in Bakken earlier? It turns out, Bakken is not a typical oil deposit. When the 21st century began, it was not economically feasible to extract Bakken's oil, but new technology has put it within reach. "The implications are already reverberating far beyond North Dakota," Edwin Dobb wrote in *National Geographic* in 2013. "Bakken-like shale formations occur across the United States, indeed, across the world. The extraction technology refined here is in effect a skeleton key that can be used to open other fossil fuel treasure chests."

fossil fuel A variety of hydrocarbons formed from the remains of dead organisms.

oil A liquid fossil fuel useful as a fuel or as a raw material for industrial products.

natural gas A gaseous fossil fuel composed mainly of simpler hydrocarbons, mostly methane.

nonrenewable resource A resource that is formed more slowly than it is used or that is present in a finite supply.

INFOGRAPHIC 1 OIL AND NATURAL GAS FOSSIL FUEL FORMATION

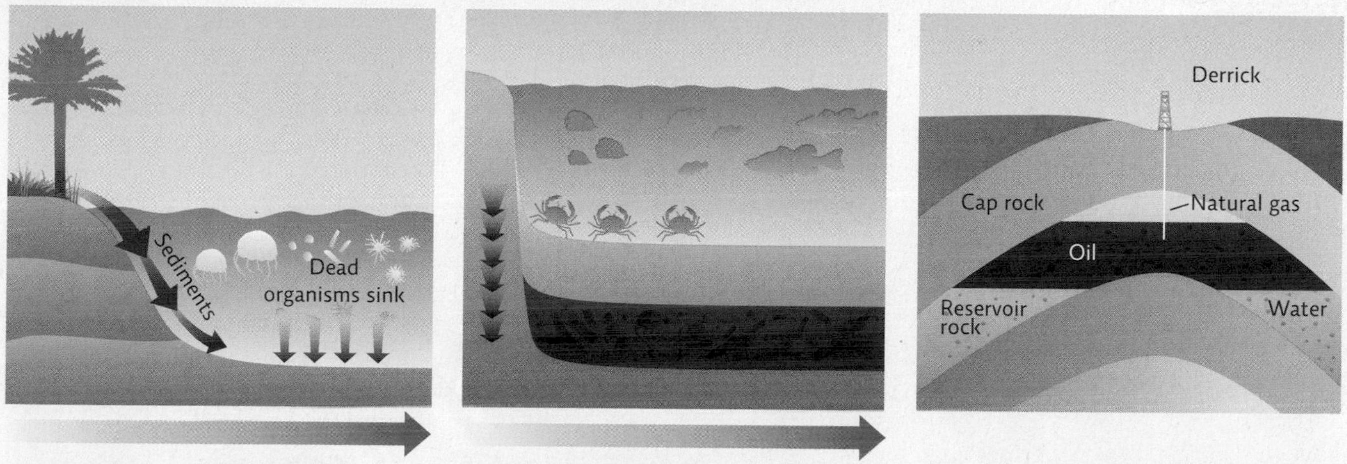

Long ago, some marine organisms died and were buried in sediment. This burial excluded oxygen, and decomposition was greatly slowed down.

As sediments accumulated, the partially decomposed buried biomass was subjected to high heat and pressure. Over the course of millions of years, it was chemically converted to oil or natural gas.

Since oil and natural gas are lighter than water, they flow upward in porous reservoir rock until stopped by a layer of dense cap rock. We tap these deposits by drilling into the porous rock reservoirs.

? How is the formation of oil and natural gas similar to and different from the formation of coal?

2 OIL AND NATURAL GAS RESERVES

Key Concept 2: The largest conventional oil and natural gas reserves are in the Middle East; however, North America has large deposits of unconventional oil and natural gas.

Oil is a liquid fossil fuel made up of hundreds of types of *hydrocarbons*, organic compounds of hydrogen and carbon. Hydrocarbons take many forms—solid, liquid, or gas—and we have developed a variety of methods for extracting all of these fossil fuels from the depths of Earth.

Oil and natural gas often occur together in formations—the lighter natural gas is found at the top of the deposit. After formation, the oil and natural gas slowly migrated up from the source rock in which they were formed into porous, sedimentary rock formations closer to the surface until they were stopped by a dense "cap rock" layer, such as granite, that prevented further migration and allowed the oil and natural gas to collect in one area. In these deposits, oil does not exist in thick black pools. Instead, if you could look down into an oil reservoir, you'd see rock. Most **crude oil** is found as tiny droplets wedged within the microscopic open spaces, or pores, inside rocks.

crude oil A mix of hydrocarbons that exists as a liquid underground; can be refined to produce fuels or other products.

proven reserves A measure of the amount of a fossil fuel that is economically feasible to extract from a known deposit using current technology.

conventional reserves Deposits of crude oil or natural gas that can be extracted by vertical drilling and pumping.

Oil and natural gas deposits, or reserves, are identified by their recovery potential. **Proven reserves** are those that hold oil or natural gas that is economically feasible to extract using current technology. **Conventional reserves** hold oil or natural gas that can be extracted with traditional oil or gas wells. Conventional oil reserves are not evenly distributed around the planet, leading to political problems among countries that have the oil, like those in the Middle East, and those that do not have enough to meet their own needs, like the United States. Alliances have been forged and wars have been fought over oil, highlighting its importance to modern society.

By many accounts, we have used up about half of the available conventional oil on the planet—an estimated 1.2 trillion barrels of oil. At current rates of extraction and use, known conventional reserves are expected to last another 50 years or so, though this number is uncertain because reports of oil reserves tend to be questionable. (Many nations keep their reserve estimates secret.)

Natural gas reserves are also finite. The Middle East has the largest conventional reserves of natural

The 800-mile Trans-Alaska pipeline delivers oil across Alaska to the southern port of Valdez.

Barry Williams/Getty Images

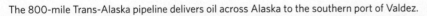

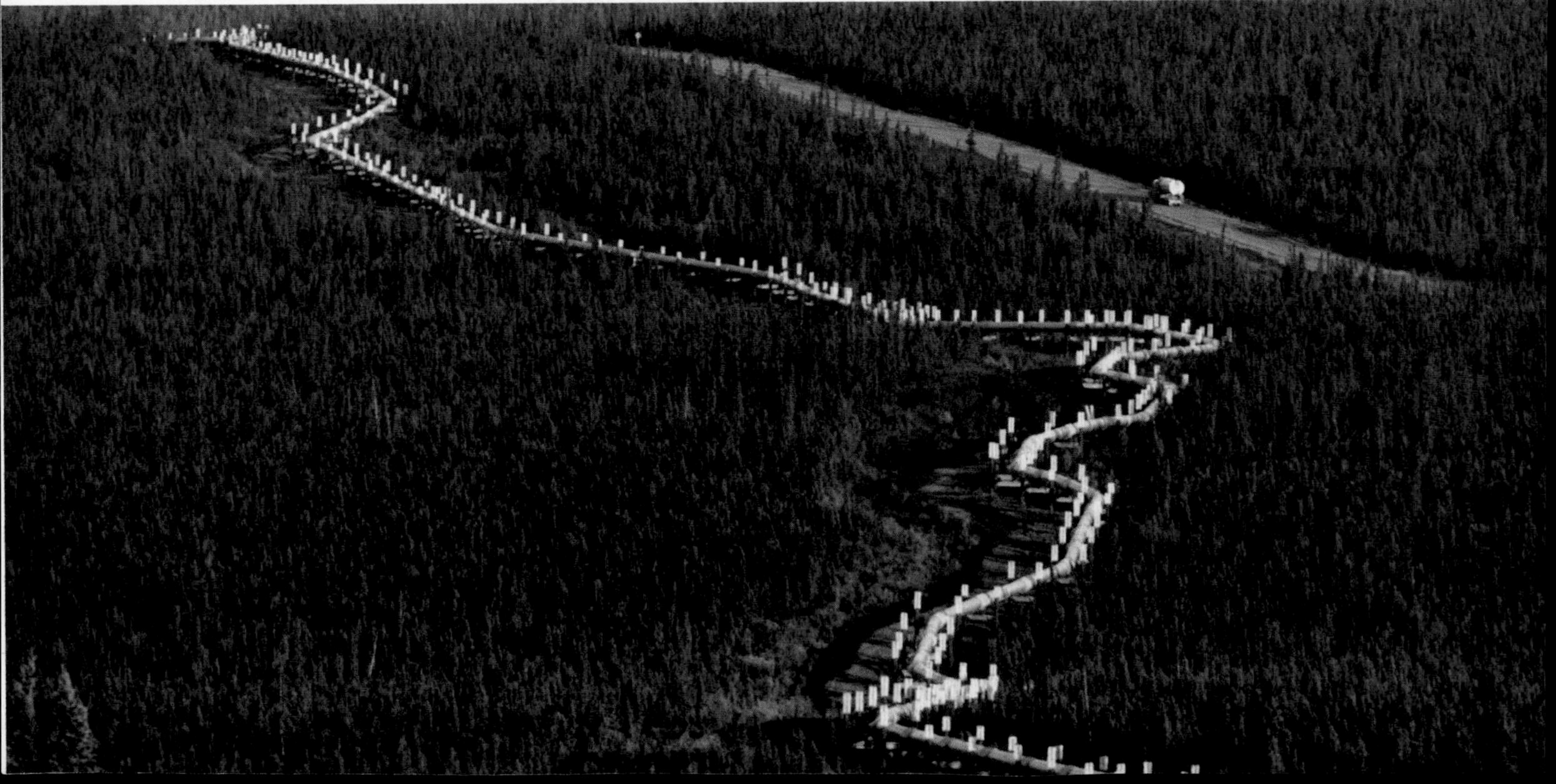

gas, with about 39% of the world's reserves; North America has only 8%. Total world conventional reserves of natural gas are expected to last 60 to 100 years at current rates of use.

However, the story does not end with conventional reserves. There are also other supplies of oil and natural gas, and our ability to access them is improving. These include **unconventional reserves**, such as Bakken oil. Bakken oil is made up almost exclusively of **tight oil**—shale oil that's trapped inside impermeable rock deposits—typically in the source rock in which it was formed (hence it is much deeper than conventional oil). Natural gas, too, is found in some shale deposits, such as the Marcellus Shale Formation that underlies much of Pennsylvania and New York. Tight oil and shale gas are of similar quality to that found in conventional deposits but are difficult to access because the oil or natural gas tends to be more widely dispersed throughout the rock. However, new technology (to be described later) is making those deposits more accessible. **INFOGRAPHIC 2**

The world runs on oil and natural gas. These fossil fuels power vehicles and provide the raw material for thousands of industrial products, and consumption continues to increase. But because we are using them much faster than they formed, they will run out. Extraction of oil and natural gas from conventional reserves is peaking—this means less can be extracted each year as time goes on. Exploiting unconventional reserves such as Bakken tight oil or Marcellus Shale natural gas will extend supplies, but not for long. Problems such as climate change, air and water pollution, and biodiversity losses are at the heart of many arguments for leaving most of that remaining oil and natural gas in the ground.

unconventional reserves Deposits of oil or natural gas that cannot be recovered with traditional oil and gas wells but may be recoverable using alternative techniques.

tight oil Light (low-density) oil in shale rock deposits of very low permeability; extracted by fracking.

INFOGRAPHIC 2 OIL AND NATURAL GAS RESERVES

Different regions of the world have different amounts of proven oil and natural gas reserves. Proven conventional reserves are shown here as bar graphs; bars represent percentage of the world total. The locations of proven unconventional reserves of shale oil and gas are shown as shaded regions on the map. The United States has a small percentage of proven conventional oil and natural gas reserves but larger amounts of unconventional reserves such as tight oil.

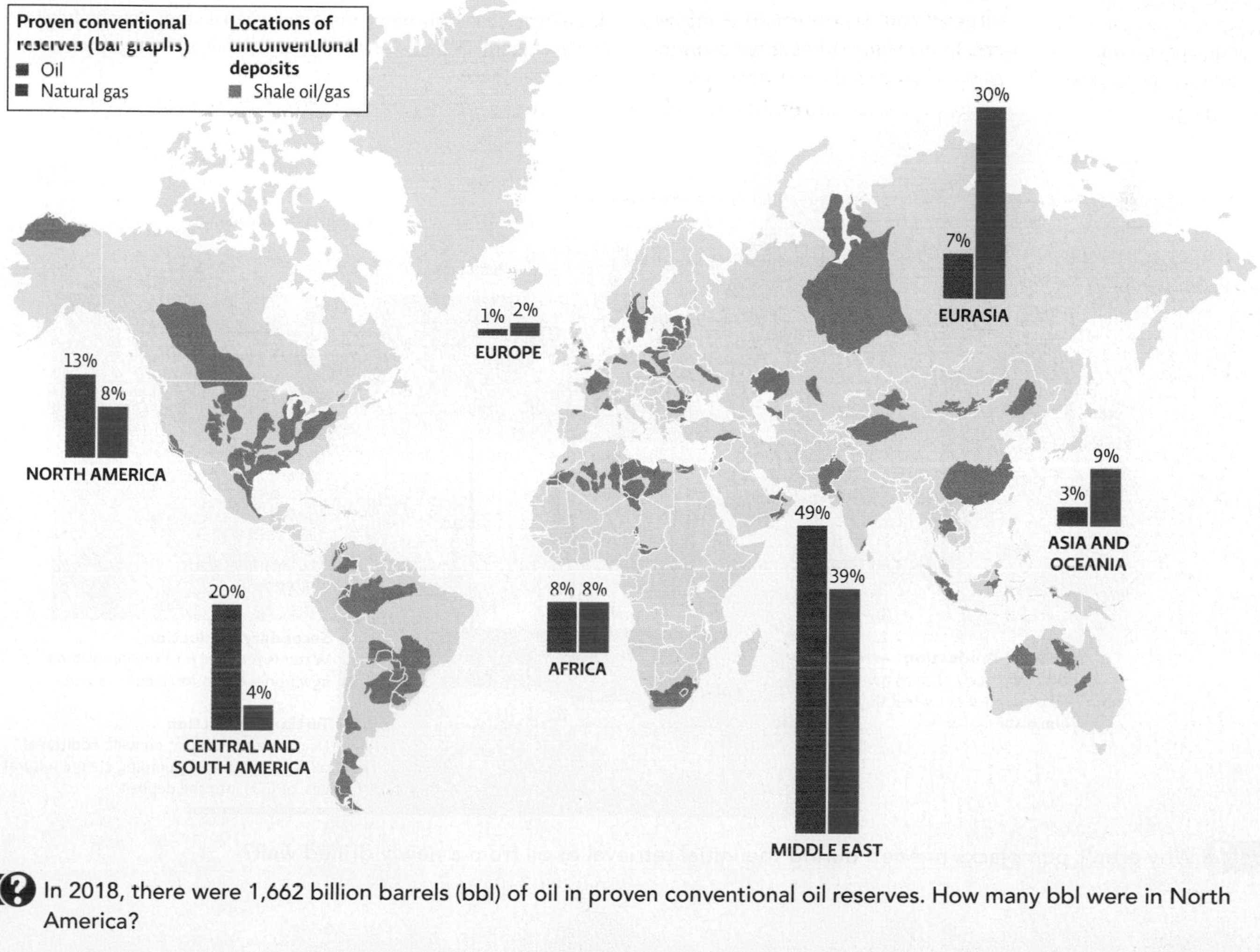

In 2018, there were 1,662 billion barrels (bbl) of oil in proven conventional oil reserves. How many bbl were in North America?

3 EXTRACTION OF CONVENTIONAL OIL AND NATURAL GAS

Key Concept 3: Oil and natural gas often occur together in rock formations and are extracted from conventional reserves using oil and gas wells.

Oil and natural gas are easier to coax out of some formations than others. Conventional oil and natural gas are the easiest to retrieve and have supplied our needs throughout the 20th century; they continue to provide most of our supply today. The fossil fuels in these deposits are extracted by drilling down into the rock formation that holds the oil and/or natural gas. For oil wells, after a new well is drilled, nature initially does most of the work. There is significant pressure on oil located deep underground from millions of tons of rocks pressing down and from Earth's heat, which causes gases around the oil and rock to expand. So when a well is first drilled, oil naturally flows upward, escaping like air gushing out of a balloon. Once the oil stops flowing freely, pumpjacks are used to pump out additional oil. This is known as *primary production*, and up to 15% of the oil can be recovered in this first phase.

Eventually, the pressure on the trapped oil will decrease, rendering pumping ineffective. So injection wells are drilled near the oil well, and water is pumped through these injection wells into the deposit. This increases the pressure on the remaining oil, forcing it upward in the deposit where it can be pumped out. During this phase of *secondary production*, an additional 20% to 40% of a reserve's oil can be recovered.

But even after maximal primary and secondary production (up to 55% of the total oil in the reserve), there is a lot of oil left in the ground. *Tertiary production* methods such as injecting steam, natural gas, or carbon dioxide gas into the reservoir allow the additional recovery of up to 20% of the reserve's oil, for a total production of up to 75% of the oil found in the reserve. **INFOGRAPHIC 3**

INFOGRAPHIC 3 HOW IT WORKS: CONVENTIONAL OIL AND NATURAL GAS WELLS

Oil and natural gas are obtained by drilling through layers of dense rock to reach the reservoir below. At first, oil easily flows due to the relief of pressure caused by the drill hole (primary production). Pumpjacks are used to mechanically pump out more when it stops flowing freely; additional secondary and tertiary production methods will allow the extraction of more oil. Because it is lighter than air, natural gas flows freely out of wells that contain only natural gas without the need for additional assistance.

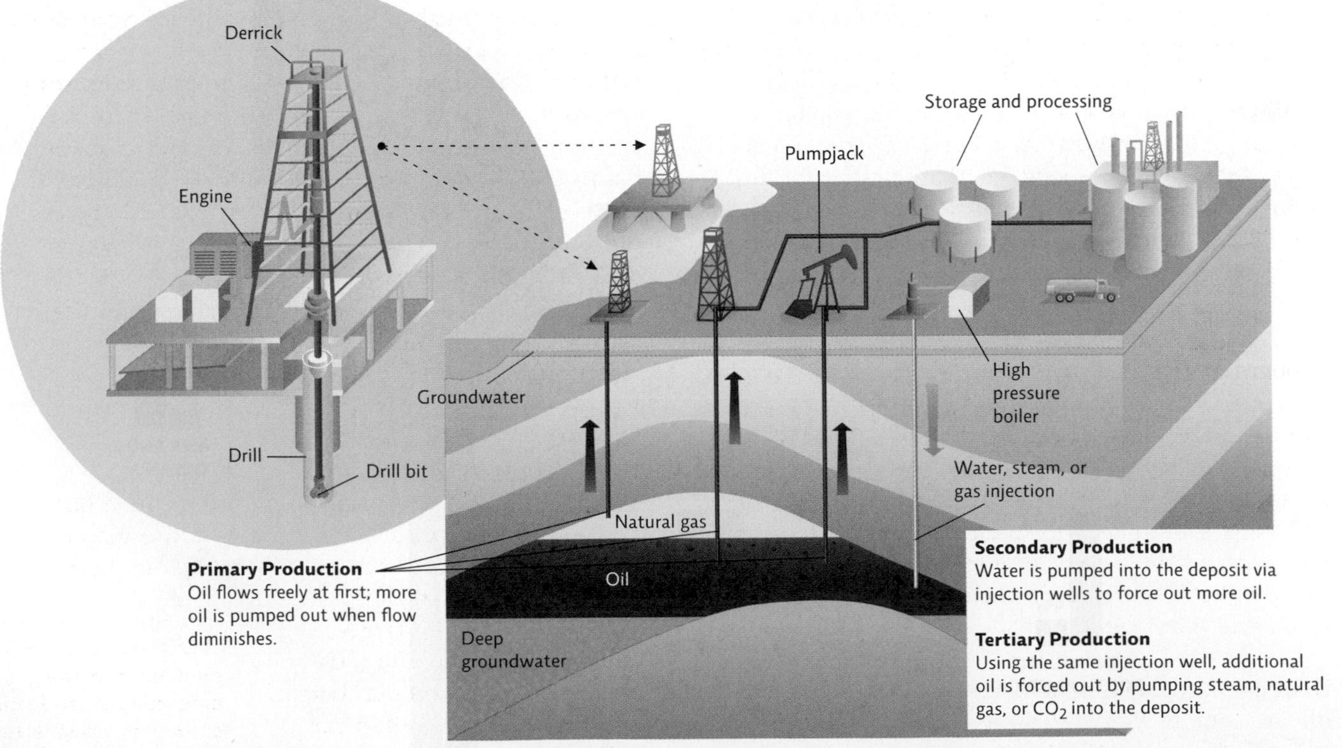

Derrick
Engine
Storage and processing
Pumpjack
Groundwater
Drill
Drill bit
High pressure boiler
Water, steam, or gas injection
Natural gas
Oil
Deep groundwater
Primary Production
Oil flows freely at first; more oil is pumped out when flow diminishes.
Secondary Production
Water is pumped into the deposit via injection wells to force out more oil.
Tertiary Production
Using the same injection well, additional oil is forced out by pumping steam, natural gas, or CO_2 into the deposit.

Why aren't pumpjacks needed during the initial retrieval of oil from a newly drilled well?

Conventional natural gas can be found in deposits that contain no oil (nonassociated gas reserves) or in deposits that contain both oil and natural gas (associated gas reserves). The extraction of natural gas is very similar to that of oil. Wells are drilled into the gas-containing formation, and natural gas flows freely up the well due to underground pressure, or it is pumped out (primary production). Secondary and tertiary methods that extract oil will also extract any natural gas that is trapped in oil, but these techniques are not used in nonassociated gas wells.

With each more intensive production method, the **energy return on energy investment (EROEI)** goes down, and the cost goes up due to the additional energy that must be expended to facilitate the process. Because fossil fuels are used to meet this higher energy need, the carbon footprint goes up, along with all the negative environmental consequences that are associated with the use of additional fossil fuels.

A comparison of the EROEI for the extraction of different energy sources helps producers and investors determine which energy sources are worth pursuing. An energy source that requires almost as much energy to tap into as it ultimately produces will be left untapped in favor of sources with higher EROEI values. In the 1950s when oil wells were first being tapped, the EROEI was close to 30:1, but as the "easy oil" was extracted with primary production and production transitioned to secondary and tertiary methods, the EROEI declined to the current average of around 10:1. (See Module 9.1 for more on EROEI.)

energy return on energy investment (EROEI) A measure of the net energy from an energy source (the energy in the source minus the energy required to get it, process it, ship it, and then use it).

4 THE TRADE-OFFS OF OIL AND NATURAL GAS

Key Concept 4: Oil and natural gas are used to create a variety of fuel and industrial products. Natural gas is considered the cleanest burning fossil fuel, but like oil, every step of acquisition and use damages the environment and impacts human health.

Modern society runs on oil and natural gas. Let's begin with a look at oil. It powers our vehicles and lubricates our machinery. It is the raw material for the plastics used to make football helmets and neonatal incubators; for the pesticides on our crops and the asphalt on our roads; for detergents, paints, solvents, and a host of other petroleum products or **petrochemicals**. The production of a desktop computer, for example, consumes 10 times the computer's weight in fossil fuels, mostly oil. A shortage like the one that occurred during the oil crisis of the 1970s can send a nation into a panic and change global politics. It is no wonder we are constantly in search of new supplies, but as they become scarcer, the extent to which we will go to extract them has some people worried.

Oil extraction has environmental consequences at every stage. To locate deposits of oil and natural gas, companies send seismic waves into the ground that bounce back to reveal the location of possible reserves. Doing this in ocean areas disorients marine wildlife; in 2008, ExxonMobil had to abandon its oil exploration off the coast of Madagascar because more than 100 whales had beached themselves, presumably because of these sonic exploratory methods.

Drilling also affects wildlife. Politicians have long debated the merits of drilling in the Arctic National Wildlife Refuge (ANWR), 20 million acres of protected wilderness. The USGS estimates that parts of ANWR located on the northern Alaskan coast could harbor between 5 and 16 billion barrels of technically retrievable crude oil and natural gas reserves. But drilling there is difficult; is expensive; and would seriously disturb the habitat of many polar species such as Arctic foxes, caribou, polar bears, and migratory birds. Because of this, drilling in some U.S. Arctic areas has been banned; in other areas, attempts to begin drilling for offshore oil have been abandoned because of the difficulty of operating in such extreme conditions. Still, in 2017, Congress added a provision to the Tax Cuts and Jobs Act of 2017 that would open the coastal plain of ANWR to oil drilling, requiring that the U.S. government offer at least two lease sales in 4 years' time. As of the end of 2019, the Bureau of Land Management was still preparing its "record of decision," a formal step that precedes lease sale preparations.

Another concern is oil spills—an unintended but frequent consequence of oil drilling and transport. Millions of gallons of oil are spilled from tankers, trains, pipelines, and ships annually. And while large-scale accidents like the 2010 *Deepwater Horizon* oil platform explosion and oil spill in the Gulf of Mexico are rare, they are

petrochemicals Distillation products from the processing of crude oil such as fuels or industrial raw materials.

John Mosier/ZUMApress.com

Boats hose down a massive fire on the oil rig *Deepwater Horizon*, 50 miles southeast of the tip of Louisiana in the Gulf of Mexico. The explosion killed 11 workers and released more than 200 million gallons of oil into the Gulf.

devastating. The *Deepwater Horizon* spill released close to 5 million barrels of oil, eventually covered 360 square miles of ocean, impacted roughly 1,100 miles of coastlines, and infiltrated and damaged coastal wetlands. The oil threatened the lives of many species and cost Gulf commercial fisheries and area tourism millions of dollars. Research shows that the impacts of that spill are still being felt in the salt marshes of the region as well as in deepwater regions near the well site. (See the *Global Case Studies* map at the end of this module for more on this oil spill.)

The hazards of oil are not limited to exploration, extraction, and transportation of oil. Burning fuels made from oil releases a variety of air pollutants that are linked to negative ecosystem and health impacts. It is also a major anthropogenic contributor to climate change. (See Module 10.1 for more on air pollution and Module 10.2 for more on climate change.) In addition, working with oil is a hazardous undertaking, even when the required precautions are taken. Workers in the petroleum chemical industry have much higher rates of cancer, rashes, heart disease, and various other health problems than the general population. This hazard extends to people living near oil refineries and petrochemical plants; rates of cancer, birth defects, headaches, and asthma in such areas exceed national averages.

Natural gas is an important alternative to oil for some fuel purposes, such as generating electricity and heat. It is made up of shorter-chain hydrocarbons (predominately methane, CH_4) and tends to have fewer impurities than oil, so when burned, fewer toxic pollutants are released. (Both oil and natural gas are cleaner than coal—the "dirtiest" fossil fuel to

ZHENGSHENG/Moment/Getty Images

Oil and natural gas are versatile fossil fuels, providing fuels and industrial products. Conventional deposits are accessed with wells, but many problems are associated with their acquisition and use.

burn.) And since natural gas has fewer impurities, it burns more efficiently and has a lower **carbon footprint**, releasing roughly 44% and 29% less CO_2 than coal or oil, respectively, per unit of heat energy. For this reason, natural gas is seen as the best fossil fuel to serve as a "bridge fuel"—one that we can use while we transition to more sustainable options. In addition, natural gas, like oil, is also the starting material for a wide variety of industrial products such as fertilizers, paints, plastics, and pharmaceuticals.

However, like oil, exploration for and extraction of natural gas are environmentally damaging. And because natural gas is highly flammable, it is dangerous to handle and ship. Shipping is accomplished with pipelines that are expensive to build and maintain. Methane leaks—from pipelines, from gas and oil wells—also significantly contribute to climate change; in the first 20 years after release, methane is almost 90 times more potent a greenhouse gas than CO_2. (After that, methane's impact relative to CO_2 decreases because CO_2 stays in the atmosphere longer.) And of course, though methane is produced today through some natural processes (such as bacterial decomposition of trash in landfills), the natural gas found in fossil fuel deposits is a nonrenewable resource with a limited and finite supply. **INFOGRAPHIC 4**

Jun Walkawa/Yomiuri Shimbun/AP Images

The sight of pedestrians wearing masks is not uncommon on the smoggy streets of Beijing where air pollution alerts are regularly issued.

carbon footprint The amount of CO_2 (and other greenhouse gases that contribute to climate change) released to the atmosphere by a person, a company, a nation, or an activity.

INFOGRAPHIC 4 THE TRADE-OFFS OF OIL AND NATURAL GAS

Oil and natural gas are versatile fossil fuels, providing fuels and industrial products, but many problems are associated with their acquisition and use.

OIL

ADVANTAGES	DISADVANTAGES
• Energy-rich fuel • Provides variety of liquid fuels that meet needs for vehicles • No ash produced when burned (no disposal issues as with coal) • Can transport via pipeline or vehicle (rail, truck, or ship) • Raw materials for a wide variety of industrial products	• Nonrenewable, finite resource • Geopolitical tensions from unequal distribution of reserves • Air pollution, including greenhouse gas emissions (extraction, processing, and burning) • Water pollution (extraction, processing, and burning) • Habitat destruction (extraction and spills) • Biodiversity loss (exploration, extraction, and spills) • Dangerous to ship • Occupational and community health hazard

NATURAL GAS

ADVANTAGES	DISADVANTAGES
• Energy-rich fuel • Versatile fuel: generates electricity, heating and cooking, vehicle fuel • Lowest air pollution and greenhouse gas emissions of the fossil fuels • No ash produced when burned (no disposal issues as with coal) • Raw material for a wide variety of industrial products	• Nonrenewable, finite resource • Methane is a potent greenhouse gas; produces CO_2 emissions when burned • Water pollution (extraction and burning) • Habitat destruction (extraction and spills) • Biodiversity loss (exploration, extraction, and spills) • Hazardous chemical • Difficult to ship

? Why is natural gas but not oil often touted as a "bridge fuel"—something to use while we transition to more sustainable options?

5 UNCONVENTIONAL OIL AND NATURAL GAS

Key Concept 5: Shale oil and natural gas deposits that can't be accessed with conventional wells can be retrieved by fracking; strip mining is used to extract tar sands oil.

As conventional reserves become depleted, we have turned more and more to unconventional sources of oil and natural gas such as the Bakken and Marcellus Shale formations. As mentioned earlier, these deposits were not previously accessible with conventional wells because the oil or natural gas, though abundant, is widely dispersed within the rock formations where it is found. But those reserves can be tapped with **fracking (hydraulic fracturing)**.

Fracking involves drilling thousands of feet straight down, deep into ancient rock beds, and then turning the drilling apparatus to drill horizontally into the gas- or oil-bearing rock formation. As drilling progresses, steel pipe casings are inserted into the wellbore, and the upper reaches of that pipe (from the surface down as far as 2,000 feet) are encased in several inches of concrete to protect groundwater and soil from contamination. After these regions of pipe are secure and protected, vertical drilling continues until the rock containing the oil or natural gas is reached. The angle of the drill bit is then slowly turned until the drilling is proceeding in a horizontal direction; this is repeated again and again to create horizontal wells in many directions (like spokes on a wheel). Horizontal wells may reach 1 mile or more in length. Steel pipes are inserted into the horizontal wellbores, and once all casings are in place, explosive charges are detonated all along the horizontal sections of the well to create holes in the pipe using a "perforating gun." This will allow the oil or natural gas to flow into the pipe. Next, between 4 and 5 million gallons of water containing a mix of sand and chemicals (some of them toxic) are injected into the well under high pressure. The water mixture enters small natural fractures in the rock, flushing out the oil or natural gas in the deposit; the sand (or other "proppant" material) holds the fractures open to aid in the retrieval of the oil or natural gas. Chemical additives serve as lubricants, corrosion inhibitors, or disinfectants. The water mixture returns to the surface, followed by the oil and/or gas as well as natural water that was trapped in the rock formation. **INFOGRAPHIC 5**

fracking (hydraulic fracturing) The extraction of oil or natural gas from dense rock formations by creating factures in the rock and then flushing out the oil or gas with pressurized fluid.

Currently, the United States leads the world in fracking, recovering enough natural gas and oil to change the global market for these fossil fuels. For example, the increased production of natural gas via fracking caused prices to fall, which made it a more attractive fuel for U.S. electricity production, significantly reducing the amount of coal used in this country. Global oil prices also fell as oil from the fracking oil fields of North Dakota and Texas entered the market.

Tar sands mining is the most destructive and energy-intensive way to acquire oil. Huge swaths of Canadian boreal forest have been destroyed and nearby bodies of water polluted by this process.

Peter Essick/Cavan/Alamy

INFOGRAPHIC 5 HOW IT WORKS: FRACKING (HYDRAULIC FRACTURING)

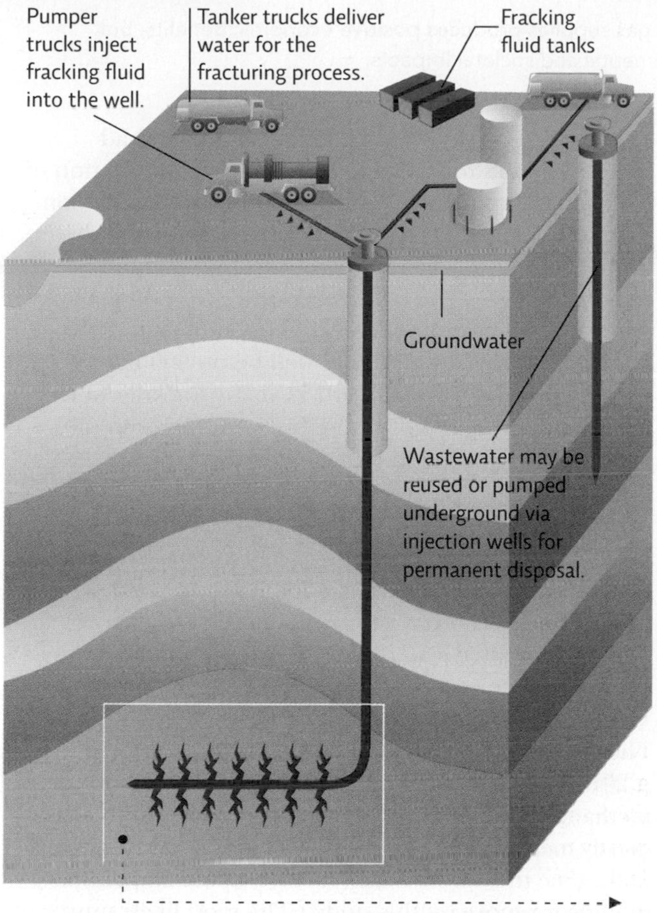

Pumper trucks inject fracking fluid into the well.
Tanker trucks deliver water for the fracturing process.
Fracking fluid tanks
Groundwater
Wastewater may be reused or pumped underground via injection wells for permanent disposal.

Suppose you are a homeowner. An oil company approaches you, offering to lease your land to install a fracking well. What questions would you have for the company as you consider what to do?

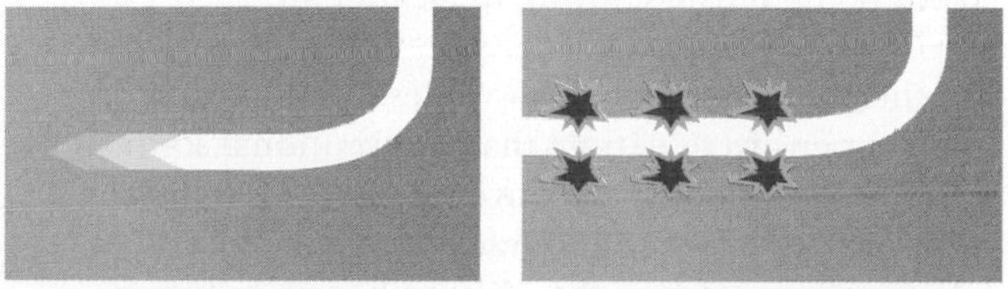

1. For deep deposits, a well is drilled down to gas-bearing rock and then extended horizontally.

2. A perforating gun blasts holes in the pipe; explosive charges then enlarge natural rock fractures.

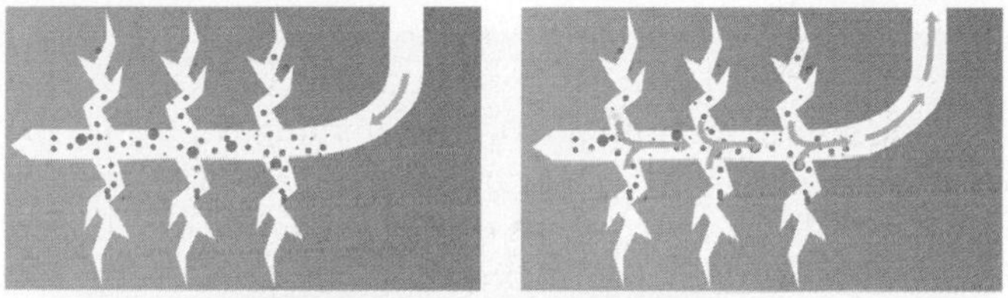

3. A slurry of sand, water, and chemicals is pumped into the rock, enlarging the fractures and holding them open so oil or gas can escape.

4. Natural gas can now flow through the more permeable, fractured rock and be extracted.

While productive, the oil and gas shale deposits aren't the only types of unconventional reserves; there's also crude *bitumen*—a heavy black oil that is often trapped in sticky, dense conglomerations of sand or clay known as **tar sands** (also known as **oil sands**). Alberta, Canada, is home to the second-largest tar sand reserve; with around 160 billion barrels of economically retrievable oil, it is the third-largest proven oil reserve in the world and represents 10% of total world proven reserves. In 2018, Alberta produced 2.9 million barrels of synthetic crude oil per day from the tar sands—96% of it shipped to the United States. Venezuela was recently recognized as having the largest tar sands deposits in the world. With 300 billion barrels of recoverable oil, its oil reserves surpass that of Saudi Arabia.

Oil is extracted from Canadian tar sands by strip mining that begins with the removal of the boreal forest above the deposit. The sticky soil is dug out and washed with copious amounts of water to separate the oil from the soil. Deeper deposits in the tar sands can be accessed with *in situ* recovery—a method that injects steam through a wellbore drilled into the deposit to melt the hardened bitumen and force it out with the rising steam (through a second well), an energy-intensive method.

Once extracted, the thick bitumen must be further processed, first by washing with copious amounts of water to separate the bitumen from the soil. It is then sent to a refinery to undergo a process known as coking during which the bitumen is heated to high temperatures to produce a lighter substance suitable for shipping. A hazardous byproduct (petcoke) produced in this process is typically stored outdoors in large piles. Dust from these piles (similar to coal dust but more hazardous because it consists of finer particles that better penetrate the lungs) can blow to nearby areas—if the refinery is close to communities, those residents can be exposed to this hazardous material. These extra processing steps are energy-intensive, giving tar sands oil the lowest EROEI of all fossil fuels (as low as 3:1).

tar sands (oil sands) Sand or clay formations that contain a heavy-density crude oil (crude bitumen); extracted by surface mining.

6 THE TRADE-OFFS OF UNCONVENTIONAL FOSSIL FUELS

Key Concept 6: Extracting unconventional oil and natural gas supplies produces positive economic benefits, but fracking and tar sands mining have serious negative environmental and societal impacts.

The upsides of the Bakken boom are not difficult to see or imagine. For one thing, such a mammoth supply of domestic oil means, for the time being at least, less reliance on foreign oil. For another, the boom has stimulated a once-declining regional economy. And more natural gas from shale formations also means less coal is being burned, giving us cleaner air.

But there are significant downsides. Because they are more expensive to extract than conventional sources, it only pays to pursue unconventional fossil fuels when prices are high. When the fracking boom took off, oil was selling at more than $100 a barrel, but soon thereafter, prices started to drop—in large part due to increased oil supplies. U.S. oil production had climbed thanks to fracking, but OPEC (Organization of Petroleum Exporting Countries) also kept production high. (Some analysts speculated this was done to drive down the price of oil and force fracking operations out of business.) The break-even price for OPEC oil is around $30 a barrel; for fracking operations in 2015, it was twice that. With oil prices well below $60 per barrel (as low as $26 in February 2016), many fracking wells were shut down and jobs were lost—the Bakken oil boom was beginning to falter.

Mike Greenlar/The Post Standard/AP Images

Fred Mayer of Candor, New York, uses a charcoal grill lighter to ignite water running from his kitchen faucet. His water has been contaminated with methane ever since a fracking operation began nearby.

New technologies deployed in 2016 increased productivity per well, making fracking more cost competitive, and fracking production began to climb again. However, this comes at the expense of jobs—automated systems are improving efficiency, decreasing the cost of fracking, and replacing workers.

The environmental costs of fracking are a concern. A 2013 study by Duke University researchers found that methane contamination of drinking water (natural gas is mostly methane) was highest in areas closest to fracking wells. (See the *Science Literacy* activity at the end of this module for more on this study.) The most likely source of the contamination is leaking pipes that deliver the gas to the surface or from improperly handled fracking water. Likewise, a 2017 study found a strong correlation between proximity to a fracking well and low birth weight—babies born to mothers who lived within 1 km (0.6 miles) of a fracking operation were 25% more likely to suffer from low birth weight and other health problems than those who lived more than 3 km away. Improved drilling methods and stricter fracking water storage requirements are decreasing those risks, industry experts say.

Fracking has also been identified as one of the major sources of the recent uptick in atmospheric methane emissions via leaks, maintenance releases, and purposeful venting in the initial stages of production from a well. Methane emissions had leveled off in the first part of the 21st century, but since 2008, they began climbing rapidly—this rate of increase will make it more difficult to achieve the international target of holding global warming to no more than 2 °C above preindustrial temperatures. (See Module 10.2.) Despite this concern, in 2019, the U.S. EPA proposed a rollback of regulations that limited methane releases from the oil and gas industry.

The high water footprint of fracking is also a concern. The volume of water required to frack a well is huge—

around 5.5 million gallons were used for each Bakken well in 2016. And Bakken wells use less water than those in other regions—that same year, each well in the Eagle Ford Formation required 8.8 million gallons; wells in Texas' nearby Permian Basin required 11 million gallons. In water-poor areas, the volume of water needed can stress local water supplies.

The production of toxic wastewater can also increase the water footprint if it contaminates the surrounding land area or surface or groundwater supplies. During fracking, much of the water injected into the ground resurfaces prior to the oil or natural gas, bringing along the injected chemicals as well as additional water known as *produced water*—naturally occurring water that can contain contaminants picked up from the rock, such as high concentrations of salt and naturally occurring radioactive materials.

Fracking wastewater contains a variety of hazardous chemicals known to be toxic, carcinogenic, or teratogenic (causing birth defects) to humans and other species. In the past, wastewater was simply disposed of by spraying it on roadways or other land surfaces. Though land application is no longer allowed (due to its demonstrated toxicity to plants), millions of gallons are spilled each year—from broken pipes or leaking storage ponds and tanker trucks. Between 2007 and 2016, roughly 4,000 wastewater spills were reported in North Dakota alone.

To deal with this wastewater, operators typically pump it back underground. In some cases, it is used in secondary production of a conventional oil well to enhance oil recovery; it may also be reused in nearby fracking wells. Much of it, however, is pumped into new wells drilled specifically for disposal (called *injection wells*). But these injection wells have their own environmental issues, most notably the triggering of earthquakes. For example, the Oklahoma City area experienced 31 earthquakes (2.0 magnitude or greater) in a 1-week period in July 2014. To put this into perspective, whereas the state experienced only 6 earthquakes between 2000 and 2008, it experienced 6,345 earthquakes between 2010 and 2013. Research published by Katie Keranen of Cornell University linked this "earthquake swarm" to nearby wastewater injection wells.

Another problem is that the amount of oil being fracked has outpaced the existing capacity to transport it. Without a pipeline to rely on, oil companies have resorted to using railways. But the sheer volume of oil to be transported is bogging down the rail system nationwide, and shipping tight oil this way is a hazardous

Ryan Remiorz/The Canadian Press/AP images

Transporting any type of oil is dangerous, but tight oil tends to be more volatile and explosive than conventional crude. In 2013, a runaway train carrying tight oil derailed in Lac-Mégantic, Quebec, destroying much of the town and killing 47 people.

undertaking. Tight oil contains natural gas dispersed throughout, making it more volatile and explosive than crude oil; it is similar to jet fuel in its combustibility. In the summer of 2013, the brakes failed on a parked, unmanned train carrying Bakken oil. The runaway train sped downhill, derailing in Lac-Mégantic, a tiny village in Quebec; the resulting crash caused an explosion when the highly combustible fuel ignited, leveling more than 40 buildings and claiming 47 lives.

Fracking operations also present a disadvantage not commonly seen with conventional drilling—the ability to drill into formations under private property without the permission of the landowner. In the United States, many landowners do not own the mineral rights to their land; mineral rights are often owned by corporations that have the right to set up drilling rigs on or near a landowner's property and frack for oil or natural gas without landowner permission. Even if a landowner does own the mineral rights to his or her property and declines to let a fracking operation set up, if the majority of neighboring landowners sign on, the reluctant landowner can be overruled.

Other unconventional sources of energy also have unique drawbacks. Tar sands extraction is the most energy- and water-intensive way to extract oil and produces 20% more greenhouse gas emissions than conventional oil and gas production. David Schindler, an ecologist at the University of Alberta in Canada, has spent years studying the effects of tar sands extraction on the water quality of the Athabasca River, which cuts through the heart of one of Alberta's biggest tar deposits and mining sites. His research showed that concentrations of toxic compounds were significantly higher in the river's tributaries located downstream of tar sands mining than they were upstream.

Processing tar sands oil is also more energy-intensive than conventional crude oil and produces more hazardous wastes—huge lakes of acidic and toxic wastewater that oil companies store at the mining sites (by-products of the refining process) are so large that they can be seen on satellite images from space. The thick crude bitumen is also difficult to ship; it will flow only in heated pipelines. The construction of the Keystone XL pipeline that will transport tar sands oil south to the Gulf Coast refineries of Texas was strongly opposed by many environmental and citizen groups (leaks and spills from other pipelines have caused significant environmental damage). It was approved by the U.S. Congress in 2015 but vetoed by President Obama. The pipeline was resurrected and approved by President Trump in 2017, though construction has been delayed by court battles.

Another controversial pipeline, the Dakota Access Pipeline, was opposed by the Standing Rock Sioux Tribe whose months-long 2016 protest grew to include thousands as they tried to prevent construction of the section that would pass near their lands, citing water quality concerns and worries it would damage sacred grounds. Soon after he took office in 2017, President Trump signed an executive order to end the environmental review of the contested leg of the pipeline; construction was completed later that year, and the pipeline now carries 500,000 gallons of Bakken crude per day. **INFOGRAPHIC 6**

INFOGRAPHIC 6 THE TRADE-OFFS OF UNCONVENTIONAL FOSSIL FUELS

Unconventional sources of fossil fuels will greatly increase supplies, but because they are harder to extract, they come with significant environmental and economic trade-offs.

ADVANTAGES	DISADVANTAGES	
• Energy-rich fuels • Can be used in the same way that conventional oil or natural gas is used • Extends worldwide supply of oil and natural gas • Abundant U.S. supply: reduces dependence on imports and stimulates local economies	• Nonrenewable, finite resource • Same air pollution when burned as with conventional oil and natural gas	
	FRACKING FOR SHALE GAS/OIL	**MINING FOR TAR SANDS OIL**
	• High water inputs • Potential for pollution of well water • Can occur without permission of property owner • Wastewater issues: hazardous and its disposal is linked to earthquakes • Dangerous to transport (tight oil) • More expensive to extract than conventional sources	• High water and energy inputs • Extensive surface water pollution • Extreme habitat destruction • Difficult to transport; pipelines must be heated • Highest carbon footprint of any fossil fuel extraction method • More expensive to extract than conventional sources

? Do the advantages of fracking or tar sands mining outweigh their disadvantages? Explain your reasoning.

7 FOSSIL FUELS AND THE FUTURE OF ENERGY

Key Concept 7: Achieving energy independence or security will require an eventual transition away from fossil fuels; conservation efforts and diversification of energy sources can help in the pursuit of these goals.

It's tempting to think of the oil made available by fracking technology as so plentiful that we can stop worrying about running out of oil for the next several lifetimes and be well on our way to achieving **energy independence**. The numbers, after all, are mind-boggling: We have more than 700 billion barrels of tight oil in Bakken, Eagle Ford, and other known reserves—and that's just in the United States. But just a tiny fraction of the total oil held in these unconventional deposits is recoverable—somewhere between 1% and 2% in the Bakken and Eagle Ford shales. "At the high end of the estimates, predicted production from Bakken and Eagle Ford together amounts to perhaps a two-year oil supply for the United States at 2011 consumption rates," says Raymond Pierrehumbert, a professor and geophysicist at the University of Chicago. "That's significant, but not a game changer."

Because fracking wells can access only a small part of the oil deposit that surrounds them, many, many wells are needed to tap an entire deposit. This diminishes the EROEI because creating each new well requires an expenditure of energy. In addition, the productivity of any given well drops off rapidly after the first couple of years and then trails off into nothing, slowly, over decades. "Tight oil is headed for a Red Queen's Race," says Pierrehumbert. "You have to keep drilling and drilling and drilling just to keep your production in the same place." (This comparison comes from Lewis Carroll's novel *Through the Looking-Glass*. When Alice commented that they seemed to be running hard but going nowhere, the Red Queen told her "it takes all the running you can do to keep in the same place.")

There is no denying that our economy and lifestyle are dependent on access to affordable energy, or **energy security**. Yet the global oil market is highly volatile and there are many reasons that our energy supplies might become unreliable or unaffordable. These include dwindling supplies,

energy independence Meeting all of one's energy needs without importing any energy.

energy security Having access to enough reliable and affordable energy sources to meet one's needs.

Ken Cedeno/Corbis/Getty Images

A fracking operation along U.S. Highway 85 just south of Watford City, North Dakota.

increasing demand, dependence on energy imports from politically unstable countries or those that might stop exporting energy resources to us for political reasons, competition from other countries, and a cartel or monopoly increasing prices or decreasing oil supplies. Many times these factors are hard to predict. In early 2020, the COVID-19 pandemic triggered sudden and drastic declines in manufacturing and transportation. As a result, the price of oil steadily fell as supply outpaced demand. Then in the midst of this global health emergency, an oil-production war between OPEC and Russia escalated (Saudi Arabia and Russia increased production in a fight for market share) and the price of oil plummeted, shaking the entire industry, especially high-priced production methods such as fracking and tar sands mining. Some analysts predict heavily leveraged operations (those heavily financed with loans) may not make it through the 2020 oil crisis. **INFOGRAPHIC 7**

To be sure, how to best meet our future energy needs is a *wicked problem*—we need abundant, reliable sources of energy to power our societies, but many of our energy choices come with significant problems. (See Module 1.1 for more on wicked problems.) Replacing fossil fuels with a cleaner, renewable energy source would mean we no longer need to worry about running out of a finite energy resource, and it would certainly cut down on air pollution, climate change, and myriad other environmental problems. A lot is at stake in this wicked problem—huge investments and even bigger profits, the health and wellbeing of human communities and ecosystems, and the need to keep society running smoothly without facing an energy shortfall. So how do we increase our energy security and reduce our dependence on fossil fuels so that we can avoid using dangerous extraction methods and polluting the planet we call home? There are many possibilities, but two important strategies are diversification and conservation.

Every sustainable energy option at our disposal comes with advantages and disadvantages. No single energy source can replace fossil fuels, but together, the wide variety of energy sources at our disposal can (see Chapter 11). We can focus efforts on acquiring a variety of domestic sources of energy (including alternative fuels such as biofuels) and importing energy from multiple suppliers. We can also shift our mobile fleets from gasoline, diesel, or natural gas to electric vehicles that run on batteries charged with non-fossil energy sources such as wind, solar, or nuclear power. At the same time, we can reduce energy needs through increased conservation and energy efficiency. Iceland, for instance, is increasing its energy security by focusing on conservation and energy independence, such that it

INFOGRAPHIC 7 ENERGY INDEPENDENCE AND SECURITY

Energy security is achieved when a nation has an affordable and reliable (uninterrupted) form of energy. If energy needs are met with national sources, the country has energy independence.

ENERGY INDEPENDENCE

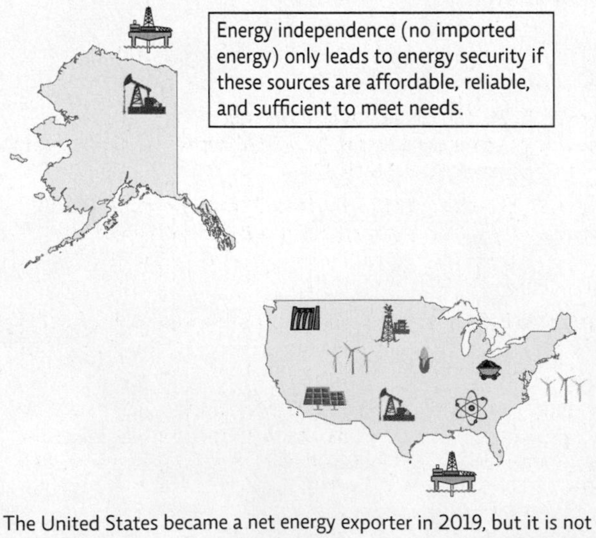

Energy independence (no imported energy) only leads to energy security if these sources are affordable, reliable, and sufficient to meet needs.

The United States became a net energy exporter in 2019, but it is not technically energy-independent because some energy, mostly crude oil, is still imported.

ENERGY SECURITY

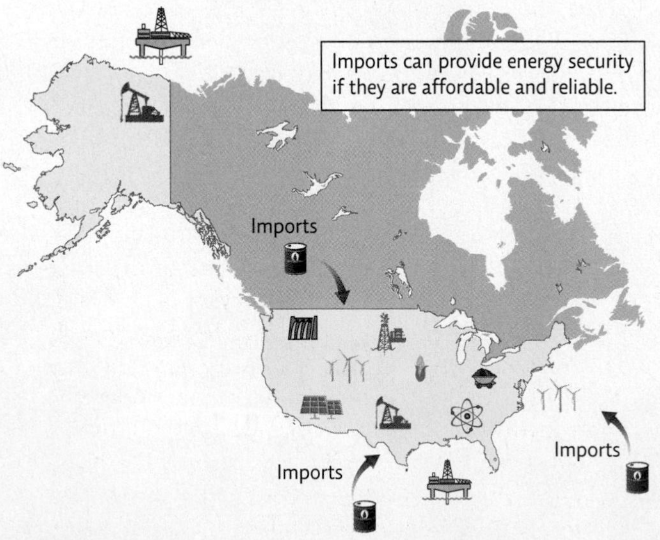

Imports can provide energy security if they are affordable and reliable.
Imports
Imports
Imports

In 2019, 77% of U.S. energy needs were met with domestic sources. The remaining 23% was predominantly imported oil.

How could the United States increase its chance of becoming energy independent if it cannot produce enough oil to meet its needs?

Andrew Cullen/Reuters/Newscom

The need for housing during the fracking boom led to a building boom, but then the oil bust came. In 2016, this apartment complex in Williston, North Dakota, offered several months of free rent, trying to entice workers who remained. If production picks up, the units may once again be in hot demand.

will meet all of its energy needs without imports by the year 2050—it already produces all its electricity with renewable energy. And in 2016, local renewables met about 85% of Sweden's energy needs; the country vows to be among the first fossil-free nations in the world.

In April 2014, the Bakken oil fields of western North Dakota reached a milestone: They yielded up their billionth barrel of crude oil. In 2018, as oil prices rose and production costs declined, the downturn in the shale fields of North Dakota and Texas began to reverse. Production increased and jobs were once again being filled. This too will be temporary—perhaps a year or two, perhaps a decade or two. Only time will tell how many more barrels remain to be harvested and how long Bakken oil will continue to flow. In the meantime, the greatest gift of Bakken may be the time it buys us to develop other domestic energy sources. "It will only do us good if we use this transitional period wisely," says Pierrehumbert. "We're in for a hard landing if we don't use our current prosperity to pave the way for a secure energy and climate future."

Select References:

Hall, C. A., et al. (2014). EROI of different fuels and the implications for society. *Energy Policy, 64,* 141–152.

Howarth, R. W. (2019). Ideas and perspectives: Is shale gas a major driver of recent increase in global atmospheric methane? *Biogeosciences, 16*(15), 3033–3046.

Jackson, R. B., et al. (2013). Increased stray gas abundance in a subset of drinking water wells near Marcellus shale gas extraction. *Proceedings of the National Academy of Sciences, 110*(28), 11250–11255.

Johnston, J. E., et al. (2016). Wastewater disposal wells, fracking, and environmental injustice in southern Texas. *American Journal of Public Health, 106*(3), 550–556.

Kelly, E. N., Schindler, D. W., et al. (2010). Oil sands development contributes elements toxic at low concentrations to the Athabasca River and its tributaries. *Proceedings of the National Academy of Sciences, 107*(37), 16178–16183.

Keranen, K. M., et al. (2014, July 25). Sharp increase in central Oklahoma seismicity since 2008 induced by massive wastewater injection. *Science, 345*(6195), 448–451. doi:10.1126/science.1255802

Kondash, A. J., et al. (2018). The intensification of the water footprint of hydraulic fracturing. *Science Advances, 4*(8), eaar5982.

GLOBAL CASE STUDIES CONSEQUENCES OF OIL AND NATURAL GAS EXTRACTION

Oil and natural gas are valuable fossil fuels that are important to modern society, but their acquisition imposes a significant environmental burden. Here are some examples of problems caused by the acquisition of these fossil fuels.

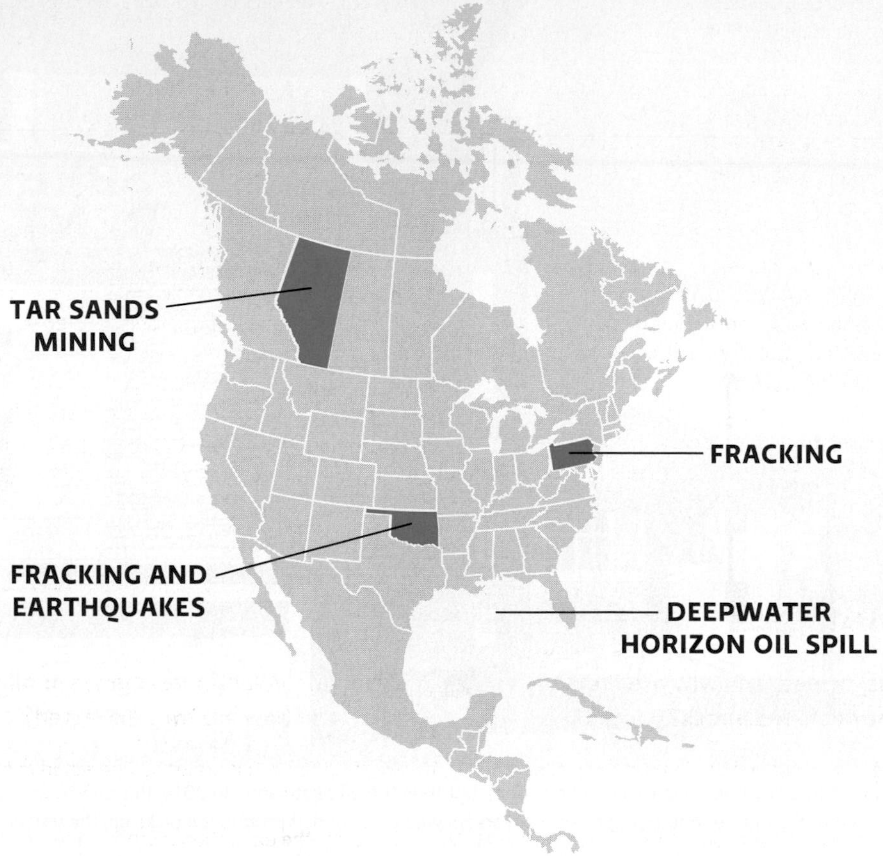

BRING IT HOME

PERSONAL CHOICES THAT HELP

Environmental and health concerns that surround our acquisition and use of fossil fuels continue to grow as we dig deeper and use more extreme methods to obtain oil and natural gas, highlighting our dependence on nonrenewable energy resources. By decreasing our oil and natural gas use, we can reduce the pressure on oil companies to pursue sources of oil that have a greater potential for environmental damage such as deepwater drilling or oil and natural gas deposits like those in the tar sands and shales.

Individual Steps

- Minimize your fuel use when driving by planning ahead to condense shopping trips and errands and reduce total miles driven and by parking as far as you can from your entrance and getting some exercise instead of wasting gas. Try not driving: See where you can safely walk, bike, or use public transport.
- Reduce your use of disposable plastics like water bottles and single-serve food containers and always recycle plastics when possible.
- Turn down your thermostat in the winter to reduce your energy use.
- Work with your school's administration to encourage public transportation and increased bike usage on your campus.
- If you live in a state where fracking occurs, research fracking regulations in your state and voice your opinion about existing or proposed regulations.

Group Action

- Organize a carpool system in your community to reduce the number of single-passenger car trips.
- Organize a screening of a documentary on energy such as *Tipping Point: The Age of the Oil and Sands, Gasland* or *Gasland 2,* or *FrackNation*. Discuss the positions presented and evaluate the evidence given in support of claims that are made. Do you detect any bias? Explain.

Jan Sonnenmair/Cavan Images

REVISIT THE CONCEPTS

- Fossil fuels (oil, natural gas, and coal) form when organic matter is buried and subjected to high temperature and pressure. Fossil fuels are important energy sources, but because we use them much faster than they form, they are considered nonrenewable resources.

- Oil and natural gas deposits vary in their recovery potential. The first deposits to be tapped became known as conventional reserves; they are extracted using oil or gas wells, but supplies are diminishing. However, unconventional deposits (those that cannot be accessed with conventional wells) also exist.

- Conventional reserves are tapped using oil and gas wells via three stages of production (primary, secondary, and tertiary). Each new stage requires more resources to implement (such as water and energy), increasing the water footprint and decreasing the energy return on energy investment (EROEI).

- Oil and natural gas are used to create a variety of fuel and industrial products. Natural gas is considered the cleanest burning fossil fuel, but like oil, every step of acquisition and use damages the environment and impacts human health. Environmental problems that arise from extraction, processing, shipping, and burning of fossil fuels include climate change, air and water pollution, and habitat destruction.

- Shale oil and gas deposits are vast but cannot be accessed with conventional wells. They can, however, be retrieved by fracking. Oil from tar sands, another unconventional deposit, is extracted using strip mining or with *in situ* steam injection wells.

- Extracting unconventional oil and natural gas supplies increases supplies and produces positive economic benefits, but the methods of extraction (fracking and tar sands mining) cause even more environmental and societal problems than conventional oil and natural gas extraction.

- Achieving energy independence or security will require an eventual transition away from fossil fuels; conservation efforts and diversification of energy sources can help in the pursuit of these goals.

ENVIRONMENTAL LITERACY Understanding the Issue

1 How are fossil fuels formed, and why are they considered nonrenewable resources?

1. Compare and contrast oil and natural gas in terms of physical properties and formation.
2. If oil and natural gas are formed from once-living organisms, why are they considered to be nonrenewable resources?

2 How are oil and natural gas reserves classified, and where are they found?

3. Distinguish among proven, conventional, and unconventional oil and natural gas reserves.
4. Why is oil the cause of geopolitical tensions around the world?

3 How are conventional oil and natural gas extracted?

5. Describe the process of conventional oil extraction, including primary production, secondary production, and tertiary production.
6. What happens to the energy return on energy invested as production transitions from primary to secondary to tertiary methods?

4 What are the trade-offs of acquiring and using conventional oil and natural gas?

7. How are oil and natural gas used in modern society?
8. Compare the environmental problems produced from extracting oil to those produced from burning oil.
9. Compare natural gas to oil in terms of advantages and disadvantages.

5 What unconventional sources of oil and natural gas exist, and how are they extracted?

10. Why is the United States so interested in extracting tight oil from shale deposits?
11. Consider the extraction methods used for these oil sources: tight oil, tar sands oil, and conventional oil. Rank them from lowest to highest EROEI. Do the same for the potential for environmental damage.
12. Explain the process of fracking. Why is fracking used for some oil and natural gas reserves and not others?
13. Explain the methods used to extract Canadian tar sands oil.

6 What are the trade-offs of pursuing unconventional oil and natural gas sources?

14. Evaluate the water footprint of fracking.
15. How is wastewater from fracking operations disposed of, and what problems are linked to its disposal?
16. Explain the environmental drawbacks of extracting tar sands oil.
17. Explain the potential risks associated with shipping light oil.

7 What obstacles stand in the way of the United States (or any nation) achieving energy independence or security?

18. What role do conservation and diversification of energy sources play in achieving energy security or independence?
19. Distinguish between the actions a country might take when pursuing energy security versus energy independence.

SCIENCE LITERACY Working with Data

Methane (CH_4), the main component of natural gas, can also normally be found in upper levels of the soil because it is produced by bacteria and can accumulate to high levels over time in some areas. Ethane (C_2H_6), however, is only created under high temperature and pressure, such as might occur deep underground at the site of shale gas deposits. Robert Jackson and colleagues suspected that methane contamination in water from wells in northeastern Pennsylvania might be linked to nearby fracking gas wells. For their study, they asked the question "Does being close to fracking gas wells increase the likelihood that the water will contain methane or ethane contamination?" They measured the amounts of methane and ethane in 141 water wells close to and far away from fracking wells. Look at the graphs below and answer the following questions.

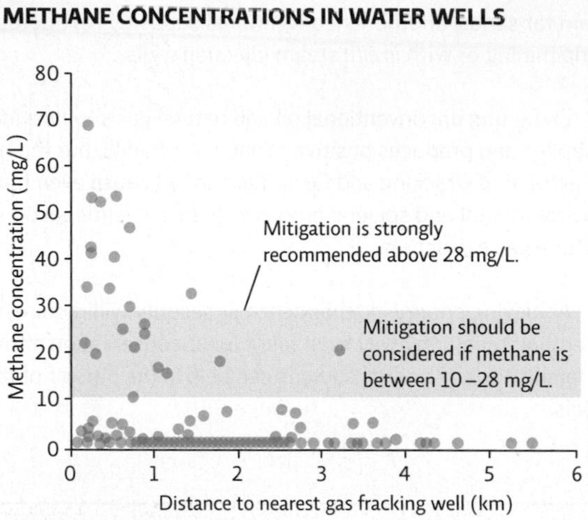

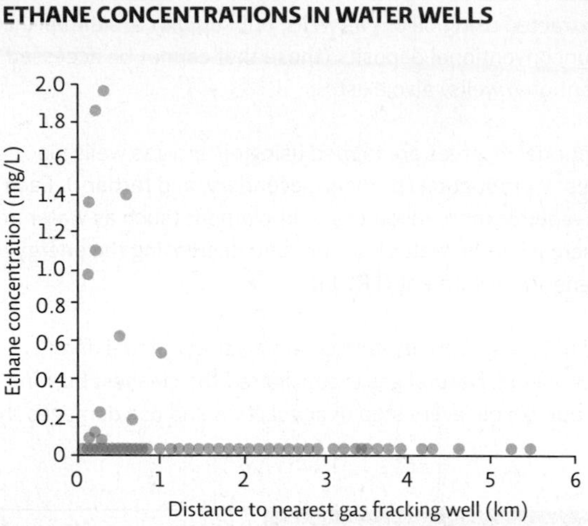

Interpretation

1. What was the hypothesis that these researchers tested in this experiment?
2. For each chemical (methane and ethane), identify the relationship between its level in well water and the water well's proximity to fracking wells. Do these relationships represent causation or correlation? Explain.
3. How many water wells had levels of methane that fell within the range where mitigation should be considered? How many fell within the range where mitigation is strongly recommended?

Advance Your Thinking

4. Do all wells that are very close to a fracking well (within 0.5 km) show evidence of methane or ethane contamination? (Are there any close-by wells with no contamination?) Does this negate the conclusion that methane and ethane contamination are due to fracking in the area?
5. Why might methane be high in water even in areas that are not close to fracking wells?
6. Why did researchers need to measure ethane in water wells to strengthen their conclusion that the water contamination is linked to fracking activity?

INFORMATION LITERACY Evaluating Information

The United States produces a lot of oil but still imports a lot from Canada, our biggest source of imported oil. For the most part, this is tar sands oil from Alberta. Transport of the mining products is accomplished through the transcontinental Keystone pipeline that currently runs from Alberta into the United States, but it does not continue all the way to refineries on the Gulf Coast. Oil companies have lobbied for years for permission to expand the pipeline by building the Keystone XL to transport oil from Canada to Texas, a request that was denied by President Barack Obama but approved by his successor, President Donald Trump.

Search for information about the Keystone XL pipeline and other proposed ways to ship tar sands crude oil. Evaluate the webpages or articles you read and answer the following questions:

1. Visit or read at least three websites or newspaper articles that advocate for the construction of the Keystone XL pipeline and three that oppose it. For each website or article, answer the following questions about the organization or authors responsible for the information you read:
   a. Who runs the website or wrote the article? Do the credentials of this individual or group make the person or group reliable or unreliable? Explain.
   b. Does this individual or group have a clear and transparent agenda? Why or why not?
   c. Do you detect any bias or logical fallacies? If so, identify them.
   d. Is this a reliable information source? Explain.
2. Briefly summarize what you believe are the most important pros and cons of the Keystone XL pipeline. Based on your analysis of the information you gathered, if you were President, would you have approved the Keystone XL pipeline or would you have rejected it? Justify your conclusions by explaining which arguments support your position and why they are valid arguments. Also explain why the counterarguments were insufficient to support the alternative conclusion.

Additional study questions are available at Achieve.macmillanlearning.com. Achieve

CHAPTER 10

Air Quality and Climate Change

Air pollution can come from a variety of sources with varied consequences that range from health problems to environmental damage to climate change. Addressing air pollution will benefit our health and the environment.

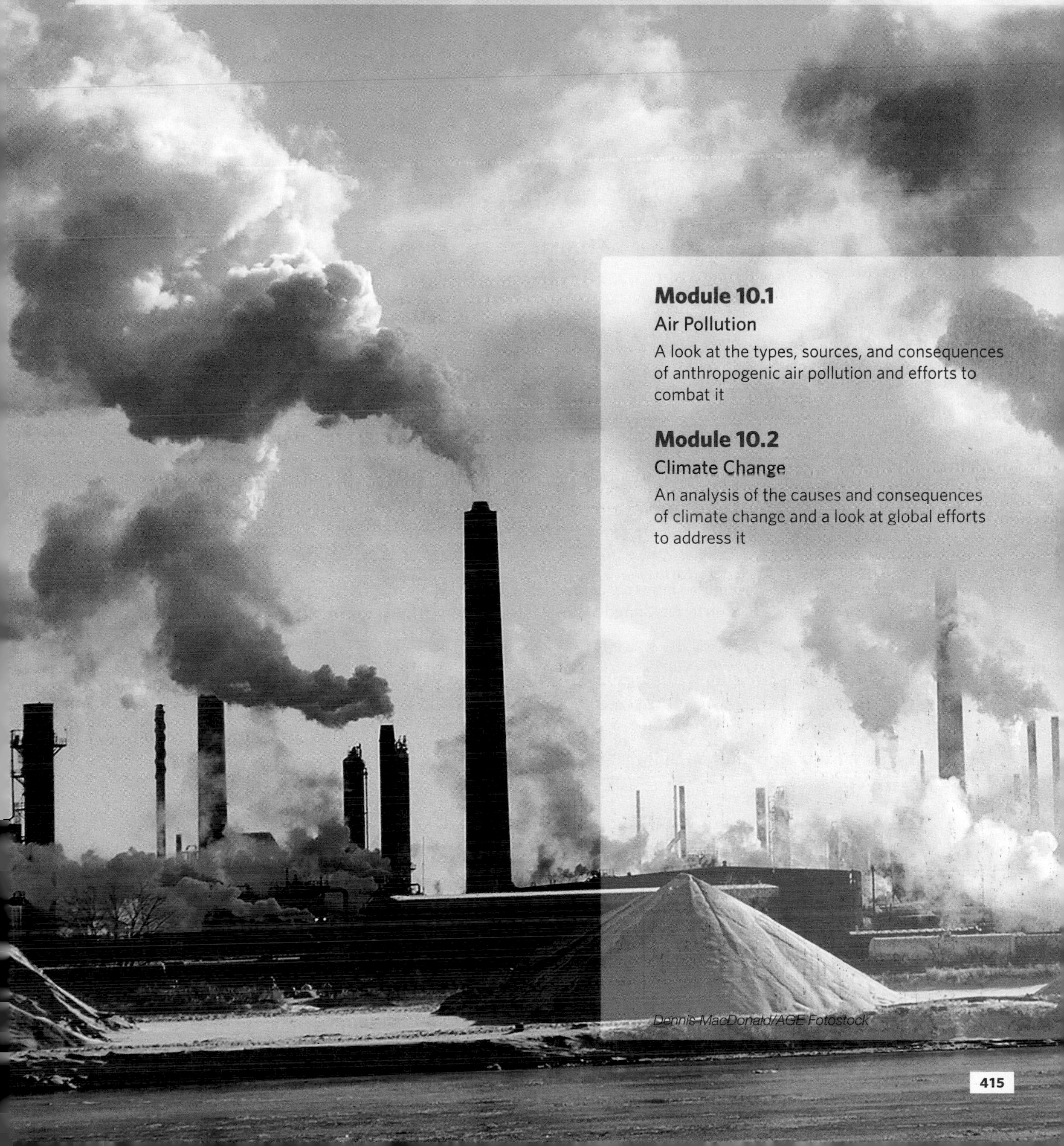

Module 10.1

Air Pollution

A look at the types, sources, and consequences of anthropogenic air pollution and efforts to combat it

Module 10.2

Climate Change

An analysis of the causes and consequences of climate change and a look at global efforts to address it

Dennis MacDonald/AGE Fotostock

Module 10.1 Air Pollution

THE YOUNGEST SCIENTISTS

Kids on the frontlines of asthma research

With more than 6 million children affected, asthma is the most common chronic illness among children in the United States.

dero2084/Alamy

After reading this module, you should be able to answer the following GUIDING QUESTIONS

1. What is air pollution, and what is its global impact?
2. What are the main types and sources of outdoor air pollution?
3. What are the health, economic, social, and ecological consequences of air pollution?
4. What are the causes and consequences of acid deposition?
5. What is the mandate of the Clean Air Act, and how is it implemented?
6. What are the main sources of indoor air pollution, and what can be done to reduce it?
7. How can outdoor air pollution be reduced, and what are the trade-offs of reducing it?

For 10-day stretches during 2003 and 2004, 45 asthmatic kids from smoggy Los Angeles County, some as young as 9 years old, carried more than books in their backpacks. The kids, from two regions of the county, wore backpacks containing small monitors that sampled the air around them continuously as they went about their daily lives—going to class, playing with friends, having dinner with their families.

Those personal air monitors are far more accurate than measurements taken at local monitoring stations, explains Ralph Delfino, an epidemiologist at the University of California, Irvine, who recruited the students to help collect the data. "A monitoring station can be many miles from where the subject lives, where they go to school, etc.—so that measurement may not represent their actual exposure very well," he says. The air monitors used in Delfino's study detected levels of harmful pollutants in the air surrounding each child.

WHERE IS LOS ANGELES, CALIFORNIA?

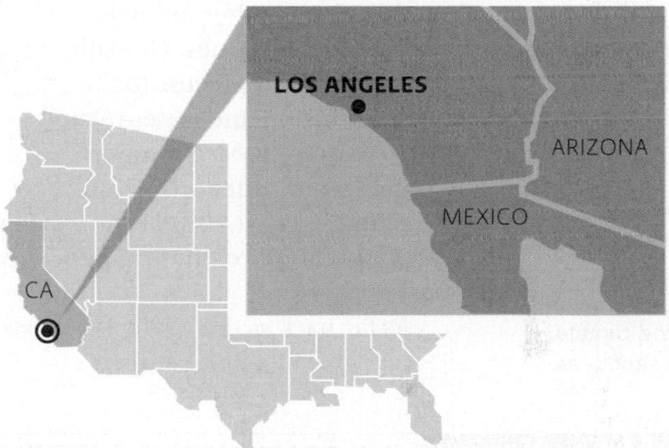

The children selected for the 10-day experiment were all currently being treated for mild to moderate asthma, and the area in which they all lived had significant vehicle air pollution. The monitors they wore measured the quantity of small particles and nitrogen dioxide (NO_2), both of which are commonly present in vehicle emissions and are known to irritate lung tissue. Exposure to these pollutants can trigger asthma symptoms such as wheezing, coughing, or shortness of breath in sensitive individuals. In addition, ten times a day, the children exhaled into a special bag that assessed their breath for nitric oxide (NO), a chemical marker of airway inflammation—a telltale symptom of asthma.

When the 10 days were over, Delfino compared the types of pollutants to which the kids were exposed with their nitric oxide levels at similar points in time, data that would allow him to paint a picture of the types and amounts of pollutants that exacerbate asthma.

AP Images/Orange County Register/Bruce Chambers

Los Angeles County in California, like many regions of southern California, owes much of its air pollution to traffic, especially diesel truck traffic that delivers the food and consumer products we use every day.

1 GLOBAL AIR POLLUTION

Key Concept 1: Air pollution is a serious problem that causes millions of deaths each year worldwide.

Researchers believe that **air pollution**—contaminants from natural sources or human activities that cause health or environmental problems—may play a key role in the recently observed rise in asthma cases that began in the 1990s and continues today in many places. In addition, air pollution has been linked to other serious health issues such as cancer, respiratory infection, and cardiovascular diseases—and it harms not only humans but also plants, other animals, and even buildings, bridges, and statues. Globally, air pollution is the fifth leading risk factor for death—higher than malnutrition and water-related health issues. In its 2019 *State of Global Air* report, the Health Effects Institute estimated that 90% of the world's population breathes polluted air.

air pollution Any material added to the atmosphere that harms living organisms, affects the climate, or impacts structures.

As far back as the 1930s, scientists recognized a link between outdoor air pollution and human illness. In 1930, for instance, 63 people died and 1,000 were sickened in Belgium when a temperature inversion—a situation that occurs when the temperature is higher in upper regions of the atmosphere than in the lower, causing pollutants to become trapped near Earth's surface—led to a sudden spike in lower atmospheric sulfur levels. In 1948, weather conditions trapped pollutants in the air of Donora, Pennsylvania, over a 4-day period, killing 20 and sickening at least 6,000. Just 4 years later, the famous "Great Smog" in London, England, is believed to have been responsible for 12,000 premature deaths. **INFOGRAPHIC 1**

Asthma is a respiratory ailment marked by inflammation and constriction of the narrow airways of the lungs. Delfino's work is important, in part because asthma is one of the most common chronic childhood diseases in developed nations, and it's a major cause of childhood

INFOGRAPHIC 1 AIR POLLUTION IS A WORLDWIDE PROBLEM

The World Health Organization (WHO) recognizes air pollution as a major threat to human health, causing as many as 7 million premature deaths worldwide (about 12.5% of the total). The vast majority of these deaths occur in low- and middle-income countries.

DEATHS FROM AIR POLLUTION (2017)

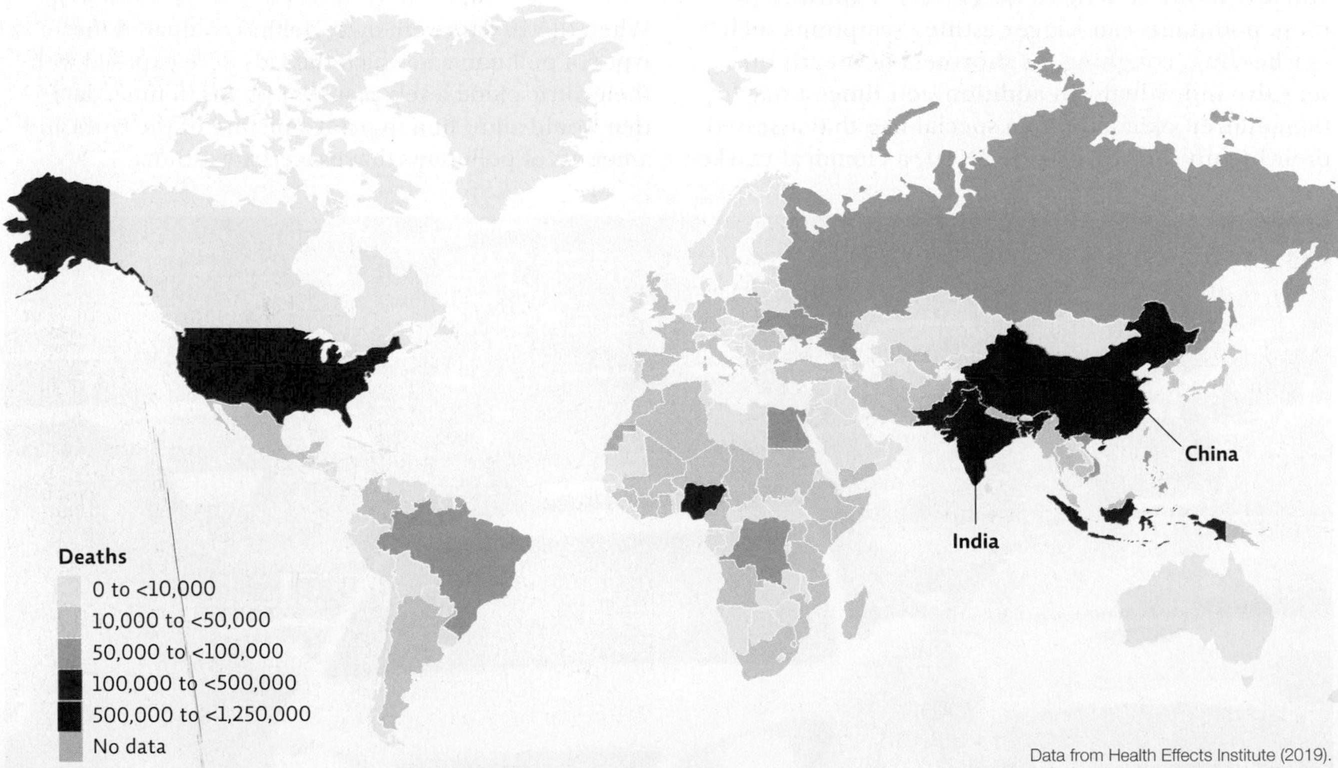

Data from Health Effects Institute (2019).

China and India accounted for roughly 50% of all air pollution deaths in 2017. Why do you suppose air pollution deaths are so high in these two countries?

disability. In the United States, for example, the prevalence of childhood asthma more than doubled from 1980 to the mid-1990s, and though it has leveled off in recent years, with 6.2 million children (8.4%) and 19 million adults (7.7%) diagnosed, it remains at historically high levels. (Because there are more adults than children, overall more adults have asthma, but a larger percentage of children have the disease.) In fact, in the United States, asthma is the leading cause of school absences and, hence, lost revenue for public schools, whose federal funding is based on attendance.

Developing nations are also seeing a rise in asthma, especially in urban centers. In all areas, asthma rates are likely underdiagnosed. Worldwide, asthma is now the most common respiratory ailment reported. More than 300 million people suffer from the disease, though with so many cases likely unreported, that number could be closer to 400 million.

2 OUTDOOR AIR POLLUTION: TYPES AND SOURCES

Key Concept 2: Outdoor air pollution is caused by natural and anthropogenic sources and includes emissions that are directly harmful (primary pollutants) and those that are converted to harmful forms (secondary pollutants). Conflicts between regions can arise when air pollution crosses borders.

In many urban areas and throughout developed nations today, much of the air pollution comes from vehicle exhaust and industry and power plant emissions. Industrial pollution is usually released as *point source pollution*—pollution that comes from an identifiable smokestack or other outlet. Point source pollution is generally easier to monitor and regulate than is *nonpoint source pollution*—pollution from dispersed or mobile sources like vehicles and lawn mowers. (See Module 6.2 for information about point source and nonpoint source water pollution.)

Outdoor air pollution includes chemicals and small particles in the atmosphere that can be natural in origin—arising from natural events like sandstorms, volcanic eruptions, or wildfires—or come from humans—such as pollution released from factories and vehicles during the combustion of fossil fuels or from burning biomass (e.g., wood, crop waste, or garbage). Of these anthropogenic sources, **primary air pollutants** are pollutants released directly from both mobile sources (such as cars) and stationary sources (such as power plants and factories).

Some primary air pollutants react with one another or with other chemicals in the air to form **secondary air pollutants**. For example, **ground-level ozone** forms when nitrogen oxides (NO and NO_2—together expressed as NO_x) released during fossil fuel combustion react with atmospheric oxygen in the presence of sunlight. It is often a component of *smog*—a term that's a combination of the words *smoke* and *fog*—a hazy air pollution that contains a variety of primary and secondary pollutants.

Another serious form of air pollution is **particulate matter (PM)**—particles or droplets small enough to remain aloft in the air for extended periods of time. Particulates are a common component in smoke and soot and are released when just about anything is burned. Although all particulates reduce visibility, it is the smallest particles—those with a diameter less than 2.5 micrometers (μm), about 1/40 the diameter of a human hair—that aggravate asthma and other chronic lung diseases and increase the risk for death.

Dust storms are a major nonpoint source of particulate matter—dust has been linked to respiratory problems since the Dust Bowl era (the 1930s) of the United States. Records from a Kansas public health report published in 1935 reported a substantial increase in respiratory problems, eye

primary air pollutants Air pollutants released directly from a mobile or stationary source.

secondary air pollutants Air pollutants formed when primary air pollutants react with one another or with other chemicals in the air.

ground-level ozone A secondary pollutant that forms when some of the pollutants released during fossil fuel combustion react with atmospheric oxygen in the presence of sunlight.

particulate matter (PM) Particles or droplets small enough to remain aloft in the air for long periods of time.

Peggy Peattie/U-T San Diego/ZUMAPRESS.com

Controlled burns of agriculture fields, like this one of an asparagus field in California, help clear land for more planting but release particulate matter (small particles) into the air, contributing to respiratory distress in sensitive individuals.

INFOGRAPHIC 2 **OUTDOOR AIR POLLUTION**

There are many sources of outdoor air pollution, both natural and anthropogenic. These sources release primary pollutants, some of which may be converted to different chemicals (secondary pollutants). Prevailing winds transport pollution that reaches the upper troposphere or stratosphere around the globe; no area is immune to air pollution. Agricultural and industrial pollutants have been found in Arctic and Antarctic air, delivered by these prevailing winds.

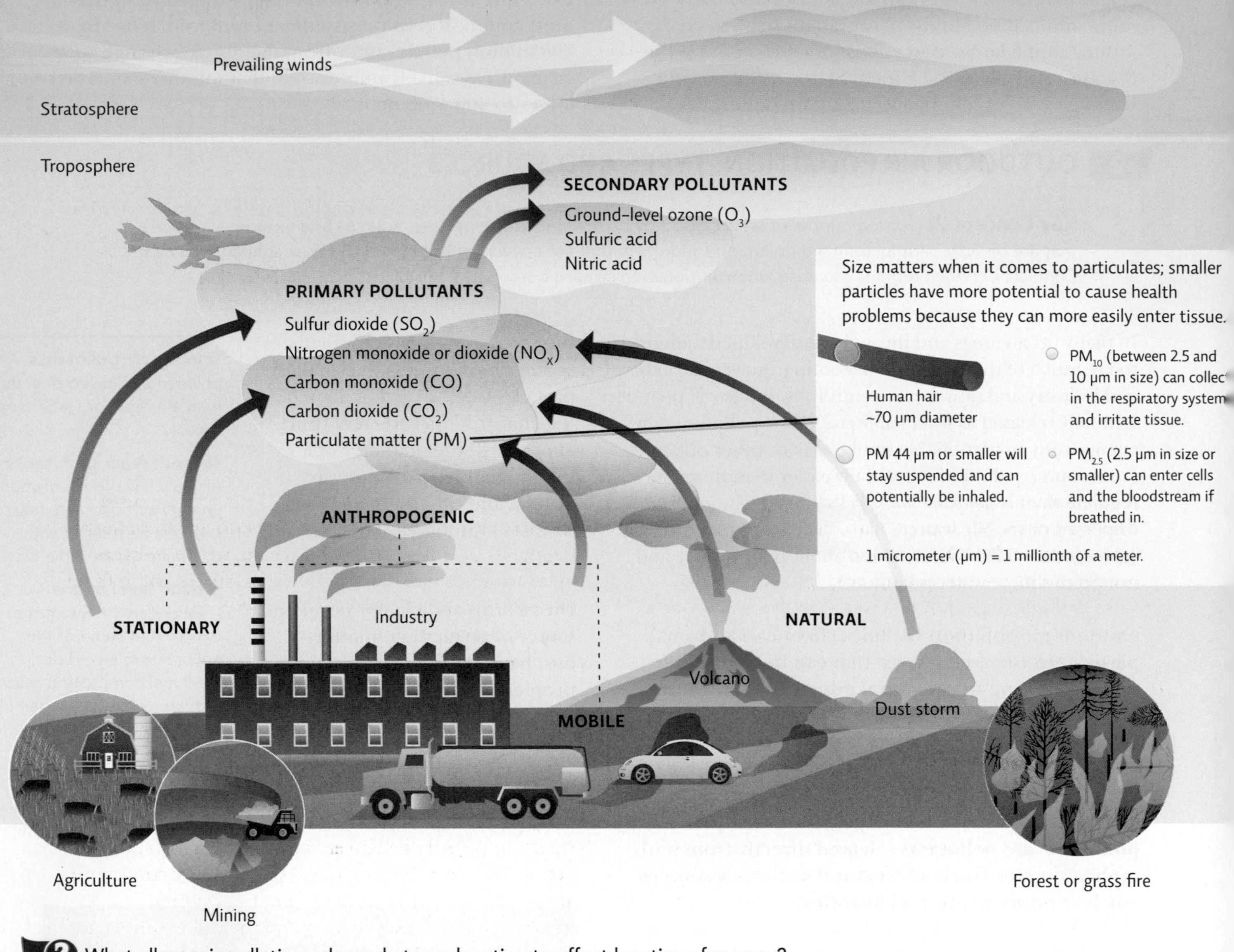
Prevailing winds
Stratosphere
Troposphere
SECONDARY POLLUTANTS
Ground-level ozone (O_3)
Sulfuric acid
Nitric acid
PRIMARY POLLUTANTS
Sulfur dioxide (SO_2)
Nitrogen monoxide or dioxide (NO_X)
Carbon monoxide (CO)
Carbon dioxide (CO_2)
Particulate matter (PM)
Size matters when it comes to particulates; smaller particles have more potential to cause health problems because they can more easily enter tissue.
Human hair ~70 μm diameter
PM_{10} (between 2.5 and 10 μm in size) can collect in the respiratory system and irritate tissue.
PM 44 μm or smaller will stay suspended and can potentially be inhaled.
$PM_{2.5}$ (2.5 μm in size or smaller) can enter cells and the bloodstream if breathed in.
1 micrometer (μm) = 1 millionth of a meter.
ANTHROPOGENIC
STATIONARY
Industry
NATURAL
Volcano
Dust storm
MOBILE
Agriculture
Mining
Forest or grass fire

What allows air pollution released at one location to affect locations far away?

infections, and other health problems following dust storm events in the state. Climate change is expected to increase the frequency of dust storms, bringing with them an increase in the incidence of the health problems they cause.

Agriculture, too, is a source of nonpoint source outdoor air pollution. Toxic pesticides sprayed on crops can become airborne and drift as far as 20 miles, confined animal feeding operations produce significant odor problems and particulate pollution (see Online Module 8.3), and cattle and rice paddies contribute to global warming by releasing the greenhouse gas methane.

Pollution can also move from ground level up into the stratosphere, a region of the upper atmosphere that contains the "ozone layer" that protects living things from some of the Sun's dangerous ultraviolet (UV) radiation. Air pollutants released by humans, such as chlorofluorocarbons (CFCs), can travel up into the stratosphere and destroy this protective ozone. (See Module 5.2 for more on stratospheric ozone depletion.) **INFOGRAPHIC 2**

Don't confuse ground-level ozone pollution with stratospheric ozone depletion. These are two very different problems, though they deal with the same molecule—O_3. Ozone in the stratosphere is a good thing, but ozone at ground level is a problem—breathing it in can directly damage the sensitive tissue of the lungs, and it is a known trigger of asthma attacks. Even plants are damaged by the corrosive action of ozone. Studies by Delfino and other researchers have shown that increases in ground-level ozone pollution have been correlated with increased hospitalizations for asthma.

In his Los Angeles County study, Delfino found that particles contained in diesel exhaust were among the worst asthma culprits. In addition, particulate levels were higher in Riverside, one of two regions he tested; the researchers concluded that Riverside had more pollution because it was downwind of the main urban areas in Los Angeles. Pollution in Riverside is an example of **transboundary pollution**—pollution that originated in one area but traveled to another.

transboundary pollution Pollution that is produced in one area but falls in a different region.

Conflicts can arise between states or nations suffering from transboundary pollution because they do not have the political ability to regulate air pollution generated outside of their borders. Even the most isolated regions on Earth are vulnerable to the effects of air pollution because atmospheric and hydrologic circulation moves pollution around the globe. For example, agricultural pesticides and industrial chemicals have been found in remote Arctic and Antarctic regions—areas far from any farms or industries.

3 CONSEQUENCES OF AIR POLLUTION

Key Concept 3: Air pollution causes environmental and health problems, damages structures, reduces visibility, and contributes to stratospheric ozone depletion and climate change. It is often worse in minority and low-income areas, raising concerns of environmental justice.

Air pollution imposes a serious health and environmental burden. In 2018, the World Health Organization (WHO) estimated that 7 million people die prematurely each year from exposure to air pollution. Respiratory ailments are common; particulates from soot and smog damage respiratory tissue and increase susceptibility to infection, particularly in children because they breathe in more air for their size than adults do and because developing tissue is more vulnerable. And damage can start before birth—maternal exposure to air pollution is linked to premature births and low birth weights.

Reducing that exposure can produce significant results. A study by Janet Currie and Reed Walker of Columbia University found that after the adoption of the E-ZPass in New Jersey and Pennsylvania (this pass allows motorists to drive through toll plazas without stopping, thus reducing air pollution from idling vehicles), the incidence of low birth weight and premature births decreased significantly (11.8% and 10.8%, respectively) for women living within a mile of the toll plaza.

Lungs are particularly vulnerable to pollution because they get so much exposure (we breathe all the time) and the tissue itself is delicate. Irritants like particles, dust, and pollen can cause the lungs to produce excess mucus to trap and expel the irritant. The lining of the airways can become inflamed; in people with asthma, the irritation may trigger muscle contractions that shrink the airway or close it off completely (an asthma attack). Particles smaller than 2.5 μm are the most dangerous because they are small enough to penetrate lung cells or enter the bloodstream, which delivers them to other body cells. If these particles come from the combustion of fossil fuels or other industrial sources, they may contain toxic substances, leading to additional problems associated with toxic exposure.

Since the cardiovascular system depends on the respiratory system to provide oxygen for the body, anything that impairs the lungs also harms the cardiovascular system, which might explain why people living in polluted areas also have higher rates of heart attacks and strokes. Cancer rates are higher in people exposed to air pollution, too; exposure to secondhand smoke and exposure to radon are the leading environmental causes of lung cancer, and exposure to smog and vehicle emissions is linked to increased risk for lung and breast cancer.

Exposure to air pollution also has a societal link. Low-income or minority areas often have some of the worst air quality. This raises questions of **environmental justice**—the concept that access to a clean, healthy environment is a basic human right (see Module 4.2). Sources of major pollution like power plants or waste incinerators are often placed in areas where residents have less ability to fight for their rights—less money, less education, little or no voice in local government. Globally,

environmental justice The concept that access to a clean, healthy environment is a basic human right.

Chang W. Lee/The New York Times/Redux

Derrick Reliford, 14, in his Bronx, New York, home. At 10 years old, he participated in the New York University study that measured his daily pollution exposure. The researchers found that students in the South Bronx who lived or attended school near a highway were exposed to more air pollution than children in other parts of the city.

exposure to and deaths from air pollution are highest in lower income populations. This also holds true in the United States. A 2018 study led by Ihab Mikati found that those living in poverty (as defined by the U.S. Census Bureau) had a 35% greater chance of living near a source of particulate pollution (e.g., a factory or power plant) than the population overall. In some cases, even when socioeconomic status is accounted for, minority communities still face more exposure to pollution than average, an example of *environmental racism* as evidenced by Mikati's findings that minorities had a 28% greater chance of living close to a pollution sources; blacks, in particular, had a 54% greater chance of doing so.

Children of low-income families are at particular risk: For example, in the Bronx, children's homes and schools are often near major roads or factories, and they typically come and go to school during rush hour, when traffic is heaviest and smog forms. More children in the South Bronx are hospitalized for asthma than anywhere else in New York State, and since many Bronx children live or attend schools adjacent to congested highways, Bronx Congressman José Serrano wondered if the two factors might be related. In 2002, he asked New York University environmental scientist George Thurston if he would be willing to conduct a study to find out. "We thought about it for a nanosecond, and then said, 'sure,'" Thurston recalls. In a study similar to Delfino's, Thurston recruited 40 South Bronx fifth-graders to tote wheeled backpacks containing personal air monitors for a month while rating their respiratory symptoms three times a day. "You rolled it, so it wasn't really that heavy," Derrick Reliford, one of the students in the study told *The New York Times*. "They were the rock stars of the class—everybody wanted to help them with the backpacks," Thurston recalls.

The children came from four different schools, two of which were close to a highway and two of which were not. Thurston found that, sure enough, the children who went to schools or lived closer to highways were exposed to more air pollution—in particular, diesel fuel exhaust—and they also had more severe respiratory symptoms. More recent research by Thurston on adults corroborated these results, finding that the closer one lived to a coal-fired power plant or major highway, the higher one's risk of dying from a heart attack.

But humans aren't the only organisms afflicted by the ill effects of air pollution. Many animals suffer the same respiratory distress as humans: All lung tissue is very vulnerable to air pollution. Gills, too, such as those on some terrestrial invertebrates, are negatively affected by air pollution. Plant tissues are also vulnerable to pollutants like smog and ozone. Exposure can reduce a leaf's ability to photosynthesize, preventing healthy growth and compromising its survival. Together with changes in soil chemistry—which can hinder plant growth—pollution damage to crops could result in dollar losses in the billions by 2030.

Pollution also damages buildings and monuments. Acids in polluted rain literally eat away at limestone and marble structures; it can etch glass and damage steel and concrete, causing billions of dollars of damage per year.

Ernie Janes/Balance/Photoshot/ZUMAPRESS.com

The image shows the degradation that has accumulated over the years, a rate of erosion that exceeds normal weathering and is likely due to acid rain.

The tourism industry is even impacted by air pollution that lowers visibility. On hazy days, for instance, it can be impossible to see across the Grand Canyon. Visibility in the Blue Ridge Mountains of North Carolina is also declining. Traditionally, summer visibility across the vistas of these mountains should be around 75 miles, but due to high levels of ground-level ozone pollution, that visibility is greatly diminished to around 15 miles or less on most days. The mountain views all but disappear in a haze of pollution. **INFOGRAPHIC 3**

INFOGRAPHIC 3 CONSEQUENCES OF AIR POLLUTION

Air pollution negatively affects the natural environment and species that live there and is strongly linked to a wide variety of human health problems. It also has impacts at the societal level, impacting the well-being of entire communities.

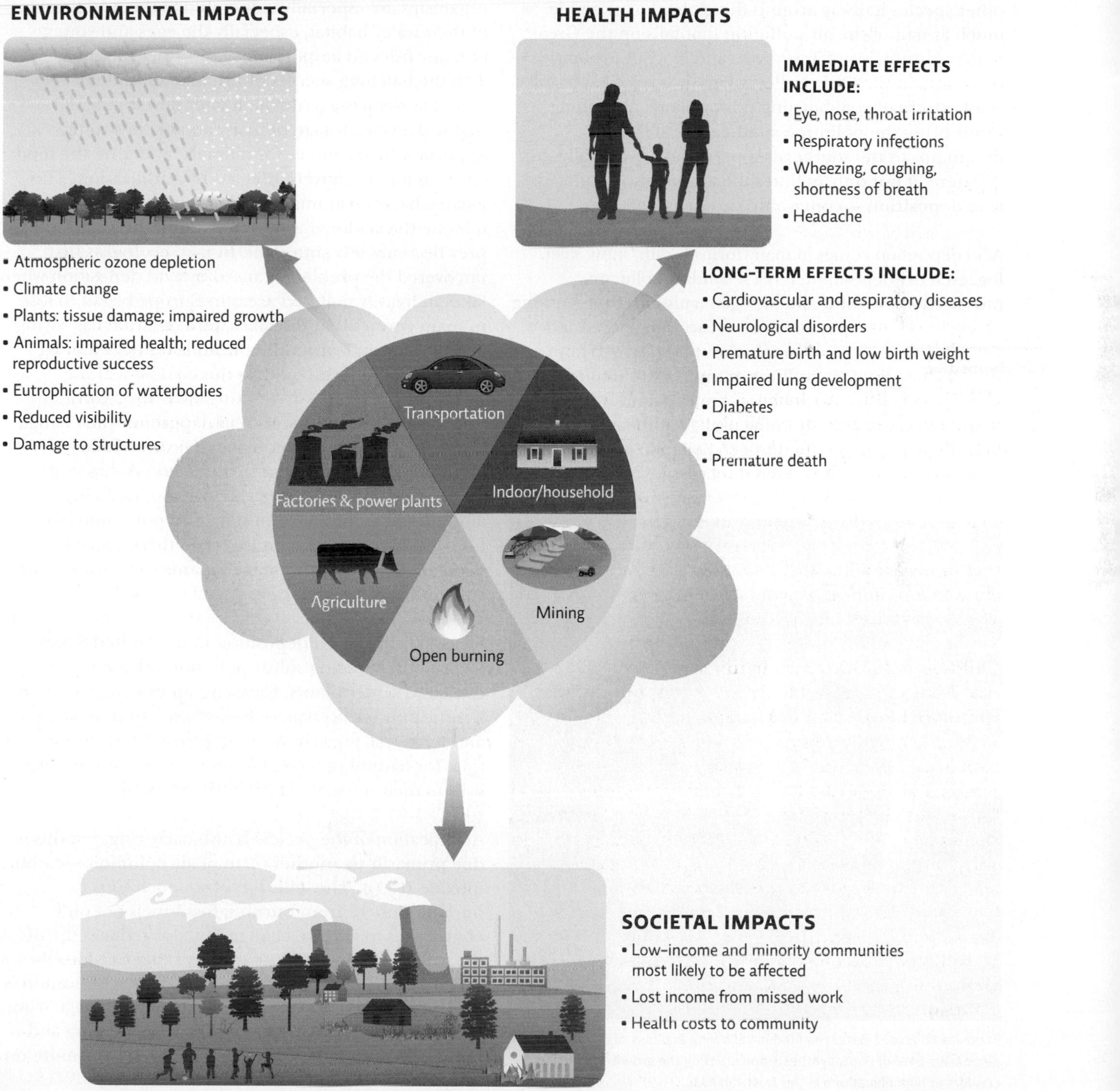

? What factors might explain why children are at a higher risk of health problems due to exposure to air pollution than adults?

4 ACIDIFICATION

Key Concept 4: Acid deposition is a secondary pollutant that results from fossil fuel burning. It lowers the pH of soil and waterbodies, which can harm plants and animals that are exposed. Anthropogenic CO_2 emissions are also increasing the acidity of the oceans with potentially devastating consequences.

One of the most problematic characteristics of air pollution is that it moves. Transboundary air pollution produced in one city can end up harming humans and other species halfway around the globe. For example, as much as half of the air pollution that falls on the Great Smoky Mountains of Tennessee and North Carolina originates in the Ohio Valley, where it is released by tall smokestacks of coal-burning power plants. Prevailing winds bring the pollution southeast, and the tall mountains in the southern Appalachians eventually stop it. There, it not only pollutes the air but also produces **acid deposition**—commonly known as acid rain.

Acid deposition comes in many forms—rain, snow, sleet, fog, even dry deposition. It is a secondary pollutant, produced when sulfur and nitrogen emissions from burning fossil fuels react with oxygen and water in the atmosphere to form sulfuric or nitric acids. (Sulfur pollution comes mainly from burning coal; the primary source of nitrogen pollution is fossil fuel–powered vehicles.)

acid deposition
Precipitation that contains sulfuric or nitric acid; dry particles may also fall and become acidified once they mix with water.

Tony Craddock/Science Source

Exposure to acid rain has resulted in yellowing and loss of needles, decreasing overall photosynthesis and stunting the growth of these conifers at high elevations in the Austrian Alps.

Acid deposition can harm aquatic organisms in affected waterbodies as well as acidify the soil, negatively impacting soil communities and plant life. Many aquatic organisms are especially vulnerable to the acidification of their water habitat, especially the eggs and young of many fish and amphibians. Lower pH can interfere with the hatching success of fish and amphibian eggs as well as the passage through later developmental stages. The larval stage of many aquatic invertebrates, important food sources for animals higher on the food chain, is also negatively affected by acidification. This means that even animals (such as adult fish) that can tolerate the acidified waters may suffer indirectly if their prey becomes less abundant. In fact, ecologists first uncovered the problems caused by acid deposition when lakes in heavily polluted areas in Europe began to lose so many species that the lakes were described as "dying." Ecologists soon learned that dead lakes were an early warning that forests could be next.

Acidification of soil due to acid deposition can change the soil chemistry and mobilize toxic metals such as aluminum, hindering plants' ability to take up water. Acids leach nutrients from the soil, too, reducing the amount of calcium, magnesium, and potassium available to plants in topsoil. Taken together, these impacts can decrease plant growth, weaken plants so they are more vulnerable to disease or pests, and even kill them.

Steps to reduce acid deposition in the United States focused on reducing sulfur pollution released from coal-fired power plants. Cleaning up those emissions has significantly reduced acid deposition, but most streams and terrestrial areas have yet to recover. It will take some time for natural processes to return these waters and soils to their normal pH. **INFOGRAPHIC 4**

Acidification of the oceans is also occurring, but this is due primarily to another form of air pollution—carbon dioxide (CO_2). The addition of so much CO_2 to our atmosphere from burning fossil fuels not only contributes to climate change but also reduces the pH of the world's oceans because much of that CO_2 dissolves in ocean water, forming carbonic acid. This acidification is projected to have wide-reaching effects that range from the weakening of seashells to impaired spawning and foraging behavior. (See Online Module 6.3 for more on the pH scale and ocean acidification.)

INFOGRAPHIC 4 ACID DEPOSITION

Burning fossil fuels releases sulfur and nitrogen oxides. These compounds react in the atmosphere to form acids. Acid rain, snow, fog, and even dry particles can fall to Earth as acid deposition, with the potential to alter the pH of lakes and soil, damaging plant and animal life.

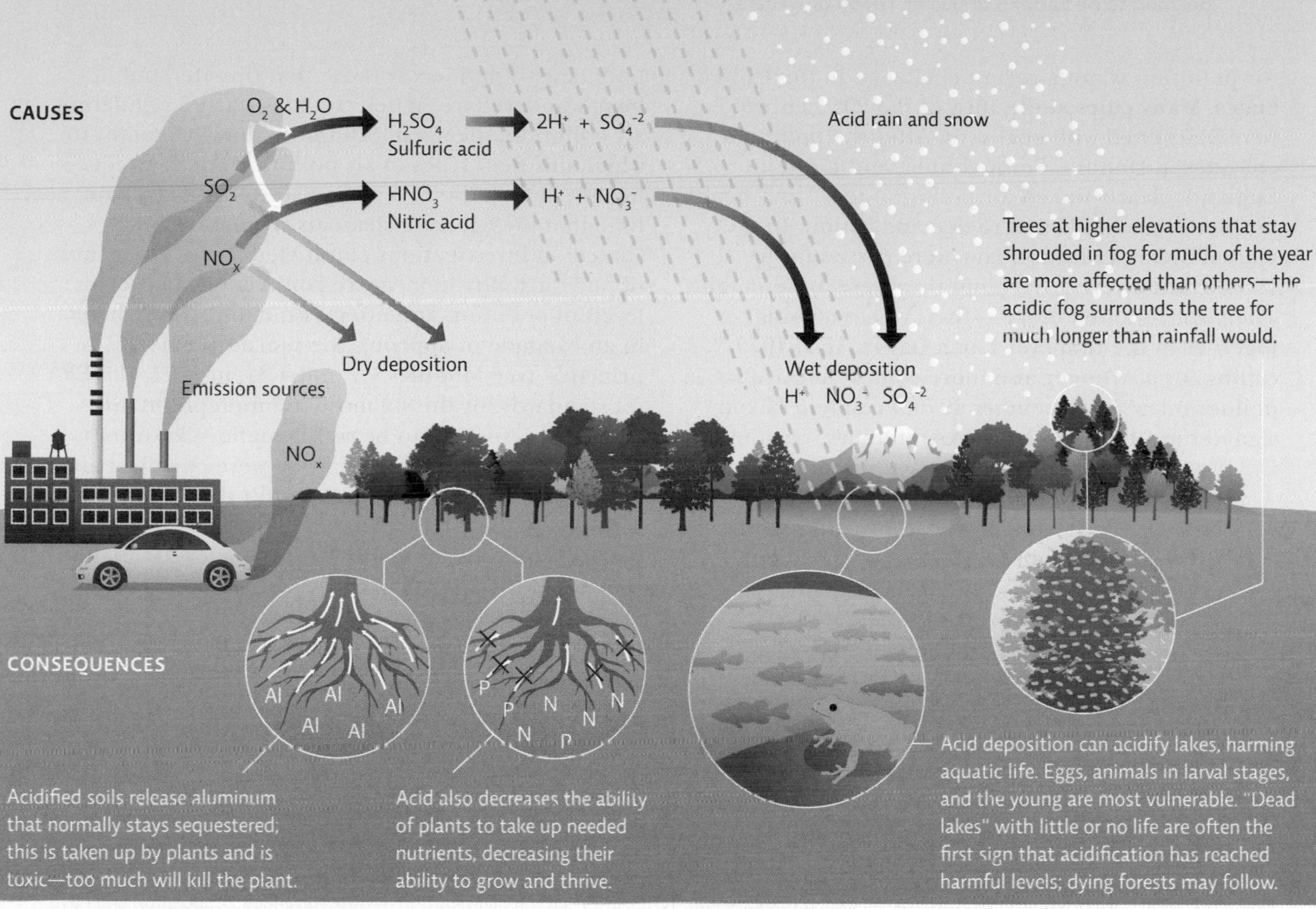

Data from National Atmospheric Deposition Program (2019).

ACID DEPOSITION HAS DECREASED

Sulfur pollution has been reduced significantly, and nitrogen pollution, shown here, is also on the decline. While the pH of U.S. precipitation has increased thanks to acid deposition policies, ecosystems have yet to recover as water and soil pHs will take longer to rebound.

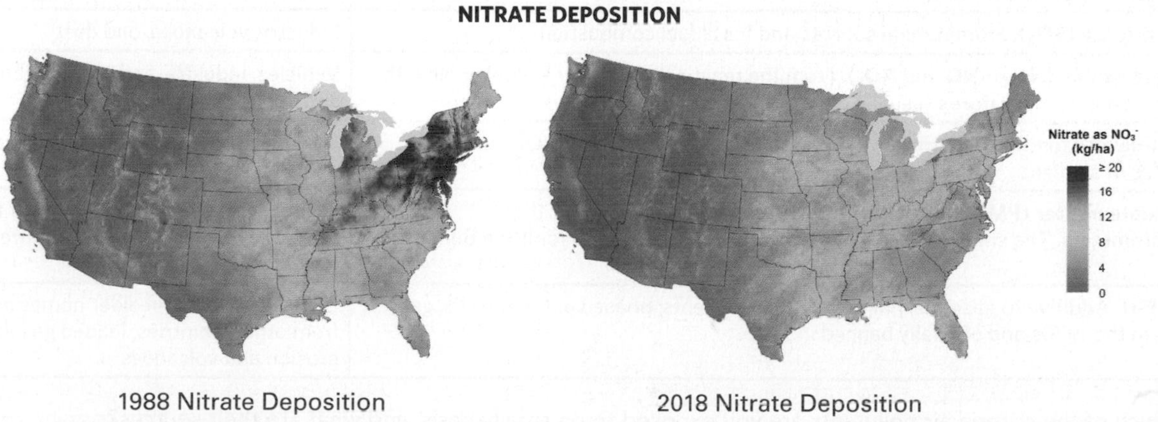

Why do you think that acid deposition from nitrogen pollution has been a bigger problem in the eastern U.S. than the western part of the country?

5 THE CLEAN AIR ACT

Key Concept 5: U.S. air pollution is regulated by the Clean Air Act, which sets standards for any pollutant deemed to be harmful to human health or welfare.

Air pollution is not a recent challenge in the United States. Many cities at the turn of the 20th century were blackened with coal soot. After the pollution episodes in Donora, Pennsylvania, and London, England, clearly linked air pollution to serious health problems, people began to demand action. In the United States, pollution laws were passed in the 1960s—they were a start, but they were not enough. The public wanted more—their voices coming together in the first-ever Earth Day in April 1970, calling for a stronger and more comprehensive air pollution law. In December 1970, President Nixon signed the **Clean Air Act (CAA)** into law, creating the U.S. Environmental Protection Agency (EPA) that same month. The CAA received major updates in 1977, 1990, and 2010.

Clean Air Act (CAA) The main U.S. law that authorizes the EPA to set standards for dangerous air pollutants and enforce those standards.

The CAA is mandated to regulate the emissions of hazardous air pollutants—defined as those that threaten public health and welfare. When the EPA started regulating air pollution, the agency did not know the extent to which different types of air pollution could affect health; in fact, the EPA administrator noted at the time that the agency's clean air regulations were "based on investigations conducted at the outer limits of our capability to measure connections between levels of pollution and effects on man." Nevertheless, in an example of applying the precautionary principle (see Modules 1.2 and 4.3), in 1971, the EPA set standards for the six most common pollutants that were suspected to be problematic—known as *criteria air pollutants.* Five of these were chemical air pollutants: carbon monoxide, sulfur dioxide (SO_2), nitrogen oxides (NO_x), lead, and ground-level ozone; standards were also set for particulate matter. **INFOGRAPHIC 5**

A landmark 1993 study published by Harvard University researcher Douglas Dockery helped firmly

INFOGRAPHIC 5 THE CLEAN AIR ACT

CLEAN AIR ACT: The primary U.S. federal law regulating air quality. Passed in its current form in 1963, it has been updated with major amendments in 1970, 1977, and 1990. The EPA is tasked with enforcing the CAA and does so by writing "rules" to lay out targets and acceptable methods states can use to comply. The first six pollutants that were targeted are known as the "criteria air pollutants." Many other air pollutants are also regulated by the CAA, including mercury and VOCs; efforts to regulate carbon dioxide have faced opposition and have been less successful.

CRITERIA AIR POLLUTANT	SOURCE
Carbon monoxide (CO). From incomplete combustion of any carbon-based fuel (and most combustion is incomplete).	Vehicles, forest fires, and volcanoes
Sulfur dioxide (SO_2). From natural sources and fossil fuel combustion.	Industry, volcanoes, and dust
Nitrogen oxides ($NO_X = NO$ and NO_2). From the reaction of nitrogen in fuel or air with oxygen at high temperatures (usually during combustion of a fuel).	Vehicles, industry, and nitrification by soil and aquatic bacteria
Ground-level ozone (O_3). Formed from reactions between NO_x and VOCs in the presence of sunlight.	Vehicles (NO_x); manufactured products and industry (VOCs)
Particulate matter (PM). Tiny airborne particles or droplets, smaller than 44 micrometers. The smaller the particle, the more dangerous it is for living things.	Combustion of any fuel or activity that produces dust, including forest fires and dust storms
Lead (Pb). Additive to gasoline, paint, and other solvents; phased out of the U.S. gas supply in the 1970s and officially banned in 1996.	Lead-based paint in older homes and from other countries; leaded gasoline; soil erosion and volcanoes

? Which of the criteria air pollutants are you exposed to on a daily basis, and what are their sources?

establish the link between air pollution and impaired health. When Dockery compared death rates in six cities before air pollution controls were put in place to death rates after the CAA was implemented in those same six cities, the data showed that as the level of small particulates ($PM_{2.5}$) decreased in the cities evaluated, the death rate decreased. A follow-up study in 2006 estimated that particulate pollution—which includes soot, ash, dust, smoke, pollen, and small, suspended droplets (aerosols)—accounted for 75,000 premature deaths per year and showed a clear dose–response effect: The higher the air pollution, the higher the risk for death.

In addition to the six criteria air pollutants, the EPA also recognizes 187 hazardous air pollutants that can have adverse effects on human health, even in small doses. These toxic substances may cause cancer or developmental defects, or they may damage the central nervous system or other body tissues. They include **volatile organic compounds (VOCs)**, a variety of chemicals that readily evaporate, entering the air as a gas. Many present a health hazard if inhaled or if sensitive tissues, such as eyes, are exposed. VOCs are produced by natural sources, such as wetlands, but the main outdoor source is fossil fuel combustion. VOCs found in household products, such as paint, carpets, and cleaners, contribute to indoor air pollution.

volatile organic compound (VOC)
A chemical that readily evaporates and is released into the air as a gas; may be hazardous.

The Los Angeles skyline is obscured by smog.

Justin Lambert/Getty Images

6 INDOOR AIR POLLUTION

Key Concept 6: Modern homes trap or are the source of many indoor air pollutants. Better ventilation and alternative building or household materials can reduce this pollution. In areas where exposure to air pollution mainly comes from poorly ventilated indoor cooking fires, indoor air pollution can be reduced with better ventilation or use of cleaner fuels.

As Delfino and others were showing, outdoor air quality substantially impacts human health. However, we breathe air indoors as well as outdoors, and indoor air quality is a growing concern among public health scientists. In fact, people living in affluent, developed nations may find that their greatest exposure to unhealthy air comes from indoors. This is because so much time is spent indoors in homes, schools, or the workplace; these areas contain many potential air pollution sources. For instance, cigarette smoke causes significant health problems that include eye, nose, and mucous membrane irritation, lung damage (which can exacerbate or cause asthma), and lung cancer. Items in our home, like paint, cleaners, and furniture, release VOCs, which can also cause health problems.

Outdoor pollutants can also find their way into our buildings. Radon is a naturally occurring radioactive gas produced from the decay of uranium in rock. It can seep through the foundations of homes and accumulate in basements; exposure to radon can cause lung cancer. Not every area has the type of rock that produces radon, but buildings constructed over areas where soil or groundwater is contaminated with VOCs, such as areas with underground chemical storage tanks, also present an infiltration risk.

Reducing indoor air pollution involves removing the sources (e.g., no carpet in the home), properly storing potential sources of air pollution (e.g., storing household chemicals properly), and providing adequate ventilation for a home. In our quest to become energy efficient, many of us might seal our homes to prevent air leaks. But if this is done, the home also needs a good system for mechanical ventilation (usually as part of the heating, ventilation, and air conditioning [HVAC] system) to ensure a good exchange of air. **INFOGRAPHIC 6**

In developing countries, where many people cook and heat with open fires, smoke and soot from burning wood, charcoal, dung, or crop waste are major sources of indoor

Indoor air pollution is a problem in many less developed regions, where cooking and heating are done indoors with wood or other combustible fuel but homes are poorly ventilated.

MATTHIEU PALEY/National Geographic Image Collection

INFOGRAPHIC 6 SOURCES OF INDOOR AIR POLLUTION

For most people, the greatest exposure to air pollution comes from being indoors. There are many sources of air pollution in a home or other building, as these structures tend to trap pollutants, keeping concentrations high. One can reduce exposure by avoiding or limiting the use of carpets, upholstered items, and furniture made with toxic glue and formaldehyde. Safer cleaners and low-VOC paints are readily available. Simple behaviors like taking off your shoes before entering the house and using a vacuum equipped with a HEPA filter will also help. Good ventilation and properly working heating and air conditioning units help keep indoor air pollutants at bay.

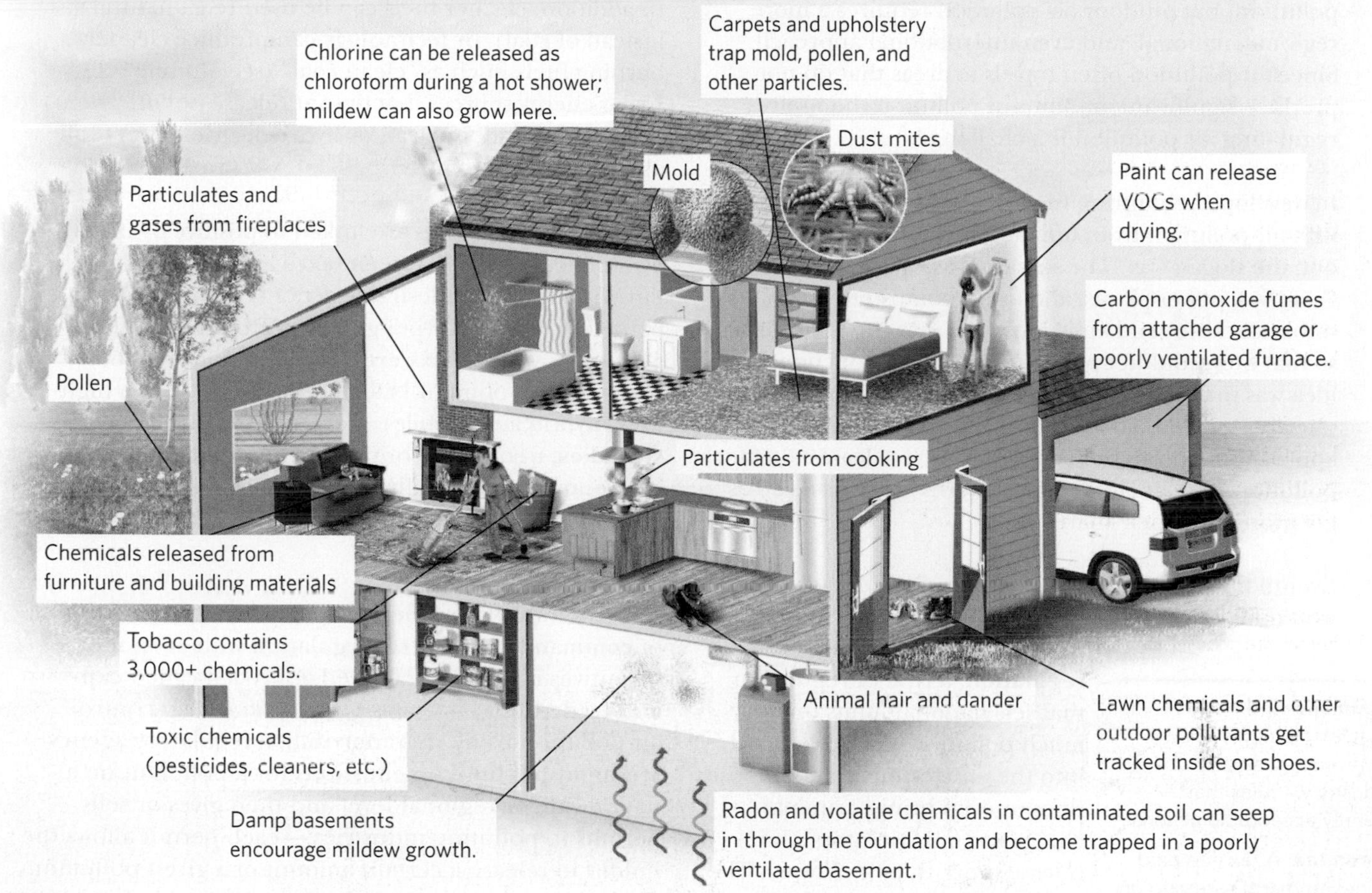
Carpets and upholstery trap mold, pollen, and other particles.
Chlorine can be released as chloroform during a hot shower; mildew can also grow here.
Dust mites
Mold
Particulates and gases from fireplaces
Paint can release VOCs when drying.
Carbon monoxide fumes from attached garage or poorly ventilated furnace.
Pollen
Particulates from cooking
Chemicals released from furniture and building materials
Tobacco contains 3,000+ chemicals.
Animal hair and dander
Lawn chemicals and other outdoor pollutants get tracked inside on shoes.
Toxic chemicals (pesticides, cleaners, etc.)
Radon and volatile chemicals in contaminated soil can seep in through the foundation and become trapped in a poorly ventilated basement.
Damp basements encourage mildew growth.

What are your main sources of indoor air pollution, and what could you do to reduce them?

pollution. Kirk Smith, a professor of environmental health at the University of California, Berkeley, has found that indoor fires increase the risk of pneumonia, tuberculosis, chronic bronchitis, lung cancer, and cataracts in adults and increase the incidence of acute lower respiratory infections in children. In addition, they are linked to a greater risk for low birth weight in babies born of women who are exposed during pregnancy. "Considering that half the world's households are cooking with solid fuels, this is a big problem," Smith says. In India, around 20% of air pollution deaths in 2017 (480,000 out of 2.4 million deaths) were linked to indoor air pollution caused by using wood or dung as cooking fuels—this number is disproportionately higher in lower income areas, where 40% of air pollution deaths were linked to household air pollution.

Addressing this type of indoor air pollution can involve improving or installing chimneys to vent smoke, installing more efficient cook stoves (this can be as simple as constructing an earthen cook stove with openings tailored to fit the pots used in cooking), or providing cleaner burning fuels such as liquid petroleum gas. Many nonprofit organizations are stepping up to meet this problem with a simple $50 solar cooker that allows people to cook food without building a fire. This technology has the added advantage of not depleting local biomass resources for fuel.

7 REDUCING OUTDOOR AIR POLLUTION

Key Concept 7: Industrial air pollution can be reduced with emission control technologies or cleaner fuels. The use of these technologies can be required by command-and-control regulations or encouraged with economic incentives, but may raise the cost of providing energy or doing business.

Solutions such as installing better ventilation systems in our homes and providing solar ovens to individuals in developing countries will help address indoor air pollution, but outdoor air pollution requires a more regional, national, and even international approach. Since air pollution often travels to areas that do not produce significant amounts of pollution themselves, regulating air pollution is a challenge.

In developed countries, the original approach to dealing with air pollution from human activities was to spread it out; the slogan was "The solution to pollution is dilution." Factories, power plants, and other point sources built tall smokestacks to send emissions high into the atmosphere so that they wouldn't pool at the site of production. The idea was that, if dispersed, the amount of pollution in any one area would be too low to cause a problem. But this approach simply doesn't work: Industry releases too much pollution, and air circulation patterns cause some areas to get more than their share.

Eventually, the U.S. government decided that regulation would be necessary. The typical approach in the 1970s was **command-and-control regulation**, a type of regulation that sets national limits on how much pollution can be released into the environment and imposes fines or even brings criminal charges against violators who release more than is allowed. The U.S. CAA, for example, sets a maximum amount of a pollutant that can be released or that can be present in ambient air. These National Air Quality Standards are an example of a *performance standard* (see Module 5.2 for more on performance standards). As a result of the CAA, the United States has seen major reductions in common air pollutants such as lead and sulfur. Indeed, since the CAA was passed in 1970, U.S. air pollution has been cut by 70%, even while the nation's economy has more than doubled.

command-and-control regulation Legislative control of an activity or industry via rules that identify acceptable actions.

green tax A tax (fee paid to government) assessed on environmentally undesirable activities.

tax credit A reduction in the tax one must pay in exchange for some desirable action.

subsidies Financial assistance given by the government to promote desired activities.

cap-and-trade program Regulations that set an upper limit for pollution, issue permits to producers for a portion of that amount, and allow producers that use or release less than their allotment to sell permits to those who exceeded their allotment.

To meet CAA-mandated performance standards, industries and power plants can use end-of-pipe solutions like scrubbers, filters, electrostatic precipitators, and catalytic converters to trap pollutants before they are released. In addition, cleaner fuels can be used (e.g., natural gas instead of coal), or technology can produce cleaner-burning fuels, such as "clean coal" (see Module 9.1). Using energy sources that do not release pollution (e.g., solar, wind, nuclear, or hydroelectric power) will also reduce air pollution.

Other regulatory tools are linked to positive or negative economic incentives. **Green taxes** are imposed on environmentally undesirable actions, such as an extra tax on low-mile-per-gallon (MPG) vehicles ("gas guzzlers"), whereas **tax credits** support actions that help the environment (such as the purchase of a high-MPG hybrid automobile). Governments also offer **subsidies**, free money or resources intended to promote environmentally friendly activities such as installing solar photovoltaic panels for the production of pollution-free electricity.

A method that combines the performance standards of command-and-control regulation with economic incentives is the market-based approach called **cap-and-trade** (also known as *permit trading*). To control air pollution using this approach, a regulatory agency sets an upper limit on emissions for a pollutant on a nationwide or regional level and then gives or sells permits to polluting industries—each permit allows the holder to release a certain amount of a given pollutant. Users that reduce their pollution emissions below what their permit allows can sell their remaining allocation to other users whose allotments were exceeded. Over time, pollution levels can be reduced as the cap—or limit—is lowered. A cap-and-trade program successfully reduced sulfur pollution from coal-fired power plants in the United States in the 1990s. A downside to cap-and-trade programs is that pollution can become concentrated in areas where individual industries choose to buy additional permits rather than reduce emissions. **INFOGRAPHIC 7**

Like many environmental laws, the CAA is regularly under attack by those who see regulation as intrusive. The fact that our air is cleaner today than it was in the 1960s—even with a larger U.S. population and more industry—is evidence that regulations are effective. Still, these improvements are costly to industry; such costs can cut into profits and are usually passed on to consumers. For this

INFOGRAPHIC 7 APPROACHES TO REDUCING AIR POLLUTION

Achieve

Many approaches can be used to lessen air pollution, including technology to reduce emissions before a fuel is burned (e.g., clean coal; see Section 7 of Module 9.1) and technology to capture emissions after a fuel is burned, as shown below (and in Infographic 7 of Module 9.1).

TECHNOLOGY

A variety of technologies, each designed to address the particular problem at hand, can be employed to reduce the release of air pollution.

Smokestack scrubbers send the emissions through a mist of water and limestone to trap contaminants and prevent their release.

TECHNOLOGY: SMOKESTACK SCRUBBER

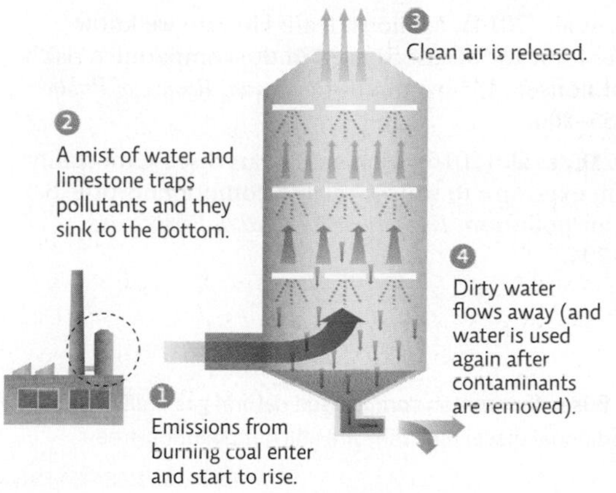
3 Clean air is released.
2 A mist of water and limestone traps pollutants and they sink to the bottom.
4 Dirty water flows away (and water is used again after contaminants are removed).
1 Emissions from burning coal enter and start to rise.

POLICY APPROACHES

Policies protect air quality by requiring or encouraging actions that reduce the production of air pollution. They include:

- Subsidies/grants: funds for technology upgrades
- Green taxes: a tax on pollution
- Penalties: fines, jail, lost contracts, etc.
- Market-driven financial incentives: cap-and-trade

Cap-and-trade is a market-driven policy that gives producers the freedom to choose how to meet the requirement and the opportunity to earn a profit in doing so.

POLICY: CAP-AND-TRADE

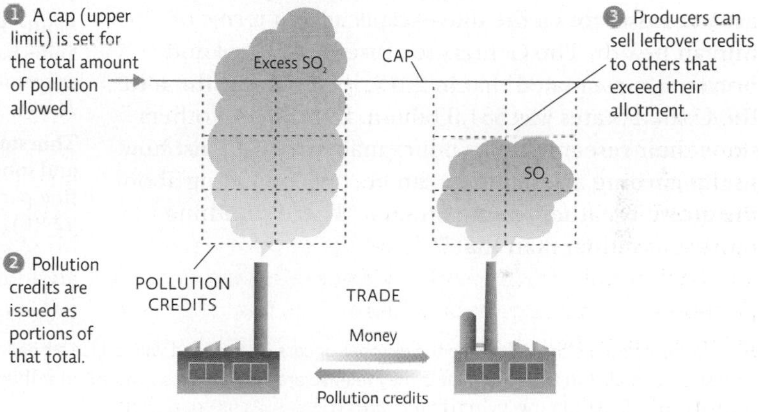
1 A cap (upper limit) is set for the total amount of pollution allowed.
Excess SO_2
CAP
3 Producers can sell leftover credits to others that exceed their allotment.
SO_2
2 Pollution credits are issued as portions of that total.
POLLUTION CREDITS
TRADE
Money
Pollution credits

How could a cap-and-trade program for sulfur pollution lead to lower pollution in one area and higher pollution in others?

reason, many individuals and groups oppose such policies, charging that the restrictions are excessive or that the government goes too far in trying to regulate emissions. However, including the cost of pollution prevention or cleanup in the cost of a product prices that product closer to its *true cost* and may actually lower the overall cost to society as health and environmental impacts decline. (See Module 5.1 for more on true cost accounting.)

While some air pollution problems have been reduced, the work of the CAA is not done; air pollution problems remain, and a new one, climate change—due in large part to carbon emissions from fossil fuel extraction and burning—threatens to be the most significant air pollution issue we will face. (See Module 10.2 for more on climate change.) In a 2007 landmark case, the U.S. Supreme Court ruled that under the CAA, the EPA had the authority to regulate emissions of greenhouse gases such as CO_2 if it could be shown that these emissions posed a threat to human health. An in-depth EPA report that evaluated the negative impacts of these emissions, most notably climate change, was released in 2009. Known as the "Endangerment Finding," the report authors concluded that greenhouse gases like CO_2 did indeed present a health threat, a finding that required the EPA to regulate these emissions. Since that 2009 report, scientific evidence has continued to mount in support of the Endangerment Finding.

The EPA's first rule that focused on regulating greenhouse gases was the Clean Power Plan (CPP), negotiated over several years and approved in 2016. The CPP's initial target was to cut carbon emissions in the power sector by 32% below 2005 levels by 2030. (Further reductions were expected to be imposed as progress was made.) The EPA estimated that reaching this target would save 3,600 lives and net the country up to $45 billion in health and climate benefits. At the time of the CPP's approval, efforts to reduce carbon emissions were already underway and the nation was on track to meet the initial targets, achieving close to a 25% reduction below 2005 levels by 2016.

In 2017, newly elected President Donald Trump ordered a reevaluation of the plan. In June 2019, the EPA replaced the CPP with a new rule—the Affordable Clean Energy (ACE) rule. According to research published in the journal *Environmental Research Letters* by Amelia Keys and colleagues, ACE, which allows most coal-fired power plants to stay online longer than the CPP would have permitted, is expected to result in little further decrease in U.S. CO_2 emissions, essentially ending the decline that began in 2005. Sulfur and nitrogen pollution that cause acid deposition will also likely increase. ACE critics argue that the seriousness of the climate crisis demands that CO_2 emission reduction targets be strengthened, not weakened. Dana Nuccitelli, writing for Yale University's *Climate Connections*, suggests that strengthening these targets would not be unduly restrictive—noting that power plants are on track to meet the CPP targets and additional progress could be reasonably expected.

Mitigating or preventing air pollution costs money, but many feel it is money well spent because it prevents far greater losses down the line—especially in terms of human health. The Centers for Disease Control and Prevention estimated that in 2013, the cost of asthma to the United States was $81.9 billion. Delfino and others hope their research helps policy makers realize just how useful curbing air pollution can be. "We're talking about the air we breathe," Thurston says. "There's nothing more communal than that."

Select References:

Delfino, R. J., et al. (2008). Personal and ambient air pollution exposures and lung function decrements in children with asthma. *Environmental Health Perspectives, 116*(4), 550–558.

Dockery, D., et al. (1993). An association between air pollution and mortality in six U.S. cities. *New England Journal of Medicine, 329*(24), 1753–1759.

Keyes, A. T., et al. (2019). The Affordable Clean Energy rule and the impact of emissions rebound on carbon dioxide and criteria air pollutant emissions. *Environmental Research Letters, 4*(14), 044018.

Nuccitelli, D. (2019, June 21). The Trump EPA strategy to undo Clean Power Plan. *YaleClimateConnects.org.* Retrieved from https://www.yaleclimateconnections.org/2019/06/the-trump-epa-strategy-to-undo-the-clean-power-plan/

Nurmagambetov, T., et al. (2018). The economic burden of asthma in the United States, 2008–2013. *Annals of the American Thoracic Society, 15*(3), 348–356.

Smith, K. R., et al. (2014). Millions dead: How do we know and what does it mean? Methods used in the comparative risk assessment of household air pollution. *Annual Review of Public Health, 35,* 185–206.

Thurston, G. D., et al. (2016). Ischemic heart disease mortality and long-term exposure to source-related components of U.S. fine particle air pollution. *Environmental Health Perspectives, 124*(6), 785–794.

Mass transit options that decrease the number of cars on the road will reduce air pollution. Buses that run on compressed natural gas emit fewer emissions overall, but the particulates they release are very small; so while better than a traditional diesel bus, they are still not pollution-free.

Pierre GLEIZES/REA/Redux

GLOBAL CASE STUDIES AIR POLLUTION IN THE UNITED STATES

Achieve

Air quality has improved in many areas of the United States thanks to the Clean Air Act, but some areas still suffer from heavy air pollution. The online Global Case Studies presented for this module profile the most polluted U.S. metropolitan area and the cleanest metropolitan area in 2019, as determined by the American Lung Association. Case studies are also presented on some recent air pollution research.

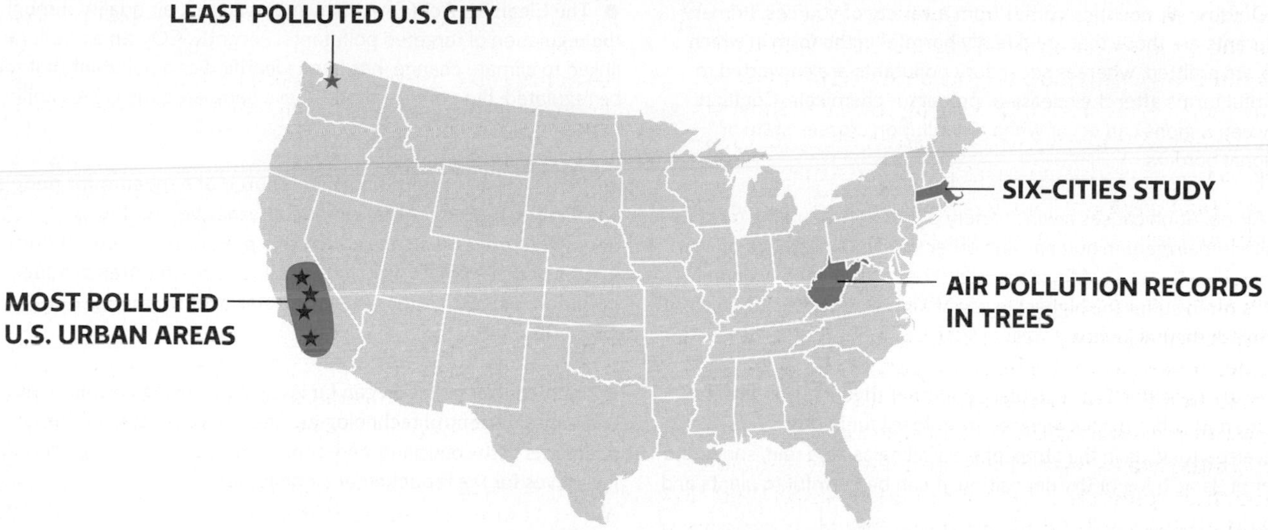

BRING IT HOME

PERSONAL CHOICES THAT HELP

Individuals can have an effect on air quality by researching the threats to their area, making appropriate behavior changes, and supporting legislation that limits the production of air pollutants.

Individual Steps

- Reduce your exposure to indoor air pollution by reducing your use of harsh cleaning products, synthetic air fresheners, vinyl products, and oil-based candles.
- Avoid outdoor exercise during poor air quality days. Go to www.airnow.gov to find the local air quality forecast.
- Buy a radon detector and carbon monoxide detector for your home to keep the household safe.
- Walk or bike more frequently if possible; look for the highest MPG vehicle that you can afford and that meets your needs if you are in the market to purchase a car or truck.

Group Action

- Organize a "car-free day" at your school, community, or workplace to reduce emissions from vehicles.
- Work with community leaders and businesses to sponsor a "free public transit" day.
- If your community does not have public transit or dedicated bike lanes, ask community leaders to investigate bringing these to your area.
- Many groups are working to improve our air quality. Find one in your region and see what issues it is addressing. For a list of national and regional organizations, go to https://insteading.com/blog/organizations-air/.

Maskot/Superstock

REVISIT THE CONCEPTS

- Air pollution comes from natural and anthropogenic sources. It damages ecosystems and is a major health concern.
- Outdoor air pollution comes from a variety of sources. Primary pollutants are those that are directly harmful in the form in which they are emitted, whereas secondary pollutants are converted to harmful forms after the release of precursor chemicals. Conflicts between regions can occur when air pollution crosses state or national borders.
- Air pollution causes a wide variety of problems ranging from health impairment in humans and other species to damage of vulnerable structures. Minorities and those of low socioeconomic status often suffer the highest levels of air pollution, raising concerns of environmental justice.
- Acid deposition is a secondary pollutant that is formed when nitrogen or sulfur oxides released from fossil fuel burning are converted to acids in the atmosphere. Falling as acid rain, snow, and sleet or as acid fog or dry deposition, it can be harmful to plants and animals that are exposed. CO_2 emissions are acidifying the oceans, with potentially devastating consequences.
- The Clean Air Act (CAA) has improved U.S. air quality through the regulation of targeted pollutants. Recently, CO_2, an air pollutant linked to climate change, has been identified as a pollutant that must be regulated, but political battles have hampered the CAA's ability to do so.
- Indoor air pollution is a major source of exposure for people worldwide. Better ventilation and alternative building or household materials can reduce this pollution in modern homes. In areas where poorly ventilated indoor cooking fires produce air pollution, using cleaner fuels and solar ovens can improve indoor air quality.
- Industrial air pollution can be lessened by using cleaner fuels and emission control technologies. The CAA can regulate these pollutants with command-and-control rules or by offering economic incentives for the reduction of air pollution.

ENVIRONMENTAL LITERACY Understanding the Issue

1 What is air pollution, and what is its global impact?

1. Worldwide, how prevalent is air pollution and how serious is it as a health problem?
2. How did air pollution events in Donora, Pennsylvania, and London, England, provide evidence that air pollution can be an immediate health threat to humans?

2 What are the main types and sources of outdoor air pollution?

3. Distinguish between primary and secondary forms of air pollution and give an example of each.
4. What is particulate matter? Identify some sources of this type of pollution where you live.
5. What is transboundary pollution, and why is it difficult to regulate?

3 What are the health, economic, social, and ecological consequences of air pollution?

6. Why are cardiovascular diseases linked to air pollution?
7. Describe a scenario in which air pollution could be an environmental justice issue.
8. How does air pollution negatively affect plants?

4 What are the causes and consequences of acid deposition?

9. In general, what kinds of actions would reduce the production of acid deposition?
10. Why are high-elevation trees often more affected by acid deposition than trees lower on a mountain?
11. Aside from health impacts, what are some of the societal and economic consequences of acid deposition?

5 What is the mandate of the Clean Air Act, and how is it implemented?

12. What is meant by the term *criteria air pollutants*? Which six pollutants are classified as criteria air pollutants?
13. What are volatile organic compounds (VOCs)? Give some examples of VOC sources.
14. Why was the EPA's initial regulation to address specific pollutants in 1971 seen as an example of the precautionary principle?

6 What are the main sources of indoor air pollution, and what can be done to reduce it?

15. The main sources of air pollution in the home differ between developed and developing countries, but they do have one factor in common. What is that factor?
16. What are some sources of indoor air pollution you might face?
17. In what part of the house is the radon that contaminates the air of some homes usually found, and where does it come from?

7 How can outdoor air pollution be reduced, and what are the trade-offs of reducing it?

18. Distinguish between these two methods of reducing air pollution: command-and-control regulations and green taxes.
19. Describe the policy of cap-and-trade. What are the advantages and disadvantages of this policy option?
20. Taking steps to reduce air pollution is often expensive. How could investments in reducing air pollution actually result in lower overall costs to the American public?
21. What was the purpose of the Clean Power Plan? How well is its replacement, the Affordable Clean Energy rule, expected to achieve the overall goal of both?

SCIENCE LITERACY Working with Data

The graph shown here indicates levels of ground-level ozone and particulate matter that exceeded national reference levels (the levels above which health or ecosystem problems occur) in areas and cities in Canada.

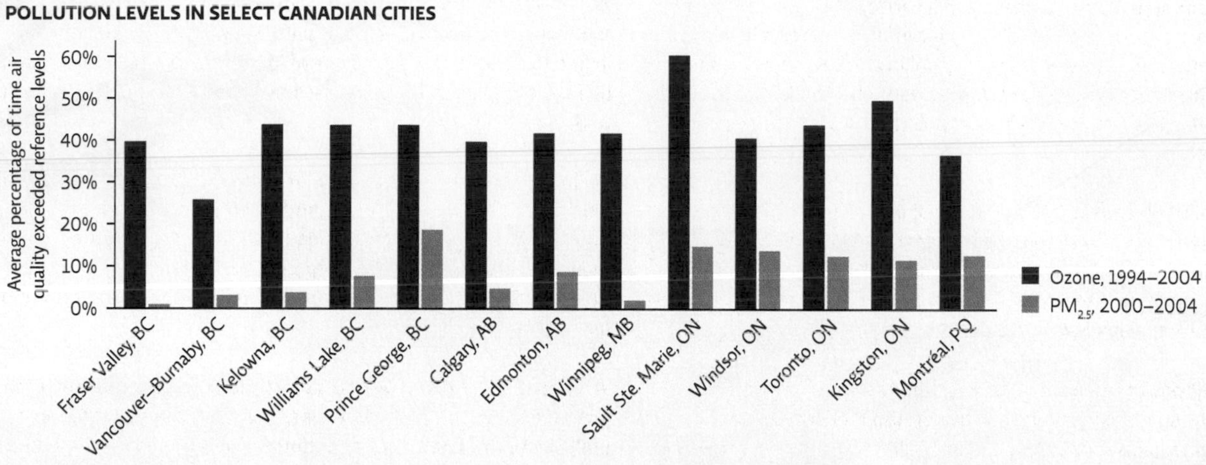

Interpretation

1. What does the *y* axis represent? Choose a city and describe the data for that city.
2. The graph presents data for $PM_{2.5}$. PM stands for *particulate matter.* The number 2.5 represents the size of the particulate in micrometers. From a health standpoint, why are the $PM_{2.5}$ values reported rather than a larger size (e.g., PM_{10})?
3. How many years of data are graphed for ozone? For particulate matter? Does the difference in the amount of time over which the data have been collected make a difference in your interpretation of these data?

Advance Your Thinking

4. Which city would be the worst for your health based on its levels of pollution? Why?
5. Based on the type of pollution present, what can you predict about the causes of pollution in Sault Ste. Marie versus Montreal? (Hint: Do some research and compare the city size, weather, and primary industries.)
6. In general, cities were out of compliance more from ground-level ozone in the time frame tested than for particulate matter. Does this mean particulate matter is less of a problem than ground-level ozone? Explain.

INFORMATION LITERACY Evaluating Information

The EPA is tasked with regulating pollutants in the United States. As part of this process, the agency collects and records data for many pollutants but not all of them. The federal EPA is assisted in this endeavor by state EPAs. However, it is impossible to collect air quality data about every locality in the United States, so most data are collected in and around cities.

Go to www.stateoftheair.org. Enter your state in the space provided and select the county closest to you with data for both ozone and particulate pollution. Note the county you chose and record these data. Then click on "Key Findings" at the top of the page and read about how the grades were calculated for each county. Finally, click on "Health Risks" and read about the specific health risks associated with both ozone and particulate matter.

Evaluate the website and work with the information to answer the following questions:

1. Determine if this is a reliable information source:
   a. Who runs the website? Do this organization's credentials make it reliable or unreliable? Explain.
   b. What is this organization's mission? How do you know this?
2. What grade did your area receive for both ozone and particulates?
   a. Based on what you read about how the grade was determined, do you feel the grading system is too lax or too strict?
   b. Why does the American Lung Association advocate for a stricter system?
3. Based on what you have learned, do you believe that the regulations of the Clean Air Act should be loosened, be tightened, or remain the same? Should more areas be monitored, or is it sufficient to monitor only large cities? Why or why not?

Additional study questions are available at Achieve.macmillanlearning.com Achieve

Module 10.2 Climate Change

CLIMATE REFUGEES

Grappling with current and future climate change

Climate change is contributing to rising sea level that is flooding the Isle de Jean Charles in Louisiana, turning its residents into climate refugees who must relocate. *Chloé Durand/Alamy*

After reading this module, you should be able to answer these GUIDING QUESTIONS

1. What is climate change, and why is it more concerning than day-to-day changes in weather?
2. What is the physical and biological evidence that climate change is occurring?
3. What is the greenhouse effect, and how are we affecting it?
4. How are atmospheric CO_2 and temperature measured, and how are they correlated?
5. Other than greenhouse gases, what factors affect climate?
6. What evidence suggests that climate change is due to human impact?
7. What are the current and potential future impacts of climate change?
8. What actions can we take to respond to a world with a changing climate?

In the spring of 2017, a team of scientists led by Matthew Hauer, then at the University of Georgia, published a study in the scientific journal *Nature Climate Change* that would make headlines across the country and raise anxiety levels in coastal city dwellers from Miami to Manhattan. The scientists had used climate forecasts and projections of population size and distribution to predict the impact that warming global temperatures and rising sea levels would have on cities in the continental United States.

The picture they painted was grim: A sea-level rise of 3 feet by the year 2100 would imperil the homes of some 4 million people. At 6 feet, the number of people displaced would top 13 million. New Orleans alone would lose 500,000 residents; New York City would lose 50,000; and more than 2.5 million people would flee Miami, the hardest-hit city of all. And it wasn't just coastal cities that would suffer. As refugees fled the rising tides, landlocked cities like Austin and Atlanta would likely be inundated with the new arrivals, all the while struggling with the effects of rising global temperatures.

Worldwide, climate-induced migrations are underway. Though it is sometimes difficult to clearly identify a single cause of migration, natural disasters and violence are the leading factors. In 2018 alone, more than 17 million people were displaced by natural disasters—the average over the past 11 years is 24 million per year according to the Internal Displacement Monitoring Center. Climate change may play a direct role in these natural disasters, but it is also seen as a threat multiplier, exacerbating problems brought on by other issues, such as increased conflict over resources made scarce by climate impacts.

The specter of *climate refugees*—people who are forced to flee their homelands because of global warming–related changes to their environment—is often viewed by those in the United States as a problem of other countries in distant parts of the world. When we talk about whole cities being swallowed by a rising ocean, we might think of the Maldives or Bangladesh, low-lying countries that have been dramatically altered by these forces in recent years. But Hauer's results painted it, very clearly, as a U.S. problem, too—and a fairly big one at that.

"These results provide the first glimpse of how climate change will reshape future population distributions," Hauer wrote. "The absence of protective measures could lead to U.S. population movements of a magnitude similar to the 20th century Great Migration of southern African Americans." (Between 1915 and 1970, some 6 million African Americans migrated from southern states to northern ones to pursue economic opportunities and to escape the segregationist South.) Already in the United States, some low-lying coastal communities are moving or contemplating relocation, including residents on the Isle de Jean Charles, a tiny spit of land about 75 miles south of New Orleans that is shrinking every year and has the distinction of being the first U.S. community to receive federal dollars to fund relocation.

WHERE IS THE ISLE DE JEAN CHARLES, LOUISIANA?

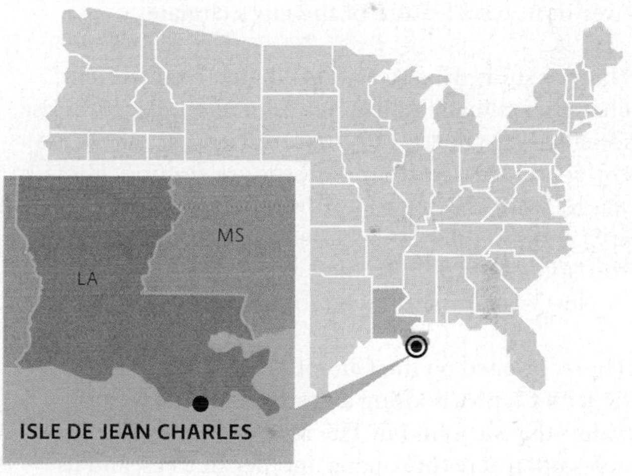

Hauer's predictions underscored a basic set of facts on which scientists have long agreed: The planet is warming at an alarming rate; that warming is due to human activity; if something isn't done to curb it, the impacts on both natural ecosystems and human communities will be severe.

But they also laid bare a string of vexing questions over which scientists and policy makers have long been puzzling (and sometimes arguing): How certain are we of what the exact impact will be? What should we do about it? And how much time do we have?

1 CLIMATE AND CLIMATE CHANGE

Key Concept 1: A change of just a few degrees in average temperature can result in a drastically different climate with negative effects on ecosystems and human societies.

To begin to understand climate change, we must first know the difference between climate and weather. **Weather** refers to the meteorological conditions in a given place on a given day, whereas **climate** refers to long-term patterns or trends. In other words, the actual temperature on any given day is part of the weather, while the range of expected values, based on the location and time of year, is the climate.

For example, Miami's climate is tropical; overall, it has hot humid summers and warm, short winters. Its average annual temperature, another characteristic of its climate, is 77°F (25°C). But the city's weather on any given day is variable. Today's temperature might be as high as 98°F or as low as 60°F. It might rain; it might be cloudy or windy. Miami often endures hurricanes in the late summer and fall. Those individual storms fall under the category of weather. But their frequency and strength over time are a feature of the city's climate.

Hauer's study tried to assess the future effects of a changing climate: As average global temperatures rise, ice sheets are melting; as that happens, sea levels are rising and local weather patterns are shifting. His task was to try to predict what those changes would mean for U.S. communities over the next century. But it turns out you don't need a crystal ball to see the future of Miami or New York. You just need to hop over to Louisiana.

There, situated on the Gulf of Mexico coast, the Isle de Jean Charles is losing a desperate battle with the waters that surround it. The island was once home to acres upon acres of banana and pecan trees, and its inhabitants—Native Americans who had lived there for centuries—used to hunt, fish, trap, and farm there. But in the past half-century, more than 98% of the land has washed away due in part to human actions such as dredging to create canals and shipping lanes that allow stronger storm surges to rush in during hurricanes and wash away land. The many levees, built to prevent flooding, also decreased the delivery of sediment that would naturally restore land lost to erosion. But an additional driver—perhaps the final nail in the coffin—may be climate change that is raising sea level and accelerating the loss of land. The rest of the island is expected to disappear in the next several years as the climate continues to change.

weather The meteorological conditions in a given place on a given day.

climate Long-term patterns or trends of meteorological conditions.

climate change Alteration in the long-term patterns and statistical averages of meteorological events.

The Isle de Jean Charles is not the only U.S. community or region threatened by rising seas—not by a long shot. Coastal communities on the Pacific and Atlantic shores are at risk—Key West, Florida; the naval base in Norfolk, Virginia; villages along the Alaskan coast; or as Jeff Goodell writes in his book, *The Water Will Come*, "any beach you've ever visited."

"The changes are underway and they are very rapid," former interior secretary Sally Jewell has warned. "We will have climate refugees."

Climate change refers to alterations in the long-term patterns and statistical averages of meteorological events. Even small shifts in climate can result in major changes to ecosystems. During the last ice age, for example, an ice sheet about a half a mile thick kissed the borders of Chicago. The average temperature at that time? Just 7°F (~4°C) colder than today.

Earth is currently undergoing climate change in the direction of warming. This global warming—the rise in average temperatures that has been measured in myriad locations around the world—is already pushing record-high temperatures even higher, creating more and longer heat waves and resulting in earlier springs and later winters. Global warming can also lead (and is leading) to extreme cold weather in some places; for example, shifting weather patterns that affect the jet stream send Arctic air farther south or for longer periods of time, resulting in some recent record cold and snow events in the United States, Europe, and Asia.

To examine these temperature changes, James Hansen and Makiko Sato of Columbia University compared the average summer temperatures for the Northern Hemisphere in recent decades to the baseline average of the decades between 1951 and 1980. In this baseline period, temperature data generated a bell curve; Hansen segregated the data into "normal," colder than normal, and hotter than normal, each occurring one-third of the time. Extreme cold or hot years were defined as those that occurred only 0.1% of the time. When he compared subsequent 10-year periods, he saw the bell curve shifting to the right. In the most recent 10-year period evaluated

(2005 to 2015), two-thirds of the yearly averages were in the hotter-than-normal range (compared to one-third in the baseline period)—15% of those moved into the extremely hot range, with many values outside of the original data set. The bell curve also gets broader and flatter with time, an indication of more variability in the climate (though some scientists attribute some of this to the uneven warming by latitude on the planet). A similar analysis of winter averages also reveals a shift toward warming, though it is not as pronounced as the summer shift. **INFOGRAPHIC 1**

The effects of global warming are not limited to temperature impacts. These changes are already setting into motion other changes that have serious consequences for life on Earth: Positional shifts in biomes (whose locations are established by temperature and precipitation [see Module 2.1]) stress already-imperiled ecosystems and their resident species, many of which are currently endangered (see Chapter 3); expanding habitats bring mosquitoes and the diseases they carry out of the tropics and subtropics into temperate regions (see Module 4.3); floods ruin crops in some areas, and drought does so in others—both of which decrease food supplies at a time when we are barely able to produce enough food to feed the world and need to increase production to feed our growing population (see Chapter 8). Other impacts (to be discussed in more detail later) include rising sea level, loss of crucial freshwater supplies (see Module 6.1), stronger storms on land and sea, and more unpredictable and variable weather in general. All of these impacts are already occurring and will only get worse as climate continues to change. For these reasons, the broader term *climate change* is more useful than *global warming* when we want to refer to all the alterations we are experiencing.

The consequences of climate change are affecting communities around the world, not just on the Isle de Jean Charles. According to a recent report by the Center for Progressive Reform, at least 17 U.S. communities are currently being forced to relocate. Arctic regions are some of the hardest hit as the area warms two to three times faster than the world average. Newtok, a small village on the northern coast of Alaska, sits on permafrost that was once rock solid but is now thawing rapidly. As the land sinks and water seeps in, the village supporting 450 or so people is disappearing. The drinking water supply was gone by the end of 2018. The airport and school will likely be gone by 2023. In 1996, the villagers voted to relocate.

INFOGRAPHIC 1 **CLIMATE CHANGE: WHY DO A FEW DEGREES MATTER?** Achieve

A shift of just under 1°C in the average global temperature is producing more frequent and extreme heat waves. Temperature changes are giving rise to other, more far-reaching effects such as changes in precipitation, ocean levels, the frequency and severity of storms, and the makeup of ecosystems across the globe—just to name a few impacts. All of these changes are already affecting life on Earth.

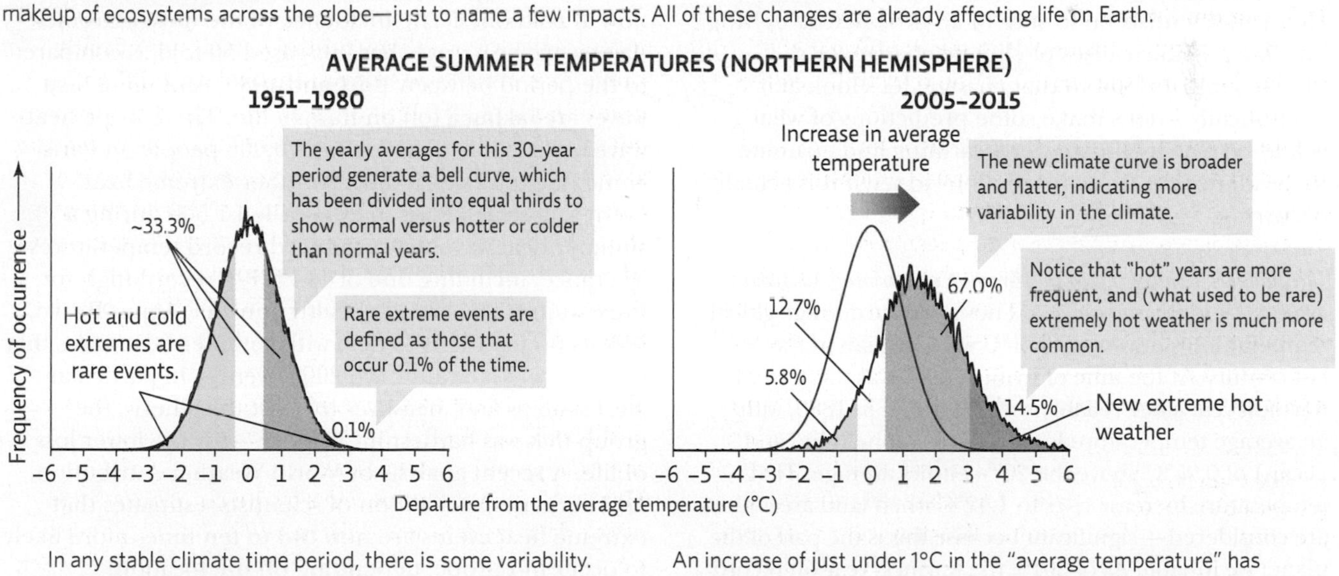

In any stable climate time period, there is some variability. Extreme hot or cold years are rare but occur with similar frequency. Here we see the distribution of average temperatures between 1951 and 1980; about a third of temperatures fall into each section of the bell curve.

An increase of just under 1°C in the "average temperature" has shifted the entire curve to the right. What used to be rare heat events become more common, and new "hot extremes" are experienced. In 2016, there were 22,470 U.S. heat records but only 4,016 cold records.

? Between 2005 and 2015, what percentage of years fell above the "normal" temperature range of the baseline time period? How does that compare to the percentage of years higher than normal from 1951 to 1980?

2 EVIDENCE FOR CLIMATE CHANGE

Key Concept 2: A warming planet should see warmer average temperatures, melting ice on sea and land, rising sea levels, and precipitation changes. All of these are currently being observed. Species' responses to climate change such as range shifts provide strong evidence that climate is changing enough to affect ecosystems.

Climate is complex, and it takes many lines of evidence to be sure that it's changing. In general, scientists look for a 30-year trend, at least, before concluding that climate has crossed the boundary of natural variability and that a new trend is emerging.

So it's no small matter to say that more than 97% of climate scientists agree that climate is now changing and that this change is in response to human activities (see Infographic 6). The evidence for this is substantial and well supported by thousands of studies from a wide variety of scientific fields (the hallmark of a well-established scientific theory; see Module 1.2). Of course, there is plenty of uncertainty over the rate at which future changes and impacts will unfold: No one can say for sure the exact year that Miami will be permanently flooded, for example. But the broad outline of consequences is not in dispute: Much of Miami will be under water at some point if we don't address climate change. An average temperature increase of 3.5° to 7°F (~2° to 4°C)—which is where we are heading—will inundate our coastlines with rising water, drive many species to extinction, and lead to significant agricultural declines.

How can scientists be so sure of all this? Before delving into the possible causes of climate change, or the human contribution to that change, let's look at it scientifically—let's make some predictions of what would be true if climate were warming and examine the evidence for those predictions to see if it is actually occurring.

If climate were changing in the direction of warming, we would expect to see temperatures rising. They are: The average global temperature is now roughly 1°C warmer than in the last century. At the time of writing, 2019 was the second warmest year since reliable records began in 1880, with an average temperature for the entire globe (land and ocean) of 0.95°C above the 20th-century average. That temperature increase rises to 1.42°C when land areas alone are considered—significant because this is the part of the planet we inhabit. As of 2019, the warmest year on record was 2016, with temperatures 0.99°C higher than the 20th-century average for land and ocean and 1.54°C higher on land areas alone. If annual temperatures are compared to preindustrial times (1850–1900), the departure from average is even greater: The global temperature over land in 2016 was more than 1.75°C higher, according to the 2019 Intergovernmental Panel on Climate Change (IPCC) special report *Climate Change and Land.*

Once the data were in, 2019 became the 43rd consecutive warmer-than-normal year, 13 years past the 30 years that climatologists require as the benchmark for the emergence of a new climate trend. Analyses by both the National Aeronautics and Space Administration (NASA) and the National Oceanic and Atmospheric Administration (NOAA) find that 2010–2019 was the warmest decade in recorded history, and the last 6 years of that decade were the warmest. According to Gavin Schmidt, director of NASA's Goddard Institute for Space Studies, this continues a trend that began in 1960—every decade since that time has been warmer than the one before.

With rising temperatures, we would expect to see changes in weather such as more precipitation in coastal areas (due to more ocean water evaporation), more extreme storms (more heat in the atmosphere and oceans means more energy in the system to fuel storms and winds), and more record temperatures. All of these are occurring: There are roughly twice as many heat records set now as cold ones. The incidence of the most extreme dangerous heat waves has increased 50-fold as compared to the period between 1951 and 1980. And these heat waves are taking a toll on human life. The Europe heat wave of 2003 killed more than 70,000 people; in Paris alone, the total was 15,000. Another extreme heat event that struck Paris in 2019 killed 1,500 during a summer heat wave that produced record temperatures in France, including one of 114.6°F, a record high for the country. The French health minister, Agnes Buzyn, credits preparations to deal with summer heat waves that were put in place after the 2003 event—in particular alert systems and measures to assist the elderly, the group that was hardest hit in 2003—for the lower loss of life. A recent analysis by World Weather Attribution, an international coalition of scientists, estimates that extreme heat events are now two to ten times more likely to occur in Europe, depending on the location.

Research by Michael Mann of Pennsylvania State University and others has linked some of these extreme events, such as storms, to alterations in the global jet stream that cause it to become stationary, keeping in place whatever weather pattern exists. Heat waves persist, rain or snow events stall over an area and dump record precipitation, or a dry spell lasts for weeks or months longer than usual. Even the catastrophic hurricane season of 2017 has a connection to climate change; though climate change didn't cause hurricanes Harvey, Irma, or Maria, scientists say it is making storms like these more powerful and thus more destructive. "We are seeing . . . remarkable changes across the planet that are challenging the limits of our understanding of the climate system. We are now in truly uncharted territory," says World Climate Research Program Director David Carlson in a 2017 World Meteorological Organization press release.

We would also expect to see ice melting if temperature were warming. We are: Ice and permafrost (a deeper soil layer that is frozen for at least 2 consecutive years) are melting, and the rate of melt is accelerating; Arctic sea ice is setting record lows for ice extent in summer and more recently in winter as well. Glaciers are melting at an increasingly rapid rate. In 1850, Glacier National Park in Montana had 150 glaciers; in 2019, only 26 remained. Greenland's ice sheet has been losing ice since 1998 at an ever-increasing rate. The response in the Antarctic is less straightforward, with some areas warming significantly and losing ice (enough to cause a noticeable increase in moss coverage, leading to a "greening" of some regions of Antarctica), while others are cooling and gaining snow. These differences are attributed to, in part, the way ocean currents are redistributing warm versus cool water.

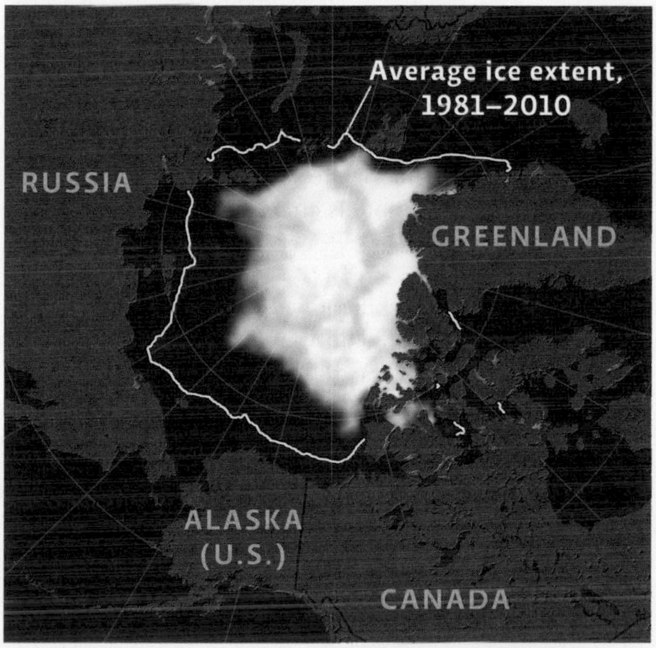

NASA Earth Observatory

Seasonal Arctic Ocean ice melt is increasing with the 2019 summer sea ice extent, shown here, the second lowest since 1979 when reliable records began. The ice is also thinner, contributing to higher melt rates.

If water is warming and land-based ice is melting, we expect to see sea level rising. We are: Water expands as it warms, and land-based ice can send meltwater to the ocean, both of which can contribute to sea-level rise. (Melting icebergs would not raise sea level since they already displace the same volume of water while floating as the meltwater they would add to the oceans.)

Globally, sea level has risen around 8 inches since 1900; almost half of that has occurred in the last 25 years, indicating that the rate of rise is increasing. In South Florida, sea-level rise is similar to the global average, but at the U.S. mid-Atlantic coast, the rise has been even greater and barrier islands there are losing ground, some even faster than the Isle de Jean Charles. And with higher sea levels come more damaging storm surges during hurricanes or violent weather and more coastal erosion. Some villages on the Alaskan coast are being destroyed by heavy wave action during storms that carves off shoreline.

Hauer's work evaluated population impacts at 3- and 6-feet sea-level increases (mid-range and high-range scenarios of future climate change as determined by Martin Vermeer of Helsinki University). This kind of rise is not out of the question—if not by 2100, then later since sea level will continue to rise, probably for centuries, even after temperatures stabilize. High-end predictions (what we are likely to see with no efforts to curb fossil fuel use or other drivers of climate change) place sea-level rise around 6 feet by 2100. Even our low-end predictions, predicated on a robust international response to address the problem, predict a sea-level rise between 2.6–4.3 feet by 2100. And as Hauer notes, long before residents may be forced to evacuate due to inundation, rising seas bring problems such as saltwater intrusion into freshwater aquifers and cropland, coastal erosion, and storm surges that reach further inland.

Unfortunately, we are not currently on track for that low-end prediction—sea-level rise will likely be higher, closer to Vermeer's mid-range scenario prediction of 3 to 4.75 feet by 2100. Indeed, in a 2016 assessment of sea-level change, Peter Clark of Oregon State University noted that our CO_2 emissions and other actions that are contributing to climate change are already dangerously close to the maximum we cannot exceed if we are to achieve the low-end scenario; unless things change drastically, and fast, that low-end scenario will soon be out of reach. Clark also pointed out that a historical analysis of sea-level rise shows that it is slow but persistent in the face of a warming climate; in the last major warming period (as Earth exited the last ice age), sea-level rise continued for thousands of years after CO_2 levels and global temperature stabilized. He writes that "twenty-first-century global average warming will [produce] a climate state not previously experienced by human civilizations." Indeed, we are already there with atmospheric CO_2 levels and temperatures higher than any human has ever seen.

So there is ample evidence of physical changes in air, on land, and at sea. (We've presented just a smattering of this physical evidence here.) The bigger question might be, does this matter—is it affecting life on Earth?

Back to our predictions: *If all of these changes were significant enough to affect species, we would expect to see clear impacts on biodiversity.* This is happening: Changes in habitats and niches that affect where species are found, which in turn can lead to broken relationships (e.g., the loss of prey or mutualistic partners), are being seen. There is ample evidence for this as well.

In general, when faced with a changing environment, a species' population can either stay put (because its members can handle the change), adapt to the change (if genetic diversity in the population allows), relocate (if possible, geographically or biologically), or face extinction. A review article by Gretta Pecl evaluated the movement of species and calls the movement of species today "the largest climate-driven global redistribution of species since the Last Glacial Maximum." These movements have consequences for human welfare as well—both directly (e.g., through the spread of disease or declines in food supply) and indirectly (through the loss of ecosystem services).

Some species are benefiting from a warmer climate. For example, bark beetles that attack trees in North America are increasing in number; their populations usually die back in the winter, helping to keep them in check and allowing the trees to recover. However, in some areas, winter temperatures don't always get cold enough to kill the beetles, allowing these insects to thrive year-round. So far, nearly 4 million acres in the western United States have suffered damage. Tropical species of mosquitoes, too, are expanding their ranges to higher latitudes. Many species of squid and octopus are also increasing in number and distribution, believed to be due to their ability to adapt to warmer temperatures even as other species (many of them competitors of the squid and octopus) decline.

However, even more species are negatively affected as their range shrinks or shifts to higher latitudes or altitudes. Unfortunately, they can't always migrate to a more suitable location because we have fragmented habitats and destroyed migratory pathways (see Module 3.2 for more on the impacts of habitat fragmentation). For species already at the upper edge of suitable habitat (top of the mountain, at the poles), there is simply nowhere else to go. But those that can relocate are on the move; scientists have documented a wide variety of terrestrial species shifting their ranges to higher latitudes or elevations—Pecl's analysis finds the average rate of movement to be 10.5 miles per decade. Marine species, too, are moving poleward at an even faster rate (45 miles per decade), or they are moving deeper into the ocean. (See the *Science Literacy* activity at the end of this module for a look at migration trends in marine species.) Even trees are "migrating" as their saplings experience higher survival rates at or just beyond the higher altitudes or latitudes of their traditional ranges.

Shifts in the timing of migration are also uncoupling species connections: Some migratory birds are arriving at summer breeding grounds at their normal time to find the insects that normally provide the main food source for their offspring have already hatched out or emerged from winter hibernation and left the area. (Many insects hatch out or emerge based on temperature cues, so a warming climate means they arrive sooner; most birds migrate in response to daylight cues, which are not affected by climate.) In addition, because there will likely be fewer predators around when these insects arrive or hatch out early, the insects can wreak havoc on plant life (both native plants and crops). The fact that so many and such a wide variety of species from all of Earth's ecosystems are responding is some of the strongest evidence we have not only that climate is changing but also that it matters.

Species differ in their ability to adjust to rapid climate changes—a reflection of their generation time, reproductive potential, the genetic diversity of the population (see Module 3.1 for more on how populations evolve in response to environmental change), and their ability to relocate, among other things. The fact that many species are endangered, with their numbers already critically low, makes them even more vulnerable. Climate is simply changing too quickly for many species to adapt. The loss of species, from habitat loss or the loss of ecosystem partners, impoverishes the ecosystem further, likely triggering the endangerment of more species in a domino effect that can escalate. This then reduces the ability of the ecosystem to provide the ecosystem services on which all species, including us, depend. **INFOGRAPHIC 2**

As humans from Alaska to Louisiana to islands of the South Pacific face migration for the same reasons—loss of suitable "habitat"—we are faced with a complicated mix of problems. In Newtok, the primary problem is money. This Alaskan village is small—just 450 people—but the costs of relocation will still be large. Most estimates put the price tag at more than $100 million. Residents know they need to move, and fast. "We just need to get out of there," Romy Cadiente, the village's relocation coordinator, told National Public Radio in 2017. "We really do, for the safety of the 450 people there."

Newtok leaders have cobbled together state and federal grants and other funding but need much more to pay for the relocation effort. A group of Newtok villagers is pushing the federal government to declare Newtok a disaster zone so that emergency funds would be available for the move. But disasters are usually declared for specific weather events, like a hurricane or flood, not for slow-moving disasters like the one Newtok faces (revealing that the differences between weather and climate are not just theoretical; they are practical too).

Despite the difficulties, Newtok is forging ahead, building a new community 9 miles south in the newly created village of Mertarvik where construction on new homes and village buildings is underway. A few homes are finished, and residents began the move in late 2019—a community split between two villages as those who remain in Newtok await construction of the final homes and hope the funds needed to finish the village and relocate their families will be forthcoming.

But even when money is available, the problems of relocation are still legion. The Isle de Jean Charles managed to secure a $48 million federal grant to move just 60 people—the very first allocation of federal money to assist climate refugees. But efforts have been fraught with bickering and uncertainty: Where should everyone go? What will become of the land they leave behind? What will become of their community and culture if they are dispersed? After 2 years of surveys, meetings, and exploration of suitable sites, a 515-acre property was purchased in Terrebonne Parish, Louisiana, some 40 miles away from the island. The community that will be constructed is now in the planning phase, but some residents are reluctant to move from the land that they feel a lifelong connection to and from homes they own free and clear. Can they afford the taxes and insurance on their new homes? Can they pay the utility bills? None of the answers come easily.

Charles Mason/The New York Times/Redux Pictures

Residents of Newtok, Alaska, walk on boardwalks that rest above flooded ground. Thawing permafrost has destabilized much of the land there; during the spring thaw, eroded soil and rising waters turn the town into an island.

INFOGRAPHIC 2 EVIDENCE FOR CLIMATE CHANGE

We know that a variety of factors can alter global temperature, but what is the evidence that temperatures have actually increased and that the climate is changing? In other words, what do we predict we would see if warming were occurring, and what do we actually see when we test those predictions?

***PREDICTION*: GLOBAL WARMING SHOULD LEAD TO WARMER TEMPERATURES.**
If climate is indeed warming, we expect to see warmer global temperatures, on average, than in the recent past (more temperature anomalies in the direction of warming).

***OBSERVATION*: TEMPERATURE IS WARMING**

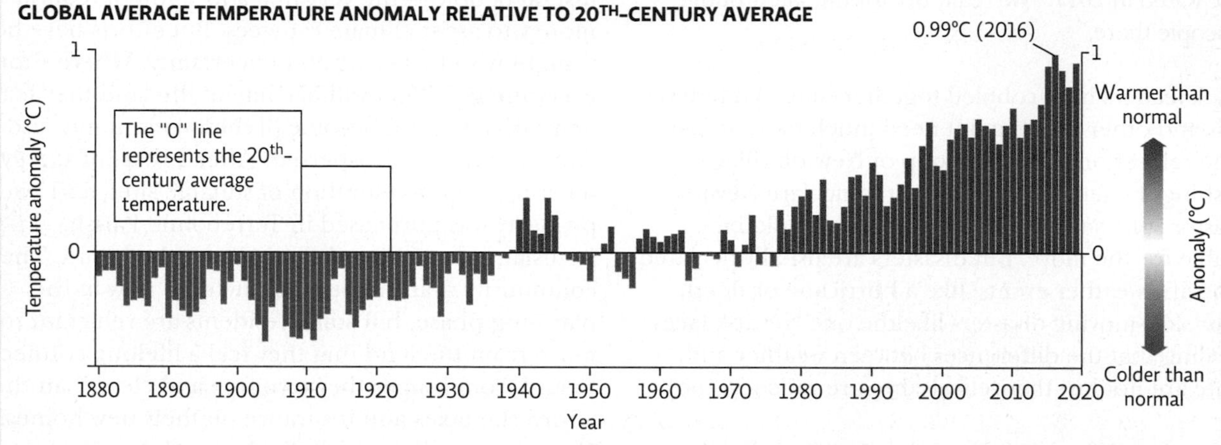

Climatologists generally look for directional changes over at least 30 years before concluding that the changes represent a trend. If you were born after 1976, you have never experienced a colder-than-average year relative to the 20th-century average.

***PREDICTION*: WARMER TEMPERATURES SHOULD LEAD TO:**

MELTING ICE
If temperatures are warming, we would expect to see more ice melt.

SEA-LEVEL RISE
We would expect to see an increase in sea level as land-based ice melts and as warmer seawater expands.

***OBSERVATIONS*: ICE IS MELTING**

SEA LEVEL IS RISING

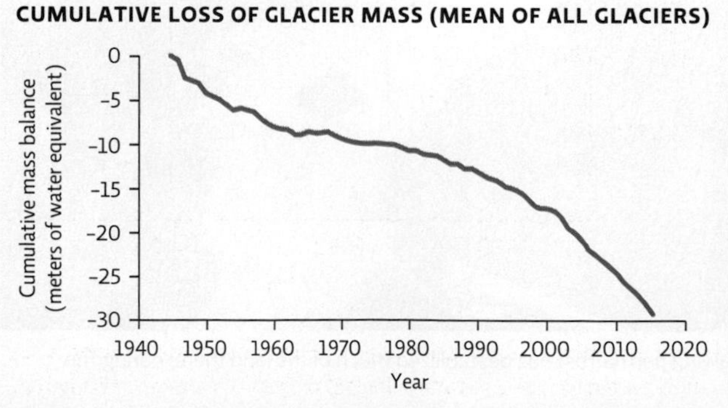

Since the middle of the 20th century, glaciers around the world have experienced a net loss of ice. Record ice melt has been observed in the Arctic and Antarctic as well. Permafrost is also thawing in high-latitude and high-altitude regions.

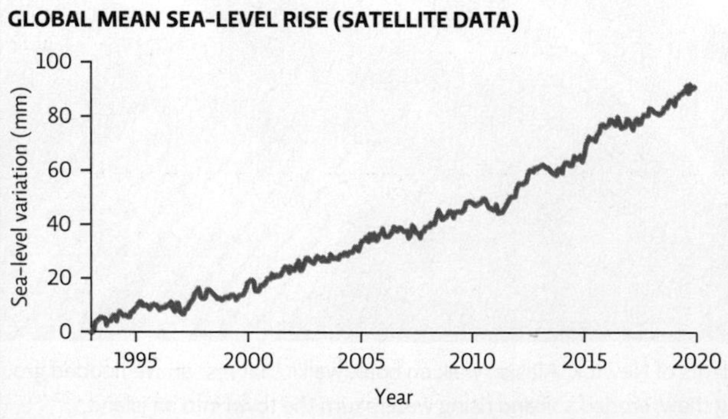

Sea level had risen ~8 inches since 1900; roughly half of that increase has come since 1993.

PREDICTION: WARMING SHOULD ALSO LEAD TO:

CHANGES IN WEATHER

OBSERVATION: PRECIPITATION IS CHANGING

EXTREME ONE-DAY PRECIPITATION EVENTS IN THE CONTINGUOUS 48 STATES, 1910–2015

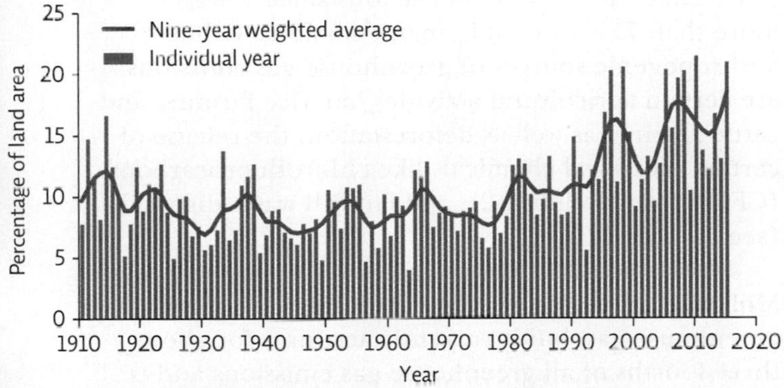

Precipitation has changed globally — some areas are wetter, and others are drier. Worldwide, the occurrence of extreme precipitation events is on the increase, even in areas that are drier than normal.

OBSERVATIONS: SEASONAL WEATHER PATTERNS ARE CHANGING

TIMING OF SPRING AND FALL FROSTS (UNITED STATES)

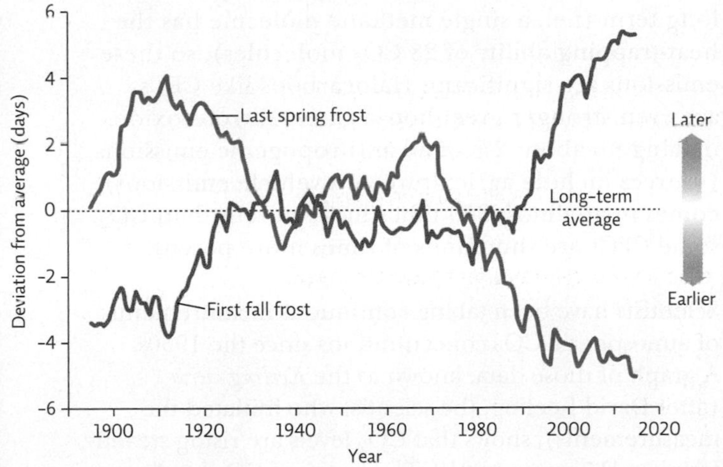

Since about 1980, the last spring frost is coming earlier and the first fall frost is coming later, extending the growing season. Note that in recent years, these first and last frost dates exceed any that were seen since 1890.

CHANGES IN BIOLOGICAL EVENTS

SPECIES ARE SHIFTING RANGES

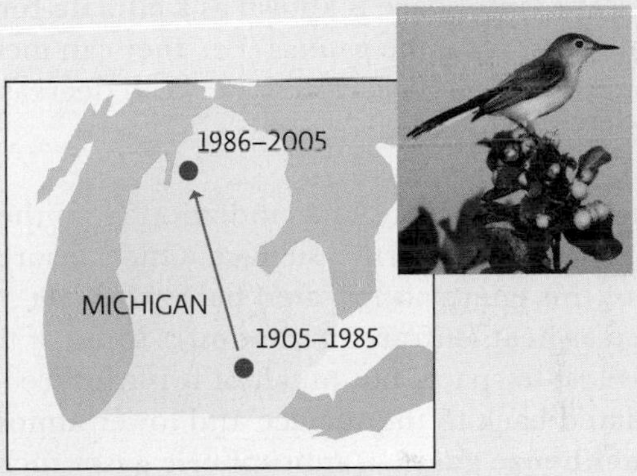

The ranges of some species are shifting. For example, the breeding range of the blue-gray gnatcatcher has shifted about 200 miles northward since 1970.

PHENOLOGICAL SHIFTS ARE BEING SEEN

LENGTH OF THE FLOWERING SEASON (COMMUNITY AVERAGE)

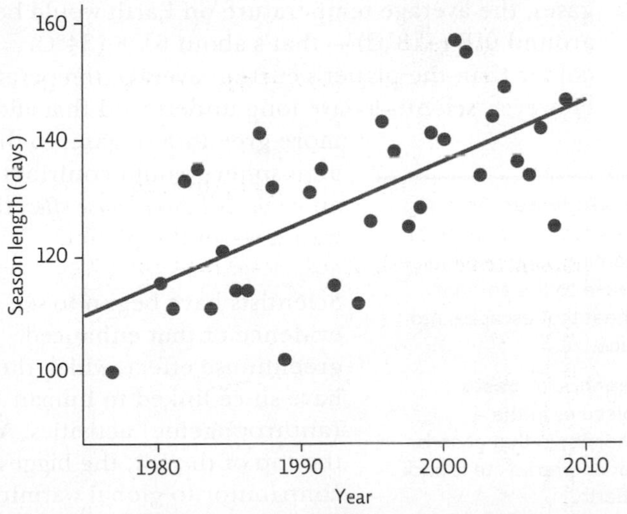

Changing climate can also change the timing of seasonal events like blooming or migration (phenology changes). A Colorado study that began in 1974 and followed 60 species showed that for the community as a whole, the length of time these flowers are in bloom has increased.

Why is it important to look at multiple lines of evidence that climate is changing rather than just evaluate global temperatures?

3 CLIMATE FORCERS: THE GREENHOUSE EFFECT AND HUMAN IMPACT

Key Concept 3: Greenhouse gases trap heat reradiated from Earth and warm the atmosphere. By adding more greenhouse gases to the atmosphere, the greenhouse effect is enhanced, warming the planet even more.

What is causing this climate change? For that matter, what kinds of things have the power to affect climate?

The Sun fuels life on Earth by providing the energy for photosynthesis (see Module 2.1), but it also fuels Earth's climate. Anything that alters the balance of incoming solar radiation relative to the amount of heat that escapes into space is known as a **climate forcer**. Climate forcers can be *positive* (i.e., they can increase warming) or *negative* (meaning they can decrease warming).

Much of the incoming solar radiation from the Sun is absorbed at the Earth's surface. Once absorbed, most of this energy is radiated back from the surface as heat (infrared radiation). Some of this heat is lost to space, but much of it is captured and reradiated back to the surface and lower atmosphere by **greenhouse gases**—atmospheric gases such as water vapor, carbon dioxide (CO_2), methane, and nitrous oxide. These are small molecules with loosely held atoms that vibrate as they absorb infrared radiation, releasing that energy as heat, which warms the atmosphere. This process, known as the **greenhouse effect**, helps warm the Earth; therefore, greenhouse gases are positive climate forcers. To be sure, the greenhouse effect is what keeps Earth habitable for humans and all current species on the planet. Without these greenhouse gases, the average temperature on Earth would be around 0°F (–18°C)—that's about 61°F (34°C) colder than the planet's current average temperature! However, scientists have long understood that adding more greenhouse gases to the atmosphere would contribute to an *enhanced greenhouse effect* that would warm the planet.

climate forcer Anything that alters the balance of incoming solar radiation relative to the amount of heat that escapes into space.

greenhouse gases Molecules in the atmosphere that absorb heat and reradiate it back to Earth.

greenhouse effect The warming of the planet that results when heat is trapped by Earth's atmosphere.

Scientists have begun to see evidence of that enhanced greenhouse effect, which they have since linked to human (anthropogenic) activities. At the top of the list, the biggest contributor to global warming is the burning of fossil fuels (coal, petroleum-based fuels, and natural gas), which releases massive quantities of greenhouse gases into the atmosphere, an action that began in earnest with the Industrial Revolution. Isotopic analysis of atmospheric CO_2 allows scientists to identify the percentage that comes from fossil fuel combustion. (There is a different ratio of two carbon isotopes [^{12}C and ^{14}C] in fossil fuel emissions as compared to CO_2 released from natural sources such as forests or soils.) This allows us to determine how much is from fossil fuel use—more than 4 gigatons (Gt): That's 400 billion metric tons since 1751, more than 75% of total human emissions. Other anthropogenic sources of greenhouse gas emissions are certain agricultural activities like rice farming and cattle rearing, as well as deforestation, the release of certain industrial chemicals like chlorofluorocarbons (CFCs) (see Module 5.2), and landfill waste disposal (see Module 5.3).

Molecule for molecule, CO_2 is not the most potent greenhouse gas, but it currently accounts for about three-fourths of all greenhouse gas emissions and is the one that is contributing most to global warming at this time. Methane, the second most abundant anthropogenic greenhouse gas, is released from rice paddies, cattle, and landfills; the largest single source is the fossil fuel industry, especially methane lost from oil and gas wells. Though methane emissions are considerably lower than CO_2 emissions, methane is around 28 times more potent than CO_2 over the long term (i.e., a single methane molecule has the heat-trapping ability of 28 CO_2 molecules), so these emissions are significant. Halocarbons like CFCs are even stronger greenhouse gases. Nitrous oxide, making up about 2% of all anthropogenic emissions (sources include agriculture and vehicle emissions), comes in at almost 300 times more potent than CO_2; some CFCs are thousands of times more potent.

Scientists have been taking continuous measurements of atmospheric CO_2 concentrations since the 1950s. A graph of those data, known as the *Keeling curve* (after David Keeling, the scientist who initiated the measurements), shows that CO_2 levels are rising steadily, topping 415 ppm in 2019. The last time CO_2 levels were this high, during the Pliocene Epoch (which ended around 3 million years ago), trees grew in Antarctica. (For perspective, atmospheric CO_2 levels have stayed around 280 ppm for much of human history.) **INFOGRAPHIC 3**

INFOGRAPHIC 3 GREENHOUSE GASES AND THE GREENHOUSE EFFECT

Achieve

Life on Earth depends on the ability of greenhouse gases in the atmosphere to trap heat and warm the planet. More greenhouse gases, however, mean more trapped heat and a warmer planet (an enhanced greenhouse effect).

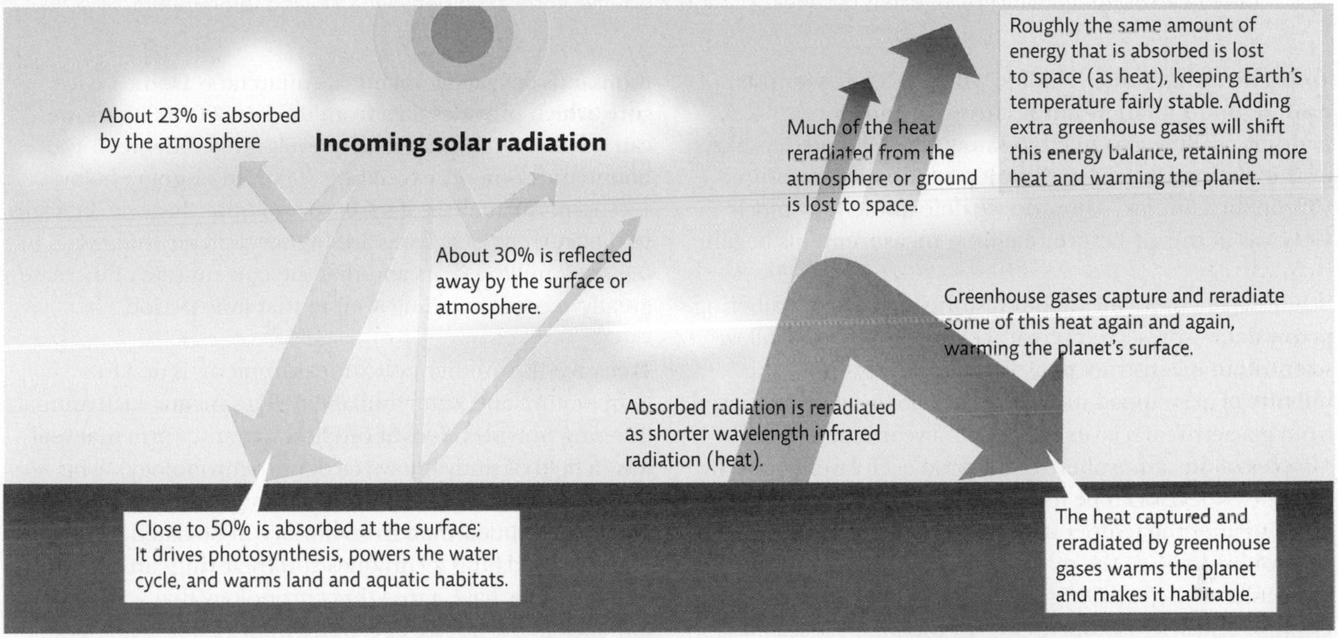
About 23% is absorbed by the atmosphere
Incoming solar radiation
Much of the heat reradiated from the atmosphere or ground is lost to space.
Roughly the same amount of energy that is absorbed is lost to space (as heat), keeping Earth's temperature fairly stable. Adding extra greenhouse gases will shift this energy balance, retaining more heat and warming the planet.
About 30% is reflected away by the surface or atmosphere.
Greenhouse gases capture and reradiate some of this heat again and again, warming the planet's surface.
Absorbed radiation is reradiated as shorter wavelength infrared radiation (heat).
Close to 50% is absorbed at the surface; It drives photosynthesis, powers the water cycle, and warms land and aquatic habitats.
The heat captured and reradiated by greenhouse gases warms the planet and makes it habitable.

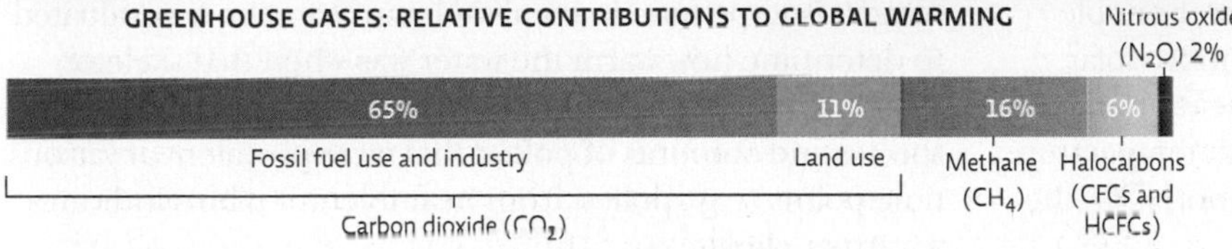
GREENHOUSE GASES: RELATIVE CONTRIBUTIONS TO GLOBAL WARMING
Nitrous oxide (N_2O) 2%
65%
11%
16%
6%
Fossil fuel use and industry
Land use
Carbon dioxide (CO_2)
Methane (CH_4)
Halocarbons (CFCs and HCFCs)

GREENHOUSE GAS	GLOBAL WARMING POTENTIAL RELATIVE TO CO_2 (OVER 100 YEARS)
Carbon dioxide	1
Methane	28
Nitrous oxide	265
Halocarbons	<1 to 13,900

Different greenhouse gases have different abilities to trap heat; their heat-trapping capacity is expressed as CO_2 equivalents (the amount of CO_2 that would produce the same warming). For example, since a molecule of methane (CH_4) traps 28 times as much heat as CO_2, one methane molecule is equivalent to 28 CO_2 molecules; one CFC-13 molecule is equivalent to 13,900 CO_2 molecules.

CO_2 is one of the greenhouse gases that is increasing. Historic CO_2 levels are estimated from ice cores like the Law Dome ice core from Antarctica; current levels have been measured directly at locations like Mauna Loa, Hawaii, since 1958.

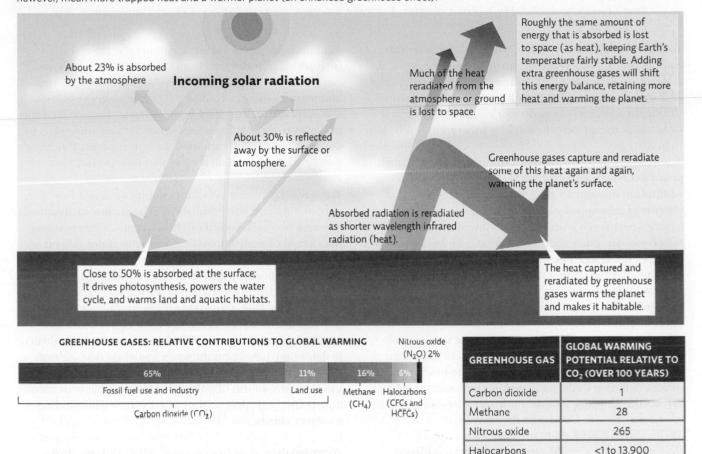

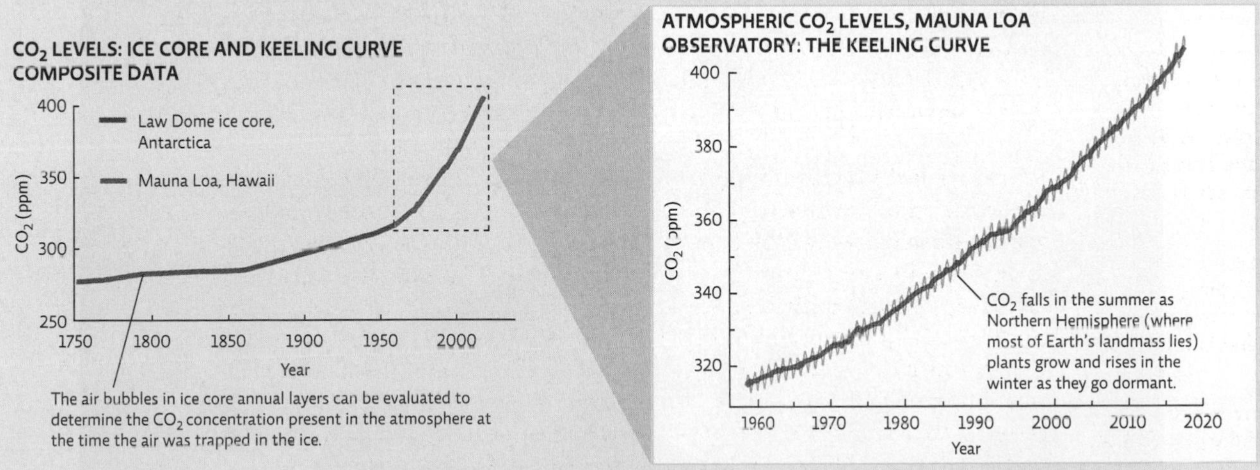
CO2 LEVELS: ICE CORE AND KEELING CURVE COMPOSITE DATA
Law Dome ice core, Antarctica
Mauna Loa, Hawaii
CO_2 (ppm)
400
350
300
250
1750
1800
1850
1900
1950
2000
Year
The air bubbles in ice core annual layers can be evaluated to determine the CO_2 concentration present in the atmosphere at the time the air was trapped in the ice.
ATMOSPHERIC CO2 LEVELS, MAUNA LOA OBSERVATORY: THE KEELING CURVE
400
380
360
340
320
CO_2 (ppm)
CO_2 falls in the summer as Northern Hemisphere (where most of Earth's landmass lies) plants grow and rises in the winter as they go dormant.
1960
1970
1980
1990
2000
2010
2020
Year

Use the concept of the greenhouse effect to explain how Earth's surface could warm up even if the Sun's output does not change.

4 TEMPERATURE AND CO_2: COLLECTING AND INTERPRETING THE DATA

Key Concept 4: A variety of methods are used to measure past and present atmospheric CO_2 levels; these levels have been rising dramatically in recent decades. All show a positive correlation between CO_2 and temperature.

To predict what future climate will look like, scientists must do more than monitor current atmospheric conditions; they also need to know what climate was like in the distant and not-so-distant past. Keeling measured CO_2 in air samples—how do we determine how much CO_2 was in the air before real-time measurements began?

Scientists uncover clues about historic climates by gathering **proxy data**—preserved physical characteristics that allow scientists to reconstruct past climates. For example, the bubbles of air trapped in annual layers of ice cores retrieved from glaciers (each layer represents a year in time) hold samples of the atmosphere at the time each layer was laid down. CO_2 levels can be measured directly, but scientists can also infer the temperature at that time by looking at the isotopic ratio of the oxygen atoms in the water (H_2O) molecules of the ice. Oxygen can exist as a light isotope (^{16}O) or a heavier isotope (^{18}O). In this analysis, scientists determine how much of the oxygen in the water sample is ^{16}O and how much is ^{18}O. Because it takes more solar energy to evaporate water that contains the heavier isotope, its concentration in the ice correlates well with temperature at the time the water sample was frozen; the more ^{18}O, the colder the climate.

proxy data Measurements that allow one to indirectly infer a value such as the temperature or atmospheric conditions in years past.

Piecing together this evidence paints a picture of the climate that persisted during each successive year, going back hundreds of thousands of years. Looking at Antarctic's "Dome C" ice core, which provides data from the last 800,000 years, we can see that atmospheric CO_2 levels ranged between two boundaries—never exceeding 300 ppm or going below 180 ppm. An analysis of ice core and other historic data tells us that current CO_2 levels are higher than anything seen in the last 2 million years and that the current rate of increase greatly exceeds anything seen in that time period.

Tree growth, another proxy measurement, is tied to temperature and water availability. This means each annual tree ring provides clues about how wet and warm that year was (a field of study known as dendrochronology). This analysis is not restricted to living trees: Trees used in ancient buildings or unearthed from burial can be dated using radiometric dating techniques and their rings analyzed. Scientists now have a tree-ring chronology that goes all the way back to the last ice age, more than 12,000 years. The oxygen isotope ratio in coral skeletons can also be evaluated to determine how warm the water was when that skeleton was laid down. And lake sediments can be evaluated for the species and amounts of pollen that were prevalent at various time points (e.g., pollen from heat-tolerant plants indicates a warmer climate).

Together, these data (and other lines of evidence) help scientists understand past climates. The ice core data, in particular, show that CO_2 and temperature are highly correlated through time; they increase and decrease in

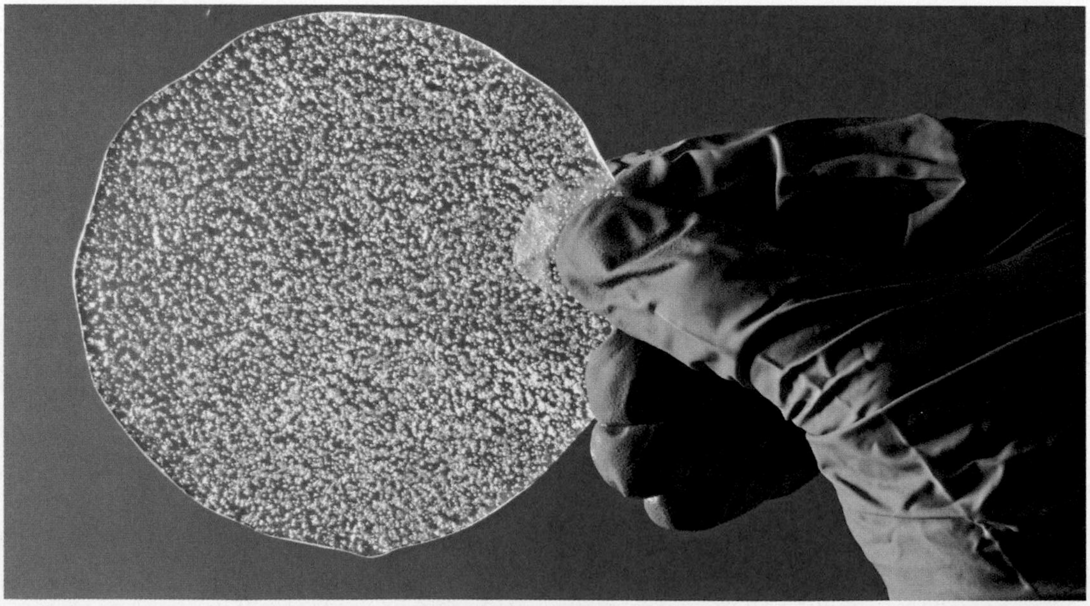
SPL/Science Source

A researcher holds a thin slice of ice from an ice core extracted from Antarctica. The bubbles in the ice contain air from long ago and can be evaluated to determine the atmospheric composition at the time this layer of ice was laid down. The water, too, can be analyzed to determine the temperature at that time.

INFOGRAPHIC 4 ATMOSPHERIC CO_2 CONCENTRATION AND TEMPERATURE OVER TIME

Data from the Vostok ice core show that temperature and CO_2 levels closely tracked one another over the last 400,000 years. Today's atmospheric CO_2 levels exceed anything in this record. Currently, temperature on Earth is still moderated by polar ice, but as that melts, temperatures will continue their upward climb.

TEMPERATURE AND ATMOSPHERIC CO_2 CONCENTRATION OVER THE PAST 400,000 YEARS (VOSTOK ICE CORE)

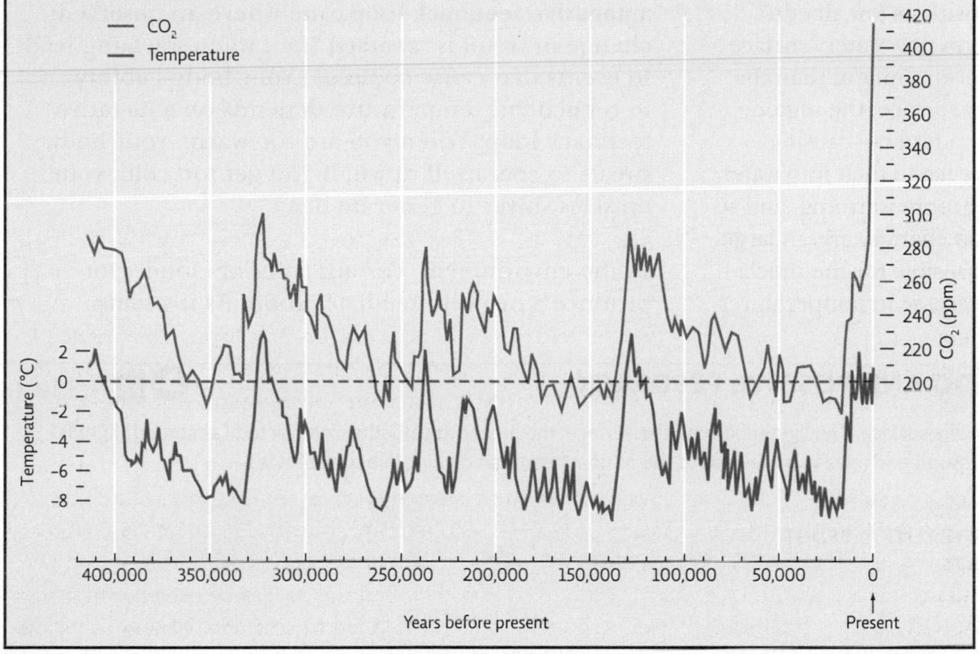

PROXY DATA

Clues about past climates can be gleaned from the use of proxy measurements.

- Ice cores have annual layers that contain bubbles of air from the past; these can be analyzed for CO_2 concentration.
- Annual layers of sediment cores can be evaluated for pollen or other climate indicators.
- Tree ring analysis reveals clues to the climate in each year of the tree's life.
- Analysis of coral allows one to determine the water temperature when the skeleton was formed.

? Estimate the range of variation for CO_2 concentrations over the last 400,000 years (ignore the recent increase in CO_2) by identifying the upper and lower values for CO_2 concentrations during that time (the peaks and the troughs). How does the current value of CO_2 concentration for 2019 compare to that range?

tandem. They also show that CO_2 can be both a cause and effect of warming temperatures. For example, if a natural event leads to warming, that warming will in turn lead to more atmospheric CO_2 (an effect) as soils and water give up more stored CO_2. Those higher CO_2 concentrations will then trigger even more warming (a cause). So regardless of whether the excessive greenhouse gases are the result of natural or human forcers, the end result is the same: Triggering warming can set into motion events that lead to even more warming. **INFOGRAPHIC 4**

Looking at the graph in Infographic 4, you might wonder why current temperature seems to be lagging behind CO_2. The CO_2 increase over the last 150 years, and especially the last 50 years, has been so rapid that it has outpaced the main factor on Earth that helps moderate our planet's temperature—polar ice. But as this ice melts, Earth's temperature will continue to rise. This ice is buying us time, but if it melts—or even if a large portion of it melts—the temperature increase that results may put Earth on a path of runaway ice melt that will be irreversible.

5 CLIMATE FORCERS OTHER THAN GREENHOUSE GASES

Key Concept 5: Greenhouse gases are not the only climate forcers. Higher albedo (reflectivity of a surface) can have a cooling effect, whereas decreasing a surface's albedo can increase warming, and this warming can escalate via positive feedback. Past climate changes are correlated with natural forcers such as the Milankovitch cycles, but these cycles do not account for current warming.

Let's revisit the concept of climate forcers. Greenhouse gases are not the only forcers. Another forcer that plays a role in warming trends is **albedo**, the ability of a surface to reflect away solar radiation. Think of it as the *reflectivity* of a surface: Light-colored surfaces, like glaciers and meadows, have a high albedo; they reflect sunlight away from the planet's surface, reducing the amount that is absorbed and then reradiated as heat. Darker surfaces like water and asphalt have low albedo: They absorb sunlight and then reradiate that solar energy back to the atmosphere as heat.

albedo The ability of a surface to reflect away solar radiation.

As surfaces with high albedo (high reflectivity) are replaced by those with low albedo, besides the planet warming, a **positive feedback loop** can be triggered—a cycle whereby an observed change in a trend is accentuated (i.e., warming causes more warming). Melting sea ice provides a good example of positive feedback: As temperatures rise, sea ice melts, and ice (with its high albedo) gives way to water (with its low albedo). Because this new watery surface absorbs more sunlight than the former icy surface, the region warms even faster—causing even more ice to melt into water, triggering more warming, and so on. Albedo changes are, in large part, responsible for the much greater increase in temperatures seen in the Arctic relative to the global average. Positive feedback loops tend to have a destabilizing effect on the environment, continually altering it as long as they are at work. (Usually, this lasts until things settle down at a new steady state; in this case, that might be when most of the ice has melted.) **INFOGRAPHIC 5A**

positive feedback loop Changes caused by an initial event that then accentuate that original event.

negative feedback loop Changes caused by an initial event that trigger events that then reverse the response.

Climate forcers also have the potential to trigger a **negative feedback loop**, one where an observed change in trend is reversed (i.e., when warming leads to events that cause cooling). Your body's ability to regulate its temperature depends on a negative feedback loop: When you are too warm, your body sweats to cool itself down; if you get too cold, your muscles shiver to generate heat.

In the environment, certain types of clouds can produce a negative feedback loop: As it warms,

INFOGRAPHIC 5A ALBEDO AND POSITIVE FEEDBACK Achieve

Albedo is a measure of the reflectivity of a surface. The lighter colored the surface, the higher the albedo. Unreflected (absorbed) light is reradiated as heat, so surfaces with a low albedo release more heat to the atmosphere than do high-albedo surfaces.

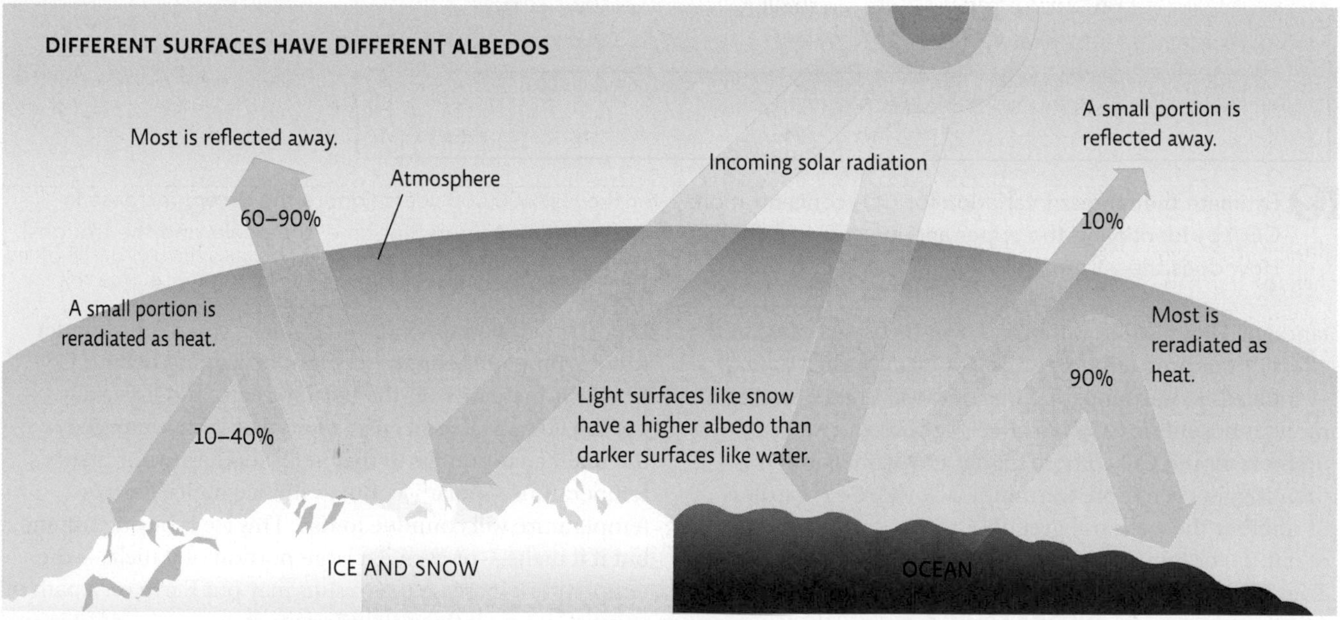

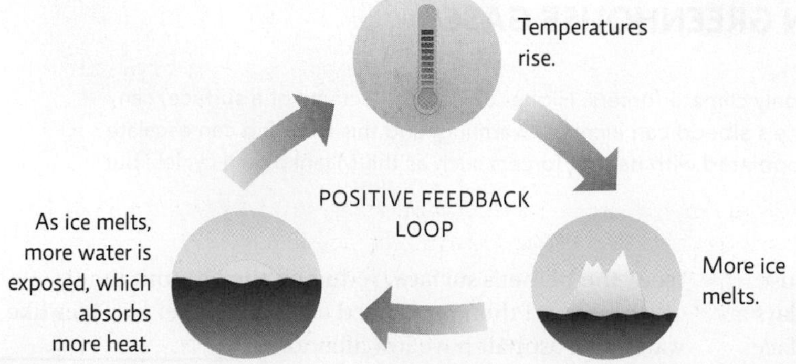

When sea ice melts, it uncovers water, which has a darker surface with a lower albedo. This activates a positive feedback loop: As the exposed water absorbs sunlight and releases more heat into the atmosphere, more ice melts and more water is exposed, which then absorbs more sunlight, releasing even more heat, and so on. Positive feedback loops represent changes that trigger additional change in the same direction (warming that triggers even more warming); the word *positive* is not meant to indicate a "positive," or beneficial, event.

One suggestion to combat global warming is to replace dark rooftops with light-colored ones. How would this help reduce warming?

more water evaporates to form clouds. If more low, thick clouds are formed, those could reflect away incoming solar radiation, thus cooling the area. As the temperature cools, less water evaporates and fewer of these clouds form. As a result, the temperature rises again, and more clouds form. And on and on. The end result of a negative feedback loop is stabilizing. In this example, if this negative feedback loop were occurring often enough, it would help stabilize the climate—but unfortunately, it is not happening enough to mitigate warming. Currently, not enough low clouds are forming to offset warming, so this particular negative feedback loop is having only a small effect on climate.

Volcanic eruptions and changes in the magnitude of solar irradiance are also considered natural forcers, as both have been known to impact climate in the past, though only over short time frames and not as strongly as greenhouse gases. Sulfur particles emitted from volcanic eruptions or from industrial activities are light in color, so they have a high albedo and act as negative forcers (they cool the planet). Darker particles, such as soot released from burning fossil fuels, are positive forcers due to their low albedo. Currently, actions that increase the albedo of the Earth's atmosphere are outweighed by positive forcers—the net effect is warming.

Scientists also have evidence that **Milankovitch cycles** (predictable long-term cycles of Earth's position in space relative to the Sun) played an important role in earlier climate change events such as the Pleistocene ice ages. During times when the Earth's position in space meant less solar irradiation reached the planet, events were triggered that initiated cooling. These events likely led to positive feedback loops that accelerated cooling (e.g., less CO_2 exchange with the atmosphere) and produced a climate several degrees colder, on average, than when Earth was closer to the Sun. During times when Earth was in a position to receive more solar energy, a warming trend could be triggered that led to a new climate. These are very long-term cycles and would not produce noticeable changes over just a few decades. Even if their effect was discernable, at present, Earth's position relative to the Sun is trending toward a cooling phase. (The shape of Earth's orbit and the tilt of its axis are not putting Earth in a position relative to the Sun that would lead to warming.) This means the Milankovitch cycles, which explain past climate changes that unfolded over long periods of time, do not correlate with current warming. **INFOGRAPHIC 5B**

In fact, none of these natural forcers—together or by themselves—account for the current warming trends.

Milankovitch cycles
Predictable variations in Earth's position in space relative to the Sun that affect climate.

INFOGRAPHIC 5B MILANKOVITCH CYCLES HELP EXPLAIN PAST CLIMATE CHANGE

Warm periods and ice ages of the past can be attributed in part to Earth's position in space relative to the Sun. Earth has three different cycles that can each have an impact on climate. The current warming we are experiencing cannot be explained by any of these cycles—Earth is currently not in a part of any cycle in which it would have greater warming.

Earth's orbit can be round or slightly elliptical; currently it is round.

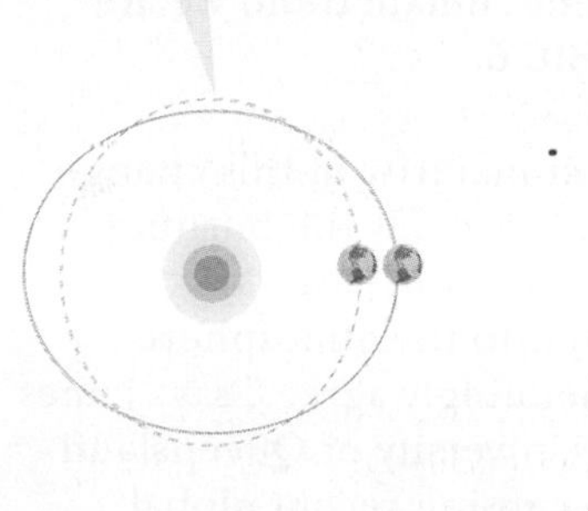

ORBITAL ECCENTRICITY

The shape of Earth's orbit around the Sun varies over a 100,000-year cycle. When it is more elliptical, climate is more variable because some seasons receive more solar radiation than normal and others receive less.

Earth's tilt is currently 23.4°, which gives us cooler summers and warmer winters than we'd have at 24.5°, which was our position 9,000 years ago. In about 32,000 years, Earth will be at 22.1°.

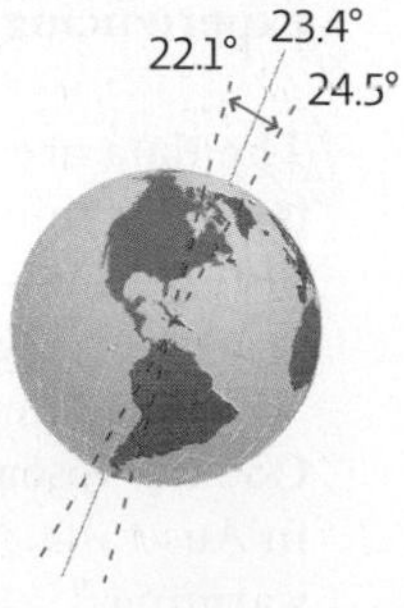

AXIAL TILT

The angle of Earth's tilt as it spins on its axis changes in a 41,000-year cycle. The greater the angle, the greater the extremes between seasons (hotter summers and colder winters).

Earth's current position is tilted toward the North Star; in about 12,000 years, the axis will point toward the star Vega.

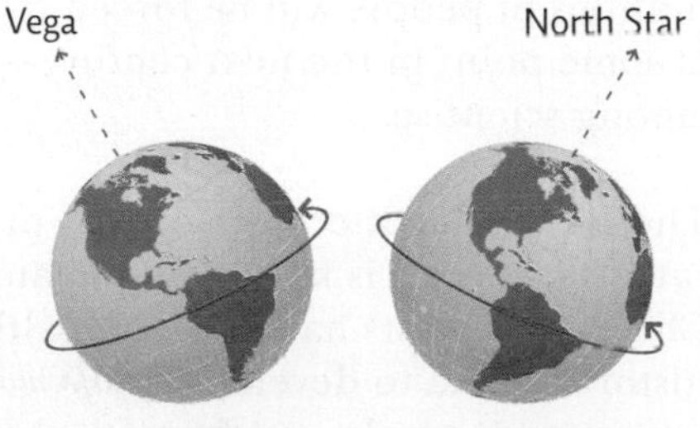

AXIAL PRECESSION

Earth "wobbles" on its axis, changing not the angle but the direction the axis points in a 20,000-year cycle. This changes the orientation of Earth to the Sun and affects the severity of the seasons. When Earth is tilted toward Vega, it is also tilted toward the Sun during summer, making summers hotter in the Northern Hemisphere.

Why do scientists conclude that Earth's axial tilt is not responsible for our current warming?

6 ATTRIBUTION: HUMAN VERSUS NATURAL CAUSES

Key Concept 6: Current warming cannot be explained without accounting for both natural and anthropogenic climate forcers.

On his laptop, Alex de Sherbinin, a climate scientist at Columbia University's Lamont Doherty Earth Observatory, can render the world's emerging disaster zones in such exquisite detail that it's possible to see exactly which homes might be destroyed by a few feet in sea-level rise. Deep red hues show where populations are densest; streaks of blue overlie the areas most likely to disappear into the ocean.

The overall picture is not pretty: With 3 feet of sea-level rise, the Ganges delta region of Bangladesh and India, home to 144 million people, will be inundated with flooding. Vietnam will lose more agricultural land than any other place on Earth. And the Sahel region of West Africa will see a doubling in the number of people facing water shortages, owing to a massive decrease in rainfall.

Climate refugee predictions (estimates of how many people will be forced to flee their homes as rising temperatures or drought make some areas inhospitable and rising seas flood out others) are also based on climate models. The most commonly cited estimates put the total number of people affected at between 50 and 200 million by the year 2050. That upper limit comes out to about 1 in every 50 people on Earth (assuming a 2050 population of 9.8 billion, the current projection).

Those specific numbers may be in dispute — no one can say for certain exactly which areas will be too hot or dry to inhabit or how much land will sink into the sea or by what year. But the bottom line — that many millions of people will be forced to flee their homes at some point in the next century — is a certainty among scientists.

The technique Sherbinin uses to produce these future scenarios is known as climate modeling. Climate scientists have used a wealth of current and historical data to develop *climate models* — computer programs (complex mathematical equations) that allow them to make future climate projections by plugging in values for temperature, CO_2 concentrations, global air circulation patterns, and so on. These models are used to see how altering the value of certain parameters (say, increasing the amount of atmospheric CO_2) might impact future climate. Their validity is tested by entering data for years past to see how well the model predictions match the climate that was seen during those years. A good match tells them they have accurately accounted for the forcers at work.

These powerful climate models allow us to determine the relative contributions of natural forcers (e.g., solar output, cloud formation, volcanic eruptions, Milankovitch cycles, etc.) and of anthropogenic forcers (e.g., burning fossil fuels, deforestation, agricultural practices, land uses) — a focus of climate science known as *attribution*. This is vitally important. By determining what the causes of climate change are, and their relative contributions, we discover which climate forcers can and should be addressed. In other words, if volcanic eruptions and the Earth's orbit were the primary drivers of climate change, there would not be much we could do about it. But if human actions are significant causes, that would be something we could address.

So what do the models say? When data for just the natural forcers are entered into the models, they match past temperatures pretty well (so we know the model is accurately accounting for these forcers), but they do not replicate the post-1960s warming we have already observed. Natural forcers alone should, according to the models, produce a fairly stable climate. Only when we consider both natural and anthropogenic forcers together do the models replicate the current climate trend we are experiencing. **INFOGRAPHIC 6**

The data are clear: The vast majority of this change is due to human activities, especially the burning of fossil fuels, and the consequent release of greenhouse gases like CO_2 into the atmosphere. "Climate scientists overwhelmingly agree," says James Cook, a researcher at the University of Queensland in Australia. "Humans are causing recent global warming."

But this is actually good news. We can alter our path and stop doing things that contribute to climate change.

INFOGRAPHIC 6 **WHAT'S CAUSING THE WARMING?**

Achie/e

Climate scientists use computer models that take into account the major factors that are known to have affected past climates in order to see what might be responsible for recent warming. Data about natural and anthropogenic factors can be fed into a computer model separately and then together to see which circumstances match up with the warming that has been observed.

COMPUTER MODELS' RECONSTRUCTION OF PAST TEMPERATURES

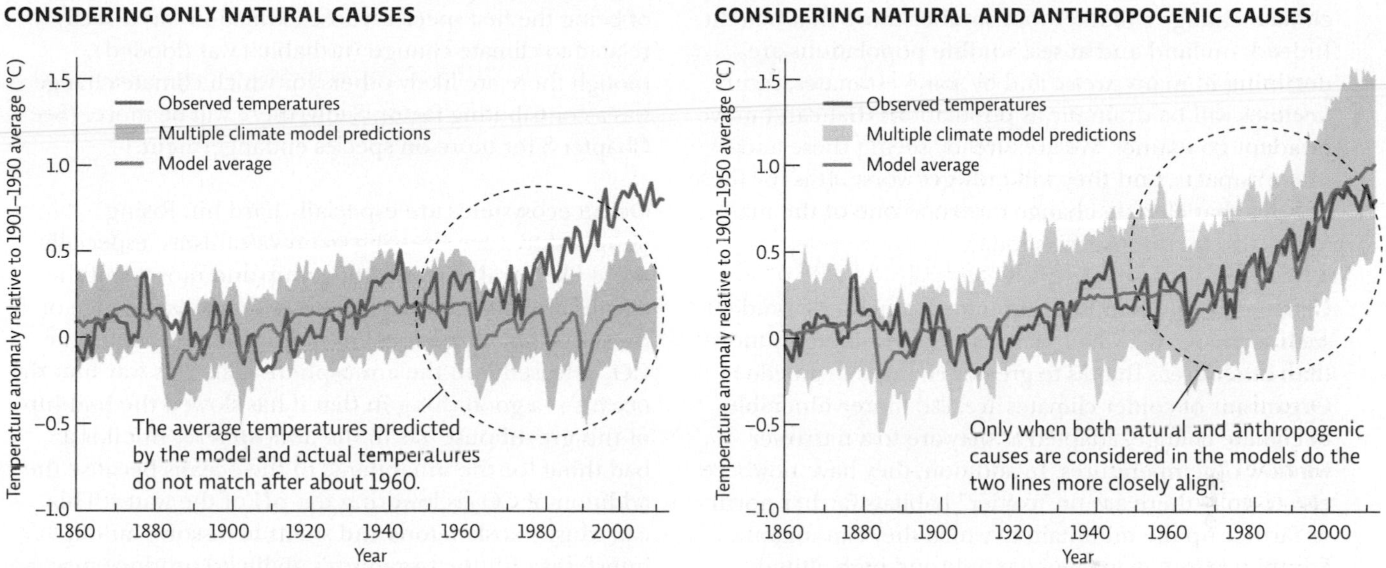

? What would you expect the line showing the model average (the purple line) in the left-hand graph of this infographic to look like, relative to the line showing observed temperatures (the blue line), if current warming could be explained by natural causes alone?

7 IMPACTS OF CLIMATE CHANGE

Key Concept 7: Climate change impacts include health and agricultural effects, biodiversity loss, more extreme storms, increased fire risk, and coastal erosion and flooding.

In some ways, the demise of communities like Newtok and Isle de Jean Charles has been slow and brutal. Even before area land was swallowed by the water, climate change was already affecting fishing and farming and the availability of clean water in the areas: Fisheries declined as a result of warming water, soil was eroded by increased storms, and clean water was contaminated or depleted by the same. And as much as anything else, communities and the cultures that they support can be wholly lost when land disappears.

These other losses illustrate something of which scientists have long been aware: Climate change won't just affect the weather. Climate change will have, and is having, environmental, economic, societal, and health consequences. For example, climate change affects water supplies—making freshwater scarce in places where droughts increase or glaciers melt away and contaminating existing water sources in places where storms and flooding become more prevalent.

Agriculture, too, is affected. While some crops or areas may benefit from the trifecta of warmer temperatures, longer growing seasons, and more CO_2 in the air, agriculture in many areas is taking a hit and productivity is declining. (While CO_2 is needed for photosynthesis, and more can boost photosynthesis in plants, other limiting nutrients that restrict growth prevent plants from taking in all the extra CO_2 we are releasing; therefore, boosted plant growth will not eliminate the CO_2 problem for us. See Module 2.1 for more on nutrient cycles and limiting growth factors.) Heat stress can hamper plant growth, as can water stress. Rainfall might decrease, and the rain that does fall could come in more powerful storms, flooding crops and washing away valuable soil. Many weed species are likely to handle the temperature and water stress of a new climate and so may proliferate, reducing harvests.

Some insect pest species may benefit from climate change as well. Pest outbreaks that harm crops or natural ecosystems, such as the bark beetle population explosion mentioned earlier, can increase in size or frequency for a number of reasons: milder winters, declines in other insect populations (the predators of those pests), or changes that allow pest species to expand their ranges to enter new territory.

However, the number of species that are harmed by climate change will likely outnumber those that benefit. Indeed, on land and at sea, wildlife populations are declining in many areas, and by some estimates, future declines will be dramatic as populations that can't move or adapt go extinct. We are already seeing these and other impacts, and they will only get worse. It is for these reasons that climate change is seen as one of the main forces threatening species today.

Climate change is most pronounced at higher latitudes such as the Arctic, which has warmed considerably more than other areas thanks to greater changes in albedo. Organisms of colder climates are also more vulnerable to climate change, adapted as they are to a narrower window of temperatures. In addition, they have nowhere else to go—there are no "colder" habitats farther north or farther up the mountain to which they can migrate. For this reason, many high-latitude and high-altitude species such as polar bears, walrus, penguins (and other polar species) as well as mountain species such as pika (a species related to the rabbit), the honeycreepers (birds) of Hawaii's mountains, and many other alpine plants and insects are threatened with extinction by climate change.

The ways that species can be affected are as varied as the species themselves. Sea-level rise that erodes and floods beaches is threatening some sea turtle species, many of whom are already endangered. Some reptiles, like turtles and alligators, are particularly vulnerable to temperature changes as the sex of an individual is affected by the incubation temperature of the egg. Warmer temperatures could produce a preponderance of females (for turtles) or males (for alligators), impacting the reproductive potential of future populations. Population surveys along the Great Barrier Reef of Australia by NOAA scientist Michael Jenson found that the population of green sea turtles on the southern section of the reef had slightly more females than males but in the population that lives in the warmer northern section of the reef, 99.8% of the young adults were females. Analysis of the data led them to conclude that there has been no successful production of males in the last two decades. The population may be all female in the near future. Since mating occurs near the beach where females were born, it is likely this population may be at a reproductive dead end.

The list of species affected goes on. A lack of moisture on the leaves of the wet forest habitat of the lemuroid ringtail possum of Queensland, Australia, is endangering this marsupial that depends on leaf moisture as its water source. The loss of sea ice threatens not only the polar bear but also its favorite prey, the ringed seal, a species that depends on the ice for reproduction (it gives birth on the ice, and young seals shelter in "snow dens" on the ice). The Bramble Cay melomys, a small rodent that inhabited only one island off the Australian coast, has the distinction of being the first species whose extinction was directly related to climate change (its habitat was flooded), though there are likely others for which climate change was a contributing factor. Sadly, there will be more. (See Chapter 3 for more on species endangerment.)

Ocean ecosystems are especially hard hit. Rising temperatures are stressing many organisms, especially those like coral that cannot get up and move. But the oceans are also experiencing another consequence of fossil fuel burning—ocean acidification. Much of the CO_2 released into the atmosphere makes its way into the oceans—a good thing in that it has slowed the buildup of this greenhouse gas in the atmosphere. But it is a bad thing for the inhabitants of the oceans because the addition of CO_2 is lowering the pH of the water. This can cause exoskeletons and shells to dissolve, and it interferes with the organism's ability to produce new shell or skeletal material—a double whammy. Rising temperatures and acidification are wreaking havoc on ocean communities around the planet; 2016 witnessed the greatest coral bleaching event ever seen in the Great Barrier Reef, serious enough to kill almost 90% of the coral in some sections of the reef.

Overall, 30% of coral were killed in the 2016 bleaching event. Coral populations take around 10 years to recover from a major bleaching event. Unfortunately, another serious bleaching event followed in 2017, killing an estimated 20% of the coral that remained. Full recovery is in question, especially if another bleaching event occurs in the near future, a likely occurrence given the state of the climate. Terry Hughes of the Australian Research Council Centre of Excellence for Coral Reef Studies says, "Climate change is not a future threat. On the Great Barrier Reef, it's been happening for 18 years." (See Online Module 6.3 for more on ocean acidification and ocean ecosystems.) Fisheries, too, are vulnerable to climate change as temperature and acidification of water alter food chains, potentially reducing the population size of many commercially important fish and shellfish species.

Human health suffers as insect-borne diseases increase in both incidence and range. Severe weather, flooding, and heat waves also imperil health. Agricultural declines can put more people at risk for undernutrition, and the lack of enough clean, fresh water can be devastating to a population. Water scarcity is already the subject of water wars (see Module 6.1) and is only expected to worsen.

Other, perhaps unexpected, effects on human health include the production of more allergy-inducing pollen by plants due to more favorable growing conditions for some or, interestingly, poor growing conditions for others (many species of plants produce an abundance of pollen when stressed, an adaptation that favors reproduction at a time when the plant might not survive to reproduce another year). Higher CO_2 levels in the atmosphere appear to be causing poison ivy plants to produce larger leaves and more of the rash-inducing urushiol oils to which so many people are allergic. **INFOGRAPHIC 7**

Another truth hidden in Hauer's Miami study is that climate change won't just harm coastal areas. Some inland cities will also become inhospitably hot, even as they absorb hundreds of thousands of refugees fleeing the coasts.

That's not to say that there won't be some "winners" in a warming world. High-latitude land in Canada and Siberia will likely become warmer and more habitable—lessening the incidences of cold-related health problems and deaths (though the melting of permafrost, the deeper ground that before the advent of climate change never thawed, is damaging roads and buildings as the ground below them becomes less supportive). Cold-weather problems overall will lessen in the northern United States and similar latitudes worldwide (think milder winters in Maine and Montana), even with the increase of aberrant cold weather in some places as the jet stream changes its course. Warmer weather during the summer months has even opened the Northwest Passage in some recent years—a long-sought-after shipping route through the Arctic Ocean—which would significantly reduce the transport time for ships that otherwise have to take the southern route through the Panama Canal. But in total, negative impacts will outweigh positive ones, especially as warming increases.

INFOGRAPHIC 7 THE IMPACTS OF CLIMATE CHANGE

Impacts from climate change are already being felt and will continue to increase and spread to other areas. The extent and severity of future impacts will depend on how quickly the world responds to climate change.

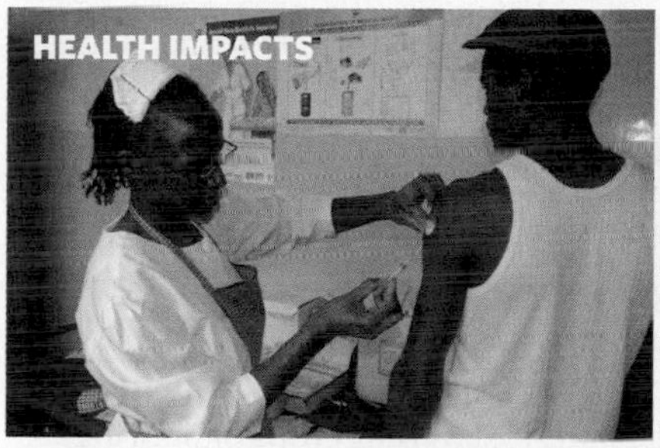

Wendy Stone/Corbis via Getty Images

Vector-borne tropical diseases spreading outside of the tropics; waterborne-disease outbreaks with flooding; more heat-related deaths.

Jim Richardson/National Geographic Image Collection

Globally, lower crop yields and more malnutrition.

U.S. Air Force photo/Alamy

More health issues (injuries and deaths) and property losses due to sea-level rise and storm events.

Paul Souders/Getty Images

Benefit some species (range expansion or loss of competitors); harm other species (habitat or mutualistic partner loss).

Sven Torfinn/Panos

Increase in land areas suffering in severe drought (weather pattern changes; loss of snowpack or glacier ice).

Layne Kennedy

Increased incidence and risk of fire, causing major property or ecosystem damage.

Which impacts concern you the most? Explain.

8 RESPONSES TO CLIMATE CHANGE

Key Concept 8: Responding to climate change will require steps that try to reduce future warming (mitigation) and steps to deal with inevitable warming (adaptation).

Back in 2009, the coming plight of climate refugees got an international audience when Mohamed Nasheed, president of the Maldives, a string of small low-lying islands in the South Pacific, signed a document that asked the nations of the world to reduce their carbon emissions. Why did this garner so much attention? He signed it underwater, wearing scuba gear, to draw attention to the fact that the very existence of his nation was at stake. At an international summit convened to negotiate an agreement on reducing carbon emissions that same year, Nasheed insisted that unless carbon emissions were kept at 350 ppm, his nation would be destined to disappear entirely. (Levels passed 400 ppm in 2016.)

Computer models like the ones used by Sherbinin can be used to predict future changes in climate, but the accuracy of the predictions depends on how well we estimate future variables such as how quickly we reduce the use of fossil fuels or what land-use changes we will pursue. Still, we know that acting now can help reduce the impact and progression of climate change. Scientists refer to efforts aimed at minimizing the extent or impact of climate change as **climate mitigation**. Mitigation includes any attempt to seriously curb the amount of greenhouse gases we are releasing into the atmosphere as well as steps to remove CO_2 from the air. Carbon capture techniques could reduce emissions (see Module 9.1), but research is also underway to develop planetary-scale technologies to mitigate climate change such as satellites that block some incoming solar radiation or giant filters that remove CO_2 from the atmosphere—approaches known as climate geoengineering.

On a national or global scale, mitigation efforts can be facilitated in a variety of ways, such as command-and-control regulations that limit greenhouse gas release or *green taxes* on pollution that is released (in this case, **carbon taxes**) as well as market-driven initiatives such as carbon *cap-and-trade* programs. Financial incentives that encourage the development and use of non-carbon fuels and more energy-efficient technology are also examples of mitigation. (See Modules 5.2 and 10.1 for more on these policy approaches.) Many industries and businesses are already stepping up with initiatives to reduce their carbon footprint, both because it is an economically sound investment and because it is the right thing to do (see Module 5.1).

climate mitigation Efforts to minimize the extent or impact of climate change.

carbon taxes Governmental fees imposed on activities that release greenhouse gases into the atmosphere.

climate adaptation Efforts to help deal with existing or impending climate change problems.

In 2004, Princeton University researchers Stephen Pacala and Robert Socolow proposed a "stabilization wedge" strategy—the multipronged implementation of currently available technology to mitigate climate change; each wedge strategy would be applied at a scale sufficient to prevent the release of 1 Gt of carbon. Strategies include transitioning away from fossil fuels (especially coal) by generating energy from low-carbon nuclear power (see Module 11.1) and sustainable sources such as solar, wind, and geothermal sources (see Module 11.2). Reforestation efforts that could increase the uptake of carbon and agricultural techniques that minimize carbon release are also strategies that will help. And not using so much energy in the first place—conservation and energy efficiency—can significantly contribute to mitigation (see Module 11.2 and Online Module 11.3). Pacala and Socolow estimate that any 8 of the 15 strategies, or "wedges," they have identified would stabilize CO_2 in the atmosphere; the longer we wait, the stronger our response would have to be—the more wedges we would need to employ.

But in places like Newtok, Alaska, and Isle de Jean Charles, Louisiana, it is likely too late for mitigation. Those residents' only choice is to adapt to the reality of a rapidly changing climate. **Climate adaptation** means responding to the climate change that has already occurred and preparing for the additional changes that are inevitable at this point. For human societies at large, that means taking steps to ensure a sufficient water supply in areas where freshwater supplies may dry up; it means planting different crops or shoring up coastlines against rising sea levels; it means preparing for heat waves and cold spells and outbreaks of infectious disease. Miami too is moving ahead with adaptation steps: installing pumps to rid city streets of water that floods during high tides and raising the level of some coastal roads and existing sea walls, just to name a couple of initiatives aimed at dealing with rising sea level. **INFOGRAPHIC 8A**

No matter which strategies we employ to confront the realities of climate change, international coordination will be essential, meaning world superpowers like the European Union, the United States, and China will have to cooperate with each other and with the world's developing nations.

So far, those efforts have been fraught. The 1992 United Nations Framework Convention on Climate Change (UNFCCC) recognized the need to address climate change, and most of the world's nations, including the United States, signed on, agreeing to cooperate. In 1997, the Kyoto Protocol, the international treaty that laid out steps to be taken, set different but specific targets for the reduction of CO_2 emissions for various countries.

INFOGRAPHIC 8A RESPONDING TO CLIMATE CHANGE

To deal with climate change, we need to pursue actions that help us adjust to current warming as well as take steps to reduce further warming.

BOTH APPROACHES WILL BE NEEDED.

ADAPTATION
Responding to the warming that has already or will inevitably occur

- Improve disease surveillance; improve sanitation in flood-prone areas.
- Plant crops to match new climate.
- Capture and conserve water.
- Erect coastal barriers to deal with sea-level rise; relocate coastal communities.
- Pursue better fire prevention management.
- Provide migration corridors for wildlife and wildlife preserves.

MITIGATION
Preventing further warming by addressing the causes of climate change

- Pursue carbon capture and sequestration.
- Use sustainable (non-fossil fuel) and nuclear energy.
- Pursue energy efficiency.
- Use waste management practices that decrease the release of methane.
- Stop deforestation; pursue reforestation projects.
- Use agricultural practices that prevent the release of methane.

MITIGATION CAN BE PURSUED USING THE "WEDGE" APPROACH.

We can take steps to curb climate change by pursuing actions that reduce greenhouse gas emissions or increase its removal from the atmosphere. Employing any 8 of 15 or more potential strategies (stabilization wedges) that each reduce CO_2 emissions by 1 billion metric tons per year over the next 50 years would stabilize emissions close to current levels. However, the longer we wait, the more wedges we will need.

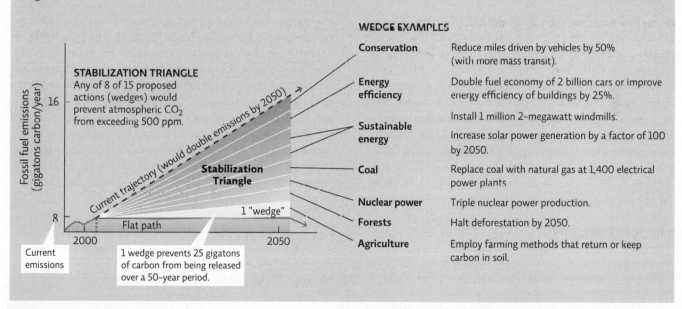

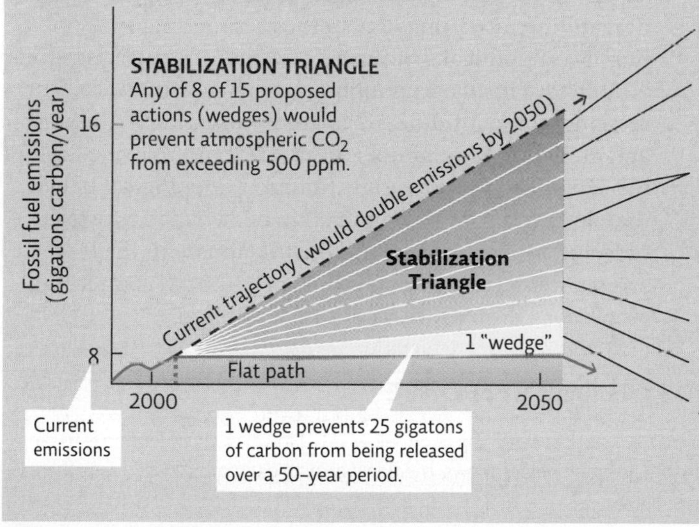

WEDGE EXAMPLES

Conservation	Reduce miles driven by vehicles by 50% (with more mass transit).
Energy efficiency	Double fuel economy of 2 billion cars or improve energy efficiency of buildings by 25%.
Sustainable energy	Install 1 million 2-megawatt windmills.
	Increase solar power generation by a factor of 100 by 2050.
Coal	Replace coal with natural gas at 1,400 electrical power plants.
Nuclear power	Triple nuclear power production.
Forests	Halt deforestation by 2050.
Agriculture	Employ farming methods that return or keep carbon in soil.

? Which stabilization wedge do you think would be the easiest to accomplish? Which would be the hardest?

The United States did not ratify the protocol, objecting because it only set reduction requirements for developed countries (which were responsible for most of the historic emissions); none were set for developing countries.

The Kyoto Protocol expired in 2012, and after several years of negotiation, it was replaced with the 2016 Paris Agreement, which allowed nations to set their own targets, with a goal of keeping warming "well below 2°C" above preindustrial levels, preferably capping warming at 1.5°C to avoid extremely dangerous impacts of climate change (e.g., minimizing sea-level rise and limiting crop and species losses to levels that are [hopefully] tolerable). We are already halfway to that 2°C (3.6°F) upper limit—Earth has already warmed about 1°C (1.8°F). The initial Paris targets will not limit warming to 2°C, but the goal is to adjust targets as needed every few years to eventually reach the 2°C (or even better the 1.5°C) goal. According to a 2018 special report by the IPCC, we have only about 12 years to reduce CO_2 emissions by 45% to have a chance of not exceeding this 1.5°C target; less aggressive action will send us to 2 degrees of warming or beyond. At our current rate of greenhouse gas emissions, we would reach 1.5°C of warming as early as 2030. And if we postpone aggressive action to 2030, the 2°C target may be out of reach. **INFOGRAPHIC 8B**

Many feel that the 2°C upper limit is too high and advocate strongly for 1.5°C as an upper limit. While half a degree of difference might not sound like much, climate models predict substantial differences in impact at 1.5°C warming versus 2.0°C. For example:

- Ice-free summers in the Arctic are 10 times more likely at 2.0°C than at 1.5°C, a loss that would be devastating to wildlife and would also decrease arctic albedo and enhance warming.
- Sea-level rise could be an additional 4 inches by 2100 at 2.0°C, exposing around 10 million more people to flooding as compared to the rise expected at 1.5°C.
- Severe drought at 1.5°C would likely affect 350 million people; at 2.0°C, that number would be closer to 410 million.
- Twice as many species would likely lose at least half of their range at 2.0°C compared to 1.5°C.
- Somewhere between 70% and 90% of tropical coral reefs could be lost at 1.5°C; at 2.0°C it is unlikely any tropical corals will survive.

Meeting the Paris Agreement goal of limiting warming to just another 0.5° or 1°C will require a dedicated and aggressive response from the international community. How aggressive? Current analysis suggests that holding the temperature increase to 1.5°C would require that we cap atmospheric anthropogenic greenhouse gases around 450 ppm. Other greenhouse gas emissions, too, need to be phased out. But even if we reduce emissions at a steady rate and reach zero emissions by 2050 (a common target), we will likely surpass 450 ppm and would need to actually remove some CO_2 from the atmosphere. This means employing some of those wedges (like reforestation and agricultural changes) that capture and sequester CO_2 and developing geoengineering technologies (giant filters) that remove atmospheric CO_2.

The Paris Agreement was signed by most nations of the world, including then–U.S. President Barack Obama. But on November 4, 2019, President Donald Trump formally submitted notification to withdraw the United States from the agreement. Exiting the agreement will take around 1 year, but actions that are dismantling Obama-era climate initiatives, which began shortly after President Trump took office in 2017, are already having an effect. A roughly 2% per year decline in greenhouse gas emissions experienced under President Obama shrank to only 0.3% due to increased emissions in both the transportation and industry sectors.

In the United States, there is a vocal minority (many with strong ties to the fossil fuel industry or politically conservative groups) that actively attacks climate science and scientists, claiming that there is insufficient evidence for climate change or that there is no evidence that it is caused by human actions. Employing many of the same tactics that stymied, for a while, action on issues such as tobacco regulation and efforts to control acid rain—dragging out long-discounted arguments, demanding more certainty before acting, spending heavily on political campaigns and lobbying efforts—their efforts have instilled enough doubt to slow the U.S response. (See Module 1.2 for more on critical thinking and analyzing arguments.) Many attribute this rejection of the scientific evidence to be fueled by ideological beliefs that favor less governmental control or by the prospect of losing money (or making less) if climate initiatives are pursued. Unfortunately, this puts the 2°C target in jeopardy—even more so the 1.5°C target.

INFOGRAPHIC 8B FUTURE WARMING DEPENDS ON OUR RESPONSE

The Paris Agreement acknowledges that all the proposed targets, even if met, will not cap warming at or below 2°C. However, the intention is to ramp up the emission reductions and strengthen these targets every 5 years or so. The graph below shows median warming estimates (this means there is a 50% chance of reaching or exceeding each value) for three response scenarios.

PROJECTED WARMING BY 2100

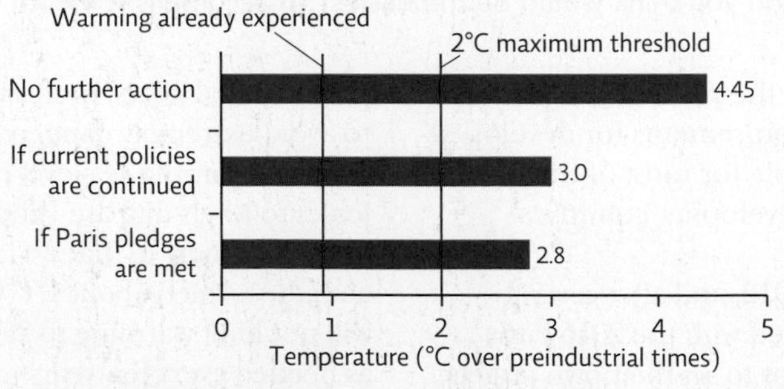

Why do you think the initial Paris pledges were set too low to meet the goal of warming no more than 2°C?

This inaction concerns people from all walks of life, especially scientists who study it closely. In a 2019 article published in the journal *Bioscience*, Oregon State University ecologist William Ripple, director of the Alliance of World Scientists, writes, "Scientists have a moral obligation to clearly warn humanity of any catastrophic threat and to 'tell it like it is.'" Recognizing this obligation, Ripple and colleagues at the Alliance of World Scientists have declared a "climate emergency," and they have invited scientists around the world to endorse their declaration by adding their signature. As of January 1, 2020, more than 13,000 had signed on.

Though, as Ripple lays out in the *Bioscience* article, the evidence for climate change is strong, there is still a startling difference between how scientists view climate change and the understanding of the general public—a difference known as the "consensus gap." Closing this gap, many believe, is key to taking real action on climate. "Disconnects between the potential threats of climate change and societal action arise from multiple factors, but changing how people perceive and conceptualize climate change is considered key to improving public engagement," writes Matthew Fitzpatrick of the University of Maryland Center for Environmental Science. To help bridge this gap, Fitzpatrick is working to translate what climate scientists understand into terms the average person can appreciate by creating an interactive map (available at https://fitzlab.shinyapps.io/cityapp/) that allows one to see which North American city has a climate today that would be expected in other cities 60 years in the future. New York City, for example, is expected to have a climate more like Jonesboro, Arkansas; Ann Arbor, Michigan, will have a warmer and wetter climate similar to Vincennes, Indiana; and Orlando, Florida's climate will be warmer and drier, more like that of Mexico City, Mexico.

Because climate change is such a complex issue, misleading the public has unfortunately been easy to do. Just claim that scientists disagree (they don't) or that the effects of climate change are nothing to worry about. (The people of Newtok, the Isle de Jean Charles, and the Maldives disagree.) Hold up a snowball in the Senate, as Senator James Inhofe famously did, and say the climate can't be warming because it snowed today (confusing *weather* with *climate*). Or even worse, dismantle or defund climate research—an action that has been compared to not taking your child's temperature because you are afraid he or she might be sick and potentially missing an opportunity to address the illness before it gets out of hand.

These and other arguments against action hurt our chances of successfully addressing this unprecedented challenge facing humanity. Even the argument that responding to climate change will hurt the economy is believed by many to be flawed because it fails to take into account the manufacturing, installation, and maintenance jobs a renewable energy industry is bringing and will bring if the world moves ahead with an ambitious renewable energy plan. Further, we have yet to measure the economic, environmental, and ethical true costs of inaction. What we do know is that people in developing countries and those in poverty will suffer the most (and they are the ones who have contributed to the problem the least). Dealing with climate change is an environmental justice issue of massive proportions.

For climate change, what is the majority conclusion? As stated in a 2017 *National Geographic* article, the scientific consensus is strong on these three points: "The world is warming. It's because of us. We're sure."

Many nations are standing by their Paris commitments, agreeing with the scientists, policy makers, and citizens who are advocating for the *precautionary principle*— choosing to act in the face of uncertainty because the stakes are so high. It remains uncertain how fast climate impacts will unfold or how bad they might be and when we might reach a tipping point (i.e., a glacial and polar ice melt that cannot be stopped) that would set into motion events that dwarf the problems with which we are currently dealing. This may even push our environment past the safe operating zones (Planetary Boundaries) that life as we know it depends on and imperil Earth's ability to sustainably support humanity (see Module 1.1 for more on Planetary Boundaries). These uncertainties arise, in part, because we continue to contribute to the problem (burning fossil fuels, cutting down forests, etc.). But what we do know is that the sooner we act, the better chance we have of successfully addressing the problem.

IR_Stone/iStock/Getty Images

An example of adapting to the threat of floods, gates on the River Thames protect central London from flooding that might occur from extremely high tides or storm surges. Gates rest flush with the riverbed when not in use and are rotated upward to form a barrier when needed. The Thames Barrier became operational in 1982, and the rate of its usage has increased in recent years as sea level rises.

Some people have referred to the impacts of climate change we are currently experiencing as our "new normal." David Wallace-Wells, a journalist and author who has written extensively on climate change, disagrees. As he explains, all of the changes Earth and human society are experiencing do not represent a "new normal" because climate is still changing and the biggest uncertainty is the human response. What happens in the future—where Earth's "new climate" stabilizes—is still a story being written, and as Wallace-Wells writes, "we are all its authors. And still writing."

Select References:

Clark, P. U., et al. (2016). Consequences of twenty-first-century policy for multi-millennial climate and sea-level change. *Nature Climate Change, 6*(4), 360–369.

Cook, J., et al. (2016). Consensus on consensus: A synthesis of consensus estimates on human-caused global warming. *Environmental Research Letters, 11*(4), 048002.

Fitzpatrick, M. C., & Dunn, R. R. (2019). Contemporary climatic analogs for 540 North American urban areas in the late 21st century. *Nature Communications, 10*(1), 614.

Hansen, J., et al. (2016). *Global Temperature in 2015.* New York: Climate Science, Awareness and Solutions, Earth Institute, Columbia University. Retrieved from http://www.columbia.edu/~jeh1/mailings/2016/20160120_Temperature2015.pdf

Hauer, M. E., et al. (2019). Sea-level rise and human migration. *Nature Reviews Earth & Environment,* 1–12.

Mann, M. E., et al. (2017). Influence of anthropogenic climate change on planetary wave resonance and extreme weather events. *Scientific Reports (Nature Publisher Group), 7,* 45242.

Ripple, W. J., et al. (2019). World scientists' warning of a climate emergency. *BioScience,* biz088. doi:10.1093/biosci/biz088.

Wallace-Wells, D. (2019). *Uninhabitable Earth: Life after Warming.* New York: Tim Duggan Books.

This 10.6-megawatt wind farm at Bada Bagh in India is an example of employing mitigation strategies to reduce the impact of climate change by displacing some fossil fuel-derived energy with a more sustainable option.

BremecR/iStock/Getty Images

GLOBAL CASE STUDIES CLIMATE CHANGE IMPACTS

Achieve

Climate change is having many and varied consequences and is constantly in the news — whether it is a new study, a new population impacted, or the responses (or lack of response) by nations of the world. Visit the Global Case Studies for this module to learn more about the effects of climate change.

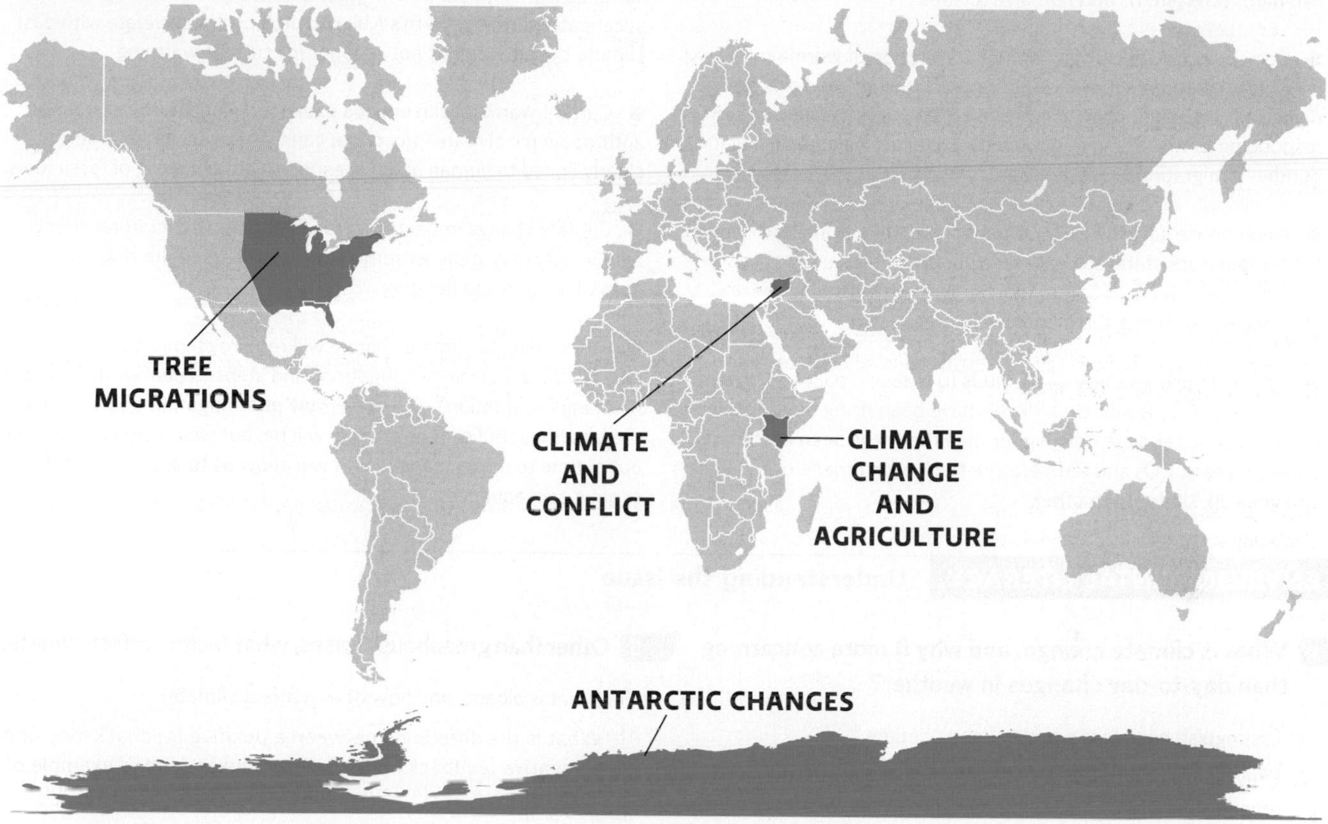

BRING IT HOME

PERSONAL CHOICES THAT HELP

The effects of climate change are already being felt by humans, other species, and ecosystems around the globe. Though significant action is needed at the city, state, and national levels, actions that individuals and community groups can make that help address climate change are also needed and will show policy makers that citizens are interested in preventing global climate change.

Individual Steps

- Do your part to reduce carbon emissions by conserving energy. Walk or ride a bike instead of driving a car. Share a ride with a coworker rather than drive alone. Negotiate with your employer to telecommute. Live close to where you work or go to school. Reduce your heating and cooling energy use and always turn off electronics and lights when not in use.
- If your utility company offers renewable energy, buy it.
- Reduce the carbon footprint of your food by decreasing the amount of feedlot-produced meat you eat. Buy your food as locally as possible to reduce energy used in transportation.
- Go to www.terrapass.com to see how you can offset CO_2 production from your car, your house, and your airplane travel.
- Write, call, or visit the offices of your elected officials and share your views about funding for research and development of clean and renewable sources of energy as well as your views on participation in the Paris Agreement. In addition, ask that they support the funding of science, especially efforts to understand and confront climate change.
- Go to https://fitzlab.shinyapps.io/cityapp/ to view an interactive map that helps users visualize climate change in North American cities. Choose any city on the map, and it will show you 60 years in the future which present-day city has a climate expected for your chosen city. Share this information with family and friends to help them understand the nature of the shift that is expected.

Group Action

- Volunteer to help build a zero-energy Habitat for Humanity home.
- Organize a community lecture on climate change with a local university expert or meteorologist as the speaker.
- Organize an event at your school or community to raise awareness about global climate change and ways to prevent it. Go to www.350.org to join a current campaign and get other program ideas.

REVISIT THE CONCEPTS

- Climate change is a serious environmental challenge. A change of just a few degrees in average temperature can result in a climate that can harm ecosystems and human societies.
- Currently climate is changing in the direction of warming, resulting in warmer average temperatures, melting land and sea ice, rising sea levels, and changing precipitation patterns. Species are responding to these changes by shifting their range or declining in number if migration is not possible.
- Greenhouse gases like CO_2 trap heat radiating from Earth, raising the temperature of the atmosphere. Our actions have enhanced this effect with the addition of more greenhouse gases to the atmosphere, warming the planet even more.
- Scientists use a variety of methods to measure past and present atmospheric CO_2 levels; these levels have been rising dramatically in recent decades and are now higher than they have been in at least 2 million years. CO_2 and temperature track each other closely — as one goes up, so does the other.
- The albedo (reflectivity) of a surface affects climate — lighter-colored surfaces have a cooling effect. Replacing light-colored surfaces with dark ones can trigger positive feedback loops that accelerate warming. Earth's Milankovitch cycles correlate with past climate cycles but does not account for current warming.
- Current warming can only be explained when both natural and anthropogenic climate forcers are considered. Recent warming is clearly linked to human actions, especially the burning of fossil fuels.
- Climate change impacts include health and agricultural effects, biodiversity loss, more extreme storms, increased fire risk, and coastal erosion and flooding.
- Responding to climate change will require steps that try to reduce future warming (mitigation) and steps to deal with inevitable warming (adaptation). The sooner we put mitigation efforts in place, the less impactful climate change will be, but we are quickly running out of time to make changes that will allow us to avoid the most serious consequences.

ENVIRONMENTAL LITERACY **Understanding the Issue**

1 What is climate change, and why it more concerning than day-to-day changes in weather?

1. Distinguish between "weather" and "climate."
2. Why is an increase of a few degrees in average global temperature more concerning than day-to-day weather changes of a few degrees?
3. In the winter of 2010, the northeastern part of the United States had several large snowstorms that resulted in record-high snowfall amounts. How does this weather fit in with the notion of global climate change?

2 What is the physical and biological evidence that climate change is occurring?

4. To what is recent sea-level rise attributed?
5. Outline the evidence for climate change. Do you feel that this evidence supports the conclusion that climate is changing? Explain.

3 What is the greenhouse effect, and how are we affecting it?

6. What are greenhouse gases, and why do we say the greenhouse effect is a good thing?
7. Identify three human actions that have led to an increase in the amount of greenhouse gases in the atmosphere. What has been the result?

4 How are atmospheric CO_2 and temperature measured, and how are they correlated?

8. How can ice cores tell us what atmospheric temperature and CO_2 levels were in the distant past?
9. What is the value of using multiple types of proxy data when reconstructing past climates?

5 Other than greenhouse gases, what factors affect climate?

10. What is albedo, and how does it affect climate?
11. What is the difference between a positive feedback loop and a negative feedback loop? Give a climate-related example of each.

6 What evidence suggests that climate change is due to human impact?

12. Explain how the accuracy of computer climate models is validated and how they are used to predict future climate based on different levels of climate forcers in the future.
13. Why do scientists conclude that current warming is largely due to anthropogenic factors?

7 What are the current and potential future impacts of climate change?

14. How might climate change help some species while hurting others? Are any species currently being helped or hurt by climate change?
15. Describe the types of problems that global climate change causes for human health. Which do you feel is likely to cause the biggest problem? Why?

8 What actions can we take to respond to a world with a changing climate?

16. Compared to preindustrial times, how much have we already warmed, and what is the Paris Agreement's target for warming? How are we currently doing with regard to meeting this goal (not exceeding this target)?
17. Distinguish between adaptation and mitigation. Why do we need both strategies?

SCIENCE LITERACY Working with Data

The location of 105 species of fish, shellfish, and other marine species in the Eastern Bering Sea and along the U.S. northeast coast was determined by tracking the location of each species from 1982 to 2015. Both latitude and depth were followed and expressed as the average center of biomass (central location for each population by weight) and shown in the graphs below. Analyze these graphs and answer the questions that follow.

CHANGE IN LATITUDE AND DEPTH OF MARINE SPECIES, 1982–2015

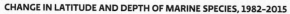

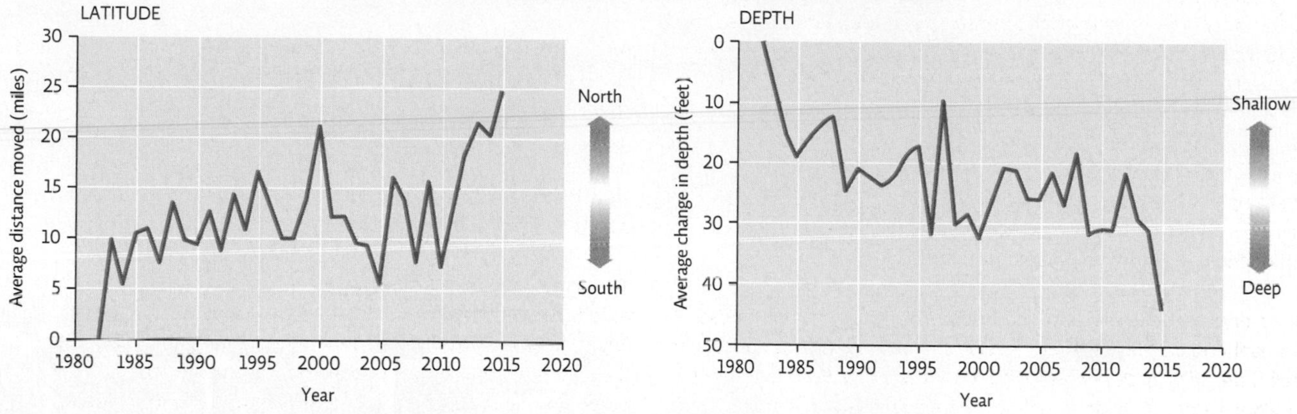

Interpretation

1. In general, what directional trend is seen over time for the average change in latitude and for depth from 1982 to 2015 for the species evaluated?
2. How much farther north were these species found in 2015 compared to 1982? How much deeper?
3. How did average latitude and depth change between 2000 and 2005? What does this suggest about water temperature in those years?

Advance Your Thinking

1. What experimental prediction would you make for change in latitude and depth if you were testing the hypothesis that warming waters were affecting the location of these species? Do the data support that hypothesis? Explain
2. Why did the researchers collect data on both latitude and depth?
3. If the researchers had only tracked one or a few species, the data would not be as compelling as it is in this study that tracked and averaged the location of 105 species. Why?

INFORMATION LITERACY Evaluating Information

Among scientists, there is broad consensus (97%) that climate change is significantly caused by human activity. Yet in a 2017 Gallup poll, only 45% of the public are worried about climate change (and this is the highest percentage in three decades). Members of the public get their information from a variety of media sources and information posted on the Internet. How can there be such a large disconnect between scientists and the public?

Go to the Global Warming Hoax page at www.globalwarminghoax.com

Evaluate the website and work with the information to answer the following questions:

1. Determine if this is a reliable information source with a clear and transparent agenda:
   a. Who runs the website? Do this person's/group's credentials make the site reliable or unreliable? Explain.
   b. What is the primary message of the website?
   c. Read the entry "Antarctic Sea Ice for March 2010 Significantly Greater than 1980." What evidence is provided in the article? Do you have any questions about the data presented? If so, what are they?

Now go to the Skeptical Science website (www.skepticalscience.com). Click on the link "Most Used Climate Myths."

2. Determine if this is a reliable information source with a clear and transparent agenda:
   a. Who runs the website? Do this person's/group's credentials make the site reliable or unreliable? Explain.
   b. What is the primary message of this website? What types of evidence does it provide to support its message?
   c. Click on the "Antarctica Is Gaining Ice" link. Read the article and compare the main point of the article to the article on the Global Warming Hoax site.
   d. Which explanation and website do you find more credible? Why?

Additional study questions are available at Achieve.macmillanlearning.com. Achieve

CHAPTER 11

Alternatives to Fossil Fuels

There are a wide variety of energy sources that can be used in place of fossil fuels. No single alternative can replace fossil fuels, but together the many alternatives can.

Module 11.1

Nuclear Power

An evaluation of the pros and cons of nuclear power as an alternative method for electricity production

Module 11.2

Sustainable Energy: Stationary Sources

A survey of sustainable energy sources for the production of electricity: solar, wind, geothermal, and hydroelectric power

Online Module 11.3

Sustainable Energy: Mobile Sources

A look at biofuels as a replacement for fossil fuel-derived transportation fuels

Online Modules are available at Achieve.macmillanlearning.com.

GIPhotoStock/Cultura/Getty Images

Module 11.1 Nuclear Power

THE FUTURE OF FUKUSHIMA

Can nuclear energy overcome its bad rep?

Police guard a checkpoint at the edge of the exclusion zone established around the crippled Fukushima Daiichi Nuclear Power Station. The sign reads "Keep Out."

AP/DAVID GUTTENFELDER/ National Geographic Creative

After reading this chapter, you should be able to answer the following GUIDING QUESTIONS

1. What are radioactive isotopes, and why are they important for nuclear power?
2. How is uranium fuel for nuclear power produced?
3. How is nuclear energy harnessed to generate electricity in a fission reactor?
4. What types of radiation are produced when isotopes decay?
5. How is the rate of decay of a radioactive atom measured?
6. What problems are associated with nuclear waste?
7. What is the history of nuclear accidents worldwide?
8. What are the advantages and disadvantages of nuclear power?

The Fukushima Daiichi Nuclear Power Station is a maze of steel and concrete perched right on Japan's Pacific coast just 150 miles north of Tokyo. Its six nuclear reactors supplied some 4.7 GW (1 gigawatt = 1 billion watts) of electric power to the country, making it one of the largest nuclear power plants in the world. On March 11, 2011, when a magnitude 9.0 earthquake struck 80 miles north of the plant, there were more than 6,000 workers inside. The quake caused a power outage, and in the darkness, chaos ensued: Men and women groped for ground that would not stabilize beneath their hands and feet, some shouting in panic as steel and concrete collided around them. When the shaking stopped, emergency lights came on, revealing a cloud of dust. But that was only the beginning of the disaster.

The earthquake had erupted beneath the ocean floor, triggering a tsunami that would arrive at the plant in two distinct waves. The first wave was not big enough to breach the 30-foot-high concrete wall that had been built between the plant and the sea. But the towering mass of water that came 8 minutes later was. The wall of water bulldozed a string of protective barriers, sent cars and trucks crashing into buildings, and eventually settled in deep black pools around the reactors themselves.

These were the strongest earthquake and largest tsunami in the country's long history. Together, they would claim more than 16,500 lives along a 250-mile stretch of coast (roughly equal to the distance between Maine and Manhattan). But as the ground steadied and the water subsided, the world's attention would quickly turn to a third disaster: the risk of nuclear meltdown at Daiichi.

WHERE IS FUKUSHIMA, JAPAN?

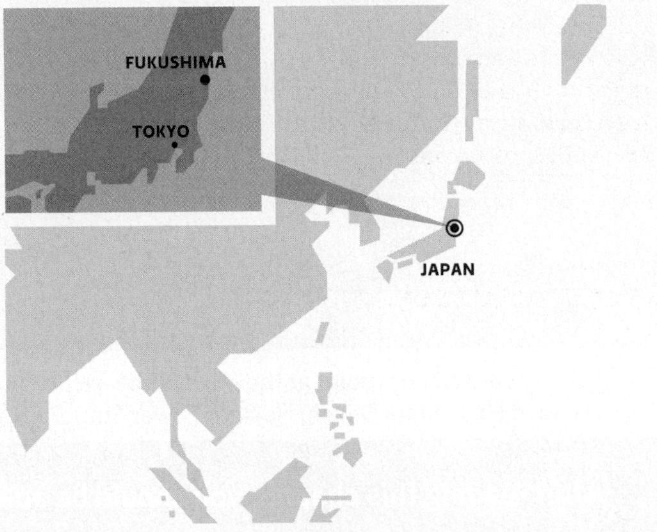

Satellite image of the damaged Fukushima Daiichi Nuclear Power Station on March 14, 2011.

Digital Globe/ABACA/Newscom

1 NUCLEAR-POWERED FUEL: RADIOACTIVE ISOTOPES

Key Concept 1: The energy contained in an atom (nuclear energy) can be harnessed to generate electricity. A radioactive isotope of uranium typically provides the fuel for nuclear power.

In some ways, electricity generated using **nuclear energy** is very similar to other forms of thermoelectric power (those that use heat to produce electricity). Just like power plants that run on oil or coal, nuclear plants use heat to boil water and produce steam, which is then used to generate electricity. The difference, really, is the source of that heat. Coal and natural gas plants create it by burning fossil fuels. In the thermonuclear production of electricity, heat is produced through a controlled nuclear reaction.

To understand nuclear power, we must delve into basic chemistry at the atomic (the atom) and subatomic (protons and neutrons) levels. All matter is made up of **atoms**— the simplest form of an element that retains the characteristics of that element. Atoms contain a nucleus where protons and neutrons are found; electrons orbit around this nucleus. A substance composed of all the same type of atoms is an **element**— hydrogen is an element, as are helium, carbon, and uranium.

The atom of any element is defined by the number of protons it possesses (known as its *atomic number*). For example, any atom that contains two protons is a helium atom; those that contain six are carbon atoms. Because electrons are so small compared to the subatomic particles in the nucleus, the atomic mass of an atom is simply the number of protons plus the number of neutrons (its *mass number*). While the number of protons in the atom of an element is unique (if the number of protons changes, the element changes), the number

nuclear energy Energy in an atom; can be released when an atom is split (fission).

atom The simplest form of an element that retains the characteristics of that element.

element A substance composed of all the same type of atoms.

PERIODIC TABLE OF ELEMENTS

Uranium

All uranium atoms have 92 protons but may have different numbers of neutrons. U-238, with 146 neutrons, is the most common form.

1 H																	2 He
3 Li	4 Be											5 B	6 C	7 N	8 O	9 F	10 Ne
11 Na	12 Mg											13 Al	14 Si	15 P	16 S	17 Cl	18 Ar
19 K	20 Ca	21 Sc	22 Ti	23 V	24 Cr	25 Mn	26 Fe	27 Co	28 Ni	29 Cu	30 Zn	31 Ga	32 Ge	33 As	34 Se	35 Br	36 Kr
37 Rb	38 Sr	39 Y	40 Zr	41 Nb	42 Mo	43 Tc	44 Ru	45 Rh	46 Pd	47 Ag	48 Cd	49 In	50 Sn	51 Sb	52 Te	53 I	54 Xe
55 Cs	56 Ba	57–71 Lantha-noids	72 Hf	73 Ta	74 W	75 Re	76 Os	77 Ir	78 Pt	79 Au	80 Hg	81 Tl	82 Pb	83 Bi	84 Po	85 At	86 Rn
87 Fr	88 Ra	89–193 Acti-noids	104 Rf	105 Db	106 Sg	107 Bh	108 Hs	109 Mt	110 Ds	111 Rg	112 Cn	113 Nh	114 Fl	115 Mc	116 Lv	117 Ts	118 Og

57 La	58 Ce	59 Pr	60 Nd	61 Pm	62 Sm	63 Eu	64 Gd	65 Tb	66 Dy	67 Ho	68 Er	69 Tm	70 Yb	71 Lu
89 Ac	90 Th	91 Pa	92 U	93 Np	94 Pu	95 Am	96 Cm	97 Bk	98 Cf	99 Es	100 Fm	101 Md	102 No	103 Lr

International Union of Pure and Applied Chemistry; December 2018.

The Periodic Table of Elements shows chemical elements arranged by their atomic number. The arrangement in rows and columns groups atoms with similar chemical properties.

of neutrons can vary. This will change the mass number. These different versions of the atom are called **isotopes**. By convention, isotopes are named according to their mass number, such as uranium-238 (with an atomic mass of 238) and uranium-235 (a uranium atom that has three fewer neutrons than U-238). **INFOGRAPHIC 1**

isotopes Atoms that have different numbers of neutrons in their nucleus but the same number of protons.

radioactive Atoms that spontaneously emit subatomic particles and/or energy.

Most isotopes are stable, meaning they do not spontaneously lose protons or neutrons. But some are **radioactive**: They are less stable and spontaneously lose mass by emitting particles (protons, neutrons, or electrons) or photons (a form of short-wavelength radiation); in the process, heat is released. Radiation is also emitted when an atom is struck by a subatomic particle, splitting the atom into one or more smaller atoms. Because they are unstable, radioactive atoms are good candidates as fuel for nuclear power because it depends on the ability of the atoms to be split, an event that releases large amounts of heat that can be captured to generate electricity.

INFOGRAPHIC 1 ATOMS AND ISOTOPES

The fundamental unit of matter is the atom. Atoms contain protons and neutrons in a central nucleus around which electrons orbit. The number of protons, the *atomic number*, is unique to each element. For example, any atom with only two protons is an atom of the element helium. The sum of the number of protons and neutrons gives an atom its *mass number*.

THE ATOM

HELIUM ATOM

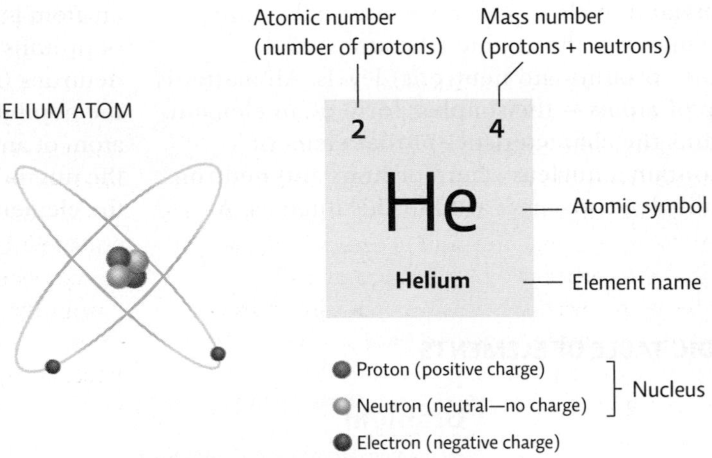

Isotopes are atoms that have the same atomic number (number of protons) but a different number of neutrons and thus a different mass number. An atom with 92 protons is uranium (U); if a uranium atom has 146 neutrons, it is U-238 (92 protons + 146 neutrons = 238). Another uranium isotope, U-235, has 143 neutrons (92 protons + 143 neutrons = 235).

ISOTOPES

URANIUM-238

URANIUM-235

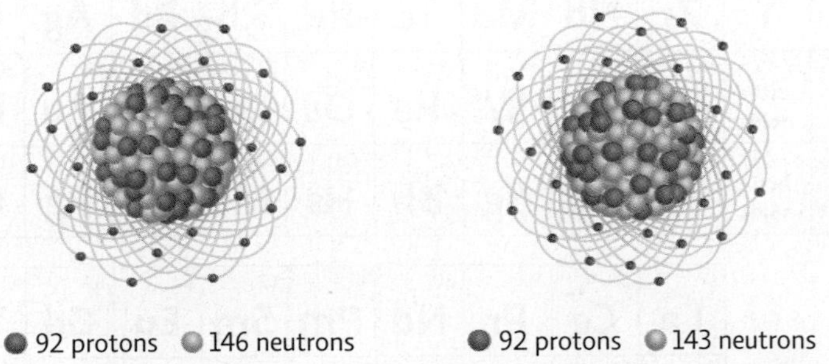

More neutrons—heavier

Fewer neutrons—lighter and less stable

? Uranium also exists as uranium-233. How many protons and neutrons does it have? Do you think it is more or less stable than U-235? Explain.

2 PRODUCTION OF NUCLEAR FUEL

Key Concept 2: The production of nuclear fuel involves mining and several processing steps, all of which generate hazardous waste.

Most nuclear reactors use the element uranium, which has several isotopes. Uranium-238 (U-238) is the most stable and is the most abundant form of uranium; it makes up roughly 99% of Earth's total supply. U-235 is a less stable form and provides the "fuel" used in a typical fission reactor. But acquiring the U-235 is not as simple as locating rock rich in uranium (uranium ore) and digging it up. (See Module 7.1 for more on mineral ores and mining.) Because U-235 makes up such a small percentage of the uranium found in uranium-rich ore, a lot of rock must be mined to eventually produce a useful fuel for the nuclear reactor.

Once the ore is dug up, several processing steps are needed. In the milling step, ore is crushed and mixed with either an acidic or alkaline solution to separate the uranium from the ore. This process creates uranium oxide, known as "yellowcake" in the industry for its yellow hue. Next, the fuel is "enriched," a process that increases the percentage of U-235 in the sample (otherwise there is not enough material in the uranium sample to participate in the nuclear reaction). In this step, the percentage of U-235 is increased from around 0.7% to around 4%. (This is nowhere near as enriched as the uranium in a nuclear bomb; that highly enriched fuel is about 90% U-235.)

The enriched uranium is formed into small pellets and stacked into 10- to 20-foot-long hollow **fuel rods**. More than 100 of these fuel rods are loaded into a framework to form a fuel assembly. Several fuel assemblies are placed into a thick-walled vessel—the reactor core—where they remain in place for a year or two before being removed and replaced with fresh fuel.

The mining and processing of uranium raise concerns about health and environmental safety. The processing of uranium, from the mining to the enrichment and production of fuel pellets, creates hazardous waste that must be dealt with. Uranium mining itself is inherently dangerous due to the radioactive material. A 2014 analysis by Benjamin Jones found that uranium miners in New Mexico who died of lung cancer had their lives cut short by almost 22 years compared to other lung cancer victims of the same age, race, and sex. Living near a uranium mining or processing operation is also linked to adverse health effects, most notably kidney and cardiovascular problems. **INFOGRAPHIC 2**

fuel rod Hollow metal cylinder filled with uranium fuel pellets for use in fission reactors.

John van Hasselt — Corbis/Getty Images

Open-pit mill tailings site in Australia. Mining and crushing uranium ore produce small-particle (sandlike consistency) waste that contains low levels of radioactive isotopes. Mill tailings waste is often stored in large open pits and covered with water to reduce windblown loss of material. There is a concern that water could overflow this pit in heavy rain years.

INFOGRAPHIC 2 NUCLEAR FUEL PRODUCTION

Uranium ore (rock that contains uranium) is mined and goes through many stages of processing to produce fuel suitable for a nuclear reactor. The process creates hazardous waste at every step.

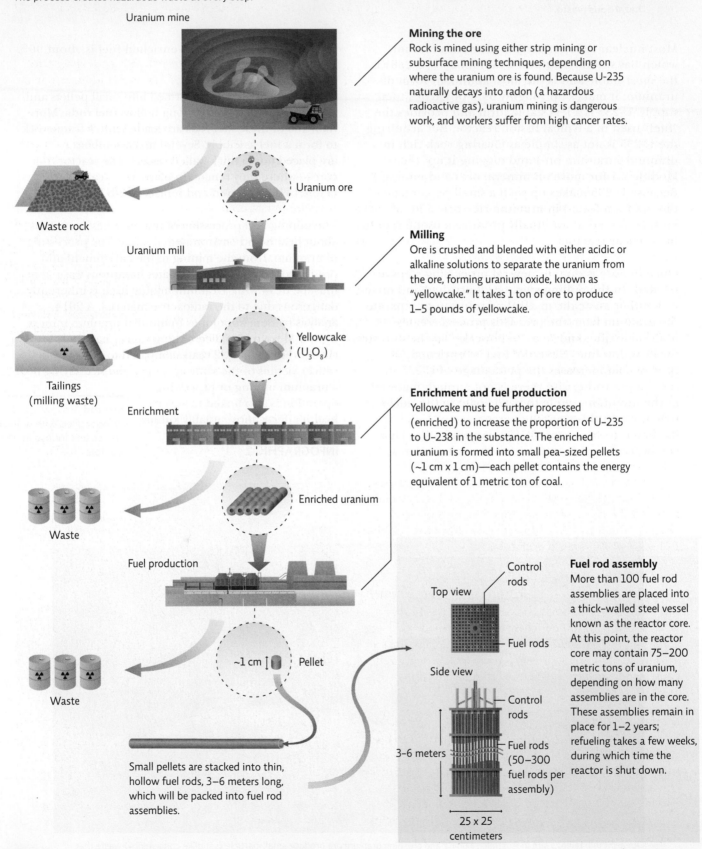

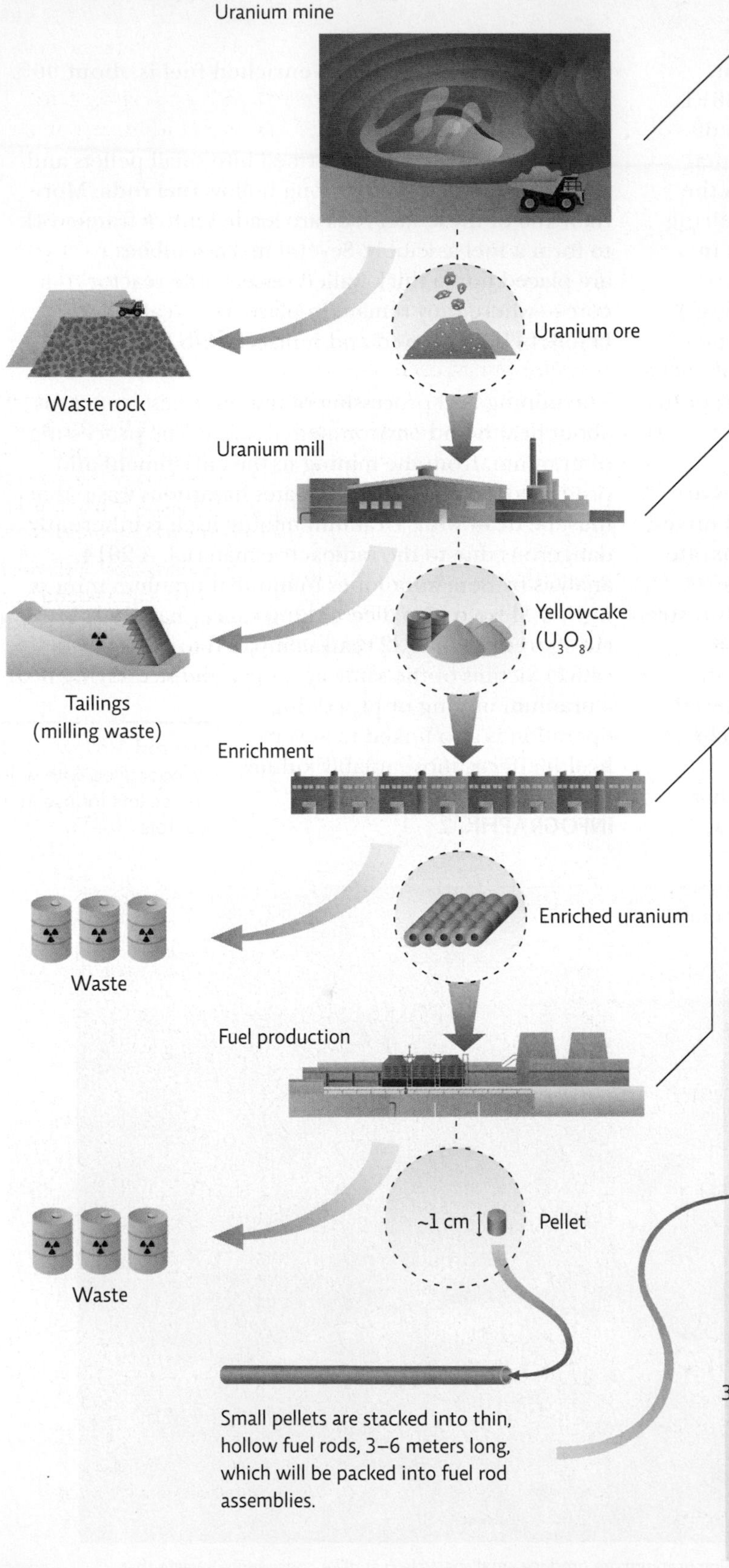

Mining the ore
Rock is mined using either strip mining or subsurface mining techniques, depending on where the uranium ore is found. Because U-235 naturally decays into radon (a hazardous radioactive gas), uranium mining is dangerous work, and workers suffer from high cancer rates.

Milling
Ore is crushed and blended with either acidic or alkaline solutions to separate the uranium from the ore, forming uranium oxide, known as "yellowcake." It takes 1 ton of ore to produce 1–5 pounds of yellowcake.

Enrichment and fuel production
Yellowcake must be further processed (enriched) to increase the proportion of U-235 to U-238 in the substance. The enriched uranium is formed into small pea-sized pellets (~1 cm x 1 cm)—each pellet contains the energy equivalent of 1 metric ton of coal.

Fuel rod assembly
More than 100 fuel rod assemblies are placed into a thick-walled steel vessel known as the reactor core. At this point, the reactor core may contain 75–200 metric tons of uranium, depending on how many assemblies are in the core. These assemblies remain in place for 1–2 years; refueling takes a few weeks, during which time the reactor is shut down.

Which stages of nuclear fuel production have radioactive material present?

3 GENERATING ELECTRICITY WITH NUCLEAR ENERGY

Key Concept 3: In a nuclear fission reactor, U-235 is bombarded with neutrons to split the atoms; this releases more neutrons, which leads to a self-perpetuating chain reaction. Heat from this reaction is used to produce steam, which turns a turbine attached to a generator to generate electricity.

Today nuclear power is generated by splitting atoms—**nuclear fission**—a process that releases a tremendous amount of heat for the production of electricity. Nuclear *fusion* reactions—the combining of two atoms to form a new one—also generate heat; the Sun is a giant fusion reactor, forming helium through the joining of two hydrogen atoms. (Research and development is underway to try to make fusion reactors a viable option.)

A nuclear fission chain reaction begins when the fuel rods (and the U-235 inside) are deliberately bombarded with neutrons. The uranium nucleus splits to form two or more lighter nuclei (hence the *fission by-products* generated are different elements from uranium and even distinct from each other if they contain a different number of protons), releasing two or three additional neutrons in the process. These newly released neutrons then hit other U-235 atoms, causing them to split and release even more neutrons, and so on, producing a chain reaction that is self-sustaining or even capable of accelerating.

Unlike the type of nuclear reaction at work in a nuclear bomb, which contains a much higher percentage of "fissile material" that sets off a massive chain reaction that is almost instantaneous, the reactions at nuclear power plants are highly controlled. **Control rods**—made of materials that absorb neutrons, such as boron or graphite—are placed in the fuel rod assembly between the fuel rods to control the speed of the reaction. They can be added (to slow down the reaction) or removed (to make it go faster).

Even controlled, this chain reaction releases a tremendous amount of heat. The heat is used to boil water, which produces steam, which turns turbines attached to a generator that creates electricity. (The electricity is produced when a metal conductor such as copper spins within a magnetic field, a process that causes the copper to release a stream of electrons [electricity].)

There are several types of fission reactors. The most common type worldwide (and in the United States) is a *pressurized water reactor* (PWR), designed so that the steam that turns the turbine is not exposed to radiation: Nuclear fission in the core heats water under pressure (like a pressure cooker); that hot (and radioactive) water enters a pipe that passes through, and thus heats, a separate container of water (which is not exposed to radiation); it is the steam from this radiation-free water that turns the turbine.

The reactors at Fukushima, however, were *boiling water reactors* (BWRs); BWRs produce steam in the reactor core itself. This means that both the steam and the turbine become radioactive in the process. The meltdown at Fukushima revealed a flaw in the BWR design: If pressure builds up in the reactor vessel, as it did in the Fukushima reactors as heat levels climbed, the corrective measure would be to vent the steam to avoid an explosion, but this would release radioactive material. Radiation would not be released if a PWR had to be vented to release excess steam since the steam itself is not radioactive. **INFOGRAPHIC 3**

nuclear fission A nuclear reaction that occurs when a neutron strikes the nucleus of an atom and breaks it into two or more parts.

control rod Cylinder that can be added to a fuel assembly to absorb neutrons and slow the fission chain reaction.

Mark Henley/Panos Pictures

The Gösgen Nuclear Plant is located within the town of Däniken, Switzerland. Steam rising from a cooling tower greets children as they walk home from school.

INFOGRAPHIC 3 HOW IT WORKS: NUCLEAR REACTORS

Achieve

Fission, or the breaking apart of atoms, begins when an atom like U-235 is bombarded with a neutron. This breaks the atom into other smaller atoms and releases free neutrons, which in turn hit other U-235 atoms, causing them to split and release more neutrons, and so on. The reaction in the fuel assembly is controlled by the insertion of control rods of nonfissionable material, which absorb some of the free neutrons.

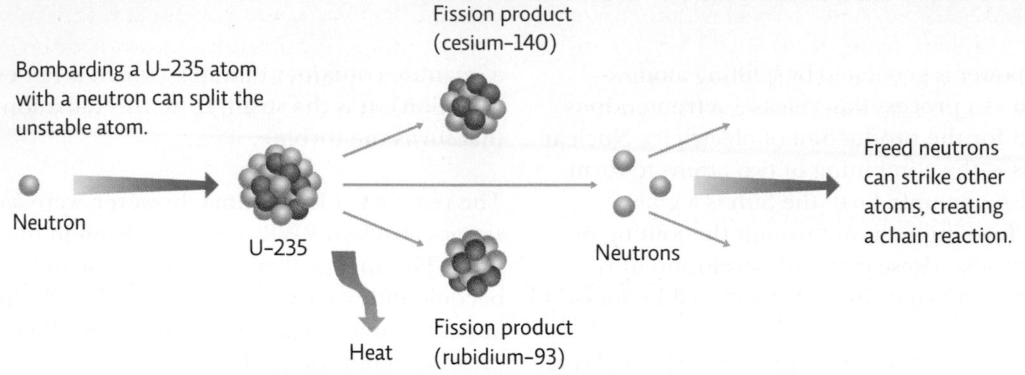

The most common type of nuclear power plant is a pressurized water reactor (PWR), but the reactor at Fukushima is an older technology: a boiling water reactor (BWR). Both designs use fuel assemblies with control rods and use water as a cooling and steam source.

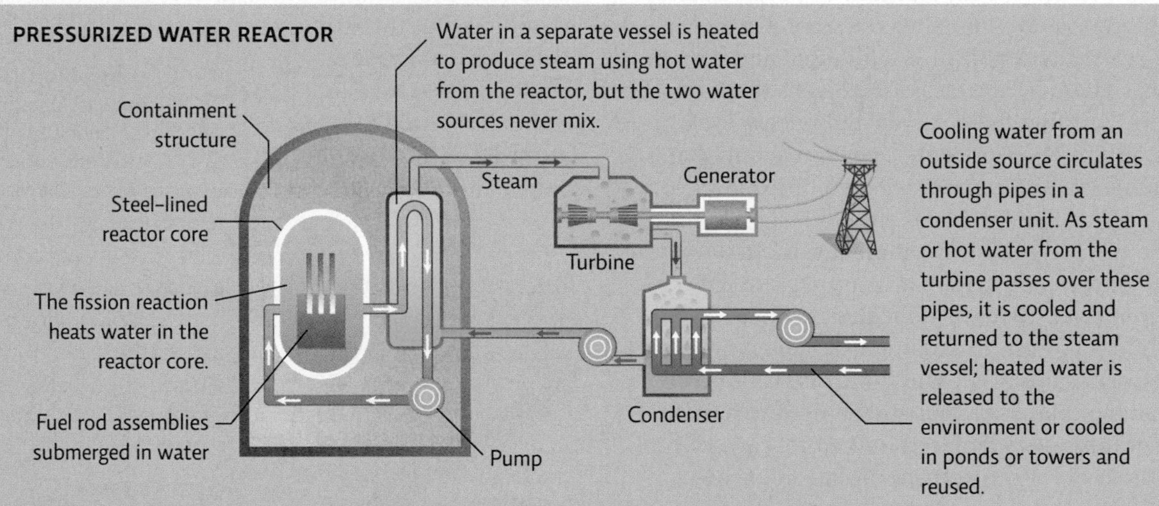

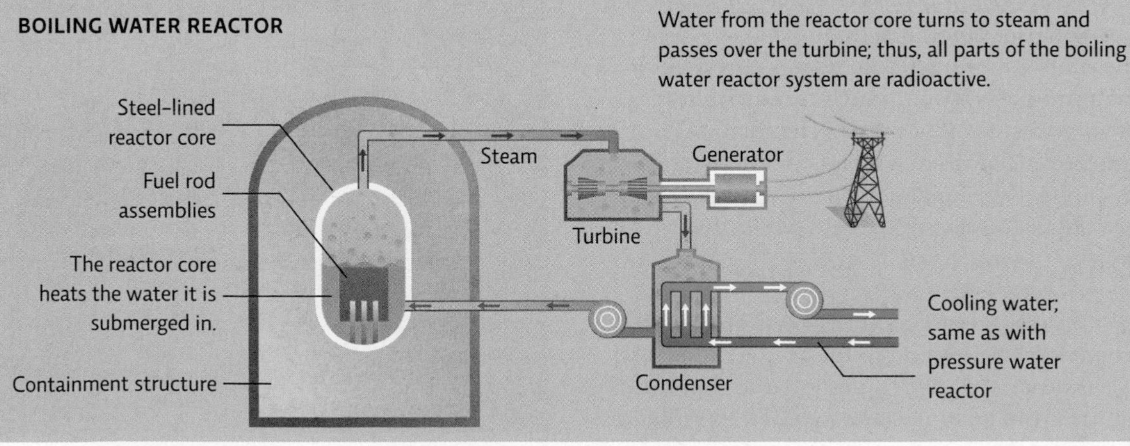

How is a nuclear reactor similar to a coal-fired power plant? How are they different?

All types of thermoelectric power take a lot of water; that's why power plants are sited near rivers and oceans. But currently nuclear power requires the most—roughly 25% more than a coal-fired power plant and 75% more than a natural gas plant. Though most of the water is not contaminated in the process, water used for cooling is returned to the source (river or ocean); releasing this warmer-than-normal water (called thermal pollution) back into the environment causes problems for species not adapted to the higher temperature.

The reason for all that water is simple: With nuclear energy, water is needed to produce steam and also to prevent the reactor from overheating. And even after the spent fuel rods are removed from the reactor, they need constant cooling because the radioactive fission products within are unstable and produce tremendous amounts of heat, creating one of the biggest challenges nuclear reactors face—storing the highly radioactive spent fuel rods.

4 TYPES OF RADIATION

Key Concept 4: Alpha, beta, and gamma radiation vary in their ability to penetrate substances; while all are dangerous, gamma radiation is the most hazardous to health.

In a fission reactor, both the nuclear fuel and the fission products produced by the chain reaction are radioactive. In fact, the fission products are more unstable and radioactive than the original U-235 fuel. On the day of the tsunami, each of the three operating reactors at the Fukushima plant held about 25,000 fuel rods. The reactors had gone through a successful emergency shutdown after the earthquake hit, but heat was still being generated by radioactive decay of the fission products in the fuel rods. Keeping the fuel cool was still vitally important. Unfortunately, the power outage had not only plunged the plant into darkness but also stopped the normal delivery of water to those reactors. (Backup generators were unfortunately located in the basement, which was flooded by the tsunami and rendered inoperable.) As Fukushima's reactor cores lost water (the heat of the fuel assemblies boiled off the water, and disabled pumps were not adding new cooling water), there was a risk of a hydrogen explosion—temperatures could climb so high that the metal cladding of the fuel rods would melt, react with water, and release explosive hydrogen gas. A steam explosion was also a risk—the buildup of tremendous pressure from the rapid production of steam could blow the top off the building.

Valiant efforts were made to get water on the reactors: Impassible roads prevented fire trucks from getting through, and helicopter attempts to drop sea water on the reactors failed. Eventually, fire trucks made their way through and, using hoses designed for jet-fuel fires, emergency responders were able to get some water to the places where it was most needed.

However, it was not enough. As the heat climbed due to lack of cooling, a steam or hydrogen explosion seemed inevitable. Radiation from damaged reactor Number 2 reached the control room; it had climbed so high that workers had to rotate out at regular intervals to avoid being poisoned with radiation.

Ionizing radiation (radiation that causes an atom to lose electrons and become "ionized" or charged) released from radioactive isotopes can take one of three forms. **Alpha** and **beta radiation** are particles. (Alpha particles are basically a helium nucleus; beta particles are electrons.) Alpha particles don't travel far and can't even penetrate paper; they do not penetrate skin but can be harmful if inhaled or ingested in food or water—raising concerns about the safety of food and dairy products produced in the Fukushima region. The much smaller beta particles can penetrate the upper layers of the skin (and can also be ingested) but are easily stopped by a thin sheet of aluminum or heavy clothing. If either type of radiation enters the body, it can cause organ damage and cancer by directly harming cell structures and DNA. (Radiation treatment for cancer is beta radiation; its goal is to kill cancer cells, but it can also kill normal cells it contacts.)

Gamma radiation is not a particle; it is a high-energy photon (an electromagnetic wave) that can easily penetrate skin—it takes thick concrete or a dense material such as lead to stop gamma rays—giving it the greatest potential to cause serious health problems such as radiation sickness, cancer, and birth defects. **INFOGRAPHIC 4**

High levels of radiation, especially gamma radiation that could not be avoided simply by wearing containment suits, quickly became a problem for Fukushima workers. On day two of the disaster, attempts to vent the steam from the reactor vessel (to avoid a steam explosion) failed due to dangerously high, and climbing, radiation levels. That afternoon the building housing reactor Number 1 exploded, sending concrete and steel debris flying. Radiation levels around the plant climbed exponentially, deterring efforts to vent the other two reactors. In the days that followed, both exploded.

alpha radiation Ionizing particle radiation that consists of two protons and two neutrons.

beta radiation Ionizing particle radiation that consists of electrons.

gamma radiation Ionizing high-energy electromagnetic waves (photons).

INFOGRAPHIC 4 IONIZING RADIATION

Three different types of ionizing radiation can be released by radioactive isotopes, and each differs in its ability to penetrate surfaces. Exposure to radiation, especially gamma rays, can damage a wide variety of body organs, leading to radiation sickness, cancer, or birth defects.

PENETRATING ABILITY

Alpha radiation Two protons and two neutrons are lost. Alpha particles can't penetrate paper or skin but can be harmful if inhaled or ingested in food or water.

Beta radiation An electron is lost. Beta particles can penetrate the upper layers of the skin but are easily stopped by a thin sheet of aluminum or very heavy clothing.

Gamma radiation Short-wavelength electromagnetic radiation is released (similar to X rays). High-energy gamma rays can only be stopped by thick or very dense material such as concrete or lead.

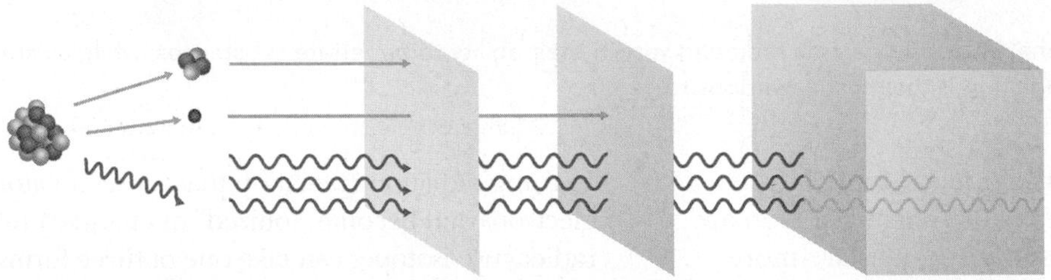

Why were the people who lived near the Fukushima power station concerned about the alpha radiation released with the reactor meltdown? After all, it cannot even penetrate paper!

5 RADIOACTIVE DECAY

Key Concept 5: Radioactive decay is measured in terms of a half-life and can vary in duration from milliseconds to billions of years for different isotopes.

The safety of nuclear power hinges on the fuel that is used and the by-products that are produced. Some of the radioactive isotopes produced in the fission reaction are more concerning than others, namely those that are highly radioactive and spontaneously release large amounts of radiation in a process known as **radioactive decay**. (Note this is not a fission reaction that splits the atom into two or more smaller atoms; it is just the spontaneous release of particle or high-energy radiation.) Many of these isotopes continue to release dangerous levels of radiation for a long time. Though the steam explosions at Fukushima did not release as much radiation as the most serious global nuclear accident (Chernobyl in 1987), the continued release of radioactive material into the surrounding area and into the oceans is uncharted territory.

radioactive decay The spontaneous loss of particle or gamma radiation from an unstable nucleus.

radioactive half-life The time it takes for half of the radioactive isotopes in a sample to decay to a new form.

The hazard presented by radioactive isotopes certainly depends on the type of radiation released (e.g., gamma radiation is more concerning and harder to avoid than alpha or beta radiation), but also how long an isotope remains radioactive is an important factor. Therefore, determining how long an isotope will be dangerously radioactive is vital information in any risk assessment.

Radioactive decay is measured in half-lives. An isotope's **radioactive half-life** is the amount of time it takes for half of the radioactive material in question to decay to a new form. After one half-life, 50% of the material will decay; in the next half-life, 50% of what's left (or 25% of the original amount) will then decay, and so on. After 10 half-lives, just 0.1% of the original radioactive material is left. **INFOGRAPHIC 5**

Some isotopes generated in the nuclear fission reaction have a very short half-life (from seconds to days or weeks), but others are more stable with half-lives in the hundreds or thousands of years. The strong gamma emitters from a fission reaction have relatively short half-lives: The half-life of radioactive iodine (I-131) is 8 days; cesium and strontium (Cs-137 and Sr-90) have half-lives of about 30 years. A general rule of thumb is that after 10 half-lives, enough radioactive material will have decayed to render the material safe. For I-131, that is 80 days; for Cs-137 and Sr-90, that is 300 years. But for isotopes with half-lives in the hundreds or thousands of years, we are looking at thousands or tens of thousands of years before the material could be considered safe.

INFOGRAPHIC 5 RADIOACTIVE DECAY

The rate of decay for a given radioactive isotope is predictable and expressed as a *half-life*—the amount of time it takes for half of the original radioactive material (*parent*) to decay to the new *daughter* material (a new isotope or even a new atom if protons are lost).

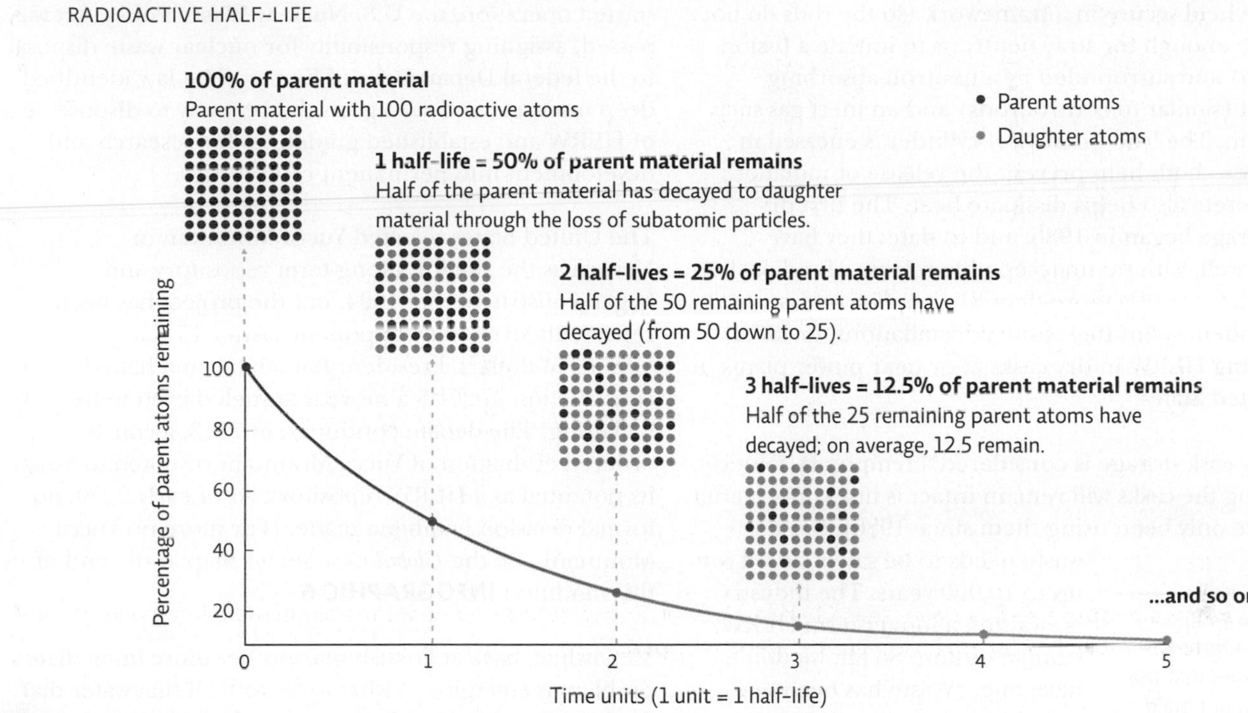

What percentage of the parent material will be left after 5 half-lives?

6 NUCLEAR WASTE

Key Concept 6: Radioactive waste is considerably more radioactive than the original fuel. We currently have no long-term storage plan for high-level radioactive waste.

At Fukushima, the reactors weren't the only problem. Experts around the world were also concerned about all the radioactive waste stored onsite at Fukushima. To understand why, it helps to know a little about radioactive waste. In general, there are two kinds. **Low-level radioactive waste (LLRW)** is material that has low amounts of radiation relative to its volume and can usually be safely buried. This includes clothing, gloves, tools, and so on, that have been exposed to radioactive material. Most LLRW isotopes decay to background levels in 100 years. The United States produces about 60,000 cubic meters (a volume equivalent to 15 million gallons) of LLRW per year, all of which is stored at just four sites—in South Carolina, Texas, Utah, and Washington.

High-level radioactive waste (HLRW) is another story. As its name suggests, it's more reactive than LLRW; in fact, because so many radioactive by-products are created in fission reactions, HLRW is considerably more radioactive than the original fuel rods—an estimated 1 million times more! It also generates tremendous heat, making storage in containers inherently dangerous—containment must successfully hold material that is both radioactive and hot. The United States produces about 2,000 metric tons of HLRW per year and has some 65,000 metric tons of the stuff in interim storage right now; this estimate includes both spent fuel rods and waste from nuclear weapon production.

The radioactive waste produced by nuclear power plants is a huge safety issue: It's extremely dangerous, there's a lot of it, and we have yet to come up with a long-term plan for disposing of it safely. Spent fuel rods are stored onsite in steel-lined pools, where short half-life isotopes can decay to safe levels. Isotopes with longer half-lives require more time to

low-level radioactive waste (LLRW) Material that has a low level of radiation for its volume.

high-level radioactive waste (HLRW) Spent nuclear reactor fuel or waste from the production of nuclear weapons that is still highly radioactive.

reach this point. After spending about 5 years in a pool, the spent fuel rods (HLRW) are moved to interim dry storage in large concrete-encased steel casks. The fuel rods inside the cask reside in a lead-lined steel cylinder; rods are held secure in a framework (so the rods do not get close enough for stray neutrons to initiate a fission reaction) and surrounded by a neutron-absorbing material (similar to control rods) and an inert gas such as helium. The lead-lined steel cylinder is encased in concrete—both help prevent the release of radiation; the concrete also helps dissipate heat. The first dry cask storage began in 1986, and to date, they have worked well, with no unacceptable release of radioactive material. Currently more than 81 sites (known as independent spent fuel storage installations, or ISFSI) are storing HLRW in dry casks at or near power plants in the United States.

Still, dry cask storage is considered a temporary solution. How long the casks will remain intact is not known (after all, we've only been using them since 1986), but the waste needs to be safely stored for up to 10,000 years. The industry is awaiting a longer-term HLRW storage option. So far, we don't have one. "Waste has been one of the biggest sore points in the debate over nuclear energy," says Charles Powers, a professor and nuclear energy scientist at Vanderbilt University. "It's the one thing we really don't have even a good theoretical solution to."

Nuclear Waste Policy Act (1982) The federal law that mandated that the federal government build and operate a long-term repository for the disposal of high-level radioactive waste.

In 1982, 25 years after the first U.S. nuclear power plant started operation, the U.S. **Nuclear Waste Policy Act** was passed, assigning responsibility for nuclear waste disposal to the federal Department of Energy. The law identified deep underground storage as the best way to dispose of HLRW and established guidelines for research and development into permanent repositories.

The United States selected Yucca Mountain in Nevada as the site for a long-term repository and began construction in 1994, but the project has been repeatedly stymied by opponents. After 15 years and billions of dollars, President Barack Obama halted construction in 2010, a move that fueled even more bickering. The debate continues; in 2013, a court-ordered evaluation of Yucca Mountain resumed to assess its potential as a HLRW repository. As of early 2020, no formal decision had been made. (For more on Yucca Mountain, see the *Global Case Studies* map at the end of this module.) **INFOGRAPHIC 6**

Meanwhile, back at Fukushima another more immediate problem is emerging—what to do with all that water that has been constantly added to the steel-lined pool where the assemblies were kept or that has been sprayed on the melted reactor cores or leaked into the basements from the groundwater below. This radioactive water is

KAZUHIRO NOGI/AFP/Getty Images

Storage tanks outside reactor Number 3 hold tritiated (radioactive) water collected from damaged reactors. The facility is running out of room to store the water and is considering releasing it into the ocean.

INFOGRAPHIC 6 RADIOACTIVE WASTE

Radioactive waste does not come just from nuclear power plants; it is also generated by industry, the medical field, research laboratories, and weapons production. All of these users produce low-level radioactive waste (LLRW); nuclear power and weapons production is responsible for almost all of the high-level radioactive waste (HLRW). This material is highly dangerous and must be safely contained for centuries.

LOW-LEVEL RADIOACTIVE WASTE DISPOSAL: BURIAL

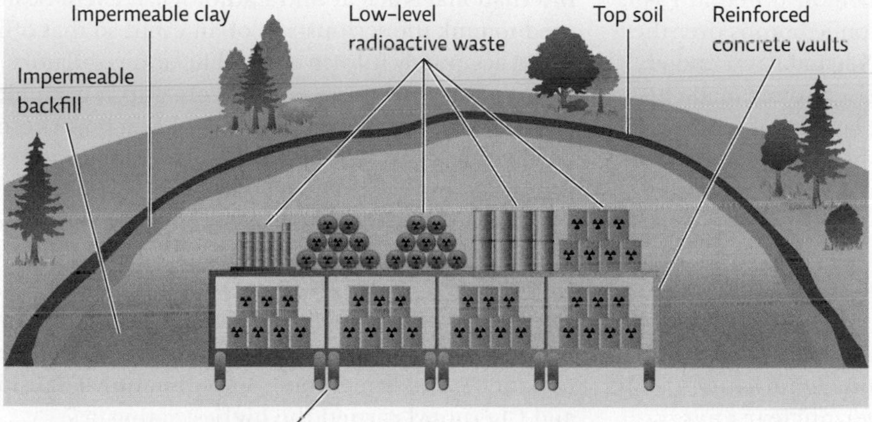

LLRW includes contaminated items such as clothing, filters, gloves, and other materials exposed to radiation. In the United States, short half-life LLRW is disposed of as regular trash after it is no longer radioactive; longer half-life LLRW is stored in casks and sent to one of four U.S. storage facilities where it is buried underground.

HIGH-LEVEL RADIOACTIVE WASTE DISPOSAL: DRY CASK STORAGE OF SPENT FUEL RODS

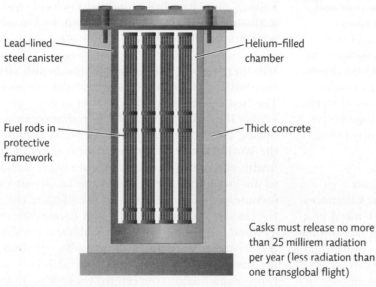

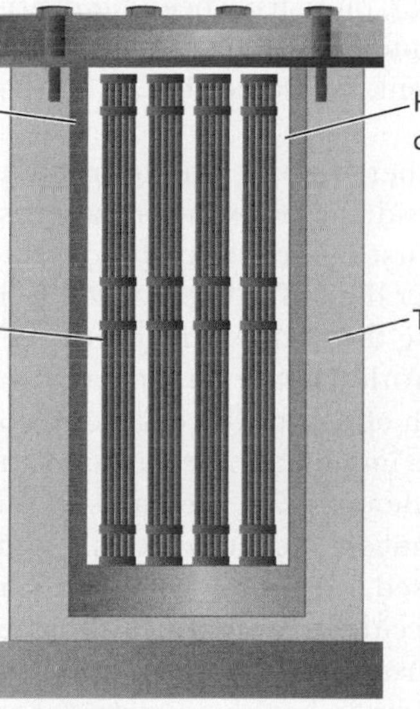

Casks must release no more than 25 millirem radiation per year (less radiation than one transglobal flight)

The United States lacks a permanent long-term storage facility for HLRW; most spent fuel rods are stored onsite at nuclear power plants in steel-lined pools. Some have been moved out of the pools and into dry casks and stored above ground.

Do you think we should continue to use nuclear power, or increase its usage, if we do not have a plan for long-term storage of HLRW?

collected, and radioactive isotopes are removed, with the exception of tritium (radioactive hydrogen). There is no way to remove tritium, so the tritiated water is stored; so far, more than 250 trillion gallons of contaminated water are currently stored in huge aboveground containers, with more containers added every day or so.

Unfortunately, some of the water escapes collection and flows into the ocean—one estimate puts it at 2,400 gallons per day. And with this water go radioactive cesium and strontium, levels of which are still detectable off the coast of Japan, indicating continued release, according to research by Maxi Castrillejo of the Autonomous University of Barcelona. No one knows what the impact of that constant release will be—we are in uncharted waters. To compound matters, Tokyo Electric says it is running out of room to store the collected water; one option is to release the stored tritiated water into the ocean. An analysis published in January 2020 by a panel convened by Japan's Ministry of Economy, Trade and Industry concludes it is safe to do so, but local fishermen and residents oppose the release.

7 NUCLEAR INDUSTRY ACCIDENTS

Key Concept 7: There have been around 100 nuclear accidents in the history of the industry; most were minor, but several caused deaths and two were severe: Chernobyl and Fukushima.

The nuclear power industry has a short history. The power of nuclear energy was first demonstrated in 1945, when the U.S. military dropped atomic bombs over the Japanese cities of Hiroshima and Nagasaki and ended World War II. In 1953, President Eisenhower made his famous "Atoms for Peace" speech, laying out a plan by which this destructive force could be harnessed for good: Instead of building bombs, we would produce cheap, reliable energy.

Worldwide, there are close to 450 nuclear power plants, accounting for about 10% of global electricity production. The United States has more nuclear reactors than any other nation, but Americans themselves have been divided over nuclear energy since the 1980s, after two infamous nuclear accidents made global headlines.

The first was a steam explosion in an experimental reactor that caused a partial meltdown (only a portion of the nuclear fuel melted) at the Three Mile Island plant near Middletown, Pennsylvania, in 1979 due to an electrical failure followed by a flurry of operator errors. This accident resulted in only minor release of radiation (below background levels). The reactor was destroyed, but an evaluation of the accident concluded that it did not cause any public health problems.

The second, caused by human error during an experiment that went awry, happened at the Chernobyl reactor in Ukraine (then part of the Soviet Union) in the spring of 1986; this accident was considerably more severe. The chain reaction escalated out of control, resulting in a full nuclear meltdown, which caused a steam explosion that released a tremendous amount of radiation over much of the western Soviet Union and Europe. The former Soviet government did not notify the public or nearby nations of the accident right away, delaying protective actions that could have reduced affected populations' exposure to the radiation. All land within a roughly 20-mile radius has been designated as an exclusion zone; entry without permission and a guide is prohibited, though some elderly residents have returned to the zone. The accident has been linked to 6,000 cases of thyroid cancer in children, and the effects of radiation exposure are still being felt. Though we may never know its full extent, low-end estimates are that the radiation will ultimately be responsible for some 94,000 premature deaths when all is said and done.

These accidents prompted the creation of the International Nuclear and Radiological Event Scale, used to rank the seriousness of an event so that officials could accurately inform the public and regulatory bodies. The scale ranks events from 1 to 7; each higher level of the scale represents a tenfold increase in severity. For "accidents" (Levels 4–7), severity is determined by the amount of radiation released. For "incidents" (Levels 1–3), severity is determined by the degree of failure of safety protocols.

There have been other incidents and accidents (fires, partial meltdowns, chemical explosions that released radiation), and some were severe, but only Fukushima and Chernobyl earned the highest rating, a 7. Thankfully, because Japanese officials moved quickly to protect people (evacuating residents and distributing iodine tablets to protect against exposure to radioactive iodine), there have been no deaths in local residents attributed to radiation exposure from the Fukushima accident. **INFOGRAPHIC 7**

Will there be any long-term physical health effects on residents and workers as a result of the accident? The first reported death occurred in 2018 when a worker died of cancer linked to radiation exposure during the accident response. A 2013 evaluation by the World Health Organization concluded that the health effects of the disaster would likely be limited to the immediate area around the facility and would include up to 7% increases in the lifetime risk for cancers such as leukemia and breast cancer in exposed individuals who were babies at the time of the accident. Mental health problems attributed to the stress of the event and its aftermath continue to represent a health concern for those who endured the disaster at this time. Residents are being allowed to reoccupy some areas (areas with acceptable levels of radiation), while other areas are only open for day visits (to clean, rebuild, and restore). Remediation of the area continues.

By early 2017, radiation in the general area around the Fukushima facility was low enough to work on demolition, though workers cannot get close to the melted reactors that still contain the as-yet-unseen melted radioactive fuel—that area remains off limits. Those fuel cores have melted into a mass that exists today at the bottom of each reactor. Robots have been sent into the reactors to examine the melted reactor

INFOGRAPHIC 7 NUCLEAR ACCIDENTS

The International Nuclear and Radiological Event Scale assigns a degree of severity to nuclear events; each subsequent level represents a tenfold increase in severity. There have been more than 100 nuclear incidents and accidents since 1952, including 20 partial or total meltdown events; only Chernobyl and Fukushima rank at the highest level of severity.

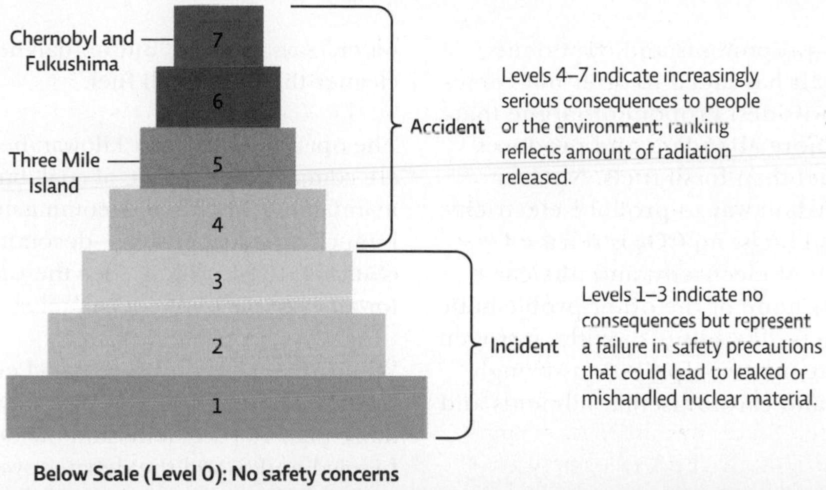
Chernobyl and Fukushima
7
6
Three Mile Island
5
4
3
2
1
Accident
Levels 4–7 indicate increasingly serious consequences to people or the environment; ranking reflects amount of radiation released.
Incident
Levels 1–3 indicate no consequences but represent a failure in safety precautions that could lead to leaked or mishandled nuclear material.
Below Scale (Level 0): No safety concerns

NOTABLE NUCLEAR EVENTS

YEAR	PLACE	EVENT LEVEL	DESCRIPTION
1952	Chalk River (Canada)	5	First reactor accident; malfunction and human error led to partial meltdown and steam explosion that released radiation.
1961	Idaho National Laboratory (U.S.)	Not determined	Failed attempt to safely restart an experimental reactor (possibly due to incorrectly inserted control rods) resulted in an uncontrolled chain reaction and explosion that killed three workers.
1979	Three Mile Island (U.S.)	5	Cooling malfunction and subsequent human error led to partial meltdown; release of radioactive water and gas occurred but was contained to the immediate area.
1986	Chernobyl (Ukraine)	7	Human error during an experiment led to a steam explosion and massive release of radiation; caused 30 deaths within a week and was, or ultimately will be, responsible for thousands of illnesses and deaths.
1987	Goiania (Brazil)	5	A radiotherapy device, stolen from an abandoned clinic where it was mistakenly left behind, was sold as scrap metal; many curious people came in contact with the colorful material inside (radioactive cesium). More than 200 were exposed to radiation, which resulted in four deaths and 20 hospitalizations.
1999	Tokaimura (Japan)	4	Worker error in the handling of radioactive material caused two deaths and resulted in local evacuations.
2011	Fukushima (Japan)	7	Earthquake and tsunami caused total meltdown and steam explosions that released large amounts of radiation.
2014	New Mexico (U.S.)	Not determined	Waste container exploded in underground bunker, releasing radiation into the atmosphere; the accident was attributed to using the wrong type of kitty litter (an absorbent material used to clean and contain chemical spills). Clean-up estimates run around $2 billion.

? Which of the accidents described above had a component of human error? What does this suggest about the safety of nuclear power?

cores—many failed to complete their mission, but one that made it down into Unit 2 (the most heavily damaged of the reactors) in 2018 sent back images that suggest at least some of the material has solidified. It has still not been determined if the radioactive material has breached the metal containment compartment that surrounds it—something that would put it in direct contact with the ground.

8 THE PROS AND CONS OF NUCLEAR POWER

Key Concept 8: The advantages (low air pollution, ample supplies) and disadvantages (radioactive hazards) of nuclear power must be weighed to determine its future role in electricity production.

Nuclear power has its proponents and opponents, and for good reasons. It has much to offer but comes with some serious downsides. Proponents argue that uranium ore is both more abundant and produces a more energy-rich fuel than fossil fuels. Nuclear energy is also a low-carbon way to produce electricity: Unlike the use of fossil fuels, no CO_2 is released during the production of electricity, and nuclear power creates virtually none of the other problematic combustion by-products, like sulfur dioxide, nitrogen oxides, and particulate matter—pollution strongly linked to respiratory and cardiovascular ailments and ecological problems.

And despite some persistent fears, research suggests that living near a nuclear power plant is actually safer than living near a coal-fired one. A 2011 study by Annette Queißer-Luft found that the risk of birth defects did not increase the closer one lived to a nuclear facility. Meanwhile, Javier García-Pérez found that in Spain, the number of cancer-related deaths does increase as proximity to coal-fired power plants increases. "You still have some environmental hazards, from mining uranium and from radioactive waste and water," says Powers. "But on balance, nuclear is far cleaner than any fossil fuel."

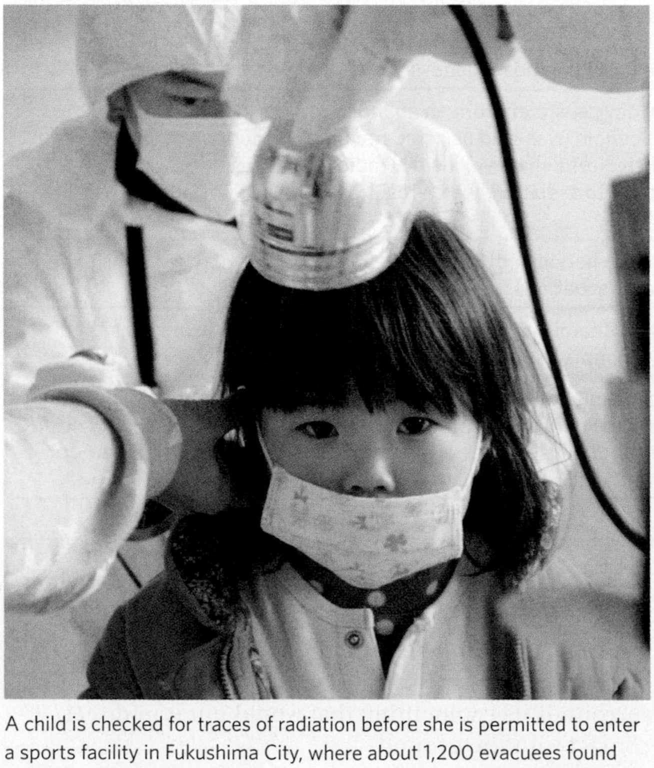

Christoph Bangert/laif/Redux

A child is checked for traces of radiation before she is permitted to enter a sports facility in Fukushima City, where about 1,200 evacuees found temporary shelter.

The operating costs per kilowatt-hour for nuclear power are comparable to those of coal, but the cost of building, maintaining, and then decommissioning nuclear plants is much more expensive—decommissioning alone could cost close to $1 trillion. (See the *Global Case Studies* map for an example of the decommissioning process.)

While there have only been two Level 7 nuclear accidents in the history of the industry, there have been more than 100 accidents and incidents ranking between Levels 1 and 6, and the debates over safety remain unresolved. Proponents point out that considering the number of existing plants and the length of time they have been operating, accidents have been exceedingly few and far between. But opponents say that such safety claims ignore three key points: the vulnerability of nuclear power plants to natural disasters (which at Fukushima led to nuclear meltdown and the release of radioactive material into the environment), the potential for nuclear fuel to be weaponized, and ways to deal with the radioactive waste. **INFOGRAPHIC 8**

The March 2011 tsunami would become the most expensive natural disaster in human history, with loss estimates skyrocketing past $1 trillion. (One estimate has it at $51 trillion, though the Japanese government's estimate is $76 billion.) In its wake, countries around the world began rethinking their plans to expand their own nuclear energy programs. Germany resolved to move up the date for its planned phase-out of all nuclear power plants by 10 years, to 2022. In the United States and other countries, safety procedures and infrastructure were examined and new protocols put in place from lessons learned at Fukushima. (For example, backup generators are no longer stored in flood-prone basements.) Consideration has also been given to the placement of nuclear power plants. Fears over a disaster at the Indian Point Power Plant, 35 miles north of Times Square in New York City (where 20 million people who live in a 50-mile radius of the plant could be at risk), have prompted officials to put into motion plans to shut it down by 2021. (See the *Global Case Studies* map for more on Indian Point.)

Given the problems associated with fossil fuels, it is likely that nuclear power will continue to have a place in our energy mix, especially with global electricity use projected to double by 2030. But in making choices about how to pursue our energy future, we

INFOGRAPHIC 8 NUCLEAR POWER: TRADE-OFFS

What role will nuclear power play in the future? Like all of our other energy options, nuclear power has advantages and disadvantages that must be weighed when making this decision.

ADVANTAGES	DISADVANTAGES
COSTS	
• Operating costs are comparable to those of a fossil fuel power plant. • The technology is available now.	• Nuclear power plants are much more expensive to build ($20 billion to build a fission reactor) and maintain than fossil fuel power plants, and even more expensive to decommission ($200 million to $1 trillion).
ELECTRICITY PRODUCTION	
• Power production can be increased or decreased to meet demand.	• Though power production can be altered to meet demand, it cannot be done as quickly as it can be done with a fossil fuel facility. • Large amounts of water are needed in the process for cooling.
SUPPLIES	
• Uranium supplies are good, and other isotopes can also be used. • Nuclear fuels are energy rich: One small uranium pellet contains the same amount of energy as 1 metric ton of coal.	• Mining and processing ore for nuclear fuel produce hazardous waste and can pollute air and water. • Some methods of nuclear power production produce radioisotopes that could be used in nuclear weapons production.
POLLUTION AND SAFETY	
• Much less pollution is released during operation at a nuclear power plant compared to a fossil fuel facility. During operation, no particulates, no sulfur or nitrogen pollution, and no CO_2 are released.	• Radioactive waste is very hazardous, and we still have no long-term plan for dealing with HLRW. • Shipping fuel or waste (by truck or rail) is also a safety concern and vehemently opposed by those who live on the transport route. • Though serious accidents are rare, they can have devastating consequences and long-term impacts when they occur.

? Which advantage of nuclear power do you consider to be its strongest advantage? Which of its disadvantages is its biggest detractor? Explain.

must consider the costs and benefits of all our energy options. A benefit analysis must evaluate how well a particular energy source meets our energy needs. A cost analysis must consider not only the monetary cost of getting kilowatts delivered to our homes but also the environmental and social costs associated with every step of the energy source's life, from acquisition to production to delivery to waste disposal. In addition, as mountaintop removal, oil spills, climate change, and nuclear meltdowns demonstrate, a risk assessment must also take place. We must answer two very crucial questions: How risky is the venture (an assessment we can do with at least some degree of accuracy), and how much risk are we willing to take?

Select References:

Burns, P. C., et al. (2012). Nuclear fuel in a reactor accident. *Science (Washington), 335*(6073), 1184–1188.

Castrillejo, M., et al. (2016). Reassessment of 90Sr, 137Cs, and 134Cs in the coast off Japan derived from the Fukushima Dai-Ichi nuclear accident. *Environmental Science and Technology, 50*(1), 173–180.

García-Pérez, J., et al. (2009). Mortality due to lung, laryngeal and bladder cancer in towns lying in the vicinity of combustion installations. *Science of the Total Environment, 407*(8), 2593–2602.

Jones, B. A. (2014). What are the health costs of uranium mining? A case study of miners in Grants, New Mexico. *International Journal of Occupational and Environmental Health, 20*(4), 289–300.

Harada, K. H., et al. (2014). Radiation dose rates now and in the future for residents neighboring restricted areas of the Fukushima Daiichi Nuclear Power Plant. *Proceedings of the National Academy of Sciences, 111*(10), E914–E923.

Queißer-Luft, A., et al. (2011). Birth defects in the vicinity of nuclear power plants in Germany. *Radiation and Environmental Biophysics, 50*(2), 313–323.

World Health Organization (WHO). (2013). *Health risk assessment from the nuclear accident after the 2011 Great East Japan Earthquake and Tsunami based on a preliminary dose estimation.* Geneva: WHO.

GLOBAL CASE STUDIES NUCLEAR POWER IN THE UNITED STATES

Achieve

In 2019, there were 58 operating U.S. nuclear power plants, 81 interim storage facilities for high-level radioactive waste (HLRW), and more than 100 other sites (reactors, industries, university labs, etc.) undergoing decommissioning. Visit these online case studies for a brief look at construction problems at a new proposed reactor (Vogtle), a power plant slated for closure (Indian Point), and another undergoing decommissioning (Zion). The final case study presents information about the proposed U.S. HLRW disposal site: Yucca Mountain in Nevada.

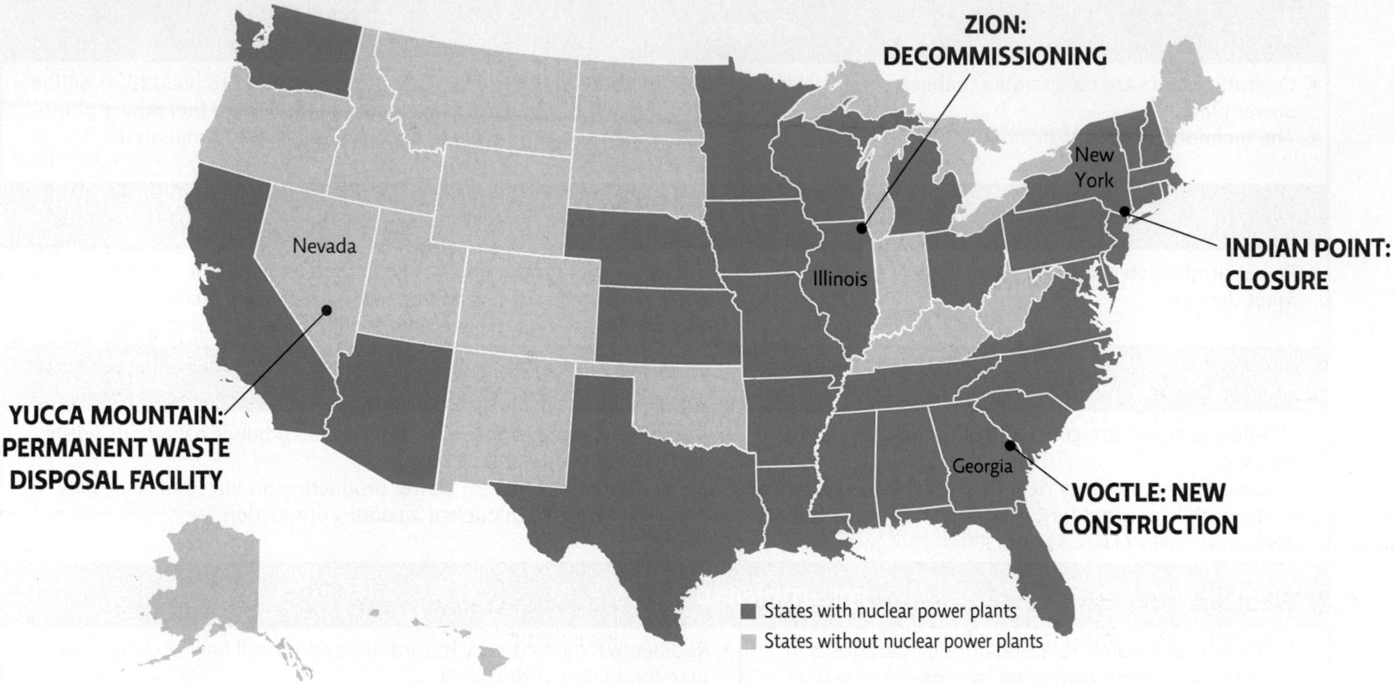

BRING IT HOME

PERSONAL CHOICES THAT HELP

Nuclear energy has been rebranded as "green energy" because it does not emit greenhouse gases. Technology has improved the safety of nuclear facilities; however, there are still safety issues and valid concerns over the long-term storage of nuclear waste. In addition, cost, national security, and uranium supplies make nuclear power a complicated energy solution.

Individual Steps

- Use the facility locator on the U.S. Nuclear Regulatory Commission website (www.nrc.gov) to find out if you have nuclear reactors where you live. If so, what type of reactor are they: PWR or BWR?
- What is your opinion about the necessity of nuclear energy in the United States? The U.S. Department of Energy website (www.energy.gov) maintains articles, updates, and the latest policy initiatives regarding nuclear energy in America. This resource can help you understand current policy, funding issues, and upcoming legislation.

Group Action

As you would expect, the policies endorsed by a particular group depend on the group's overall view of nuclear energy. To see two examples, check out the following sites and see how they compare:

- Visit the Nuclear Energy Institute (www.nei.org), which is a pronuclear organization, to see proposed legislation regarding increasing our current nuclear energy production.
- Visit the Nuclear Energy Information Service (www.neis.org), which is a nonprofit antinuclear organization committed to a nuclear-free future.

AP Photo/Toby Talbot, File

REVISIT THE CONCEPTS

- Nuclear energy can be harnessed to generate electricity. Typically, U-235, a radioactive isotope of uranium, provides the fuel for nuclear power.

- The production of U-235 fuel begins with mining of uranium ore followed by several processing steps that increase the proportion of U-235 in the rock. Each step generates hazardous waste.

- The energy of U-235 is released by splitting atoms (nuclear fission). In the reactor, U-235 is bombarded with neutrons to split the uranium atoms into two or more smaller atoms; this process releases more neutrons, which leads to a self-perpetuating chain reaction. Heat from this reaction is used to boil water, producing steam that turns a turbine attached to a generator, which generates electricity.

- Alpha and beta radiation result from the release of subatomic particles, and while dangerous, they are less dangerous than the high-energy gamma radiation, which can more easily penetrate the body.

- Radioactive decay is measured in terms of the time it takes for half of the material to decay into a new isotope or atom (half-life).

- Radioactive waste is considerably more radioactive than the original U-235 fuel. Low-level radioactive waste is buried under highly controlled conditions. We currently store most high-level radioactive waste onsite at the power plant and as yet have no long-term storage plan for it.

- In the short history of nuclear power there have been around 100 nuclear accidents; most were minor, but some of these did cause deaths. The two most severe events, Chernobyl and Fukushima, earned the highest severity rating on the International Nuclear and Radiological Event Scale.

- While nuclear power has some noteworthy advantages, including low air pollution and ample fuel supplies, its disadvantages (most notably radioactive hazards) are significant and, like all energy sources, must be weighed to determine whether or how much to pursue it.

ENVIRONMENTAL LITERACY Understanding the Issue

1 What are radioactive isotopes, and why are they important for nuclear power?

1. How does the generation of electricity at a nuclear power facility compare to other thermoelectric power plants, such as one that runs on coal?
2. Define atomic number and mass number and explain how these differ between isotopes of the same element.
3. What does it mean for an isotope to be radioactive? Are all isotopes radioactive?

2 How is uranium fuel for nuclear power produced?

4. What hazards are associated with uranium mining and processing?
5. What is the purpose of the enriching step of nuclear fuel production?
6. Describe the steps of fuel preparation from pellet to fuel assembly.

3 How is nuclear energy harnessed to generate electricity in a fission reactor?

7. Explain how a nuclear fission reaction is initiated in a fission reactor and how it proceeds to produce a chain reaction.
8. What is the purpose of the control rods in a nuclear reactor?
9. Why would one be more concerned about the venting of steam from a boiling water reactor than a pressure water reactor?

4 What types of radiation are produced when isotopes decay?

10. What is ionizing radiation, and why is it dangerous?
11. Distinguish among alpha, beta, and gamma radiation in terms of their makeup, penetrating power, and hazard posed to living things.

5 How is the rate of decay of a radioactive atom measured?

12. What is radioactive decay, and how does it relate to the concept of a half-life?
13. After four half-lives, about how much radioactive parent material is left?
14. In a nuclear fuel assembly, even if the fission reaction could be completely halted, significant heat would still be produced. Explain why.

6 What problems are associated with nuclear waste?

15. Define and give examples of low-level radioactive waste (LLRW).
16. How are spent fuel rods currently disposed of in the United States?
17. Why is nuclear waste seen as a major, yet unresolved problem for the nuclear industry?

7 What is the history of nuclear accidents worldwide?

18. On the International Nuclear and Radiological Event Scale, how do the different levels compare to each other in terms of seriousness?
19. Why were the health and environmental impacts so much more severe at Chernobyl than at Fukushima?

8 What are the advantages and disadvantages of nuclear power?

20. Barring a major accident, is it more or less hazardous to your health to live near a nuclear reactor than a coal-fired power plant? Explain.
21. From an economic standpoint, which type of electricity production — nuclear or fossil fuel — is less expensive? Explain. Remember to include a full life cycle analysis in your answer.

SCIENCE LITERACY Working with Data

A life cycle evaluation of greenhouse gas emissions, a metric that includes every step of energy production from acquiring raw materials to decommissioning facilities, is a useful way to compare the carbon footprint of various energy sources. The following graph shows one life cycle estimate that compares the greenhouse gas emissions of various non-fossil fuel energy sources for electricity production. (For comparison, the life cycle greenhouse gas emissions for coal are more than 1,000 g of CO_2-equivalent/kWh.)

CARBON FOOTPRINTS OF NON-FOSSIL FUEL ENERGY SOURCES

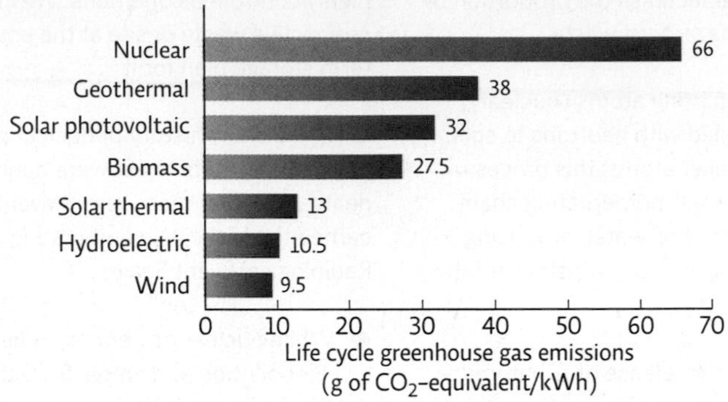

Interpretation

1. What does the length of each bar represent?
2. Based on the graph, how does the carbon footprint of nuclear power compare to other industries?
3. Compare the carbon footprint of wind (the lowest greenhouse gas emitter) and geothermal (the second-highest emitter) to that of nuclear power by calculating the percentage of CO_2-equivalent greenhouse gases released by each compared to nuclear. (**Hint:** Divide the total shown for the energy source in question by the total for nuclear power and then multiple by 100.)

Advance Your Thinking

4. Use these data to draw a conclusion that answers this question: If a country's goal is to reduce greenhouse gas release to be no more than half that of nuclear power, which energy sources should be pursued?
5. Since no CO_2 is released to generate the heat of nuclear fuel (nothing is burned), what do you think contributes to the relatively higher release of greenhouse gases from nuclear power compared to other non-fossil fuel sources?

INFORMATION LITERACY Evaluating Information

The Nuclear Regulatory Commission (NRC) is tasked with regulating nuclear power in the United States and "protecting people and the environment." The members of the commission "formulate policies, develop regulations governing nuclear reactor and nuclear material safety, issue orders to licensees, and adjudicate legal matters" (www.nrc.gov/about-nrc/organization/commfuncdesc.html). The NRC maintains a detailed website with information about nuclear power plants and nuclear materials.

Go to the NRC's Facilities page (www.nrc.gov/info-finder/region-state). Find your state in the list of states and territories.

Evaluate the website and work with the information to answer the following questions:

1. Determine if this is a reliable information source:
   a. Who runs the website? Do this person's/group's credentials make the information reliable or unreliable?
   b. Is the information on the website up to date? Explain.
2. Choose a U.S. state to evaluate and record the name of that state. How many nuclear power reactors are active in this state? How many are being decommissioned?
3. Click on the links for an active nuclear power reactor in the state you are evaluating and look at the information provided via the other links on its main page.
   a. Write down information about its age, intended expiration date, type of reactor in use (PWR or BWR), and location. Find the location on a map and record the closest large body of water.
   b. Look at safety information (e.g., inspections, accidents, safety performance) about this reactor and evaluate its safety record.
   c. Look over the "Enforcement Actions" link to evaluate the types of actions that were taken (including severity and penalties assessed) and briefly summarize them.
4. Consider the information you have gathered and do a basic risk assessment analysis for one of the sites.
   a. Do you think the facility is a danger to the environment? To human health? Why or why not?
   b. Is the facility located in an area that may have natural disasters? If so, what are they? If a natural disaster were to occur, predict some outcomes.

Additional study questions are available at Achieve.macmillanlearning.com. Achieve

Module 11.2 Sustainable Energy: Stationary Sources

MOVING TOWARD A RENEWABLE ENERGY FUTURE

Chicago tackles renewable energy

Solar panels line the roof of Shedd Aquarium in Chicago, Illinois. *Jeremy Graham/dbimages/Alamy*

After reading this module, you should be able to answer these GUIDING QUESTIONS

1. What are the characteristics of a sustainable energy source, and what types are commonly used?
2. What are the current and projected roles of renewables in global energy production?
3. How can wind energy be captured to generate electricity, and what are the pros and cons of these methods?
4. How can solar energy be captured and used, and what are the pros and cons of these methods?
5. How can geothermal energy be captured and used, and what are the pros and cons of these methods?
6. How can the power of water be captured and used, and what are the pros and cons of these methods?
7. What roles do conservation and energy efficiency play in helping us meet our energy needs sustainably?
8. What economic adjustments and technological advances are needed for renewables to become a viable replacement for fossil fuels?

Chicago, Illinois, has a long history of responding and adapting to the needs of the times. In 1871, residents had to rebuild the Midwestern city after a fire destroyed more than 3 miles of the city center, killing 300 and leaving more than 100,000 Chicagoans homeless. In 1900, engineers reversed the flow of the Chicago River to protect the cleanliness of Lake Michigan, the city's drinking water source. And after a 1995 heat wave killed more than 700 people, the city overhauled its emergency response efforts to prevent similar weather-related catastrophes in the future. Now Chicago has chosen to pursue another challenge—committing to a 100% clean energy future.

It's a big undertaking. In 2017, when less than 4% of the region was powered by renewable energy sources, it became the largest city in the United States to pledge to power its public buildings with renewable energy by the year 2025. Two years later the city expanded its plan, promising to power all of Chicago's buildings with renewable power by 2035 and to use electricity to power Chicago Transit Authority buses by 2040. The city is also leading the way in energy equity with its attempts to reduce energy burdens and improve energy affordability for low-income residents. "To all those who put out the lie that somehow you can't grow the economy and improve the environment, come to Chicago—we've got something we've got to show you," said former Chicago mayor Rahm Emanuel in 2019. "If you're going to be a global city, you have to be a global leader."

WHERE IS CHICAGO?

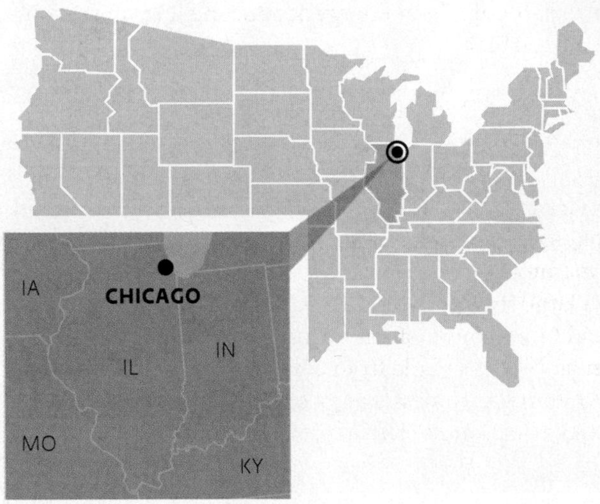

Chicago is just one of many big cities and smaller communities around the world taking steps to wean themselves from fossil fuels. The path a particular community takes varies, a reflection of its needs and the energy sources at hand. In each case, there is much to be learned from the successes and setbacks of these trailblazers—lessons others can use in their quest toward a sustainable energy future.

1 CHARACTERISTICS OF SUSTAINABLE ENERGY SOURCES

Key Concept 1: Sustainable energy sources are those that meet our needs and are readily replenished while producing impacts we, and ecosystems, can live with long term.

Chicago used to be just like most other cities, fully dependent on fossil fuels: coal, oil, and natural gas. In the United States and in other nations, the fossil fuel industry is heavily subsidized, making it hard for non-fossil energy sources to get a foothold. But in 2007, Chicago took a big step toward creating a future powered by **renewable energy**—energy from sources that are replenished over short time scales or that are perpetually available—when the Illinois state legislature enacted the Illinois Power Agency Act, which pledged that 25% of the state's power would come from renewable sources by 2025. Then in 2012, Chicago Mayor Rahm Emanuel announced another proposal to improve the energy efficiency of the city's buildings by 20% over 5 years, and he also called for the installation of 10 additional megawatts (MW) of renewable energy on city properties. Finally, in 2019, the city announced its plan to power all of Chicago's buildings with "clean power" (includes renewables such as solar, wind, and hydroelectric as well as nuclear power) by 2035. While Chicago may just be beginning its clean energy transition in earnest, many other communities, cities, and even entire countries are well on their way. But Chicago's ambitions are noteworthy—if this densely populated urban center can do it, Chicago's actions may provide a roadmap that other large cities can follow.

renewable energy Energy that comes from an infinitely available or easily replenished source.

To qualify as a **sustainable energy** source, the energy must be renewable—because energy cannot be recycled or reused, living organisms need to depend on an energy source that is readily replenished. (See Infographic 6 in Module 1.1 for more on the characteristics of a sustainable ecosystem.) In addition, this renewable energy source must be used at a rate equal to or less than its replacement rate. And finally, to be sustainable, the acquisition and use of the energy must have a low enough environmental impact that it could be used for the long term. Fossil fuels, our main fuel source at present, meet none of these requirements. They cause significant environmental damage when they are extracted, processed, and burned, including being a major contributor to climate change (see Chapters 9 and 10). Indeed, climate change is one of the biggest environmental challenges we face today, and addressing it will require that we "decarbonize" our energy production and move away from fossil fuels.

The earliest renewable energy source used by humans is **biomass energy**—energy that is harnessed when biological material is burned. Around the world, many people still depend on wood, animal waste, or crop residue as a fuel for cook stoves and heat. Biomass can also be burned in waste incinerators or power plants to produce heat or electricity on a large scale (see Module 5.3), and it can be converted to liquid or gaseous biofuels for use in homes or vehicles—in Illinois, biofuels made from corn and soybean oil are the predominant biomass energy source. In fact, in 2019, biomass became the leading form of renewable energy in the state. Again, biofuels are only considered renewable if the rearing or harvesting of biomass for fuels does not exceed the rate at which it is replaced by nature. (For more on biofuels, see Online Module 11.3.)

While biomass is useful fuel, it has a major drawback—there isn't enough land to grow all the food and fuel crops we would need if we turned to biofuels for a large part of society's energy supply. This limitation makes biomass less likely to contribute significantly as a renewable fuel in the future, especially as a stationary source of energy for electricity production. Four other energy options that are useful for electricity production are *solar, wind, geothermal,* and *hydropower.* (While this module focuses on stationary sources of renewable energy, Module 11.3 discusses renewable energy sources for mobile uses such as cars and planes.)

Chicago and its surrounding suburbs have begun investing in solar energy, thanks in part to the Future Energy Jobs Act, a state bill enacted in 2016 by the Illinois General Assembly that subsidizes renewable energy and has established a program that funds new solar energy systems in low-income communities. Each renewable energy source has its own advantages and disadvantages. Each of these will be defined and discussed later in this module. **INFOGRAPHIC 1**

sustainable energy Energy from renewable sources that are used no faster than replenished and whose use has an acceptable environmental impact.

biomass energy Energy from biological material such as plants (wood, charcoal, crops) and animal waste.

INFOGRAPHIC 1 SUSTAINABLE ENERGY SOURCES

Sustainable energy sources are those that are readily replenished and that have impacts that are acceptable; they only remain sustainable if we use them at or below the rate of replenishment.

SOLAR POWER
The energy of the Sun can be trapped in a variety of ways. More solar energy arrives daily than we could possibly use.

WIND POWER
Energy in the motion of air can be captured to generate electricity.

HYDROPOWER
The energy of moving water can be tapped in a variety of ways to produce electricity or other services.

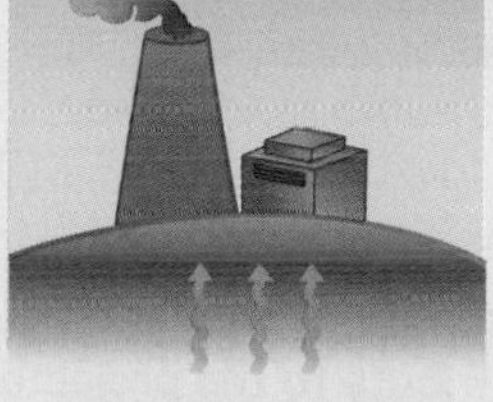

GEOTHERMAL POWER
The heat of the Earth can be tapped to make electricity or to provide heating or cooling for buildings.

BIOMASS
Chemical energy in biological materials (crops, waste, wood, etc.) can be released when the biomass is burned. (See Online Module 11.3.)

Which of these renewable energy options are a direct or indirect form of solar energy?

2 CURRENT AND PROJECTED USE OF SUSTAINABLE ENERGY SOURCES

Key Concept 2: Fossil fuels are the leading fuel for electricity production. The use of renewables is rising, but so is the use of fossil fuels, due to a rise in energy demand.

Human societies depend on energy, and for modern societies, this energy comes predominately from fossil fuels. Since the Industrial Revolution, world energy use has been on the rise, remarkably so since the mid-20th century when it began to increase exponentially. In 2019, fossil fuels such as coal, oil, and natural gas accounted for 63% of U.S. utility-scale electricity production (for global production, this value is around 64%), while renewables have increased from around 7% of production in 1990 to 17.5% in 2019—a time period that also saw a 20% increase in overall electricity production. (Nuclear power provides the rest of U.S. electricity.)

The use of renewables is slowly climbing, but so is the demand for energy. U.S. electricity production from renewables is expected to more than double by 2050 (compared to 2018). But this would account for just an increase to 31% of usage because, by 2050, we will be using more electricity than we did in 2018. By 2050, the single major source of electricity is still projected to be natural gas (which surpassed coal in 2015).

Still, it is a start. In the near future, wind and solar power will see the biggest increases as costs come down for these technologies and the benefits of using carbon-free energy sources mount. This transition is being facilitated by technological improvements, leading to more efficient systems and cheaper materials. And more and more groups are developing renewable projects, leading to stronger competition between providers who hope to become industry leaders.

But transitioning to a fossil fuel–free economy will not be easy. As we will see, each of our renewable energy options comes with its own set of trade-offs, and no single source can replace fossil fuels. However, together, their advantages are significant—and in many cases, they complement each other in ways that diminish their disadvantages. Whether or how quickly we transition away from fossil fuels will depend on society's assessment of the costs of using fossil fuels and our willingness to put our money and resources into deploying and developing renewable options. **INFOGRAPHIC 2**

INFOGRAPHIC 2 RENEWABLE ENERGY USE

According to the U.S. Energy Information Administration, in 2019, renewable sources of energy contributed around 17.5% of U.S. energy production. By 2050, energy production by renewable energy sources is projected to more than double, but will account for only about 30% of the total because total usage is also expected to increase over that time.

U.S. ELECTRICITY PRODUCTION BY SOURCE, 2019

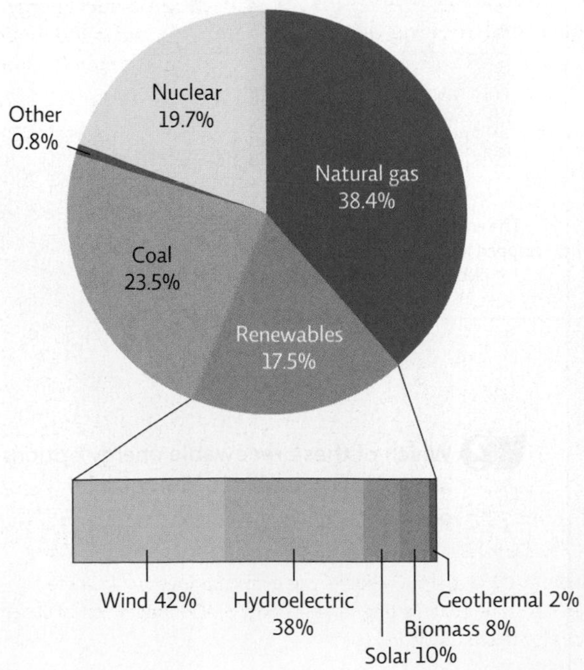

RENEWABLE ELECTRICITY GENERATION

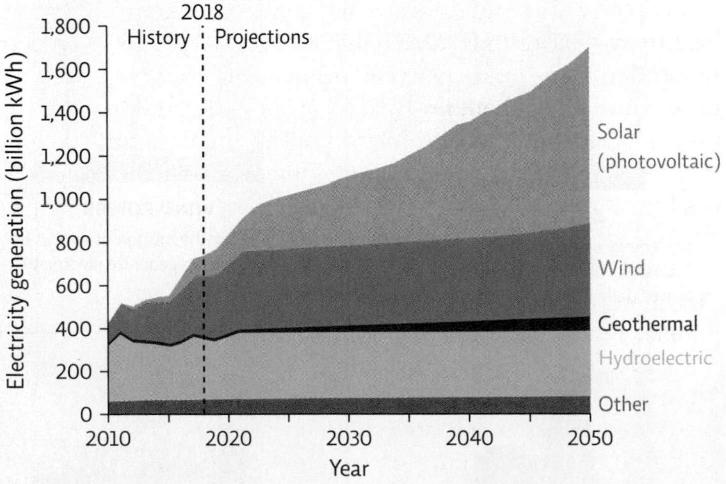

Estimate the percentage each renewable energy source is expected to contribute to the projected 1,700 billion kWh of electricity production in 2050. (For help interpreting an area graph, see online Appendix 2.)

3 WIND POWER

Key Concept 3: Wind power is a cost-effective way to generate electricity. It has a low environmental impact but is an intermittent energy source. In addition, turbines are dangerous to wildlife, and some consider them eyesores.

Wind power, energy contained in the motion of air across Earth's surface, is actually an indirect form of solar energy; it results from the difference in temperature between different regions of Earth, such as the poles and the equator, causing air to move from cooler regions to warmer regions. Although Chicago is called the "Windy City," it does not generate any wind power itself, but many other parts of Illinois do. In fact, the state of Illinois is ranked sixth in the nation in installed wind energy capacity. Chicago faces the same problem of many densely populated urban areas—the challenge of installing large-scale renewable energy within its city limits. Acquiring *renewable energy credits* (paying others to install renewable energy elsewhere) is a big part of its plan.

Wind power has long been used to power ships and to grind grain or pump water using windmills. But now, we can also use wind to generate electricity; spinning blades turn a shaft inside a generator. It's similar to how coal or nuclear power plants create electricity, but rather than using steam to turn a turbine, wind does the job.

Wind power has the advantage of being a pollution-free technology that can be used on a large scale (via huge turbines) or a small scale (via small turbines suitable for a single home). In some especially windy locations, *wind farms* containing dozens of turbines are cropping up (there are now 50 wind farms in Illinois). In the United States, Texas is by far the leading producer of wind power with wind turbines that generate enough electricity—more than 20,000 megawatts (MW)—to power 5 million homes per year. Illinois, too, could make enough wind energy by 2030 to power 7.2 million average U.S. homes, according to the U.S. Department of Energy. The career of a wind turbine technician has been touted as the fastest-growing energy job in the United States. In windy areas, wind is also seen as the most cost-effective renewable today, sometimes producing electricity at a cheaper per-kilowatt-hour rate than coal or natural gas.

But wind energy isn't perfect and usually needs to be paired with other energy sources to meet needs. First, wind is often intermittent—in most places, it stops and starts irregularly, not producing a steady stream of power. And large wind turbines are not cheap; onshore turbines can cost $1 million, and offshore ones can cost up to $5 million each. (Less expensive, smaller wind turbines can be purchased by homeowners for personal use.) Beyond cost, wind turbines can create noise, and some people view them as eyesores (although the same could be said for power lines and transformer stations, which we are all now very used to). Some Illinois counties have tightened wind project restrictions, too, creating a complicated approval process.

Turbines can also have a negative impact on wildlife due to habitat loss and collision with blades. Hundreds of thousands of bats are killed each year by wind turbines in the United States, some by collision and others by simply flying too close to the rapidly spinning blades—the sudden change in atmospheric pressure can be enough to cause internal hemorrhaging and death.

Wind turbines are also a threat to birds—an estimated 570 million birds are killed by wind turbines annually in the United States. While that is a large number, it is far less than the number of birds killed each year by flying into windows (600 million) or those killed by domestic cats (up to 2.6 billion). Still, to decrease the risk from windmills, engineers now avoid placing them in known migratory flight paths or close to areas frequented by birds of prey such as eagles. Technological solutions being tested include auditory or visual deterrents, such as the installation of ultrasonic "boom boxes" (to deter bats, which depend on echolocation to navigate) or strategically placed predator models (such as owls) to discourage birds from flying too close. New turbine designs aimed at reducing wildlife impacts are also being developed. **INFOGRAPHIC 3**

wind power Energy contained in the motion of air across Earth's surface.

Ron and Patty Thomas/iStock/Getty Images

This windfarm in southern California near Palm Springs is one of the largest in the state, contributing to the almost 6,000 megawatts of wind power produced statewide in 2019.

INFOGRAPHIC 3 HOW IT WORKS: WIND TURBINES

Wind turbines can be large or small. All work by the same principle: Spinning blades turn a shaft inside a generator and produce electricity.

ADVANTAGES	DISADVANTAGES
• Abundant in some areas. • Pollution-free. • Low environmental impact and carbon footprint. • Effective at large and small scales. • Lowest-cost renewable energy source. • Creates jobs.	• Wind is unpredictable and intermittent. • Not all areas are windy enough to support it. • Wildlife (birds and bats) are killed by windmills. • Considered an eyesore by some. • Usually must be paired with other energy sources to meet needs.

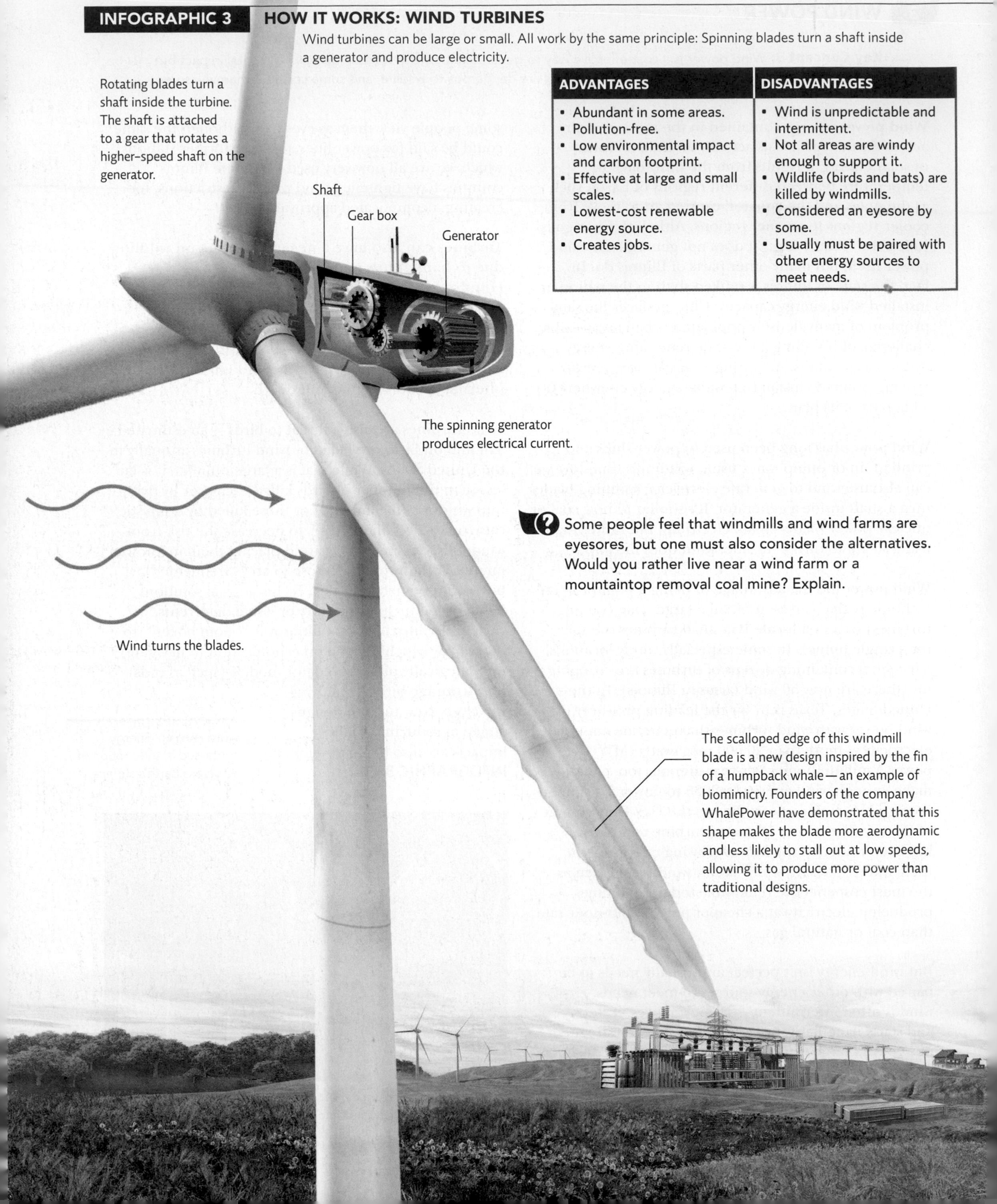

Rotating blades turn a shaft inside the turbine. The shaft is attached to a gear that rotates a higher-speed shaft on the generator.
Shaft
Gear box
Generator
The spinning generator produces electrical current.
Wind turns the blades.
The scalloped edge of this windmill blade is a new design inspired by the fin of a humpback whale—an example of biomimicry. Founders of the company WhalePower have demonstrated that this shape makes the blade more aerodynamic and less likely to stall out at low speeds, allowing it to produce more power than traditional designs.

? Some people feel that windmills and wind farms are eyesores, but one must also consider the alternatives. Would you rather live near a wind farm or a mountaintop removal coal mine? Explain.

4 SOLAR POWER

Key Concept 4: Solar energy can be harnessed in many ways. It is pollution-free but is expensive and less productive at higher latitudes, on cloudy days, and at night.

The most abundant sustainable energy source is the one that powers the planet—the Sun. Each day Earth receives a staggering amount of energy from the Sun (more than 10,000 times the amount of energy we use). The Sun meets many energy needs: It warms our homes and provides light—but new technologies allow more effective use, and even storage, of solar energy.

Solar power is energy harnessed from the Sun in the form of heat or light, and it can be used in two ways: through active or passive technologies. In many communities, homes are dotted with **photovoltaic (PV) cells**, also called solar panels. PV cells are **active solar technologies** that use mechanical equipment to convert solar energy directly into electricity. If just 4% of the world's deserts were covered in PV cells, it would supply all of the world's electricity needs. PV panels can be set up in large-scale installations (such as solar farms), as arrays on a rooftop or as a single panel that powers a light in a remote region, or even just as a solar cell on the face of a watch or in a small calculator.

Illinois's Future Energy Jobs Act calls for 400 MW of community solar projects to be developed by 2030. Since 2009, the city has installed enough solar to offset 1 billion kilowatt-hours of electricity that formerly came from coal. One way in which Chicago is promoting solar is through community solar gardens: Residents who can't afford to or can't install solar panels in their homes or apartments can still purchase a portion of their electricity produced by community solar installations.

When it comes to solar power, California is ahead of most states with policies that promote solar power. In fact, California leads the nation in solar generation with more than 37% of total U.S. production in 2019, and it is pushing for more, passing a law that will require all new builds of single-family homes and other buildings three stories tall or less to install solar panels or to tap into community solar installations. The mandate went into effect on January 1, 2020.

The production of electricity can also be done on a large scale using *solar-thermal systems*, another active solar technology. There are many types of solar "power plants." One early design consists of large arrays of huge parabolic (curved) mirrors that capture solar energy and focus it on pipes that contain a fluid that can be heated to high temperatures. These high temperatures are used to boil water to produce steam that turns a turbine attached to a generator, producing electricity. Newer technologies may vary the way the solar energy is captured and focused, but they operate on the same general principle—heating a fluid to produce steam to generate electricity.

Less expensive alternatives to PV cells or solar thermal systems are **passive solar technologies**. A greenhouse, for example, captures light and heat without any electronic or mechanical assistance. Many energy-conscious homes are designed with passive solar energy in mind, incorporating strategically oriented windows to maximize sunlight in a room and dark-colored walls or floors to absorb that light and heat the home.

Because solar energy is abundant and its capture is conceptually simple, safe, and clean, with no noise or moving parts, it is the most popular member of the renewable energy club. Today hundreds of thousands of buildings around the world are powered by PV cells.

But like any energy source, solar has its trade-offs. Start-up costs can be high, though PV cells are becoming cheaper every day—their price dropped 80% between 2010 and 2020. However, it can still take a homeowner or business owner many years to save enough in reduced power bills to offset the cost of installation, a metric known as **payback time**. In the United States, on average, the installation cost is recouped in 6 to 8 years, payback time is 3 to 5 years in sunny places like the desert Southwest or in places where electricity costs are higher, such as the Northeast.

Intermittency is a bigger problem. Sunlight is available for only roughly half of each day—and even less at higher latitudes where areas only see a few hours a day of pale sunlight in the winter. (However, in those places, solar power is quite useful in the summertime when production can go on for 18 hours or more. This makes solar panels popular in places like Alaska, where many homes are built "off the grid.") **INFOGRAPHIC 4**

Diversification is a hallmark of a sustainable energy future: Because sustainable energy sources have strengths and weaknesses, no single source will likely meet the needs of any particular community and certainly not those of the entire world. But together, each can provide a unique contribution. For example, solar is productive in daylight hours, whereas wind tends to blow harder at night; thus, these two renewable sources complement each other in many places. But in some regions, it makes more sense to tap other sources of renewable energy.

solar power Energy harnessed from the Sun in the form of heat or light.

photovoltaic (PV) cell A technology that converts solar energy directly into electricity.

active solar technology Mechanical equipment for capturing, converting, and sometimes concentrating solar energy into a more usable form.

passive solar technology A technology that captures solar energy (heat or light) without any electronic or mechanical assistance.

payback time The amount of time it takes to save enough money in operation costs to pay for equipment.

INFOGRAPHIC 4 SOLAR ENERGY TECHNOLOGIES TAKE MANY FORMS

There are many ways to capture and use solar energy. Passive solar homes are constructed in a way that maximizes solar heating potential.

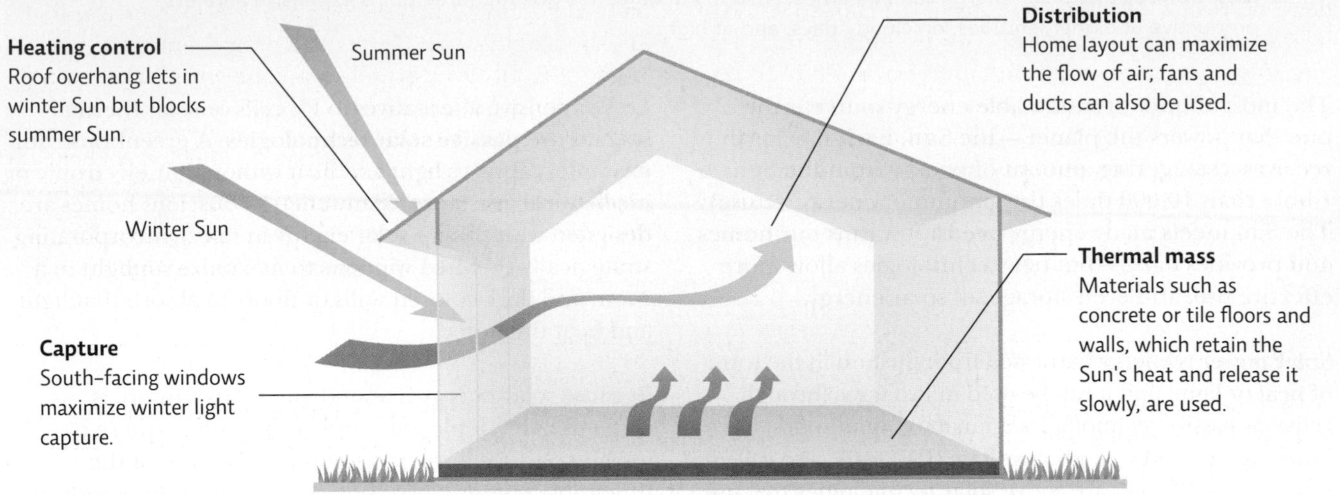
Heating control
Roof overhang lets in winter Sun but blocks summer Sun.
Summer Sun
Winter Sun
Distribution
Home layout can maximize the flow of air; fans and ducts can also be used.
Thermal mass
Materials such as concrete or tile floors and walls, which retain the Sun's heat and release it slowly, are used.
Capture
South-facing windows maximize winter light capture.

Active solar technologies use mechanical equipment to capture solar energy and covert it to a usable form. Photovoltaic (PV) cells convert sunlight to electricity. They are made of a semiconductor material, such as silicon, that will lose electrons when light strikes the surface, creating an electrical current.

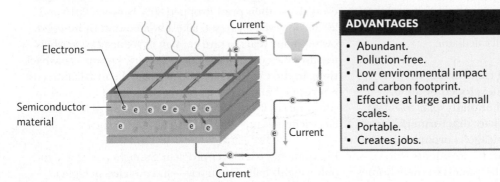

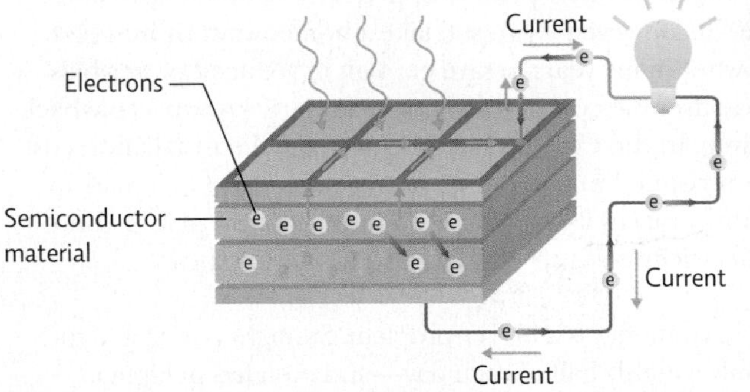
Current
Electrons
Semiconductor material
Current
Current

ADVANTAGES	DISADVANTAGES
• Abundant. • Pollution-free. • Low environmental impact and carbon footprint. • Effective at large and small scales. • Portable. • Creates jobs.	• Location matters; shady and high-latitude areas get less winter sunlight. • Little or no generation on cloudy days or at night. • Expensive compared to other renewable energy sources.

In terms of reducing fossil fuel use, why should you make your home as energy efficient as possible before installing active solar technologies?

5 GEOTHERMAL POWER

Key Concept 5: Geothermal energy can be captured on a small scale to lower home heating and cooling costs or on a large scale to produce electricity or heat.

Unlike Chicago, some communities are fortunate enough to be located near sources of **geothermal energy**. A tremendous amount of heat is produced deep in Earth from radioactive decay of isotopes (see Section 2 in Module 7.1). This is the same heat that bubbles hot springs and causes geysers to erupt. The heat in geothermal "hot spots" (areas where molten rock, or magma, is located close to the surface) can be tapped by **geothermal power plants** to generate electricity. Steam from water injected into geothermal wells is used to spin turbines and produce electricity. Captured steam can also be piped to a nearby community to provide heat or hot water.

geothermal energy Heat stored underground, contained in either rocks or fluids.

geothermal power plant A large-scale facility that captures steam produced from Earth's internal heat to turn turbines, which generate electricity.

Geothermal heat pumps (also called ground-source heat pumps) offer a very different way to trap and use this heat. Common in southern Illinois, more and more geothermal heat pumps are being installed in new builds and in restorations of older Chicago buildings, including historic Unity Temple in the Chicago suburb of Oak Park. These systems, used in more than half a million homes around the world, do not generate electricity—they reduce our use of it.

If you have ever been in a cave, you might have been struck by the constant temperature within. It could be 100°F outside on a blistering summer day or 0°F in the dead of winter, yet it always feels like a cool spring day inside a cave. This is because ground temperature is amazingly constant—around 55°F year-round. To tap into this temperature using heat pumps, fluid-filled pipes are buried; the fluid, which takes on that ground temperature, is then pumped into the home, where a heat pump captures the heat within the fluid, using it to maintain the desired temperature in the home much more efficiently than a traditional heating, air conditioning, and ventilation (HVAC) system. In the winter, the heat pump operates as if it is only 55°F outside, and in the summer, this system provides natural cooling. Such systems are fairly expensive to install ($20,000 to $25,000 on average) but result in lower monthly energy bills than conventional heating and cooling systems. Areas with more extreme climates save enough money in monthly heating and cooling bills to offset the cost of installation in as little as 5 years. **INFOGRAPHIC 5**

geothermal heat pump A system that captures the steady 55°F underground temperature to heat or cool a building.

INFOGRAPHIC 5 GEOTHERMAL ENERGY CAN BE HARNESSED IN A VARIETY OF WAYS

The high temperatures found underground in some regions can be tapped to generate electricity. Geothermal heat can also be piped directly to communities to provide heat or hot water. Geothermal energy can also be tapped using ground-source heat pumps, reducing the cost to heat and cool individual buildings.

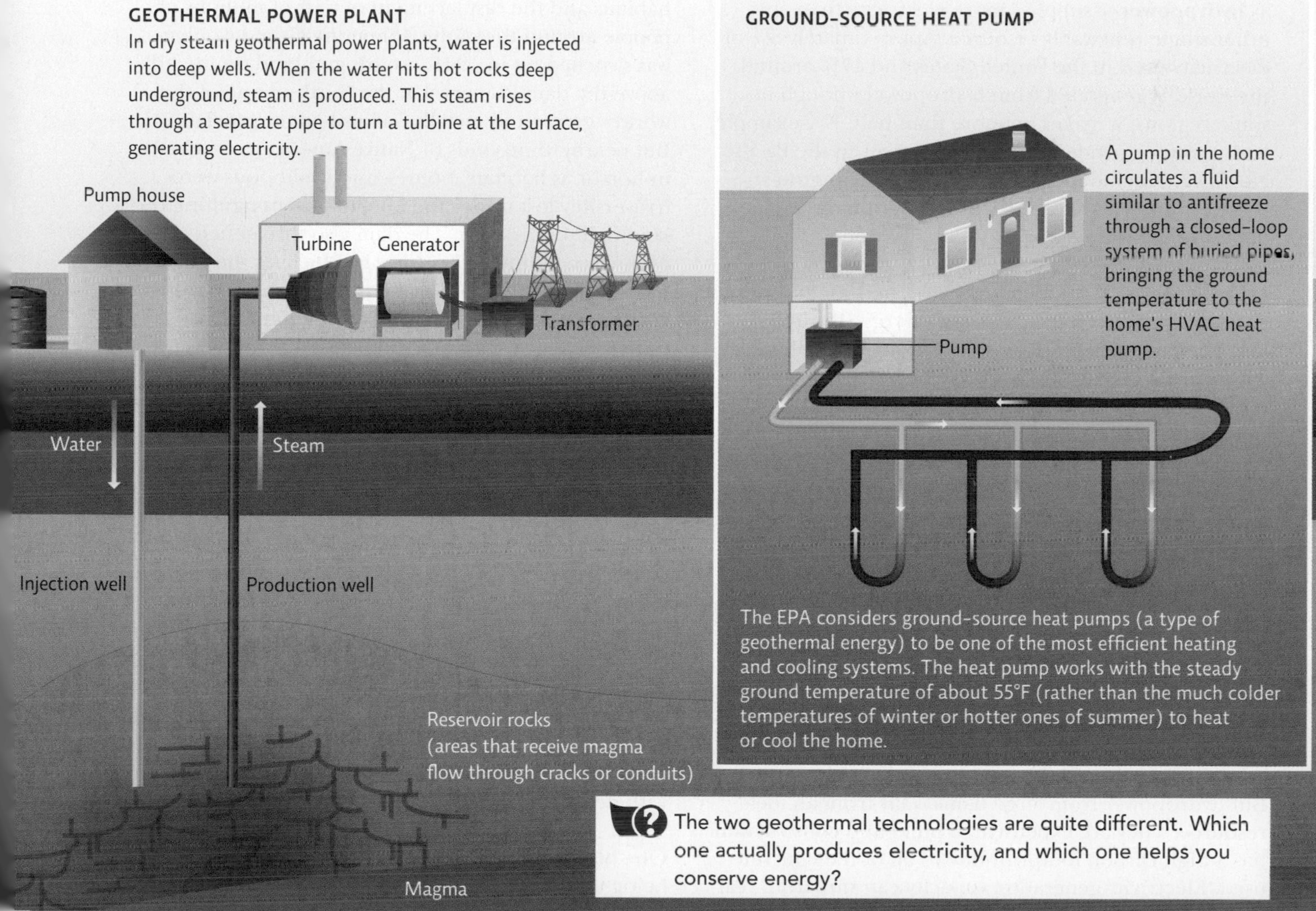
GEOTHERMAL POWER PLANT
In dry steam geothermal power plants, water is injected into deep wells. When the water hits hot rocks deep underground, steam is produced. This steam rises through a separate pipe to turn a turbine at the surface, generating electricity.
Pump house
Turbine
Generator
Transformer
Water
Steam
Injection well
Production well
Reservoir rocks (areas that receive magma flow through cracks or conduits)
Magma
GROUND-SOURCE HEAT PUMP
A pump in the home circulates a fluid similar to antifreeze through a closed-loop system of buried pipes, bringing the ground temperature to the home's HVAC heat pump.
Pump
The EPA considers ground-source heat pumps (a type of geothermal energy) to be one of the most efficient heating and cooling systems. The heat pump works with the steady ground temperature of about 55°F (rather than the much colder temperatures of winter or hotter ones of summer) to heat or cool the home.

The two geothermal technologies are quite different. Which one actually produces electricity, and which one helps you conserve energy?

ADVANTAGES	DISADVANTAGES
• Pollution-free. • Low environmental impact and carbon footprint. • Can be used at large or small scales to produce electricity or to capture heat. • Geothermal heat pumps can be installed anywhere and will reduce monthly heating and air conditioning costs.	• Geothermal power plants are expensive to build and are suitable only in some areas. • Geothermal heat pumps are more expensive to install than traditional HVAC systems. • Usually must be paired with other energy sources to meet needs.

This type of renewable energy is becoming more popular: It has a low carbon footprint, and its use does not generate toxic air pollution. It also is dependable—it has none of the intermittency issues of wind or solar.

Harnessing geothermal power for electricity production can be difficult, however. While geothermal power plants are reliable and efficient, their potential is entirely dependent on location. But drilling is expensive and difficult to do because of the depths one must drill to tap into Earth's heat. In early 2007, an earthquake of magnitude 3.4 on the Richter scale shook the city of Basel, Switzerland. The quake was attributed to a geothermal mining project in the area, and the project was halted.

6 HYDROPOWER

Key Concept 6: Hydroelectric dams are the most common way to harness the power of water, but they permanently alter the environment and have other drawbacks.

Humans have harnessed the power of falling water for thousands of years, ever since early civilizations used waterwheels at grist mills to grind grain into flour. Today energy produced from moving water—known as **hydropower**—supplies more electricity than any other single renewable resource. Approximately 8% of electricity used in the United States and 17% around the world is generated from hydropower, though in some regions, it makes up more than half. For example, hydropower provides much of the power to the Pacific Northwest, including 90% of Seattle, Washington's power. Large-scale hydroelectric power plants at giant dams are the source of most of that power.

Hydropower is abundant and clean, and no greenhouse gases are produced to access the energy (though large emissions result from the construction of dams). There are a variety of ways to harness the energy of moving water—from using ocean waves and the tides to capturing energy released from variations in ocean temperatures (ocean thermal energy conversion)—but the most common way to generate hydropower is with dams. The Grand Coulee Dam on the Columbia River in Washington State is the largest electrical power producer in the United States, with a total generating capacity of 6,809 MW; average production is 21 billion kWh per year—enough for 4.2 million households. The reservoir created by the dam, an artificial lake that pools behind the structure, stretches some 150 miles to the Canadian border. Dams can provide recreation and flood control, too.

But hydropower from large dams is far from an ideal resource. They are expensive to build and construction has a high carbon footprint due to all of the concrete used. Electricity generation capacity can vary from year to year based on rainfall amounts, and dams that restrict water flow can curtail water supplies to downstream areas. Further, in hot climates, huge amounts of water evaporate from reservoirs.

hydropower Energy produced from moving water.

In addition, dams are responsible for the loss of major habitats and the displacement of tens of millions of people around the globe. When the Grand Coulee was switched on in 1941, crowds gathered on the hill above the dam to marvel at the "birth of one of the world's greatest waterfalls," a local paper proclaimed. But nearby, thousands of Native Americans watched in horror as habitats, homes, and livelihoods were irreversibly lost under the pool of water engulfing the land behind the dam. The dam also threatened a staple of their economy and culture by blocking the migratory path of the area's iconic salmon populations; salmon runs upstream of the dam were completely wiped out. A variety of methods can be used to address this problem, including fish ladders (a structure that creates a passageway along which the fish can swim to move up and over the dam), trucks that move fish from one side of the dam to the other, and fish elevators that transport fish to the top of the dam. **INFOGRAPHIC 6**

Not all hydropower systems have the same impact as a giant dam. A *run-of-the-river hydroelectric* system, often a low stone or concrete wall, doesn't block the water; it merely directs some of the flowing water past a turbine and thus generates electricity from the natural flow of a river, which is less disruptive to the river ecosystem since the area behind the dam is not flooded. But energy production is dependent on the flow of a river at any particular time, so such a system is suitable only for rivers with dependable flow rates year-round.

One of the newest hydroelectric technologies, currently being used in Portland, Oregon, is the placement of small turbines inside large water pipes. Tapping the flow of this water (drinking water for the city) provides electricity without disrupting fish migrations or destroying habitat.

INFOGRAPHIC 6 HARNESSING THE POWER OF WATER

The energy of moving water can be captured in a variety of ways to produce electricity.

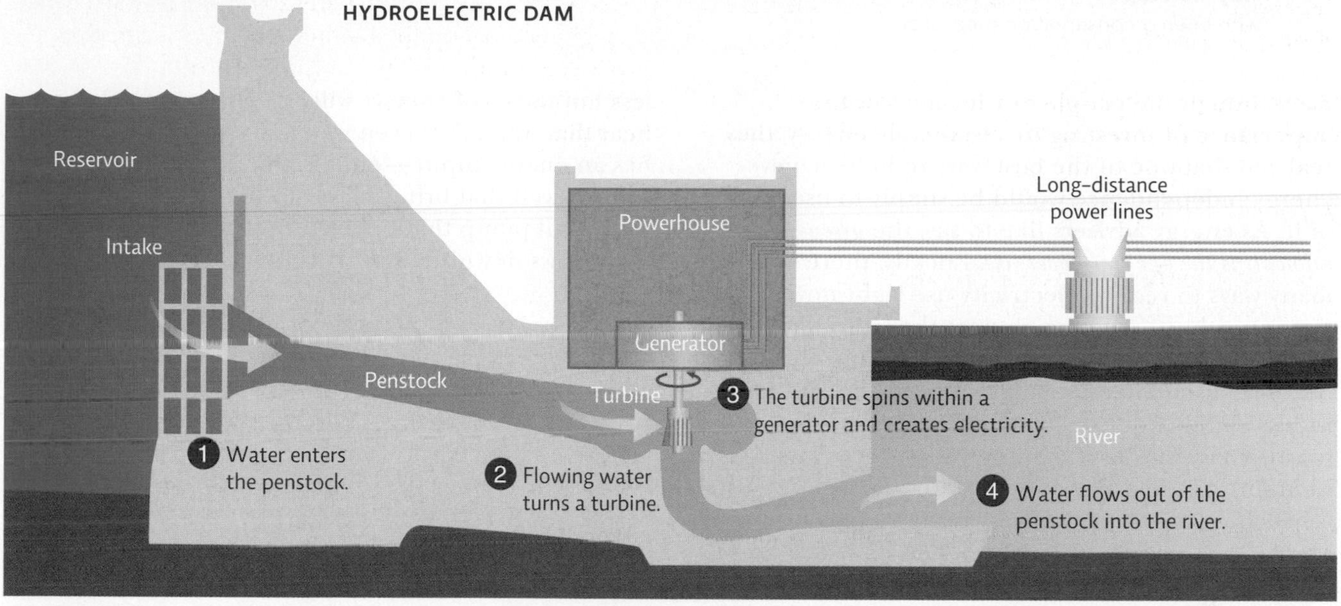
HYDROELECTRIC DAM
Reservoir
Intake
Powerhouse
Long-distance power lines
Generator
Penstock
Turbine
1 Water enters the penstock.
2 Flowing water turns a turbine.
3 The turbine spins within a generator and creates electricity.
River
4 Water flows out of the penstock into the river.

The most common way to harness the power of water is with a hydroelectric dam. When a large river is dammed, a reservoir is formed behind the dam. The buildup of water in the reservoir creates an enormous amount of pressure; water diverted from the top of the dam flows through long pipes, called penstocks, to turbines below. As the water rushes past the curved blades of the turbines, they spin and generate electricity.

ADVANTAGES	DISADVANTAGES
• Pollution-free. • Low environmental impact and carbon footprint from operation. • Large dams offer recreation and flood control advantages. • Small run-of-the-river systems can generate electricity without damming a river. • Tremendous power potential remains to be harnessed from the oceans.	• Large dams are expensive to build and permanently damage habitats and displace communities. • Production capacity varies seasonally and from year to year based on rainfall. • Usually must be paired with other energy sources to meet needs. • High evaporative water loss from reservoirs in hot, arid climates. • Blocks upstream migration of fish.

Do you feel the energy production and recreational/flood control benefits of large hydroelectric dams outweigh the permanent ecological and community destruction they cause? Explain your position.

Fish ladders and other technologies are used to try to help migratory fish such as salmon get past a dam that has blocked a river.

U.S. Army Corps of Engineers

7 THE ROLE OF CONSERVATION AND ENERGY EFFICIENCY

Key Concept 7: Meeting our energy needs with renewable sources becomes more likely when we pair renewables with energy conservation measures.

Even though the people of Chicago saw the importance of investing in sustainable energy, they realized that one of the best ways to help achieve energy independence would be simply to use less of it. As energy advisers like to say, the *greenest kilowatt is the one you never use.* Luckily, there are many ways to reduce electricity use right now. **Conservation**—making choices that result in less energy consumption—is a vital part of our quest to become sustainable users of energy. Conservation means simply not using energy if we don't need it as well as using energy more efficiently when we do use it.

Much of conservation involves changing behavior. For example, lighting typically accounts for about 25% of the average home's electric bill, and simply turning off the lights when you leave a room can make a big difference. Other options include studying in a room with ample natural light instead of a dark corner that requires artificial light. Unplugging phone chargers when not in use or setting your computer to "go to sleep" after a short idle period will also reduce the drain of "energy vampires"—devices that use electricity even when not in use.

Embracing **energy efficiency** is also a way to reduce energy use. All kinds of electrical devices are being made to perform the same tasks with less energy—more efficient lightbulbs and home appliances can reduce one's electricity bill enough to pay for themselves in a short period of time. Energy-efficiency advances are rapidly being made. For example, today LED lightbulbs represent the most efficient sources of light for your home, using one-third of the energy that a comparable compact fluorescent bulb would use or one-thirtieth the amount of energy needed to power an incandescent bulb. They also last longer, reducing the embedded cost of producing and disposing of more bulbs. Energy efficiency is a big part of Chicago's plan. Its Smart Lighting Program has replaced almost 200,000 energy-hungry sodium city streetlights with LED bulbs.

conservation Efforts that reduce waste and increase efficient use of resources.

energy efficiency A measure of the amount of energy needed to perform a task; higher efficiency means less energy is wasted.

Changes that allow you to save water also save energy. Using less hot water, of course, will save energy needed to heat that water, but even water from your cold tap has an energy input—either from the pump attached to your well that brings water to the surface or the industrial pump that pumps it into a water tower that then flows downhill into your home.

Before embarking on any energy-efficiency home improvements, a homeowner should conduct or commission an energy audit to see what actions would provide the biggest energy return on investment. For example, it makes no sense to invest in solar panels if your home is poorly insulated and losing more energy than the panels produce. **INFOGRAPHIC 7**

In Chicago, Mayor Emanuel focused on ways to improve energy efficiency in city buildings. As part of the Retrofit Chicago Energy Challenge, building owners and property managers have been encouraged to commit to reduce energy use by 20% over 5 years; as of February 2019, participating teams had achieved a nearly 15% energy reduction across a total of 87 city buildings, saving over $10.6 million in annual energy costs. "From Rogers Park to Trumbull Park, big buildings around the city are showing they can work together to reduce emissions, save money and put people to work," Mayor Emanuel said. (See Module 4.2 for more on green building.)

Ale-ks/iStock/Getty Images

Simple home improvements, such as installing insulation, can make a big difference in your energy bill. Request an energy audit from your local electricity provider for suggestions on which projects will give you the most return on your investment.

INFOGRAPHIC 7 SAVING ENERGY

Energy conservation is about wise use: using less and using energy more efficiently. Some steps can be taken immediately, with no investment; others require money and time to implement. Whatever you can do will not only reduce your energy use but also reduce your energy bill and the pollution generated from producing that energy.

NO-COST WAYS TO SAVE ENERGY	STEPS THAT COST MONEY BUT HAVE QUICK PAYBACK TIMES	MORE EXPENSIVE OPTIONS WITH LONGER PAYBACK TIMES
• Turn off computers and monitors and unplug phone chargers when not in use. • Lower the thermostat on the water heater to 120°F; turn it off if you will be away for several days. • In the winter, open curtains on south-facing windows to let in heat and light during the day; close them at night to reduce heat loss. • Close the fireplace damper when not in use to prevent loss of heat. • Take short showers instead of baths to reduce hot water use. • Wash only full loads of dishes and clothes; wash clothes with cold water.	• The first step is to have an energy audit performed to identify where your home is energy-inefficient. (Go to www.energy.gov/energysaver/energy-saver for more information.) • Install attic and wall insulation to meet recommendations for your area. • Weatherstrip and caulk door and window frames; check heating ducts for leaks and seal if needed. • Insulate hot-water pipes and water heater. • Install a programmable thermostat to automatically adjust the temperature when you are not at home or are sleeping. • Replace old lightbulbs with the more energy-efficient LED bulbs. • Replace home appliances with Energy Star-rated appliances.	• Replace older windows with high-efficiency windows that restrict loss of heating or cooling while allowing plenty of light. • Replace exterior doors with energy-efficient varieties. • Replace heating, ventilation, and air conditioning (HVAC) system with an energy-efficient model; consider a ground-source heat pump. • Install a high-efficiency on-demand (tankless) or geothermal water heater. • If you live in a hot climate, install a light-colored roof to reflect solar radiation.

? Which of the suggestions for saving energy could be done by a student living in a dormitory or rental apartment?

8 THE WAY FORWARD: MEETING ENERGY NEEDS SUSTAINABLY

Key Concept 8: Pricing energy sources according to their true costs and finding ways to store and more efficiently deliver electricity are needed to make renewables a more viable option. We can best meet local energy needs if we use a variety of renewable technologies that fit each locale and decentralize energy production.

The ideal sustainable energy system of the future needs to have a number of characteristics. It needs to produce cheap, dependable, carbon-free (net-zero emissions) electricity or, in locales where electricity isn't an option, carbon-neutral fuels. Replacing internal combustion engines with electric or biofuel-powered motors is also needed as well as efficient and low- or no-emission manufacturing. Finally, integration between sectors is needed so that gains in one sector are not lost by failures in others.

One recurring criticism of renewable energy technologies is that they are more expensive than fossil fuel methods. Even though sunlight, wind, geothermal heat, and water are free, constructing and installing the solar cells, wind turbines, underground pipes, and hydroelectric systems needed to harness their energy is not. But perhaps we should not be asking why renewables are so expensive but instead ask why fossil fuels are so cheap. We need to consider *all* the costs—environmental, social, and economic—of these technologies to fairly compare them. Because fossil fuels, led by coal, have the highest external (environmental and health) costs, when the *true cost* per kilowatt hour is estimated, renewables come out far ahead of fossil fuels. (See Module 5.1 for more on true cost accounting.)

The need for true cost pricing may become less of an issue as prices for renewables continue to drop. Wind and solar are already comparable in price to fossil fuels (and in some places, they are cheaper). By 2020, all renewable energy sources used for commercial production were, on average, equal in price to or cheaper than fossil fuel electricity.

While renewable technology is improving and becoming more diversified every year, there is a key hurdle that must be surmounted if renewables are to replace fossil fuels—the ability to store excess energy for later use. This is especially important for the two leading renewables, solar and wind, because their production of electricity is intermittent—sometimes you would have more than you needed, and other times there would not be enough. But using a variety of renewable options in the right combinations will allow us to better meet energy needs as the strength of one approach helps to offset the weaknesses of another.

Diversification may also help to *decentralize* the production and delivery of energy, which will allow the system to better meet the needs of local communities and will give community members the incentives to make sustainable energy work. One policy tool that supports this approach is the **feed-in tariff**, which has been used in Germany to great success. With these tariffs, individuals and groups that install solar or wind power and sell it to the utility (send it out over the grid) earn guaranteed profits from that investment—they are paid a fixed price for the energy they produce over a period of time (such as 5 or 10 years). This tool provides property owners with a stable incentive to encourage investments in renewable energy. In the United States, for example, Vermont offers a feed-in tariff for rooftop solar; in 2019, solar owners were guaranteed that they would be paid more for the electricity than it would cost to buy that electricity from the utility—in this case, for every kWH of electricity they provided, they earned retail plus an additional 2 cents.

Unfortunately, electricity itself can't be stored—it is sent across wires when generated and immediately used. This problem leads to a balancing act that power plant operators must deal with every day: They must match electricity production with consumption. The output of fossil fuel plants, and to a lesser extent nuclear power plants, can be scaled up or down as needed, but for renewables, what is needed is a cost-effective way to convert excess electrical production into a storable form of energy. This is a big obstacle to overcome, but we are making great strides. One novel solution being used in some areas is to use extra energy—say, electricity produced by solar power during the day—to pump water uphill to a storage reservoir. At night, the water can flow downhill past turbines that generate electricity. The next day the water is pumped uphill again—a simple solution, but one that only works in areas with a suitable water supply and terrain conducive to such a system. Other, more generally applicable methods are needed, too.

The most appealing option may be to convert electricity to chemical energy and store it in a battery. But to work on a large scale—one that works for a city or nation—we need battery arrays that efficiently store excess energy and readily give it up on demand. Batteries would offer a faster and more seamless way to juggle fluctuating energy flows than altering electricity production in a power plant could ever do. Several companies are putting major resources into the development and manufacture of battery technologies geared to solving the problem of storing electricity.

Prices for large-scale batteries are dropping rapidly: The cost of utility-scale lithium-ion batteries, for instance, has fallen 76% since 2012. And a new large-scale solar array plus battery project in Kern County, California, is projected to produce electricity that costs less than fossil fuel–generated electricity. Of course, batteries come with their own environmental problems, such as mining for battery components (see Module 7.1). With all approaches, trade-offs have to be considered and addressed.

The development of a **smart grid** is also part of the renewable future wish list. The present-day electricity grid consists of the power plants and power lines that crisscross our communities and nations, delivering electricity at the flip of a switch. This is described as a one-way communication—you "ask" for power, and it is delivered. But that grid is aging and frequently unable to cope with the electricity demands placed on it, leading to brownouts, blackouts, and power surges.

Turning the existing infrastructure into a smart grid means improving efficiency and smoothing out the flow of electricity—avoiding power surges or brownouts that occur when demand doesn't match production. This is accomplished by a better sharing of information—the grid becomes a "two-way conversation" with buildings like your home sending information back to the power company, allowing it to better anticipate electricity needs. "Smart meters" allow you (or your home) to schedule some energy use at low-demand times and automatically detect problems and quickly restore delivery of power.

As part of its sustainability initiative, the city of Chicago has partnered with its electricity provider, Commonwealth Edison, to install more than 4 million smart meters in area homes and businesses at a cost of around $3 a month, a fee tacked on to the electric bill. Using the information collected by the smart meter and managing energy demands (e.g., running a dishwasher at night when energy prices are lowest) can more than pay back that fee, managers say.

schmidt-z/E+/Getty Images

Many communities are pursuing sustainable energy to meet part or all of their energy needs, showing that a renewable energy future is within our grasp.

feed-in tariff Fixed payments, usually higher than the retail price, made to owners of renewable energy production equipment for electricity they send to the grid.

smart grid A modernized network that provides electricity to users in a way that automatically optimizes the delivery of electricity.

INFOGRAPHIC 8 A RENEWABLE ENERGY FUTURE

The transition away from fossil fuels to renewables for electricity production will require some changes in the way we choose energy sources (true cost accounting and diversification) as well as developing technological solutions to logistical problems.

TRUE COST PRICING

Factoring in the health and environmental costs of fossil fuel energy raises its costs above renewables, making renewables the most cost-effective choice.

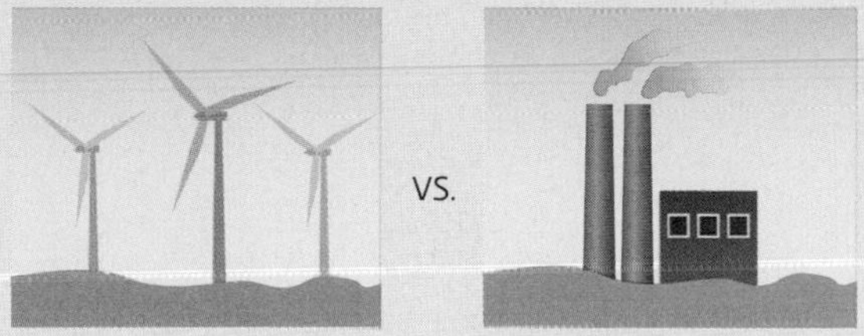

DIVERSIFY AND DECENTRALIZE

A variety of renewable options allows the strengths of one to offset the weaknesses of another and supports decentralization, which can better meet local needs.

ENERGY STORAGE

Finding cost-effective ways to store excess electricity can address the problem of intermittent production.

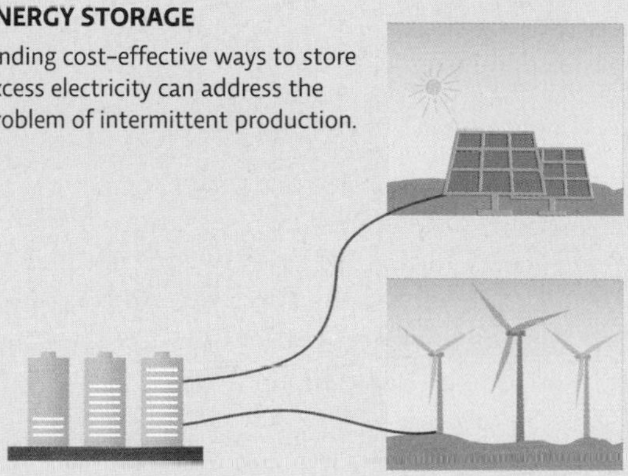

SMART GRID

The power grid needs to be updated and enhanced to better match electricity supply with demand and to handle the fluctuating power supply of renewables.

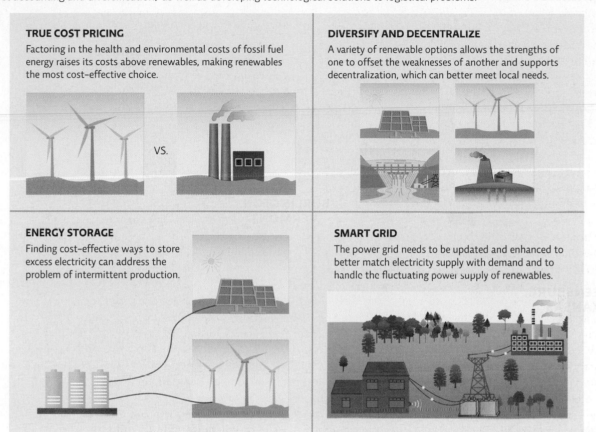

How would the development of dependable storage of electricity complement the operation of a smart grid?

As necessary as it is, a smart grid won't come cheaply. The Electric Power Research Institute estimates the price tag for a fully implemented U.S. smart grid at close to half a trillion dollars. The benefits, however, should be four times that. **INFOGRAPHIC 8**

In its quest to become more energy-neutral, the city of Chicago has focused its efforts on a variety of renewable technologies and conservation measures. Along with other renewable energy and conservation leaders—cities such as Eugene, Oregon, and Seattle, Washington, and countries such as Iceland, Denmark, and Costa Rica—Chicago is helping to blaze a trail others can follow and following in the footsteps of other cities that are already meeting their electricity needs with 100% renewable energy such as Burlington, Vermont, and Aspen, Colorado. Smaller cities like Greensburg, Kansas, destroyed by a tornado in 2007, are showing that they too can be renewable energy cities. This city of fewer than 800 residents gets all its electricity from renewable energy. (See the *Global Case Studies* map at the end of this module for more on some of these success stories.)

The next phase of Chicago's transition will involve replacing the city's gas-powered buses with electric buses by 2040. Indeed, Chicago was one of only four U.S. cities rated as having "exemplary" energy plans in a 2020 report published by the nonprofit American Council for an Energy-Efficient Economy. "We are taking concrete steps towards ensuring Chicago remains a resilient city that can build on the strengths of our past, while adapting to the challenges of the future," wrote Mayor Emanuel in 2019. "We are committed to reinvention."

Select References:

City of Chicago. (2019). *Resilient Chicago: A plan for inclusive growth and a connected city.* Retrieved from https://resilient.chicago.gov

Service, R. F. (2019). Solar plus batteries is now cheaper than fossil power. *Science, 365*(6449), 108.

U.S. Energy Information Administration. (2019). *Annual Energy Outlook 2019 with projections to 2050.* #AEO2019. Washington, DC: U.S. Energy Information Administration.

York, D., & Jarrah, J. (2020). *Community resilience planning and clean energy initiatives: A review of city-led efforts for energy efficiency and renewable energy.* Report U2002. Washington, DC: American Council for an Energy-Efficient Economy.

GLOBAL CASE STUDIES GREEN ENERGY

Achieve

Chicago is not the only community pursuing energy neutrality. Read about how other communities, and even a nation (Costa Rica), are pursuing sustainable energy.

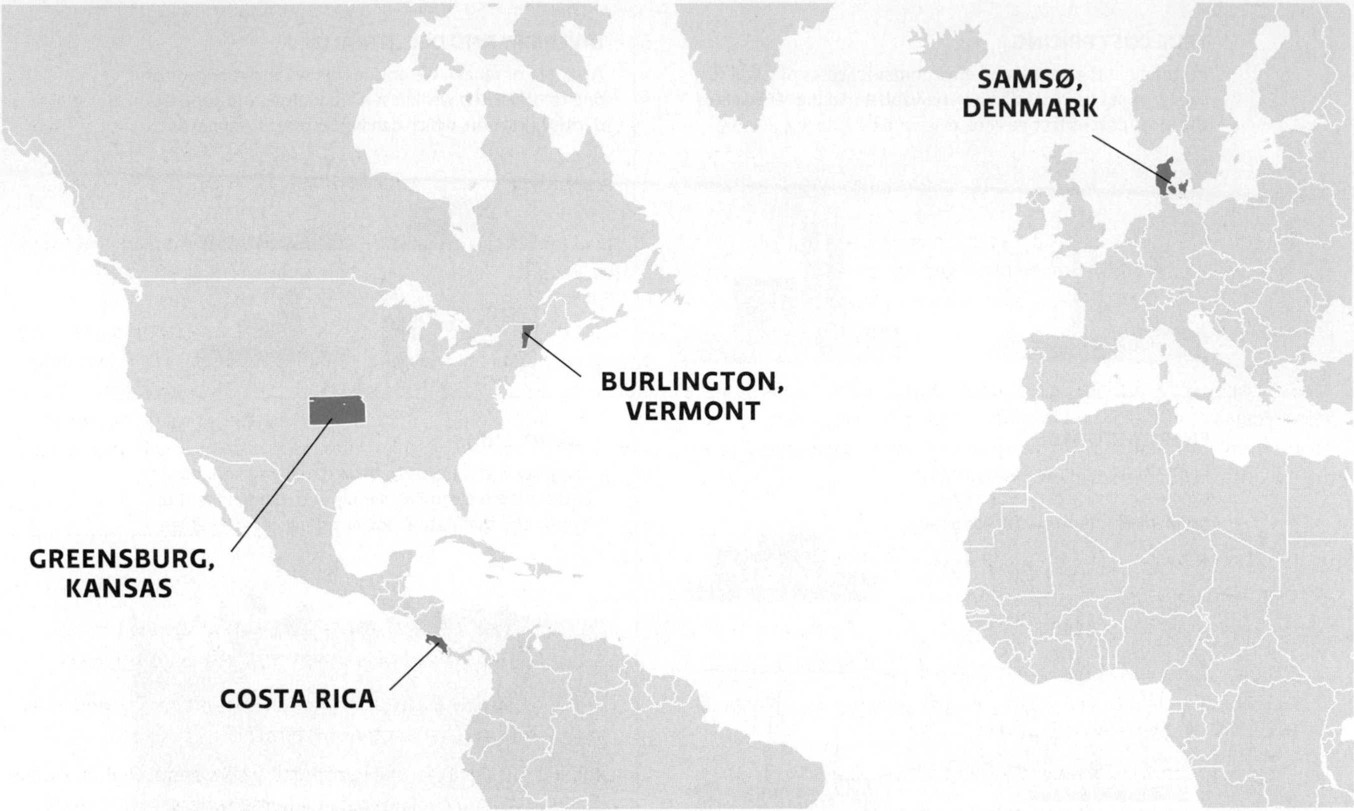

BRING IT HOME

PERSONAL CHOICES THAT HELP

A key component to developing a sustainable society is using renewable energy. Its use decreases harmful impacts of mining associated with nonrenewable energy and also reduces the amount of air and water pollution produced when the fuel is processed, transported, and burned. The efficiency and availability of renewable energy are rapidly increasing as new technologies emerge.

Individual Steps

- Contact your energy provider to see if you can purchase a percentage of your energy from a renewable source or enroll in a community solar or wind project. Also ask if smart meters are available in your area and request that one be installed if possible.
- Regardless of your home's energy source, make sure you are using energy efficiently. Review Infographic 7 and visit www.energy.gov/energysaver/energy-saver for ideas.
- If you have a smartphone, download the PVme app to see how many solar panels you would need to meet your household energy needs.
- A major barrier to using solar energy for many people is the cost. For information on tax incentives, rebates, and other programs that make using renewable energy easier, check out www.dsireusa.org or look into the growing trend of leasing solar panels.

Group Action

- Ask your local electricity provider if there are any local community solar or wind energy projects in your area and consider investing in a project. Learn about opportunities to invest in solar energy projects for lower-income communities or how to start one in your own community from the nonprofit Solar United Neighbors, at www.solarunitedneighbors.org.
- Visit the EPA's Power Profiler website at https://www.epa.gov/energy/power-profiler#/ to see how much renewable energy is used to generate electricity in your region. Ask your electricity provider about its plan to increase renewables and advocate for more clean energy to be part of its electricity portfolio.
- Contact your state and federal legislators and ask them to support or sponsor legislation that provides financial incentives for the purchase of renewable technologies.

REVISIT THE CONCEPTS

- Like natural ecosystems, human societies should rely on sustainable energy sources — those that are readily replenished while producing impacts that we can live with.
- Fossil fuels, the leading fuel for electricity production, are not renewable, and their use generates considerable pollution and environmental damage. The use of renewables is slowly rising, but this increase cannot keep up with rising demand. This means the use of fossil fuels, especially natural gas, is also increasing.
- The power of wind can be harnessed to generate electricity that is less expensive than other renewables and has a low environmental impact. Drawbacks include wind's intermittent nature and hazards that turbines pose to wildlife.
- Solar energy can be harnessed using passive and active solar technologies. It has a low environmental footprint, and operation is pollution-free. However, it can be expensive and is less productive at higher latitudes, on cloudy days, and at night.
- The heat of the Earth, geothermal energy, can be captured to lower home heating and cooling costs using a geothermal heat pump or to generate utility-scale electricity or hot water in a geothermal power plant.
- Electricity can be produced by harnessing the power of moving water, most commonly by damming rivers. While large dams have significant environmental drawbacks, other methods that use smaller run-of-the river systems or tap ocean power are available.
- Meeting our energy needs without fossil fuels will require that we pair renewables with energy conservation measures that reduce demand.
- Transitioning away from fossil fuels becomes more feasible if we price energy sources according to their true costs and find ways to store and more efficiently deliver electricity. Diversifying our energy sources and decentralizing production can better meet local needs and make it easier to provide community members with incentives to make sustainable energy work.

ENVIRONMENTAL LITERACY Understanding the Issue

1 What are the characteristics of a sustainable energy source, and what types are commonly used?

1. Why does society (or a living organism) need sources of energy that are renewable (readily replenished)?
2. Distinguish between renewable energy and sustainable energy.
3. Why are fossil fuels not a sustainable energy source?

2 What are the current and projected roles of renewables in global energy production?

4. Currently how important are fossil fuels in the production of electricity?
5. Which renewable energy sources are expected to increase the most in the next few decades, and which will see little change? What might account for these trends?

3 How can wind energy be captured to generate electricity, and what are the pros and cons of these methods?

6. Identify the advantages of wind power. Which do you feel is this technology's greatest strength? Explain.
7. Identify the disadvantages of wind power. Which do you feel is the most problematic, and how could it be addressed?
8. Compare the production of electricity by a wind turbine and the production of electricity by heating water to produce steam (thermoelectric production).

4 How can solar energy be captured and used, and what are the pros and cons of these methods?

9. Distinguish between passive and active solar technologies and give an example of each.
10. Which aspect of solar power do you feel is its greatest strength, and which its greatest weakness?
11. What is the concept of "payback time," and how can it be useful in deciding what types of conservation measures to pursue?

5 How can geothermal energy be captured and used, and what are the pros and cons of these methods?

12. How do geothermal power plants generate electricity, and what is the main limitation for their placement?
13. Explain how a geothermal heat pump works and how it can lower both heating and cooling costs for homeowners.

6 How can the power of water be captured and used, and what are the pros and cons of these methods?

14. How does a hydroelectric dam generate electricity?
15. Which disadvantage of large dams can the smaller run-of-the-river dams avoid?
16. How were Native Americans and wild salmon populations adversely affected by the construction of the Grand Coulee Dam?

7 What roles do conservation and energy efficiency play in helping us meet our energy needs sustainably?

17. With regard to electricity use, *energy efficiency* can be considered to be a subset of the larger energy-saving category known as *conservation*. Explain why this is true and give an example that supports your explanation.
18. What is an "energy vampire," and what can you do to reduce your usage of this energy?
19. Explain the value of an energy audit for homeowners who want to lower energy use.

8 What economic adjustment and what technological advances are needed for renewables to become a viable replacement for fossil fuels?

20. Explain the importance of the ability to store electricity for renewable energy to be a reliable source of electricity.
21. Why is diversification such an important part of a renewable energy future?

SCIENCE LITERACY Working with Data

One criticism of wind turbines is the toll they take on wildlife, especially birds and bats. Look at the data provided below to examine the impact of wind turbines and potential actions that could reduce impact.

A: NIGHTLY ACTIVITY PATTERNS OF BATS

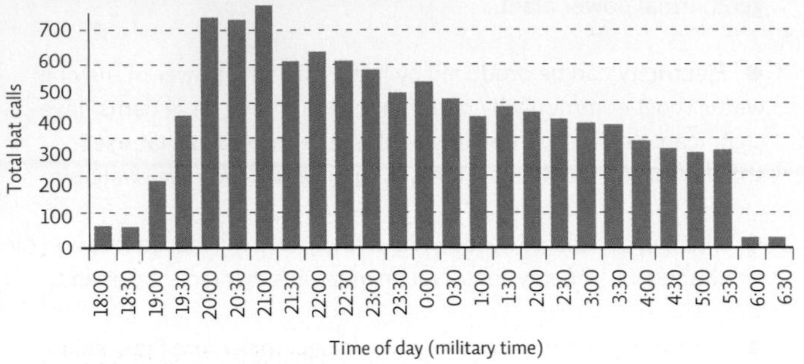

B: HUMAN-RELATED CAUSES OF BIRD MORTALITY (CANADA AND THE UNITED STATES)

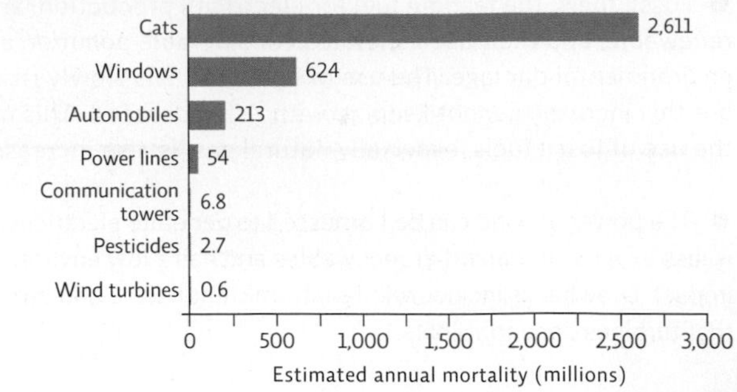

Interpretation

1. What question were the researchers asking for each of the two studies represented here (Graph A and Graph B)?
2. Based on the number of bat calls recorded for one population of bats (Graph A), when are these bats most active? (**Hint:** 18:00 is 6 P.M.)
3. Based on Graph B, what is likely to be the biggest threat to birds who live in large cities? To those who live in rural areas?

Advance Your Thinking

4. Based on Graph A, during what hours of the day should wind turbines be turned off to protect bats? How might this impact wind energy production?
5. Identify three actions that would do the most to decrease bird mortality. Which do you think would be the easiest to implement? Explain.
6. If a wind farm was installed close to you — on the outskirts of your city or in a nearby rural area — how do you think the impact of that wind farm would compare to other threats faced by birds in that area? Explain.

INFORMATION LITERACY Evaluating Information

Renewable energy resources have become a priority around the world as we grapple with the cost and limited supply of fossil fuels, not to mention the effect on the environment resulting from their use. How do we know which energy source is appropriate to invest in with both our time and money?

To learn more about renewable energy options, use the search engine of your choice and enter the key words "Renewable Energy."

Select a website that you think presents reliable, unbiased information, such as a government website. Explore the website and evaluate it by answering Questions 1a and 1b. After answering these questions, go back to your original search page and choose a website from a potentially biased source, such as an energy company or a political group. Explore and evaluate this second website and work with the information presented to answer the same questions in Question 1. Continue on to answer Questions 2 and 3.

1. Evaluate each website:
   a. Identify the name of the website and provide its URL. Who runs this website? Do the credentials of this organization make it reliable or unreliable? Explain.
   b. Does the website offer evidence for claims, cite information sources, or offer links to other information sources? Explain and give examples that support your evaluation.
2. Compare these two information sources:
   a. Is the information from the two websites consistent, or are there some contradictions? Explain and give examples that support your evaluation.
   b. Does one website have more detail, evidence, or links to scientific sources than the other? Is one easier to understand than the other or more useful to the average person? Explain and give examples that support your evaluation.
3. After completing your evaluation and comparison of the websites, write a paragraph that addresses these questions:
   a. Can an industry or political group (or any group with a vested interest in an issue) be a trusted information source, or is the information it presents always suspect?
   b. Are scientific or otherwise unbiased information sources always trusted information sources, or might the information they present be suspect?

Additional study questions are available at Achieve.macmillanlearning.com. Achieve

GLOSSARY

A

abiotic The nonliving components of an ecosystem, such as rainfall and mineral composition of the soil. (Module 2.1)

acid deposition Precipitation that contains sulfuric or nitric acid; dry particles may also fall and become acidified once they mix with water. (Module 10.1)

acid mine drainage Water that flows past exposed rock in mines and leaches out sulfates. These sulfates react with the water and oxygen to form acids (low-pH solutions). (Modules 7.1, 9.1)

acidification The lowering of the pH of a solution. (Module 6.3)

active solar technology Mechanical equipment for capturing, converting, and sometimes concentrating solar energy into a more usable form. (Module 11.2)

acute effect Adverse reaction that occurs very rapidly after exposure to a toxic substance has occurred. (Module 4.3)

adaptive management A plan that allows room for altering strategies as new information becomes available or as the situation itself changes. (Modules 5.2, 7.3)

additive effects Exposure to two or more chemicals that has an effect equivalent to the sum of their individual effects. (Module 4.3)

age structure The percentage of the population that is distributed into various age groups. (Module 4.1)

agroecology A scientific field that considers the area's ecology and indigenous knowledge and favors agricultural methods that protect the environment and meet the needs of local people. (Module 8.2)

air pollution Any material added to the atmosphere that harms living organisms, affects the climate, or impacts structures. (Module 10.1)

albedo The ability of a surface to reflect away solar radiation. (Module 10.2)

alleles Variants of genes that account for the diversity of traits seen in a population. (Module 3.1)

alpha radiation Ionizing particle radiation that consists of two protons and two neutrons. (Module 11.1)

annual crops Crops that grow, produce seeds, and die in a single year and must be replanted each season. (Module 8.2)

antagonistic effects Exposure to two or more chemicals that has a lesser effect than the sum of their individual effects would predict. (Module 4.3)

Anthropocene A proposed new geologic epoch that is marked by modern human impact. (Module 1.1)

anthropocentric worldview A human-centered view that assigns intrinsic value only to humans. (Module 1.1)

applied science Research whose findings are used to help solve practical problems. (Module 1.1)

aquaculture Fish farming; the rearing of aquatic species in tanks, ponds, or ocean net pens. (Module 8.4)

aquifer An underground, permeable region of soil or rock that is saturated with water. (Module 6.1)

artificial selection A process in which humans decide which individuals breed and which do not in an attempt to produce a population of plants or animals with desired traits. (Module 3.1)

atmosphere The blanket of gases surrounding Earth. (Module 5.2)

atom The simplest form of matter that retains the characteristics of that element. (Module 11.1)

B

background rate of extinction The average rate of extinction that occurred before the appearance of humans or that occurs between mass extinction events. (Module 3.1)

benthic macroinvertebrates Easy-to-see (not microscopic) arthropods such as insects that live on the stream bottom. (Module 6.2)

beta radiation Ionizing particle radiation that consists of electrons. (Module 11.1)

bioaccumulation The buildup of a substance in the tissues of an organism over the course of its lifetime. (Module 4.3)

biocentric worldview A life-centered approach that views all life as having intrinsic value, regardless of its usefulness to humans. (Module 1.1)

biodegradable Capable of being broken down by living organisms. (Module 5.3)

biodiesel A diesel-type fuel made from animal or vegetable oils and fats. (Module 11.3)

biodiversity The variety of life on Earth; it includes species, genetic, and ecological diversity. (Modules 1.1, 3.2)

biodiversity hotspot An area that contains a large number of endangered endemic species. (Module 3.2)

bioethanol An alcohol fuel made from crops in a process of fermentation and distillation. (Module 11.3)

biofuel Solid, liquid, or gaseous fuel produced from biological material. (Module 11.3)

biological assessment The process of sampling an area to see what lives there as a tool to determine how healthy the area is. (Module 6.2)

biomagnification The increased concentration of substances in the tissue of animals at successively higher levels of the food chain. (Module 4.3)

biomass Material from living or recently living organisms or their by-products. (Module 11.3)

biomass energy Energy from biological material such as plants (wood, charcoal, crops) and animal waste. (Module 11.2)

biome One of many distinctive types of ecosystems determined by climate and identified by the predominant vegetation and organisms that have adapted to live there. (Module 2.1)

biosphere The sum total of all of Earth's ecosystems. (Module 2.1)

biotic The living (organic) components of an ecosystem, such as the plants and animals and their waste (dead leaves, feces). (Module 2.1)

biotic potential (*r*) The maximum rate at which the population can grow due to births if each member of the population survives and reproduces. (Module 2.2)

boreal forest Coniferous forest found at high latitudes and altitudes characterized by low temperatures and low annual precipitation. (Module 7.2)

bottleneck effect The situation that occurs when population size is drastically reduced, leading to the loss of some genetic variants and resulting in a less diverse population. (Module 3.1)

bottom-up regulation Population sizes in a community are limited primarily by availability of resources that enhance growth and survival of organisms lower on the food chain. (Module 2.2)

bycatch Nontarget species that become trapped in fishing nets and are usually discarded. (Module 8.4)

C

canopy The upper layer of a forest formed where the crowns (tops) of the majority of the tallest trees meet. (Module 7.2)

cap-and-trade program Regulations that set an upper limit for pollution, issue permits to producers for a portion of that amount, and allow producers that use or release less than their allotment to sell permits to those who exceeded their allotment. (Modules 5.2, 10.1)

carbon capture and sequestration (CCS) The process of removing carbon from fuel combustion emissions or other sources and storing it to prevent its release into the atmosphere. (Module 9.1)

carbon cycle Movement of carbon through biotic and abiotic parts of an ecosystem via photosynthesis and cellular respiration as well as in and out of other reservoirs such as oceans, soil, rock, and atmosphere. (Module 2.1)

carbon footprint The amount of CO_2 (and other greenhouse gases that contribute to climate change) released to the atmosphere by a person, company, nation, or activity. (Modules 4.2, 8.2, 8.3)

carbon sequestration The storage of carbon in a form that prevents its release into the atmosphere. (Module 7.3)

carbon sink An area such as a forest, ocean sediment, or soil, where accumulated carbon does not readily reenter the carbon cycle. (Module 7.2)

carbon taxes Governmental fees imposed on activities that release greenhouse gases into the atmosphere. (Module 10.2)

carrying capacity (*K*) The maximum population size that a particular environment can support indefinitely; for human populations, it depends on resource availability and the rate of per capita resource use by the population. (Modules 2.2, 4.1)

cash crops Food and fiber crops grown to sell for profit rather than for use by local families or communities. (Module 8.1)

cause-and-effect relationship An association between two variables that identifies one (the effect) occurring as a result of or in response to the other (the cause). (Module 1.2)

cellular respiration The process in which all organisms break down sugar to release its energy, using oxygen and giving off CO_2 as a waste product. (Module 2.1)

cellulosic ethanol Bioethanol made by breaking down cellulose in plants. (Module 11.3)

childhood mortality rate The number of children under 5 years of age that die per every 1,000 live births in that year. (Module 4.1)

chronic effect Adverse reaction that happens only after repeated long-term exposure to low doses of a toxic substance. (Module 4.3)

circular economic system A production system in which the product is returned to the resource stream when consumers are finished with it or is disposed of in such a way that nature can decompose it. (Module 5.1)

cisgenic organism An organism that received DNA from a close relative, DNA that could have been acquired via traditional breeding. (Module 8.1)

citizen suit provision A provision that allows a private citizen to sue, in federal court, a perceived violator of certain U.S. environmental laws, such as the Clean Air Act, in order to force compliance. (Module 5.2)

Clean Air Act (CAA) The main U.S. law that authorizes the EPA to set standards for dangerous air pollutants and enforce those standards. (Module 10.1)

clean coal A liquid or gaseous product produced by removing some contaminants contained in coal so that the resulting fuel releases less pollution when burned. (Module 9.1)

Clean Water Act (CWA) U.S. federal legislation that regulates the release of point source pollution into surface waters and sets water quality standards for those waters. It also supports best management practices to reduce nonpoint source pollution. (Module 6.2)

clear-cut Timber-harvesting technique that cuts all trees in an area. (Module 7.2)

climate Long-term patterns or trends of meteorological conditions. (Module 10.2)

climate adaptation Efforts to help deal with existing or impending climate change problems. (Module 10.2)

climate change Alteration in the long-term patterns and statistical averages of meteorological events. (Module 10.2)

climate forcer Anything that alters the balance of incoming solar radiation relative to the amount of heat that escapes into space. (Module 10.2)

climate mitigation Efforts to minimize the extent or impact of climate change. (Module 10.2)

clumped distribution A distribution in which individuals are found in groups or patches within the habitat. (Module 2.2)

coal A fossil fuel that is formed when plant material is buried in oxygen-poor conditions and subjected to high heat and pressure over a long time. (Module 9.1)

coevolution A special type of natural selection in which two species each provide the selective pressure that determines which traits are favored by natural selection in the other. (Module 3.1)

collapsed fishery A fishery in which annual catches fall below 10% of their historic high; stocks can no longer support a fishery. (Module 8.4)

command-and-control regulation Legislative control of an activity or industry via rules that identify acceptable actions. (Modules 5.2, 10.1)

commensalism A symbiotic relationship between individuals of two species in which one benefits from the presence of the other but the other is unaffected. (Module 2.3)

communicable disease A disease that can spread from one person to another. (Module 4.3)

community All the populations (plants, animals, and other species) living and interacting in an area. (Module 2.1)

community ecology The study of all the populations (plants, animals, and other species) living and interacting in an area. (Module 2.3)

competition Species interaction in which individuals are vying for limited resources. (Module 2.3)

composting Allowing waste to biologically decompose in the presence of oxygen and water, producing a soil-like mulch. (Module 5.3)

concentrated animal feeding operation (CAFO) A situation in which meat or dairy animals are raised in confined spaces, maximizing the number of animals that can be reared in a small area. (Module 8.3)

conservation Efforts that reduce waste and increase efficient use of resources. (Module 11.2)

conservation biology The science concerned with preserving biodiversity. (Module 3.3)

conservation genetics The scientific field that relies on species' genetics to inform conservation efforts. (Module 3.3)

consumer An organism that obtains energy and nutrients by feeding on another organism. (Modules 2.1, 2.3)

control group The group in an experimental study to which the test group's results are compared; ideally, the control group will differ from the test group in only one way. (Module 1.2)

control rod Cylinder that can be added to a fuel assembly to absorb neutrons and slow the fission chain reaction. (Module 11.1)

convention An international agreement that represents a position on an issue and identifies general goals that the signing countries agree to pursue. (Module 5.2)

Convention on International Trade in Endangered Species of Wild Fauna and Flora (CITES) An international treaty that regulates the global trade of selected species. (Module 3.3)

conventional reserves Deposits of crude oil or natural gas that can be extracted by vertical drilling and pumping. (Module 9.2)

convergent plate boundary A place where tectonic plates are moving toward each other. (Module 7.1)

coral Colonial marine animal that secretes a hard outer shell in which it lives and is mutualistically dependent on an algal partner. (Module 6.3)

coral bleaching A stress response in a coral in which the mutualistic algal partner is expelled; this weakens and even can kill the coral if it is not recolonized soon. (Module 6.3)

coral reef A large underwater structure formed by colonies of tiny animals (corals) that produce a calcium carbonate exoskeleton that over time builds up; commonly found in shallow, warm, tropical seas. (Module 6.3)

correlation Two things occurring together but not necessarily having a cause-and-effect relationship. (Module 1.2)

cradle-to-cradle Refers to management of a resource that considers the impact of its use at every stage, from raw material extraction to final disposal or recycling. (Module 5.1)

critical thinking Skills that enable individuals to logically assess information, reflect on that information, and reach their own conclusions. (Module 1.2)

crude oil A mix of hydrocarbons that exists as a liquid underground; can be refined to produce fuels or other products. (Module 9.2)

D

dam A structure that blocks the flow of water in a river or stream. (Module 6.1)

debt-for-nature-swaps Arrangements in which a wealthy nation forgives the debt of a developing nation in return for a pledge to protect natural areas in that developing nation. (Module 3.3)

decomposers Organisms such as bacteria and fungi that break organic matter all the way down to constituent atoms or molecules in a form that plants can take back up. (Module 2.3)

deforestation Net loss of trees in a forested area. (Module 7.2)

demographic factors Population characteristics such as birth rate that influence changes in population size and composition. (Module 4.1)

demographic transition A theoretical model that describes the expected drop in once-high population growth rates as economic conditions improve the quality of life in a population. (Module 4.1)

denitrification Conversion of nitrate to molecular nitrogen (N_2). (Module 2.1)

density-dependent factors Factors, such as predation or disease, whose impact on a population is influenced by the size of that population. (Module 2.2)

density-independent factors Factors, such as a storm or an avalanche, whose impact on a population is not related to population size. (Module 2.2)

dependent variable The variable in an experiment that is evaluated to see if it changes due to the conditions of the experiment. (Module 1.2)

desalination The removal of salt and minerals from seawater to make it suitable for consumption. (Module 6.1)

desertification The process that transforms once-fertile land into desert. (Module 7.3)

desired fertility The ideal number of children an individual indicates he or she would like to have. (Module 4.1)

detritivores Consumers (including worms, insects, and crabs) that eat dead organic material. (Module 2.3)

discounting the future Giving more weight to short-term benefits and costs than to long-term ones. (Module 5.1)

dissolved oxygen (DO) The amount of oxygen in the water. (Module 6.2)

divergent plate boundary A place where tectonic plates are moving away from each other. (Module 7.1)

domestic water use Indoor and outdoor water used by households and small businesses. (Module 6.1)

dose-response study An experiment that tests the strength of the effect produced at different doses of a substance. (Module 4.3)

E

earthquake A sudden shaking of the ground caused by movement of tectonic plates at a plate boundary or an intraplate fault, as well as volcanic activity. (Module 7.1)

ecocentric worldview A system-centered view that values intact ecosystems, not just the individual parts. (Module 1.1)

ecolabeling Providing information about how a product is made and where it comes from. Allows consumers to make more sustainable choices and support sustainable products and the businesses that produce them. (Module 5.1)

ecological diversity The variety within an ecosystem's structure, including many communities, habitats, niches, and trophic levels. (Module 3.2)

ecological footprint The land area needed to provide the resources for and assimilate the waste of a person or population. (Modules 4.1, 5.1)

ecological succession Progressive replacement of plant (and then animal) species in a community over time due to the changing conditions that the plants themselves create (more soil, shade, etc.). (Module 2.3)

economics The social science that deals with the production, distribution, and consumption of goods and services. (Module 5.1)

ecosystem All of the organisms in a given area plus the physical environment in which and with which they interact. (Module 2.1)

ecosystem conservation A management strategy that focuses on protecting an ecosystem as a whole in an effort to protect the species that live there. (Module 3.3)

ecosystem services Essential ecological processes that make life on Earth possible. (Modules 3.2, 5.1, 7.2)

ecotones Regions of distinctly different physical areas that serve as boundaries between different communities. (Module 2.3)

ecotourism Low-impact travel to natural areas that contributes to the protection of the environment and respects the local people. (Modules 3.3, 7.2)

edge effect The change in species diversity that occurs due to the different conditions that either attract or repel certain species at an ecotone. (Module 2.3)

effluent Wastewater discharged into the environment. (Module 6.2)

electricity The flow of electrons (negatively charged subatomic particles) through a conductive material (such as wire). (Module 9.1)

electrolysis Processing step that uses an electric current to separate metal from ore. (Module 7.1)

element A substance composed of all the same type of atoms. (Module 11.1)

emergent layer The region where a tree that is taller than the canopy trees rises above the canopy layer. (Module 7.2)

empirical evidence Information detected with the senses—or with equipment that extends our senses. (Module 1.2)

empirical science A scientific approach that investigates the natural world through systematic observation and experimentation. (Module 1.1)

endangered species Species at high risk of becoming extinct. (Module 3.2)

Endangered Species Act (ESA) The primary federal law that protects biodiversity in the United States. (Module 3.3)

endemic Describes a species that is native to a particular area and is not naturally found elsewhere. (Module 3.1, 3.2)

energy The capacity to do work. (Module 9.1)

energy efficiency A measure of the amount of energy needed to perform a task; higher efficiency means less energy is wasted. (Module 11.2)

energy flow The one-way passage of energy through an ecosystem. (Module 2.1)

energy independence Meeting all of one's energy needs without importing any energy. (Module 9.2)

energy return on energy investment (EROEI) A measure of the net energy from an energy source (the energy in the source minus the energy required to get it, process it, ship it, and then use it). (Modules 9.1, 9.2)

energy security Having access to enough reliable and affordable energy sources to meet one's needs. (Module 9.2)

environment The biological and physical surroundings in which any given living organism exists. (Module 1.1)

environmental economics New theory of economics that considers the long-term impact of our choices on people and the environment. (Module 5.1)

environmental ethic The personal philosophy that influences how a person interacts with his or her natural environment and thus affects how one responds to environmental problems. (Module 1.1)

environmental health The branch of public health that focuses on factors in the natural world and the human-built environment that impact the health of populations. (Module 4.3)

environmental impact statement (EIS) A document outlining the positive and negative impacts of any federal action that has the potential to cause environmental damage. (Module 5.2)

environmental justice The concept that access to a clean, healthy environment is a basic human right. (Modules 4.2, 10.1)

environmental literacy A basic understanding of how ecosystems function and of the impact of our choices on the environment. (Module 1.1)

environmental policy A course of action adopted by a government or an organization that is intended to improve the natural environment and public health or reduce human impact on the environment. (Module 5.2)

Environmental Protection Agency (EPA) The federal agency responsible for setting policy and enforcing U.S. environmental laws. (Module 5.2)

environmental science An interdisciplinary field of research that draws on the natural and social sciences and the humanities in order to understand the natural world and our relationship to it. (Module 1.1)

epidemiologist A scientist who studies the causes and patterns of disease in human populations. (Module 4.3)

erosion The movement of broken-down rock, soil, and other materials from one location to another. (Module 7.1)

estuary A region where a river empties into the ocean. (Module 6.3)

eutrophication A process in which excess nutrients in aquatic ecosystems feed biological productivity, ultimately lowering the oxygen content in the water. (Module 6.2)

evolution Differences in the gene frequencies within a population from one generation to the next. (Module 3.1)

e-waste Unwanted computers and other electronic devices such as discarded televisions and cell phones. (Modules 5.3, 7.1)

exclusive economic zones (EEZs) Zones that extend 200 nautical miles from the coastline of any given nation, where that nation has exclusive rights over marine resources, including fish. (Module 8.4)

experimental prediction A statement that identifies what is expected to happen if the hypothesis being tested is correct. (Module 1.2)

experimental study Research that manipulates a variable in a test group and compares the response to that of a control group that was not exposed to the same variable. (Module 1.2)

exponential growth The kind of growth in which a population becomes progressively larger each breeding cycle; produces a J curve when plotted over time. (Module 2.2)

external cost A cost associated with a product or service that is not taken into account when a price is assigned to that product or service but rather is passed on to a third party who does not benefit from the transaction. (Module 5.1)

extinction The complete loss of a species from an area; may be local (gone from an area) or global (gone for good). (Module 3.1)

extirpation Locally extinct in one geographic area but still found elsewhere. (Module 3.1)

exurbs Towns beyond the immediate suburbs whose residents commute into the city for work. (Module 4.2)

F

fair trade A certification program whose products are made in ways that are environmentally sustainable and socially beneficial (e.g., fair wages, good working conditions). (Module 5.1)

feed conversion ratio The amount of edible food that is produced per unit of feed input. (Module 8.3)

feed-in tariff Fixed payments, usually higher than the retail price, made to owners of renewable energy production equipment for electricity they send to the grid. (Module 11.2)

feedstocks Biomass sources used to make biofuels. (Module 11.3)

fertilizer A natural or synthetic mixture that contains nutrients that is added to soil to boost plant growth. (Module 8.1)

fisheries The industry devoted to commercial fishing or the places where fish are caught, harvested, processed, and sold. (Module 8.4)

fishing down the food chain The harvest of fish at lower trophic levels once fish stocks at higher trophic levels become depleted. (Module 8.4)

flagship species The focus of public awareness campaigns aimed at generating interest in conservation in general; usually an interesting or charismatic species, such as the giant panda or tiger. (Module 3.3)

food chain A simple, linear path starting with a plant (or other photosynthetic organism) that identifies what each organism in the path eats. (Module 2.3)

food desert A locale where access to affordable, fresh, and nutritious food is limited or nonexistent. (Module 8.1)

food miles The distance a food travels from its site of production to the consumer. (Module 8.2)

food security Having physical, social, and economic access to sufficient safe and nutritious food. (Module 8.1)

food self-sufficiency The ability of an individual nation to grow enough food to feed its people. (Module 8.1)

food sovereignty The ability of an individual nation to control its own food system. (Module 8.1)

food web A linkage of all the food chains together that shows the many connections in the community. (Module 2.3)

forest An ecosystem made up primarily of trees and other woody vegetation. (Module 7.2)

forest ecosystem management (FEM) A system that focuses on managing the forest as a whole rather than for maximizing yields of a specific product. (Module 7.2)

forest floor The lowest level of the forest, containing herbaceous plants, fungi, leaf litter, and soil. (Module 7.2)

fossil fuel A variety of hydrocarbons formed from the remains of dead organisms. (Modules 9.1, 9.2)

founder effect The situation that occurs when a small group with only a subset of the larger population's genetic diversity becomes isolated and evolves into a different population, missing some of the traits of the original. (Module 3.1)

fracking (hydraulic fracturing) The extraction of oil or natural gas from dense rock formations by creating fractures in the rock and then flushing out the oil or gas with pressurized fluid. (Module 9.2)

fuel crop A crop specifically grown to be used to produce biofuels. (Module 11.3)

fuel rod Hollow metal cylinder filled with uranium fuel pellets for use in fission reactors. (Module 11.1)

G

gamma radiation Ionizing high-energy electromagnetic waves (photons). (Module 11.1)

gene frequencies The assortment and abundance of particular variants of genes relative to each other within a population. (Module 3.1)

genes Stretches of DNA, the hereditary material of cells, that each direct the production of a particular protein and influence an individual's traits. (Module 3.1)

genetic diversity The heritable variation among individuals of a single population or within a species as a whole. (Modules 3.1, 3.2, 8.1)

genetic drift The change in gene frequencies of a population over time due to random mating that results in the loss of some gene variants. (Module 3.1)

genetically modified organism (GMO) Organism that has had its genetic information modified to give it desirable characteristics such as pest or drought resistance. (Module 8.1)

geothermal energy Heat stored underground, contained in either rocks or fluids. (Module 11.2)

geothermal heat pump A system that captures the steady 55°F underground temperature to heat or cool a building. (Module 11.2)

geothermal power plant A large-scale facility that captures steam produced from Earth's internal heat to turn turbines, which generate electricity. (Module 11.2)

grassland A biome that is predominately grasses due to low rainfall, grazing animals, and/or fire. (Module 7.3)

green building Construction and operational designs that promote resource and energy efficiency and provide a better environment for occupants. (Module 4.2)

green business Doing business in a way that is good for people and the environment. (Module 5.1)

green city A city designed to improve environmental quality and social equity while reducing its overall environmental impact. (Module 4.2)

Green Revolution A plant-breeding program in the mid-1900s that dramatically increased crop yields and paved the way for mechanized, large-scale agriculture. (Module 8.1)

green space A natural area such as a park or an undeveloped landscape containing grass, trees, or other vegetation in an urban area, usually set aside for recreational use. (Module 4.2)

green tax A tax (a fee paid to the government) assessed on environmentally undesirable activities (e.g., a tax per unit of pollution emitted). (Modules 5.2, 10.1)

greenhouse effect The warming of the planet that results when heat is trapped by Earth's atmosphere. (Module 10.2)

greenhouse gases Molecules in the atmosphere that absorb heat and reradiate it back to Earth. (Module 10.2)

greenwashing Claiming environmental benefits about a product when the benefits are actually minor or nonexistent. (Module 5.1)

ground-level ozone A secondary pollutant that forms when some of the pollutants released during fossil fuel combustion react with atmospheric oxygen in the presence of sunlight. (Module 10.1)

groundwater Water found underground trapped in soil or porous rock. (Module 6.1)

growth factors Resources individuals need to survive and reproduce that allow a population to grow in number. (Module 2.2)

H

habitat The physical environment in which individuals of a particular species can be found. (Module 2.3)

habitat fragmentation The destruction of part of an area that creates a patchwork of suitable and unsuitable habitat areas that may exclude some species altogether. (Module 3.2)

hazardous waste Waste that is toxic, flammable, corrosive, explosive, or radioactive. (Module 5.3)

high-level radioactive waste (HLRW) Spent nuclear reactor fuel or waste from the production of nuclear weapons that is still highly radioactive. (Module 11.1)

holistic planned grazing Grazing livestock in a way that mimics wild grazers by grazing intensively on a small section of pasture before moving to another. (Module 7.3)

hydropower Energy produced from moving water. (Module 11.2)

hypothesis A possible explanation for what we have observed that is based on some previous knowledge. (Module 1.2)

hypoxia A situation in which a body of water contains inadequate levels of oxygen, compromising the health of many aquatic organisms. (Module 6.2)

I

igneous rock Rock that forms when molten rock cools and solidifies. (Module 7.1)

incinerators Facilities that burn trash at high temperatures. (Module 5.3)

independent variable The variable in an experiment that a researcher manipulates or changes to see if the change produces an effect. (Module 1.2)

indicator species A species that is particularly vulnerable to ecosystem perturbations, and that, when we monitor it, can give us advance warning of a problem. (Module 2.3)

industrial agriculture Farming methods that rely on technology, synthetic chemical inputs, and economies of scale to increase productivity and profits. (Module 8.1)

inferences Conclusions drawn based on observations. (Module 1.2)

infill development The development of empty lots within a city. (Module 4.2)

infiltration The process of water soaking into the ground. (Module 6.1)

information literacy The ability to find and evaluate the quality of information. (Module 1.2)

instrumental value An object's or species' worth, based on its usefulness to humans. (Module 3.2)

integrated pest management (IPM) The use of a variety of methods to control a pest population, with the goal of minimizing or eliminating the use of chemical toxins. (Module 8.2)

internal cost A cost—such as for raw materials, manufacturing costs, labor, taxes, utilities, insurance, or rent—that is accounted for when a product or service is evaluated for pricing. (Module 5.1)

intragenic organism An organism whose own DNA has been edited. (Module 8.1)

intrinsic value An object's or species' worth, based on its mere existence; it has an inherent right to exist. (Module 3.2)

invasive species A non-native species (a species outside its range) whose introduction causes or is likely to cause economic or environmental harm or harm to human health. (Module 3.1)

IPAT model An equation (I = P × A × T) that measures human impact (I), based on three factors: population (P), affluence (A), and technology (T). (Module 5.1)

isotopes Atoms that have different numbers of neutrons in their nucleus but the same number of protons. (Module 11.1)

K

keystone species A species that impacts its community more than its mere abundance would predict, often altering ecosystem structure. (Module 2.3)

***K*-selected species** Species that have a low biotic potential and that share characteristics such as long life span, late maturity, and low fecundity; generally show logistic population growth. (Module 2.2)

L

landscape conservation An ecosystem conservation strategy that specifically identifies a suite of species, chosen because they use all the vital areas within an ecosystem; meeting the needs of these species will keep the ecosystem fully functional, thus meeting the needs of all species that live there. (Module 3.3)

landslide The sudden movement of unstable rock or soil material down a slope due to the force of gravity, often triggered by heavy rain or an earthquake. (Module 7.1)

LD50 (lethal dose 50%) The dose of a substance that would kill 50% of the test population. (Module 4.3)

leachate Water that carries dissolved substances (often contaminated) that can percolate through soil. (Module 5.3)

Leadership in Energy and Environmental Design (LEED) A certification program that awards a rating (standard, silver, gold, or platinum) to buildings that include environmentally sound design features. (Module 4.2)

life-history strategies Biological characteristics of a species, such as life span and fecundity, that influence how quickly a population can potentially increase in number. (Module 2.2)

linear economic system A production model that is one way: inputs are used to manufacture a product, and waste is discarded. (Module 5.1)

lithosphere The rigid outer layer of Earth made up of the crust and the hard uppermost layer of mantle. (Module 7.1)

logical fallacies Arguments that attempt to sway the reader without using reasonable evidence. (Module 1.2)

logistic growth The kind of growth in which population size increases rapidly at first but then slows down as the population becomes larger; produces an S-shaped curve when plotted over time. (Module 2.2)

low-level radioactive waste (LLRW) Material that has a low level of radiation for its volume. (Module 11.1)

M

malnutrition A state of poor health that results from inappropriate caloric intake (too many or too few calories) or is deficient in one or more nutrients. (Module 8.1)

marine calcifiers Organisms that make a hard calcium-based shell or exoskeleton. (Module 6.3)

marine protected areas (MPAs) Discrete regions of ocean that are legally protected from various forms of human exploitation. (Module 8.4)

marine reserves Restricted areas where all fishing is prohibited and absolutely no human disturbance is allowed. (Module 8.4)

matter cycles Movement of life's essential chemicals or nutrients through an ecosystem. (Module 2.1)

maximum sustainable yield (MSY) The amount that can be harvested without decreasing the yield in future years. (Modules 7.2, 8.4)

media literacy The ability to evaluate digital sources of information. (Module 1.2)

metal A malleable substance that can conduct electricity; usually found in nature as part of a mineral compound. (Module 7.1)

metamorphic rock Rock that forms when existing rock in Earth's crust is transformed by high heat and pressure. (Module 7.1)

Milankovitch cycles Predictable variations in Earth's position in space relative to the Sun that affect climate. (Module 10.2)

mine tailings The waste of mining operations; includes mill tailings, the finely ground rock left over from processing mineral ores. (Module 7.1)

mineral A naturally occurring chemical compound that exists as a solid with a predictable, three-dimensional, repeating structure. (Module 7.1)

minimum viable population The smallest number of individuals that would still allow a population to be able to persist or grow, ensuring long-term survival. (Module 2.2)

mining The extraction of natural resources from the ground. (Module 7.1)

monoculture A farming method in which a single variety of one crop is planted, typically in rows over huge swaths of land, with large inputs of fertilizer, pesticides, and water. (Module 8.1)

mountaintop removal (MTR) A surface mining technique that involves using explosives to blast away the top of a mountain to expose the coal seam underneath; the waste rock and rubble are deposited in a nearby valley. (Modules 7.1, 9.1)

Multiple-Use Sustained-Yield Act U.S. legislation (1960) mandating that national forests be managed in a way that balances a variety of uses. (Module 7.2)

municipal solid waste (MSW) Everyday garbage or trash (solid waste) produced by individuals or small businesses. (Module 5.3)

mutualism A symbiotic relationship between individuals of two species in which both parties benefit. (Module 2.3)

N

National Environmental Policy Act (NEPA) A 1969 U.S. law that established environmental protection as a guiding policy for the nation and required that the federal government take the environment into consideration before taking action that might affect it. (Module 5.2)

natural capital The wealth of resources on Earth. (Module 5.1)

natural gas A gaseous fossil fuel composed mainly of simpler hydrocarbons, mostly methane. (Module 9.2)

natural interest Readily produced resources that we could use and still leave enough natural capital behind to replace what we took. (Module 5.1)

natural selection The process by which organisms best adapted to the environment (the fittest) survive to reproduce, leaving more offspring than less well-adapted individuals. (Module 3.1)

negative feedback loop Changes caused by an initial event that trigger events that then reverse the response. (Module 10.2)

niche The role a species plays in its community, including how it gets its energy and nutrients, what habitat requirements it has, and with which other species and parts of the ecosystem it interacts. (Module 2.3)

niche generalist A species who occupies a broad niche because it can utilize a wide variety of resources. (Module 2.3)

niche specialist A species with very specific habitat or resource requirements that restrict where it can live. (Module 2.3)

nitrification Conversion of ammonia to nitrate (NO_3^-). (Module 2.1)

nitrogen cycle A continuous series of natural processes by which nitrogen passes from the air to the soil to organisms and then returns back to the air or soil. (Module 2.1)

nitrogen fixation Conversion of atmospheric nitrogen into a biologically usable form, carried out by bacteria found in soil or via lightning. (Module 2.1)

NOAEL (no-observed-adverse-effect level) The highest dose where no adverse effect is seen. (Module 4.3)

noncommunicable disease (NCD) An illness that is not transmissible between people; not infectious. (Module 4.3)

nondegradable Incapable of being broken down under normal conditions. (Module 5.3)

nonpoint source pollution Runoff that enters the water from overland flow. (Module 6.2)

nonrenewable resource A resource that is formed more slowly than it is used or that is present in a finite supply. (Modules 1.1, 9.2)

nuclear energy Energy in an atom; can be released when an atom is split (fission). (Module 11.1)

nuclear fission A nuclear reaction that occurs when a neutron strikes the nucleus of an atom and breaks it into two or more parts. (Module 11.1)

Nuclear Waste Policy Act (1982) The federal law that mandated that the federal government build and operate a long-term repository for the disposal of high-level radioactive waste. (Module 11.1)

O

observational study Research that gathers data in a real-world setting without intentionally manipulating any variable. (Module 1.2)

oil A liquid fossil fuel useful as a fuel or as a raw material for industrial products. (Module 9.2)

open dumps Places where trash, both hazardous and nonhazardous, is simply piled up. (Module 5.3)

open-pit mining A surface mining method that extracts rock or mineral from a pit excavated for that purpose. (Module 7.1)

ore mineral/ore A rock deposit that contains economically valuable amounts of metal-bearing minerals. (Module 7.1)

organic agriculture Farming that does not use synthetic fertilizer, pesticides, GMOs, or other chemical additives like hormones (for animal rearing). (Module 8.2)

overburden The rock and soil removed to uncover a mineral deposit during surface mining. (Modules 7.1, 9.1)

overexploited fisheries More fish are taken than is sustainable in the long run, leading to population declines. (Module 8.4)

overgrazing Too many herbivores feeding in an area, eating the plants faster than they can regrow. (Module 7.3)

overpopulated The number of individuals in an area exceeds the carrying capacity of that area. (Module 4.1)

ozone (O_3) A molecule made up of three oxygen atoms. (Module 5.2)

ozone-depleting substances Chemicals that break down stratospheric ozone. (Module 5.2)

P

parasitism A symbiotic relationship between individuals of two species in which one benefits and the other is negatively affected. (Module 2.3)

particulate matter (PM) Particles or drop lets small enough to remain aloft in the air for long periods of time. (Module 10.1)

passive solar technology A technology that captures solar energy (heat or light) without any electronic or mechanical assistance. (Module 11.2)

pathogen An infectious agent that causes illness or disease. (Module 4.3)

payback time The amount of time it takes to save enough money in operation costs to pay for equipment. (Module 11.2)

peer review A process whereby researchers' work is evaluated by outside experts to determine whether it is of a high enough quality to publish. (Module 1.2)

perennial crops Crops that do not die at the end of the growing season but live for several years, which means they can be harvested annually without replanting. (Module 8.2)

performance standards The levels of pollutants allowed to be present in the environment or released over a certain time period. (Modules 5.2, 6.2)

persistence A measure of how resistant a chemical is to degradation. (Module 4.3)

pesticide A natural or synthetic chemical that kills or repels plant or animal pests. (Module 8.1)

pesticide resistance The ability of a pest to withstand exposure to a given pesticide; the result of natural selection favoring the survivors of an original population that was exposed to the pesticide. (Module 8.1)

petrochemicals Distillation products from the processing of crude oil such as fuels or industrial raw materials. (Module 9.2)

phosphorus cycle A series of natural processes by which the nutrient phosphorus moves from rock to soil or water to living organisms and back to soil. (Module 2.1)

photosynthesis The chemical reaction performed by producers that uses the energy of the Sun to convert carbon dioxide and water into sugar and oxygen. (Module 2.1)

photovoltaic (PV) cell A technology that converts solar energy directly into electricity. (Module 11.2)

pioneer species Species that move into an area during early stages of succession. (Module 2.3)

placer mining The mining of underwater sediments (e.g., streambeds) for minerals. (Module 7.1)

point source pollution Pollution from wastewater treatment plants or industrial sites, such as that from discharge pipes or smokestacks. (Module 6.2)

policy A formalized plan that addresses a desired outcome or goal. (Module 1.2)

political lobbying Contacting elected officials in support of a particular position; some professional lobbyists are highly organized, with substantial financial backing. (Module 5.2)

polyculture A farming method in which a mix of different species are grown together in one area. (Modules 8.2, 8.3)

population All the individuals of a species that live in the same geographic area and are able to interact and interbreed. (Modules 2.1, 2.2)

population density The number of individuals per unit area. (Module 2.2)

population distribution The location and spacing of individuals within their range. (Module 2.2)

population dynamics Changes over time in population size and composition. (Module 2.2)

population growth rate The change in population size over time that takes into account the number of births and deaths as well as immigration and emigration numbers. (Modules 2.2, 4.1)

population momentum The tendency of a young population to continue to grow even after birth rates drop to replacement fertility (two children per couple). (Module 4.1)

positive feedback loop Changes caused by an initial event that then accentuate that original event. (Module 10.2)

potable Water that is clean enough for consumption. (Module 6.1)

potency The dose size required for a chemical to cause harm. (Module 4.3)

precautionary principle Acting in a way that leaves a safety margin when the data are uncertain or severe consequences are possible. (Modules 1.2, 4.3)

predation Species interaction in which one individual (the predator) feeds on another (the prey). (Module 2.3)

primary air pollutants Air pollutants released directly from a mobile or stationary source. (Module 10.1)

primary source Information source that presents original data or firsthand information. (Module 1.2)

primary succession Ecological succession that occurs in an area where no ecosystem existed before (e.g., on bare rock with no soil). (Module 2.3)

producer An organism that converts solar energy to chemical energy via photosynthesis. (Modules 2.1, 2.3)

pronatalist pressure Factor that increases the desire to have children. (Module 4.1)

protected areas Geographic spaces on land or at sea that are recognized, dedicated, and managed to achieve long-term conservation of nature. (Module 3.3)

protocol A document that sets precise goals and targets. (Module 5.2)

proven reserves A measure of the amount of a fossil fuel that is economically feasible to extract from a known deposit using current technology. (Modules 9.1, 9.2)

proxy data Measurements that allow one to indirectly infer a value such as the temperature or atmospheric conditions in years past. (Module 10.2)

public health The science that deals with the health of human populations. (Module 4.3)

R

radioactive Atoms that spontaneously emit subatomic particles and/or energy. (Module 11.1)

radioactive decay The spontaneous loss of particle or gamma radiation from an unstable nucleus. (Module 11.1)

radioactive half-life The time it takes for half of the radioactive isotopes in a sample to decay to a new form. (Module 11.1)

random distribution A distribution in which individuals are spread out over the environment irregularly, with no discernible pattern. (Module 2.2)

range The geographic area where a species or one of its populations can be found. (Module 2.2)

rangeland Grassland used for grazing of livestock. (Module 7.3)

range of tolerance The range, within upper and lower limits, of a limiting factor that allows a species to survive and reproduce. (Module 2.1)

recirculating aquaculture system (RAS) A method used to rear fish indoors in tanks that filter and recirculate the water. (Module 8.4)

reclamation The process of restoring a damaged natural area to a less damaged state. (Modules 7.1, 9.1)

recycle The fourth of the four Rs of waste reduction: Return items for reprocessing into new products. (Module 5.3)

reduce The second of the four Rs of waste reduction: Make choices that allow you to use less of a resource by, for instance, purchasing durable goods that will last or can be repaired. (Module 5.3)

refuse The first of the four Rs of waste reduction: Choose not to use or buy a product if you can do without it. (Module 5.3)

renewable energy Energy that comes from an infinitely available or easily replenished source. (Modules 1.1, 11.2)

replacement fertility The rate at which children must be born to replace the previous generation. (Module 4.1)

reservoir An artificial lake formed when a river is impounded by a dam. (Module 6.1)

resilience The ability of an ecosystem to recover when it is damaged or perturbed. (Module 2.3)

resistance factors Things that directly (predators, disease) or indirectly (competitors) reduce population size. (Module 2.2)

Resource Conservation and Recovery Act (RCRA) The federal law that regulates the management of solid and hazardous waste. (Modules 4.3, 5.3)

resource partitioning The use of different parts or aspects of a resource by different species rather than direct competition for exactly the same resource. (Module 2.3)

restoration ecology The science that deals with the repair of damaged or disturbed ecosystems. (Module 2.3)

reuse The third of the four Rs waste reduction: Use a product more than once for its original purpose or for another purpose. (Module 5.3)

riparian areas The land areas close enough to a body of water to be affected by the water's presence (e.g., areas where water-tolerant plants grow) and that affect the water itself (e.g., provide shade). (Module 6.2)

risk assessment The systematic process of weighing the risks and benefits of an undertaking to determine whether to take action. (Module 4.3)

rock Solid aggregate of one or more minerals that occurs in a variety of configurations. (Module 7.1)

rock cycle The process in which rock is constantly made and destroyed. (Module 7.1)

rotational grazing Moving animals from one pasture to the next in a predetermined sequence to prevent overgrazing. (Module 7.3)

***r*-selected species** Species that have a high biotic potential and that share other characteristics, such as short life span, early maturity, and high fecundity. (Module 2.2)

S

Safe Drinking Water Act (SDWA) Federal law that protects public drinking water supplies in the United States. (Module 6.1)

saltwater intrusion The inflow of ocean (salt) water into a freshwater aquifer that happens when an aquifer has lost some of its freshwater stores. (Module 6.1)

sanitary landfills Disposal sites that seal in trash at the top and bottom to prevent its release into the atmosphere; the sites are lined on the bottom, and trash is dumped in and covered with soil daily. (Module 5.3)

science A body of knowledge (facts and explanations) about the natural world and the process used to get that knowledge. (Module 1.2)

scientific method The procedure scientists use to empirically test a hypothesis. (Module 1.2)

secondary air pollutants Air pollutants formed when primary air pollutants react with one another or with other chemicals in the air. (Module 10.1)

secondary source Information source that presents and interprets information solely from primary sources. (Module 1.2)

secondary succession Ecological succession that occurs in an ecosystem that has been disturbed but not rendered lifeless. (Module 2.3)

sedimentary rock Rock that forms when fragments of mineral or biological origin are deposited, accumulate, and are compacted and cemented. (Module 7.1)

sediments Fragments of mineral, rock, or organic material. (Module 7.1)

selective harvesting Timber-harvesting technique that cuts only the highest-value trees; the remaining trees reseed the plot. (Module 7.2)

selective pressure A nonrandom influence that affects who survives or reproduces. (Module 3.1)

service economy A business model whose focus is on leasing and caring for a product in the customer's possession rather than on selling the product itself (i.e., selling the *service* that the product provides). (Module 5.1)

shelterbelts A stand of trees that blocks the wind and thus decreases soil erosion. (Module 7.3)

shelterwood harvesting Timber-harvesting technique that cuts all but the best trees, which reseed the plot and are then harvested. (Module 7.2)

single-species conservation A management strategy that focuses on protecting one particular species. (Module 3.3)

sinks Abiotic or biotic components of the environment that serve as storage places for cycling nutrients. (Module 2.1)

smart grid A modernized network that provides electricity to users in a way that automatically optimizes the delivery of electricity. (Module 11.2)

smart growth Strategies that help create walkable communities with lower environmental impacts. (Module 4.2)

smelting Mineral processing step in which the material is melted at high temperatures and mixed with chemicals that separate the mineral from the rock. (Module 7.1)

social traps Decisions by individuals or groups that seem good at the time and produce a short-term benefit but that hurt society in the long run. (Module 1.1)

soil A complex ecosystem of mineral and organic material, including living organisms such as bacteria, invertebrates, and fungi, that supports the growth of plants and is, in turn, affected by those plants. (Module 8.2)

soil erosion The process in which soil is moved from one location to another, most often by wind and water. (Module 8.2)

solar power Energy harnessed from the Sun in the form of heat or light. (Module 11.2)

solid waste Any material that humans discard. (Module 5.3)

solubility The ability of a substance to dissolve in a water- or fat-based liquid or gas. (Module 4.3)

solution mining Extraction method where desired minerals are dissolved by a liquid injected into the deposit, then pumped back out and purified. (Module 7.1)

species A group of plants or animals that have a high degree of similarity and can generally only interbreed among themselves. (Module 2.1)

species diversity The variety of species, including how many are present (richness) and their abundance relative to each other (evenness). (Modules 2.3, 3.2)

species evenness The relative abundance of each species in a community. (Module 2.3)

species richness The total number of different species in a community. (Module 2.3)

statistics The mathematical evaluation of experimental data to determine how likely it is that any difference observed is due to the variable being tested. (Module 1.2)

stormwater runoff Water from precipitation that flows over the surface of the land. (Modules 6.2, 7.2)

stratosphere A layer of atmosphere that lies directly above the troposphere. (Module 5.2)

strip mining A surface mining method that accesses fossil fuel or mineral resources from deposits close to the surface on level ground, one section at a time. (Modules 7.1, 9.1)

subduct The movement of one tectonic plate below another at a convergent plate boundary. (Module 7.1)

subsidies Financial assistance given by the government to promote desired activities. (Modules 5.2, 10.1)

subsurface mines Sites where tunnels are used to access underground fossil fuel or mineral resources. (Modules 7.1, 9.1)

suburban sprawl Low-population-density developments that are built outside of a city. (Module 4.2)

surface mining A form of mining that involves removing soil and rock that overlay a mineral deposit close to the surface in order to access that deposit. (Modules 7.1, 9.1)

surface water Any body of water found above ground, such as oceans, rivers, and lakes. (Module 6.1)

sustainable Capable of being continued indefinitely. (Modules 1.1, 5.1)

sustainable agriculture Farming methods that can be used indefinitely because they do not deplete resources, such as soil and water, faster than they are replaced. (Module 8.2)

sustainable development Development that meets present needs without compromising the ability of future generations to do the same. (Module 1.1)

sustainable energy Energy from renewable sources that are used no faster than replenished and whose use has an acceptable environmental impact. (Module 11.2)

sustainable fishery A fishery that ensures that fish stocks are maintained at healthy levels, the ecosystem is fully functional, and fishing activity does not threaten biological diversity. (Module 8.4)

sustainable grazing Practices that allow animals to graze in a way that keeps pastures healthy and allows grasses to recover. (Module 7.3)

symbiosis A close biological or ecological relationship between two species. (Module 2.3)

synergistic effects Exposure to two or more chemicals that has a greater effect than the sum of their individual effects would predict. (Module 4.3)

T

take-back program Program that allows a consumer, once they are finished with a product, to return it to the manufacturer that made it. (Module 5.1)

tar sands (oil sands) Sand or clay formations that contain a heavy-density crude oil (crude bitumen); extracted by surface mining. (Module 9.2)

tax credit A reduction in the tax one must pay in exchange for some desirable action. (Modules 5.2, 10.1)

tectonic plates Rigid pieces of Earth's lithosphere that move above the asthenosphere. (Module 7.1)

temperate forest Forest found in areas that have four seasons and a moderate climate, receive 30 to 60 inches of precipitation per year, and may include evergreen and deciduous conifers and broadleaf trees. (Module 7.2)

tertiary source Information source that uses information from at least one secondary source. (Module 1.2)

test group The group in an experimental study that is manipulated such that it differs from the control group in only one way. (Module 1.2)

theory A widely accepted explanation of a natural phenomenon that has been extensively and rigorously tested scientifically. (Module 1.2)

threatened species Species that are at risk for extinction; various threat levels have been identified, ranging from "least concern" to "extinct." (Module 3.3)

tight oil Light (low-density) oil in shale rock deposits of very low permeability; extracted by fracking. (Module 9.2)

top-down regulation Population sizes in a community are limited primarily by predation from organisms at the top of the food chain. (Module 2.2)

total fertility rate (TFR) The number of children the average woman has in her lifetime. (Module 4.1)

total maximum daily loads The maximum amount of a pollutant allowed to enter a waterbody so that the waterbody will meet water quality standards. (Module 6.2)

toxic substance/toxic A substance that causes damage when it contacts, or enters, the body. (Module 4.3)

Toxic Substances Control Act (TSCA) The primary federal law governing chemical safety. (Module 4.3)

trade-offs The imperfect and sometimes problematic responses that we must at times choose between when addressing complex problems. (Module 1.1)

tragedy of the commons The tendency of an individual to abuse commonly held resources in order to maximize his or her own personal interest. (Modules 1.1, 8.4)

transboundary pollution Pollution that is produced in one area but falls in a different region. (Module 10.1)

transboundary problem A problem that extends across state and national boundaries; pollution that is produced in one area but falls in or reaches other states or nations. (Module 5.2)

transform plate boundary A place where two tectonic plates slide side to side relative to each other. (Module 7.1)

transgenic organism An organism that contains genes from another species. (Module 8.1)

transpiration The loss of water vapor from plants. (Module 6.1)

triple bottom line The combination of the environmental, social, and economic impacts of our choices. (Modules 1.1, 5.1)

trophic cascade Top-down effects from the presence or absence of a top predator that propagate all the way down a food chain to the ecosystem's plant communities. (Module 2.2)

trophic levels Feeding levels in a food chain. (Module 2.3)

tropical forest Forest found in equatorial areas with warm temperatures year-round and high rainfall; some have distinct wet and dry seasons, but none has a winter season. (Module 7.2)

troposphere The lowest level of the atmosphere. (Module 5.2)

true cost The sum of both external and internal costs of a good or service. (Module 5.1)

tsunami A series of high, long, fast-moving water waves caused by the displacement of a large volume of water by an underwater earthquake, landslide, or volcanic eruption, usually in the Pacific Ocean but also in other ocean areas and in large lakes. (Module 7.1)

U

ultraviolet (UV) radiation High-energy radiation that is harmful to living things. (Module 5.2)

unconventional reserves Deposits of oil or natural gas that cannot be recovered with traditional oil and gas wells but may be recoverable using alternative techniques. (Module 9.2)

undergrazing Grazing too few animals on a grassland to maintain its ecological integrity. (Module 7.3)

understory The smaller trees, shrubs, and saplings that live in the shade of the forest canopy. (Module 7.2)

uniform distribution A distribution in which individuals are spaced evenly, perhaps due to territorial behavior or mechanisms for suppressing the growth of nearby individuals. (Module 2.2)

urban areas Densely populated regions that include cities and the suburbs that surround them. (Module 4.2)

urban flight The process of people leaving an inner-city area to live in surrounding areas. (Module 4.2)

urbanization The migration of people to large cities; sometimes also defined as the growth of urban areas. (Module 4.2)

U.S. Farm Bill Legislation that deals with many aspects of the production and sale of farm-raised commodity crops. (Module 8.3)

U.S. Surface Mining Control and Reclamation Act Federal law that mandates coal mines be restored to their approximate original condition after closure. (Module 9.1)

V

vector An agent that transmits a pathogen to an organism. (Module 4.3)

vein Distinct region within a rock that contains a mineral ore. (Module 7.1)

volatile organic compound (VOC) A chemical that readily evaporates and is released into the air as a gas; may be hazardous. (Module 10.1)

volcano An opening (vent) through which lava, gases, and other material escape from beneath Earth's crust, often accumulating to form a mountain or hill. (Module 7.1)

W

wastewater Used and contaminated water that is released after use by households, businesses, industry, or agriculture. (Module 6.1)

wastewater treatment The process of removing contaminants from wastewater to make it safe enough to release into the environment. (Module 6.1)

water cycle The movement of water through various water compartments such as surface waters, atmosphere, soil, and living organisms. (Module 6.1)

water footprint The water consumed by a given group (that is, person or population) or appropriated and/or polluted by industry to produce products or energy. (Modules 6.1, 8.3)

water pollution The addition of any substance to a body of water that might degrade its quality. (Module 6.2)

water scarcity Not having access to enough clean water. (Module 6.1)

water table The uppermost water level of the saturated zone of an aquifer. (Module 6.1)

water wars Political conflicts over the allocation of water sources. (Module 6.1)

waterborne disease An infectious disease acquired through contact with contaminated water. (Module 4.3)

watershed The land area surrounding a body of water over which water such as rain can flow and potentially enter that body of water. (Module 6.2)

watershed management Management of what goes on in an area around streams and rivers. (Module 6.2)

weather The meteorological conditions in a given place on a given day. (Module 10.2)

weathering The breakdown of rock by physical or chemical forces. (Module 7.1)

wind power Energy contained in the motion of air across Earth's surface. (Module 11.2)

Z

zero-population growth The absence of population growth; occurs when birth rates equal death rates. (Module 4.1)

zoonotic disease An infectious disease of animals that can be transmitted to humans. (Module 4.3)

zooxanthellae Mutualistic photosynthetic algal partner of a coral polyp; each provides nutrients that the other needs. (Module 6.3)

triple bottom line The combination of the environmental, social, and economic impacts of our choices. (Module [illegible])

trophic cascade Top-down effects from the presence or absence of a top predator that propagate all the way down a food chain to [illegible] the ecosystem's plant communities. (Module [illegible])

trophic levels Feeding levels in a food chain. (Module [illegible])

tropical forest Forest found near the equator [illegible] temperatures [illegible] year round [illegible] [illegible]. (Module [illegible])

troposphere The lowest layer of the atmosphere. (Module [illegible])

true cost The sum of both the internal and the external costs of a good or service. (Module [illegible])

tsunami A series of waves [illegible] water where [illegible] the displacement of a large volume of water [illegible] [illegible] large [illegible]. (Module [illegible])

U

ultraviolet (UV) radiation [illegible] energy [illegible] than visible light [illegible]. (Module [illegible])

unconventional reserves Deposits of oil and fossil fuels that cannot be extracted with [illegible] but only [illegible]. (Module [illegible])

undernourished [illegible] to maintain [illegible]. (Module [illegible])

understory The smaller trees, shrubs, and plants that live in the shade of the forest canopy. (Module [illegible])

uniform distribution A distribution in which individuals are evenly spaced [illegible] of [illegible] the growth of nearby individuals. (Module [illegible])

urban areas Densely populated regions [illegible] and the suburbs that surround them. (Module [illegible])

urban flight The process of people leaving an urban city area to live in suburban or rural areas. (Module [illegible])

urbanization The migration of people to [illegible], sometimes also defined as the growth of urban areas. (Module [illegible])

U.S. Farm Bill Legislation that deals with many aspects of the production and sale of [illegible] commodity crops. (Module [illegible])

U.S. Surface Mining Control and Reclamation Act Federal law that mandates that mines be restored to their approximate original condition after mining. (Module [illegible])

V

vector An agent that transmits a pathogen to a new host. (Module [illegible])

vein [illegible] within a rock that contains [illegible]. (Module [illegible])

volatile organic compound (VOC) A chemical that readily evaporates [illegible] [illegible]. (Module [illegible])

volcano A [illegible] through which [illegible] gases [illegible] from beneath Earth's crust [illegible]. (Module [illegible])

W

wastewater Used and contaminated water that is released after use by households, businesses, industry, or agriculture. (Module [illegible])

wastewater treatment The process of removing contaminants from wastewater to make it safe enough to release into the environment. (Module [illegible])

water cycle The movement of water [illegible] such [illegible]

[illegible] waters, atmosphere, land, and organisms. (Module [illegible])

water footprint The water consumption of a person, city, or population [illegible] water consumed and/or polluted by industry to produce products or energy. (Modules [illegible])

water pollution The addition of any substance to a body of water that [illegible] its quality. (Module [illegible])

water scarcity Not having access to enough clean water. (Module [illegible])

water table The uppermost water level of the saturated zone of an aquifer. (Module [illegible])

water wars Armed conflicts over the allocation of water sources. (Module [illegible])

waterborne disease An infectious disease acquired through contact with contaminated water. (Module [illegible])

watersheds The areas of land surrounding [illegible] bodies of water [illegible] such as [illegible] [illegible] and [illegible] that body of water. (Module [illegible])

watershed management [illegible] [illegible]. (Module [illegible])

weather The short-term [illegible] in [illegible]. (Module [illegible])

weathering The breakdown of rocks by physical, chemical, or biological forces. (Module [illegible])

wind power Energy generated by the motion of air across Earth's surface. (Module [illegible])

Z

zero population growth The absence of population growth; [illegible] birth rates and death rates. (Module [illegible])

zoonotic disease A disease that [illegible] animals [illegible] to humans. (Module [illegible])

zooxanthellae Mutualistic photosynthetic algae [illegible] coral polyps [illegible] [illegible]. (Module [illegible])

INDEX

Note: Page numbers followed by f indicate figures, infographics, photographs, and associated captions. Italicized page numbers indicate online module information.

A

C

E

F

H

I

M

N

P

Q

R

S

T

V

W

Y

Z